Lecture Notes in Mechanical Engineering

Lecture Notes in Mechanical Engineering (LNME) publishes the latest developments in Mechanical Engineering—quickly, informally and with high quality. Original research reported in proceedings and post-proceedings represents the core of LNME. Volumes published in LNME embrace all aspects, subfields and new challenges of mechanical engineering. Topics in the series include:

- Engineering Design
- Machinery and Machine Elements
- Mechanical Structures and Stress Analysis
- Automotive Engineering
- Engine Technology
- Aerospace Technology and Astronautics
- Nanotechnology and Microengineering
- Control, Robotics, Mechatronics
- MEMS
- Theoretical and Applied Mechanics
- Dynamical Systems, Control
- Fluid Mechanics
- Engineering Thermodynamics, Heat and Mass Transfer
- Manufacturing
- Precision Engineering, Instrumentation, Measurement
- Materials Engineering
- Tribology and Surface Technology

To submit a proposal or request further information, please contact the Springer Editor of your location:

China: Dr. Mengchu Huang at mengchu.huang@springer.com
India: Priya Vyas at priya.vyas@springer.com
Rest of Asia, Australia, New Zealand: Swati Meherishi at swati.meherishi@springer.com
All other countries: Dr. Leontina Di Cecco at Leontina.dicecco@springer.com

To submit a proposal for a monograph, please check our Springer Tracts in Mechanical Engineering at http://www.springer.com/series/11693 or contact Leontina.dicecco@springer.com

Indexed by SCOPUS. The books of the series are submitted for indexing to Web of Science.

More information about this series at http://www.springer.com/series/11236

Sandip Kumar Saha · Mousumi Mukherjee
Editors

Recent Advances in Computational Mechanics and Simulations

Volume-II: Nano to Macro

Springer

Editors
Sandip Kumar Saha
School of Engineering
Indian Institute of Technology Mandi
Mandi, Himachal Pradesh, India

Mousumi Mukherjee
School of Engineering
Indian Institute of Technology Mandi
Mandi, Himachal Pradesh, India

ISSN 2195-4356 ISSN 2195-4364 (electronic)
Lecture Notes in Mechanical Engineering
ISBN 978-981-15-8314-8 ISBN 978-981-15-8315-5 (eBook)
https://doi.org/10.1007/978-981-15-8315-5

This Springer imprint is published by the registered company Springer Nature Singapore Pte Ltd.
The registered company address is: 152 Beach Road, #21-01/04 Gateway East, Singapore 189721, Singapore

Foreword

Selected papers, presented in the 7th International Congress on Computational Mechanics and Simulation (ICCMS 2019) held under the aegis of Indian Association for Computation Mechanics (IndACM), and organized by Indian Institute of Technology Mandi, are compiled into two book volumes. The advances made in recent years in mechanics of materials and structures are reported in the first volume. It gives me immense pleasure to write the foreword for the second volume that reports the advances in nano-to-macro mechanics and application of computational techniques in emerging areas.

There are 55 chapters, thematically arranged and subdivided into seven subsections in the second volume. The first section, having six chapters, is dedicated to recent advances in nanomechanics. These chapters report new developments in understanding the behavior of nanoshells and nanotubes. As we gain more clarity on the behavior of nanoscale materials and elements, their application scope, especially in medical use, will widen. With use of supercomputer and advanced simulation techniques, we are progressing toward the use of digital twins of the human body to monitor and forecast medical conditions. It is, therefore, necessary to understand the mechanics of various parts of the human body. The second section focuses on the advancements in biomechanics. The five chapters in this section are dedicated to the application of computational mechanics and simulation techniques in the modeling of bone, tissue, veins, etc.

The third section, with nine chapters, deals with the recent developments in solid mechanics and finite element method (FEM). Studies related to fatigue life estimation of large complex systems, like turbo generators, use of extended finite element method (XFEM) in creep life estimation and analysis of advanced honeycomb structures are presented here. The 13 chapters in section four present the advancements in computational fluid dynamics (CFD) and transport phenomena. From the use of meshless methods in simulating flow and the transport process to modeling bio-inspired motions are in the focus of at least five chapters. Another five chapters focus on the turbulent and non-Newtonian fluid flow and associated areas. The remaining three chapters discuss different forces on hypersonic blunt body, acoustic analysis and effect of laser on natural convection.

The fifth section presents the recent advancements in modeling and simulation of multiscale-multiphysics problems. There are six chapters in this section highlighting various ranges of applications. This section also stresses upon the simulation of multi-scale systems with advanced materials such as piezoelectric materials and magneto-rheological materials. The sixth section has seven chapters dedicated to recent advances and new applications of numerical techniques and algorithms. These include solution of Levy-type plates, time integration schemes for nonlinear problems, algorithms for simulating geomaterial response, solution of bivariate population balance equation, etc.

The last section, comprises nine chapters and presents the recent applications of computational mechanics and techniques in emerging areas. Some observations presented herein on the use of X-ray microtomographic images of concrete for phase segmentation, design of nonlinear energy harvesting systems, battery module for electric vehicles and dynamic analysis of launch vehicle structures truly showcase the wide and futuristic applications of computational mechanics and simulation techniques. There is no doubt that computational mechanics is going to play a driving role to utilize the true potential of ever-increasing computer capabilities for solving twenty-first century's complex real-life problems.

I hope the readers will enjoy reading the book volume and get some high-quality food for thought leading to future advancements of this field. The editors have done a commendable job in bringing such a wide range of excellent research articles under one umbrella. I must congratulate them for their sincere efforts.

Tarun Kant, Ph.D., FNAE, FASc, FNA, FNASc
Professor Emeritus
Department of Civil Engineering
Indian Institute of Technology Bombay
Powai, Mumbai, India

Preface

Indian Institute of Technology Mandi (IIT Mandi) successfully hosted the 7th International Congress on Computational Mechanics and Simulation (ICCMS 2019). This prestigious biennial event was attended by over 250 delegates from all across the world. This unique congress has exceptional reputation for showcasing the latest developments in various fronts of computational mechanics—from theories to applications. In the modern era, the scientific exploration has gained a new height with the advancements in mathematical and computational methods. Its significant impact can further be realized through the possibilities of performing engineering analysis and design tasks which could have not been managed by manual efforts earlier. Computational Mechanics emphasizes the development of mathematical models representing physical phenomena and applies modern computing methods to analyze these phenomena. Essentially, being interdisciplinary in nature, it thrives on the fields related to physics, mechanics, mathematics and computer science, and encompasses applying numerical methods to various problems in science and engineering. Recent advances in the field of Computational Mechanics have generated considerable interest among researchers and practicing engineers to gain more knowledge and insight into the various aspects of modeling, analysis and design. The principles of Computational Mechanics can be effectively applied to rational design of engineering components under various extreme and complex loading conditions. Within the aforementioned general framework, Computational Mechanics is currently being used in a broad range of applications including civil, offshore, aerospace, automotive, naval, biomechanics and nuclear structures. Keeping such multidisciplinary aspects in mind, this book volume presents the recent advancements which took place in the field of Computational Mechanics by compiling selected papers presented in the ICCMS 2019.

About 215 technical papers were presented during the three days' congress by researchers and practitioners from a wide spectrum of research background. In addition, there were 15 plenary speeches, 10 keynote speeches and four invited talks delivered by eminent researches in respective areas. All the submitted articles went through a rigorous three-level review process. The review comments were shared with the authors to comply and improve. The articles were accepted for presentation only after receiving positive recommendations from the reviewers. It was indeed an enthralling experience to assimilate all the outstanding research papers from the multidisciplinary domains. However, based on the reviewers' feedback and recommendations, only 105 papers were selected for compilation as book volumes. We are delighted that the state-of-the-art technical articles are finally coming up in the form of two exclusive book volumes by Springer titled *Recent Advances in Computational Mechanics and Simulations*. The technical articles are categorically presented as 105 chapters distributed in two volumes. Volume-I is broadly dedicated to the use of computational mechanics and simulations at the level of *Materials to Structures*; and Volume-II broadly focuses on computational mechanics at *Nano to Macro* levels and its applications in emerging areas. Nevertheless, there are several chapters that encompass multiple frontiers of the computational mechanics. Therefore, drawing a fine boundary while distributing the articles in two volumes was really challenging.

The wholehearted support received from the organizing committee members and colleagues at IIT Mandi in organizing the ICCMS 2019 is truly appreciated. We are grateful to Prof. Tarun Kant, Founding President of Indian Association of Computational Mechanics (IndACM), and Prof. Vasant Matsagar, Executive Secretary of IndACM, for their mentoring in the overall organization of the congress. We would like to thank the authors who contributed the exceptional technical articles for the congress. We express our sincere gratitude to all the reviewers for their valuable time and painstaking efforts in reviewing the articles. We are indebted to Prof. Tarun Kant for writing the foreword of this volume. The list of acknowledgements will be incomplete without mentioning the passionate efforts put in by our postgraduate students at various levels of the organization. Without their active support and devotion, it would not have been possible to organize an event of this stature. We thank Springer for accepting our proposal and Ms. Swati Meherishi, Editorial Director, and her team for continuous support in successfully bringing out these book volumes.

Mandi, Himachal Pradesh, India
May 2020

Sandip Kumar Saha
Mousumi Mukherjee

7th International Congress on Computational Mechanics and Simulation (ICCMS 2019)

Under the auspices of the Indian Association for Computational Mechanics (IndACM), International Congress on Computational Mechanics and Simulation (ICCMS) is organized biennially. The Association was founded on January 1, 2000, and has about 200 life members. The previous congresses (ICCMS) were held at IIT Bombay (2015, 2009), CSIR-SERC Chennai (2014), IIT Hyderabad (2012), IIT Guwahati (2006) and IIT Kanpur (2004). Computational Mechanics, being an interdisciplinary domain of mechanics, involving mathematical models of physical real-life problems and their solutions, the congress is typically attended by delegates with background in civil engineering, mechanical engineering, aerospace engineering, materials engineering/science, physics, mathematics and other disciplines.

The 7th edition of ICCMS was organized by Indian Institute of Technology Mandi (IIT Mandi), during 11–13 December 2019, at the serene Kamand Valley located in the lap of the Great Himalayas. The congress hosted 15 plenary speakers, 10 keynote speakers and four invited speakers from, Australia, India, Japan, UK and USA. More than 215 technical papers were presented by the delegates from across the world in addition to the plenary, keynote and invited presentations. The technical papers were invited under 14 different subdomains, such as biomechanics; computational fluid dynamics and transport phenomena; computational geomechanics and geotechnics; natural materials; computational structural dynamics; constitutive modeling of materials; composites and multifunctional materials; fracture and failure mechanics; interfaces, contacts and interactions; multiscale and multiphysics problems, simulation; numerical methods and algorithms in engineering and science; simulation and analysis under accidental and extreme loadings; structural health monitoring; vibration control; structural mechanics, materials and engineering; uncertainty quantification, reliability analysis; and application of computational techniques in other areas. The technical presentations were conducted in six parallel sessions during the three days of the congress.

Along with the regular technical sessions, International Society for Soil Mechanics and Geotechnical Engineering (ISSMGE), in association with ICCMS 2019, organized a day-long Mini-Symposium of ISSMGE's Technical committee TC-105 on GeoMechanics from Micro to Macro. This comprised seven invited lectures and four regular presentations by young researchers. The objective of this mini-symposium was to promote the micromechanics-based modeling and research within the geomechanics community of India.

After seven successful occasions, the 8th edition of ICCMS will be organized by Indian Institute of Technology Indore in 2021.

Organizing Committee (ICCMS 2019)

Patron

Prof. Timothy A. Gonsalves, Director, IIT Mandi

Mentor

Prof. Tarun Kant, Founding President, IndACM, Professor Emeritus, IIT Bombay, Visiting Distinguished Professor, IIT Mandi

Core Organizing Committee

Dr. Sandip Kumar Saha, Assistant Professor, School of Engineering, IIT Mandi (Convenor)
Dr. Mousumi Mukherjee, Assistant Professor, School of Engineering, IIT Mandi
Dr. Maheshreddy Gade, Assistant Professor, School of Engineering, IIT Mandi
Dr. Venkata Uday Kala, Assistant Professor, School of Engineering, IIT Mandi
Dr. Rajesh Ghosh, Assistant Professor, School of Engineering, IIT Mandi
Dr. Subhamoy Sen, Assistant Professor, School of Engineering, IIT Mandi
Dr. Gaurav Bhutani, Assistant Professor, School of Engineering, IIT Mandi
Dr. Rajneesh Sharma, Assistant Professor, School of Engineering, IIT Mandi
Dr. Kaustav Sarkar, Assistant Professor, School of Engineering, IIT Mandi

Contents

Recent Advances in Nanomechanics

Buckling of Pressurized Nano-Sized Spherical Shell 3
Manjur Alam and Sudib Kumar Mishra

On Flexure of Shear Deformable Isotropic Rectangular Nanobeams 17
Kedar S. Pakhare, P. J. Guruprasad, and Rameshchandra P. Shimpi

Rolling and Sliding Resistance as Carbon Nanotubes are Driven on a Graphene Sheet 27
Avirup Sircar and Puneet Kumar Patra

Surface Stress-induced Degradation of Electrochemical Performance of Cylindrical Silicon Anode Particles in Li-ion Batteries 41
Amrita Sengupta, Sourav Das, and Jeevanjyoti Chakraborty

Molecular Dynamics Simulation of Single-Wall Carbon Nanotube Aluminum Composite 53
Pramod Rakt Patel, Sumit Sharma, and S. K. Tiwari

Low Velocity Impact Behavior of Carbon Nanotubes Reinforced Aluminum Foams 63
Y. M. Chordiya and M. D. Goel

Recent Advances in Biomechanics

Effect of Interfacial Crack on the Prediction of Bone–Cement Interface Failure of Cemented Acetabular Component 75
Ajay Kumar, Rajesh Ghosh, and Rajeev Kumar

Effect of Occlusion Percentage and Lesion Length on Stenosed Coronary Artery: A Numerical Study 87
Supratim Saha, T. Purushotham, and K. Arul Prakash

Model-Based Simulation of Surface Electromyography Signals and Its Analysis Under Fatiguing Conditions Using Tunable Wavelets 99
Lakshmi M. Hari, S. Edward Jero, G. Venugopal, and S. Ramakrishnan

Fractal Order Poro-elastic Model for Modelling Biphasic Tissue and Tissue-Like Materials 109
Shib Sundar Banerjee, Arockiarajan Arunachalakasi, and Ramakrishnan Swaminathan

Quantification of Brain Retraction Using Visco-hyperelastic Framework for Image-Guided Neurosurgical Applications 121
Abhilash Awasthi, Suryanarayanan Bhaskar, Umesh Gautam, and Sitikantha Roy

Recent Trends in Solid Mechanics and FEM

Calculation of NSIFs and Shape Factors of Four-Point Bend Specimens Containing Sharp V-Notches 133
Mirzaul Karim Hussain and K. S. R. K. Murthy

Tera-Hertz Wave Propagation in Non-classical Beams Using Spectral Finite Element Method 141
K. Sainath and S. Narendar

Fem Simulations for Fatigue Life Estimation of Big Turbo-generator Shaft During Various Fault Disturbances Under Active Control 153
Tarun Kumar, Rajeev Kumar, and S. C. Jain

Complete Creep Life Prediction Using Continuum Damage Mechanics and XFEM 169
V. B. Pandey, I. V. Singh, and B. K. Mishra

Numerical Study of Coupled Elasto-Plastic Hydrogen Diffusion at Crack Tip Using XFEM 177
Anjali Jha, I. V. Singh, B. K. Mishra, Ritu Singh, and R. N. Singh

Role Played by Grain Boundaries in Plastic Deformation of Polycrystalline Metals: A Discrete Dislocation Dynamics Study 187
Tawqeer Nasir Tak, Aditya Prakash, Indradev Samajdar, and P. J. Guruprasad

Effective Properties of Two-Scale Viscoelastic Composites 195
Vivek Singh, Jayram Desai, and Vikranth Racherla

Thermo-Mechanical Analysis of Shape Memory Polymer Composite 205
L. Bhola, P. M. Mujumdar, and P. J. Guruprasad

Energy Absorption Characteristics of Balsa Wood Filled Aluminum Honeycomb Structures Subjected to Axial Quasi-static Loading 217
Nadeem Ahmad, Devendra K. Dubey, and Rohit Sankrityayan

Recent Advances in Computational Fluid Dynamics

Simulation of Flow and Transport Process—Scope of Meshless Methods 235
Aatish Anshuman and T. I. Eldho

Computational Modelling of Stochastic Buffeting of Fin-Like Structures 249
Rahul Sundar, Avisha Ghorpade, Jithin Jith, Sayan Gupta, and Sunetra Sarkar

Investigating the Dynamical Behaviour of Dipteran Flight-Inspired Flapping Motion Using Immersed Boundary Method-Based FSI Solver 259
Chhote Lal Shah, Dipanjan Majumdar, and Sunetra Sarkar

Probing into the Efficacy of Discrete Forcing Immersed Boundary Method in Capturing the Aperiodic Transition in the Wake of a Flapping Airfoil 271
Dipanjan Majumdar, Chandan Bose, and Sunetra Sarkar

Numerical Investigation of Switching of a Jet Generated by a Foil Pitching in Still Fluid 283
Chayanit Nigaltia and Sachin Y. Shinde

Studying the Channel Confluence Hydraulics Using Eddy Viscosity Models and Reynolds Stress Model 295
Abhishek K. Pandey, Pranab K. Mohapatra, and Vikrant Jain

Computational Modelling of Turbulent Flows Using an Adaptive Mesh Finite Element Method: A Benchmarking Study 307
Naman Agarwal and Gaurav Bhutani

Finite Element Computational Modelling of Non-Newtonian Fluids Using Anisotropic Mesh Adaptivity 323
Neeraj Kr. Singh and Gaurav Bhutani

Computational Study of Shear Flow Past Square Cylinder with Horizontal Control Plate 339
Ashwani and Rajendra K. Ray

Numerical Study of Shear Flow Past an Inclined Square Cylinder with Vertical Control Plate 351
Rishabh Saxena and Rajendra K. Ray

Numerical Analysis of the Pressure, Temperature, and Aerodynamic Forces on Hypersonic Blunt Hemispherical Shaped Body 363
Krishnakumar V. Patel and Prem R. Patel

Interior Acoustic Analysis of Rectangular-Shaped Rigid Cavity with Opening 375
Subhankar Pramanik, Sreyashi Das, and Arup Guha Niyogi

Numerical Investigation of Effect of Laser on Natural Convection in Two-Dimensional Geometry 385
G. Chanakya and Pradeep Kumar

Recent Advances in Multiscale–Multiphysics Problems

Finite Deformation of a Dielectric Cylindrical Actuator: A Continuum Mechanics Approach 401
Deepak Kumar, Subrat Kumar Behera, and Somnath Sarangi

Numerical Studies for Generalized Modified Polarization Saturation (PS) Model in Piezoelectric Media 419
Kuldeep Sharma and Sandeep Singh

Thermal Conduction in One-Dimensional Φ^4 Chains with Colliding Particles 429
Sankhadeep Bhattacharyya and Puneet Kumar Patra

Sequence of Hydrodynamic Phenomena During the Interactions of Drop and Bubble in Vertical Conduit 443
Subhav Chauhan and Parmod Kumar

Investigating Fluid-Structure Interaction Behavior of a Chord-Wise Flexible Filament in the Wake of a Bluff Body Using Partitioned Strong Coupling Approach 453
Chandan Bose, Rajanya Chatterjee, Sayan Gupta, and Sunetra Sarkar

Optimal Design of Rotary Magneto-Rheological Drum Brake for Transfemoral Prosthesis 465
Radhe Shyam Tak Saini, Hemantha Kumar, Sujatha Chandramohan, and Sujatha Srinivasan

Recent Advances in Numerical Techniques

Single Variable New First-Order Shear Deformation Plate Theory: Numerical Solutions of Lévy-Type Plates Using Fourth-Order Runge-Kutta Technique 477
Himanshu Sawhney, Kedar S. Pakhare, Rameshchandra P. Shimpi, P. J. Guruprasad, and Yogesh M. Desai

Investigation of Some Recently Proposed Explicit Time Integration Schemes for Nonlinear Problems 487
Abhijeet Singh, Rishiraj K. Thakur, Vishal Agrawal, and Sachin S. Gautam

Performance Evaluation of Some Novel Composite Time Integration Schemes for Dynamic Problems 499
Jasti Mahesh Kumar, Vishal Agrawal, and Sachin Singh Gautam

Development of Efficient Stress Return Algorithms for Simulating Geomaterial Response 509
Divyanshu Kumar Lal and Arghya Das

Kernel Gradient Free Smoothed Particle Hydrodynamics for Transient Boundary Value Problems 519
K. C. Charan and Siva Prasad AVS

A New Edge-Based Meshless Scheme for High-Speed Inviscid and Viscous Flows 527
H. S. Krishna

A Numerical Framework for the Solution of Bivariate Population Balance Equation—Model Implementation and Verification 539
Deepak Kumar Singh and Gaurav Bhutani

Application of Computational Mechanics in Emerging Areas

Application of Novel Radial Thresholding Method for the Segmentation of Different Phases from X-Ray Microtomographic Images of Concrete 555
Vishwas Chhimpa, Subhasis Pradhan, and Sudhirkumar V. Barai

Design of a Nonlinear Energy Harvesting Dynamic Vibration Absorber 563
Soumi Bhattacharyya and Shaikh Faruque Ali

Harvesting Energy from a Series of Harvesters 573
Mohammad Reyaz Ahmad Vali and Shaikh Faruque Ali

Design of a Flow Control Device Using a Special Class of Hybrid Symmetric Bistable Laminates in Clamped Boundary Condition 587
Aghna Mukherjee, Akash Mundwaik, Shaikh Faruque Ali, and Arunachalkasi Arockiarajan

Dynamic Stability of Spent Fuel Trays Stack Submerged in Water Pool Incorporating Coupled Fluid Structure Interaction 597
Binu Kumar, R. S. Singh, O. P. Singh, G. R. Reddy, K. M. Singh, and N. Gopala Krishnan

Frequency Domain Based Robust Flutter Analysis of Swept Back Wing Using μ Method 609
A. Arun Kumar and Amit Kumar Onkar

Optimal Parameters Identification of Quarter Car Simulink Model for Better Ride Comfort and Road Holding 623
N. P. Puneet, Abhinandan Hegale, Hemantha Kumar, and K. V. Gangadharan

Parametric Study of Lithium-ion Battery Module for Electric Vehicle Application 637
R. Umesh, Basant Singh Sikarwar, Ayush Goyal, and Sachin Singh Gautam

An Efficient Implementation of Lanczos Method for Dynamic Analysis of Launch Vehicle Structures 649
P. Deepak, P. V. Anil Kumar, and R. Neetha

About the Editors

Sandip Kumar Saha is an Assistant Professor in the School of Engineering at the Indian Institute of Technology Mandi (IIT Mandi). He obtained his Ph.D. from IIT Delhi, India and postdoctoral work in University of Canterbury, New Zealand. His research interests include computational structural dynamics, performance based earthquake engineering, uncertainty modeling in dynamical systems, multi-hazard (earthquake, wind, fire, blast etc.) protection of structures.

Mousumi Mukherjee is an Assistant Professor in the School of Engineering at the Indian Institute of Technology Mandi (IIT Mandi), India. She received her Ph.D. in civil engineering from Indian Institute of Technology Kanpur, India and pursued post-doctoral work in the School of Civil, Environmental and Mining Engineering at the University of Adelaide, Australia. Her research interests include geomechanics, constitutive modeling of frictional material, soil instability analysis, modeling of debris flow, numerical modelling of large deformation problems pertinent to geotechnical engineering.

Recent Advances in Nanomechanics

Buckling of Pressurized Nano-Sized Spherical Shell

Manjur Alam and Sudib Kumar Mishra

1 Introduction

Since the invention of carbon nanotube and a variant of nanostructures, the interest in nanotechnology has been flourishing rapidly. This has led to arbitrary miniaturization of many devices and systems, which are widely acclaimed in a variety of critical infrastructural applications. Key to the success is the exceptional thermo-electro-mechanical properties of various nanostructures, that have attracted the attention of the researchers' community. Understanding the mechanical responses are important for economic and reliable design of nanostructures. This requires the evaluation of stress, strains and their functions. Because of their characteristic dimensions, the mechanics of nanostructures/nanodevices are significantly dominated by the molecular forces and their interactions, which needs to be adequately incorporated in the analysis. The Molecular dynamics (MD) simulations [1] are traditionally used for this. In instances, even the quantum–mechanical description of such interactions is also included [2]. However, the MD simulation becomes prohibitively exhaustive to probe length scales of engineering significance to prevent their usage in routine analysis/design of nanostructures. As an alternative, the theory of continuum solid mechanics has been extended by Eringen et al. [3, 4] by accommodating the long-range molecular interactions, referred as nonlocal theory. In these theories, the stress is not only related to the strains but their gradients as well, described by the Laplacian of the field variable.

Since the concept of nonlocal elasticity is introduced by Eringen et al. [3, 4], several aspects of molecular mechanics have been reported in literature using this theory. Eringen [5] et al. showed that the stress singularity at the crack tip for Griffith

M. Alam (✉) · S. Kumar Mishra
Indian Institute of Technology, Kanpur, UP 208016, India
e-mail: manjur@iitk.ac.in

S. K. Saha and M. Mukherjee (eds.), *Recent Advances in Computational Mechanics and Simulations*, Lecture Notes in Mechanical Engineering,
https://doi.org/10.1007/978-981-15-8315-5_1

crack is eliminated in nonlocal media. The solution of the screw dislocation using nonlocal theory showed good agreement with the experimental observations [6]. The concern of these investigations is material instability. It is important to note that, with increasingly smaller dimensions of nanostructures, their strength gets higher [7, 8]. Whereas improved strength permits slender nanostructures, this also increases their vulnerability to geometric instability, such as buckling. Therefore, geometric instability of nanostructures attracts considerable attention in recent times.

Peddieson et al. [9] provided a nonlocal formulation of the Euler–Bernoulli beam theory for its potential application in nanotechnology. Zhang et al. [10] studied buckling of an axially compressed cylindrical shell based on the nonlocal continuum theory. Wang et al. [11] indicated a resemblance between the effect of transverse shear deformation and small length scale in reducing the buckling load. Reddy [12] presented beam theories based on Eringen's differential constitutive model to show that the nonlocal effect decreases critical loads. Pradhan et al. [13] studied the effect of nonlocality on the buckling analysis of bi-axially compressed single-layered Graphene using nonlocal theory. The effect of shear deformation is also illustrated [14]. Hashemi et al. [15] investigated buckling of bi-axially loaded nonlocal Mindlin plates. Zhang et al. [16] calibrated the parameters of nonlocal model from the critical load.

It is apparent from the review that the nonlocal elasticity theory has been used most extensively in order to study the buckling and post-buckling behavior of different nanostructures, such as beams and plates. Shen and co-investigators [17–19] presented the size dependency of buckling and post-buckling of shear deformable shell model under axial, radial and torsional loading. The buckling instability of cylindrical nonlocal shell (such as CNT) has also been recently studied by Ghavanloo et al. [20]. However, otherwise, research works on buckling of spherical shell is scanty and needs further exploration. In fact, several systems at nanoscale can be modelled as thin-walled spherical shells. The most common example is that of Fullerene, which has been approximated as spherical shell by the previous investigators [21, 22]. The viral capsid is also modelled as spherical shell by Zhang and Ru [23] and Ru [24]. However, these studies do not take the nonlocal interactions into account. Although due to their characteristic lengths, nonlocal interactions are expected to be significant. Therefore, it is important to study the buckling instability of these structures under external pressures, which remains unexplored. An analysis of nonlocal spherical shell for buckling will be a significant add on to the literature. This article derives the basic equations governing the buckling of pressurized nano spherical shell. The shallow shell segment assumptions are made use of, in order to significantly reduce the complexity of the problem. The equations are formulated using nonlocal differential constitutive equations in conjunction with the shallow segment shell theory. An expression for the critical pressure for buckling is derived in close form on solving these equations. The effect of parametric variations of important parameters is also demonstrated.

2 Nonlocal Model for a Spherical Shell

2.1 *Nonlocal Elasticity*

According to Eringen's nonlocal elastic field theory [3–6], the stress at a reference point x is a functional of a strain field at every point $x^{'}$ in the body. This observation accords the atomic theory of lattice dynamics and experimental observations on phonon dispersion, details of which may be obtained elsewhere [4, 25, 26]. In a limiting case, as the effect of strains at points other than x is neglected, the classical local theory of elasticity is reclaimed. For homogeneous, isotropic elastic solid, the stress–strain relationship is expressed as

$$\sigma_{ij}(x) = \int_{v} \alpha\left(\left|x - x'\right|, \tau\right)\sigma_{ij}^{cl}\left(x'\right)dv\left(x'\right) \tag{1}$$

In which $\sigma_{ij}(x)$ is the nonlocal stress tensor and $\sigma_{ij}^{cl}(x^{'})$ is the classical stress tensor at $x^{'}$ which is related to the linear strain tensor $\varepsilon_{ij}(x^{'})$ at any point $x^{'}$ in the body. The kernel function $\alpha\left(\left|x - x^{'}\right|, \tau\right)$ is the nonlocal modulus with an argument $\left|x - x^{'}\right|$ as the distance, and $\tau = \left(\frac{e_0 a}{l}\right)$ is a constant with $\left(\frac{a}{l}\right)$ as the characteristic length ratio of the nanostructure and e_0 is a material constant, which is determined by matching the atomic dispersion curve of lattice dynamics with experiments [25]. In these expressions, a is an internal characteristic length (e.g. lattice parameter, intergranular distance) and l is the external characteristic length (e.g. wavelength, crack length or width of localization band). The nonlocal modulus acquires maximum value at $x = x'$ and attenuates with $\left|x - x^{'}\right|$. It is observed from the constitutive equation that in evaluating the stress at a point, the kernel function takes the contribution of strains from the entire region, over which the integration is performed. In general, this results into a set of integro-partial differential equations which are quite involved. In order to overcome this difficulty, Eringen [6] proposed a simpler set of constitutive equations in differential form as

$$\left(1 - e_0^2 a^2 \nabla^2\right)\sigma_{ij} = C_{ijkl}\varepsilon_{kl} \tag{2}$$

In which, σ_{ij} and ε_{ij} are the nonlocal stress and strain tensor, C_{ijkl} are the elastic constants and ∇ is the Laplacian operator. Equation (2) is the equation that differentiates the nonlocal elasticity theory from the classical (local) theory.

2.2 *Nonlocal Shallow Shell Equations*

The characteristics wavelength in a sphere due to buckling are small in comparison to the radius of the shell. With this as the key assumption, the nonlinear equations of shallow shell are used here for the analysis. This simplification was extensively employed by Hutchinson [27] and Reissner [28, 29] in the analysis of spherical shells. A segment will be called shallow if the ratio of its height to base diameter becomes less than (1/8), however, the results obtained based on such theory are also applicable to shells which are not necessarily shallow but the stresses are effectively restricted to shallow zones [28]. A complete sphere is shown in Fig. 1, out of which, a shallow segment (S_0) is taken out. The analysis is performed by considering this shallow section, where the buckled pattern is duplicated many times, thereby adequately representing the buckled scenario. The Cartesian coordinate system is adopted, in which the x and y-axis are chosen to be on the base of the section (S_0) and the z-axis align with the direction normal to the base, as illustrated in Fig. 1 a and 1.b. The sphere has a radius (R) and the thickness (h) as shown in Fig. 1a.

The displacements along x and y are denoted as u and v and the normal displacement along z is denoted as w. Subsequently, the kinematic relation for the shell can be written as

$$\varepsilon_x = \frac{\partial u}{\partial x} - \frac{w}{R} + \frac{1}{2}\left(\frac{\partial w}{\partial x}\right)^2; \ \varepsilon_y = \frac{\partial v}{\partial y} - \frac{w}{R} + \frac{1}{2}\left(\frac{\partial w}{\partial y}\right)^2;$$
$$\gamma_{xy} = \frac{\partial u}{\partial y} + \frac{\partial v}{\partial x} + \left(\frac{\partial w}{\partial x}\right)\left(\frac{\partial w}{\partial y}\right) \tag{3}$$

$$\chi_x = -\frac{\partial^2 w}{\partial x^2}; \ \chi_y = -\frac{\partial^2 w}{\partial y^2}; \ \chi_{xy} = -\frac{\partial^2 w}{\partial x \partial y} \tag{4}$$

The expressions in Eqs. (3–4) are similar to that used in plate bending theory, except the ones with curvature terms. The von-Karmann nonlinear terms are also included. This can be justified by the results from buckling of cylindrical shell, in which, the shell gets subdivided into independent shallow panels in each buckling modes. The shell is considered to be thin, for which the normal stress (σ_z), the corresponding normal strain (ε_z) and the shear stresses $(\sigma_{xz}, \sigma_{yz})$ and strains $(\gamma_{xz}, \gamma_{yz})$ are considered to be negligibly small.

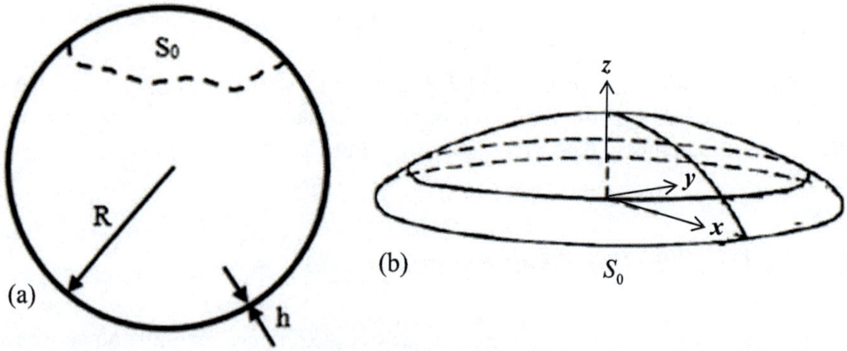

Fig. 1 **a** A spherical shell with a Shallow section, **b** the Shallow section illustrated in detail

The resultant direct and shear stresses are denoted as (N_x, N_y, N_{xy}) in the respective directions, designated by the subscripts. Further, M_x, M_y and M_{xy} denotes the resultant moments in the respective directions. The equations for the forces are moments can be obtained by integrating Eq. (2), which provides the following equations.

$$N_x - e_0^2a^2\left(\frac{\partial^2 N_x}{\partial x^2} + \frac{\partial^2 N_x}{\partial y^2}\right) = C(\varepsilon_x + \nu\varepsilon_y) \tag{5.1}$$

$$N_y - e_0^2a^2\left(\frac{\partial^2 N_y}{\partial x^2} + \frac{\partial^2 N_y}{\partial y^2}\right) = C(\varepsilon_y + \nu\varepsilon_x) \tag{5.2}$$

$$N_{xy} - e_0^2a^2\left(\frac{\partial^2 N_{xy}}{\partial x^2} + \frac{\partial^2 N_{xy}}{\partial y^2}\right) = \frac{(1-\nu)}{2}C\gamma_{xy} \tag{5.3}$$

The moment–curvature relations can be expressed as

$$M_x - e_0^2a^2\left(\frac{\partial^2 M_x}{\partial x^2} + \frac{\partial^2 M_x}{\partial y^2}\right) = -D\left(\frac{\partial^2 w}{\partial x^2} + \nu\frac{\partial^2 w}{\partial y^2}\right) \tag{6.1}$$

$$M_y - e_0^2a^2\left(\frac{\partial^2 M_y}{\partial x^2} + \frac{\partial^2 M_y}{\partial y^2}\right) = -D\left(\frac{\partial^2 w}{\partial y^2} + \nu\frac{\partial^2 w}{\partial x^2}\right) \tag{6.2}$$

$$M_{xy} - e_0^2a^2\left(\frac{\partial^2 M_{xy}}{\partial x^2} + \frac{\partial^2 M_{xy}}{\partial y^2}\right) = -D(1-\nu)\frac{\partial^2 w}{\partial x\partial y} \tag{6.3}$$

where the constants C and D are expressed as

$$C = \frac{Eh}{(1-\nu^2)};\ D = \frac{Eh^3}{12(1-\nu^2)}; \tag{7}$$

in which the symbols E and ν are respectively the elastic modulus and poisons ratio for the material considered. The system of differential equations for equilibrium of nonlocal shallow shell takes the form

$$\frac{\partial N_x}{\partial x} + \frac{\partial N_{xy}}{\partial y} = 0 \tag{8.1}$$

$$\frac{\partial N_{xy}}{\partial x} + \frac{\partial N_y}{\partial y} = 0 \tag{8.2}$$

$$\frac{\partial Q_x}{\partial x} + \frac{\partial Q_y}{\partial y} + N_x\left(\frac{1}{R} + \frac{\partial^2 w}{\partial x^2}\right) + N_y\left(\frac{1}{R} + \frac{\partial^2 w}{\partial y^2}\right) + 2N_{xy}\left(\frac{\partial^2 w}{\partial y\partial x}\right) + p = 0 \tag{8.3}$$

$$Q_x - \frac{\partial M_x}{\partial x} - \frac{\partial M_{xy}}{\partial y} = 0 \tag{9.1}$$

$$Q_y - \frac{\partial M_{xy}}{\partial x} - \frac{\partial M_y}{\partial y} = 0 \tag{9.2}$$

These equations of equilibrium can be reduced further into two equations, one equation of equilibrium and one of compatibility. It may be noted that the theory of plate bending is reclaimed by the present formulation, in which, the radius of curvature $R = \infty$. The stress functions are defined as

$$N_x = \frac{\partial^2 f}{\partial y^2}; \quad N_y = \frac{\partial^2 f}{\partial x^2}; \quad N_{xy} = -\frac{\partial^2 f}{\partial y \partial x}; \tag{10}$$

A differential equation for the stress function f can be obtained by using the compatibility of deformation using the set of Eq. (3) as

$$\begin{aligned}
&\frac{\partial^2 N_x}{\partial y^2} - e_0^2 a^2\left(\frac{\partial^4 N_x}{\partial y^2 \partial x^2} + \frac{\partial^4 N_x}{\partial y^4}\right) - \nu\left[\frac{\partial^2 N_y}{\partial y^2} - e_0^2 a^2\left(\frac{\partial^4 N_y}{\partial y^2 \partial x^2} + \frac{\partial^4 N_y}{\partial y^4}\right)\right] + \frac{\partial^2 N_y}{\partial x^2} \\
&\quad - e_0^2 a^2 \frac{\partial^4 N_y}{\partial x^4} - e_0^2 a^2 \frac{\partial^4 N_y}{\partial y^2 \partial x^2} - \nu\left[\frac{\partial^2 N_x}{\partial x^2} - e_0^2 a^2\left(\frac{\partial^4 N_x}{\partial x^4} + \frac{\partial^4 N_x}{\partial y^2 \partial x^2}\right)\right] \\
&\quad - (2 + 2\nu)\left[\frac{\partial^2 N_{xy}}{\partial y \partial x} - e_0^2 a^2\left(\frac{\partial^4 N_{xy}}{\partial y \partial x^3} + \frac{\partial^4 N_{xy}}{\partial y^3 \partial x}\right)\right] \\
&\quad = Eh\left[\left(\frac{\partial^2 w}{\partial y \partial x}\right)^2 - \left(\frac{\partial^2 w}{\partial x^2}\right)\left(\frac{\partial^2 w}{\partial y^2}\right)\right] - \frac{Eh}{R}\left[\left(\frac{\partial^2 w}{\partial x^2}\right) + \left(\frac{\partial^2 w}{\partial y^2}\right)\right]
\end{aligned} \tag{11}$$

Combining Eqs. (10) and (11), one obtains

$$\nabla^4 f - e_0^2 a^2 \nabla^6 f = \frac{Eh}{R}\left[R\left(\frac{\partial^2 w}{\partial y \partial x}\right)^2 - R\left(\frac{\partial^2 w}{\partial x^2}\right)\left(\frac{\partial^2 w}{\partial y^2}\right) - \left(\frac{\partial^2 w}{\partial x^2}\right) - \left(\frac{\partial^2 w}{\partial y^2}\right)\right] \tag{12}$$

Equation (12) is the first equation for the buckling problem of nonlocal shallow spherical shell. The other equation is obtained from the equilibrium. Substituting the moment equilibrium Eqs. (9.1–9.2) into the force equilibrium (8.1, 8.2, 8.3), one attains

$$\frac{\partial^2 M_x}{\partial x^2} + 2\frac{\partial^2 M_{xy}}{\partial x \partial y} + \frac{\partial^2 M_y}{\partial y^2} + N_x\left(\frac{1}{R} + \frac{\partial^2 w}{\partial x^2}\right) + N_y\left(\frac{1}{R} + \frac{\partial^2 w}{\partial y^2}\right) + 2N_{xy}\frac{\partial^2 w}{\partial x \partial y} + p = 0 \tag{13}$$

On substituting the moment–curvature relation from Eqs. (6.1, 6.2, 6.3) into the Eq. (13), it appears

$$\begin{aligned}
&-D\,\nabla^2 w + N_x\left(\frac{1}{R} + \frac{\partial^2 w}{\partial x^2}\right) + N_y\left(\frac{1}{R} + \frac{\partial^2 w}{\partial y^2}\right) - 2N_{xy}\left(\frac{\partial^2 w}{\partial y \partial x}\right) \\
&\quad + p - e_0^2 a^2\left[\left(\frac{\partial^2 N_x}{\partial x^2}\right)\left(\frac{1}{R} + \frac{\partial^2 w}{\partial x^2}\right) + 2\frac{\partial N_x}{\partial x}\left(\frac{\partial^3 w}{\partial x^3}\right) + N_x\left(\frac{\partial^4 w}{\partial x^4}\right)\right. \\
&\quad \left. + \left(\frac{\partial^2 N_x}{\partial y^2}\right)\left(\frac{1}{R_1} + \frac{\partial^2 w}{\partial x^2}\right) + 2\frac{\partial N_x}{\partial y}\left(\frac{\partial^3 w}{\partial y \partial x^2}\right) + N_x\left(\frac{\partial^4 w}{\partial y^2 \partial x^2}\right)\right] \\
&\quad - e_0^2 a^2\left[\left(\frac{\partial^2 N_y}{\partial x^2}\right)\left(\frac{1}{R} + \frac{\partial^2 w}{\partial y^2}\right) + 2\frac{\partial N_y}{\partial x}\left(\frac{\partial^3 w}{\partial y^2 \partial x}\right) + N_y\left(\frac{\partial^4 w}{\partial y^2 \partial x^2}\right)\right.
\end{aligned}$$

$$+\left(\frac{\partial^2 N_y}{\partial y^2}\right)\left(\frac{1}{R}+\frac{\partial^2 w}{\partial y^2}\right)+2\left(\frac{\partial N_y}{\partial y}\right)\left(\frac{\partial^3 w}{\partial y^3}\right)+N_y\left(\frac{\partial^4 w}{\partial y^4}\right)\Bigg]$$
$$-2e_0^2a^2\Bigg[-\left(\frac{\partial^2 N_{xy}}{\partial x^2}\right)\left(\frac{\partial^2 w}{\partial y\partial x}\right)+2\left(\frac{\partial N_{xy}}{\partial x}\right)\left(\frac{\partial^3 w}{\partial y\partial x^2}\right)+N_{xy}\left(\frac{\partial^4 w}{\partial y\partial x^3}\right)$$
$$+\left(\frac{\partial^2 N_{xy}}{\partial y^2}\right)\left(\frac{\partial^2 w}{\partial y\partial x}\right)+2\left(\frac{\partial N_{xy}}{\partial y^2}\right)\left(\frac{\partial^2 w}{\partial y\partial x}\right)+2\left(\frac{\partial N_{xy}}{\partial y}\right)\left(\frac{\partial^3 w}{\partial y^2\partial x}\right)$$
$$+N_{xy}\left(\frac{\partial^4 w}{\partial y^3\partial x}\right)\Bigg]=0 \tag{14}$$

Introducing the stress functions (10) into Eq. (14), results in the second equation governing the buckling

$$-D\,\nabla^2 w+\left(\frac{\partial^2 f}{\partial y^2}\right)\left(\frac{1}{R}+\frac{\partial^2 w}{\partial x^2}\right)+\left(\frac{\partial^2 f}{\partial x^2}\right)\left(\frac{1}{R}+\frac{\partial^2 w}{\partial y^2}\right)-2\left(\frac{\partial^2 f}{\partial y\partial x}\right)\left(\frac{\partial^2 w}{\partial y\partial x}\right)$$
$$+p-e_0^2a^2\Bigg[\left(\frac{\partial^4 f}{\partial y^2\partial x^2}\right)\left(\frac{1}{R}+\frac{\partial^2 w}{\partial x^2}\right)+2\left(\frac{\partial^3 f}{\partial y^2\partial x}\right)\left(\frac{\partial^3 w}{\partial x^3}\right)+\left(\frac{\partial^2 f}{\partial y^2}\right)\left(\frac{\partial^4 w}{\partial x^4}\right)$$
$$+\left(\frac{\partial^4 f}{\partial y^4}\right)\left(\frac{1}{R}+\frac{\partial^2 w}{\partial x^2}\right)+2\left(\frac{\partial^3 f}{\partial y^3}\right)\left(\frac{\partial^3 w}{\partial y\partial x^2}\right)+\left(\frac{\partial^2 f}{\partial y^2}\right)\left(\frac{\partial^4 w}{\partial y^2\partial x^2}\right)\Bigg]$$
$$-e_0^2a^2\Bigg[\left(\frac{\partial^4 f}{\partial x^4}\right)\left(\frac{1}{R}+\frac{\partial^2 w}{\partial y^2}\right)+2\left(\frac{\partial^3 f}{\partial x^3}\right)\left(\frac{\partial^3 w}{\partial y^2\partial x}\right)+\left(\frac{\partial^2 f}{\partial x^2}\right)\left(\frac{\partial^4 w}{\partial y^2\partial x^2}\right)$$
$$+\left(\frac{\partial^4 f}{\partial y^2\partial x^2}\right)\left(\frac{1}{R}+\frac{\partial^2 w}{\partial y^2}\right)+2\left(\frac{\partial^3 f}{\partial y\partial x^2}\right)\left(\frac{\partial^3 w}{\partial y^3}\right)+\left(\frac{\partial^2 f}{\partial x^2}\right)\left(\frac{\partial^4 w}{\partial y^4}\right)\Bigg]$$
$$-2e_0^2a^2\Bigg[-\left(\frac{\partial^4 f}{\partial y\partial x^3}\right)\left(\frac{\partial^2 w}{\partial y\partial x}\right)-2\left(\frac{\partial^3 f}{\partial y\partial x^2}\right)\left(\frac{\partial^3 w}{\partial y\partial x^2}\right)-\left(\frac{\partial^2 f}{\partial y\partial x}\right)\left(\frac{\partial^4 w}{\partial y\partial x^3}\right)$$
$$-\left(\frac{\partial^4 f}{\partial y^3\partial x}\right)\left(\frac{\partial^2 w}{\partial y\partial x}\right)-2\left(\frac{\partial^3 f}{\partial y^2\partial x}\right)\left(\frac{\partial^3 w}{\partial y^2\partial x}\right)-\left(\frac{\partial^2 f}{\partial y\partial x}\right)\left(\frac{\partial^4 w}{\partial y^3\partial x}\right)\Bigg]=0 \tag{15}$$

It can be observed from both the Eqs. (11) and (15) equations that, if the nonlocal parameter e_0a is made zero, the equations for the classical shell can be regained.

3 Solution of the Equation Governing Nonlocal Buckling

The equations derived for a shallow spherical shell segment is solved herein. The spherical shell is subjected to a constant external pressure p and the state of stress prior to buckling is approximated to be the membrane state of stress, given as.

$N_x^0 = N_y^0 = -\frac{pR}{2}$ and $N_{xy}^0 = 0$(16).

The inward displacement, prior to buckling is given as

$$w^0 = -\frac{(1-\nu)pR^2}{2Eh} \tag{17}$$

In the light of pre-buckled state of stress and deformation, the stress function and the displacement may be written as

$$f = -\left(\frac{pR}{4}\right)(x^2 + y^2) + F \tag{18.1}$$

$$w = -\frac{(1-\nu)pR^2}{2Eh} + W \tag{18.2}$$

In these expressions F and w are zero prior to buckling. In order to find out the critical buckling pressure, at which, the bifurcation from the pre-buckling state occurs, a linear buckling analysis is needed. The equations for linear buckling analysis can be obtained by substituting Eqs. 18.1 and 18.2 into Eqs. 12 and 15. Subsequent linearizing with respect to F and W furnishes.

$$-D\nabla^4 W + \frac{1}{R}\nabla^2 F - \frac{pR}{2}\nabla^2 W - e_0^2 a^2\left(\frac{1}{R}\nabla^4 F - \frac{pR}{2}\nabla^4 W\right) = 0 \tag{19}$$

$$\nabla^4 F - e_0^2 a^2 \nabla^6 F + \frac{Eh}{R}\nabla^2 W = 0 \tag{20}$$

The problem, therefore, is reduced to a coupled set of homogeneous eigenvalue problem in F and W. Only for specific values of p, the equations will have non-trivial solution. A periodic solution can be taken of the following form

$$W = cos\left(\tfrac{\kappa_x x}{R}\right)cos\left(\tfrac{\kappa_y y}{R}\right) \tag{21.1}$$

$$F = Bcos\left(\frac{\kappa_x x}{R}\right)cos\left(\frac{\kappa_y y}{R}\right) \tag{21.2}$$

where κ_x and κ_y are the wavenumbers in x and y directions respectively. The buckled modes will have characteristic wavelengths (λ_x, λ_y), which are assumed to be much smaller than the characteristic dimension of the shallow shell segment (S_0). Substituting 21.1 and 21.2 into the Eqs. 19 and 20, the critical load for buckling is found to be

$$p = \frac{\left(\frac{2Eh}{R}\right)}{\left\{1 + \left(\frac{e_0 a}{R}\right)^2\left(\kappa_x^2 + \kappa_y^2\right)\right\}}\left[\frac{1}{\left(\kappa_x^2 + \kappa_y^2\right)} + \frac{\left(\kappa_x^2 + \kappa_y^2\right)}{q_0^4}\right] \tag{22}$$

The respective value of the coefficient (B) is obtained as

$$B = \frac{EhR}{\left\{1 + \left(\frac{e_0 a}{R}\right)^2\left(\kappa_x^2 + \kappa_y^2\right)\right\}\left(\kappa_x^2 + \kappa_y^2\right)} \tag{23}$$

in which, the parameter (q_0) is given as

$$q_0^4 = 12\left(1 - \nu^2\right)\left(\frac{R}{h}\right)^2 \tag{24}$$

Minimizing p with respect to κ_x and κ_y provides the critical load for a nonlocal shallow spherical shell

$$p_{cr} = \frac{4EhR}{\left(e_0^2 a^2 q_0^2 + \sqrt{e_0^4 a^4 q_0^4 + R^4}\right) q_0^2} \tag{25}$$

which is associated with any combination of wavenumber κ_x and κ_y satisfying

$$\kappa_x^2 + \kappa_y^2 = \frac{\left(e_0^2 a^2 q_0^2 + \sqrt{e_0^4 a^4 q_0^4 + R^4}\right) q_0^2}{R^2} \tag{26}$$

Equation (25) gives the expression for critical buckling load of the nano-spherical shell when subjecting to uniform external pressure. It is clear from the expressions (22–26) that for vanishing nonlocal scale parameter$a$, the results reduce to classical solution. However, the solution obtained here is based on a shallow shell representation, the validity of this assumption is supported by large wavenumbers κ_x and κ_y, or small wavelengths (λ_x, λ_y), in comparison to unity i.e. the buckling wavelengths should be small compared with the radius (R) of the sphere. It is apparent from Eq. (26) that κ_x and κ_y is of order of q_0 and hence adequately large. If either of the wavenumber κ_x or κ_y becomes zero, the shallow shell representation of a complete sphere is also valid. This case is analogous to the condition of buckling of cylinders under axial pressure, where the shallow shell equations accurately represent the axially symmetric mode.

4 Numerical Illustration and Discussion

It is obvious from the derivation that the nonlocal scale effect reduces the buckling load, which is non-conservative, if not adequately taken care of while designing. In order to illustrate the effect of nonlocality on the critical buckling load, a load parameter λ is defined as the ratio of classical (local) buckling load to the nonlocal buckling load of the spherical shell, following the literature [10]. A spherical C_{60} fullerene structure of thickness $h = 0.0665$ nm is considered with a radius $R = 0.355$ nm, Poisson's ratio (ν) 0.2 and an average bond length $a = 0.14$ nm. Following Eringen [6], a value of $e_0 = 0.39$ for the nonlocal material constant is adopted for illustration.

It is well established in the literature that the scale parameter ($e_0 a$) affects different modes of buckling differently, with a more pronounced effect on the higher modes.

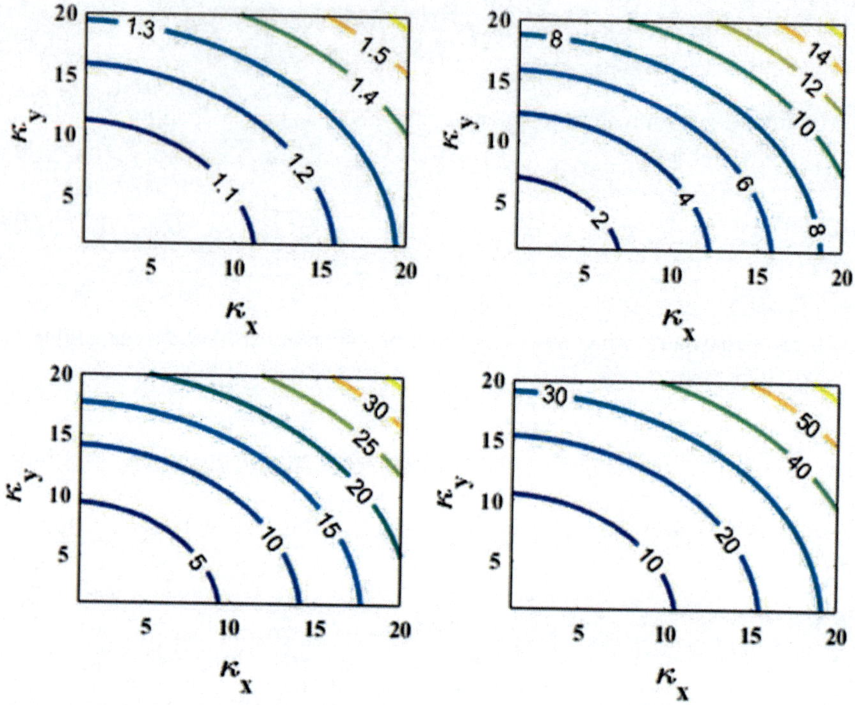

Fig. 2 The contour of the load parameter (λ) for varying combinations of wavenumbers, with the nonlocal parameter (e_0a) as **a** zero **b** 0.05 **c** 0.075 and **d** 0.10

Contour plots of this load parameter (λ) with varying combinations of the wavenumbers (κ_x, κ_y) are shown in Fig. 2a–d, for different values of the nonlocal length scale parameter. It is clearly observed from these figures (Fig. 2a–d) that for a particular value of the nonlocal parameter, the load parameter λ increases substantially for increasingly higher modes. Further, the rate of increase becomes more pronounced for higher value of the nonlocal length scale. The increase in load parameter λ essentially indicates the respective fall in the critical loads for increased degree of nonlocality, dictated by the length scale parameter.

It is also observed that the combination of wave numbers needs to satisfy a critical relationship (Eq. 26) for being qualified for a legitimate buckling mode. The relationship is observed to be affected by the nonlocal length scale parameter. By setting this nonlocal parameter to be zero, Eq. (26) reclaims the one for classical ("local") shell. Therefore, the effect of length scale parameter on the wavenumber combination is shown in Fig. 3. This relation basically forms a circle, as also shown in this figure. It is seen that with increasing length scale parameter (and thus increasing nonlocality), the radius of the circle increases. This implies that the number of wavenumbers becomes higher for increasing nonlocality. This also means that the respective wavelengths become smaller, which further justifies the representation of nonlocal spherical shell with a mere shallow segment.

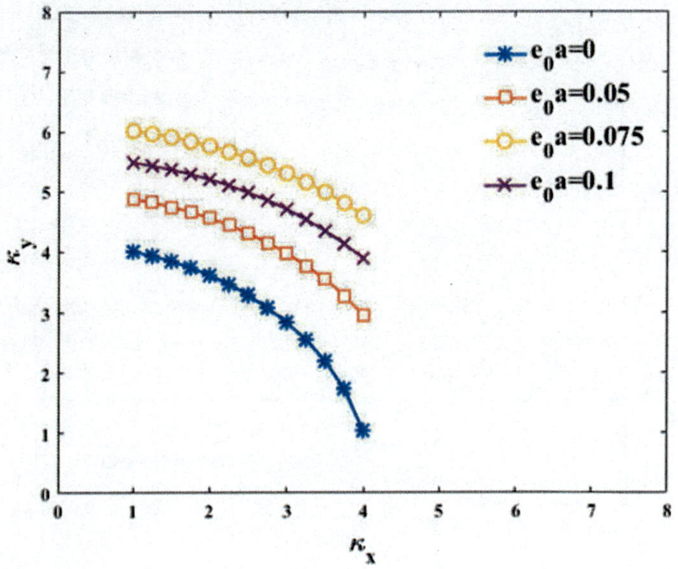

Fig. 3 Wavenumber combinations for minimum buckling load with different length scale parameters

Another parameter, having a profound influence on the buckling of shell is the ratio of the shell thickness to its radius $\left(\frac{h}{R}\right)$. With this in view, the parametric variation of the decrease in the critical load with the nonlocal length scale parameter is also studied with respect to the varying $\left(\frac{h}{R}\right)$ ratio. This is presented in Fig. 4 . The variation of the load parameter λ seems to be significantly get affected by the $\left(\frac{h}{R}\right)$ ratio. It is important to note that for higher values of $\left(\frac{h}{R}\right)$ ratio, the value of load parameter λ reduces, i.e. the nonlocality induced reduction of critical load, becomes less prominent in thicker shells. Thus, the effect of nonlocality counteracts the effect of increasing $\left(\frac{h}{R}\right)$ ratio, i.e. the thickness. On the other hand, the effect of nonlocality on reducing the critical load magnitude gets amplified for increasingly thinner shell.

5 Conclusion

Buckling of nano-sized spherical shell subjected to external pressure is presented herein based on the nonlocal theory. A simplified shallow shell model of the spherical shell is considered to significantly simplify the analysis. The governing equations and their linearized version are developed. On solving, a closed-form expression for the critical load is derived. The expression for the critical pressure involves the nonlocal length scale as a parameter. The salient effects are demonstrated using a numerical example. It is found that the nonlocality effect reduces the critical buckling load. The influence of nonlocality gets amplified for higher modes of buckling

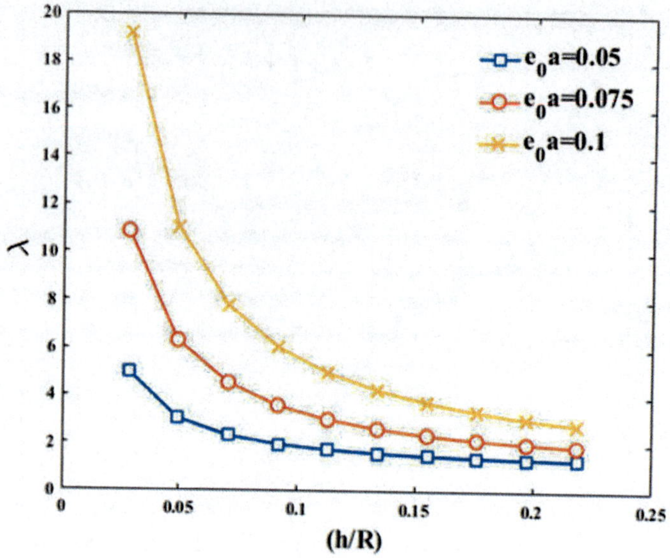

Fig. 4 Critical load parameter for varying (h/R) ratio, shown for different length scale parameters

and increasingly thinner shell. The critical relationship that must be satisfied by the wavenumbers of a buckling mode, is also shown to be modified by nonlocality. With increasing nonlocality, an increased number of wavenumbers must be accommodated in nonlocal shallow shell segment, which also results in increasingly smaller wavelengths. This fact provides additional support to the simplifying assumptions of shallow shell segment in order to study the buckling of nonlocal spherical shell.

References

1. Rapaport, D.C.: The Art of Molecular Dynamics Simulation. Cambridge University Press, Cambridge, United Kingdom (2004)
2. Kresse, G., Hafner, J.: Ab initio molecular dynamics simulation of the liquid-metal-amorphous-semiconductor transition in germanium.Phys. Rev. B **49**, 14251 (1994)
3. Eringen, A.C., Edelen, D.G.B.: On nonlocal elasticity. Int. J. Eng. Sci. **10**(3), 233–248 (1972)
4. Eringen, A., Wegner, J.L.: J. Nonlocal Contin. Field Theor. (2003)
5. Eringen, A.C., Speziale, C.Z., Kim, B.S.: Crack-tip problem in non-local elasticity. J. Mech. Phys. Solids **25**, 339–355 (1977)
6. Eringen, A.C.: On differential equations of nonlocal elasticity and solutions of screw dislocation and surface waves. J. Appl. Phys. **54**, 4703–4710 (1983)
7. Huang, Y., Zhang, F., Hwang, K.C., Nix, W.D., Pharr, G.M., Feng, G.: A model of size effects in nano-indentation. J. Mech. Phys. Solids **54**(8), 1668–1686 (2006)
8. Wang, W., Zhong, Y., Lu, K., Lu, L., McDowell, D.L., Zhu, T.: Size effects and strength fluctuations in nanoscale plasticity. Acta Mater. **60**(8), 3302–3309 (2012)
9. Peddieson, J., Buchanan, G.R., McNitt, R.P.: Application of nonlocal continuum models to nanotechnology. Int. J. Eng. Sci. **41**, 305 (2003)

10. Zhang, Y.Q., Liu, G.R., Wang, J.S.: Small-scale effects on buckling of multi-walled carbon nanotubes under axial compression. Phys. Rev. B **70**, 205430 (2004)
11. Wang, C.M., Zhang, Y.Y., Ramesh, S.S., Kitipornchai, S.: Buckling analysis of micro- and nano-rods/tubes based on nonlocal Timoshenko beam theory. J. Phys D. Appl. Phys. **39**, 3904–3909 (2006)
12. Reddy, J.N.: Nonlocal theories for bending, buckling and vibration of beams. Int. J. Eng. Sci. **45**, 288–307 (2007)
13. Pradhan, S.C., Murmu, T.: Small scale effect on the buckling of single-layered graphene sheets under biaxial compression via nonlocal continuum mechanics. Comput. Mater. Sci. **47**, 268–274 (2009)
14. Pradhan, S.C.: Buckling of single-layer graphene sheet based on nonlocal elasticity and higher-order shear deformation theory, Phys. Lett. Sect. a Gen. At. Solid State Phys. **373**, 4182–4188 (2009)
15. Hashemi, S.H., Samaei, A.T.: Buckling analysis of micro/nanoscale plates via nonlocal elasticity theory. Phys. E Low-Dimens. Syst. Nanostructures **43**, 1400–1404 (2011)
16. Zhang, Z., Wang, C.M., Challamel, N.: Eringen's length scale coefficient for buckling of nonlocal rectangular plates from microstructured beam-grid model. Int. J. Solids Struct. **51**(25), 4307–4315 (2014)
17. Shen, H.: Nonlocal shear deformable shell model for post-buckling of axially compressed microtubules embedded in an elastic medium. Biomech. Model. Mechanobiol. **9**, 345–357 (2010)
18. Shen, H., Zhang, C.: Nonlocal shear deformable shell model for post-buckling of axially compressed double-walled carbon nanotubes embedded in an elastic matrix. J. Appl. Mech. **77**, 1–12 (2010)
19. Shen, H.: Nonlocal shear deformable shell model for torsional buckling and post-buckling of microtubules in thermal environments. Mech. Res. Commun. **54**, 83–95 (2013)
20. Ghavanloo, E., Fazelzadeh, S.A., Sohrabpour, S.: Buckling analysis of nonlocal anisotropic thin-walled cylindrical shells subjected to combined loading. Journal of Engineering Mechanics **142**(12), 06016009 (2016)
21. Ahmad, S.: Continuum elastic model of fullerenes and the sphericity of the carbon onion shells. J. Chem. Phys. **116**, 3396–3400 (2002)
22. Omidi, M.J., ShayanMehr, M., Rafiee, R.: A study on equivalent spherical structure of Buckyball-C_{60} based on continuum shell model. Lat. Am. J. Solids Struct. **13**, 1016–1029 (2016)
23. Zhang, L., Ru, C.Q.: Imperfection sensitivity of pressured buckling of biopolymer spherical shells. Phys. Rev. E **93**, 062403 (2016)
24. Ru, C.Q.: Buckling of empty spherical viruses under external pressure. J. Appl. Phys. **105**, 124701 (2009)
25. Eringen, A.C.: Linear theory of nonlocal elasticity and dispersion of plane waves. Int. J. Eng. Sci. **10**(5), 425–435 (1972)
26. Zhang, P., Jiang, H., Huang, Y., Geubelle, P.H., Hwang, K.C.: An atomistic based continuum theory for carbon nanotubes, analysis of fracture nucleation. J. Mech. Phys. Solids **52**, 977–998 (2004)
27. Hutchinson, J.W.: Imperfection sensitivity of externally pressurized spherical shells. J. Appl. Mech., ASME 49–55 (1967)
28. Reissner, E.: Stresses and small displacements of shallow spherical shells. I, Report, Massachusetts Institute of Technology, USA, pp. 80–85
29. Reissner, E.: Stresses and small displacements of shallow spherical shells, II, Report, Massachusetts Institute of Technology, USA, pp. 279–300

On Flexure of Shear Deformable Isotropic Rectangular Nanobeams

Kedar S. Pakhare, P. J. Guruprasad, and Rameshchandra P. Shimpi

1 Introduction

Size-dependent effects which are insignificant in case of macroscopic structures become significant for small-scale structures such as nanobeams and nanoplates. Classical continuum theories do not take into account such size-dependent effects (Eltaher et al. [1]). Nonlocal elasticity theory of Eringen (references [2, 3]) has brought to light the importance of size-dependent effects with regard to small-scale structures. According to Eringen's theory, stress at a point in an elastic continuum depends on strains at that point as well as strains at all other points in continuum; hence the name nonlocal elasticity.

Effects of transverse shear with regard to beam deformation are insignificant for slender beams but become prominent for shear deformable beams. In literature, various first-order shear deformation beam theories (*FSDT*) and higher order shear deformation beam theories (*HSDT*) are present which capture these effects of shear in beam deformation (Ghugal and Shimpi [4]). It should be noted that, assumed displacement functions of *FSDT* result in constant transverse shear strain through the beam thickness. Whereas, assumed displacement functions of *HSDT* result in more realistic non-linear variation of transverse shear strain through the beam thickness. In comparison with *FSDT*, *HSDT* generally have increased number of independent unknowns and require specification of increased number of boundary conditions at each beam end.

To accurately predict mechanical behavior of shear deformable nanobeams, it is essential to take into account not only size-dependent effects but also beam transverse shear deformation effects. In this paper, newly developed single variable new first-order shear deformation nonlocal beam theory (*SVNBT*) of Pakhare et al. [5] is used for performing flexure of shear deformable isotropic rectangular nanobeams under the action of sinusoidally distributed transverse load. *SVNBT* involves only

K. S. Pakhare (✉) · P. J. Guruprasad · R. P. Shimpi
Department of Aerospace Engineering, IIT Bombay, Mumbai 400 076, India
e-mail: kedar188200@gmail.com

S. K. Saha and M. Mukherjee (eds.), *Recent Advances in Computational Mechanics and Simulations*, Lecture Notes in Mechanical Engineering,
https://doi.org/10.1007/978-981-15-8315-5_2

one governing differential equation which is obtained by utilizing beam gross equilibrium equations. Unlike nonlocal Timoshenko beam theory (*NTBT*) which involves two unknown functions (Wang et al. [6]), *SVNBT* involves only one unknown function. Effects of nonlocal parameter of Eringen and beam thickness-to-length ratio on profiles of non-dimensional beam transverse displacement and on maximum non-dimensional beam transverse displacement for simply supported, clamped, and cantilever isotropic rectangular nanobeams under the action of sinusoidally distributed transverse load by utilizing *SVNBT* are presented.

2 Theoretical Formulation

Governing differential equation of *SVNBT* (Pakhare et al. [5]) is as follows:

$$E\,I\,\frac{\mathrm{d}^4 w_b}{\mathrm{d}x^4} + K\,\frac{\mathrm{d}^2 q}{\mathrm{d}x^2} - q = 0 \tag{1}$$

In this theory, total beam transverse displacement is given as follows:

$$w = w_b - \frac{h^2\,(1+\mu)}{6\,\kappa}\,\frac{d^2 w_b}{\mathrm{d}x^2} \tag{2}$$

Beam geometry, co-ordinate system, symbols appearing in Eqs. (1) and (2) are the same as those of Shimpi et al. [7], κ is shear correction factor.

For beam under the action of sinusoidally distributed transverse load:

$$q = q_o \sin\left(\frac{\pi\,x}{L}\right) \tag{3}$$

where q_o is amplitude of applied transverse load.

Substituting Eq. (3) in Eq. (1) and solving the resulting expression, general solution of w_b is obtained which is as follows:

$$\begin{aligned} w_b = {} & \frac{q_o\,(\pi^2 L^2 K + L^4)}{\pi^4\,E\,I}\sin\left(\frac{\pi\,x}{L}\right) + C_1\left(\frac{x}{L}\right)^3 \\ & + C_2\left(\frac{x}{L}\right)^2 + C_3\left(\frac{x}{L}\right) + C_4 \end{aligned} \tag{4}$$

where in Eq. (4); C_1, C_2, C_3 and C_4 are arbitrary integration constants whose values can be found out by utilizing boundary conditions imposed on beam ends.

Beam Boundary Conditions

Pakhare et al. [5] have also stated physically meaningful beam boundary conditions. For the problem under consideration, these conditions with regard to one of the beam ends assumed at $x = 0$ for illustrative purpose are as follows:

Simply supported boundary conditions:

$$\left[w_b - \frac{h^2\,(1+\mu)}{6\,\kappa}\frac{d^2 w_b}{dx^2}\right]_{x=0} = 0 \tag{5}$$

$$\left[\frac{d^2 w_b}{dx^2}\right]_{x=0} = 0 \tag{6}$$

Clamped boundary conditions:

$$\left[w_b - \frac{h^2\,(1+\mu)}{6\,\kappa}\frac{d^2 w_b}{dx^2}\right]_{x=0} = 0 \tag{7}$$

$$\left[\frac{d w_b}{dx}\right]_{x=0} = 0 \tag{8}$$

Free end boundary conditions:

$$\left[\frac{d^2 w_b}{dx^2}\right]_{x=0} = 0 \tag{9}$$

$$\left[E\,I\,\frac{d^3 w_b}{dx^3} + K\,\frac{dq}{dx}\right]_{x=0} = 0 \tag{10}$$

Boundary conditions are chosen from Eq. (5) through (10) for each beam end in case of simply supported, clamped, and cantilever nanobeams.

3 Numerical Results and Discussion

Tables 1 through 3 present maximum non-dimensional beam transverse displacement ($\overline{w}_{\max}$) for the problem under consideration obtained by utilizing *SVNBT* along with corresponding results of *NTBT* (Wang et al. [6]) for simply supported, clamped and cantilever nanobeams, respectively, for various values of nonlocal parameter of Eringen (K) and beam thickness-to-length ratio (h/L). It should be noted that, $\overline{w}_{\max}$ obtained using *NTBT* is calculated by the present authors by utilizing the work reported by Wang et al. [6].

Figure 1 through 3 present profiles of non-dimensional beam transverse displacement ($\overline{w}$) for the problem under consideration obtained by utilizing *SVNBT* for simply supported, clamped, and cantilever nanobeams, respectively, for various values of K and h/L.

It should be noted that beam ends $x = 0$ and $x = L$ are simply supported in the case of simply supported nanobeam (Table 1 and Fig. 1), beam ends $x = 0$ and $x = L$ are clamped in the case of clamped nanobeam (Table 2 and Fig. 2), beam end $x = 0$ is clamped and $x = L$ is free in the case of cantilever nanobeam (Table 3 and Fig. 3).

Maximum non-dimensional beam transverse displacement ($\overline{w}_{\max}$) obtained by utilizing *SVNBT* matches exactly with corresponding results obtained by utilizing *NTBT* in the case of simply-supported, clamped and cantilever nanobeams (Tables 1 through 3 respectively) with nanobeam under the action of sinusoidally distributed transverse load for various values of nonlocal parameter of Eringen ($K = 0, 1, 2$ and 3) and beam thickness-to-length ratio ($h/L = 0.01, 0.05, 0.10$ and 0.15). It should also be noted that, *SVNBT* contains only one unknown function as opposed to *NTBT* which contains two unknown functions.

Following points should be noted with regards to profiles of non-dimensional beam transverse displacement ($\overline{w}$) presented in Figs. 1 through 3:

1. For simply-supported and clamped nanobeams under the action of sinusoidally distributed transverse load (Figs. 1 and 2 respectively), increase in nonlocal parameter of Eringen as well as beam thickness-to-length ratio cause increase in $\overline{w}$.

Table 1 $\overline{w}_{\max} = (100\ w_{\max}\ E\ I)/(q_o\ L^4)$ for simply supported nanobeam, $\mu = 0.3$

Theory	$\overline{w}_{\max}$ with $K = 0$			
	$h/L = 0.01$	$h/L = 0.05$	$h/L = 0.10$	$h/L = 0.15$
Present$	1.02686	1.03318	1.05294	1.08587
NTBT$ [6]	1.02686	1.03318	1.05294	1.08587
$\overline{w}_{\max}$ with $K = 1$				
Theory	$h/L = 0.01$	$h/L = 0.05$	$h/L = 0.10$	$h/L = 0.15$
Present$	1.12821	1.13516	1.15686	1.19304
NTBT$ [6]	1.12821	1.13516	1.15686	1.19304
$\overline{w}_{\max}$ with $K = 2$				
Theory	$h/L = 0.01$	$h/L = 0.05$	$h/L = 0.10$	$h/L = 0.15$
Present$	1.22956	1.23713	1.26078	1.30021
NTBT$ [6]	1.22956	1.23713	1.26078	1.30021
$\overline{w}_{\max}$ with $K = 3$				
Theory	$h/L = 0.01$	$h/L = 0.05$	$h/L = 0.10$	$h/L = 0.15$
Present$	1.33090	1.33910	1.36471	1.40738
NTBT$ [6]	1.33090	1.33910	1.36471	1.40738

$Shear correction factor of 5/6 is utilized

Table 2 $\overline{w}_{\max} = (100\ w_{\max}\ E\ I)/(q_o\ L^4)$ for clamped nanobeam, $\mu = 0.3$

Theory	$\overline{w}_{\max}$ with $K = 0$			
	$h/L = 0.01$	$h/L = 0.05$	$h/L = 0.10$	$h/L = 0.15$
Present$	0.220573	0.226896	0.246653	0.279583
NTBT$ [6]	0.220573	0.226896	0.246653	0.279583
$\overline{w}_{\max}$ with $K = 1$				
Theory	$h/L = 0.01$	$h/L = 0.05$	$h/L = 0.10$	$h/L = 0.15$
Present$	0.242343	0.249289	0.270997	0.307176
NTBT$ [6]	0.242343	0.249289	0.270997	0.307176
$\overline{w}_{\max}$ with $K = 2$				
Theory	$h/L = 0.01$	$h/L = 0.05$	$h/L = 0.10$	$h/L = 0.15$
Present$	0.264113	0.271683	0.295341	0.334770
NTBT$ [6]	0.264113	0.271683	0.295341	0.334770
$\overline{w}_{\max}$ with $K = 3$				
Theory	$h/L = 0.01$	$h/L = 0.05$	$h/L = 0.10$	$h/L = 0.15$
Present$	0.285882	0.294077	0.319685	0.362364
NTBT$ [6]	0.285882	0.294077	0.319685	0.362364

$Shear correction factor of 5/6 is utilized

Table 3 $\overline{w}_{\max} = (100\ w_{\max}\ E\ I)/(q_o\ L^4)$ for cantilever nanobeam, $\mu = 0.3$

Theory	$\overline{w}_{\max}$ with $K = 0$			
	$h/L = 0.01$	$h/L = 0.05$	$h/L = 0.10$	$h/L = 0.15$
Present$	7.38600	7.40587	7.46794	7.57139
NTBT$ [6]	7.38600	7.40587	7.46794	7.57139
$\overline{w}_{\max}$ with $K = 1$				
Theory	$h/L = 0.01$	$h/L = 0.05$	$h/L = 0.10$	$h/L = 0.15$
Present$	7.06769	7.08756	7.14963	7.25308
NTBT$ [6]	7.06769	7.08756	7.14963	7.25308
$\overline{w}_{\max}$ with $K = 2$				
Theory	$h/L = 0.01$	$h/L = 0.05$	$h/L = 0.10$	$h/L = 0.15$
Present$	6.74938	6.76925	6.83132	6.93477
NTBT$ [6]	6.74938	6.76925	6.83132	6.93477
$\overline{w}_{\max}$ with $K = 3$				
Theory	$h/L = 0.01$	$h/L = 0.05$	$h/L = 0.10$	$h/L = 0.15$
Present$	6.43107	6.45094	6.51301	6.61646
NTBT$ [6]	6.43107	6.45094	6.51301	6.61646

$Shear correction factor of 5/6 is utilized

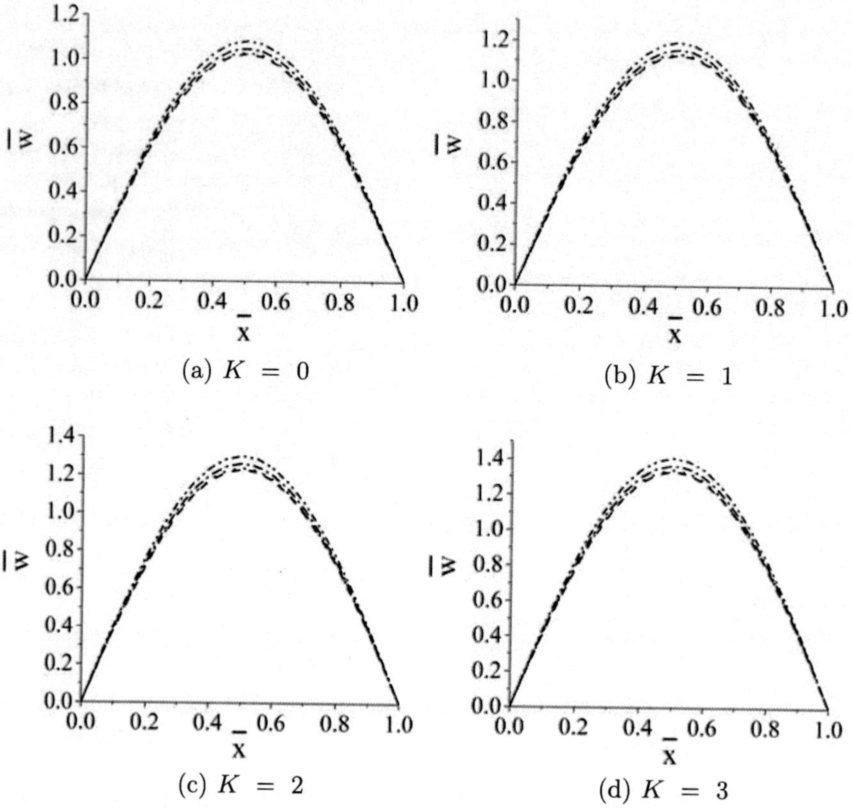

Fig. 1 For simply-supported nanobeam, non-dimensional beam transverse displacement ($\overline{w} = (100\ w\ E\ I)/(q_o\ L^4)$) versus non-dimensional beam axial location ($\overline{x} = x/L$) for various values of nonlocal parameter of Eringen (K) and beam thickness-to-length ratio (h/L); --- for $h/L = 0.01$, ··· for $h/L = 0.05$, -·- for $h/L = 0.10$ and -··- for $h/L = 0.15$

2. For cantilever nanobeam under the action of sinusoidally distributed transverse load (Fig. 3); increase in nonlocal parameter of Eringen for a given beam thickness-to-length ratio causes reduction in $\overline{w}$. Whereas, increase in beam thickness-to-length ratio for a given nonlocal parameter of Eringen causes increase in $\overline{w}$.

4 Concluding Remarks

In this paper, newly-developed single variable new first-order shear deformation nonlocal beam theory (*SVNBT*) is utilized for performing flexure of shear deformable isotropic rectangular nanobeams under the action of sinusoidally distributed trans-

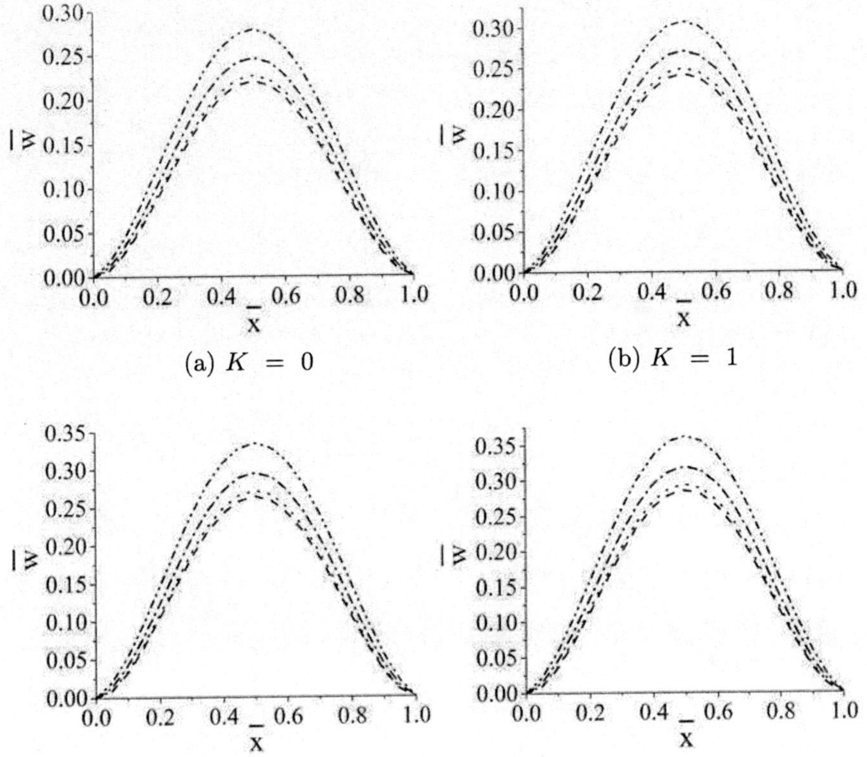

(a) $K = 0$ (b) $K = 1$

(c) $K = 2$ (d) $K = 3$

Fig. 2 For clamped nanobeam, non-dimensional beam transverse displacement ($\overline{w} = (100\ w\ E\ I)/(q_o\ L^4)$) versus non-dimensional beam axial location ($\overline{x} = x/L$) for various values of nonlocal parameter of Eringen (K) and beam thickness-to-length ratio (h/L); - - - for $h/L = 0.01$, · · · for $h/L = 0.05$, - · - for $h/L = 0.10$ and - ·· - for $h/L = 0.15$

verse load. Effect of nonlocal parameter of Eringen and beam thickness-to-length ratio on profiles of non-dimensional beam transverse displacement and maximum non-dimensional beam transverse displacement for thick isotropic nanobeams with various beam fixity conditions is presented. Obtained results are compared with corresponding results of nonlocal Timoshenko beam theory to demonstrate the effectiveness of *SVNBT*. These results and comparison prove that the flexural analysis of shear deformable isotropic rectangular nanobeams carried out using easy to use *SVNBT* is reliable for a range of beam fixity conditions, nonlocal parameter and beam thickness-to-length ratio.

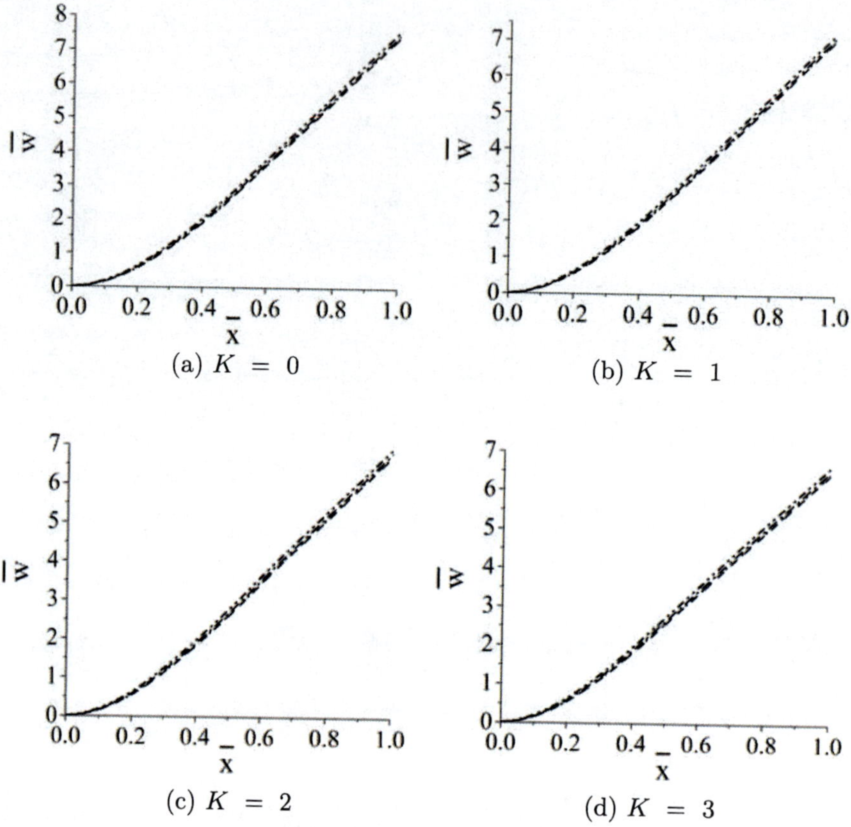

Fig. 3 For cantilever nanobeam, non-dimensional beam transverse displacement ($\overline{w} = (100\ w\ E\ I)/(q_o\ L^4)$) versus non-dimensional beam axial location ($\overline{x} = x/L$) for various values of nonlocal parameter of Eringen (K) and beam thickness-to-length ratio (h/L); - - - for $h/L = 0.01$, · · · for $h/L = 0.05$, - · - for $h/L = 0.10$ and - ·· - for $h/L = 0.15$

References

1. Eltaher, M.A., Khater, M.E., Emam, S.A.: A review on nonlocal elastic models for bending, buckling, vibrations, and wave propagation of nanoscale beams. Appl. Math. Model. **40**, 4109–4128 (2016). https://doi.org/10.1016/j.apm.2015.11.026
2. Eringen, A.C.: Nonlocal polar elastic continua. Int. J. Eng. Sci. **10**, 01–16 (1972). https://doi.org/10.1016/0020-7225(72)90070-5
3. Eringen, A.C.: Linear theory of nonlocal elasticity and dispersion of plane waves. Int. J. Eng. Sci. **10**, 425–435 (1972). https://doi.org/10.1016/0020-7225(72)90050-X
4. Ghugal, Y.M., Shimpi, R.P.: A review of refined shear deformation theories for isotropic and anisotropic laminated beams. J. Reinf. Plast. Comp. **20**, 255–272 (2001). https://doi.org/10.1177/073168401772678283
5. Pakhare, K.S., Guruprasad, P.J., Shimpi, R.P.: A single-variable first-order shear deformation nonlocal theory for the flexure of isotropic nanobeams. J. Braz. Soc. Mech. Sci. Eng. **42**, 42 (2020). https://doi.org/10.1007/s40430-019-2128-6

6. Wang, C.M., Kitipornchai, S., Lim, C.W., Eisenberger, M.: Beam bending solutions based on nonlocal Timoshenko beam theory. J. Eng. Mech. **134**, 475–481 (2008). https://doi.org/10.1061/(ASCE)0733-9399(2008)1343A6(475)
7. Shimpi, R.P., Shetty, R.A., Guha, A.: A single variable shear deformable nonlocal theory for transversely loaded micro- and nano-scale rectangular beams. Sādhanā **43**, 73 (2018). https://doi.org/10.1007/s12046-018-0852-8

Rolling and Sliding Resistance as Carbon Nanotubes are Driven on a Graphene Sheet

Avirup Sircar and Puneet Kumar Patra

1 Introduction

Two objects in physical contact with each other display interesting phenomena such as friction, wear, and adhesion at multiple scales. With the increasing popularity of nanoscale electro-mechanical devices, understanding the properties of two nanoscale objects in "contact" with each other is gaining prominence. However, research on nanoscale devices continues to challenge our knowledge on wear and tribology [1–5] and is not a mere extension of continuum scale mechanics. This is because as the length scale of a device decreases from micro to nano, surface forces such as adhesion, friction, and viscous forces that are proportional to area become a thousand times larger than the body forces, such as inertia, which depend on the volume [5]. Additionally, physical "contact" becomes more prominent, making the objects particularly vulnerable to adhesion. At nanoscale, atomic interactions such as chemical bond formation [6, 7] and Van der Waals force across the interfaces [7, 8] play a crucial role. The interaction between the two objects governs the nature of motion—whether the objects roll, slip, or roll with slip [2, 9].

One of the most important nanoscale materials to have come up in the last two decades is Carbon nanotubes (CNTs). Having cylindrical shape [8], they possess exceptionally high elastic modulus [10, 11], thermal conductivity [12], electrical conductivity [13], etc., making them suitable for future nanoscale devices [14, 15] such as mass spectrometers, drug delivery devices, high density data storage systems, etc. A proper understanding of the workings of these nanoscale devices require a better knowledge of energy dissipation mechanism. Nanotribology of CNTs and its applications have been a subject of several research works.

Multi-walled CNTs, put together concentrically to form CNT bearings, have been studied through experiments [16]. With separately imposed fixed and free boundary conditions on the outer nanotube shell, and rotation on the inner nanotube, the

A. Sircar · P. Kumar Patra (✉)
Indian Institute of Technology Kharagpur, Kharagpur 721302, India
e-mail: puneet.patra@civil.iitkgp.ac.in

S. K. Saha and M. Mukherjee (eds.), *Recent Advances in Computational Mechanics and Simulations*, Lecture Notes in Mechanical Engineering,
https://doi.org/10.1007/978-981-15-8315-5_3

frictional characteristics have been studied using molecular dynamics simulations [17]. It was found that friction forces increase linearly with the rotation speed. At higher temperatures, thermal vibrations increase, causing frequent "collisions" between atoms, and hence, more transfer of orderly kinetic energy into disorderly thermal motion, resulting in enhancement of frictional forces [17]. Conversely, friction was found to increase with decreasing inter-tube spacing, since the tighter spacing increases the stiffness of the van der Waals interaction between the atoms of adjacent tubes. The frictional characteristics of capped single-walled CNTs (SWCNTs) used as the tip of atomic force microscope have been studied using molecular dynamics simulations [18]. As the tip moves over a graphitic surface, numerical investigations of the lateral and normal forces of interactions reveal that the coefficient of friction varies with the distance between the tip and surface along with the layers of graphene present in the graphitic surface. The variable frictional characteristics of the capped SWCNTs may be utilized to replace the standard Silicon-based tip of atomic force microscopy to perform high resolution surface probing.

However, whether a CNT rolls or slides on the graphitic surface, when pushed, is yet to be well understood. Using molecular dynamics at constant energy ensemble, Buldum and Lu [7] suggested that rolling of CNTs is energetically more favorable than sliding *only* in certain locking positions under suitable external forcing conditions. Depending on the driving force a CNT may very well exhibit sliding and a combination of sliding and rolling (rolling with slipping). Experimentally, similar observations have been made [19, 20] by Falvo et. al. where stick-slip rolling have been observed for some driving forces. The origin of this stick-slip rolling and transition from sliding to rolling is yet to be understood. Surface effects also tend to influence the characteristics of motion. Consider the case, where a CNT lies on top of a graphitic surface. As the CNT moves, commensuration of the two contacts determine the nature of motion—CNTs tend to get locked to commensurate positions and, as a result, the motion changes from sliding to stick-rolling while *maintaining* commensuration just like a rack and pinion gear arrangement. The effect of commensurate contact on rolling/sliding was further elucidated by Miura et. al [21] in their experimental study involving multi-walled CNTs (MWCNTs) moving over graphitic surface. Despite these promising findings, experiments with CNTs rolling on graphitic surfaces have revealed that the energy dissipation in nanoscale rolling is often higher than that in sliding motion when the contacting surfaces are incommensurate [22–26]. In addition to the geometrical considerations, the relative motion of CNT-graphitic system depends on the interaction of outermost layer of CNT with the graphitic substrate. The interaction energy comprises short-range attraction due to chemical bonding, very short-range repulsion due to electronic orbital overlap [27] and attractive longer range Casimir forces [28], all of which may be modeled through a Lennard-Jones type potential [29].

In this manuscript, we revisit the motion of a CNT on a graphitic substrate to establish the different regimes of nanotribology—sliding, rolling, and sliding with rolling—using molecular dynamics simulations with Tersoff potential for C-C interactions of both CNT and graphene and Lennard-Jones potential for cross C-C interaction of CNT and graphene. The center of mass (CoM) of CNT is attached to a fictitious spring whose other end is pulled at a constant speed. Defining frictional forces as the component of interaction force between the CNT and graphene along the direction of motion, we specifically try to answer the following questions—(i) how do frictional forces evolve, as the rate of doing external work by the pulling agent increases, (ii) is there a regime where stick-roll behavior is observed regardless of effects of commensuration, and (iii) does the temperature of the environment has any bearing on the nature of motion and magnitude of frictional forces? A new dynamics-based scheme has been proposed to identify the different regimes of motion.

This manuscript is organized as follows: the next section details the simulation methodology adopted in the study, subsequent to which the important results are highlighted. Lastly, we conclude this study with several questions that have arisen from this study.

2 Simulation and Modeling

Usually experiments involving nanotribology have been performed with one of the three setups: (i) nanoindentation experiments where one surface slides over the other in order to estimate sliding friction as the atomic structures evolve [30], (ii) experiments where two surfaces move with respect to each other in presence of sandwiched lubricant molecules for estimating the resistance to the relative motion [31], and (iii) experiments where a fluid moves around a nanoscale system and the drag force is measured [32]. Motivated by these experiments, a different computational experiment is used here to understand nanotribology.

A pictorial representation of the problem studied in this manuscript is shown in Fig. 1. The figure depicts a graphene sheet modeled in 3 dimensions as brown dots, on top of which an SWCNT, shown in blue dots, is kept. The SWCNT has a chirality of (10,0) and length of 10 Å, while the graphene sheet is 500 Å × 10 Å. The CoM of SWCNT is attached with a fictitious spring, shown in black dashed line, whose free end is pulled with a velocity, v in the y direction. The spring constant is taken as

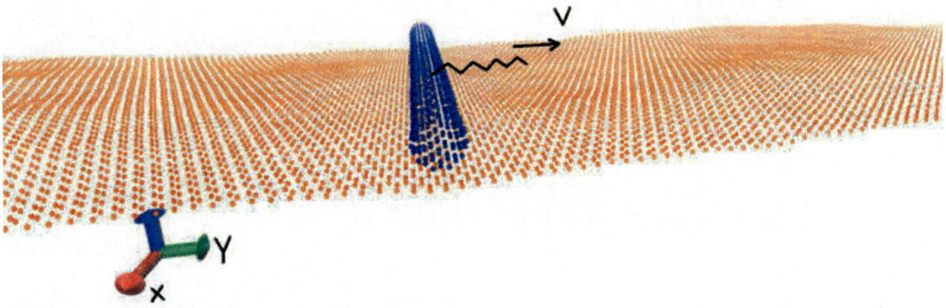

Fig. 1 Pictorial representation of the problem studied: an SWCNT is connected to a fictitious spring which is pulled with a constant velocity v. The SWCNT moves on top of a graphene sheet

100 eV/Å^2. As the spring is pulled, the atoms of SWCNT experience a force in the y direction. However, its motion on the graphene sheet (in the y direction) is resisted due to its interaction with the atoms of the graphene sheet. For the remainder of this study, the force of resistance along the y direction is treated as the friction force between the SWCNT and the graphene sheet. In order to make a uniform comparison, the results of friction presented later are with respect to lattice units, where the lattice constant of graphene = 2.456 Åhas been used.

The interaction between the C-C atoms of both SWCNT and graphene sheet have been modeled using the three-body Tersoff-like potential [33]. Although several potentials are available for modeling the C-C interactions [34–37], we have chosen the Tersoff-like potential owing to its wide usage in the molecular dynamics community [38]. In fact, the Tersoff potential has been a preferred choice for investigating nanoscale friction in CNTs by various researchers [39–42]. Further, the Tersoff-like potential is valid across a wide range of temperature, making possible a meaningful comparison of nanotribology regimes as the temperature changes. Mathematically, the Tersoff potential can be expressed as

$$E = \sum_i E_i = \frac{1}{2} \sum_{i \neq j} \sum_j \phi(r_{ij}),$$
$$\phi(r_{ij}) = f_c(r_{ij})[f_R(r_{ij}) + b_{ij} f_A(r_{ij})], \tag{1}$$

where E is the total potential energy of the system and E_i is the potential energy of the ith atom and ϕ is the potential energy between the ith and jth atoms.

In Eq. 1, r_{ij} is the distance between the ith and the jth atoms, b_{ij} is the bond-order function, f_c the cutoff function that ensures nearest-neighbor interaction, f_R the repulsive pair potential, and f_A the attractive pair potential. The mathematical forms of these individual functions are as follows:

$$f_c(r_{ij}) = \begin{cases} 1 & \forall r_{ij} < P_{ij} \\ \frac{1}{2} - \frac{1}{2}\sin(\frac{\pi}{2}\frac{r_{ij}-R_{ij}}{D_{ij}}) & \forall P_{ij} < r_{ij} < Q_{ij} \\ 0 & \forall r_{ij} > Q_{ij} \end{cases}$$

here, $P_{ij} = R_{ij} - D_{ij}$, $Q_{ij} = R_{ij} + D_{ij}$

The cutoff function is continuous and goes from 1 to 0 smoothly across R_{ij} from P_{ij} to Q_{ij}. R_{ij} is chosen so as to include only the first-neighbor shell for most structures of interest.

$$f_R(r_{ij}) = Ae^{-\lambda_1 r_{ij}}, \; f_A(r_{ij}) = -Be^{-\lambda_2 r_{ij}},$$
$$b_{ij} = (1 + \beta^n \zeta_{ij}^n)^{-\frac{1}{2n}},$$
$$\zeta_{ij} = \sum_{k \neq i,j} f_C(r_{ik}) g(\theta_{ijk}) \exp[\lambda_3^3 (r_{ij} - r_{ik})^3],$$
$$g(\theta_{ijk}) = 1 + c^2/d^2 - c^2/[d^2 + (h - \cos\theta_{ijk})^2)], \tag{2}$$

where θ_{ijk} represents the angle between the bonds $i-j$ and $i-k$. Depending upon the system being simulated, the parameters in the equation take different values. In the current simulation the parameters are taken according to the paper by Lindsay and Broido [38]. The simulations by these parameters have been seen to agree with the experimental results and from first principles, which motivated us to use this [38].

While the Tersoff potential accounts for the interaction within the SWCNT and graphene sheet, the cross-interaction between the C-C atoms of SWCNT and graphene sheet is modeled using the two-body Lennard-Jones (LJ) potential. The 12-6 LJ potential:

$$\phi_{LJ}(r_{ij}) = 4\epsilon \left[\left(\frac{\sigma}{r_{ij}} \right)^{12} - \left(\frac{\sigma}{r_{ij}} \right)^{6} \right], \tag{3}$$

has been adopted in the present study owing to its widespread use in modeling the van der Walls forces between two nanoscale objects [39–42]. Here, ϵ is the depth of the potential well and equals 0.002411 eV. σ denotes the distance between any two atoms such that $\phi_{LJ}(r_{ij}) = 0$, and equals 3.4 Å.

All simulations have been performed using the free to use software LAMMPS (Large-scale Atomic/Molecular Massively Parallel Simulator) [43]. Each simulation run begins with a conjugate gradient-based potential energy minimization. This step ensures that the entire system reaches a minimum potential energy state. Meaningful friction results can *only* be obtained if the system reaches a proper minimized energy configuration. Subsequent to energy minimization, 500,000 time steps, each of 1 fs have been run in constant temperature and pressure (NPT) ensemble with temperature, $T = 1$K and 300K, and pressure equalling 1 bar. In NPT ensemble, the simulation box is allowed to change, ensuring that the system does not become unstable in later stages. Post NPT relaxation, periodic boundary conditions have been imposed in the system along the x and y directions. A constant force of $0.0001 eV/$Å is applied on each atom of the SWCNT in the z direction to mimic normal reactive force that arise in macroscopic systems. In order to prevent translation of the system under the external force, the total linear momentum of the graphene sheet is kept at zero. Under this setup, 10,000,000 time steps, each of 1 fs, have been performed in the constant volume and temperature (NVT) ensemble. Two different values of $T = 1$K and 300K have been imposed for understanding the temperature dependence of frictional forces. Unlike in NPT, we employ a Langevin [44, 45] thermostat at this step, and not one of the deterministic thermostats [46, 47] for temperature control. Once the system is thermalized, the free end of the spring is displaced continuously with a speed v until the CoM of SWCNT reaches a specific location on the graphene sheet. Three different values of v (in Å$/fs$) have been used in the study for understanding the effect of frictional forces on rate at which work is done by the spring: $v = 0.005, 0.5$ and 1.0. All coordinate-based results have been normalized with respect to a lattice unit, $a = 2.456$ Å, which essentially represents the lattice constant of graphene.

3 Results and Discussion

At continuum scales, distinct regimes of friction can be typically observed as an object achieves steady-state motion starting from rest—initial regime where there is resistance to motion due to static friction force, intermediate regime when the motion initiates but is yet to reach steady state, and lastly the steady-state regime where the resistance is provided by kinetic friction. Defining the different regimes of friction in the problem investigated is somewhat tricky, as there is no clear transition. Figure 2 shows the variation of frictional forces with different values of v and T. In none of the different cases of v and T, one can observe distinct regime of static, intermediate, and kinetic friction.

As the temperature of the SWCNT-graphene system increases, thermal vibration effects start to become more prominent. The ordered kinetic energy imparted to the SWCNT by the external work due to the fictitious spring is converted to disordered thermal energy of both SWCNT and graphene sheet by the Langevin thermostat. As a result, at times, the atoms of SWCNT and graphene come very close to each other causing a sudden rise in friction values at higher temperature. A similar reason may be attributed to the increase in absolute value of the frictional forces at higher temperature vis-á-vis lower temperature.

Dynamics at low temperatures and low pulling velocities is significantly different from those at higher velocities and temperatures. A zoomed-in view of the frictional force profile shown in Fig. 2a is depicted in Fig. 3a. In the steady-state regime, the frictional force almost linearly decreases to prevent the motion of SWCNT as the free

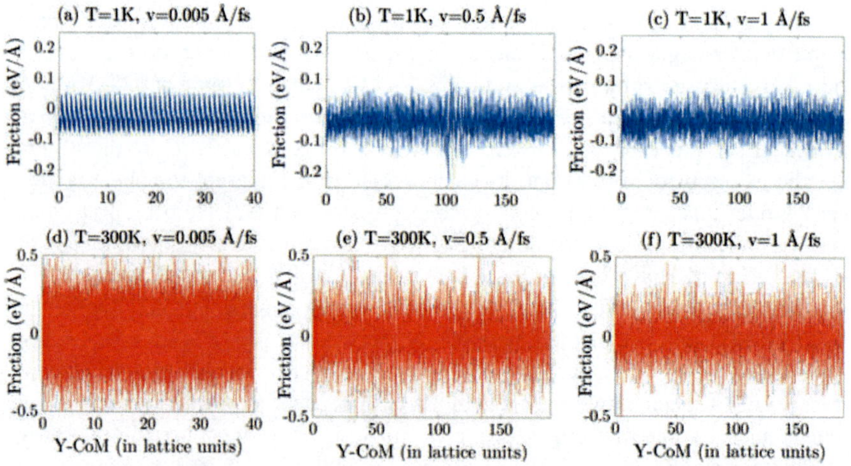

Fig. 2 Variation of friction force at 1K (top figures) and at 300K (bottom figures) for different pulling velocities, v. The frictional force is sensitive to both v and T. While one observes stick-rolling in figure (**a**), pure rolling in figure (**b**), thermal effects dominate in figure (**d**). For figures (**c**) and (**f**), friction occurs due to sliding of SWCNT over graphene

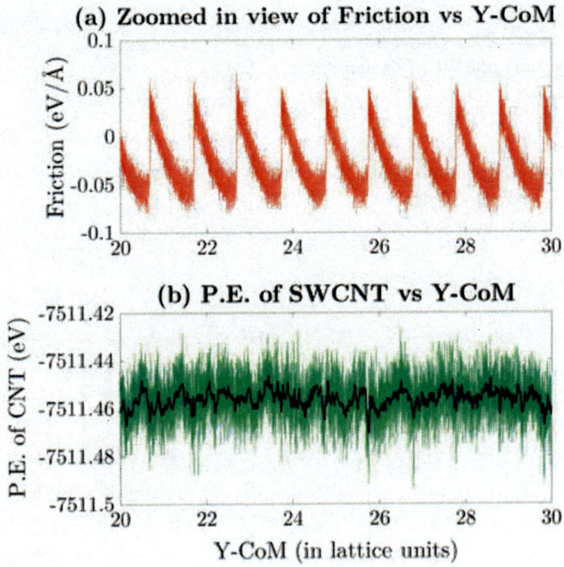

Fig. 3 (**a**) Zoomed-in view of friction versus $Y - CoM$ at T = 1K showing evidence of stick-roll motion of SWCNT at low temperatures and $v = 0.005\text{Å}/fs$, and (**b**) Variation of potential energy with $Y - CoM$ (smoothed average indicated in black line) depicting that each sudden change in frictional force is associated with a drop in potential energy

end of the spring moves. This is reminiscent of rolling about the bottom-most atom of SWCNT in contact with the graphene sheet, like a rack and pinion arrangement. The gradual decrease is followed by a sudden increase in friction force indicating a slip in the motion—a new atom takes the bottom-most position at this instant. Thus, the SWCNT undergoes a stick-roll motion. The evolution of potential energy of SWCNT is indicated in Fig. 3b in green, and the moving average of potential energy with 150 sample points is depicted in black line. Each slip event is associated with a sudden dip in potential energy establishing that the SWCNT jumps to a local potential energy minimum configuration.

To identify the deviation of the dynamics of the SWCNT from pure rolling, a new methodology has been designed which we detail next. Consider a spherical ring moving on a flat surface. In pure rolling, the trajectory of a point on the spherical ring follows a cycloid with the distance between the two peaks and troughs equalling $\pi \times D_{cyc}$. For our cases, where the ring comprises a cross-section of SWCNT, D_{cyc} corresponds to the diameter of the SWCNT and equals 3.915 Å. Figure 4 shows the evolution of Z and Y coordinates of a single particle of a SWCNT when $v = 0.005\text{Å}/fs$: (a) for 1K and (b) for 300K. At finite temperature, the graphene sheet tends to vibrate causing undulations in the geometry. As a consequence, the SWCNT tends to also move in the Z direction. In order to negate this behavior, the the Z coordinate of the particle is plotted with respect to the CoM of the ring whose one of the constituents is the chosen particle. On top of $Y - Z$ plot, a perfect cycloid motion is plotted in black dashed line. Notice that the plot at 1K completely coincides with a normal cycloid which suggests that the SWCNT undergoes pure rolling motion. The SWCNT remains "stuck" to the graphene through its bottom-most point about which the remaining atoms of SWCNT rotate. Subsequently, a new atom takes the bottom-

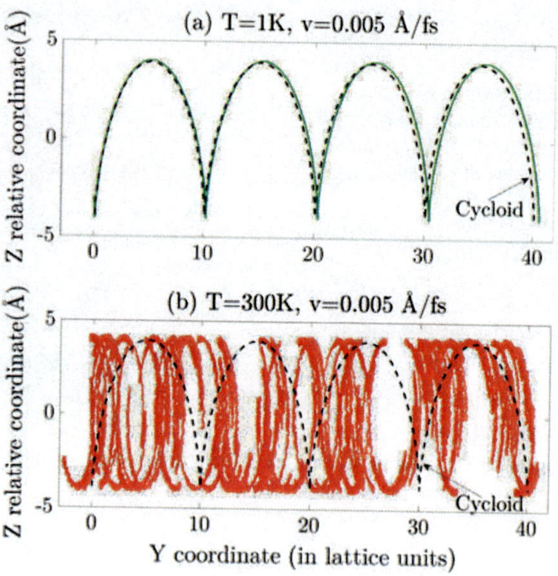

Fig. 4 Evolution of Y and relative Z coordinates of a particle of SWCNT when the spring is pulled by $v = 0.005\text{Å}/fs$ for (**a**) 1K temperature and (**b**) 300K temperature. A perfect cycloid is indicated in dashed black line

most position, and similar dynamics occur. Figure 4b plots the $Y - Z$ coordinate in a similar manner at 300K. Notice the difference from the cycloid. The motion is dominated by thermal vibration of the underlying graphene sheet because of which stick-roll dynamics is not observed.

In Fig. 5, the steady-state $Y - Z$ plot of a single atom of SWCNT, when pulled with $v = 0.5\text{Å}/fs$, is shown at two temperatures: (a) at 1K and (b) at 300K. The corresponding cycloids are shown in the black dashed line. At 1K there is rolling *with* slipping which gradually turns into pure rolling and like in the case of $v = 0.005\text{Å}/fs$ at $T = 1K$, the deviation of dynamics from a cycloid is small for Fig. 5a. There is a subtle difference between the two, however, the friction profile does not indicate any stick-roll phenomenon, rather, there is a hint of slipping for the bottom-most particle. At 300K, on the other hand, the dynamics is rolling *with* slipping. This is evident from the loops formed by the atom as it reaches bottom-most position. The competing effects of driving force due to the fictitious spring, thermal vibrations and frictional force due to graphene cause slipping, and therefore, one observes a significant deviation from a cycloid.

The dynamics is completely different for $v = 1\text{Å}/fs$. Friction profile is shown in Fig. 2c and f indicates that stick-roll or pure roll motion do not occur. In order to ascertain this, the $Y - Z$ evolution of a single atom of SWCNT is shown in Fig. 6. At 1K (see Fig. 6a), the green line depicting the motion of the atom is almost a flat line, indicating that the SWCNT gets simply dragged across the graphene sheet. At 300K (see Fig. 6b), though, there is a hint of rotation as can be judged from the height difference in the relative Z coordinate of the atom. The SWCNT undergoes a very interesting motion in this case—at first the nanotube rolls followed by rolling and sliding.

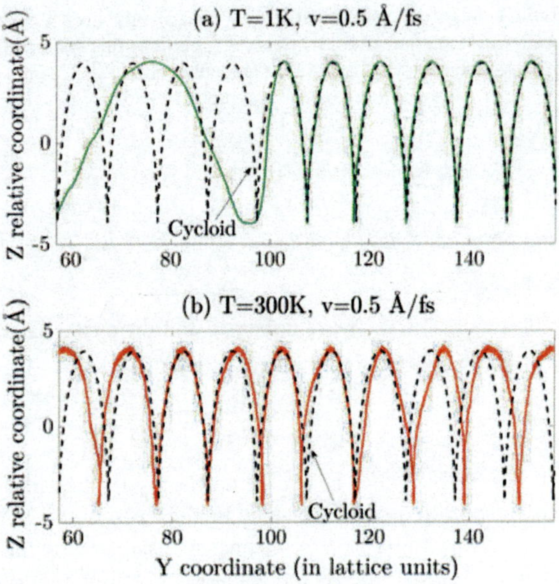

Fig. 5 Evolution of Y and relative Z coordinates of a particle of SWCNT when the spring is pulled by $v = 0.5\text{Å}/fs$ for (**a**) 1K temperature and (**b**) 300K temperature. A perfect cycloid is indicated in dashed black line

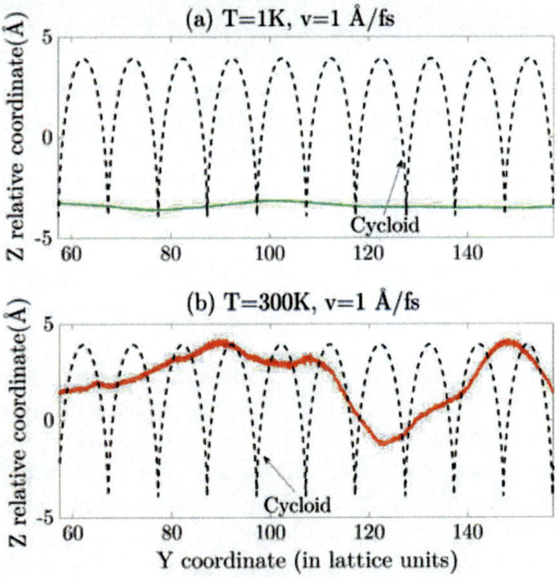

Fig. 6 Evolution of Y and relative Z coordinates of a particle of SWCNT when the spring is pulled by $v = 1\text{Å}/fs$ for (**a**) 1K temperature and (**b**) 300K temperature. A perfect cycloid is indicated in dashed black line. Both figures (**a**) and (**b**) suggest that the SWCNT gets dragged (slides) along the graphene sheet. At 300K, however, there is a small component of rotation along with sliding

Table 1 Work done per unit distance traveled by the CoM along the Y direction, $\langle W \rangle_d$, as evaluated using Eq. 4. All velocities, v are in $Å/fs$. Notice that stick-rolling motion ($v = 0.005Å/fs$, $T = 1K$) is associated with the smallest amount of energy dissipation at 1K, suggesting that it is possible to engineer low-friction regime by appropriately driving the nanotube

T(K)	v = 0.005	v = 0.5	v = 1
1	−0.0261	−0.0392	−0.0393
300	0.0067	−0.0069	−0.0027

Let us define the work done per unit distance, $\langle W \rangle_d$, by the friction force as

$$\langle W \rangle_d = \frac{\sum F_f \Delta Y_{CoM}}{\sum \Delta Y_{CoM}}, \tag{4}$$

As the SWCNT does not travel uniformly for the different cases, the normalization with respect to distance traveled by the CoM in the Y direction provides a tool for a meaningful comparison of the energy dissipated by the frictional forces. The results are indicated in Table 1. Notice that stick-rolling motion ($v = 0.005Å/fs$, $T = 1K$) is associated with the smallest amount of energy dissipation at 1K, suggesting that it is possible to engineer low-friction regime by appropriately driving the nanotube. Surprisingly enough, the case of $v = 1Å/fs$, $T = 300K$ also has a very low amount of energy dissipation. At high T, thermal effects are dominant, and coupled with large v, the SWCNT does not get enough time to relax to an energy minimum state, i.e., its Z coordinate is such that the force due to interaction is small, and as a consequence, $\langle W \rangle_d$ is also very small.

4 Conclusion

This manuscript studies the tribological characteristics of an SWCNT moving over a graphene sheet at two different temperatures using molecular dynamics simulations. There are several differences between the tribological characteristics of the nanoscale problem studied here and an equivalent continuum scale problem. For example, unlike in the continuum scale where friction always opposes the direction of motion, at nanoscale, friction may act in the direction of motion. The direction of friction force is a complex interplay between the position of SWCNT (and as a consequence its interaction with graphene), the force exerted by the spring and the effect of thermal vibrations. When the CNT approaches a local minimum potential energy configuration, the van der Waals forces are directed toward the local minimum energy configuration, causing the frictional forces to "aid" the motion. On the other hand, as the nanotube is pulled away from a local minimum energy configuration, the frictional forces oppose the motion as the van der Waals forces continue to remain directed toward the local minimum energy configuration. Further, defining uniquely the static and kinetic friction forces is not always possible.

The dynamics of SWCNT on graphene is significantly influenced by the controlling temperature, which in turn, influences the temporal and spatial evolution of frictional forces. At very low temperatures and low pulling velocities, we observe stick-roll phenomenon in the SWCNT, and as a consequence, the frictional force suddenly changes sign from being negative to positive. Such phenomenon is typically absent at relatively higher temperature and very low pulling velocities. In here, the thermal vibration of the graphene dominates and determines the motion of SWCNT. However, one can still experience pure rolling at high temperatures if the velocity of pulling is increased. At very high pulling velocities, the steady-state motion of SWCNT is such that it simply gets dragged across the graphene sheet. Typically pure rolling is associated with lower energy dissipation. Our results suggest that for every temperature, there is a pulling velocity where the SWCNT experiences pure (or nearly pure) rolling, corresponding to which the energy dissipation is small. As a result, one can engineer a regime of superlubricity where frictional losses are very small by tuning the pulling velocity at each temperature.

However, it must be noted that a more comprehensive study is in order to determine if it is possible to reach superlubricity regime by controlling the driving mechanism of SWCNT over graphene. Determining the temperature and velocity at which dynamics changes can help in creating better nanoelectromechanical devices. While at continuum scale, friction is typically associated with wear and material loss, at nanoscale such studies are still at a nascent stage, and we hope that our study will foster research in this domain. Lastly, we would like to highlight that the present setup may be extended to calculating Peierls-Nabarro barrier by looking at the temperature at which the SWCNT moves without any external influence.

Acknowledgements Support for the research provided in part by Indian Institute of Technology Kharagpur under grant DNI is gratefully acknowledged.

References

1. Vettiger, P., Brugger, J., Despont, M., Drechsler, U., Dürig, U., Häberle, W., et al.: Ultrahigh density, high-data-rate NEMS-based AFM data storage system. Microelectron. Eng. **46**(1–4), 11–17 (1999)
2. Martin, F.J., Grove, C.: Microfabricated drug delivery systems: concepts to improve clinical benefit. Biomed. Microdevices **3**(2), 97–108 (2001)
3. Hornbeck, L.J.: Bistable deformable mirror device. Spat. Light. Modul. Appl. (1988)
4. Gupta, B., Chevallier, J., Bhushan, B.: Tribology of ion bombarded silicon for micromechanical applications. J. Tribol. **115**(3), 392–399 (1993)
5. Bhushan, B.: Nanotribology and nanomechanics: an introduction. Springer Science & Business Media (2008)
6. Mo, Y., Turner, K.T., Szlufarska, I.: Friction laws at the nanoscale. Nature **457**(7233), 1116 (2009)
7. Buldum, A., Lu, J.P.: Atomic scale sliding and rolling of carbon nanotubes. Phys. Rev. Lett. **83**(24), 5050 (1999)

8. Luan, B., Robbins, M.O.: The breakdown of continuum models for mechanical contacts. Nature **435**(7044), 929 (2005)
9. Buldum, A., Ciraci, S.: Controlled lateral and perpendicular motion of atoms on metal surfaces. Phys. Rev. B **54**(3), 2175 (1996)
10. Yu, M.F., Lourie, O., Dyer, M.J., Moloni, K., Kelly, T.F., Ruoff, R.S.: Strength and breaking mechanism of multiwalled carbon nanotubes under tensile load. Science **287**(5453), 637–640 (2000)
11. Filleter, T., Bernal, R., Li, S., Espinosa, H.: Ultrahigh strength and stiffness in cross-linked hierarchical carbon nanotube bundles. Adv. Mater. **23**(25), 2855–2860 (2011)
12. Pop, E., Mann, D., Wang, Q., Goodson, K., Dai, H.: Thermal conductance of an individual single-wall carbon nanotube above room temperature. Nano Lett. **6**(1), 96–100 (2006)
13. Hong, S., Myung, S.: Nanotube Electronics: a flexible approach to mobility. Nat. Nanotechnol. **2**(4), 207 (2007)
14. Rapoport, L., Bilik, Y., Feldman, Y., Homyonfer, M., Cohen, S., Tenne, R.: Hollow nanoparticles of WS 2 as potential solid-state lubricants. Nature **387**(6635), 791 (1997)
15. Schwarz, U.D., Zwörner, O., Köster, P., Wiesendanger, R.: Quantitative analysis of the frictional properties of solid materials at low loads. I. Carbon compounds. Phys. Rev. B **56**(11), 6987 (1997)
16. Bourlon, B., Glattli, D.C., Miko, C., Forró, L., Bachtold, A.: Carbon nanotube based bearing for rotational motions. Nano Lett. **4**(4), 709–712 (2004)
17. Cook, E.H., Buehler, M.J., Spakovszky, Z.S.: Mechanism of friction in rotating carbon nanotube bearings. J. Mech. Phys. Solids **61**(2), 652–673 (2013)
18. Liu, P., Zhang, Y.: A theoretical analysis of frictional and defect characteristics of graphene probed by a capped single-walled carbon nanotube. Carbon **49**(11), 3687–3697 (2011)
19. Falvo, M.R., Steele, J., RM Taylor, I.I., Superfine, R.: Gearlike rolling motion mediated by commensurate contact: Carbon nanotubes on HOPG. Phys. Rev. B **62**(16), R10665 (2000)
20. Falvo, M., Taylor Ii, R., Helser, A., Chi, V., Brooks Jr., F.P., Washburn, S., et al.: Nanometre-scale rolling and sliding of carbon nanotubes. Nature **397**(6716), 236 (1999)
21. Miura, K., Takagi, T., Kamiya, S., Sahashi, T., Yamauchi, M.: Natural rolling of zigzag multi-walled carbon nanotubes on graphite. Nano Lett. **1**(3), 161–163 (2001)
22. Overney, R., Takano, H., Fujihira, M., Paulus, W., Ringsdorf, H.: Ansiotropy in friction and molecular stick-slip motion. Phys. Rev. Lett. **72**(22), 3546 (1994)
23. Liley, M., Gourdon, D., Stamou, D., Meseth, U., Fischer, T., Lautz, C., et al.: Friction anisotropy and asymmetry of a compliant monolayer induced by a small molecular tilt. Science **280**(5361), 273–275 (1998)
24. Sokoloff, J.: Theory of energy dissipation in sliding crystal surfaces. Phys. Rev. B **42**(1), 760 (1990)
25. Harrison, J., White, C., Colton, R., Brenner, D.: Molecular-dynamics simulations of atomic-scale friction of diamond surfaces. Phys. Rev. B **46**(15), 9700 (1992)
26. So, M., Jacobsen, K.W., Stoltze, P., et al.: Simulations of atomic-scale sliding friction. Phys. Rev. B **53**(4), 2101 (1996)
27. Salem, L.: The forces between polyatomic molecules. II. Short-range repulsive forces. Proc. R. Soc. Lond. Ser. Math. Phys. Sci. **264**(1318), 379–391 (1961)
28. Hertlein, C., Helden, L., Gambassi, A., Dietrich, S., Bechinger, C.: Direct measurement of critical Casimir forces. Nature **451**(7175), 172 (2008)
29. Girifalco, L., Lad, R.: Energy of cohesion, compressibility, and the potential energy functions of the graphite system. J Chem Phys **25**(4), 693–697 (1956)
30. Buldum, A., Ciraci, S., Batra, I.P.: Contact, nanoindentation, and sliding friction. Phys. Rev. B **57**(4), 2468 (1998)
31. Braun, O.: Simple model of microscopic rolling friction. Phys. Rev. Lett. **95**(12), 126104 (2005)
32. Tang, W., Advani, S.G.: Drag on a nanotube in uniform liquid argon flow. J. Chem. Phys. **125**(17), 174706 (2006)
33. Tersoff, J.: New empirical approach for the structure and energy of covalent systems. Phys. Rev. B **37**(12), 6991 (1988)

34. Brenner, D.W., Shenderova, O.A., Harrison, J.A., Stuart, S.J., Ni, B., Sinnott, S.B.: A second-generation reactive empirical bond order (REBO) potential energy expression for hydrocarbons. J. Phys.: Condens. Matter **14**(4), 783 (2002)
35. Marks, N.: Generalizing the environment-dependent interaction potential for carbon. Phys. Rev. B **63**(3), 035401 (2000)
36. Van Duin, A.C., Dasgupta, S., Lorant, F., Goddard, W.A.: ReaxFF: a reactive force field for hydrocarbons. J. Phys. Chem. A **105**(41), 9396–9409 (2001)
37. Los, J., Fasolino, A.: Intrinsic long-range bond-order potential for carbon: performance in Monte Carlo simulations of graphitization. Phys. Rev. B **68**(2), 024107 (2003)
38. Lindsay, L., Broido, D.: Optimized Tersoff and Brenner empirical potential parameters for lattice dynamics and phonon thermal transport in carbon nanotubes and graphene. Phys. Rev. B **81**(20), 205441 (2010)
39. Suekane, O., Nagataki, A., Mori, H., Nakayama, Y.: Static friction force of carbon nanotube surfaces. Appl. Phys. Express **1**(6), 064001 (2008)
40. Chen, Y., Yang, J., Wang, X., Ni, Z., Li, D.: Temperature dependence of frictional force in carbon nanotube oscillators. Nanotechnology **20**(3), 035704 (2008)
41. Barreiro, A., Rurali, R., Hernández, E.R., Moser, J., Pichler, T., Forro, L., et al.: Subnanometer motion of cargoes driven by thermal gradients along carbon nanotubes. Science **320**(5877), 775–778 (2008)
42. Servantie, J., Gaspard, P.: Rotational dynamics and friction in double-walled carbon nanotubes. Phys. Rev. Lett. **97**(18), 186106 (2006)
43. Plimpton, S.: Fast parallel algorithms for short-range molecular dynamics. J. Comput. Phys. **117**(1), 1–19 (1995)
44. Grønbech-Jensen, N., Hayre, N.R., Farago, O.: Application of the G-JF discrete-time thermostat for fast and accurate molecular simulations. Comput. Phys. Commun. **185**(2), 524–527 (2014)
45. Dünweg, B., Paul, W.: Brownian dynamics simulations without Gaussian random numbers. Int. J. Mod. Phys. C **2**(03), 817–827 (1991)
46. Evans, D.J., Holian, B.L.: The nose-hoover thermostat. J. Chem. Phys. **83**(8), 4069–4074 (1985)
47. Patra, P., Bhattacharya, B.: A deterministic thermostat for controlling temperature using all degrees of freedom. J. Chem. Phys. **140**(6), 064106 (2014)

Surface Stress-induced Degradation of Electrochemical Performance of Cylindrical Silicon Anode Particles in Li-ion Batteries

Amrita Sengupta, Sourav Das, and Jeevanjyoti Chakraborty

1 Introduction

With growing demands for high energy storage systems, researchers are trying to replace the traditionally used graphite with silicon as the anode material in the widely used lithium-ion batteries. Although Si has a theoretical specific capacity ten times higher than that of graphite, it has not been accepted commercially because of its large volumetric expansion and contraction on lithiation and delithiation, respectively. This cylic change in volume during charging and discharging of the battery lead to mechanical failure of the anode particles and eventually capacity fade of the battery. Using nanostructured Si anode particles helps circumvent these problems satisfactorily.

For large anode particles (>100 nm) the surface to volume ratio is negligible and the surface stresses are negligible as compared to the bulk stresses. But, as we reduce the dimension of the particles, surface stresses become significant. Gibbs [1], Shuttleworth [2], Gurtin and Murdoch [3] among others [4–8], had set up the mathematical framework to calculate surface stresses originating at solid surfaces. As an application of these theoretical frameworks in modeling nanostructured anode particles, pioneering works have been done by Cheng and Verbrugge [9], Yang [10], Deshpande et al. [11], and Hao et. al. [12]. Subsequent studies [13–15] have further investigated the effect of surface stresses on different battery parameters. All the above studies catering to small deformation cases, fail to determine the surface stresses involved in the large volumetric expansion of Si particles during lithiation.

In the present work, we model a single annular/hollow cylindrical-shaped Si anode particle in the nano-scale range, undergoing large volumetric expansion during lithiation. The novelty of this work lies in deducing the surface stress expression for the particle which is consistent with finite deformation. We adopt the chemo-mechanical model for cylindrical Si anode particle from earlier works [16–21]. We examine how

A. Sengupta (✉) · S. Das · J. Chakraborty
Department of Mechanical Engineering, Indian Institute of Technology, Kharagpur 721302, India
e-mail: asengupta.nitd@gmail.com

S. K. Saha and M. Mukherjee (eds.), *Recent Advances in Computational Mechanics and Simulations*, Lecture Notes in Mechanical Engineering,
https://doi.org/10.1007/978-981-15-8315-5_4

surface stresses affect the diffusion-induced stresses (DISs) during the period of lithiation/charging. The rest of the paper is organized as follows: Problem Formulation in Section II, Results and Discussion in Section III, and Conclusion in Section IV.

2 Problem Formulation

2.1 *Model Set up*

The silicon anode particle is modeled as a hollow circular cylinder of inner radius R_i, outer radius R_o, and length L_0 in the reference configuration. Standard cylindrical coordinate system (r, θ, z) is considered. The diffusion of lithium into and out of the anode particle during lithiation (charging) and delithiation (discharging) causes deformation of the particle. The DISs caused by the inhomogeneous lithium concentration, further affects the diffusion process. This coupled phenomenon has been formulated following a previous work [18] involving cylindrical silicon anode particles undergoing finite deformation during lithium insertion. The mechanical properties and parameters used in the model are given in Table 1.

Table 1 Mechanical properties and parameters used in the model

Material property or parameter	Value
A_0, parameter used in activity constant	-29549 Jmol^{-1} [17]
B_0, parameter used in activity constant	-38618 Jmol^{-1} [17]
D_0, diffusivity of Si	1×10^{-16}m^{2}s^{-1} [22]
$\dot{d}_0$, characteristic strain rate for plastic flow in Si	1×10^{-3} s^{-1} [17]
Y_0, modulus of elasticity of pure Si	90.13 [23] GPa
m, stress exponent for plastic flow in Si	4 [16]
R_g, universal gas constant	8.314 JK^{-1} mol^{-1}
T, temperature	300 K
V_m^{Si}, molar volume of Si	1.2052×10^{-5} m^3 mol^{-1} [17]
x_{max}, maximum concentration of of Li in Si	4.4
α, coefficient of diffusivity	0.18 [24]
η, coefficient of compositional expansion	0.2356 [17]
η_E, rate of change of modulus of elasticity with concentration	-0.1464 [23]
ν, Poisson's ratio of Si	0.28 [17]
σ_f, initial yield stress of Si	0.49 ± 0.08 GPa [25]
λ_s, surface Lamé constant	3.5 Nm^{-1} [6]
μ_s, surface Lamé constant	-6.23 Nm^{-1} [6]

2.2 Multiplicative Decomposition of Deformation Gradient

Let the material points in the reference configuration be given by $X \subset B_0$ such that, $X = r\mathbf{e}_r + z\mathbf{e}_z$, where $(\mathbf{e}_r, \mathbf{e}_\theta, \mathbf{e}_z)$ are the direction vectors in the radial, circumferential, and axial displacements, respectively. After deformation, $\boldsymbol{x} \subset B$ denotes the corresponding points in the current configuration, such that $\boldsymbol{x} = \Phi(X)$, where Φ is the deformation map connecting X to $\boldsymbol{x}$. The displacement field is described as $\boldsymbol{u}(X) = \Phi(X) - X$ such that, $\boldsymbol{u} = u\mathbf{e}_r + v\mathbf{e}_\theta + w\mathbf{e}_z$ where (u, v, w) are the displacements in the radial, circumferential, and axial displacements, respectively. Considering axisymmetric lithium diffusion and axisymmetric deformation, we assume only radial growth of the particle, which renders $v \equiv 0$. The total deformation $\Phi(X)$ of the particle as a result of lithium diffusion can be considered as a sequence of stress-free transformation (represented by $\mathbf{F}^*$) to an intermediate state, followed by an elastic transformation ($\mathbf{F}^{\mathrm{e}}$) to the final state [17, 26]. Using multiplicative decomposition of deformation gradient, the total deformation ($\mathbf{F}$) from the initial state to the final state can be expressed as

$$\mathbf{F} = \mathbf{F}^{\mathrm{e}}\mathbf{F}^*, \qquad \mathbf{F}^* = \mathbf{F}^{\mathrm{SF}}\mathbf{F}^{\mathrm{p}}, \tag{1}$$

where $\mathbf{F}^{\mathrm{SF}}$ is the stress-free deformation gradient, $\mathbf{F}^{\mathrm{p}}$ is the plastic deformation gradient and $\mathbf{F}^{\mathrm{e}}$ is the elastic deformation gradient. Similarly, the total surface deformation gradient $\widehat{\mathbf{F}}$ can also be decomposed as

$$\widehat{\mathbf{F}} = \widehat{\mathbf{F}}^p \cdot \widehat{\mathbf{F}}^e \cdot \widehat{\mathbf{F}}^{\mathrm{SF}}. \tag{2}$$

2.3 Influence of Surface Stress

According to Gibbs, [1] the variation of excess free energy of the interface on the area element dA should be equal to the reversible work required to elastically stretch this pre-existing surface element. Mathematically,

$$\delta(\widehat{J}^* w_s)\mathrm{d}A_0 = (\widehat{\mathbf{T}} : \delta\widehat{\mathbf{E}})\mathrm{d}A_0. \tag{3}$$

Here, w_{s} is the interfacc energy in the current configuration per unit area $\mathrm{d}A^*$ of the intermediate state, [17] $\widehat{J}^* w_{\mathrm{s}} =: W_{\mathrm{s}}$ is the interface energy per unit area $\mathrm{d}A_0$ in the reference configuration and $\widehat{J}^*$ is the ratio between $\mathrm{d}A^*$ (in intermediate configuration) and $\mathrm{d}A_0$ (in reference configuration). The second Piola-Kirchhoff surface stress $\widehat{\mathbf{T}}$ is the stress conjugate of the Green-Lagrange surface strain tensor $\widehat{\mathbf{E}}$. For a linear elastic isotropic material w_{s} is

$$w_s = w_0 + \frac{1}{2}\left[\lambda_s (E^e_{\gamma\gamma})^2 + 2\mu_s E^e_{\alpha\beta}E^e_{\beta\alpha}\right], \tag{4}$$

where w_0 is the residual surface energy, $E^e_{\alpha\beta}$ denotes elastic part of the Lagrangian surface strain. The subscripts α, β, γ are indices which can take values of θ and z only (since cylindrical surface is on the $\theta - z$ surface).

In case of finite deformation, the constitutive relation of the interface can be derived from Eq. 3. The first Piola-Kirchoff (PK1) stress for surface can be obtained by [27]

$$\widehat{\boldsymbol{\sigma}}^{\mathbf{0}} = \frac{\partial(\widehat{J}^* w_s)}{\partial \widehat{\mathbf{F}}}, \tag{5}$$

where $\widehat{\mathbf{F}}$ is the surface deformation gradient. Note that w_s is a function of elastic strains only. So, w_s can be expressed as a function of invariants $\widehat{J}^e_1$ and $\widehat{J}^e_2$ where $\widehat{J}^e_1 = \mathrm{tr}(\widehat{\mathbf{F}}^{\mathbf{e}})$ (i.e., trace of $\widehat{\mathbf{F}}^{\mathbf{e}}$) and $\widehat{J}^e_2 = \widehat{J}^e = \det(\widehat{\mathbf{F}}^{\mathbf{e}})$. Using Eq. 4 in Eq. 5 and simplifying, we have the Piola-Kirchoff surface stress in the circumferential direction as

$$\widehat{\sigma}^0_\theta = (J^c)^{1/3} \left[\frac{\lambda_z}{F^e_\theta} w_s + \left(\lambda_z \frac{\partial w_s}{\partial \widehat{J}^e_1} + \lambda_z F^e_z \frac{\partial w_s}{\partial \widehat{J}^e_2} \right) \right]. \tag{6}$$

From the analysis of the mechanical equilibrium of a surface,

$$\boldsymbol{\sigma}^{\mathbf{0}} \cdot \mathbf{N} = \widehat{\mathrm{Div}}\widehat{\boldsymbol{\sigma}}^{\mathbf{0}}, \tag{7}$$

where $\mathbf{N}$ is the outward unit normal vector to the surface, $\boldsymbol{\sigma}^{\mathbf{0}}$ denotes continuous extension of second-order stress tensor in the bulk body, to the surface, and $\widehat{\mathrm{Div}}\widehat{\boldsymbol{\sigma}}^{\mathbf{0}} := \mathrm{Div}\widehat{\boldsymbol{\sigma}}^{\mathbf{0}} \cdot \widehat{\mathbf{I}}$ is the surface divergence of $\widehat{\boldsymbol{\sigma}}^{\mathbf{0}}(\widehat{X})$ defined on the surface [12] ($\widehat{\mathbf{I}} = \mathbf{I} - \mathbf{N} \otimes \mathbf{N} = \mathbf{I} - \mathbf{e}_r \otimes \mathbf{e}_r$ = refers to the surface unit tensor in both the reference and current configurations). This generalized equation under the special case of a cylindrical particle, owing to its axisymmetric properties, simplifies to give

$$\sigma^0_r \mathbf{e}_r = \frac{\widehat{\sigma}^0_\theta}{R_i} \mathbf{e}_r, \qquad \sigma^0_r \mathbf{e}_r = -\frac{\widehat{\sigma}^0_\theta}{R_o} \mathbf{e}_r \tag{8}$$

Equation 8 actually states the boundary conditions which are described in the next section.

2.4 *The Chemo-Mechanical Model*

The set of partial differential equations governing the coupled phenomenon of diffusion-induced stress and stress enhanced diffusion is given below

$$\frac{1}{V^{\mathrm{Si}}_m} \frac{\partial c}{\partial t} = -\frac{\partial J_r}{\partial r} - \frac{J_r}{r}, \tag{9}$$

$$\frac{\partial \sigma_r^0}{\partial r} + \frac{\sigma_r^0 - \sigma_\theta^0}{r} = 0. \tag{10}$$

Equation 9 governs the diffusion of lithium into the cylindrical silicon particle. Here $c = x/x_{\max}$ gives the non-dimensional concentration of lithium in silicon particle [18], and J_r is the flux of Li. Equation 10 is the radial component of the mechanical equilibrium equation in the reference configuration in terms of PK1 stresses. Radius r varies from R_i to R_o. The PK1 stresses are expressed in terms of the Lagrangian strain components E_r, E_θ and E_z as

$$\sigma_r^0 = J^c \frac{Y(c)}{(1+\nu)(1-2\nu)}[(1-\nu)E_r^e + \nu(E_\theta^e + E_z^e)]\frac{2E_r^e + 1}{1 + \partial u/\partial r}, \tag{11a}$$

$$\sigma_\theta^0 = J^c \frac{Y(c)}{(1+\nu)(1-2\nu)}[(1-\nu)E_\theta^e + \nu(E_r^e + E_z^e)]\frac{2E_\theta^e + 1}{1 + u/r}, \tag{11b}$$

$$\sigma_z^0 = J^c \frac{Y(c)}{(1+\nu)(1-2\nu)}[(1-\nu)E_z^e + \nu(E_\theta^e + E_r^e)]\frac{2E_z^e + 1}{1 + \partial w/\partial z}. \tag{11c}$$

Here $Y(c) = Y_0(1 + \eta_E x_{\max} c)$ is the concentration-dependent modulus of elasticity [18]. The radial (u) and axial (w) displacements of the cylindrical particle are related to the elastic strain components as

$$\mathbf{E}^e = \frac{1}{2}[(\mathbf{F}^e)^T \cdot \mathbf{F}^e - \mathbf{I}] = \mathrm{diag}(E_r^e, E_\theta^e, E_z^e). \tag{12}$$

The plastic stretches in the radial and circumferential directions evolve as

$$\frac{\partial \lambda_{r,\theta}}{\partial t} = \sqrt{\frac{3}{2}}\lambda_{r,\theta}\dot{d}_0 \left(\frac{\sigma_{\text{eff}}}{\sigma_f} - 1\right)^m \frac{\tau_{r,\theta}}{\sqrt{\tau_r^2 + \tau_\theta^2 + \tau_z^2}} H\left(\frac{\sigma_{\text{eff}}}{\sigma_f} - 1\right), \tag{13}$$

where $\dot{d}_0$ is the characteristic strain rate, σ_f is the yield stress of silicon, and H is the Heaviside step function. τ_r, τ_θ, τ_z are the non-dimensionalized deviatoric parts of the Cauchy stress tensor. The effective stress in Eq. 13 is given by

$$\sigma_{\text{eff}} = \sqrt{\frac{3}{2}}\sqrt{\tau_r^2 + \tau_\theta^2 + \tau_z^2}. \tag{14}$$

The flux of Li in Eq. 9 is given as

$$J_r = -\frac{D}{R_g T}\frac{c}{V_m^{\text{Si}}}\frac{\partial \mu}{\partial r}. \tag{15}$$

Here D is the diffusivity of Li in Si, J_r is a function of the chemical potential μ which can be further decomposed into stress-independent and stress-dependent parts [17]. The initial and boundary conditions corresponding to Eq. 9 are

$$c(r,0)=0,\quad J_r(R_i,t)=0,\quad J_r(R_o,t)=\underbrace{J_0(1-c)}_{\text{Charging}}\text{ or }\underbrace{-J_0c}_{\text{Discharging}}. \tag{16}$$

For Eq. 10, we have two boundary conditions:

$$\sigma_r^0(R_i)=\sigma_r^0|_{S_i}=\frac{\widehat{\sigma}_\theta^0}{R_i},\quad \sigma_r^0(R_o)=\sigma_r^0|_{S_0}=-\frac{\widehat{\sigma}_\theta^0}{R_o}. \tag{17}$$

For large particles, the reference radius R_o is large and hence $\widehat{\sigma}_\theta^0/R_{i,o}\to 0$, which is same as the conventional free-surface boundary condition, $\sigma_r=0$.

2.5 Non-dimensionalized Governing Differential Equations

The governing equations as a result of non-dimensionalization [18] are

$$\frac{\partial c}{\partial\tilde{t}}=-\frac{\partial\tilde{J}_r}{\partial\tilde{r}}-\frac{\tilde{J}_r}{\tilde{r}}, \tag{18a}$$

$$\frac{\partial\tilde{\sigma}_r^0}{\partial\tilde{r}}+\frac{\tilde{\sigma}_r^0-\tilde{\sigma}_\theta^0}{\tilde{r}}=0, \tag{18b}$$

$$\frac{\partial\lambda_r}{\partial\tilde{t}}=\sqrt{\frac{3}{2}}\lambda_r\dot{d}_0\frac{R_o^2}{D_0}\left(\frac{\tilde{\sigma}_{\text{eff}}}{\tilde{\sigma}_f}-1\right)^m\frac{\tilde{\tau}_r}{\sqrt{\tilde{\tau}_r^2+\tilde{\tau}_\theta^2+\tilde{\tau}_z^2}}H\left(\frac{\tilde{\sigma}_{\text{eff}}}{\tilde{\sigma}_f}-1\right), \tag{18c}$$

$$\frac{\partial\lambda_\theta}{\partial\tilde{t}}=\sqrt{\frac{3}{2}}\lambda_\theta\dot{d}_0\frac{R_o^2}{D_0}\left(\frac{\tilde{\sigma}_{\text{eff}}}{\tilde{\sigma}_f}-1\right)^m\frac{\tilde{\tau}_\theta}{\sqrt{\tilde{\tau}_r^2+\tilde{\tau}_\theta^2+\tilde{\tau}_z^2}}H\left(\frac{\tilde{\sigma}_{\text{eff}}}{\tilde{\sigma}_f}-1\right). \tag{18d}$$

The initial and boundary conditions are

$$c(\tilde{r},0)=0,\quad \tilde{J}_r\left(\frac{R_i}{R_o},\tilde{t}\right)=0,\quad \tilde{J}_r(1,\tilde{t})=\underbrace{\tilde{J}_0(1-c)}_{\text{charging}}\text{ or }\underbrace{-\tilde{J}_0c}_{\text{discharging}}. \tag{19}$$

$$\tilde{\sigma}_r^0\left(\frac{R_i}{R_o},\tilde{t}\right)=\frac{\widehat{\sigma}_\theta}{R_i}\frac{w_0V_m^{\text{Si}}}{R_gT},\quad \tilde{\sigma}_r^0(1,\tilde{t})=-\frac{\widehat{\sigma}_\theta}{R_o}\frac{w_0V_m^{\text{Si}}}{R_gT} \tag{20}$$

$$\lambda_r(\tilde{r},0)=1\quad\text{and}\quad\lambda_\theta(\tilde{r},0)=1. \tag{21}$$

Equations 18(a)–(d) along with initial and boundary conditions 19, 20, and 21 are solved numerically using COMSOL Multiphysics 5.3a.

Results and Discussions

The results are discussed for an axially unconstrained case where the cylindrical Si particles are free of any physical constraints in the axial direction, thus

$$2\pi \int_0^1 \tilde{\sigma}_z^0 \tilde{r}\, d\tilde{r} = 0. \tag{22}$$

Throughout the discussion, we have referred to the State of Charge or SOC of the electrode particle. We define SOC as the average concentration of lithium within the Si particle, with respect to its reference configuration. Mathematically, SOC = $\left(\int_0^{R_o} c2\pi r\, dr\right) / \left(\int_0^{R_o} 2\pi r\, dr\right)$.

2.6 *Effect of Surface Stress on DIS-trends*

Figure 1 shows the trends of Cauchy stresses σ_r, σ_θ, σ_z versus $\tilde{r}$ with and without considering surface stresses, at various time instants $\tilde{t}$ for $\tilde{J}_0 = 0.1$ and $R_o = 50$ nm. It is observed that with surface stresses considered, the plots for the radial stresses in Fig. 1a shift in the negative direction, but the span or range of stress magnitudes remain intact. This trend is absent in circumferential and axial stress plots, where the tensile stress becomes more tensile and compressive stress becomes more compressive in an annular particle. When we compare the σ_θ trends for an annular particle with a solid cylindrical particle (Fig. 1d and e), we observe the qualitative difference. For the solid case, the σ_θ shifts in the negative direction (Fig. 1d), unlike the annular case. Here, we define $\Delta\sigma_{r,\theta,z} = (\sigma_{r,\theta,z})_{\text{w/o--SS}} - (\sigma_{r,\theta,z})_{\text{w/SS}}$, i.e., difference in stresses calculated without and with considering surface stresses. In Fig. 1e, we plot $\Delta\sigma_\theta$ along $\tilde{r}$ for solid and hollow cylindrical particles at $\tilde{t} = 0.05$, to understand how surface stress affects them differently. While in the solid particle $\Delta\sigma_\theta$ is throughout positive denoting a negative shift, the $\Delta\sigma_\theta$ for the annular particle is negative near the inner surface and positive near the outer surface (compare Fig. 1b and d). This difference stems from the fact that in case of a solid particle, there is no deformation at the center of the particle, whereas for the annular particle, the inner surface is subjected to deformation as well as surface stress. It is important to note that the axial stresses (Fig. 1c) are negligibly affected by the consideration of surface stresses. Furthermore, similar to circumferential stresses, the magnitudes of axial stresses at different time instants in Fig. 1c are higher when compared to the case without surface stress. The evolution of the stresses are extensively discussed in a previous work by Chakraborty et. al. [18] With increase in SOC, the inhomogeneity in lithium concentration within the silicon particle decreases and hence the stresses (radial, circumferential and axial) die down at higher SOCs.

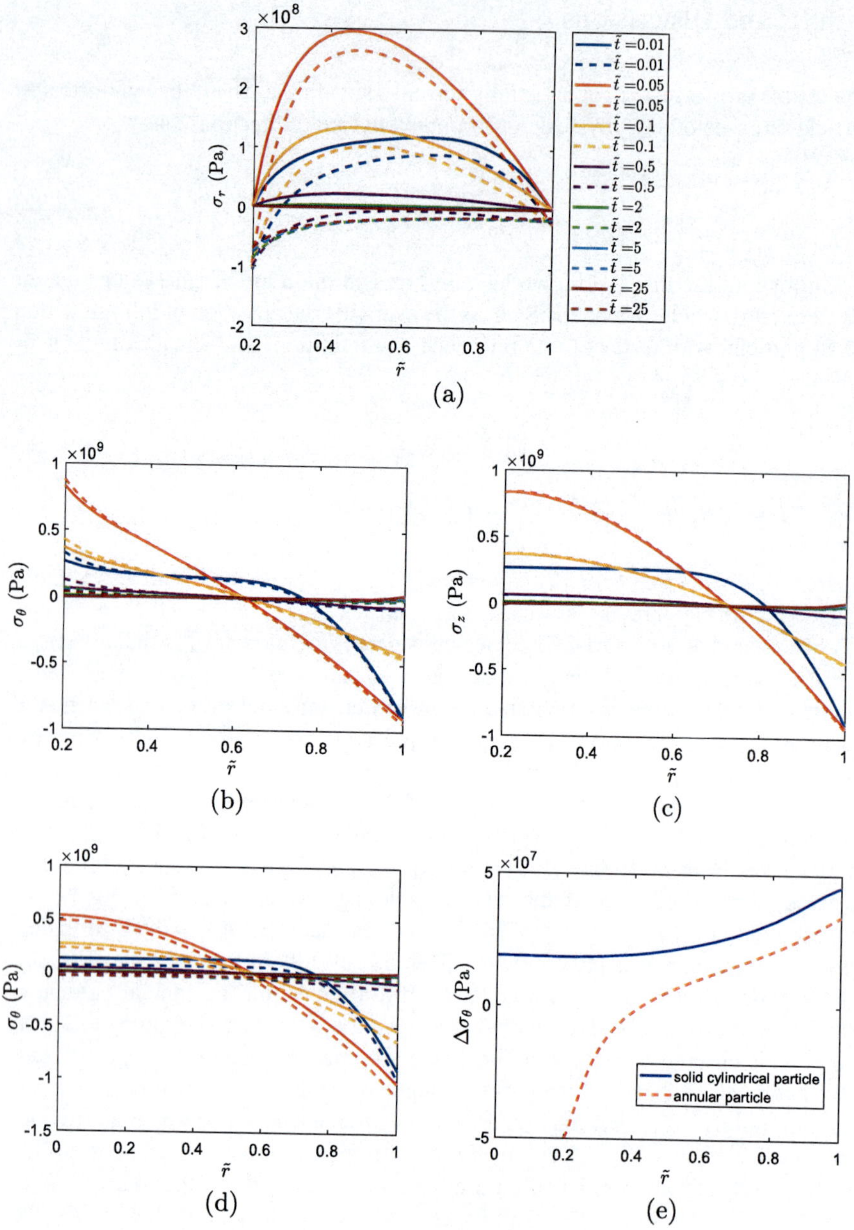

Fig. 1 Variation of **a** σ_r, **b** σ_θ, and **c** σ_z with $\tilde{r}$ for $\tilde{J}_0 = 0.1$, $R_o = 50$ nm, and $R_i/R_0 = 0.2$ ((–): no surface stress, (- -): surface stress is present); **d** variation of σ_θ with $\tilde{r}$ for solid cylindrical particle of radius $R_o = 50$ nm, **e** variation of $\Delta\sigma_\theta$ in solid and hollow cylindrical particles at time $\tilde{t} = 0.05$

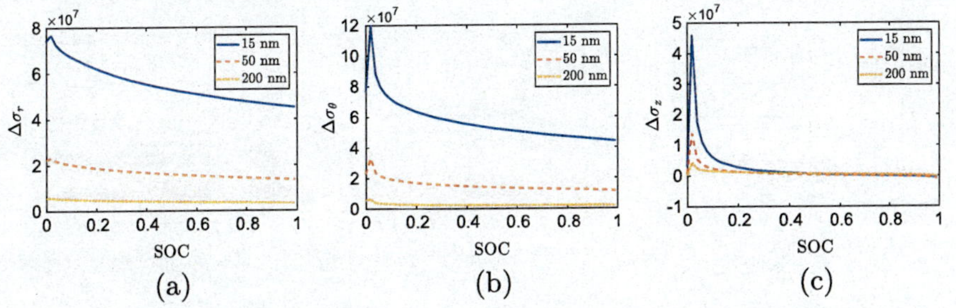

Fig. 2 Difference in stresses (with and without surface stresses) in **a** radial, **b** circumferential, and **c** axial directions at the surface, plotted against SOC for various R_o values. Influx rate is kept at $\tilde{J}_0 = 0.1$ and the ratio $R_i/R_0 = 0.2$

2.7 *Dependence of Surface Stresses on Size of the Particle*

The stress trends in Fig. 1 are generated for a particle of initial outer radius R_o=50 nm. As R_o is decreased, keeping the R_i/R_0 ratio constant at 0.2, the differences in stress magnitudes with and without surface stress become significant. These differences become negligible at higher outer radii (>200 nm). To analyze how R_o affects the surface stress, we plot $\Delta\sigma_r$, $\Delta\sigma_\theta$, $\Delta\sigma_z$ at outer surface ($\tilde{r} = 1$) against SOC, for $R_o = 15$, 50 and 200 nm in Fig. 2, keeping $\tilde{J}_0$ fixed at 0.1 for all the silumations. It is observed that $\Delta\sigma_{r,\theta,z}$ increases as R_o is decreased.

2.8 *Dependence of Surface Stress on C-Rate*

Surface stress generated at the anode surface is controlled by a number of factors (Eq. 6). It is observed that for a particle of outer radius R_o, if we increase the C-rate, consequently the rate of influx $\tilde{J}_0$ increases. To understand the effect of C-rate on surface stresses, we plot $\Delta\sigma_r$, $\Delta\sigma_\theta$, $\Delta\sigma_z$ at surface ($\tilde{r} = 1$) against SOC (Fig. 3), for C-rates = 0.3C, 3C, and 30C, keeping R_o fixed at 50 nm. At the initial stages of lithiation, the difference in stresses increases with an increase in C-rate. However with an increase in SOC, the concentration gradient at the outer surface decreases faster in case of 30 C than for 0.3C. Hence, the surface stress decreases with increasing C-rate, at higher SOCs.

2.9 *Effect of Surface Stresses on Plastic Stretches*

Here we analyze the effect of surface stress on plastic stretches (λ_r, λ_θ, λ_z) for a range of yield stress values [25]. In Fig. 4 we plot (a) $\lambda_r - 1$, (b) $\lambda_\theta - 1$ and (c)

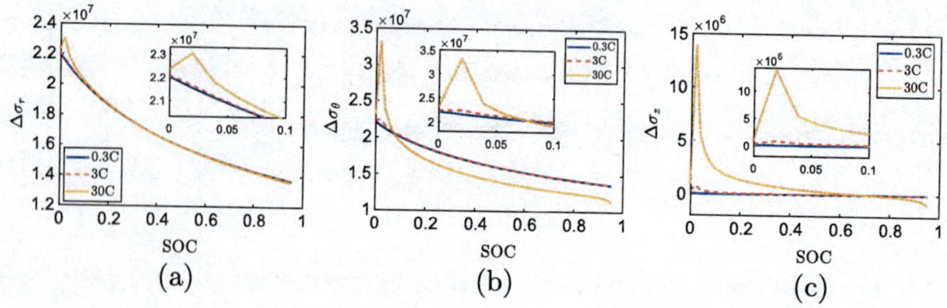

Fig. 3 **a** $\Delta\sigma_r$, **b** $\Delta\sigma_\theta$, and **c** $\Delta\sigma_z$ at $\tilde{r}=1$, plotted against SOC for various C-rates. The initial outer radius is $R_o = 50$ nm and $R_i/R_0 = 0.2$

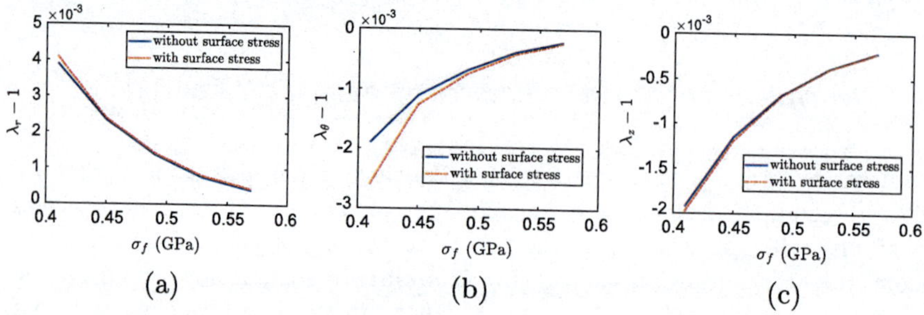

Fig. 4 **a** $\lambda_r - 1$, **b** $\lambda_\theta - 1$, and **c** $\lambda_z - 1$ at $\tilde{r}=1$, plotted against σ_f when SOC = 1. Influx rate $\tilde{J}_0 = 0.1$, initial outer radius $R_o = 50$ nm and $R_i/R_0 = 0.2$

$\lambda_z - 1$ with and without surface stress, at $\tilde{r} = 1$ (outer surface) and SOC = 1, against σ_f. For all the cases, $R_o = 50$ nm and $\tilde{J}_0 = 0.1$. It is observed that the diffrerence in stretches decreases with increasing σ_f. Also, the magnitude of plastic stretches is higher with surface stresses, than without.

3 Conclusion

A mathematical framework, consistent with finite deformation is established to study the effects of surface stresses in nanostructured, hollow cylindrical silicon anode particle under charging conditions of a lithium-ion battery. We analyze the effects for an axially unconstrained case. The consideration of surface stresses shifts the radial diffusion-induced stresses in the negative direction. Also, the magnitude of circumferential and axial stresses increase when surface stresses are taken into account for an annular particle, which is in contrast to a solid cylindrical particle that experiences similar shift like the radial stress. The influence of surface stress depends upon various factors such as initial particle size and the influx rate. It is observed that with decrease in particle size below $R_o = 50$ nm, the surface stress becomes most prominent. Again, with an increase in the influx rate or C-rate for a particle of fixed initial size, the surface stress enhances. Further, the plastic stretches are affected when surface stress is considered. We conclude that while surface stresses are insignificant at radii of curvature above 100 nm and low C-rates of 0.3C, they play a crucial role in determining how bulk stresses evolve for $R_o < 50$ nm and C-rate > 3C.

References

1. Gibbs, J.W.: The Scientific Papers of J. Willard Gibbs, vol. 1. Green and Company, Longmans (1906)
2. Shuttleworth, R.: Proceedings of the physical society. Sect. A **63**(5), 444 (1950)
3. Gurtin, M.E., Murdoch, A.I.: Arch. Ration. Mech. Anal. **57**(4), 291 (1975)
4. Miller, R.E., Shenoy, V.B.: Nanotechnology **11**(3), 139 (2000)
5. Cammarata, R.C.: Prog. Surf. Sci. **46**(1), 1 (1994)
6. Sharma, P., Ganti, S., Bhate, N.: Appl. Phys. Lett. **82**(4), 535 (2003)
7. Papastavrou, A., Steinmann, P., Kuhl, E.: J. Mech. Phys. Solids **61**(6), 1446 (2013)
8. McBride, A., Javili, A., Steinmann, P., Bargmann, S.: J. Mech. Phys. Solids **59**(10), 2116 (2011)
9. Cheng, Y.T., Verbrugge, M.W.: J. Appl. Phys. **104**(8), 083521 (2008)
10. Yang, F.: J. Appl. Phys. **108**(7), 073536 (2010)
11. Deshpande, R., Cheng, Y.T., Verbrugge, M.W.: J. Power Sources **195**(15), 5081 (2010)
12. Hao, F., Gao, X., Fang, D.: J. Appl. Phys. **112**(10), 103507 (2012)
13. Stein, P., Zhao, Y., Xu, B.X.: J. Power Sources **332**, 154 (2016)
14. Liu, W., Shen, S.: Acta Mech. **229**(1), 133 (2018)
15. Lu, Y., Zhang, P., Wang, F., Zhang, K., Zhao, X.: Electrochim. Acta **274**, 359 (2018)
16. Bower, A.F., Guduru, P.R., Sethuraman, V.A.: J. Mech. Phys. Solids **59**(4), 804 (2011)
17. Cui, Z., Gao, F., Qu, J.: J. Mech. Phys. Solids **60**(7), 1280 (2012)
18. Chakraborty, J., Please, C.P., Goriely, A., Chapman, S.J.: Int. J. Solids Struct. **54**, 66 (2015)
19. Zhao, K., Pharr, M., Cai, S., Vlassak, J.J., Suo, Z.: J. Am. Ceram. Soc. **94**, s1 (2011)
20. Neogi, S., Chakraborty, J.: J. Appl. Phys. **124**(15), 154302 (2018). https://doi.org/10.1063/1.5052236
21. Sengupta, A., Chakraborty, J.: Acta Mechanica, pp. 1–21 (2019). https://doi.org/10.1007/s00707-019-02550-4
22. Liu, X.H., Zheng, H., Zhong, L., Huang, S., Karki, K., Zhang, L.Q., Liu, Y., Kushima, A., Liang, W.T., Wang, J.W., et al.: Nano Lett. **11**(8), 3312 (2011)
23. Rhodes, K., Dudney, N., Lara-Curzio, E., Daniel, C.: J. Electrochem. Soc. **157**(12), A1354 (2010)
24. Haftbaradaran, H., Song, J., Curtin, W., Gao, H.: J. Power Sources **196**(1), 361 (2011)
25. Bucci, G., Nadimpalli, S.P., Sethuraman, V.A., Bower, A.F., Guduru, P.R.: J. Mech. Phys. Solids **62**, 276 (2014)
26. Rodriguez, E.K., Hoger, A., McCulloch, A.D.: J. Biomech. **27**(4), 455 (1994)
27. Huang, Z., Wang, J.X.: Acta Mech. **182**(3–4), 195 (2006)

Molecular Dynamics Simulation of Single-Wall Carbon Nanotube Aluminum Composite

Pramod Rakt Patel, Sumit Sharma, and S. K. Tiwari

1 Introduction

Carbon Nanotube (CNT) has been used as a reinforcing material for aluminum (Al) due to its novel material properties [1–13]. Al has been used as the matrix material due to its good mechanical properties. The uniform dispersion of CNTs is the key factor for the enhanced material properties of CNT–Al composites. Various studies [1–17] have been conducted on CNT–Al composites experimentally. But there is a need of more studies at nanoscale for further improving properties of CNT–Al composites. A few studies have been performed at nanoscale using computational techniques [18–20]. Ab initio methods and Molecular Dynamics (MD) simulation are the two nanoscale computational simulation techniques that have been used to predict the properties of CNT–Al composites. The ab initio method has been used to predict electron densities, energies, and other properties by solving the electronic Schrodinger equation based on the quantum chemistry, while the MD simulation method is useful to understand molecular failure mechanisms and to predict mechanical properties of composite material [21]. Xiao, Hou, et al. [9] used the domain coupling method to study the elastic modulus and fracture strength of CNT-based aluminum composites. In the present study, MD simulations have been used to study the effect of length, diameter, volume fraction, and type of CNTs on Young's modulus and stress–strain behavior of CNT–Al composite.

P. R. Patel · S. Sharma (✉) · S. K. Tiwari
B. R. Ambedkar National Institute of Technology Jalandhar, Jalandhar 144011, India
e-mail: sharmas@nitj.ac.in

S. K. Saha and M. Mukherjee (eds.), *Recent Advances in Computational Mechanics and Simulations*, Lecture Notes in Mechanical Engineering,
https://doi.org/10.1007/978-981-15-8315-5_5

2 Materials and Methods

In this study, pure aluminum has been used as the matrix with Young's modulus of 69 GPa [2], and SWCNTs have been used as reinforcement, which are formed by rolling a graphene sheet into a cylinder along a (m, n) lattice vector. The (m, n) determines the diameter and chirality (degree of twist) of tube. To study the effect of diameter, volume fraction and the type of SWCNT on Young's modulus and ultimate tensile strength of SWCNT–Al composite, armchair (m = n), and zig-zag (n = 0) CNTs of different diameter were inserted centrally in a periodic aluminum simulation cell of size 40.495 X 40.495 X 40.495$Å^3$. Further, to study the effect of the length of CNTs on the mechanical properties, a (6, 6) armchair SWCNT with different length (29.5, 36.9, 46.7, 54.1, 61.5 Å) was inserted centrally in an Al cell of cross-sectional area 40.495 X 40.495 $Å^2$. Accordingly, the length of the cell was decided. Condensed Phase Optimized Molecular Potential for Atomistic Simulation Studies (COMPASS), a common force field, has been used for describing the inter and intra-molecular atomic interactions.

2.1 *MD Modeling of CNT–Al Composite*

Firstly, a structure of pure Al was imported as a 3D atomistic document from the inbuilt library of Biovia Materials Studio (MS) 2017. Then, an Al super-cell was generated with a super-cell range of 10 X 10 X 10 and a cell volume of 6405.5 $Å^3$ containing 4631 atoms as shown in Fig. 1. Then a periodic single-walled CNT was built with a super-cell range of 1 X 1 X 17 for armchair and 1 X 1 X 10 for zig-zag CNT, as shown in Fig. 2a and b, respectively. Further, the CNT was inserted centrally by removing Al atoms from the specified space in such a way that the interatomic spacing between CNT and Al didn't exceed 3.02 Å, as shown in

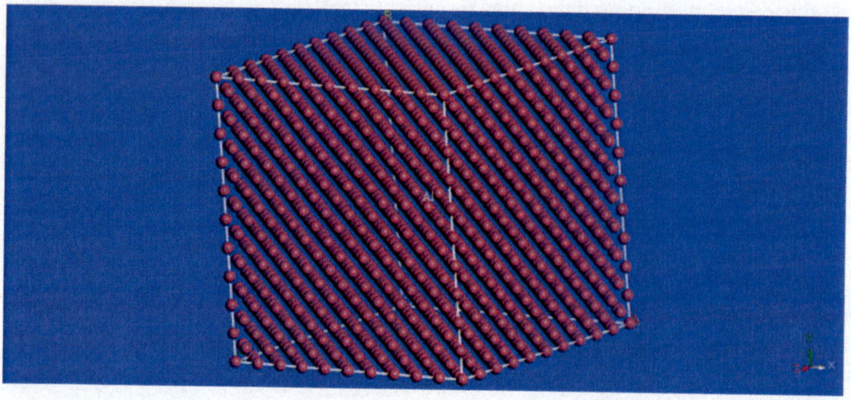

Fig. 1 Al super-cell of 10 X 10 X 10 range

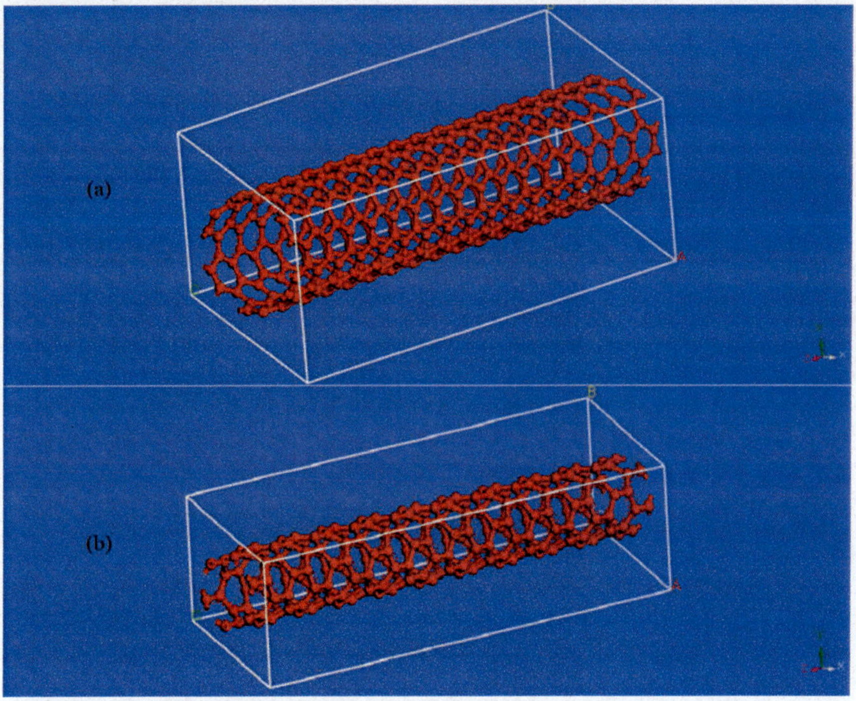

Fig. 2 Super-cell of (**a**) armchair SWCNT (6, 6) and (**b**) zig-zag (6, 0) SWCNT

Figs. 3, 4, and 5. Generally, there are four steps involved in equilibration process and mechanical property calculation of the system: (i) energy optimization (ii) geometry optimization (iii) dynamics (iv) mechanical property. The energy optimization is used to provide stability to the cell by attaining desired coordinates. In this study, energy

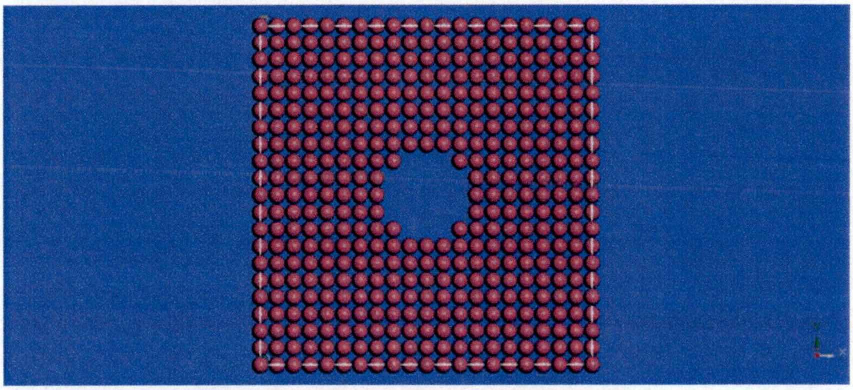

Fig. 3 Al super-cell with central cavity to insert armchair (6, 6) SWCNT

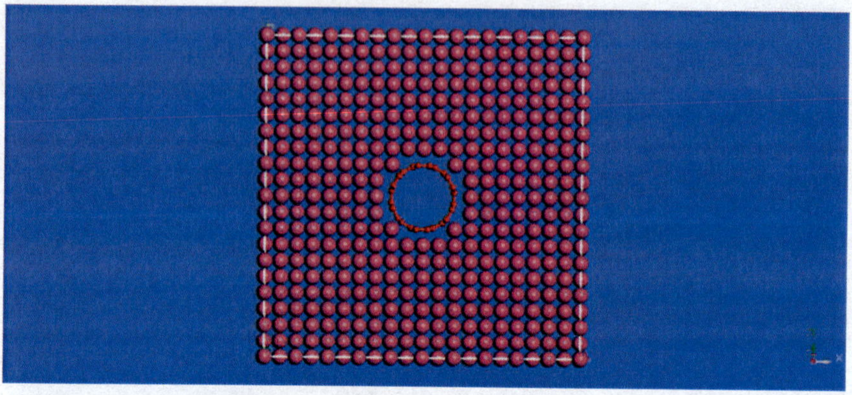

Fig. 4 Front view of SWCNT reinforced Al composite

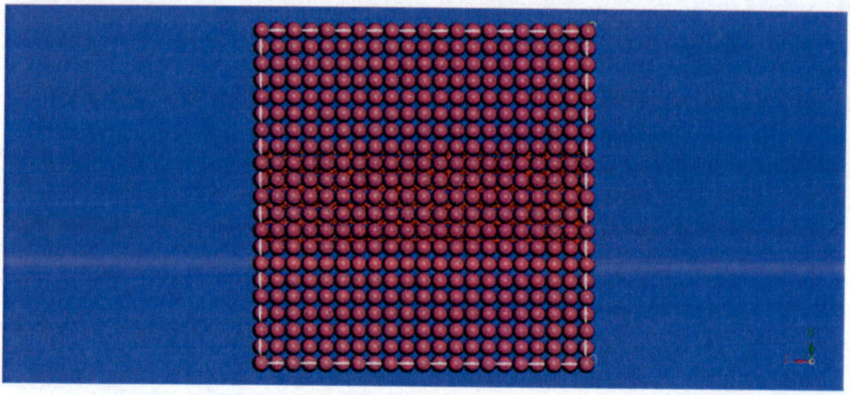

Fig. 5 Side view of SWCNT reinforced Al composite

was minimized using the "Smart algorithm" until the change in energy between subsequent steps was less than 1×10^{-4} kcal/mol. Geometry optimization is used to obtain the proper density and low residual stresses.

The steps of geometry optimization and dynamics were executed over the file generated from the previous steps respectively. The parameters for geometry optimization, dynamics, and calculation of mechanical properties have been listed in Tables 1, 2, and 3.

2.2 *MD Simulation for Young's Modulus*

After dynamics run, a trajectory file has been obtained on which mechanical property calculation was performed using a constant strain approach, where a constant strain

Table 1 Geometry optimization parameters for SWCNT–Al composites [22]

S. No	Parameter	Value
1	Algorithm	Smart
2	Quality convergence tolerance	Fine
3	Energy convergence tolerance	10^{-4} kcal/mol
4	Force convergence tolerance	0.005 kcal/mol/Å
5	Displacement convergence tolerance	5×10^{-5} Å
6	Maximum number of iterations	50,000

Table 2 Dynamics run parameters for SWCNT–Al composites [22]

S. No	Parameter	Value
1	Ensemble	NVE
2	Initial velocity	Random
3	Temperature	298 K
4	Time step	1 fs
5	Total simulation time	100 ps
6	Number of steps	100,000
7	Frame output every	50,000 steps
8	Collision ratio	1
9	Energy deviation	5×10^{12} kcal/mol

Table 3 Mechanical properties simulation parameters for SWCNT–Al composites [22]

S. No	Parameter	Value
1	Number of steps for each strain	6
2	Maximum strain	0.001
4	Algorithm	Smart
5	Maximum number of iterations	5000
6	Forcefield	COMPASS

of 0.001 was applied. The effective elastic moduli can be determined directly from the Virial theorem, in which the expression of the stress tensor in a microscopic system is given as the function of atomic coordinates and interatomic forces [23], which is shown in succeeding text:

$$\sigma = -\frac{1}{V_0}\left[\left(\sum_{i=1}^{n} m_i\left(v_i v_i^T\right)\right) + \left(\sum_{i<j} r_{ij} f_{ij}^T\right)\right] \quad (1)$$

The relative position of particles changes due to the application of stress and expressed quantitatively via strain tensor:

$$\varepsilon_{ij} = \begin{bmatrix} \varepsilon_{11} & \varepsilon_{12} & \varepsilon_{13} \\ \varepsilon_{21} & \varepsilon_{22} & \varepsilon_{23} \\ \varepsilon_{31} & \varepsilon_{32} & \varepsilon_{33} \end{bmatrix} \quad (2)$$

To calculate the axial Young's modulus (E_{11}) the atoms are displaced by $u_1 = \varepsilon_{11}^0$. The average strain and stress are

$$\varepsilon_{11} = \varepsilon_{11}^0 \quad (3)$$

$$\sigma_{11} \neq 0, \text{anyother} \sigma_{ij} = 0 \quad (4)$$

Through MD simulation the longitudinal elastic modulus, E11, can be calculated as

$$E_{11} = \frac{\sigma_{11}}{\varepsilon_{11}^0} \quad (5)$$

In other simulation runs, the load was applied either in transverse or shear direction.

2.3 *MD Simulation for Ultimate Tensile Strength*

A script has been written using the "Perl Scripting Module" in Biovia Materials Studio (2017), to obtain the stress–strain behavior of CNT–Al composite for which trajectory document obtained from dynamics was used as an input file. After successful execution of the Perl script, a stress–strain data was obtained in a tabulated form.

3 Results and Discussion

The evaluated Young's modulus was found to lie between the upper and lower limits predicted by Choi et al. [20] and George et al. [24] for CNT volume fraction ranging from 1 to 4%. From Fig. 6, it could be observed that the maximum value of Young's modulus (130.37 GPa) of CNT–Al composite occurred at the volume fraction of 11.75% for zig-zag SWCNT. Whereas for armchair SWCNT reinforced Al composite, the maximum value of Young's modulus (91.68 GPa) was found at a volume fraction 8.81%.

Young's modulus of zig-zag SWCNT–Al composite is higher as compared to armchair SWCNT–Al composite at the same volume fraction due to the fact that zig-zag SWCNT possess slight higher Young's modulus than armchair SWCNT [25].

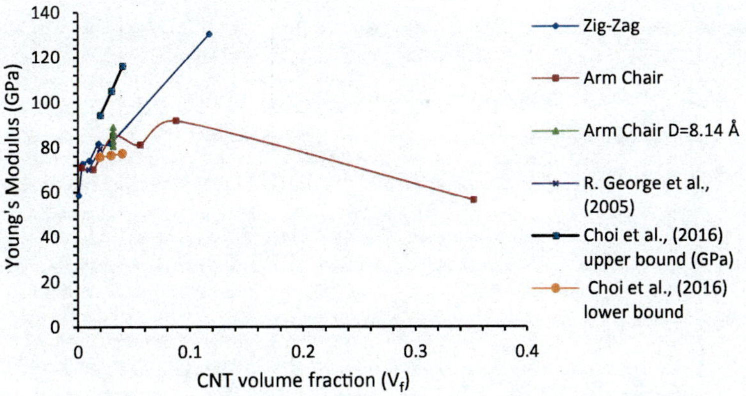

Fig. 6 Variation of Young's modulus of CNT–Al composite with SWCNT volume fraction

From Fig. 6, it could be inferred that Young's modulus for armchair CNT reinforced Al composite increased with an increase in the volume fraction of CNTs up to a certain limit (8.81%) and after that it was found to decrease. This may be due to the reason that at a low volume fraction there is a low chance of agglomeration of CNT while non-uniform dispersion may occur at higher volume fraction. Also the simulation has been performed to study the effect of armchair (6, 6) CNT length at a constant volume fraction on Young's modulus, and it has been observed that the effects were insignificant. The fact behind this behavior is that the fiber axial stress increases with increase in CNT length, hence Young's modulus remains approximately the same [22].

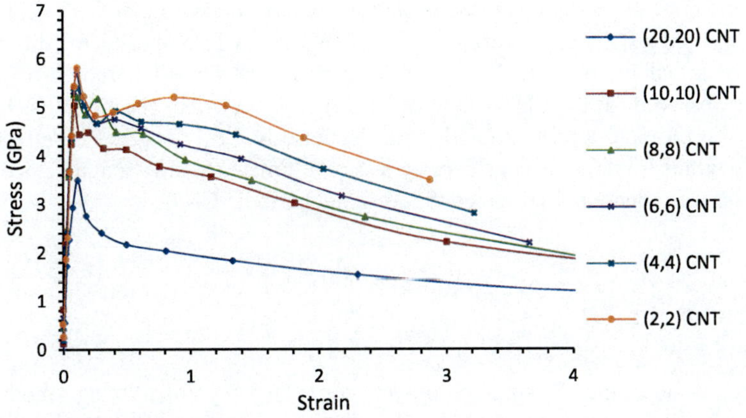

Fig. 7 Stress–strain curve of CNT–Al composite with various armchair CNTs for a constant length of CNT ($L = 40.495$ Å)

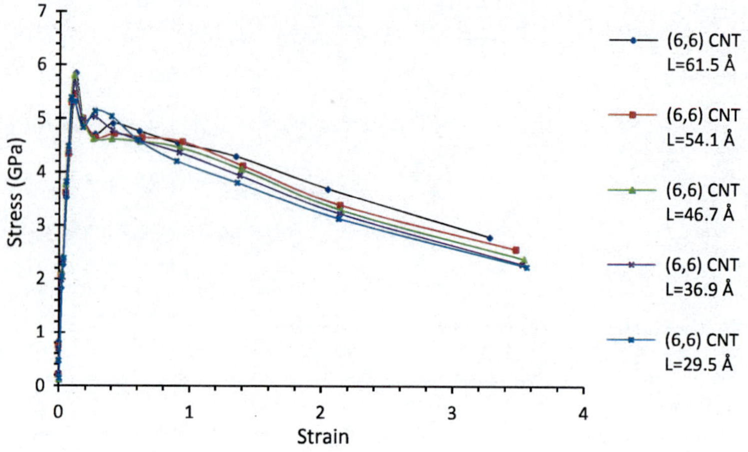

Fig. 8 Stress–strain curve of CNT–Al composite with various lengths of armchair CNTs for a fixed CNT diameter ($d = 8.14$ Å)

Figure 7 depicts the stress–strain behavior of various armchair CNT reinforced Al composite. It could be seen that with an increase in the diameter of the CNTs the ultimate tensile strength of CNT–Al composite kept on decreasing.

The maximum ultimate tensile strength of 5.81 GPa was achieved by armchair (2, 2) CNT reinforced Al composite. On the other hand, with an increase in diameter of the CNTs, from armchair (10, 10) to armchair (20, 20) CNT reinforced Al composite, a degradation of ultimate tensile strength in comparison to pure Al was observed [20]. This is due to the fact that fracture is dominated by extrinsic factors most likely consist of defects and atomic vacancies, since the crack length indicates a linear relationship with diameter of tube which is responsible for fracture/weakening of CNT [26].

Figure 8 shows the stress–strain behavior of CNT–Al composite for different lengths of armchair (6, 6) CNT. The maximum ultimate tensile strength of 5.81 GPa was observed for the CNT length of 61.5 Å, and the minor reduction has been observed in ultimate tensile strength with decrease in CNT length. The effect of CNT's length is found to be considerably less on ultimate tensile strength, since the simulation was conducted for the continuous and periodic CNT.

4 Conclusion

In the present study, the MD simulations were conducted to predict Young's modulus and ultimate tensile strength of CNT–Al composites. Young's modulus increased considerably with increase in CNT reinforcement. This was observed only for a small volume fraction of the CNT ($V_f = 11.75\%$). For the higher volume fraction of CNTs, the increase in Young's modulus was found to be insignificant. The ultimate tensile strength of CNT–Al composite was found to be higher for CNTs having longer length and smaller diameter. The tensile strength CNT–Al composite was found to decrease when CNTs of short length and larger diameter were used as reinforcement. This could be attributed to the lattice structure change and micro-void nucleation.

The interpretation of this study can be used by the researchers to develop lightweight Al composites for various automobiles and structural applications. This study could be extended to predict the mechanical properties of Al composites reinforced with CNTs of various types such as functionalized, defective, and coated CNTs.

References

1. Kurita, H., Kwon, H., Estili, M., Kawasaki, A.: Multiwalled carbon nanotube aluminum matrix composites prepared by combination of hetero-agglomeration method, spark plasma sintering and hot extrusion. Mater. Trans. **52**, 1960–1965 (2011)
2. Kwon, H., Leparoux, M.: Hot extruded carbon nanotube reinforced aluminum matrix composite materials. Nanotechnology **23**, 415701 (2012)
3. Laha, T., Chen, Y., Lahiri, D., Agarwal, A.: Tensile properties of carbon nanotube reinforced aluminum nanocomposite fabricated by plasma spray forming. Compos. A Appl. Sci. Manuf. **40**, 589–594 (2009)
4. Lee, W., Jang, S., Kim, M.J., Moon, J.M.: interracial interactions and dispersion relations in carbon-aluminium nanocomposite systems. Nanotechnology **19**, 285701 (2008)
5. Majid, M., Majzoobi, G., Noozad, G.A., Reihani, A., Mor-tazavi, S., Gorji, M.: Fabrication and mechanical properties of MWCNTs-reinforced aluminum composites by hot extrusion. Rare Metals **31**, 372–378 (2012).
6. Song, H.Y., Zha, X.W.: Influence of nickel coating on the interfacial bonding characteristics of carbon nanotube aluminum composites. Comput. Mater. Sci. **49**, 899–903 (2010)
7. Stein, J., Lenczowski, B., Frety, N., Anglaret, E.: Mechanical reinforcement of a high-performance aluminium alloy AA5083 with homogeneously dispersed multi-walled carbon nanotubes. Carbon **50**, 2264–2272 (2012).
8. Tokunaga, T., Kaneko, K., Horita, Z.: Production of aluminum matrix carbon nanotube composite using high pressure torsion. Mater. Sci. Eng. A **490**, 300–304 (2008)
9. Xiao, S., Hou, W.: Studies of nanotube based aluminum composites using the bridging domain coupling method. Int. J. Multiscale Comput. Eng. **5**, (2007).
10. Hu, Z., Zhang, J., Yan, Y., Yan, J., Sun, T.: Molecular dynamics simulation of tensile behavior of diffusion bonded Ni/Al nanowires. J. Mech. Sci. Technol. **27**, 43–46 (2013)
11. Kutana, A., Giapis, K.: Transient deformation regime in bending of single walled carbon nanotubes. Phys. Rev. Lett. **97**, 245501 (2006)
12. Munilla, J., Castro, M., Carnicero, A.: Surface effects in atomistic mechanical simulations of Al nanocrystals. Phys. Rev. B **80**, 024109 (2009).
13. Shazed, M., Suraya, A., Rahmanian, S., Salleh, M.M.: Effect of fibre coating and geometry on the tensile properties of hybrid carbon nanotube coated carbon fibre reinforced composite. Mater. Des. **54**, 660–669 (2014)
14. Farrash, S.M.H., Shariati, M., Rezaeepazhand, J.: The effect of carbon nanotube dispersion on the dynamic characteristics of unidirectional hybrid composites: An experimental approach. Compos. B Eng. **122**, 1–8 (2017)
15. Lee, S.E., Park, S.H.: Enhanced dispersion and material properties of multiwalled carbon nanotube composites through turbulent Taylor-Couette flow. Compos. A Appl. Sci. Manuf. **95**, 118–124 (2017)
16. Wang, J., Pui, D.Y.: Dispersion and filtration of carbon nanotubes (CNTs) and measurement of nanoparticle agglomerates in diesel exhaust. Chem. Eng. Sci. **85**, 69–76 (2013)

17. Liao, J.Z., Tan, M.J., Sridhar, I.: Spark plasma sintered multi-wall carbon nanotube reinforced aluminum matrix composites. Mater. Des. **31**, S96–S100 (2010)
18. Silvestre, N., Faria, B., Lopes, J.N.C.: Compressive behavior of CNT reinforced aluminum composites using molecular dynamics. Compos. Sci. Technol. **90**, 16–24 (2014)
19. Simoes, S., Viana, F., Reis, M.A., Vieira, M.F.: Influence of dispersion/mixture time on mechanical properties of Al–CNTs nanocomposites. Compos. Struct. **126**, 114–122 (2015)
20. Bong Kyu Choi: Gil Ho Yoon, Seungjun Lee: molecular dynamics studies of CNT-reinforced aluminum composites under uniaxial tensile loading. Compos. B **91**, 119–125 (2016)
21. Winey, J., Kubota, A., Gupta, Y.: A thermodynamic approach to determine accurate potentials for molecular dynamics simulations: thermoelastic response of aluminum. Modell. Simul. Mater. Sci. Eng. **17**, 055004 (2009)
22. Sharma, S., Chandra, R., Kumar, P., Kumar, N.: Molecular dynamics simulation of functionalized SWCNT–polymer composites. J. Compos. Mater. **0**(0), 1–16 (2016).
23. Sharma, S.: Molecular dynamics simulation of nanocomposites using Biovia Materials Studio, LAMMPS and Gromacs, 1st edn. Matthew Deans, Oxford, United Kingdom-Elsevier (2019)
24. George, R., Kashyap, K.T., Rahul, R., Yamdagni, S.: Strengthening in carbon nanotube/aluminium (CNT/Al) composites. Scripta Mater. **53**, 1159–1163 (2005)
25. Liang, Y., Han, Q., Xin, H.: Elastic properties of carbon nanotubes. J. Comput. Theor. Nanosci. **10**, 1061–1071 (2013)
26. Takakura, A., Beppu, Ko., Nishihara, T., Fukui, A., Kozeki, T., Namazu, T., Miyauchi, Y., Itami, K.: Strength of carbon nanotubes depends on their chemical structures. Nat. Commun. **10**, 3040 (2019)

Low Velocity Impact Behavior of Carbon Nanotubes Reinforced Aluminum Foams

Y. M. Chordiya and M. D. Goel

1 Introduction

From the previous reports on road accident in India, it is observed that there has been a tremendous amount of increase in number of accidents over the past years [1]. Hence, a lot of research is being done in this area, i.e., to increase the crashworthiness of the vehicles because vehicular occupant safety is the prime concern while designing the vehicles. The crash boxes are the devices which are filled in with different materials and are impact tested to get energy absorption response of the different materials. One of the best suitable materials to be used as energy absorber is foams, and it is because of their lightweight structure and their impressive compressive properties. The reason can be attributed to the fact that internal structure of the foam has pores present within them, so when they are subjected to impact it leads to cellular rearrangement of the foam which ultimately leads to energy absorption. Also, the stress–strain curve of the foam (Fig. 1) has a plateau region present which is the region in which maximum energy absorption of the foam takes place [2]. Further, foams find their use in a wide range of applications like packaging, automobiles, safety guards, packaging, safety guards, blast lining materials, and helmets [3–7]. Hence, the material used for in this investigation is carbon nanotubes reinforced aluminum foam either.

There are many tests available for impact testing but the most widely used test is drop weight impact hammer machine test. It is used by many researchers in the past for testing different materials like foams, graphite-fiber-reinforced composite, hybrid fiber engineered cementitious composite, concrete [8–12]. Further, many variations in these tests had been suggested by earlier researchers [13–17]. Considering its simplicity and application, drop weight impact hammer test is test simulated in

Y. M. Chordiya · M. D. Goel (✉)
Visvesvaraya National Institute of Technology (VNIT), Nagpur 440 010, India
e-mail: mdgoel@apm.vnit.ac.in

S. K. Saha and M. Mukherjee (eds.), *Recent Advances in Computational Mechanics and Simulations*, Lecture Notes in Mechanical Engineering,
https://doi.org/10.1007/978-981-15-8315-5_6

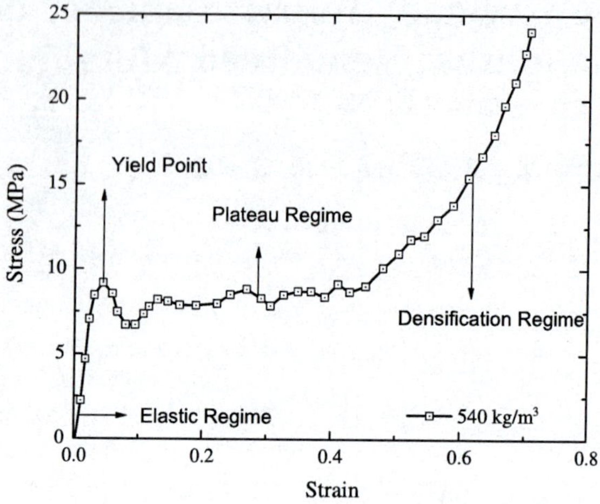

Fig. 1 A Quasi-static stress–strain curve for carbon nanotubes reinforced aluminum foams for density of 540 kg/m^3 [2]

the present investigation. The commonly used foams include aluminum cenosphere syntactic foam, polymeric foams, closed-cell aluminum foam, and polymeric foams [18–23]. However, in this study, carbon nanotubes (CNTs) reinforced aluminum foam is used. In this foam, CNTs are added to the molten aluminum matrix using the liquid metallurgy method, and it is observed that CNTs addition improves the strength of the foam composite [18].

Drop weight impact hammer is an optimum option for conducting impact testing but it requires a lot of manpower, and it is also evident from the literature review that use of numerical simulation is very scarce. Hence, numerical model is prepared in LS-DYNA® for numerical simulation of drop weight impact hammer test. Herein, the hammer is modeled using bilinear material model and foam is modeled using crushable foam material model. In this study, in addition to the effect of drop height and effect of density of foam, the effect of skin is also investigated. Moreover, a comparative study is done based on the parameters such as reaction force–time history, displacement–time history, and energy absorption for all the models developed in this study.

2 Finite Element Modeling and Material Properties

Numerical simulation of drop weight impact hammer test is done by preparing a model in LS-DYNA®, wherein the dimensions of the hammer are 720.2 mm length and 155 mm diameter and the dimensions of foam are 100 mm length and 80 mm diameter. The FE model is said to be validated from Fig. 2a and for further details about validation the author's earlier investigation can be referred [20, 24]. In this study, hammer and skin around the foam are modeled using MAT_003 (MAT_ PLASTIC_KINEMATIC) material model of LS-DYNA® material library

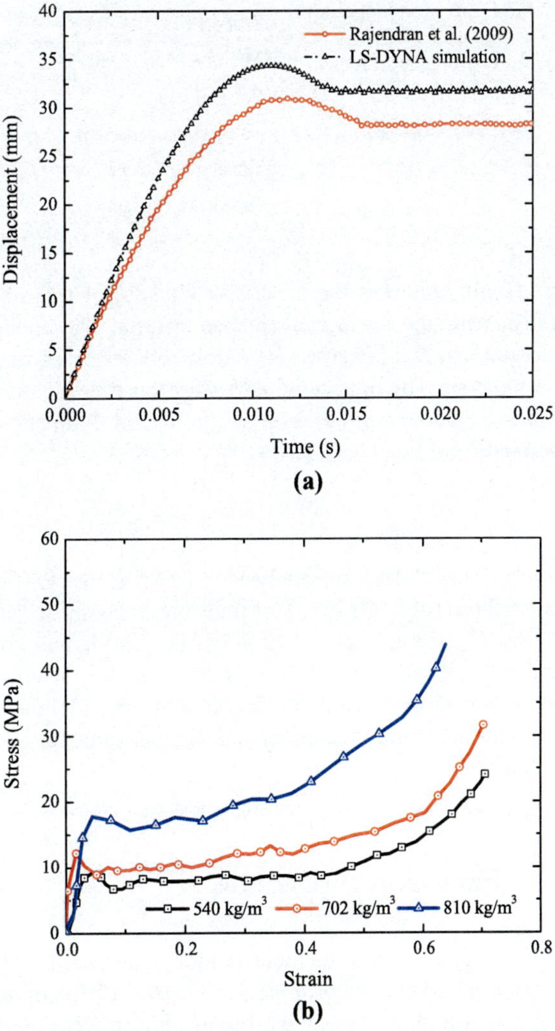

Fig. 2 **a** Variation of displacement–time history of foam in comparison with results reported by Rajendran et al. [7] and **b** Quasi-static stress–strain curve for carbon nanotubes reinforced aluminum foams for three densities [2]

[25]. Hammer is modeled using the eight noded hexahedral solid elements, whereas skin is modeled using shell element. The MAT_003 material model card requires density (ρ), modulus of elasticity (E), Poisson's ratio (μ), yield stress (σ_y), and tangent modulus (E_t) to simulate the behavior of material. The corresponding values for hammer are 7800 kg/m^3, 210 GPa, 0.3, 230 MPa, 800 MPa, respectively, and for the skin, these are 2700 kg/m^3, 70 GPa, 0.3, 364 MPa, 700 MPa, respectively. Foam is modeled using MAT_063 (MAT_ CRUSHABLE_ FOAM) material model and eight noded hexahedral solid elements of LS-DYNA® material library [25]. The properties of the foam for all the three densities are reported in Table 1, and stress–strain curve of the foam is reported in Fig. 2b. The mesh size chosen for hammer

Table 1 Material properties of carbon nanotubes reinforced aluminum foams [2]

Property	Values		
Young's Modulus, E (MPa)	250	900	520
Poisson's ratio, υ	0.3	0.3	0.3
Damping co-efficient	0.3	0.3	0.3
Tension cutoff, p_t	2	2	2
Density, ρ (kg/m^3)	540	702	810

is 20 mm, whereas the mesh size chosen for foam and skin is chosen as 10 mm considering the mesh convergence criteria. The nodes between foam and skin are merged together because separation between foam and skin is not desirable during the analysis. The impact velocities for this investigation are 6.26, 7.67, and 8.85 m/s and these correspond to drop height of 2 m, 3 m, and 4 m, respectively, and can be derived from Eq. 1.

$$v = \sqrt{2gh} \tag{1}$$

The time chosen for analysis is 0.025 s, and the results are extracted at a time interval (Δt) of 0.00125 s. This time interval is chosen such that it satisfies the relation $\Delta t < l/C_L$. Here, l is the length of smallest division of sample and C_L is speed of longitudinal wave which travels through the material. In the present simulation, automatic surface to surface contact criteria is defined between top surface of foam and bottom surface of hammer and clamped boundary condition is given at the bottom of foam.

3 Results and Discussions

In this study, a finite element model is prepared in LS-DYNA® and parameters considered herein are drop height, density of foam, and effect of skin. This study comprises a total of eighteen models and the comparison is done on the basis of parameters such as reaction force–time history and displacement–time history. The deformed shape of the foam at a time 0.015 s is reported in Table 2 for all the models considered herein. It can be observed from the deformed shape of foam that increase in drop velocity leads to increase in deformation in the foam and increase in density leads to reduction in the deformation in the foam. The effect of skin is observed for foam of all the densities and it leads to reduction in deformation of the foam. It can also be observed that deformed patterns of the foams for models without skin follow a buckling type of deformation. Whereas, for model with skin, concertina mode of deformation is observed for foam with density 540 and 702 kg/m^3, but for foam with density 810 kg/m^3 only the top and bottom layer of the foam folds, whereas the other part bulges out. So, for this foam density model, hammer is imparted a velocity of 12 m/s and the deformation pattern is observed to be concertina, but it is observed

Table 2 Deformed shape of foam at time $t = 0.015$ s for different densities

Density (kg/m^3)	Deformed Shape					
	Velocity = 6.26 m/s		Velocity = 7.67 m/s		Velocity = 8.85 m/s	
	Without Skin	With Skin	Without Skin	With Skin	Without Skin	With Skin
540						
702						
810						

that the model with skin and without skin model gave comparable results, i.e., if the strain rate is increased then the effect of skin does not matter much. The peak stress values of model with skin and without skin are compared, and it is observed that presence of skin reduces stress on foam by 4–30%.

Figure 3a depicts the displacement–time history, and it can be observed that increase in drop height leads to increase in displacement and increase in density leads to reduction in displacement. The effect of skin reduces the deformation of foam and this reduction can be observed for foam of all three densities considered herein. The peak displacement comparison is done for model with skin and without skin for velocity of 7.67 m/s, and it is observed that model with skin results in 22.26, 27.71, 16.05% lower peak displacement response in comparison with model without skin for density 540 kg/m^3, 702 kg/m^3, and 810 kg/m^3, respectively.

Figure 3b depicts the reaction force–time history, and it can be observed that increase in drop height and density leads to increase in reaction force. The presence of skin leads to significant amount of increase in the reaction force for all the three densities of foam considered in this investigation. Eventually, up to time duration of 0.01 s, the model with skin gives higher reaction force in comparison with model without skin. Whereas, in some cases model without skin showed more reaction force in comparison with model with skin with all other parameters being same.

The peak reaction force comparison is done for model with skin and without skin for velocity of 7.67 m/s, and it is observed that model with skin results in 7.65, 17.3, 13.04% higher peak reaction force response in comparison with model without skin for density 540 kg/m^3, 702 kg/m^3, and 810 kg/m^3, respectively. It is interesting

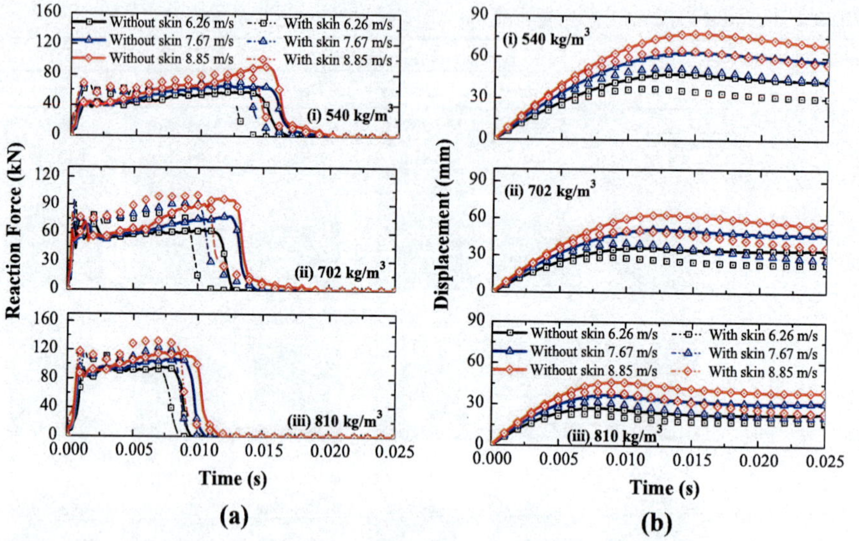

Fig. 3 **a** Reaction force–time history under different drop velocities for foam with density (i) 540 kg/m^3 (ii) 702 kg/m^3 (iii) 810 kg/m^3 and **b** Displacement–time history under different drop velocities for foam with density (i) 540 kg/m^3 (ii) 702 kg/m^3 (iii) 810 kg/m^3

to note that there is a plateau region present in almost all the reaction force–time histories and the displacement corresponding for that particular time duration goes on increasing. This means that in this region there is almost constant force for an increase in displacement which is the basic principle of energy absorption. Hence, these foams are good energy absorbers.

The area under the force–displacement curve gives us the energy absorbed by the foam and it is calculated by Eq. 2.

$$E = \int_{0}^{u} F du \tag{2}$$

Figure 4 shows force–displacement variation for foam of all the densities considered herein, and it observed that the curve increases up to a certain point and then it retraces back. For a particular foam density and the skin configuration the trajectory followed is same. Based on Eq. 2, energy absorption is calculated and reported in Table 3. It can be observed from the table that increase in drop velocity leads to increased energy absorption, increase in density leads to increase in energy absorption and presence of skin accounts an appreciable increase in energy absorption for foam models with skin in comparison with model without skin. The comparison for energy absorption is done for model with skin and without skin for velocity of 7.67 m/s, and it is observed that model with skin results in 14.57, 13.42 17.97% more energy

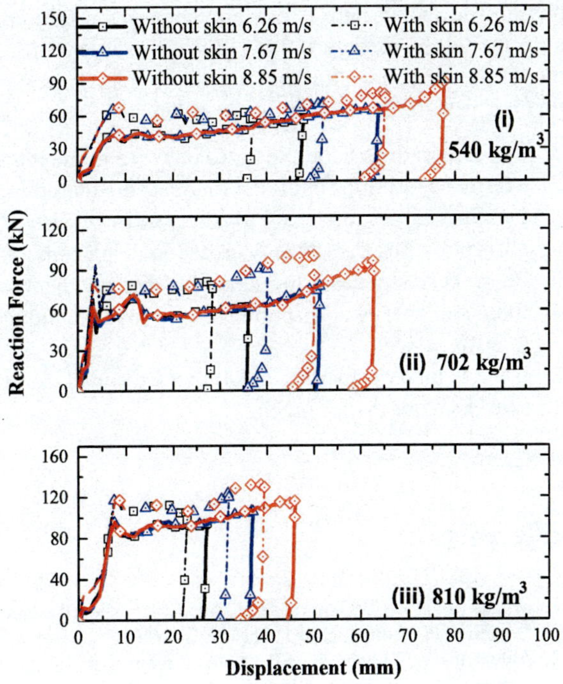

Fig. 4 Variation of force with displacement under different drop velocities for foam with density (i) 540 kg/m³ (ii) 702 kg/m³ (iii) 810 kg/m³

Table 3 Energy absorbed by foam of different densities under different impact velocities

Properties	Energy absorption (J)			
Density (kg/m³)	Provision of skin	Velocity = 6.26 m/s	Velocity = 7.67 m/s	Velocity = 8.85 m/s
540	No	2006.24	3023.84	4013.79
	Yes	2651.83	3464.58	4785.43
702	No	2050.64	3104.62	4150.07
	Yes	2706.4	3521.35	4826.84
810	No	2080.81	3150.6	4218.62
	Yes	2807.4	3176.8	5000.45

absorption in comparison with model without skin for density 540 kg/m³, 702 kg/m³, and 810 kg/m³, respectively.

4 Conclusions

The basic aim of this study was to investigate the effect of drop height, density of foam, and effect of skin on energy absorption characteristics of foam. The material chosen for testing here is carbon nanotubes reinforced aluminum foams. For this, a FE model is prepared in LS-DYNA®, wherein bilinear material model is used to model the hammer, skin, and crushable foam material model is used to model the foam. Based on the study following conclusions can be deduced

1. Increase in drop height leads to increase in the reaction force, displacement, and energy absorption for all the models considered in the present investigation.
2. Increase in density leads to increase in reaction force and increase in energy absorption but it leads to decrease in displacement.
3. Presence of skin leads to decrease in displacement but leads to increase in reaction force and energy absorption. Also, it leads to reduction in stress on foam from 4 to 30%.
4. If the velocity imparted to hammer exceeds a certain value then the presence of skin does not matter much.

References

1. Ministry of Road Transport and Highways Transport Research Wing: Road Accidents in India. Government of India, New Delhi, pp. 2–7 (2016)
2. Aldoshan, A., Khanna, S.: Effect of relative density on the dynamic compressive behavior of carbon nanotube reinforced aluminum foam. Mater. Sci. Eng. A. **689**, 17–24 (2017)
3. Barr, B., Bouamrata, A.: Development of a repeated dropweight impact testing apparatus for studying fibre reinforced concrete materials. Composites **19**, 453–466 (1988)
4. Bowles, K.: The correlation of low-velocity impact resistance of graphite-fiber-reinforced composites with matrix properties. In: Composite Materials: Testing and Design (Eighth Conference). ASTM International, 100 Barr Harbor Drive, PO Box C700, West Conshohocken, PA 19428–2959, pp. 124-124–19 (1988)
5. Gupta, N., Pinisetty, D., Shunmugasamy, V.C.: Reinforced Polymer Matrix Syntactic Foams. Springer International Publishing, Cham (2013)
6. Pinnoji, P.K., Mahajan, P., Bourdet, N., Deck, C., Willinger, R.: Impact dynamics of metal foam shells for motorcycle helmets: experiments and numerical modeling. Int. J. Impact Eng. **37**, 274–284 (2009)
7. Rajendran, R., Moorthi, A., Basu, S.: Numerical simulation of drop weight impact behaviour of closed cell aluminium foam. Mater. Des. **30**, 2823–2830 (2009)
8. Banthia, N., Mindess, S., Bentur, A., Pigeon, M.: Impact testing of concrete using a drop-weight impact machine. Exp. Mech. **29**, 63–69 (1989)
9. Roesset, J.M., Kausel, E., Cuellar, V., Monte, J.L., Valerio, J.: Impact of weight falling onto the ground. J. Geotech. Eng. **120**, 1394–1412 (1994)
10. Zhang, J., Maalej, M., Quek, S.T.: Performance of hybrid-fiber ECC blast/shelter panels subjected to drop weight impact. J. Mater. Civ. Eng. **19**, 855–863 (2007)
11. Sahu, S., Mondal, D.P., Cho, J.U., Goel, M.D., Ansari, M.Z.: Low-velocity impact characteristics of closed cell AA2014-SiCp composite foam. Compos. Part B Eng. **160**, 394–401 (2019)
12. Zhang, W., Chen, S., Liu, Y.: Effect of weight and drop height of hammer on the flexural impact performance of fiber-reinforced concrete. Constr. Build. Mater. **140**, 31–35 (2017)
13. Bhattacharya, S., Krishnamurthy, K.C., Rajendran, R., Prem Sai, K., Basu, S.: Impact studies on structural components using a free-flight drop tower. Exp. Tech. **30**, 52–58 (2006)
14. Goyal, S., Buratynski, E.K.: Methods for realistic drop-testing. Int. J. Microcircuits Electron. Packag. **23**, 45–52 (2000)

15. Gunawan, L., Dirgantara, T., Putra, I.S.: Development of a dropped weight impact testing machine. Int. J. Eng. Technol. **11**, 120–126 (2011)
16. Browne, A.L., Johnson, N.L.: Dynamic crush tests using a "free-flight" drop tower: theory. Exp. Tech. **26**, 43–46 (2002)
17. Taheri-Behrooz, F., Shokrieh, M.M., Abdolvand, H.R.: Designing and manufacturing of a drop weight impact test machine. Eng. Solid Mech. **1**, 69–76 (2013)
18. Aldoshan, A., Mondal, D.P., Khanna, S.: Dynamic high temperature compression of carbon nanotubes reinforced aluminum foams. J. Dyn. Behav. Mater. **3**, 1–11 (2017)
19. Aymerich, F., Marcialis, P., Meili, S., Priolo, P.: An instrumented drop weight machine for low velocity impact testing. Trans. Built Environ. **34**, 53–58 (1997)
20. Chordiya, Y.M., Goel, M.D.: Study of impact behavior of polymeric foams using numerical simulation of drop weight impact. In: Proceedings of National Conference: Advanced Structures, Materials and Methodology in Civil Engineering (ASMMCE–2018), pp. 227–233. NIT Jalandhar, Jalandhar (2018). Last accessed 03–04th Nov 2018
21. Goel, M.D., Mondal, D.P., Yadav, M.S., Gupta, S.K.: Effect of strain rate and relative density on compressive deformation behavior of aluminum cenosphere syntactic foam. Mater. Sci. Eng. A. **590**, 406–415 (2014)
22. Goel, M.D., Matsagar, V.A., Gupta, A.K.: Blast resistance of stiffened sandwich panels with aluminum cenosphere syntactic foam. Int. J. Impact Eng. **77**, 134–146 (2015)
23. Matsagar, V.A.: Computing stress and displacement response of composite plates under blast. Disaster Adv. **7**, 23–38 (2014)
24. Chordiya, Y.M., Goel, M.D.: Low velocity impact behaviour of aluminium cenosphere syntactic foam. In: Proceedings of International Conference: Proceedings of Ninth International Conference of Materials Processing and Characterization (ICMPC-2019), Gokuraju Rangraju Institute of Engineering and Technology (GRIET), Hyderabad (2019). Last accessed 08–10 March 2019
25. LS-DYNA®: KEYWORD USER'S MANUAL. Livermore Software Technology Corporation, Livermore, California, U.S.A (2019)

Recent Advances in Biomechanics

Effect of Interfacial Crack on the Prediction of Bone–Cement Interface Failure of Cemented Acetabular Component

Ajay Kumar, Rajesh Ghosh, and Rajeev Kumar

1 Introduction

The total number of hip revision was 106,320 between "April 2003 and December 2017"; out of that, approximately 45% of hip revision was caused due to the aseptic loosening [1]. Failure of the cemented acetabular component is promising to be begun by mechanical reasons, and it is related to its design, material properties of the implant, interface strength, and fracture properties [2]. One of the leading causes of loosening is interface debonding, which would rise due to the micro-crack generation at the interface. The presence of holes and cavities in the cement or interface results in the initiation of the cracks and finally causes failure or fracture [3]. The cement has further chances of debonding under tensile loading due to the effects of material discontinuity and the crack at the cement–implant interface on interface failure [4]. The bone–cement interface is more responsible for cup failure than the cement–implant interface [5]. Some of the researchers investigated the fatigue failure analysis and crack growth consideration in the cement mantle of acetabular replacements and concluded that the crack propagates linearly in the radial direction until the bone–cement interface [6, 7]. Earlier published studies investigated the interfacial fracture behavior and fracture toughness of the bone–cement interface under mixed-mode loading conditions considered both experimental and numerical approaches, and they provide some primary understanding of the cement fixation technique for joint replacements [8, 9]. However, they did not consider accurate geometry of bone and loading condition and the impact of interfacial crack on the bone–cement interface failure of the cemented acetabular component. Interfacial crack and its impact on the interface failure of the cemented acetabular component is still vague. In this

A. Kumar (✉) · R. Ghosh · R. Kumar
Indian Institute of Technology Mandi, Kamand, Mandi 175005, Himachal Pradesh, India
e-mail: D16033@students.iitmandi.ac.in

S. K. Saha and M. Mukherjee (eds.), *Recent Advances in Computational Mechanics and Simulations*, Lecture Notes in Mechanical Engineering,
https://doi.org/10.1007/978-981-15-8315-5_7

present study, an attempt has been made to understand the impact of the interfacial crack on the bone–cement interface failure of the cemented acetabular component in terms of the mixed-mode Stress Intensity Factor (SIF). To the authors' knowledge, there is a shortage of data related to the impact of interfacial crack on the mixed-mode SIF and potential causes of bone–cement interface failure of the cemented acetabular component by considering the realistic geometry of a bone and corresponding forces and boundary conditions. The goal of this study is to determine the impact of interfacial crack on the SIF at the bone–cement interface of the cemented acetabular component and predict the potential causes of failure of the cemented acetabular component due to THR.

2 Materials and Methods

2.1 3D FEM Model Generation and Material Properties

Three-Dimensional (3D) FE model of the implanted pelvis was generated based on Computed Tomography (CT) dataset of a 62-year-old female patient of weighing 70 kg. FE model created in this study was tested and validated with a previously published research through the use of alike loading and boundary conditions [10–12]. The development procedure of the FE model was described in detail in an earlier published study [12]. The cup was placed at an anteversion angle of 20° and an inclination angle of 40° [13]. For the present study, Polymethyl Methacrylate (PMMA) was considered as the cement material. Uniform cement mantle thickness (3 mm) was considered for the FE model of implanted pelvic bone. Ultra-High Molecular Weight Polyethylene (UHMWPE) was considered for the acetabular implant, and cobalt-chromium-molybdenum (Co-Cr-Mo) was considered for the spherical head. The material properties of cortical bone, cancellous bone, cement, and implants are presented in earlier published research [12, 14–17]. The cortical bone was assumed elastic, isotropic, and homogeneous having Young's Modulus of 17 GPa, Poisson's ratio of 0.3, and a density of 1.73 gm.cm^{-3}. Young's Modulus of the UHMWPE (polymer) acetabular cup is considered as 1.174 GPa, and the Poisson's ratio is considered as 0.4 [12, 14–17]. Young's modulus and Poisson's ratio for cement mantle (PMMA) were considered as 2 GPa and 0.33, respectively. Ten nodded tetrahedral elements were considered in the FE Modeling of implanted pelvic bone. The FE model of the implanted pelvic bone model (Fig. 1) was solved using FEM-based ANSYS FE software v 17 (ANSYS, Inc., PA, USA).

The loading conditions were analyzed similarly to the data given in the earlier published literature [18]. All the forces and boundary conditions were considered as the earlier published studies [10, 19]. The maximum value of the reaction force of the hip joint was found for 13% of the gait cycle, and corresponding hip joint force and muscle forces were considered in this present analysis. These forces were calculated according to body weight and also mentioned in earlier published studies

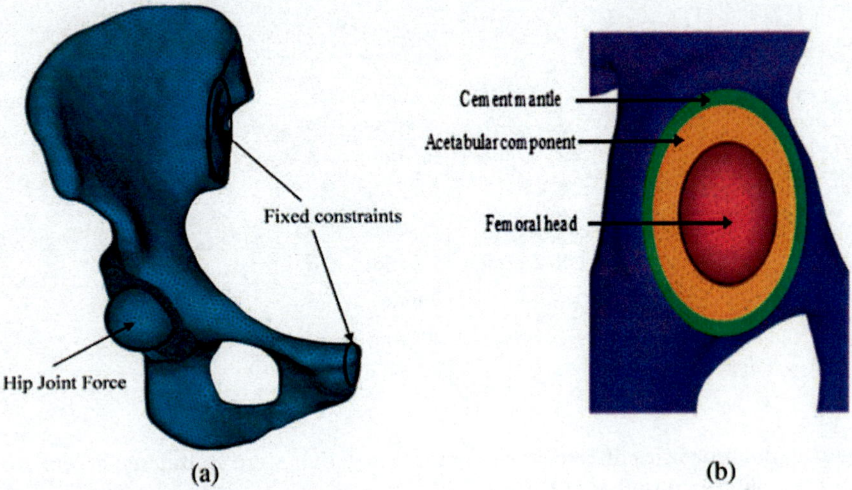

Fig. 1 Finite element model of implanted hemipelvis

[10–12]. Surface-to-surface contact elements (six nodes) were considered to implement contact analysis between the cup and the femoral head, and it was taken to be frictionless [10]. The bonded condition was considered for the bone–cement interface.

2.2 Crack Analysis

To investigate the impact of crack at the interface on the bone–cement interface failure of the cemented acetabular cup, four anatomic locations (superior, inferior, anterior, and posterior) were considered for the current analysis. To understand the impact of interfacial crack at the bone–cement interface, the Two Dimensional (2D) analysis was performed by considering Y–Z (Coronal plane) and X–Z (Sagittal plane) planes. The average stress values (x, y, and z component nodal stresses) obtained from realistic FE models were assigned at the boundary of the rectangular sections. Considering those values of average stresses as boundary loading conditions (mixed-mode loading conditions) (Table 1). There will be less chance of edge crack generation in the Y–Z plane, and thus we considered the edge crack only in the X–Z plane at the bone–cement interface for the FE model to understand the impact of interfacial crack on bone–cement interface failure. Edge and center crack were modeled at the interface (bone–cement) and analyzed using the Element Free Galerkin Method (EFGM). EFGM analysis was implemented to compute the mixed-mode SIF and to minimize the computational time as compared to FEM. The size of the crack was considered the same for both the cases of center and edge crack model. For fracture analysis,

Table 1 Nature of the stress components obtained through FEM analysis was considered as the mixed-mode loading conditions to simulate mixed-mode SIF analysis using EFGM

Analysis	Anatomic locations	Mode-I load (Opening load) (σ)	Mode-II load (Shear load) (τ)
Y–Z plane	Superior	σ_z	σ_y
	Inferior	σ_z	σ_y
	Anterior	σ_y	σ_z
	Posterior	σ_y	σ_z
X–Z plane	Superior	σ_z	σ_x
	Inferior	σ_z	σ_x
	Anterior	σ_x	σ_z
	Posterior	σ_x	σ_z

the material properties of cement and implant materials were considered as similar in earlier published studies [12, 14–17].

In recent years, EFGM has become one of the critical Meshless Methods (MMs) to solve the fracture problem [20–22]. Moving Least Squares (MLS) interpolation is used in the EFGM to compute shape functions. Lagrange's multipliers are used to enforce the boundary conditions as the shape function derived from (MLS) does not follow the Kronecker delta property.

The bi-material interfacial crack problem contains both weak (material discontinuity) and strong discontinuity (crack). Material discontinuity is modeled using the jump function approach [23]. A standard intrinsic enrichment criterion was considered to model the crack-tip stress fields [23]. A modified interaction integral approach was considered to determine of mixed-mode SIF for the bone–cement interfacial crack problem [23]. In this analysis, a domain of 4 × 8 mm has been used for simulation with the edge and center crack of 2 mm. 6 Gauss quadrature and 50 × 100 nodes were considered for simulation after a mesh convergence study. A MATLAB (MATLAB_R2016b; Math Works, United States) script was developed to implement EFGM.

3 Results

Mode-I and mode-II SIF (K_I and K_{II}) at the bone–cement interface of the cemented acetabular component are presented in Figs. 2, 3, and 4. It is shown by the results that mixed-mode SIF in the anterior location at the bone–cement interface is more as compared to other locations (Figs. 2, 3, and 4).

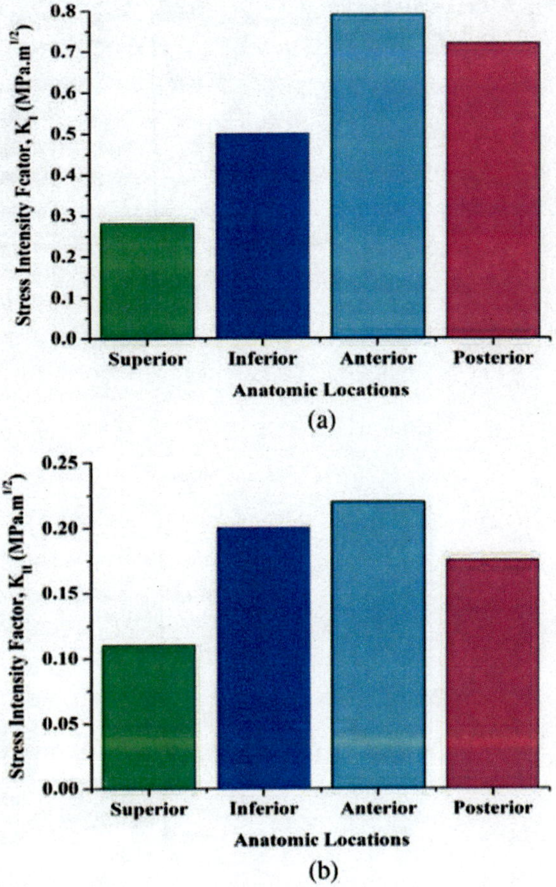

Fig. 2 Comparison of Stress Intensity Factor (SIF) **a** mode-I, and **b** mode-II for Y–Z plane (coronal plane) with the center crack in all anatomic locations (superior, inferior, anterior, and posterior)

3.1 Impact of the Center Crack in the Coronal Plane (Y–Z Plane)

The highest value of K_I and K_{II} at the anterior location was found to be 0.79 and 0.22 MPa.m$^{1/2}$ at the bone–cement interface (Fig. 2). The highest value of SIF (K_I and K_{II}) at the superior location was found to be 0.282 and 0.11 MPa.m$^{1/2}$ at the bone–cement interface. The lowest value of K_I and K_{II} was found to be in a superior location as compared to other anatomic locations (Fig. 2). The inferior and posterior locations have an intermediate amount of K_I and K_{II} at both the interfaces as compared to superior and anterior locations. The value of mixed-mode SIF was found to be positive, and the value of K_I is higher than K_{II}. Hence there will be more chance of opening mode of failure in the Y–Z plane with center crack (Fig. 2).

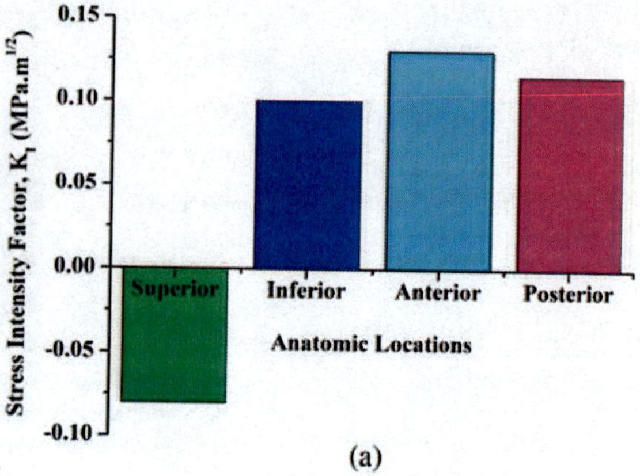

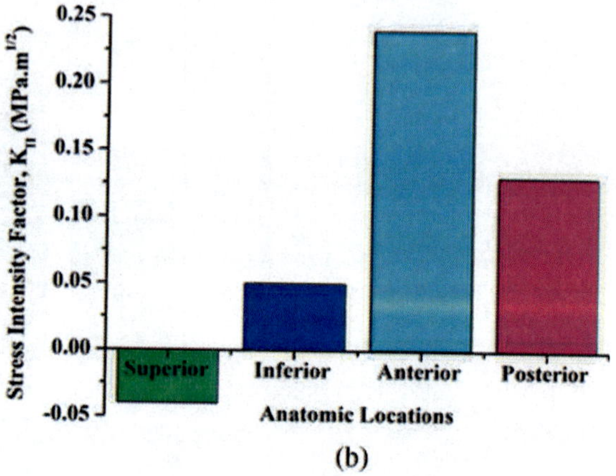

Fig. 3 Comparison of Stress Intensity Factor (SIF) **a** mode-I, and **b** mode-II for X–Z plane (Sagittal Plane) with the center crack in all anatomic locations (superior, inferior, anterior, and posterior

3.2 Impact of the Center Crack in the Sagittal Plane (X–Z Plane)

The highest value of K_I and K_{II} were found in the anterior location at the bone–cement interface.

The highest value of K_I and K_{II} at the anterior location was found to be 0.13 and 0.24 MPa.m$^{1/2}$ at the bone–cement interface. The lowest and negative values of K_I and K_{II} were found to be at the superior location as compared to any other locations for both the interfaces (Fig. 3). The highest value of K_I and K_{II} at the superior location was found to be −0.083 and −0.045 MPa.m$^{1/2}$ at the bone–cement interface. Results at the anterior location indicated that the value of K_{II} is more as compared to $K_{I,}$ and the chance of shear failure (mode-II) is more as compared to opening mode (mode-I) (Fig. 3). In the superior location, the value of K_I and K_{II} for

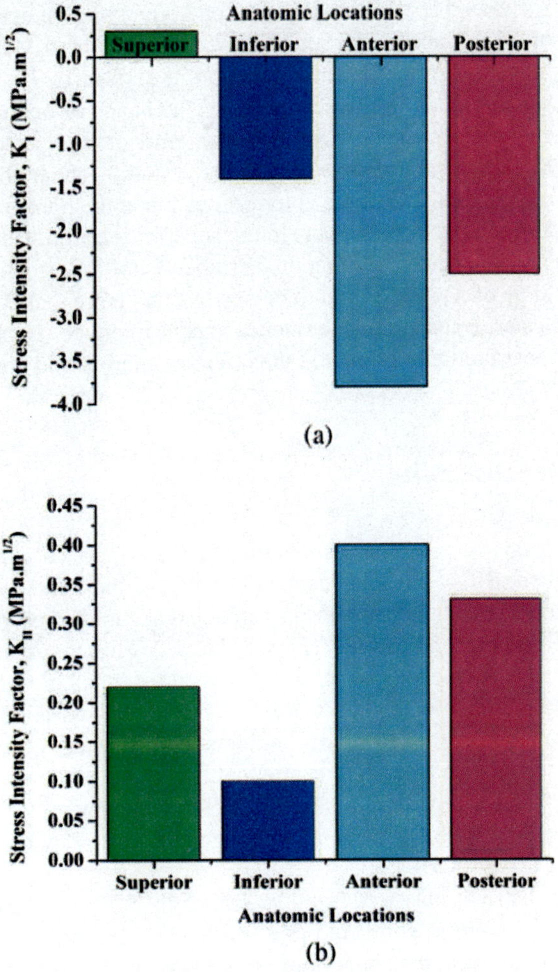

Fig. 4 Comparison of Stress Intensity Factor (SIF) **a** mode-I, and **b** mode-II for X–Z plane (Sagittal Plane) with edge crack in all anatomic locations (superior, inferior, anterior, and posterior)

the bone–cement interface was found to be negative (Fig. 3). Hence, there are very fewer chances of failure due to the opening of the crack (Fig. 3).

3.3 Impact of an Edge Crack in the Sagittal Plane (X–Z Plane)

Edge crack at the X–Z plane has a more substantial impact on K_I and K_{II} as compared to center crack in all the anatomic locations and at the bone–cement interface of the cemented acetabular component (Figs. 3 and 4). For edge crack, the highest value of K_I was found in the superior location, whereas the highest value of K_{II}

was found in the anterior location at the bone–cement interface. The highest value of K_I in the superior location was found to be 0.31 MPa.m$^{1/2}$ at the bone–cement interface. In the superior location, the value of K_I at the bone–cement interface was found to be positive and have a chance of opening mode (mode-I) of failure. However, other locations (inferior, anterior, and posterior) have negative values of K_I due to compressive load fields (Fig. 4a). The highest value of K_{II} was found to be 0.4 MPa.m$^{1/2}$ at the anterior location at the bone–cement interface. The value of K_I is found to be highest in the superior location. Hence, the superior location has more chance of opening mode (mode-I) failure as compared to other locations. The value of K_{II} was found to be positive and more than K_I in the inferior, anterior, and posterior location at the bone–cement interface. Hence, there will be a chance of shear failure (mode-II) at the inferior, anterior, and posterior locations (Fig. 4).

4 Discussion

In this research paper, we studied the impact of interfacial crack on the mixed-mode SIF at the bone–cement interface of the cemented acetabular component. The maximum and positive values of stress and the corresponding mixed-mode SIF were considered in this study (Figs. 2, 3 and 4). The maximum positive value of K_I and K_{II} at the bone–cement interface was obtained for all models are within the limit of interfacial fracture toughness (K_{IC}) of cancellous bone and bone–cement (PMMA) under mode-I and mixed-mode loading conditions [24–26]. One of the earlier published studies stated that the fracture toughness of the bone–cement interface under mode-I loading was within the limit of 0.5–0.8 MPa.m$^{1/2}$ [24]. Though, the fracture toughness of the bone–cement interface under mixed-mode loading was within the limit of 0.11–1.6 MPa.m$^{1/2}$ [25, 26].

It can be observed from the analysis of SIF (Y–Z plane) with the center crack that K_I is more as compared to K_{II} (Fig. 2). This indicates that in the Y–Z plane with center crack, the chances of failure would be dominated by opening mode (Fig. 2). For SIF analysis with the center crack in the Y–Z plane, the highest value of SIF was observed at the anterior location as compared to the other anatomic locations. Hence, the anterior location was more prone to failure as compared to other locations (Figs. 2, 3 and 4). The influence of center crack at X–Z plane indicated that mode-II loading or failure plays a crucial role at the anterior location, where K_{II} (shear mode) was found to be more as compared to K_I at the bone–cement interface (Fig. 3). However, the value of K_{II} is more as compared to K_I, which indicates that chances of the shear mode of failure are more as compared to opening mode. The inferior location also has the positive value of K_I at the bone–cement interface. However, the value is less than the anterior location. In the analysis of edge crack in the X–Z plane, in the superior location, the K_I is positive and more as compared to K_{II} at the bone–cement interface (Fig. 4). This indicated that the chances of failure in superior location would be dominated by opening (mode-I) mode. Although, having a very high negative value of K_I, anterior have the maximum chance of shear failure due to the higher value of K_{II} as compared to other anatomic locations and the fact that bone–cement interface is more compliance in tension than compression [27]. The main reason behind this is due to the increase in stresses in the bone, cement, and the interface of the cemented acetabular component in anterior location as compared to other anatomic locations (superior, inferior, and posterior). The high value of stress and SIF at the bone–cement interface could increase the risk of failure of the cemented acetabular component.

There are a few limitations and assumptions in this study. We do not take into account the aspects like tissue porosity, cement line, and microstructure changes to predict the mixed-mode Stress Intensity Factor (SIF) at the bone–cement interface. The cancellous bone was assumed to be linear, elastic, isotropic, and heterogeneous. In the case of numerical analysis, Linear Elastic Fracture Mechanics (LEFM) was used to perform the interfacial SIF analysis, and it may underestimate the total fracture parameters of bone because it did not account for the energy required for plastic deformations. However, the bone cement (PMMA) is generally brittle, so there will be less chance of plastic deformation around the interfacial crack tip.

5 Conclusions

The main aim of this study is to understand the influence of interfacial crack on bone–cement interface failure of the cemented acetabular component. Mode-I and mode-II SIF (K_I and K_{II}) at the superior, inferior, anterior, and posterior locations were identified by considering center and edge crack at the bone–cement interface. The following conclusions may be drawn from the present analysis.

- The anterior location has the highest value of mixed-mode SIF and more likely to fail as compared to other anatomic locations (superior, inferior, and posterior).
- The present analysis concluded that in the X–Z plane (Sagittal plane), the chances of interface failure are more for the generation of edge crack as compared to center crack.

6 Conflict of Interest

None.

Acknowledgements We acknowledge the Department of Science and Technology (DST), India (Grant No: ECR/2016/000023) for financial support. The authors would like to thank IIT Mandi for supporting this study.

References

1. National Joint Registry, 2018. 15th Annual Report–National Joint Registry for England and Wales, pp. 1–220 (2018)
2. Huiskes, R.: Mechanical failure in total hip arthroplasty with cement. Curr. Orthop. **7**, 239–247 (1993)
3. Haghpanahi, M., Ghomashchi, H.: Cement-implant interface fracture failure by crack initiation due to interface cavity stress concentration. Int. J. Eng. Sci. **91**, 173–186 (2008)
4. Mann, K.A., Bhashyam, S.: Mixed-mode fracture toughness of the cobalt-chromium alloy/polymethylmethacrylate cement interface. J. Orthop. Res. **17**(3), 321–328 (1999)
5. Sih, G.C., Moyer, E.T., Berman, A.T.: Analytical modeling of bone-cement interface and failure prediction. Eng. Fract. Mech. **14**(4), 779–787 (1981)
6. Zant, N.P., Wong, C.K.Y., Tong, J.: Fatigue failure in the cement mantle of a simplified acetabular replacement model. Int. J. Fatigue **29**(7), 1245–1252 (2007)
7. Benbarek, S., Bachir Bouiadjra, B.A., El Mokhtar, B.M., Achour, T., Serier, B.: Numerical analysis of the crack growth path in the cement mantle of the reconstructed acetabulum. Mater. Sci. Eng. C **33**(1), 543–549 (2013)
8. Wang, J.Y., Tozzi, G., Chen, J., Contal, F., Lupton, C., Tong, J.: Bone-cement interfacial behaviour under mixed mode loading conditions. J. Mech. Behav. Biomed. Mater. **3**(5), 392–398 (2010)
9. Tong, J.: Interfacial fracture toughness of synthetic bone-cement interface. Key. Eng. Mater. **312**, 251–256 (2006)
10. Ghosh, R., Pal, B., Ghosh, D., Gupta, S.: Finite element analysis of a hemi-pelvis: the effect of inclusion of cartilage layer on acetabular stresses and strain. Comput. Methods Biomech. Biomed. Eng. **18**(7), 697–710 (2015)
11. Ghosh, R., Gupta, S., Dickinson, A., Browne, M.: Experimental validation of finite element models of intact and implanted composite hemipelvises using digital image correlation. J. Biomech. Eng. **134**(August), 1–9 (2012)
12. Ghosh, R.: Assessment of failure of cemented polyethylene acetabular component due to bone remodeling: a finite element study. J. Orthop. **13**(3), 140–147 (2016)
13. Lewinnek, G.E., Lewis, J.L., Tarr, R., Compere, C.L., Zimmerman, J.R.: Dislocation after total hip-replacement arthroplasties. J. Bone Jt. Surg. **60**(2), 217–220 (1978)
14. Ghosh, R., Pratihar, D.K., Gupta, S.: Towards the optimal design of an uncemented acetabular component using genetic algorithms. Eng. Optim. 1–15 (2014)
15. Ghosh, R., Gupta, S.: Bone remodelling around cementless composite acetabular components: the effects of implant geometry and implant–bone interfacial conditions. J. Mech. Behav. Biomed. Mater. **32**, 257–269 (2014)
16. Coultrup, O.J., Hunt, C., Wroblewski, B.M., Taylor, M.: Computational assessment of the effect of polyethylene wear rate, mantle thickness, and porosity on the mechanical failure of the acetabular cement mantle. J. Orthop. Res. **28**(5), 565–570 (2010)
17. Dalstra, M., Huiskes, R., Odgaard, A., van Erning, L.: Mechanical and Textural Properties of Pelvic Trabecular Bone. J Biomech **26**(4–5), 523–535 (1993)
18. Dalstra, M., Huiskes, R.: Load transfer across the pelvic bone. J. Biomech. **28**(6), 715–724 (1995)
19. Zant, N.P., Heaton-Adegbile, P., Hussell, J.G., Tong, J. : In Vitro fatigue failure of cemented acetabular replacements: a hip simulator study. J. Biomech. Eng. **130**(2), 021019 (2008)
20. Belytschko, T., Organ, D., Krongauz, Y.: A coupled finite element-element-free galerkin method. Comput. Mech. **17**(3), 186–195 (1995)
21. Belytschko, T., Krongauz, Y., Organ, D., Fleming, M., Krysl, P.: Meshless methods: an overview and recent developments. Comput. Methods Appl. Mech. Eng. **139**(1), 3–47 (1996)
22. Fleming, M., Moran, B.: Enriched element-free Galerkin methods for crack tip fields. Int. J. Numer. Methods Eng. **40**, 1483–1504 (1997)
23. Pant, M., Singh, I.V., Mishra, B.K.: Evaluation of mixed mode stress intensity factors for interface cracks using EFGM. Appl. Math. Model. **35**(7), 3443–3459 (2011)

24. Chiang, F.T., Hung, J.P.: Investigation of the fracture characteristics of the interfacial bond between bone and cement: experimental and finite element approaches. J. Mech. Sci. Technol. **24**(6), 1235–1244 (2010)
25. Kumar, A., Sanjay, D., Mondal, S., Ghosh, R., Kumar, R.: Influence of interface crack and non-uniform cement thickness on mixed-mode stress intensity factor and prediction of interface failure of cemented acetabular cup. Theoretical and Applied Fracture Mechanics 107, 102524 (2020). https://doi.org/10.1016/j.tafmec.2020.102524
26. Wang, X., Agrawal, C.M.: A mixed mode fracture toughness test of bone-biomaterial interfaces. J. Biomed. Mater. Res. **53**, 664–672 (2000)
27. Janssen, D., Mann, K.A., Verdonschot, N.: Finite element simulation of cement-bone interface micromechanics: a comparison to experimental results. J. Orthop. Res. **27**(10), 1312–1318 (2009)

Effect of Occlusion Percentage and Lesion Length on Stenosed Coronary Artery: A Numerical Study

Supratim Saha, T. Purushotham, and K. Arul Prakash

1 Introduction

Atherosclerosis is a disease state that causes about 20 million deaths worldwide. CHD-related demise in India increased from 17% in 2001–2003 to 23% in 2010–2013 [1]. Therefore, any rapid and accurate approach to the diagnosis of coronary artery defects will improve its succession planning, along with clinical therapy in the coronary artery. Nonetheless, most cases of CHD fall between extremely mild and very severe cases. It perplexes the situation for physicians to assess the safest choice in this case. Fractional flow reserve (FFR) is the ratio of coronary stenosis pressures used as a clinical measure to evaluate the functional severity [2]. FFR is defined as the ratio of distal pressure (P_d) to proximal pressure (P_p) of the stenotic lesion. This can be written as

$$FFR = \frac{P_d}{P_p} \quad (1)$$

Diagnostic procedure for patients to assess severity of the stenosis is invasive. The process requires a wire probe to be inserted into a patient blood vessel and sent to stenosis site for examination. In the end, pressure values are measured across the occluded artery to measure the severity of the occlusion. A computed tomography angiogram (CTA) obtains 3D images of patient's beating heart and the larger blood vessels associated with the human body. This 3D image is reproduced and computational calculations are performed on the generated 3D artery model to estimate the FFR value. While 3D modelling predicts FFR value similar to that achieved using the intrusive wire insertion process, the time needed for computing is immense because

S. Saha · T. Purushotham · K. A. Prakash (✉)
Indian Institute of Technology, Chennai 600036, India
e-mail: arulk@iitm.ac.in

S. K. Saha and M. Mukherjee (eds.), *Recent Advances in Computational Mechanics and Simulations*, Lecture Notes in Mechanical Engineering,
https://doi.org/10.1007/978-981-15-8315-5_8

it involves a realistic stenotic lesion and also considers the incorporation of vessels as a lumped model at the terminal end in the 3D computational domain which is due to the computing expenses in a overall scenario [3]. The numerical simulation is done for implementing a full arterial tree with the stenotic lesion as one of the blood vessels, and 1D simulation with reasonably good accuracy can predict clinical parameters at minimal numerical expenses. The partial differential equations which govern the flow of blood through the blood vessel were given by Sherwin et al. [4]. A complete 1D arterial circulation model by incorporating the systemic circulation dynamics is given by Mynard and Nithiarasu with the coronary flow circulation [5]. They integrated a cardiac model into the arterial flow and performed simulations. The majority of earlier studies used a 3D model for FFR calculation. Nonetheless, 3D simulations require much computational time, whereas 1D simulations require comparatively less computational time and cost. Iguchi et al. [6] clinically studied the effect of lesion span on functional severity for intermediate coronary lesions. They found that lesion length has physiological importance in coronary lesions of intermediate-grade coronary. Several authors have clinically studied the impact of lesion size on the severity of stenosis but no substantial numerical analysis has been reported. Detailed analysis of the haemodynamic variables is expected due to the impact of the occlusion percent and the size of the lesion combined concurrently. A 1D model and calculations were conducted from the arterial network. The combined effects of occlusion percentage and span of the lesion on haemodynamic parameters are investigated in this study. The geometric effect of occlusion on FFR is primarily investigated and the correlation between FFR and geometric attributes is generated.

2 Methodology

2.1 Numerical Results

The artery is regarded to be an elastic cylindrical conduit. For continuity and momentum conservation, the 1D equations (as provided in Sherwin et al. [4]) are

$$\frac{\partial A}{\partial t} + \frac{\partial (Au)}{\partial x} = 0 \tag{2}$$

$$\frac{\partial u}{\partial t} + u\frac{\partial u}{\partial x} + \frac{1}{\rho}\frac{\partial p}{\partial x} - \frac{f}{\rho A} = 0 \tag{3}$$

where A represents the area of the cross-section, the mean velocity is u, the pressure inside the artery is p, the density of the blood is $\rho \approx 1060\ \text{kg/m}^3$ and the friction force is f per unit length. For modelling the friction force term, a steady, laminar and Poiseuille flow is assumed.

Equations (2) and (3) contain three unknown parameters. The system of equations is closed by incorporating an equation relating the pressure to the cross-sectional area based on the vessel's elasticity (as shown in Formaggia et al. [7], Olufsen et al. [8]) by

$$p = p_{ext} + \beta\left(\sqrt{A} - \sqrt{A_0}\right) \tag{4}$$

where p_{ext} is the ambient tissue pressure, A_0 is the region where the transmural pressure is nil (i.e. $p = p_{ext}$) and the vessel's material properties are accounted for β. w_1 and w_2 are roots of a typical characteristics system of hyperbolic equations. w_1 and w_2 are characteristics of the hyperbolic system of partial differential equations corresponding to the 1D model. Dependent variables (A) and (u) are derived from the values of forward (w_1) characteristics and backward characteristics (w_2) as

$$A = \frac{(w_1 - w_2)^2}{1024}\left(\frac{\rho}{\beta}\right)^2 \tag{5}$$

$$u = \frac{1}{2}(w_1 + w_2) \tag{6}$$

For numerical simulation, the Locally Conservative Galerkin (LCG) approach is used. Mynard et al. [5] implemented the LCG numerical strategy, in which each element is regarded as a subdomain with its own elemental boundaries.

2.2 *Modelling of the Arterial Tree*

The whole arterial framework coupling systemic and coronary flow is developed and is demonstrated in Fig. 1. The blood vessel and their properties that are utilized for carrying out the simulation are acquired from the works of Mynard et al. [5]. The left endocardial artery is modelled as a left anterior descending artery in addition to all other segments to be healthy [9].

2.3 *Computational Domain of the Stenosed Section*

The geometry of the Left Anterior Descending Artery (LAD) is given in Fig. 2. The stenosed section is modelled based on Eq. (7) and it is represented as

$$y = \left[\frac{\varepsilon(L_{seg} - x)}{L_{seg}}\right] - \left[\frac{R}{2}S_0\left\{1 - \cos\left(2\pi\left(\frac{x - L_m - \frac{L_s}{2}}{L_s}\right)\right)\right\}\right] \tag{7}$$

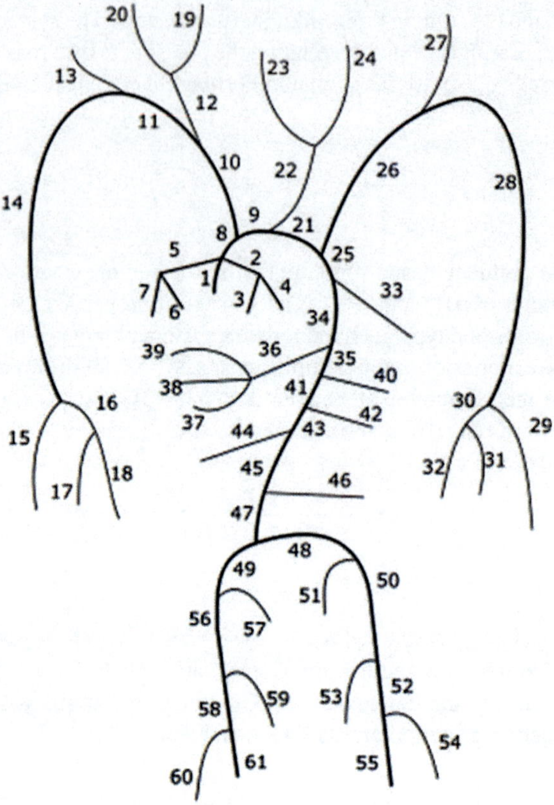

Fig. 1. 1D blood vessel network of prime arteries is utilized from the works of Mynard et al. [5]

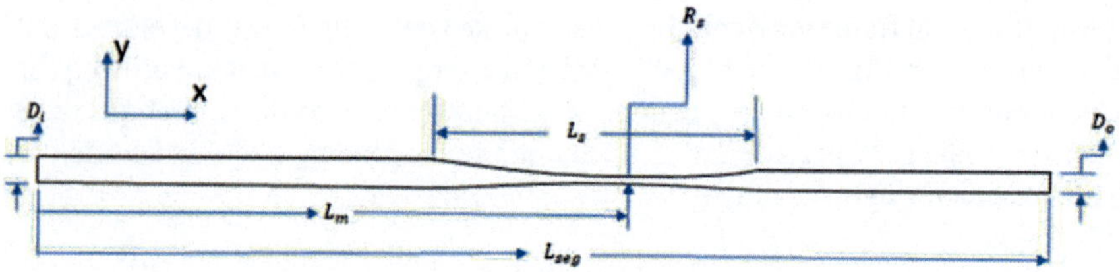

Fig. 2 Geometry with label

where

$$\left\{L_m - \frac{L_s}{2} \le x \le L_m - \frac{L_s}{2}\right\}, \tag{8}$$

$$S_0 = \frac{\text{severity}}{100}, e = \frac{(D_i - D_o)}{2} \tag{9}$$

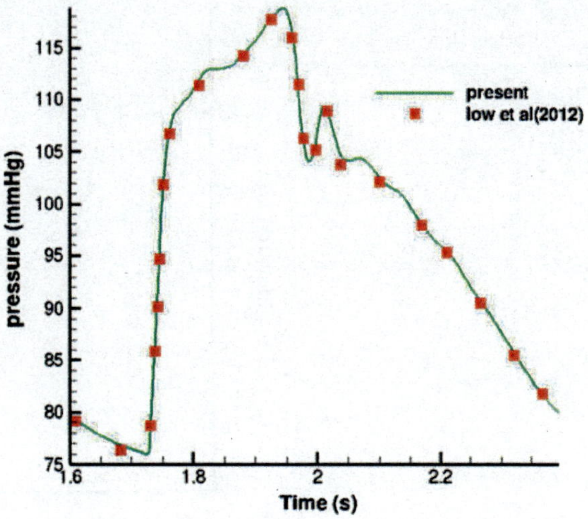

Fig. 3 Comparison of right carotid artery pressure with results reported by Low et al. [10]

L_{seg} is the size of the blood vessel segment, R_s is the radius at the stenotic lesion, L_m is the distance from the start of the arterial segment to the location where the flow cross-section is maximum constricted, L_s is the span of the blood vessel having stenosis, D_i is the diameter of the inlet, D_o is the diameter of the exit and severity or occlusion percentage is the percentage of blockage that occurred in the blood vessel, i.e. any value ranging from 0 to 100.

2.4 Validation of Results

The results obtained from present study are compared numerical results generated by Low et al. [10] in Figs. 3 and 4. The right carotid artery is the artery used for validation. Under normal conditions and heart function, the pattern of pressure waveform and flow waveform is found to be similar [11].

2.5 Grid Independence Test

Three grid sizes having nodes of 50, 100 and 150 are used in the calculations. For the three grid sizes in Fig. 5, the axial velocity value variation along the span of the artery is examined. It is observed that both 100 grid sizes along with 150 gride sizes generate identical numerical results. Thus, for simulation, 100 nodes per artery are chosen.

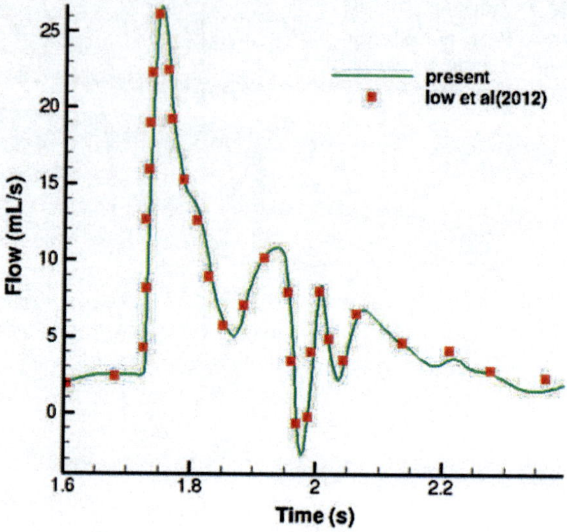

Fig. 4 Comparison of right carotid artery flow rate with results reported by Low et al. [10]

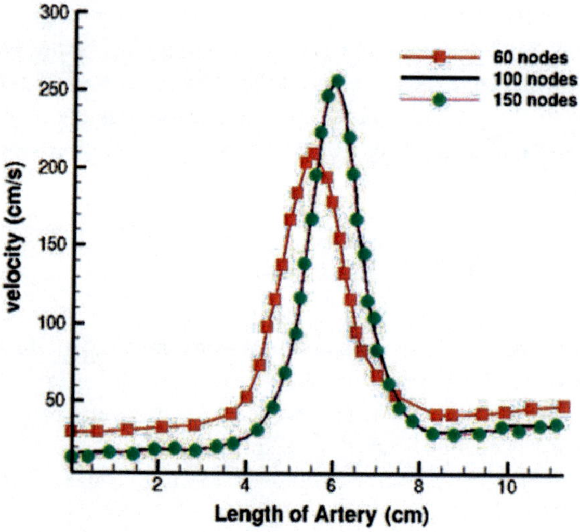

Fig. 5 Velocity change for the grid independence check along the axis of the artery (70% blockage)

3 Results and Discussion

3.1 *Waveforms of Flow Variables*

Figures 6 and 7 demonstrate the flow waveform found in the occluded coronary artery for different scenarios of occlusion percent (OP) and lesion length (LL). In the systolic phase of heart cycle, a flow increase is found for situations of the diseased state.

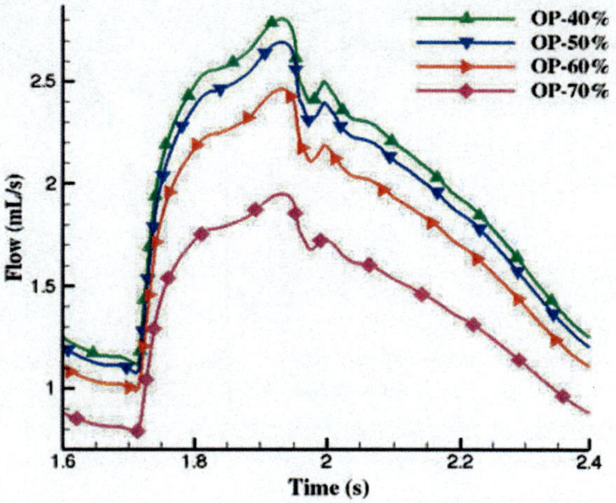

Fig. 6 Variation of the arterial flow rate for different percent blockages having same length of lesion (5 cm) percent blockage 70

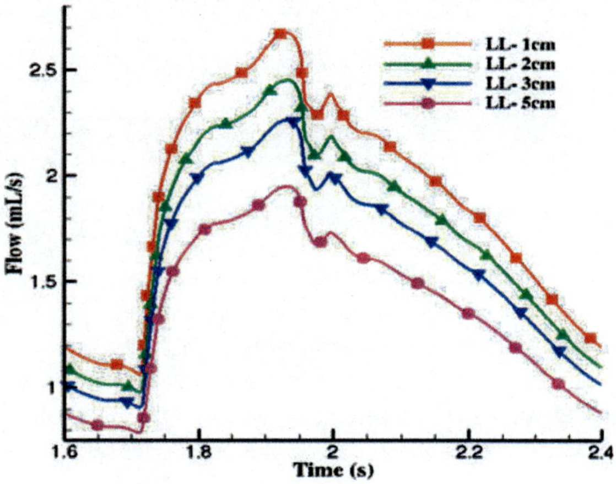

Fig. 7 Variation of the arterial flow rate for different lengths of a lesion having percent blockage 70

The increase of flow during the systole phase of the heart cycle is evident, as reported by Rammos et al. [12]. In the literature, this trend is referred to as the systolic flow rise. With the rise in percent blockage of stenosis, the wave reflection increases. As the amount of blockage increases, the velocity magnitude based on the characteristics should diminish. So, the rate of flow declines with accretion in occlusion. For the same percent blockage, the similar occurrence is observed with the span of the lesion as illustrated in Fig. 7. This finding demonstrates that not only percent blockage is a significant parameter but also the span of the lesion enact a vital role in the artery of the intermediate level.

The flowrate waveform shows a more significant trend for higher percent blockage due to lesion length, as shown in Figs. 7 and 8. The percent decrease of flow rate for 70% blockage is almost four times compared to 50% blockage for the same increase in lesion length. As the percent blockage increases, the influence of lesion

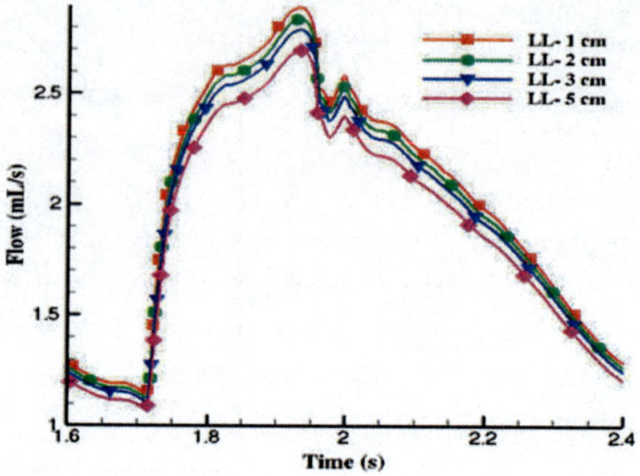

Fig.8 Variation of the arterial flow rate for different lengths of a lesion having percent blockage 50

Table 1 FFR variation with lesion length for 70% blockage

Lesion length (cm)	FFR
1	0.93053
2	0.88932
3	0.85397
4	0.82311
5	0.79564

length is more significant in hemodynamic parameters. This trend is similarly found for pressure waveform also. For same blockage percentage, FFR decreases with the increase in lesion length as shown in Table 1.

The value of the maximum u_b increases with increasing occlusion percentage, but the values of maximum u_a and maximum u_c decrease as shown in Fig. 9. As the occlusion percentage rises, the cross-section becomes thinner at the midspan of

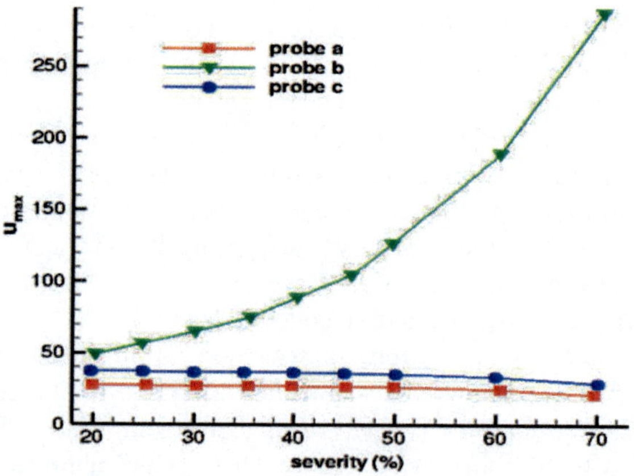

Fig.9 Maximum velocity variation with blockage percent at different probes

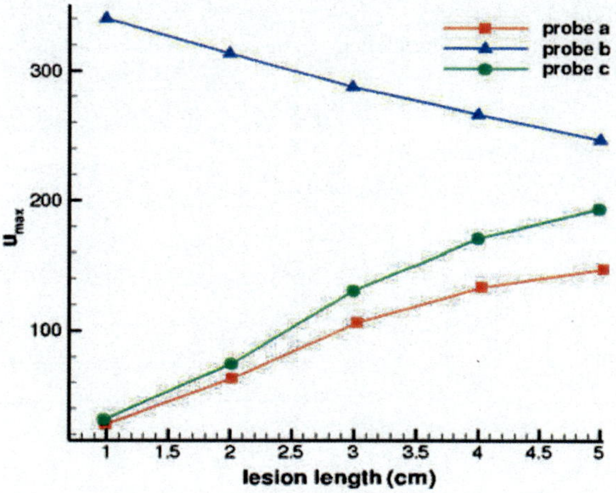

Fig.10 Maximum velocity variation with lesion length at different probes

the stenosed flow section causing an increase in the backward travelling wave (w_2), resulting decrease in the value of velocity ($u \propto (w_1 + w_2)$). Therefore, maximum u_a decreases with increase in severity. Probe c is located at the end of diverging portion in the stenosed section, and hence the velocity values decrease from probe b till c. The maximum u_c values also show a similar trend shown by maximum u_a values. The cross-sectional area increases till c and then it tapers downstream, and this tapering cross-section causes an increase in w_2 and reduces the velocity value at location c. Similarly, the value of maximum u_b decreases with increasing lesion length unlike the value of maximum velocity at probe a and probe c following the same principles of forward and backward travelling waves as demonstrated in Fig. 10.

A correlation between FFR and geometric factors is proposed based on the findings of numerical computation by the statistical analysis. The FFR data obtained using numerical computation and correlation are in fairly good agreement and can be utilized to predict FFR, rather than performing the numerical simulation.

Instead of carrying out numerical simulation for predicting FFR

$$
\begin{aligned}
FFR = a + b\ln L^* + cb^* + d(\ln L^*)^2 + e(b^*)^2 + \mathrm{f}(b^*)\ln\left(L^*\right) \\
+ g((\ln L^*))^3 + h(b^*)^3 + i(b^*)^2\ln L^* + jb^*(\ln L^*)^2
\end{aligned}
$$

where L * is the ratio of the span of the lesion to the span of the arterial section and b * is the occlusion in terms of percent blockage. It is perceived that the R^2 value corresponding to FFR is 0.9894 for correlation model having a confidence of 99%. Table 2 shows the value of the function's correlation coefficients. Figure 11 indicates the effectiveness of the FFR regression model, where the parity is posed between the values predicted by the correlation model and numerical simulation. It is demonstrated that the deviations were estimated between the numerical computation values and the regression values, and the deviation from the estimated value is well within ±2.

Table 2 Values of coefficients of the correlation function

Coefficient	Value
a	0.93053
b	0.88932
C	0.85397
d	0.82311
e	0.79564
f	0.88932
g	0.88932
h	0.88932
i	0.88932
j	0.88932

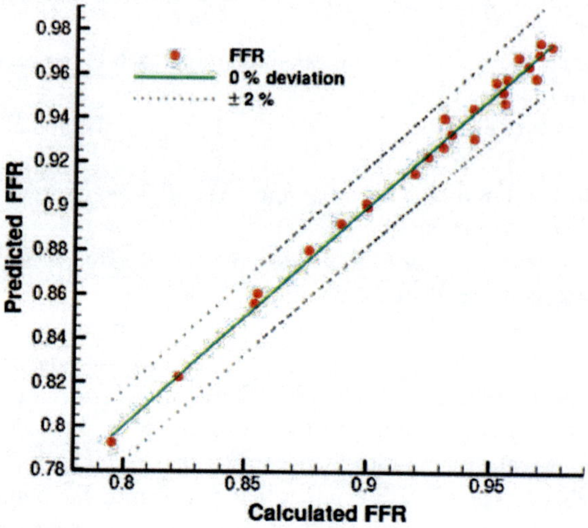

Fig. 11 Parity plot between numerical result and correlation having a 99% confidence for estimating FFR

4 Conclusions

1. The flow circulation that is happening inside the artery is affected by characteristics which are forward and backward.
2. In the stenosed artery, the changes within systolic flow are found as opposed to healthy state.
3. The FFR values drop consistently with the increment of severity and lesion unlike the pressure drop across the stenosis region.
4. The lesion span has a functional seriousness in the severity of stenotic lesion.
5. The influence of lesion length on the haemodynamic parameter becomes more significant for higher percent blockage.

6. A regression analysis of FFR is performed, and it is in good reasonable agreement, and the results of the suggested correlation may be used instead of actual numerical computation.
7. Better calculation of the FFR values cans be accomplished by treating terminal vessels as Windkessel elements in comparison to the model of resistance considered in this study.

References

1. Gupta, R., Mohan, I., Narula, J.: Trends in coronary heart disease epidemiology in India. Ann. Global Health **82**(2), 307–315 (2016)
2. Tonino, P.A., De Bruyne, B., Pijls, N.H., Siebert, U., Ikeno, F., vant Veer, M., Klauss, V., Manoharan, G., Engstrm, T., Oldroyd, K.G., Ver Lee, P.N.: Fractional flow reserve versus angiography for guiding percutaneous coronary intervention. New England J. Medi. **360**(3), 213–224 (2009)
3. Koo, B.K., Erglis, A., Doh, J.H., Daniels, D.V., Jegere, S., Kim, H.S., Dunning, A., DeFrance, T., Lansky, A., Leipsic, J., Min, J.K.: Diagnosis of ischemia-causing coronary stenoses by noninvasive fractional flow reserve computed from coronary computed tomographic angiograms: results from the prospective multicenter DISCOVERFLOW (Diagnosis of Ischemia-Causing Stenoses Obtained Via Noninvasive Fractional Flow Reserve) study. J. Am. Coll. Cardiol. **58**(19), 1989–1997 (2011)
4. Sherwin, S.J., Formaggia, L., Peiro, J., Franke, V.: Computational modelling of 1D blood flow with variable mechanical properties and its application to the simulation of wave propagation in the human arterial system. Int. J. Numeri. Methods Fluids **43**(67), 673–700 (2003)
5. Mynard, J.P., Nithiarasu, P.: A 1D arterial blood flow model incorporating ventricular pressure, aortic valve and regional coronary flow using the locally conservative Galerkin (LCG) method. Commun. Numer. Methods Eng. **24**(5), 367–417 (2008)
6. Iguchi, T., Hasegawa, T., Nishimura, S., Nakata, S., Kataoka, T., Ehara, S., Hanatani, A., Shimada, K., Yoshiyama, M.: Impact of lesion length on functional significance in intermediate coronary lesions. Clin. Cardiol. **36**(3), 172–177 (2013)
7. Formaggia, L., Nobile, F., Quarteroni, A., Veneziani, A.: Multiscal modelling of the circulatory system: a preliminary analysis. Comput. Vis. Sci. **2**(2–3), 75–83 (1999)
8. Olufsen, M.S., Peskin, C.S., Kim, W.Y., Pedersen, E.M., Nadim, A., Larsen, J.: Numerical simulation and experimental validation of blood flow in arteries with structured-tree outflow conditions. Ann. Biomed. Eng. **28**(11), 1281–1299 (2000)
9. Saha, S., Purushotham, T., Prakash, K.A.: Comparison of fractional flow reserve value of patient-specific left anterior descending artery using 1D and 3D CFD analysis. Int. J. Adv. Eng. Sci. Appl. Math. **20**, 1–10 (2020 Feb)
10. Low, K., van Loon, R., Sazonov, I., Bevan, R.L., Nithiarasu, P.: An improved bascline model for a human arterial network to study the impact of aneurysms on pressureflow waveforms. Int. J. Numer. Methods Biomed. Eng. **28**(12), 1224–1246 (1998)
11. Saha, S., Purushotham, T., Prakash, K.A.: Numerical and experimental investigations of Fractional Flow Reserve (FFR) in a stenosed coronary artery. In: E3S Web of Conferences 2019, EDP Sciences, vol. 128, p. 02006 (2019)
12. St. Rammos, K., Koullias, G.J., Pappou, T.J., Bakas, A.J., Panagopoulos, P.G., Tsangaris, S.G.: Computer model for the prediction of left epicardial coronary blood ow in normal, stenotic and bypassed coronary arteries, by single or sequential grafting. Cardiovas. Surgery **6**(6), 635–648 (1998)

Model-Based Simulation of Surface Electromyography Signals and Its Analysis Under Fatiguing Conditions Using Tunable Wavelets

Lakshmi M. Hari, S. Edward Jero, G. Venugopal, and S. Ramakrishnan

1 Introduction

Muscle fatigue is a neuromuscular condition during which the muscle cannot produce adequate force to perform a particular task. The muscle force production depends on the availability of oxygen, nutrition, and regulated blood supply, and lack of these can cause fatigue in a muscle [1, 2]. Fatigue analysis is an important part of diagnosis in any neuromuscular disease, sports medicine, and ergonomics [3, 4]. Among various fatigue assessment techniques available, Surface Electromyography (sEMG) is the most commonly used method [5]. It is non-invasive, reliable, and requires less medical training.

The signal depends on several factors such as firing rate, number of motor units, types of motor units, conduction velocity, muscle fibers, and recruitment pattern of motor units [6]. The sEMG signals are generally noisy, non-stationary, and multi-component in nature. For the better understanding about the muscular system and for more accurate analysis, synthetic sEMG signals are generated by many researchers using different models.

The influence and effect of various physiological parameters such as structure of muscle, volume conductor, motor unit recruitment and firing rate, synchronization of motor units, and fiber diameter have been analyzed and evaluated using different models [5–12]. The effect of external parameters such as crosstalk, noise, and percentage of maximum voluntary contraction [10] has also been addressed

L. M. Hari (✉) · S. E. Jero · S. Ramakrishnan
Department of Applied Mechanics, Indian Institute of Technology Madras, Chennai 600036, India
e-mail: lakshmimhari@gmail.com

G. Venugopal
Department of Instrumentation and Control Engineering, N.S.S. College of Engineering Palakkad, APJ Abdul Kalam Technological University, Kerala, India

S. K. Saha and M. Mukherjee (eds.), *Recent Advances in Computational Mechanics and Simulations*, Lecture Notes in Mechanical Engineering,
https://doi.org/10.1007/978-981-15-8315-5_9

by researchers. Synthetic signals are generated for many muscles such as biceps brachii, tibialis anterior, dorsal interosseous, rectus abdominis, and other skeletal muscles [6, 7, 12, 13].

Large number of features have been used to relate progression of fatigue with amplitude and frequency variations. Root Mean Square (RMS), mean, mean frequency, median frequency, and Integrated EMG (IEMG) are some of the commonly used amplitude- and frequency-based features [14–16].

Studies show that both time and frequency features fail to represent non-stationary property of the signal. Time frequency analysis has been gaining importance in signal processing and interpretation over the past few years, because it is capable of providing considerable information about the nonstationarity [17, 18]. In the proposed work, tunable Q-factor wavelets are used on generated synthetic signal to increase the time frequency resolution [19]. Tunable Q-Wavelet Transform (TQWT) has found a wide range of applications in biomedical signal analysis due to the ability to tune parameters to match the oscillatory behavior of the signals [20].

The objective of this study is to evaluate an adopted sEMG model during isometric contraction of the biceps brachii muscle using a set of amplitude-based features in time domain and time frequency domain.

2 Methods

Synthetic sEMG signals for nonfatigue and fatigue conditions are generated using the selected mathematical model. Statistical time-domain features such as RMS value, Mean Absolute Value (MAV), Variance (VAR), and IEMG are extracted in this work to examine the behavior of the signal. For the time frequency analysis of the generated signal, TQWT is used. This transform decomposes the signal to different energy subbands. Feature values are extracted from each decomposed frequency subbands using the abovementioned amplitude-based features.

Further, percentage difference in extracted features between nonfatigue and fatigue signal in each subband is analyzed. The conventional time-domain feature analysis method and TQWT performance are compared in terms of percentage difference of feature values between nonfatigue and fatigue signals.

2.1 *Synthetic sEMG Generation Model*

The components of the adopted synthetic sEMG model used in this work are current source, volume conductor, motor unit recruitment, and firing behavior functions [8]. The current source function is expressed as [11]

$$I_m(t) = CA\lambda^2(\lambda(vt - z_e))(6 - 6\lambda(vt - z_e) + \lambda^2(vt - z_e)^2)e^{-\lambda(vt-z_e)} \quad (1)$$

where C is a constant which depends on axoplasm conductivity σ_i, muscle fiber diameter d, and conduction velocity υ; z_e is the distance between neuromuscular junction and electrode; A is the amplitude factor which varies with amplitude of Motor Unit Action Potential (MUAP); and λ is the scaling factor which varies with duration of the MUAP.

This source function is used to generate MUAP for nonfatigue and fatigue signals. MUAP amplitude and duration change with A and λ values. A range of values adopted from [13] is used for λ and A to generate the signal.

Volume conductor function used in this study is proposed by [9]. The major factors in this model are radial conductivity σ_r, longitudinal conductivity σ_z, radial distance between muscle fiber and electrode r_f, and distance between source and sink currents b. This function is expressed mathematically as

$$\phi(t) = \frac{1}{4\pi\sigma_r}\left[\frac{1}{\sqrt{r_f^2 \cdot \frac{\sigma_z}{\sigma_r} + (\upsilon t - z_e)^2}} - \frac{1}{\sqrt{r_f^2 \cdot \frac{\sigma_z}{\sigma_r} + (\upsilon t - z_e + b)^2}}\right] \quad (2)$$

Motor Unit (MU) recruitment and firing pattern function in this model are adopted from [10]. It depends on two parameters excitatory drive (E) and Recruitment Threshold Excitation (RTE). RTE is calculated as

$$\mathrm{RTE}(i) = e^{ai} \quad (3)$$

where i is the index of MU, $a = \frac{\ln RR}{n}$, n is the total number of MUs, and RR is the range of desired threshold value. E drive is calculated as

$$\mathrm{E}(i) = \mathrm{RTE}_n + \frac{\mathrm{PFR}(i) - \mathrm{MFR}}{g_e} \quad (4)$$

where RTE_n is the highest excitation threshold, $\mathrm{PFR}(i)$ is the peak firing rate for each MU, MFR is minimum firing rate, g_e is the excitatory drive–firing rate relationship. The firing rate of ith MU is given as

$$\mathrm{FR}(i) = g_e(\mathrm{E}(i) - \mathrm{RTE}(i)) + \mathrm{MFR} \quad (5)$$

sEMG signals are generated by convolving the expressions for the current source function, volume conductor function, and recruitment and firing behavior function for each MU and by summing it for the entire MU pool. The signal is sampled at a frequency of 1 kHz. The percentage variation of λ and A from nonfatigue to fatigue for a fast fatigable muscle is given as 174 ± 45 and 47 ± 24, respectively [13]. Biceps brachii is a fast fatigable muscle which is the muscle of interest in this study. Generated synthetic signal is based on needle EMG values, and convolution of this values with volume conductor function can be considered as sEMG signal.

Numerical values used for model parameters are taken from the literature reported by [9, 10].

2.2 Features

A set of most commonly used amplitude-based features such as Root Mean Square value (RMS), Mean Absolute Value (MAV), Variance (VAR) and Integrated EMG (IEMG) is used to analyze the energy characteristics of the signal.

RMS value is used to represent the energy variations in a signal. It directly depends on the amplitude of the signal. It is mathematically expressed as [14]

$$\mathrm{RMS} = \sqrt{\frac{1}{N}\sum_{i=1}^{N} x_i^2} \tag{6}$$

where N is the length of the signal and x_i represents the sEMG signal. IEMG is the summation of amplitude of a signal. It represents the force generated by the muscle during contraction [16].

$$\mathrm{IEMG} = \sum_{i=1}^{N} |x_i| \tag{7}$$

where N is the length of the signal and x_i represents the sEMG signal. MAV represents the muscle force and gives the mean of the amplitude values of a sEMG signal [15] as

$$\mathrm{MAV} = \frac{1}{N}\sum_{i=1}^{N} |x_i| \tag{8}$$

where N is the length of the signal and x_i represents the sEMG signal. VAR is the power index of a sEMG signal. It is generally calculated by the mean of the square of deviation in the signal. In EMG, mean is approximately equal to zero. Thus, here variance is calculated by [15]

$$\mathrm{VAR} = \frac{\sum_{i=1}^{N} (x_i)^2}{(N-1)} \tag{9}$$

where N is the length of the signal and x_i represents the sEMG signal.

2.3 *Tunable Q-Wavelet Transform (TQWT)*

TQWT is a discrete-time wavelet transform with three tunable parameters, namely, Q-factor (Q), the number of subbands (J), and the redundancy or the over sampling rate (r). These factors can be tuned according to the oscillatory behavior of the signal to which it is applied. The TQWT is a combination of two channel filter bank operations, a low-pass filter with scaling factor α, and high-pass filter with scaling factor β [19].

Certain conditions have to be followed for setting abovementioned parameters: First condition is that scaling parameters must satisfy $0 < \alpha < 1$ and $0 < \beta \leq 1$. Second condition is that $\alpha + \beta > 1$ for the perfect reconstruction of the signal by oversampling filter bank.

Q-factor is the tunable factor which depends on α and β values. It is mathematically expressed as [19, 20]

$$Q = \frac{2 - \beta}{\beta} \tag{10}$$

Redundancy or oversampling rate is estimated by a factor

$$r = \frac{\beta}{1 - \alpha} \tag{11}$$

Maximum of number of decomposition level J depends on the length of a signal N. It is expressed as

$$J_{\max} = \frac{\log(\beta N/8)}{\log(1/\alpha)} \tag{12}$$

3 Results and Discussion

Synthetic sEMG signals are generated for both fatigue and nonfatigue conditions. MUAP duration and amplitude values to generate the signal are adopted from [13]. Table 1 shows the range of MUAP duration (λ) and amplitude (A) values for both conditions. Fifty signals are generated for nonfatigue and fatigue conditions

Table 1 Range of amplitude and duration values of nonfatigue and fatigue signals

	Amplitude (mV)	Duration (mS)
Nonfatigue	0.2–1.3	6.9–19.3
Fatigue	0.15–0.36	21.15–23.40

each. Signals are generated for the duration of 2 s. Representative signals for both conditions are shown in Fig. 1a and b, respectively.

The generated sEMG amplitude varies from −80 to 90 mV for nonfatigue signal and −60 to 65 mV for fatigue signal. It shows that amplitude reduced in the course of fatigue as compared to nonfatigue signal. This might be due to the loss of number of MUs involved in the activity as the fatigue progresses.

Table 1 shows the range of values A and λ during nonfatigue condition and after 4 min of fatigue in fatigue test which is adopted from [13].

The amplitude-based features such as RMS, VAR, IEMG, and MAV are performed on the synthetic signal in time domain. Table 2 illustrates the feature values in both conditions.

It is observed that all features have higher values in the case of nonfatigue signals as compared to fatigue. This change in amplitude might be due to loss in number of motor units involved in the activity in the course of fatigue. Amplitude of the MUAP

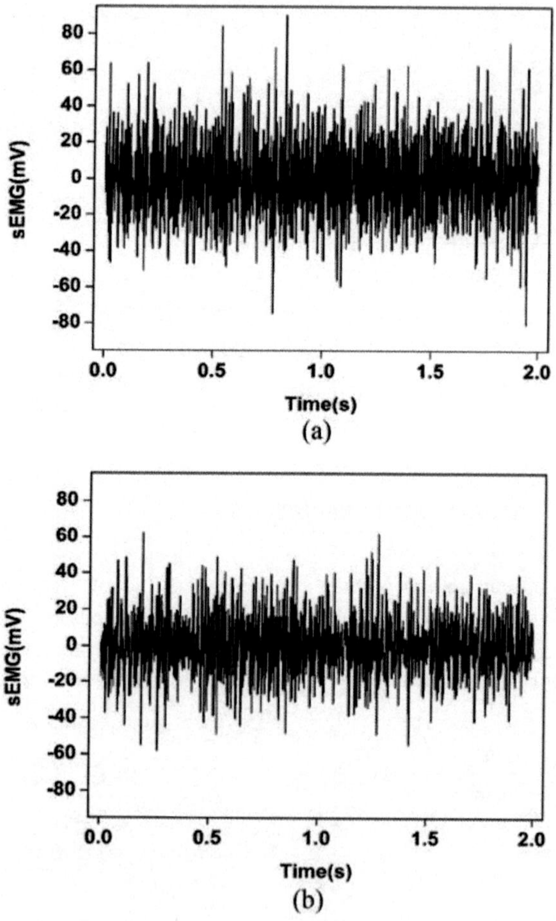

Fig. 1 Representative synthetic sEMG signals during **a** nonfatigue, **b** fatigue conditions

Table 2 Feature values in time domain

Features	Nonfatigue	Fatigue
RMS	19.82	13.41
Variance	1.58e+02	77.90
IEMG	3.15e+04	2.09e+04
MAV	15.36	10.19

will be higher than nonfatigue at the beginning of fatigue test due to large number of MU recruitment. It is found that amplitude decreased till the end of fatigue test and duration of MUAP prolonged [13]. Similar response is obtained with the generated synthetic signal using proposed model. Statistical significance of the features is tested using Wilcoxon signed rank test. All the features are found highly significant with p value < 0.001.

TQWT parameter values are chosen to satisfy the conditions mentioned in Sect. 2.3 ($Q = 1.22$, $r = 3$ and $J_{max} = 15$). Amplitude-based features used in time domain are analyzed for each subband which have been obtained from fatigue and nonfatigue signals after TQWT analysis. Mean of the feature values for all the 50 signals in each subband is calculated for both conditions. Figure 2 shows the feature values obtained in each subband for both nonfatigue and fatigue conditions.

It is found that nonfatigue signal has higher amplitude for all the feature values as compared to fatigue signal which is similar to time-domain response. This might be due to the temporary chemical imbalances in muscles. This is in agreement with existing literature on experimentally obtained sEMG signals [13]. All the features give clear distinction between both conditions during high-frequency subbands. Based on the results obtained from Wilcoxon signed rank test, each feature is found highly significant with p value < 0.001 in high-frequency subbands.

Percentage difference in feature values from nonfatigue signal to fatigue signal is estimated to extract the frequency distribution in the signal as illustrated in Fig. 3. It is observed from Fig. 3 that there is a percentage decrease in feature values in high-frequency subbands and a percentage increase is found in low-frequency subbands.

It is inferred that fatigue signal has more amplitude than nonfatigue in lower frequency subbands. It is observed that during fatigue, frequency is shifted toward the lower region of spectrum. This might be due to the synchronization of motor unit recruitment during fatigue and also due to the increase in duration of MUAP with the progression of fatigue.

Time -domain analysis and TQWT are compared in terms of percentage variation in Table 3. It is found that percentage decrease is higher in TQWT analysis during high-frequency subband. All the features perform better in time frequency domain as compared to conventional time domain. Conventional time-domain method does not provide any information about the frequency components present in the signal.

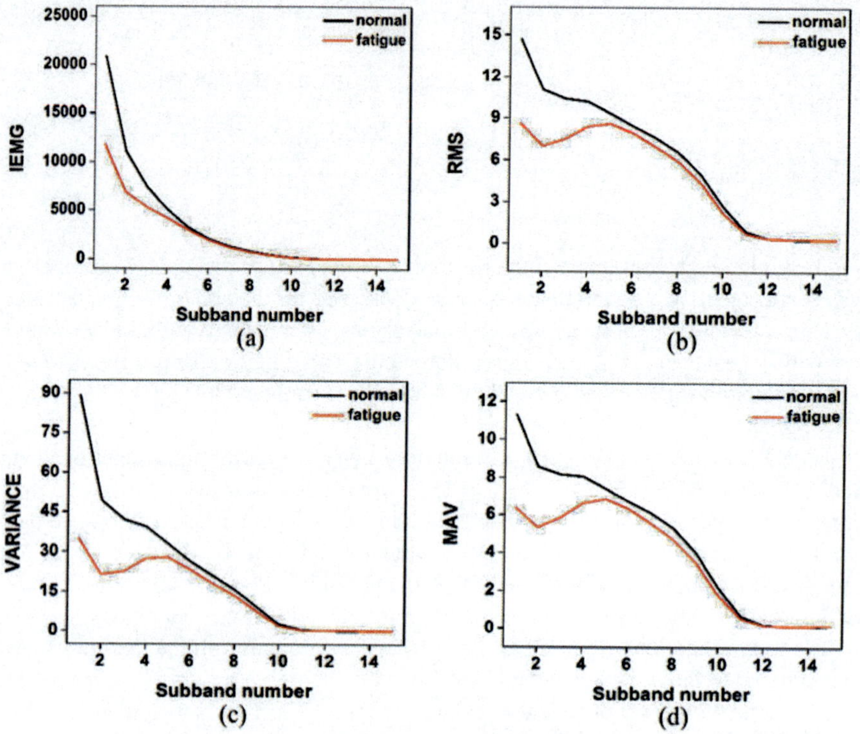

Fig. 2 Analysis of synthetic sEMG signal under nonfatigue and fatigue conditions in time frequency domain using amplitude-based features **a** IEMG, **b** RMS, **c** VAR, **d** MAV

Fig. 3 Percentage variation in feature values from nonfatigue to fatigue signal in each subband

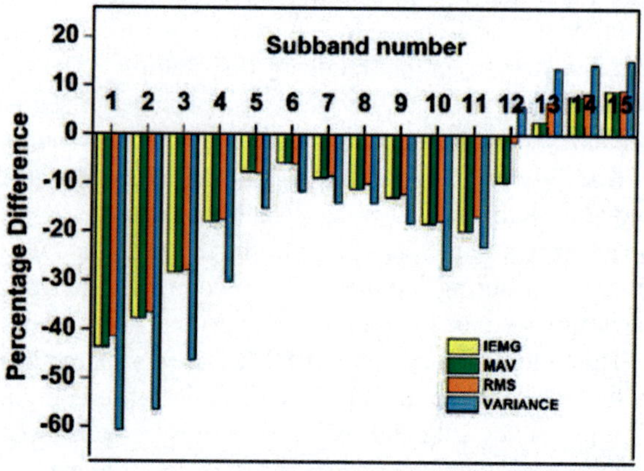

Table 3 Comparison of percentage variation in feature values of time domain and TQWT

Features	Time domain (%)	TQWT (%)
IEMG	−33.65	−43.67
RMS	−32.34	−41.44
VAR	−50.69	−60.77
MAV	−33.66	−43.67

4 Conclusion

This study adopted values for amplitude and duration of MUAP to generate synthetic sEMG signal using a mathematical model. From the results obtained, it is found that the proposed model is able to provide a realistic signal and it is capable of representing the sEMG signal for the biceps brachii muscle. This model can be used to extract characteristics of the signal in detail using different features and signal processing tools. The synthetic signal generated by the model shows that the fatigue signal is not able to maintain the force required to perform the activity from the reduced value of IEMG feature. The proposed amplitude-based features are able to differentiate between fatigue and nonfatigue signals. This model can be recommended for the analysis of neuromuscular diseases. TQWT performs better than time-domain analysis with higher percentage difference. Percentage difference estimated in each subband showed the shift in frequency toward lower region of spectrum during fatigue. This method can be adopted for fatigue analysis in experimentally acquired sEMG signals.

References

1. Cifrek, M., Medved, V., Tonković, S., Ostojić, S.: Surface EMG based muscle fatigue evaluation in biomechanics. Clin. Biomech. **24**(4), 327–340 (2009)
2. Merletti, R., Parker, P.A.: Electromyography: physiology, engineering, and non-invasive applications. Wiley, New Jersey, USA (2004)
3. Zwarts, M.J., Bleijenberg, G., Van Engelen, B.G.M.: Clinical neurophysiology of fatigue. Clin. Neurophysiol. **119**(1), 2–10 (2008)
4. Massó, N., Rey, F., Romero, D., Gual, G., Costa, L., Germán, A.: Surface Electromyography Applications. Apunt. Med. L'Esport. **45**(166), 127–136 (2010)
5. Cao, H., Boudaoud, S., Marin, F., Marque, C.: Surface EMG-force modelling for the biceps brachii and its experimental evaluation during isometric isotonic contractions. Comput. Methods Biomech. Biomed. Eng. **18**(9), 1014–1023 (2015)
6. Venugopal, G., Deepak, P., Ghosh, D.M., Ramakrishnan, S.: Generation of synthetic surface electromyography signals under fatigue conditions for varying force inputs using feedback control algorithm. Part H: J. Eng. Medi. **231**(11), 1025–1033 (2017)
7. Petersen, E., Rostalski, P.A.: Comprehensive mathematical model of motor unit pool organization, surface electromyography, and force generation. Front. Physiol. (2019)
8. Stegeman, D.F., Blok, J.H., Hermens, H.J., Roeleveld, K.: Surface EMG models: properties and applications. J. Electromyogra. Kinesiol. **10**(5), 313–326 (2000)

9. Fuglevand, A.J., Winter, D.A., Patla, A.E., Stashuk, D.: Detection of motor unit action potentials with surface electrodes: influence of electrode size and spacing. Biol. Cybern. **67**, 143–153 (1992)
10. Fuglevand, A.J., Winter, D.A., Patla, A.E.: Models of recruitment and rate coding organization in motor-unit pools. J. Neurophysiol. **70**(6), 2470–2488 (1993)
11. Wheeler, K.A., Kumar, D.K., Shimada. H.: An accurate bicep muscle model with sEMG and muscle force outputs. J. Med. Biol. Eng. **30**(6), 393–398 (2010)
12. Rannou, F., Nybo, L., Andersen, J.E., Nordsborg, N.B.: Monitoring muscle fatigue progression during dynamic exercise. Med. Sci. Sports Exerc. **51**(7), 1498–1505 (2019)
13. Celichowski, J., Grottel, K., Rakowska, A.: Changes in motor unit action potentials during the fatigue test. Acta. Neurobiol. Exp. **51**, 145–155 (1991)
14. Caesarendra, W., Tjahjowidodo, T.: A review of feature extraction methods in vibration-based condition monitoring and its application for degradation trend estimation of low-speed slew bearing. Machines **5**(4), 21 (2017)
15. Phinyomark, A., Limsakul, C., Phukpattaranont, P.: A novel feature extraction for robust EMG pattern recognition. J. Comput. **1**(1), 71–80 (2009)
16. Phinyomark, A., Phukpattaranont, P., Limsakul, C.: Feature reduction and selection for EMG signal classification. Expert Syst. Appl. **39**, 7420–7431 (2012)
17. Bonato, P., Roy, S.H., Knaflitz, M., De Luca, C.J.: Time-frequency parameters of the surface myoelectric signal for assessing muscle fatigue during cyclic dynamic contractions. IEEE Trans. Biomed. Eng. **48**(7), 745–753 (2001)
18. Karthick, P.A., Ramakrishnan, S.: Surface electromyography based muscle fatigue progression analysis using modified B distribution time–frequency features. Biomed. Signal Process. Control **26**, 42–51 (2016)
19. Selesnick, I.W.: Wavelet transform with tunable Q-factor. IEEE Trans. Signal Process. **59**(8), 3560–3575 (2011)
20. Abdel-Ghaffar, E.A.: Effect of tuning TQWT parameters on epileptic seizure detection from EEG signals. In: Proceedings of IEEE 12th International Conference on Computer Engineering and Systems (ICCES), pp. 47–51. Egypt (2017).

Fractal Order Poro-elastic Model for Modelling Biphasic Tissue and Tissue-Like Materials

Shib Sundar Banerjee, Arockiarajan Arunachalakasi, and Ramakrishnan Swaminathan

1 Introduction

Since 1980s, the common approach is to analyse the mechanical characteristics of soft tissue to consider it to be an ensemble of solid and fluid constituents and then devise a poro-elastic (PE) model to capture the interaction of these two phases in the tissue matrix [1–3]. Various soft tissues, e.g. brain [4], kidney and liver [5], have been characterized under the assumptions of PE substrate. A time-dependent stress, strain and fluid pressure production is reported in soft tissues which is attributed to the flow dynamics of the interstitial fluid [6]. Varieties of other biomaterials such as hydrogels are also similarly comprised of aqueous and solid constituents and exhibit time-dependent mechanical response [7]. Rate dependency in biological materials has been addressed in two competent theories, namely, viscoelasticity and poro-elasticity. Such materials, when subjected to compressive deformation, can undergo a conformational change at microstructural level which gives rise to viscous effect, while the movement of solvent results in PE behaviour [8]. Studies have observed that PE performs better at explaining responses in quasi-static and intermediate deformation rates while viscoelasticity is more relevant at fast strain rates [4]. Moreover, Chandran and Barocas [9] concluded from a microstructural study that poro-elastic processes are more likely to contribute in low and intermediate strain rates with network viscoelasticity being significant only in long-time relaxation phenomena.

Preliminary poro-elastic formulations of tissues exploited the classical theory by Biot in a finite deformation scenario [2, 10]. Standard practice is to adopt both theory of porous media and mixture theory to design an effective framework for a biphasic constitutive law. The solid skeleton is considered as a deformable skeleton with a

S. S. Banerjee (✉) · A. Arunachalakasi · R. Swaminathan
Indian Institute of Technology, Madras, Chennai 600036, India
e-mail: banerjeeshib2@gmail.com

S. K. Saha and M. Mukherjee (eds.), *Recent Advances in Computational Mechanics and Simulations*, Lecture Notes in Mechanical Engineering,
https://doi.org/10.1007/978-981-15-8315-5_10

governing hyperelastic strain energy function. Neo-Hookean- or Mooney–Rivlin-type models are suitable choices for reduced computational cost [11–13]. Pore fluid is often assumed to be non-viscous and barotropic, as the biphasic system in concern is perfused by mostly water [14]. Standard Fickian diffusion is assumed for the pressure-induced fluid transport in the body. Dynamics of fluid is often hypothesized to be obeying Darcy's law, i.e. there exists a linear relationship between interstitial fluid velocity and inherent pressure gradient [15].

Recent literatures have demonstrated deviations from Fickian diffusion in many real-world problems across the domains ranging from geophysics to biophysics. Such departure from idealized behaviours have been encountered in fields of atomistic energy diffusion and heat diffusion in complex medium, and has constituted a special class of problem named as anomalous diffusion problem [16]. Heterogeneous biological tissues and hydrogels trigger trapping, tortuosity and turbulence during diffusion because of the complex topology of the effective pore space lead to anomaly in Fickian diffusion [17, 18]. Previously, a number of corrective measures either in flux vector (Brinkman correction) or nonlinear, Forchheimer-type, corrective terms of the state variables of the problems have been introduced to mitigate the deviations observed [19]. Statistical scale invariance in microstructural domain of real heterogeneous materials such as tissues and hydrogels has been reported, which might induce anomaly in diffusion [20]. In a physical sense, the inertial effect during the pore fluid flow might not be restricted to a local length scale Δx and an integer-order spatial derivative cannot correctly capture the contribution of this effect. Fractional derivatives, however, are non-local operators which have found applicability in such scenarios [21].

Despite immense success in modelling nonlinearity, fractional calculus has several shortcomings due to the convolutional integral on Mittag–Leffler functional embedded in its definition. This gives rise to singularities, violates ideal properties of integer-order derivatives and increases the computational cost. An alternative definition, namely, fractal derivative, a local derivative based on Hausdorff metric space has been introduced by Chen [16] which can effectively mitigate these shortcomings while preserving nonlinear properties. Fractal derivative diffusion has been introduced for biological tissues in recent years [22]. Moreover, if the pore space is fractal in nature, the matrix can be considered to be non-fractal [23] while the interface between solid matrix and pore space can also be a fractal [24].

The aim of the work is to amend the classical poro-elasticity for heterogeneous biological media by taking non-Fickian diffusion into consideration. Previously Ostoja-Starzewski [25] has introduced the idea of fractal media in poro-elasticity. A reformulation of mass conservation and fluid percolation at fractal length scale has been proposed in the current study. The Fractal Poro-Hyperelastic (FPHE) model along with its numerical application has been developed on that premise. A u-p representation of the model has been generated. The model has also been validated against literature data.

2 Theory

2.1 Modelling Framework

The material is assumed to be a fluid saturated porous hyperelastic body. The total Cauchy stress (σ_{ij}) generated in the body under application of imposed deformation is comprised of the Cauchy stress relevant to the porous hyperelastic matrix ($\sigma_{ij}^{'}$) and isotropic pore fluid pressure (p). The effective stress formulation can be written as

$$\sigma_{ij} = \sigma_{ij}^{'} - p\delta_{ij} \tag{1}$$

where δ_{ij} is the Kronecker delta. This relation (1) is a limiting case of Biot's theory where the porous skeleton has relatively higher compressibility as compared to interstitial fluid. Early studies in the field of tissue mechanics have assumed full incompressibility of tissues, but recent studies have shown limited but non-zero volume deformability of fluid-rich systems such as cartilage, liver and hydrogel [26].

The reference coordinates of the undeformed skeleton are assumed to be $\mathbf{X} = X_i(i = 1, 2, 3)$ which, under an applied strain, moves to a current coordinate $\mathbf{x} = x_i(i = 1, 2, 3)$. The deformation gradient, $\mathbf{F}$, can therefore be defined as

$$\mathbf{F} = F_{ij} = \frac{\partial x_i}{\partial X_j} \tag{2}$$

The corresponding finite strain can be expressed in terms of right Cauchy Green tensor $\mathbf{C}$ and the definition is as follows: $\mathbf{C} := \mathbf{F}^T\mathbf{F}$. The compressibility can be measured using volume Jaccobian, J, defined as $J := \det(\mathbf{F})$. Then the distortional part of the deformation gradient can be computed as $\overline{\mathbf{F}}= J^{-1/3}\mathbf{F}$. The modified first and second invariant can be written as $\overline{I}_1 = J^{-2/3}I_1$ and $\overline{I}_2 = J^{-4/3}I_2$, where I_1 and I_2 are the two preliminary strain invariants, respectively. The Helmholtz free energy of compressible porous hyperelastic skeleton is a function of the modified invariants, i.e. $U = U(\overline{I}_1, \overline{I}_2, J)$. If λ_1, λ_2 and λ_3 are the principal stretch components, then the corresponding Cauchy stress can be defined as

$$\sigma_{ii}^{'} = \frac{\lambda_i}{J}\frac{\partial U}{\partial \lambda_i} \tag{3}$$

In the proposition to capture the complex topology of porous media, Chen [16] has employed the idea of Hausdorff distance and a scaling transform to form a new metric space. Under assumption of fractal invariance and equivalence, the metric space can be defined as $\Delta\hat{x} = \Delta x^q$, where q is the order of fractal. The derivative on this space can be defined as

$$\frac{df(x)}{d^q x} = \frac{1}{qx^{q-1}} \frac{df(x)}{dx} \tag{4}$$

The mass balance equation for this space can be rewritten under the modified approach in mixture theory in a fractal configuration space $\hat{\mathbf{x}}$ under assumption of scalability of linear domain into fractal reference volume unit as follows:

$$\frac{\partial \rho_\alpha}{\partial t} + \nabla^\beta (\rho_\alpha \hat{\mathbf{v}}_\alpha) = 0, \alpha = s, f \tag{5}$$

where t is the diffusion time, β is the order of fractality in the media, ∇^β is fractal laplacian operator, ρ_α is fractional density of the respective phase α and s, f stand for solid and fluid phases, respectively. $\hat{\mathbf{v}}_\alpha$ is the fractal velocity of the annotated phase and can be defined as $\frac{d\hat{\mathbf{x}}}{dt}$.

The modified spatial fractal Darcy's law [21] can be written as

$$\varphi(\hat{\mathbf{v}}_f - \hat{\mathbf{v}}_s) = -\frac{\kappa_\beta}{\eta} \nabla^\beta p \tag{6}$$

where φ is porosity of the media, η is dynamic viscosity, p is interstitial pore pressure and κ_β is fractal order permeability.

Solving (5) and (6) simultaneously one can get the following relation between displacements ($\mathbf{u} = u_i$) and intrinsic pressure (p), in a purely axial loading condition:

$$\frac{\partial u_i}{\partial t} = \frac{1}{\beta^2 x_i^{2(\beta-1)}} \frac{\kappa_\beta}{\eta} \frac{\partial p}{\partial x_i} \tag{7}$$

Considering a one-dimensional confined compression scenario with unidirectional deformation $\mathbf{u} = [u(t)\ 0\ 0]$, one can define the deformation gradient as follows:

$$F = \begin{bmatrix} 1 + \frac{\partial u(t)}{\partial X} & 0 & 0 \\ 0 & 1 & 0 \\ 0 & 0 & 1 \end{bmatrix} \tag{8}$$

where X_1 is simply represented as X. Corresponding volume Jaccobian turns out to be $J = \left(1 + \frac{\partial u(t)}{\partial X}\right)$.

A compressible Mooney–Rivlin energy function is considered for the solid hyperelastic media:

$$U = c_{10}\left(\overline{I}_1 - 3\right) + c_{01}\left(\overline{I}_2 - 3\right) + \frac{1}{D_1}(J-1)^2 \tag{9}$$

where c_{10}, c_{01} and D_1 are constitutive parameters. For the abovementioned deformation case, the resulting Cauchy stress in the skeleton can be written as follows:

$$\sigma'_{xx} = \frac{4}{3}c_{10}\left\{\left(1+\frac{\partial u}{\partial X}\right)^{1/3} - \left(1+\frac{\partial u}{\partial X}\right)^{-5/3}\right\}$$
$$+\frac{4}{3}c_{01}\left\{\left(1+\frac{\partial u}{\partial X}\right)^{-1/3} - \left(1+\frac{\partial u}{\partial X}\right)^{-7/3}\right\} + \frac{2}{D_1}\frac{\partial u}{\partial X} \quad (10)$$

2.2 Formulation for Ramp and Hold and Creep Conditions

The momentum balance equation in case of a fractal space can be formulated as follows:

$$\nabla^{\beta}\hat{\mathbf{v}}_c = \nabla^{\beta}\left((1-\varphi)\hat{\mathbf{v}}_s + \varphi\hat{\mathbf{v}}_f\right) = 0 \quad (11)$$

where the cumulative fractal velocity of the biphasic mixture, $\hat{\mathbf{v}}_c$, satisfies the properties of a solenoid vector field. Essentially, (11) necessitates spatial invariance of $\hat{\mathbf{v}}_c$ across the body. Now, by the boundary condition at $x = 0$, both the apperent Euclidean and fractal velocities for each phase are zero (no permeation), which leads to $\hat{\mathbf{v}}_c = 0$. The uniformity of cumulative fractal velocity will require the value to be zero at any x. This reveals the relationship between the fluid and solid phase velocity ($\hat{\mathbf{v}}_f$ and $\hat{\mathbf{v}}_s$, respectively) as follows:

$$(1-\varphi)\hat{\mathbf{v}}_s + \varphi\hat{\mathbf{v}}_f = 0$$
$$\Rightarrow \mathbf{v}_f = \left(1 - \frac{1}{\varphi}\right)\mathbf{v}_s \quad (12)$$

In case of the dynamic compression, the momentum balance equation for a biphasic mixture in macroscale can be written as follows:

$$\rho_s\frac{\partial \mathbf{v}_s}{\partial t} + \rho_f\frac{\partial \mathbf{v}_f}{\partial t} + \nabla\left(\boldsymbol{\sigma}' - p\mathbf{I}\right) = 0 \quad (13)$$

This expression can be solved for x direction, and one can arrive at a relation solely based on the deformation u and total stress σ'_{xx} as follows:

$$\left(\rho_s + \rho_f\left(1 - \frac{1}{\varphi}\right)\right)\frac{\partial^2 u}{\partial t^2} + \left(\frac{\rho_f}{\varphi^2}\frac{\partial \varphi}{\partial t} - \frac{\eta}{\kappa_\beta}\beta^2 x^{2(\beta-1)}\right)\frac{\partial u}{\partial t} = -\frac{\partial \sigma'_{xx}}{\partial x} \quad (14)$$

For creep condition, the inertial effect can be neglected and the problem consequently reduces to a much simpler PDE with the following configuration:

$$\frac{\eta}{\kappa_\beta}\beta^2 x^{2(\beta-1)}\frac{\partial u}{\partial t} = \frac{\partial \sigma'_{xx}}{\partial x} \tag{15}$$

These Eqs. (14) and (15) are to be solved for a given boundary condition in order to find out the temporal variation of the deformation u along x direction. It is noteworthy that setting $\beta = 1$ will result into transformation of the abovementioned model into a Poro-Hyperelastic (PHE) category.

3 Methodology

3.1 Numerical Solution Scheme

A nonstandard finite difference scheme has been developed to solve the dynamic problem as described in (14) [27]. The equation can be written in the following format:

$$P(t)\frac{\partial^2 u}{\partial t^2} + Q(t)\frac{\partial u}{\partial t} = R(t) \tag{16}$$

With $P(t) = \left(\rho_s + \rho_f\left(1 - \frac{1}{\varphi}\right)\right)$, $Q(t) = \left(\frac{\rho_f}{\varphi^2}\frac{\partial\varphi}{\partial t} - \frac{\eta}{\kappa_\beta}\beta^2 x^{2(\beta-1)}\right)$ and $R(t) = -\frac{\partial \sigma'_{xx}}{\partial x}$. The corresponding difference equation can be written in terms of the following substitution:

$$\begin{aligned}\frac{\partial^2 u}{\partial t^2} &\approx \frac{u_m^{k+1} - 2u_m^k + u_m^{k-1}}{(\Delta t)^2}\\ \frac{\partial u}{\partial t} &\approx (1-2\epsilon)\left(\frac{u_m^{k+1} - u_m^k}{\Delta t}\right) + 2\epsilon\left(\frac{u_m^{k+1} - u_m^{k-1}}{2\Delta t}\right)\end{aligned} \tag{17}$$

where m is the spatial position, k is the time instance and Δt is the time step. The first time derivative u has been considered as a combination of forward and central difference and ϵ is the weighting factor. Now (16) can be rewritten as a difference equation by employing (17), and to eliminate the dependence of u_m^{k-1} on the temporal solution, one can find this necessary condition after a little algebraic manipulation that:

$$\epsilon_m^k = \frac{1}{\Delta t}\frac{P_m^k}{Q_m^k} \tag{18}$$

Hence, we can write (16) in the form as follows:

$$\left(\frac{P_m^k}{(\Delta t)^2}+\frac{Q_m^k}{\Delta t}(1-\epsilon_m^k)\right)u_m^{k+1}-\left(\frac{2P_m^k}{(\Delta t)^2}+\frac{Q_m^k}{\Delta t}(1-2\epsilon_m^k)\right)u_m^k=Q_m^k \tag{19}$$

which can be solved further in an explicit scheme.

For (15) a central finite difference scheme is adequate. The stability of numerical approach is ensured by optimally choosing $\Delta t/(\Delta x)^2$ where Δt and Δx are time step and spatial difference, respectively. If sample height is h and permeation of water is only allowed through the surface at $X = h$, then the boundary conditions for a confined compression case can be set as follows for a ramp and hold conditions:

(i) $u(X,t) = 0$ at $X = 0$ for $t \geq 0$
(ii) $u(X,t) = 0$ at $t = 0$ for $X \in (0,h)$
(iii) $\frac{\partial u}{\partial t}\big|_{X=h} = \beta$ for $0 \leq t \leq t_{ramp}$ and 0 for $t > t_{ramp}$

where β is the rate of deformation imposed and t_{ramp} is the ramping time. For a creep scenario, the third condition can be replaced with $\sigma'_{xx}\big|_{X=h} = \sigma_0$ where σ_0 is the constant creep load.

3.2 *Parameter Optimization*

A two-phase genetic algorithm-based search programme has been developed for parameter optimization, for a crude search of a global minima of squared fitting error in first phase, and fine-tuning the parameters in the second. The methodology is validated against test values. The model has been validated against the data of Knapp et al. [28].

4 Results

The fractal order β has been observed to considerably affect the sensitivity to strain rates at any loading case in FPHE model. A number of numerical experiments are attempted to understand the salient features of the proposed formulation. For the purpose of numerical simulation, the parameters chosen are $c_{10} = 0$ Pa, $c_{01} = 10.8$ Pa, $D_1 = 6 \times 10^{13}$ Pa^{-1}, $\frac{\kappa_\beta}{\eta} = 9.88 \times 10^{-7}$ unit and $m = 1$. The fractal order β is varied from 0.8 to 1.2. The test protocols are kept similar to that of Knapp et al. [28].

The cylindrical collagen gel sample is of a height of 5 mm, which is operated under confined configuration. The first case is of a ramp and hold experiment where a ramp strain of 10% is applied gradually over a period of 120 s, which is followed by a hold phase.

The second case is of a creep test where a load of 0.43 Pa is applied on the gel sample. Both of the simulation results are shown in Fig. 1. As it is visible, increased β reflects in increased stiffness, and consequent decrease in compliance

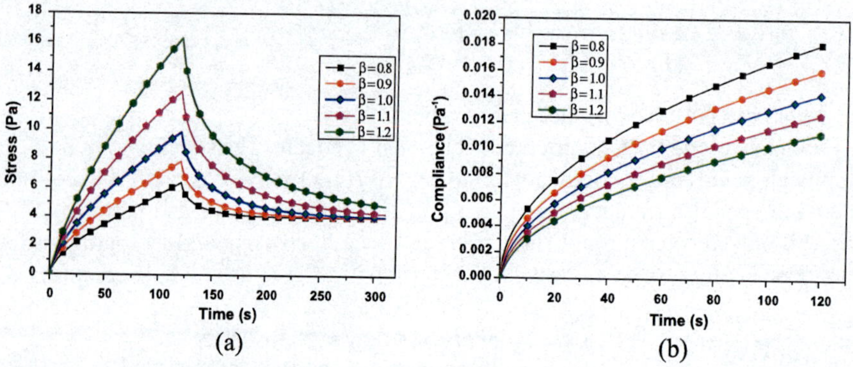

Fig. 1 Numerical simulation with different fractal orders (β) for two different test scenarios: **a** Ramp and hold and **b** Creep test

of the material. Figure 1a shows that the peak stress attained by ramp loading is increased by increasing β, but the long-term residual stress at hold phase converges for all β at ample time interval. Figure 1b shows that the equilibrium compliance in creep is higher for lower β.

Figure 2 illustrates the variation of the pore pressure at the non-permeable surface in the bottom for different test cases. Increased β has resulted in reduced order percolation rate which reflects in the following behaviour. Figure 2a shows the peak pore pressure attained to be higher in case of a higher β. Also increased fractal order resulted in sharper drop of pressure in the transition from ramp to hold and lengthened the decay time during hold phase. Figure 2b also reflects the same result as higher β leads to slower attainment of equilibrium.

The two-phase genetic algorithm is validated against a synthetic data in ramp and hold case. The optimization algorithm is able to predict the parameters from

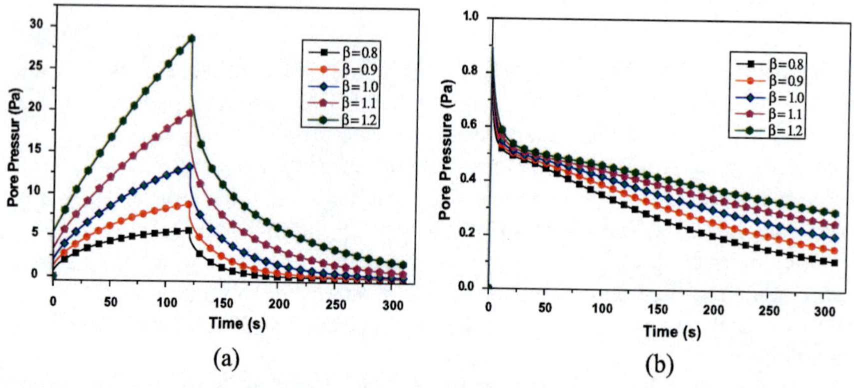

Fig. 2 Evolution of fluid pressure at non-permeable bottom surface for different fractal orders (β) in two different test scenarios: **a** Ramp and hold and **b** Creep test

Table 1 Input parameters and predicted parameters found from two-step GA optimization

Parameters	c_{10} (Pa)	c_{01} (Pa)	D_1 (Pa^{-1})	$\frac{K\beta}{\eta}$ (unit)	m	β
Test	10	12	$5 \times 10^{+13}$	1.2×10^{-6}	3	1.2
Predict	9.47	10.52	$5.34 \times 10^{+13}$	9.63×10^{-7}	3.15	1.16

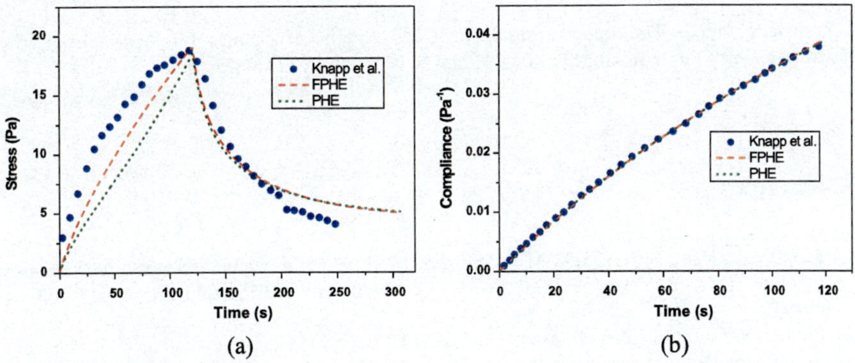

Fig. 3 FPHE and PHE model fitting with the Knapp et al. data in two different protocols: **a** Ramp and hold and **b** Creep test

synthetic data with an error of 2% in fitting with the model. The correlation of the synthetic data and predicted data has been shown to be $R^2 = 0.99$ with $p < 0.001$. The corresponding input parameters and predicted parameters can be seen in Table 1.

The fitting algorithm has been employed to fit the literature data as obtained from Knapp et al. The original data and fitted curves are shown in Fig. 3. Figure 3b shows an excellent fit with the creep data with a R^2 of 0.99. Figure 3a shows a satisfactory fit with a square-rooted error of 4%. An underestimation of stress level can be observed in the ramp phase, but the ramp peak has been identified satisfactorily. PHE model underestimates the load level throughout the ramping phase with a 6% deviation in the prediction of the peak load. The overall squared error obtained is 7% for PHE. Model performed better capturing the behaviour in the hold phase. Figure 3b shows a satisfactory fit for both FPHE and PHE models, and the reason might be absence of manifestation of rate dependency in material response in quasi-static deformation history.

5 Conclusion

In this work, an attempt has been made to accommodate for non-Fickian diffusion in the premise of finite poro-elasticity. A spatially fractal ordered Darcy's law is taken into account and the balance equations are reformulated in terms of fractal metric space. A parameter optimization scheme based on two-phase GA is also proposed. The model has been demonstrated to be sensitive to the fractal order, which might indicate the relation between the index β and the order of the pore structure. The model has been shown to characterize biphasic biological materials more accurately. Thus, the model can be helpful in simulating tissues and biomaterials for various applications in surgical or physiological relevance.

Acknowledgements The support from the SMC lab, Applied Mechanics, IIT Madras, especially of Mr. Ratnadeep Pramanik for his valuable inputs, is gratefully acknowledged.

References

1. Mow, V.C., Kuei, S.C., Lai, W.M., Armstrong, C.G.: Biphasic creep and stress relaxation of articular cartilage in compression: theory and experiments. J. Biomech. Eng. **102**(1), 73–84 (1980)
2. Simon, B.R., Liable, J.P., Pflaster, D., Yuan, Y., Krag, M.H.: A poroelastic finite element formulation including transport and swelling in soft tissue structures. J. Biomech. Eng. **118**(1), 1–9 (1996)
3. Islam, M.T., Righetti, R.: An analytical poroelastic model of a spherical tumor embedded in normal tissue under creep compression. J. Biomech. **89**, 48–56 (2019)
4. McGarry, M.D.J., Johnson, C.L., Sutton, B.P., Georgiadis, J.G., Van Houten, E.E.W., Pattison, A.J., Weaver, J.B., Paulsen, K.D.: Suitability of poroelastic and viscoelastic mechanical models for high and low frequency MR elastography. Med. Phys. **42**(2), 947–957 (2015)
5. Cardoso, L., Cowin, S.C.: Role of structural anisotropy of biological tissues in poroelastic wave propagation. Mech. Mater. **44**, 174–188 (2012)
6. Leiderman, R., Barbone, P.E., Oberai, A.A., Bamber, J.C.: Coupling between elastic strain and interstitial fluid flow: ramifications for poroelastic imaging. Phys. Med. Biol. **51**(24), 6291 (2006)
7. Castro, A.P.G., Laity, P., Shariatzadeh, M., Wittkowske, C., Holland, C., Lacroix, D.: Combined numerical and experimental biomechanical characterization of soft collagen hydrogel substrate. J. Mater. Sci. Mater. Med. **27**(4), 79 (2016)
8. Hu, Y., Suo, Z.: Viscoelasticity and poroelasticity in elastomeric gels. Acta Mech. Solida Sin. **25**(5), 441–458 (2012)
9. Chandran, P.L., Barocas, V.H.: Microstructural mechanics of collagen gels in confined compression: poroelasticity, viscoelasticity, and collapse. J. Biomech. Eng. **126**(2), 152–166 (2004)
10. Simon, B.R.: Multiphase poroelastic finite element models for soft tissue structures. Appl. Mech. Rev. **45**(6), 191–218 (1992)
11. Baldit, A., Ambard, D., Cherblanc, F., Royer, P.: Mechanical behaviour of annulus fibrosus tissue: identification of a poro-hyper-elastic model from experimental measurements. Comput. Methods Biomech. Biomedi. Eng. **16**(sup1), 280–281 (2013)
12. Townsend, M.T., Sarigul-Klijn, N.: Updated Lagrangian finite element formulations of various biological soft tissue non-linear material models: a comprehensive procedure and review. Comput. Methods Biomech. Biomedi. Eng. **19**(11), 1137–1142 (2016)
13. Castro, A.P., Yao, J., Battisti, T., Lacroix, D.: Poroelastic modeling of highly hydrated collagen hydrogels: experimental results vs. numerical simulation with custom and commercial finite element solvers. Front. Bioeng. Biotechnol. **6**, 142 (2018)
14. Gajo, A.: A general approach to isothermal hyperelastic modelling of saturated porous media at finite strains with compressible solid constituents. Proc. R Soc. A Mathemat. Phys. Eng. Sci. **466**(2122), 3061–3087 (2010)

15. Crevacore, E., Di Stefano, S., Grillo, A.: Coupling among deformation, fluid flow, structural reorganisation and fibre reorientation in fibre-reinforced, transversely isotropic biological tissues. Int. J. Non-Linear Mech. **111**, 1–13 (2019)
16. Chen, W.: Time–space fabric underlying anomalous diffusion. Chaos, Solitons Fractals **28**(4), 923–929 (2006)
17. Norris, D.G., Niendorf, T., Leibfritz, D.: Healthy and infarcted brain tissues studied at short diffusion times: the origins of apparent restriction and the reduction in apparent diffusion coefficient. NMR Biomed. **7**(7), 304–310 (1994)
18. Kihara, T., Ito, J., Miyake, J.: Measurement of biomolecular diffusion in extracellular matrix condensed by fibroblasts using fluorescence correlation spectroscopy. PLoS ONE **8**(11), e82382 (2013)
19. Deseri, L., Zingales, M.: A mechanical picture of fractional-order Darcy equation. Commun. Nonlinear Sci. Numer. Simul. **20**(3), 940–949 (2015)
20. Balankin, A.S.: Toward the mechanics of fractal materials: mechanics of continuum with fractal metric. https://arxiv.org/abs/1409.5829 (2014)
21. Chang, A., Sun, H., Zhang, Y., Zheng, C., Min, F.: Spatial fractional Darcy's law to quantify fluid flow in natural reservoirs. Phys. A **519**, 119–126 (2019)
22. Liang, Y., Allen, Q.Y., Chen, W., Gatto, R.G., Colon-Perez, L., Mareci, T.H., Magin, R.L.: A fractal derivative model for the characterization of anomalous diffusion in magnetic resonance imaging. Commun. Nonlinear Sci. Numer. Simul. **39**, 529–537 (2016)
23. Dathe, A., Thullner, M.: The relationship between fractal properties of solid matrix and pore space in porous media. Geoderma **129**(3–4), 279–290 (2005)
24. Neimark, A.: A new approach to the determination of the surface fractal dimension of porous solids. Phys. A **191**(1–4), 258–262 (1992)
25. Ostoja-Starzewski, M.: Towards poroelasticity of fractal materials. In: Continuous Media with Microstructure, pp. 157–164. Springer, Berlin, Heidelberg (2010)
26. Zhong, H., Peters, T.: A real time hyperelastic tissue model. Comput. Methods Biomech. Biomedi. Eng. **10**(3), 185–193 (2007)
27. Mickens, R.E., Jordan, P.M.: A positivity-preserving nonstandard finite difference scheme for the damped wave equation. Numeri. Methods Partial Different. Eqs. Int. J. **20**(5), 639–649 (2004)
28. Knapp, D.M., Barocas, V.H., Moon, A.G., Yoo, K., Petzold, L.R., Tranquillo, R.T.: Rheology of reconstituted type I collagen gel in confined compression. J. Rheol. **41**(5), 971–993 (1997)

Quantification of Brain Retraction Using Visco-hyperelastic Framework for Image-Guided Neurosurgical Applications

Abhilash Awasthi, Suryanarayanan Bhaskar, Umesh Gautam, and Sitikantha Roy

1 Introduction

In neurosurgery, brain retraction technique has become popular in the field of image-guided (IG) procedure for intracranial operations such as in brain tumor, cerebral aneurysms deep into the brain, cerebral hematoma, etc. Retraction provides accessibility and adequate exposure to the operative field, which are otherwise hard to access. During retraction, the brain tissue near a fissure is split open using a metallic strip called spatula (see Fig. 1). The pressure applied to the spatula affects the brain tissue in contact with it which causes temporary or permanent deformation of adjacent brain tissue and partial or total closure of blood vessels, thus impairing the oxygen supply to the brain [1]. This subsequently leads to several local brain contusions, which limit the reliability of the IG neurosurgical systems. To obtain an effective exposure of operative field without any damage to tissue, factors such as magnitude and direction of brain retraction pressure (BRP), amount of exposure, speed of retracting, etc. must be administered carefully. It is nearly impossible for a new practitioner in neurosurgery to master these skills. Because of the softness of brain tissue with marginal blood supply, a small variation in BRP could be fatal. IG neurosurgical procedures help neurosurgeons in pre-planning and training well before time and help in minimizing the risk of tissue damage. Therefore, there is a need for surgical training in this field to enhance the precision, efficiency, and reliability of the procedure.

A. Awasthi · U. Gautam · S. Roy (✉)
Department of Applied Mechanics, Indian Institute of Technology Delhi, Hauz Khas, New Delhi 110016, India
e-mail: sroy@am.iitd.ac.in

S. Bhaskar
Department of Neurosurgery, All India Institute of Medical Sciences Jodhpur, Jodhpur 342037, Rajasthan, India

S. K. Saha and M. Mukherjee (eds.), *Recent Advances in Computational Mechanics and Simulations*, Lecture Notes in Mechanical Engineering,
https://doi.org/10.1007/978-981-15-8315-5_11

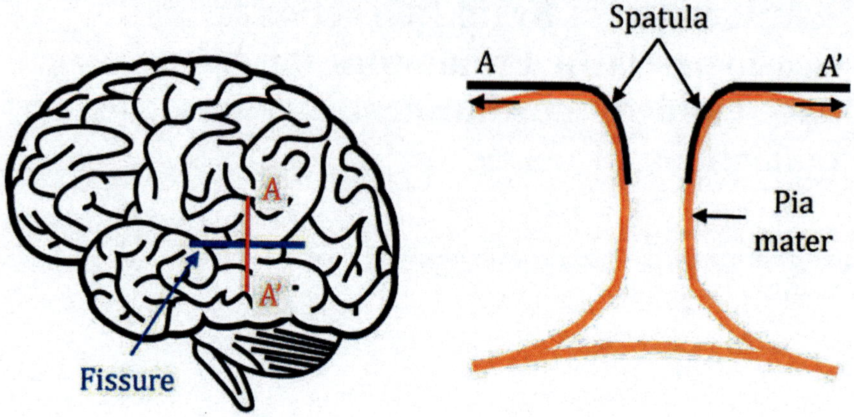

Fig. 1 Schematic of brain retraction with the help of spatulas by taking advantages of anatomical corridors of the brain called fissures

Neuronavigation systems rely on the accurate registration of anatomical details with the help of imaging techniques. The registration errors caused by the intraoperative brain shift, retraction, or resection often hamper the accuracy of IG navigation systems. With the help of computational modeling, registration errors can be avoided. However, their accuracy increases proportionally with the level of computing power, thus incurring a high computational cost. Brain retraction is the first and inevitable procedure of neurosurgery and was studied extensively in the recent past. Most of the studies on brain retraction remained focused on enhancing the accuracy, efficiency, and reliability of IG neurosurgical systems [2–4].

Although brain retraction is indispensable during many intracranial procedures, this comes with a cost of brain tissue getting damaged. There have been experimental studies on the brain tissue injury following brain retraction [5–7], but little effort has been made to study it using a computational model. Availability of a computational model helps in determining the gauged retraction parameters preoperatively that will minimize the risk of any cerebral injury such as cerebral ischemia. The model could further be used to study the postoperative effects on brain tissue following retraction. An anatomically correct numerical model with accurate material description to study these complications is still lacking in the literature. The present work is an attempt to bridge this gap by using a three-dimensional (3D) finite element (FE) model of the human brain in the visco-hyperelastic framework. The study aims at demonstrating the potential of an anatomically and constitutively correct 3D FE model of the human brain to predict the stress relaxation following a retraction, which eventually prevents it from injuries. The model could further be coupled with the theory of continuum damage mechanics to enhance its capabilities.

2 Materials and Methods

Due to several difficulties encountered during the experimental testing of many soft tissues such as the brain, the computer models are providing a great advantage. With the help of imaging techniques, a computer model of any complex shape can be generated with almost similar anatomical features as well as its constituents. For the clinical methods, a stack of images in Digital Imaging and Communications in Medicine (DICOM) format is used for the 3D model reconstruction. Image segmentation using image processing tools is used for differentiating different parts of the model and finally converted into voxels. A similar procedure has also been followed in the present study. The brain retraction problem has been defined as a boundary value problem (BVP) in the non-linear elasticity framework. The essential components for posing a complete BVP are a geometric/mathematical model, constitutive relation/assumptions, and boundary conditions (BC). A complete description of BVP with its components, which eventually will be solved for the field quantities has been discussed in next subsections.

2.1 Geometric Modeling

Several anatomically correct 3D FE models that were developed in the past [8] remained focused on Traumatic Brain Injury, an injury to the brain resulting from the application of sudden or impact load on the head. Linear as well as angular acceleration or deacceleration from these impacts, causes the relative motion to occur between different parts of the brain having different densities, hence causing damage to the brain at the neuronal level. For surgical procedures such as brain retraction, the speed at which forces act on the brain is medium in the range (1–10 mm/min). In the past few decades, the effects of brain retraction on animal brains have been studied experimentally as well as clinically. Recently, few studies used a numerical model to study the deformed shape of the brain due to retraction and used it to register the real-time deformation to a computer model intra-operatively. However, the prime focus of these studies was to enhance the accuracy of IG neurosurgical procedures. Due to higher computational cost associated with such systems and need for real-time deformation mapping, makes them hard to execute. Also, the associated higher computational cost resulted in the use of relatively simple (2D and 3D) computer models of the brain in those studies without mimicking its actual geometry. All these factors consequently lead to the requirement of examining the effects of retraction on brain tissue numerically by simulating the real-time/actual procedure with the help of an anatomically correct numerical model.

A 3D human brain model was segmented and reconstructed from the T1 weighted magnetic resonance (MR) image data of the human head in DICOM format. The DICOM images of an average adult subject were acquired using 1.5T Philips Achieva MRI scanner with 1 mm thick slice interval at All India Institute of Medical Sciences,

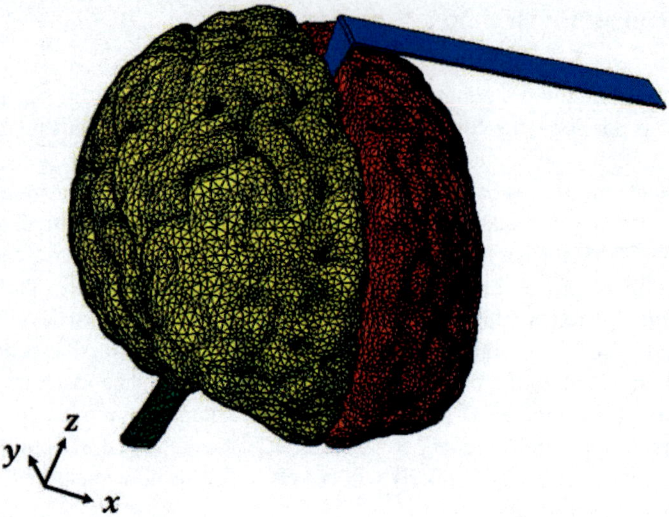

Fig. 2 Schematic of assembled 3D FE brain model with left hemisphere in red color, right hemisphere in yellow color, brain stem in green color, and spatula in blue color

New Delhi, India. A computer-aided design model of the human brain was developed from the stack of acquired MR images. The different parts of the brain, such as left and right hemispheres, brain stem, etc. (see Fig. 2) were identified and segmented based on their grayscale value using a custom software ScanIP from Simpleware [9]. The geometry of spatula in contact with brain tissue was tapered with 5 mm width at the bottom. The brain model was meshed using C3D4 (4-node linear tetrahedral continuum element) element type while the spatula was meshed using C3D8H element type (8-node linear hybrid brick element). The FE model in total consists of 249,270 number of elements (both C3D4 and C3D8H) and 57,816 number of nodes.

2.2 *Constitutive Model*

Most of the studies on retraction considered the brain tissue to be linearly elastic or poroelastic. However, these models, which are quite simple and have their solution procedures already optimized, are applicable to numerical models undergoing small deformations only. Usually surgery-related loads result into large deformations in the tissue. This type of behavior can only be captured using the non-linear elastic framework. Also, brain tissue is very soft and exhibits hyperelasticity and viscoelasticity as its primary characteristics. Accordingly, the non-linear, incompressible, isotropic material response of brain was modeled using a time-dependent polynomial strain energy density function (W) given by [10]:

Table 1 Coefficients for visco-hyperelastic material model

Hyperelastic parameters	Viscoelastic parameters
$C_{100} = C_{010} = 263\,\text{Pa}$	$g_1 = 0.27259$ & $\tau_1 = 0.83924\,\text{s}$
$C_{200} = C_{020} = 491\,\text{Pa}$	$g_2 = 0.25741$ & $\tau_1 = 9.60894\,\text{s}$
$C_{110} = 0$	

$$W(t) = \int_0^t \left\{ \sum_{i+j=1}^{N} C_{ij0} \left[1 - \sum_{k=1}^{n} g_k \left(1 - \exp\left(\frac{-(t-\tau)}{\tau_k} \right) \right) \right] \right\} \cdots \tag{1}$$

$$\cdots \times \frac{d}{d\tau} \left[(\overline{I}_1 - 3)^i (\overline{I}_2 - 3)^j \right] d\tau$$

where C_{ij0} is the instantaneous relaxation modulus, g_k is the dimensionless relaxation modulus, t represents current time, and τ_k are the characteristic times denoting running time variable $(0 \le \tau \le t)$. $\overline{I}_1$ and $\overline{I}_2$ are the invariants of left Cauchy-Green strain tensor (**B**) given by

$$\overline{I}_1 = \lambda_1^2 + \lambda_2^2 + \lambda_3^2 \qquad and \qquad \overline{I}_2 = \lambda_1^{-2} + \lambda_2^{-2} + \lambda_3^{-2}$$

where λ_i with $i = 1, 2, 3$ are the principal stretches in three principal directions. Stretch is defined as the ratio between the deformed and undeformed length ($\lambda = L/L_0$) of a line element. From the second law of thermodynamics for an isothermal process, the Cauchy stress tensor (**T**) can be expressed as

$$\mathbf{T} = \frac{1}{J} \frac{\partial W(t)}{\partial \mathbf{F}} : \mathbf{F}^T \tag{2}$$

where **F** is the deformation gradient and $J = det(\mathbf{F})$ is the relative volume change.

A two-term polynomial strain energy density function ($N = 1, 2$) and two-term Prony series ($k = 1, 2$) were considered for the present case. The values for the hyperelastic and Prony series parameters were obtained from unconfined compression test on brain tissue [10], values of which are given in Table 1. The spatula is assumed to be made up of steel and was modeled as linearly elastic material with Young's modulus and Poisson's ratio as 200 GPa and 0.3, respectively. Combining constitutive relation with the balance laws (from 1st and 2nd law of thermodynamics) gives the field equations which are then solved using appropriate BC. The BC and FE implementation of BVP are discussed in the next subsection.

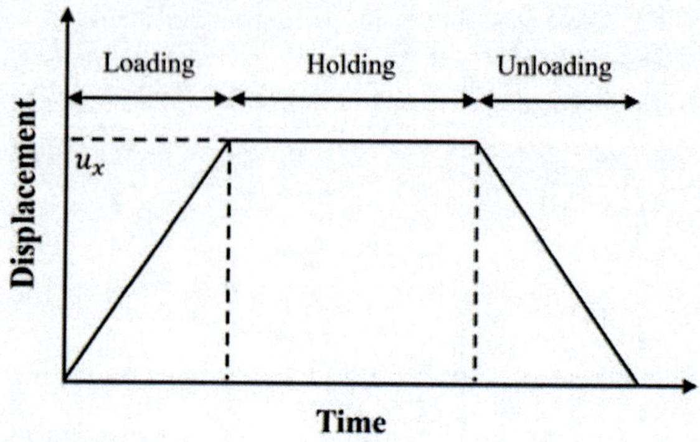

Fig. 3 Time-dependent BC on spatula was divided into three-time steps: loading till displacement u_x, holding at same displacement for 30 min, and then unloading

2.3 *Finite Element Simulation*

The last and essential part of a complete BVP is the BC. Based on expert neurosurgeon's advice, it was hypothesized that for interhemispheric retraction, the right hemisphere and brain stems remain fixed in their position while the left hemisphere is retracted. Hence for the simulation purpose, the right hemisphere and brain stem were kept as fixed. The deformable left hemisphere was kept traction free. Time-dependent displacement BC were prescribed on the spatula by dividing the total loading time into three steps, viz., loading, holding, and unloading steps. The schematic description of these steps is given in Fig. 3.

Initially, the spatula was displaced in $x-$direction by prescribing a displacement (u_x) of 5 mm at a loading speed of 5 mm/min. After loading step, the same displacement value ($u_x = 5$ mm) was kept constant for 30 min describing the holding period. Finally, the spatula was slowly unloaded to its initial position with the same speed as that for the loading step. The holding period is the time during which whole surgery is performed. Also, the tool-tissue interaction between spatula and brain tissue has been modeled by specifying a non-penetrating frictionless surface to surface contact.

The solution to the BVP has been obtained using the FE method and implemented using commercially available FE solver ABAQUS 6.13/Standard [11]. The 3D model took approximately 1 h to solve for the field quantities using a Windows 7 operating system with 16 GB RAM. The BRP during the surgery can be computed by dividing the reaction force on the spatula by the area of contact. Real-time monitoring of BRP could help in preventing the brain tissue against any damage such as contusions or hemorrhage. The variation of BRP throughout different loading steps has been explained and discussed in the next section.

3 Results and Discussion

Over the past few decades, there has been a tremendous rise in the use of computational models for mimicking the experimental procedures. This is implemented to make the surgical procedures less-invasive, thereby minimizing the risk of injury to soft tissues. Also, the validity of these numerical models becomes critical, requiring the representation of actual physical processes. In an attempt to model the real-time brain retraction procedure, anatomically correct 3D FE model was used to predict the effects of retraction. A full-field solution using the visco-hyperelastic material model in non-linear elasticity framework was compared with the experimental data [5]. The experiments consisted of the retraction procedure on the temporal lobe of the dog brain using a strain gauge spatula with pressure transducer on its front surface, in-vivo. The brain tissue was retracted to expose the operative field. After achieving sufficient exposure, the spatula was fixed in the same position, and attenuation of BRP was recorded until a plateau value reached. Similar BC were also applied to our model, and resulted variation in BRP was studied.

The variation in reaction force due to the displacement BC applied to spatula as the time progresses is given in Fig. 4. The black curve shows different loading modes during the deformation of brain tissue. The interesting and valuable information that could be extracted from these plots is the value of peak BRP and a drop in its value during the holding period. This attenuation of BRP is responsible for relaxing the brain tissue from stresses and hence minimizing the chances of tissue damage during brain retraction.

The excess of peak BRP results in secondary brain injuries such as focal contusions, cortical incisions, brain hemorrhage, or other neurological disorders [12]. The attenuation of BRP during the holding period (see Fig. 5) after obtaining sufficient exposure continues until reaching a plateau at 30% of initial retraction pressure. The decrease in BRP is attributed to the drainage of cerebrospinal fluid (CSF) from the subarachnoid spaces and ventricular system. This further accounts for the viscoelastic nature of brain tissue and demonstrate its relaxation behavior. Excessive and frequently employed brain retraction results in higher BRP values, thereby putting the whole surgical procedure at risk. The level of BRP and its attenuation does not solely depend on the drainage of CSF but also get affected by the relative position of the head, the exposure required, the physical condition of the brain, etc. The level of BRP also gets amplified from abnormal brain conditions such as hypotension or hypoxemia [13].

Over the past few decades, several researchers have provided threshold for brain retraction using different methods such as mean and local retraction pressure, force measurement method, cerebral pulsation method, brain electrical activity monitoring, blood flow monitoring, among others. However, a general consensus for these methods or threshold parameters provided by them is still not available at present [1, 12, 14]. Also, most of the brain injuries following retraction are not evident immediately postoperatively, thereby limiting the use of clinical or imaging techniques for their prevention. The use of patient-specific computer head models, coupled with various sensor technologies and haptic feedback devices for tool-tissue interaction, will help to improve the reliability of surgical simulations. These simulations can then be used for the surgical guidance of neurosurgeon or realistic training of the practitioners.

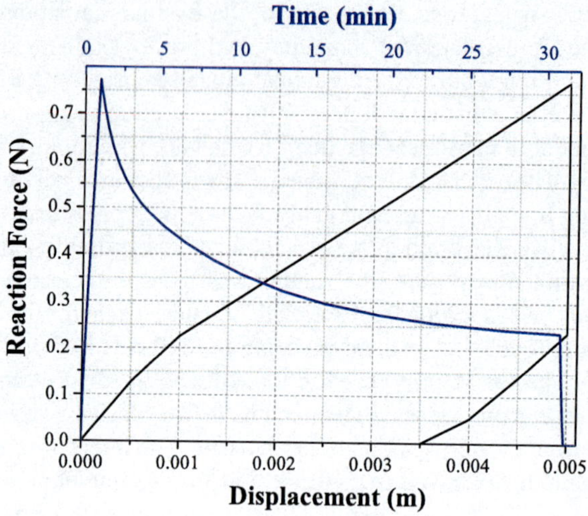

Fig. 4 Reaction force on spatula versus, (*i*) displacement (black curve), and (*ii*) time (blue curve). The trend of the curve shows different loading modes during the deformation process

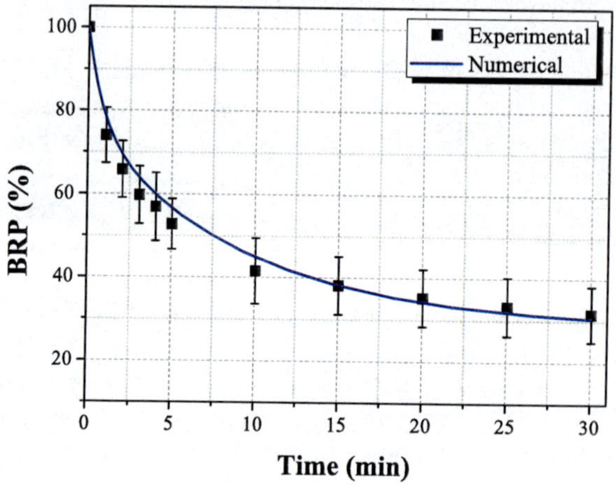

Fig. 5 Comparison of attenuation in BRP during the holding period between experimental data and results from present model

4 Conclusions

A 3D FE model of the human brain was devised to investigate the effects of deep brain retraction. The FE brain model was segmented and reconstructed using MRI data of the human head. A visco-hyperelastic material model was used to model the brain tissue constitutively. The loading history was divided into three steps, viz., loading, holding, and unloading; mimicking the actual surgical procedure. The attenuation of BRP during the holding period (the period during which whole surgery is performed) was compared with the available experimental data. The BRP decreases to 30% of the initial BRP after 30 min. The results could help in understanding the deformation behavior of brain tissue while retracting and could further be used for preventing and minimizing the risk of brain tissue getting injured. The present model could further be used to study the advantages of intermittent retraction as compared to continuous one or the use of multiple retractors instead of one, etc. Finally, the present model can also be used for the computation of model-updated intraoperative images for neurosurgical image-guidance in minimally invasive surgeries.

Acknowledgements The authors would like to thank Indian Institute of Technology (IIT) Delhi for supporting this study.

References

1. Zhong, J., Dujovny, M., Perlin, A.R., Perez-Arjona, E., Park, H.K., Diaz, F.G.: Brain retraction injury. Neurol. Res. **25**(8), 831–838 (2003). https://doi.org/10.1179/016164103771953925
2. Sase, K., Fukuhara, A., Tsujita, T., Konno, A.: GPU-accelerated surgery simulation for opening a brain fissure. Robomech. J. **2**(1), 17 (2015). https://doi.org/10.1186/s40648-015-0040-0
3. Li, P., Wang, W., Zhang, C., An, Y., Song, Z.: In vivo investigation of the effectiveness of a hyper-viscoelastic model in simulating brain retraction. Scienti. Rep. **6**, 28654 (2016). https://doi.org/10.1038/srep28654
4. Lamprich, B.K., Miga, M.I.: Analysis of model-updated MR images to correct for brain deformation due to tissue retraction. In: Medical Imaging 2003: Visualization, Image-Guided Procedures, and Display, vol. 5029, pp. 552–560. International Society for Optics and Photonics (2003). https://doi.org/10.1117/12.480217
5. Hongo, K., Kobayashi, S., Yokoh, A., Sugita, K.: Monitoring retraction pressure on the brain: an experimental and clinical study. J. Neurosur. **66**(2), 270–275 (1987). https://doi.org/10.3171/jns.1987.66.2.0270
6. Rosenrn, J.: The risk of ischaemic brain damage during the use of self-retaining brain retractors. Acta neurologica Scandinavica. Supplementum **120**, 1–30 (1989). https://doi.org/10.1111/j.1600-0404.1989.tb08017.x
7. Kaido, T., Nakase, H., Nagata, K., Otsuka, H., Sakaki, T.: Intermittent isometric exposure prevents brain retraction injury under venous circulatory impairment. Neurolo. Res. **23**(7), 739–744 (2001). https://doi.org/10.1179/016164101101199261
8. Dixit, P., Liu, G.R.: A review on recent development of finite element models for head injury simulations. Arch. Comput. Methods Eng. **24**(4), 979–1031 (2017). https://doi.org/10.1007/s11831-016-9196-x
9. Simpleware ScanIP: 3D image visualization and processing software. http://www.simpleware.com/software/scanip/
10. Miller, K.: Constitutive model of brain tissue suitable for finite element analysis of surgical procedures. J. Biomech. **32**(5), 531–537 (1999). https://doi.org/10.1016/S0021-9290(99)00010-X
11. ABAQUS 6.13. ABAQUS/CAE User's Manual, Dassault Systems, USA (2013)
12. Andrews, R.J., Bringas, J.R.: A review of brain retraction and recommendations for minimizing intraoperative brain injury. Neurosurgery **33**(6), 1052–1064 (1993). https://doi.org/10.1097/00006123-199312000-00014
13. Albin, M.S., Bunegin, L., Bennett, M.H., Dujovny, M., Jannetta, P.J.: Physiopathological responses to graded brain retractor pressure under induced hypotension. Proc. Am. Assoc. Neurol. Surg. **23** (1976)
14. Laha, R.K., Dujovny, M., Rao, S., Barrionuevo, P.J., Bunegin, L., Hellstrom, H.R., Albin, M.S., Taylor, F.H.: Cerebellar retraction: significance and sequelae. Surg. Neurol. **12**(3), 209–215 (1979)

Recent Trends in Solid Mechanics and FEM

Calculation of NSIFs and Shape Factors of Four-Point Bend Specimens Containing Sharp V-Notches

Mirzaul Karim Hussain and K. S. R. K. Murthy

1 Introduction

Structures or machine components often contain sharp V-notches. The singular stress field exists in the vicinity of the notch tip of sharp V-notches. In linear elastic fracture mechanics, Williams [1] suggested that the singular stress field can be stated in the form of $Kr^{\lambda-1}f_{ij}(\theta)$, here K is the notch stress intensity factor (NSIF), r is the radial distance from the notch tip, $\lambda - 1$ is the order of stress singularity, and $f_{ij}(\theta)$ are the angular functions of stress components. Among the various brittle fracture criteria developed, the more popular and widely employed criteria are based on the critical values of NSIF [2–5].

In past decades, many techniques have been developed by researchers to compute the NSIFs. Gross and Mendelson [6] proposed the boundary collocation method (BCM) to determine the NSIFs of sharp V-notches. Chen [7] developed a body force method (BFM) to evaluate the NSIFs for specimens under mixed mode (I/II) tensile and in-plane bending loading conditions. Moreover, Ju and Chung [8] and Liu et al. [9] used finite element (FE) stresses, and utilizing the least-squares method determined the NSIFs. These least-squares methods [8, 9] require higher order Williams coefficients and/or very fine meshes for attaining the accurate values of NSIFs. In another research, Ayatollahi and Nejati [10] proposed an over deterministic method using the FE displacements to calculate the NSIFs along with the higher order terms of Williams coefficients. Recently, a collocation technique and a point substitution technique were proposed by authors [11, 12] to obtain the mixed mode (I/II) NSIFs using the FE displacements along the notch flanks.

M. K. Hussain (✉) · K. S. R. K. Murthy
Department of Mechanical Engineering, Indian Institute of Technology Guwahati, Guwahati 781039, India
e-mail: mirzaul@iitg.ac.in

S. K. Saha and M. Mukherjee (eds.), *Recent Advances in Computational Mechanics and Simulations*, Lecture Notes in Mechanical Engineering,
https://doi.org/10.1007/978-981-15-8315-5_12

It is worth mentioning here that, unlike the availability of popular quarter-point elements (QPEs) in the crack problems, no such singular elements are currently available for the sharp V-notched problems. Moreover, even though the displacement components are more accurate than the stress components in the FE method and the notch flank displacements are the most accurate in the entire domain, there are not many popular methods available for the determination of NSIFs using notch flank FE displacements. One of the reasons for neglecting the advantages of the notch flank displacement may be the presence of rigid body displacements in the displacement components which are absent in stress components. Therefore, the method proposed by authors [11], the rotational components in the displacements are not neglected; rather, nicely negotiated using a simple formula. Thus, the main thrust of the present work is to demonstrate the efficacy and capabilities of the collocation technique proposed by authors [11] in terms of accurate estimation of the NSIFs of a four-point bend (FPB) specimens with sharp V-notches. The FPB specimens with sharp V-notches are one of the widely used specimens to study the fracture of brittle materials.

In this research, the mode I and mode II NSIFs of the FPB are determined under various loading and geometry conditions using the collocation technique. The dimensionless notch shape factors and the mixing ratios are also presented for all the examples. The results obtained in the present investigation are then compared with the available solutions [13].

2 Theoretical Background

Consider a homogeneous elastic 2D geometry containing a sharp V-notch having notch angle $\gamma (= 180 - 2\alpha)$ as shown in Fig. 1. Considering the singular terms and the constant displacement terms, the displacement field at any pint point $P(r, \theta)$ near to the notch tip O under any arbitrary in-plane loading can be given as [8]

$$u = \frac{\kappa + 1}{2G} A_0 + \frac{A_1}{2G} r^{\lambda_1^I} \left\{ \left(\kappa + \lambda_1^I \cos 2\alpha + \cos 2\alpha\lambda_1^I \right) \cos \lambda_1^I \theta - \lambda_1^I \cos\left(\lambda_1^I - 2 \right)\theta \right\}$$

Fig. 1 Notch geometry

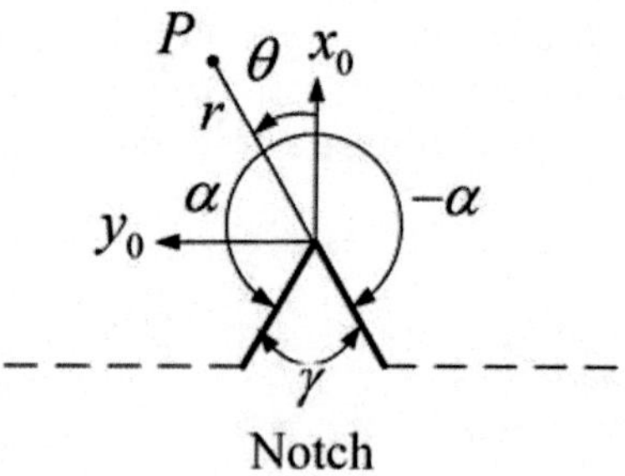

$$+\frac{B_1}{2G}r^{\lambda_1^{II}}\left\{\left(-\kappa-\lambda_1^{II}\cos 2\alpha+\cos 2\alpha\lambda_1^{II}\right)\sin\lambda_1^{II}\theta+\lambda_1^{II}\sin\left(\lambda_1^{II}-2\right)\theta\right\}-\frac{\kappa+1}{2G}B_2 r\sin\theta$$

$$v=\frac{A_1}{2G}r^{\lambda_1^{I}}\left\{\left(\kappa-\lambda_1^{I}\cos 2\alpha-\cos 2\alpha\lambda_1^{I}\right)\sin\lambda_1^{I}\theta+\lambda_n^{I}\sin\left(\lambda_1^{I}-2\right)\theta\right\}+\frac{\kappa+1}{2G}B_0$$

$$+\frac{B_1}{2G}r^{\lambda_1^{II}}\left\{\left(\kappa-\lambda_1^{II}\cos 2\alpha+\cos 2\alpha\lambda_1^{II}\right)\cos\lambda_1^{II}\theta+\lambda_1^{II}\cos\left(\lambda_1^{II}-2\right)\theta\right\}+\frac{\kappa+1}{2G}B_2 r\cos\theta \quad (1)$$

where A_0, A_1, B_0, B_1, and B_2 are Williams' coefficients, A_0 and B_0 are coefficients corresponding to the rigid body translation and B_2 is coefficient corresponding to the rigid body rotations, Kolosov constant κ is equal to $(3-\nu)/(1+\nu)$ for plane stress and $3-4\nu$ for plane strain conditions, $G=E/2(1+\nu)$ is the shear modulus, ν and E are the Poisson's ratio and Young's modulus, respectively, and λ_1^I and λ_1^{II} are the modes I and II eigenvalues, respectively.

Similar to the crack problems, mode I and mode II NSIFs can be defined as [6]

$$\begin{aligned}K_I&=\sqrt{2\pi}\lambda_1^I\left(1+\lambda_1^I-\lambda_1^I\cos 2\alpha-\cos 2\alpha\lambda_1^I\right)A_1\\K_{II}&=\sqrt{2\pi}\lambda_1^{II}\left(-1+\lambda_1^{II}-\lambda_1^{II}\cos 2\alpha+\cos 2\alpha\lambda_1^{II}\right)B_1\end{aligned} \quad (2)$$

It can be shown that [11], the notch opening displacement (NOD) can be written as

$$\Delta v=v^I_{\theta=+\alpha}-v^I_{\theta=-\alpha}=2v^I_{\theta=+\alpha}=2\frac{A_1}{2G}r^{\lambda_1^I}\begin{bmatrix}\left(\kappa-\lambda_1^I\cos 2\alpha-\cos 2\alpha\lambda_1^I\right)\sin\lambda_1^I\alpha\\+\lambda_1^I\sin\left(\lambda_1^I-2\right)\alpha\end{bmatrix}$$

$$=2A_1C_1r^{\lambda_1^I} \quad (3)$$

Similarly, the notch sliding displacement (NSD) can be obtained as [11]

$$\Delta u=u^{II}_{\theta=+\alpha}-u^{II}_{\theta=-\alpha}=2u^{II}_{\theta=+\alpha}=2\frac{B_1}{2G}r^{\lambda_1^{II}}\begin{bmatrix}\left(-\kappa-\lambda_1^{II}\cos 2\alpha+\cos 2\alpha\lambda_1^{II}\right)\sin\lambda_1^{II}\alpha\\+\lambda_1^{II}\sin\left(\lambda_1^{II}-2\right)\alpha\end{bmatrix} \quad (4)$$

$$-2\frac{\kappa+1}{2G}B_2 r\sin\alpha=2B_1C_2r^{\lambda_1^{II}}+2B_2C_3r$$

where C_1, C_2, and C_3 are constants which depend upon the geometry of the notch and material property.

Considering N number of nodes are considered along the notch flanks (as shown in Fig. 2a), and the residual R_I between the analytical and the FE NOD can be obtained as

$$R_I=\sum_{j=1}^{N}\left[\ln\left(\Delta v_j^{FE}\right)-\ln(A_1)-\ln(2C_1)-\lambda_1^I\ln\left(r_j\right)\right]^2 \quad (5)$$

For the minimum value of R_I, the partial differentiation of R_I with respect to $\ln(A_1)$ should be equals to zero

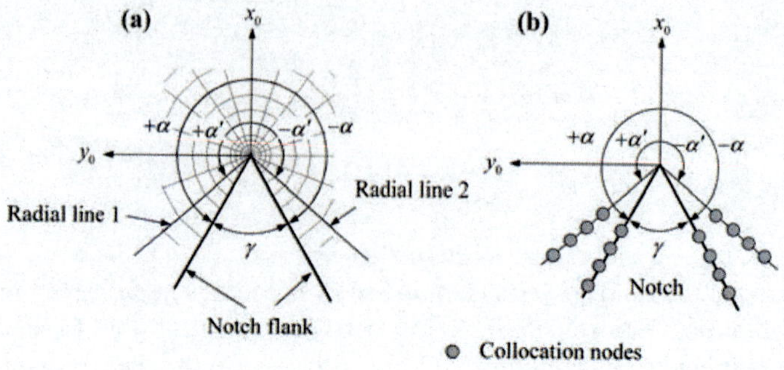

Fig. 2 **a** Selection of the collocation nodes and **b** typical mesh arrangement around the notch tip

$$\frac{\partial R_I}{\partial \ln(A_1)} = 0 \tag{6}$$

From (6), the Williams' constant A_1 can be obtained as [11]

$$A_1 = \frac{\exp\left[\frac{1}{N}\sum_{j=1}^{N}\ln\left(\Delta v_j^{FE} r_j^{-\lambda_1^I}\right)\right]}{2C_1} \tag{7}$$

The NSD (Eq. (4)) contains two unknown Williams constants (B_1 and B_2) and therefore, computation of B_1 is not straight forward. Hence, two additional radial lines (radial line 1 and radial line 2 as shown in Fig. 2b) along $\theta = +\alpha'$ and $\theta = -\alpha'$ apart from the notch flanks ($\theta = +\alpha$ and $\theta = -\alpha$) are considered. Similar to mode I, using effective NSD ($\Delta u_{eff,j}^{FE}$) at N number of nodes along the notch flanks and the radial lines 1 and 2, B_1 can be obtained as [11]

$$B_1 = \frac{\exp\left[\frac{1}{N}\sum_{j=1}^{N}\ln\left(\Delta u_{eff,j}^{FE} r_j^{-\lambda_1^{II}}\right)\right]}{2C_4} \tag{8}$$

Using the Williams constants the mode I and mode II NSIFs can be obtained from Eq. (2). The NSIFs can be normalized using the following equation [13]

$$F_I = (2\pi a)^{\lambda_1^I - 1}\frac{2L}{P_1}K_I \quad \text{and} \quad F_{II} = (2\pi a)^{\lambda_1^{II} - 1}\frac{2L}{P_1}K_{II} \tag{9}$$

Moreover, the state of mixity can be defined by an auxiliary parameter notch mode mixity ratio parameter (M^e), which can be defined as [13]

$$M^e = \frac{2}{\pi}\tan^{-1}\left(\frac{F_I}{F_{II}}\right) \tag{10}$$

3 Numerical Results and Discussions

A four-point bend (FPB) specimen is considered for the present analysis, as shown in Fig. 3a. The geometry and loading conditions for the FPB are shown in Fig. 3a. The length and width of the specimen are $L = 80$ and $w = 10$, respectively. The point load $P = 10$ is applied as shown in Fig. 3. The notch length to width ratio $a/w = 0.4$ is considered.

The finite element solution is completed in ANSYS. Eight-noded quadratic elements are used in the entire domain, and the notch tip elements are collapsed to form a spider web pattern at the notch tip. The typical FE mesh used for the FPB specimen is shown in Fig. 3b. The enlarged notch tip elements are shown in Fig. 3c.

At first, the results for the FPB for the notch angle $\gamma = 30°$ obtained using the collocation method [11] are validated with the results of Ref. [13]. In Ref [13], the shape factors were determined using an overdeterministic technique [10]. For the validation the following configurations were considered: $L_1/L = L_2/L = 0.45$, $L_4/L = 0.3$, and $L_3/L = 0.3$, 0.4, 0.5, 0.6, 0.7, 0.8, and 0.9. The present and published results are shown in Fig. 4. From Fig. 4, it can be seen that both mode I and mode II shape factors using the present approach are in good agreement with the published results [13]. After that, some new results for the shape factors with the new configurations, $L_1/L = L_2/L = 0.6$, $L_4/L = 0.45$, and $L_3/L = 0.45$, 0.5, 0.6, 0.7, 0.8, and 0.9 are plotted in Figs. 5 and 6 for $\gamma = 60°$ and 90°, respectively. Moreover, the mixity ratio parameters (M^e) are also plotted in Figs. 5c and 6c. It can be seen that when $L_3/L = L_4/L$, the specimen will be under pure mode I condition and as L_4 increases the specimen will be under mixed-mode loading conditions. It has

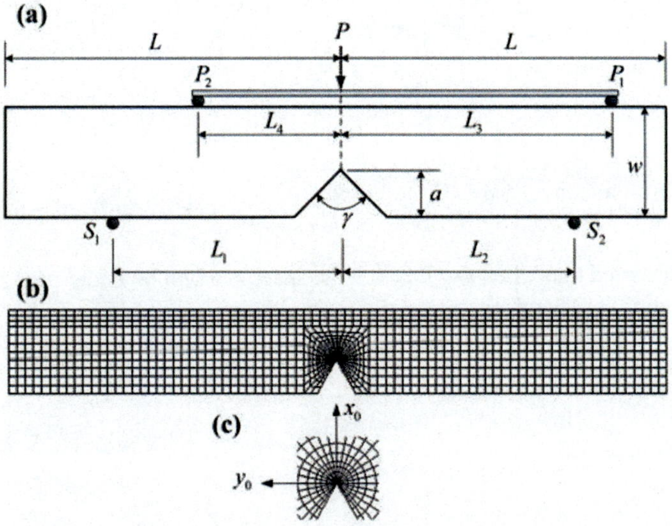

Fig. 3 **a** FPB specimen, **b** typical FE mesh used for the FPB specimen, and **c** typical mesh arrangement around the notch tip

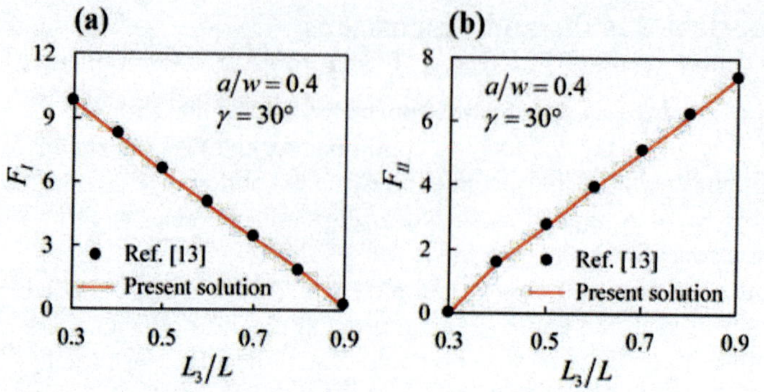

Fig. 4 Validation of shape factors for **a** F_I and **b** F_{II} for $\gamma = 30°$

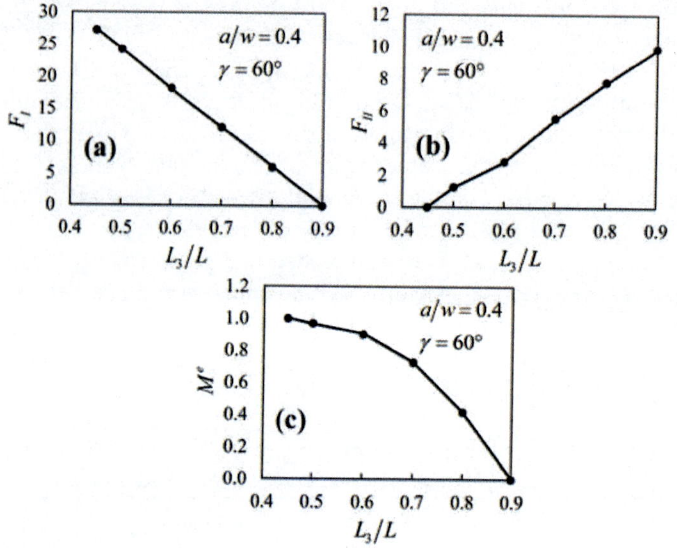

Fig. 5 Results for **a** F_I, **b** F_{II} and **c** M^e for $\gamma = 60°$

been also observed that when $L_3/L = 0.9$ the specimen will be under pure mode II loading conditions. Therefore, the mixity ratio (M^e) is 1 for $L_3/L = L_4/L = 0.45$, and the ratio of L_3/L changes, the value of M^e decreases and becomes equal to zero at $L_3/L = 0.9$.

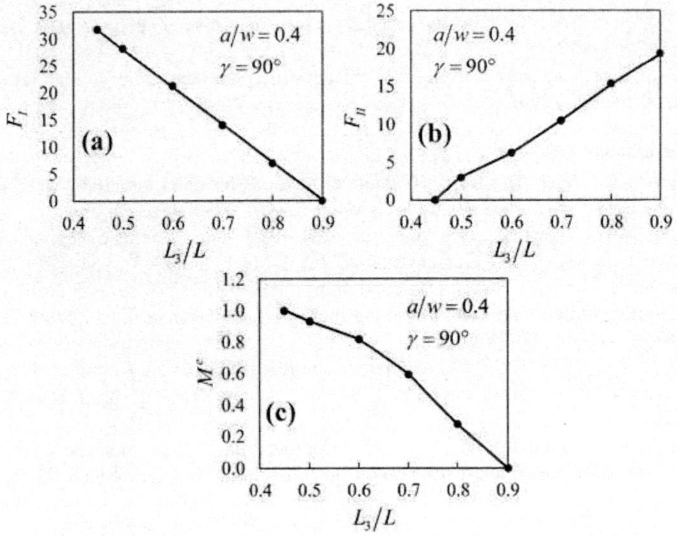

Fig. 6 Results for **a** F_I, **b** F_{II} and **c** M^e for $\gamma = 90°$

4 Conclusions

In this paper, the efficacy of an FE based NSIF extraction technique has been demonstrated by determining the NSIFs of four-point bending specimens. The notch flank opening and sliding finite element (FE) displacements are considered for determining the NSIFs. The rigid body displacements are nicely bypassed to evaluate the mixed mode (I/II) NSIFs. The NSIFs obtained using the present technique are found to be in good agreement with the available solutions. The mixed mode mixity ratio parameters are also determined for various loading positions from pure mode I to pure mode II. The results verify that the present collocation technique is simple, straightforward, and easy to be implemented in the available FE code.

References

1. Williams, M.L.: Stress singularities resulting from various boundary conditions in angular corners of plates in extension. J. Appl. Mech. **19**(4), 526–528 (1952)
2. Dunn, M.L., Suwito, W., Cunningham, S.: Fracture initiation at sharp notches: correlation using critical stress intensities. Int. J. Solids Struct. **34**(29), 3873–3883 (1997)
3. Seweryn, A.: Brittle fracture criterion for structures with sharp notches. Eng. Fract. Mech. **47**(5), 673–681 (1994)
4. Lazzarin, P., Livieri, P.: Notch stress intensity factors and fatigue strength of aluminium and steel welded joints. Int. J. Fatigue **23**(3), 225–232 (2001)
5. Fischer, C., Fricke, W., Rizzo, C.M.: Review of the fatigue strength of welded joints based on the notch stress intensity factor and SED approaches. Int. J. Fatigue **84**, 59–66 (2016)

6. Gross, B., Mendelson, A.: Plane elastostatic analysis of V-notched plates. Int. J. Fract. Mech. **8**(3), 267–276 (1972)
7. Chen, D.H.: Stress intensity factors for V-notched strip under tension or in-plane bending. Int. J. Fract. **70**, 81–97 (1995)
8. Ju, S.H., Chung, H.Y.: Accuracy and limit of a least-squares method to calculate 3D notch SIFs. Int. J. Fract. **148**(2), 169–183 (2007)
9. Liu, Y., Wu, Z., Liang, Y., Liu, X.: Numerical methods for determination of stress intensity factors of singular stress field. Eng. Fract. Mech. **75**(16), 4793–4803 (2008)
10. Ayatollahi, M.R., Nejati, M.: Determination of NSIFs and coefficients of higher order terms for sharp notches using finite element method. Int. J. Mech. Sci. **53**(3), 164–177 (2011)
11. Hussain, M.K., Murthy, K.S.R.K.: Calculation of mixed mode (I/II) stress intensities at sharp V-notches using finite element notch opening and sliding displacements. Fati. Fract. Eng. Mater. Struct. **42**(5), 1130–1147 (2019)
12. Hussain, M.K., Murthy, K.S.R.K.: A point substitution displacement technique for estimation of elastic notch stress intensities of sharp V-notched bodies. Theoret. Appl. Fract. Mech. **97**, 87–97 (2018)
13. Ayatollahi, M.R., Dehghany, M., Kaveh, Z.: Computation of V-notch shape factors in four-point bend specimen for fracture tests on brittle materials. Arch. Appl. Mech. **83**(3), 345–356 (2014)

Tera-Hertz Wave Propagation in Non-classical Beams Using Spectral Finite Element Method

K. Sainath and S. Narendar

1 Introduction

The present work is motivated by the growing attention toward the development of the futuristic nanoscale devices. Design, modeling and analysis of such devices under the umbrella of molecular dynamics methodologies and computations is quite cumbersome and costly. Due to this, the modified continuum theories are used to predict the accurate results which will give an idea about the result trends for a particular model [1, 2].

In the literature, nonlocal elasticity theory [2–6] is very popular to model the nanoscale elements to study their behavior. Many researchers have applied this theory to nanostructures to analyze the vibration, buckling or wave dispersion behavior [7, 8] at nanoscale [9]. These nanostructures are the primary elements in the nanoscale devices. So, the present paper deals with the dynamic wave behavior of two types of non-classical beams, i.e., Euler-Bernoulli and Timoshenko beams. To analyze the dynamic wave behavior of these beams, spectral finite element method [7] is used to obtain the non-classical dynamic stiffness matrix and the variation of the elements of dynamic stiffness with respect to frequency are studied in detail.

K. Sainath (✉) · S. Narendar
Defence Research and Development Laboratory, Hyderabad 500058, TS, India
e-mail: sainathaero88@gmail.com

S. Narendar
e-mail: narendar@drdl.drdo.in

S. K. Saha and M. Mukherjee (eds.), *Recent Advances in Computational Mechanics and Simulations*, Lecture Notes in Mechanical Engineering,
https://doi.org/10.1007/978-981-15-8315-5_13

2 Mathematical Formulation

2.1 The Theory of Non-classical Elasticity

The theory of non-classical elasticity was formally initiated by the papers of Erigen [3–6]. This takes in to account the remote action of forces between atoms. As per this theory, the stress state at a reference point **x** in the body is regarded to be dependent not only on the strain state at **x** but also on the strain states at all other points $\mathbf{x}'$ of the body as

$$\sigma_{ij,j} = 0 \tag{1}$$

$$\sigma_{ij}(\mathbf{x}) = \int_V \alpha(|\mathbf{x} - \mathbf{x}'|, \xi) C_{ijkl} \varepsilon_{jk}(\mathbf{x}') dV(\mathbf{x}'), \qquad \forall x \in V \tag{2}$$

$$\varepsilon_{ij} = \frac{1}{2}(u_{i,j} + u_{j,i}) \tag{3}$$

where C_{ijkl} is the elastic modulus tensor of classical isotropic elasticity, σ_{ij} and ε_{ij} are stress and strain tensors, respectively, and u_i is the displacement vector. $\alpha = \alpha(|\mathbf{x} - \mathbf{x}'|, \xi)$ is nonlocal modulus. $|\mathbf{x} - \mathbf{x}'|$ is the Euclidean distance, and $\xi = \frac{e_0 a}{\ell}$, where e_0 is nonlocal parameter.

3 Non-classical Beams

As mentioned in introduction, we will use two different beam models, namely, the Bernoulli-Euler model and Timoshenko beam models, to study the dynamic wave behavior in these non-classical beams.

3.1 Non-classical Euler-Bernoulli Beam Model

The governing equation of the non-classical Euler-Bernoulli beam model is given as [6]

$$EI\omega_{,xxxx} + \rho A\ddot{\omega} - (e_0 a)^2 \ddot{\omega}_{,xx} = 0 \tag{4}$$

where $\omega = \omega(x, t)$ is the flexural deflection, ρ is the mass density, A is the cross-sectional area, EI is the bending rigidity of the beam structure and $e_0 a$ is the nonlocal scaling parameter. Here () denotes derivative with respect to time(t) and ($_{,}x$) is the derivative with respect to x. It is observed that if the internal length scale a is identically zero, then the classical Euler-Bernoulli beam model is recovered.

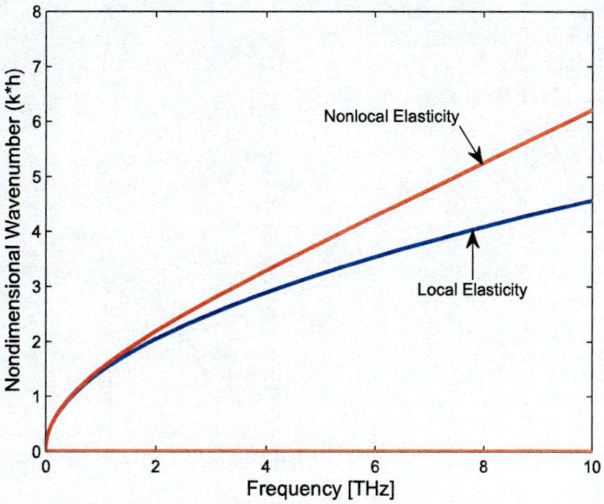

Fig. 1 Spectrum curves (Wavenumber dispersion), for a nonlocal beam obtained from both local and nonlocal Euler-Bernoulli beam theories

The time variable can be eliminated from the governing partial differential equation of the non-classical beam (Eq. (4)) by using the Fourier transformation,

$$\omega(x, t) = \sum_{n=1}^{N} \widehat{W}(x, \omega_n) e^{i\omega_n t} \tag{5}$$

where ω_n the circular frequency of the nth sampling point and N is the Nyquist frequency.

For non-trivial solution of the amplitude, this implies that

$$EIk^4 - \rho A\omega^2 (e_0 a)^2 k^2 - \rho A\omega^2 = 0 \tag{6}$$

this is known as the dispersion equation for the assumed non-classical Euler-Bernoulli beam. The wavenumbers are obtained by solving the dispersion equation (6). The wavenumber (k) variation with the frequency (ω) is shown in Fig. 1 for both the classical and non-classical beams. The dispersion equation gives four wavenumber solutions. Out of these four wavenumbers two are purely real and the other two are purely imaginary. The real part gives rise to the propagating component of the wave while the imaginary part gives rise to the spatially damped mode of the wave. From Eq. (6) it is obvious that, there is no possibility for a cut-off frequency for these flexural modes, above which the spatially damped mode turns to be propagative. The relation between the wavenumber k and frequency ω is called the spectrum relation and it is a nonlinear function of frequency and is clearly seen in Fig. 1. The non-classical elasticity predicts the higher wavenumbers than the classical elasticity. This leads to the lower group speeds ($C_g = \mathrm{Re}(\partial k/\partial\omega)$) in non-classical elasticity as shown in Fig. 2. The complete dynamic wave dispersion in the non-classical Euler-Bernoulli beam is obtained from the dynamic stiffness.

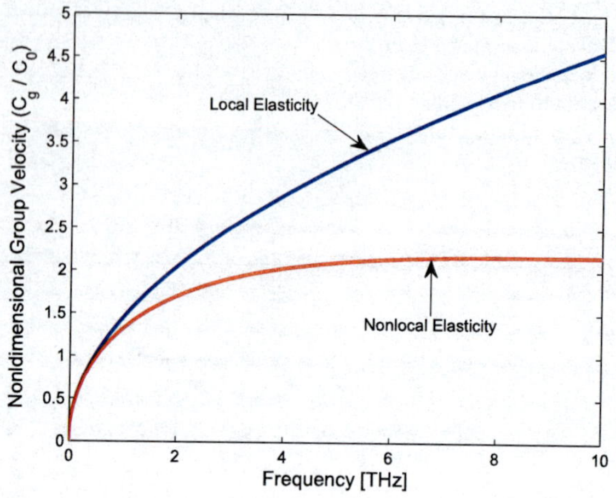

Fig. 2 Dispersion curves (group speed dispersion), for a nonlocal beam obtained from both local and nonlocal Euler-Bernoulli beam theories

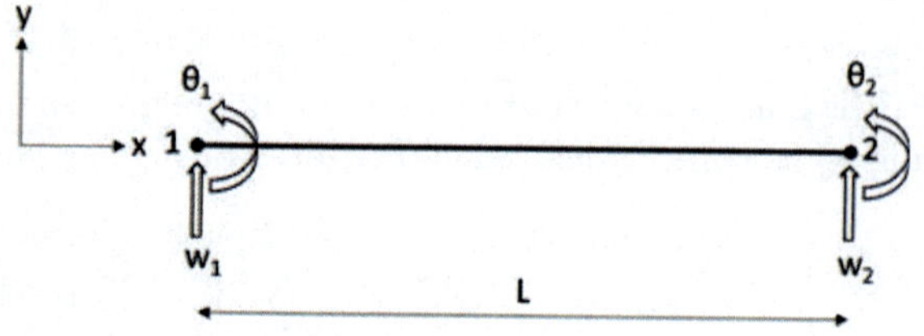

Fig. 3 Schematic of Euler-Bernoulli spectral beam element

3.2 *Computation of Dynamic Stiffness Matrix for Euler-Bernoulli Beam*

Since the spectral element can be very long then the ability to compute the response between nodes is necessary. This can be done conveniently using the dynamic stiffness matrix. Thus the total solution for the displacement field to be considered as:

$$\omega(x,t) = (\overline{A}e^{ik_1x} + \overline{B}e^{ik_2x} + \overline{C}e^{ik_3x} + \overline{D}e^{ik_4x})e^{i\omega t} \tag{7}$$

Now that the representation of the displacement field has been established from the governing partial differential equation, the dynamic stiffness can be formulated. The dynamic stiffness is formulated using a direct method. both the nodal displacements and nodal forced can be written in terms of the unknown coefficients $(\overline{A}, \overline{B}, \overline{C}, \overline{D})$. These relations can be used to form an expression for the dynamic stiffness matrix. the spectral beam element developed will use two nodes with two DOF, transverse displacement (ω_1, ω_2) and rotation (θ_1, θ_2) at each node as shown in Fig. 3.

Writing the displacements and rotations at the nodes in matrix form gives

$$\begin{Bmatrix} \omega(0,t) \\ \theta(0,t) \\ \omega(L,t) \\ \theta(L,t) \end{Bmatrix} = \begin{Bmatrix} \omega_1 \\ \theta_1 \\ \omega_2 \\ \theta_2 \end{Bmatrix} e^{i\omega t} = [D] \begin{Bmatrix} \overline{A} \\ \overline{B} \\ \overline{C} \\ \overline{D} \end{Bmatrix} e^{i\omega t} \tag{8}$$

for the case of the Euler-Bernoulli beam elements, $[D]$ is a 4×4 matrix and has the form

$$[D] = \begin{bmatrix} 1 & 1 & 1 & 1 \\ ik_1 & ik_2 & ik_3 & ik_4 \\ e^{ik_1 L} & e^{ik_2 L} & e^{ik_3 L} & e^{ik_4 L} \\ ik_1 e^{ik_1 L} & ik_2 e^{ik_2 L} & ik_3 e^{ik_3 L} & ik_4 e^{ik_4 L} \end{bmatrix} \tag{9}$$

writing the shears and moments at the nodal locations in matrix form gives

$$\begin{Bmatrix} V(0,t) \\ -M(0,t) \\ -V(L,t) \\ M(L,t) \end{Bmatrix} = \begin{Bmatrix} V_1 \\ M_1 \\ V_2 \\ M_2 \end{Bmatrix} e^{i\omega t} = [F] \begin{Bmatrix} \overline{A} \\ \overline{B} \\ \overline{C} \\ \overline{D} \end{Bmatrix} e^{i\omega t} \tag{10}$$

For the case of Euler-Bernoulli beam element, $[F]$ has the following form

$$[F] = \begin{bmatrix} -ik_1^3 & -ik_2^3 & -ik_3^3 & -ik_4^3 \\ k_1^2 & k_2^2 & k_3^2 & k_4^2 \\ ik_1^3 e^{ik_1 L} & ik_2^3 e^{ik_2 L} & ik_3^3 e^{ik_3 L} & ik_4^3 e^{ik_4 L} \\ -ik_1^2 e^{ik_1 L} & -ik_2^2 e^{ik_2 L} & -ik_3^2 e^{ik_3 L} & -ik_4^2 e^{ik_4 L} \end{bmatrix} \tag{11}$$

The transverse displacement and rotation evaluated at the nodes in Eq. (11) form the vector of nodal DOF $\{d\} = \{\omega_1\ \theta_1\ \omega_2\ \theta_2\}^T e^{i\omega t}$. The shear force and bending moments are evaluated at the nodes in Eq. (10) from the vector of nodal forces $\{f\} = \{V_1\ M_1\ V_2\ M_2\}^T e^{i\omega t}$. The dynamic stiffness matrix $[K_{dyn}]$ relates the nodal DOF $\{d\}$ to the nodal loads $\{f\}$ as $\{f\} = [K_{dyn}]\{d\}$. The dynamic stiffness matrix can be solved from Eqs. (8) and (10) to be $[K_{dyn}] = [F][D]^{-1}$.

The dynamic stiffness matrix is symmetric and includes transverse inertial effects. The dynamic stiffness matrix was developed through a direct method using the nodal displacements and forces. the displacement field that the nodal displacements and forces were based on was the exact solution of governing PDE. Because the dynamic stiffness is based on the exact solution to the governing PDE, the nodal displacements and forces obtained through the use of the dynamic stiffness will be exact. The dynamic stiffness elements variation with frequency is shown in Figs. 4 and 5. Due to space limitation the dynamic stiffness for diagonal terms of K_{11}, K_{44} and off-diagonal terms of K_{13}, K_{24} are shown in the results. The variation of the diagonal stiffness elements with frequency are shown in Fig. 4 for classical and non-classical solutions. A slight good match in the dynamic stiffness obtained for very low frequencies, i.e., <1 THz between classical and non-classical solutions. A tremendous variation in the dynamic stiffness is seen in the non-classical solutions at higher frequencies.

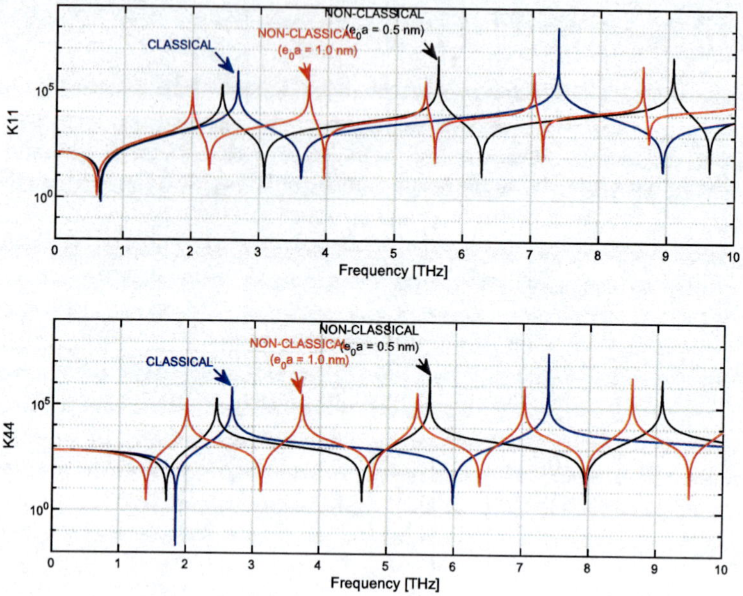

Fig. 4 Dynamic stiffness (K_{11}, K_{44}) behavior of non-classical Euler-Bernoulli beam at lower frequencies

As the nonlocal scale coefficient increases the amplitude of the dynamic stiffness response reduces and also this effect introduces extra number of peaks within the given frequency range. This effect cannot be captured in classical solutions. Similar variations are observed in the off-diagonal stiffness elements and are shown in Fig. 5. These variations in dynamic stiffness elements are directly related to the dynamic response of the non-classical beams.

3.3 Non-classical Timoshenko Beam Model

The governing equation of the non-classical Timoshenko beam model is given as [6]

$$GA\kappa\left(\psi_{,x} - \omega_{,xx}\right) + \rho A\left(\ddot{w} - (e_0 a)^2 \ddot{\omega}_{,xx}\right) = 0 \tag{12}$$

$$EI\psi_{,xx} + GA\kappa\left(\omega_{,x} - \psi\right) - \rho I\left(\ddot{\psi} - (e_0 a)^2 \ddot{\psi}_{,xx}\right) = 0 \tag{13}$$

where $G = \frac{E}{2(1+\nu)}$ is shear modulus of the beam, ν is the Poisson's ratio, κ is the shear adjustment coefficient, vary with the cross section of the beam [5], I is the moment of inertia of the cross section of the beam, A is the cross-sectional area, and

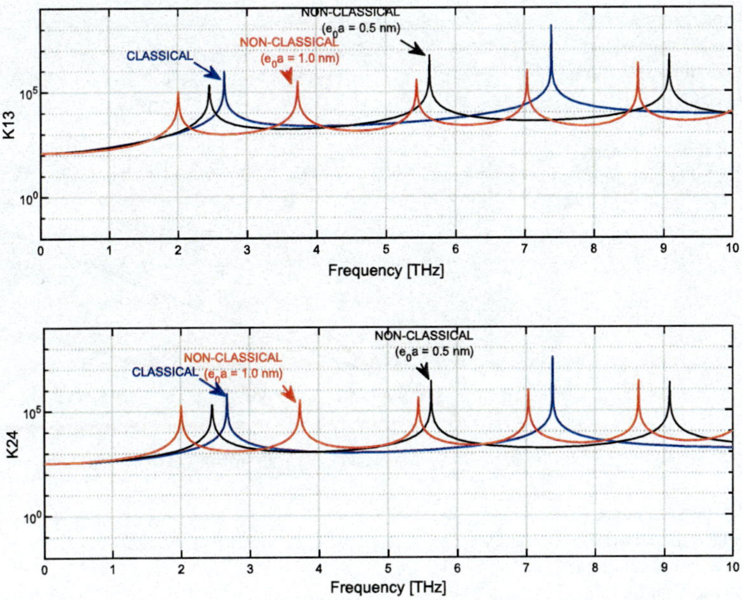

Fig. 5 Dynamic stiffness (K_{13}, K_{24}) behavior of non-classical Euler-Bernoulli beam at lower frequencies

ρ is the mass density of the beam. It is observed that if the internal length scale a is identically zero, then the local Timoshenko beam model is recovered.

The time variable can be eliminated from the governing partial differential equation of the nonlocal beam Eqs. (12-13) by using the Fourier transformation,

$$w\,(x,t) = \sum_{n=1}^{N} \widehat{W}(x,\omega_n)e^{i\omega_n t}, \quad \psi\,(x,t) = \sum_{n=1}^{N} \widehat{\Psi}(x,\omega_n)e^{i\omega_n t} \tag{14}$$

The wavenumbers and hence the group speeds are solved from resulted characteristic equation by using Polynomial Eigenvalue Problem (PEP) [7].

Multiplying the temporal and spatial components and combining coefficients give the complete solution to be

$$\begin{aligned} \omega(x,t) &= (\overline{A}R_1e^{ik_1x} + \overline{B}R_2e^{ik_2x} + \overline{C}R_3e^{ik_3x} + \overline{D}R_4e^{ik_4x})e^{i\omega t} \\ \psi(x,t) &= (\overline{A}e^{ik_1x} + \overline{B}e^{ik_2x} + \overline{C}e^{ik_3x} + \overline{D}e^{ik_4x})e^{i\omega t} \end{aligned} \tag{15}$$

The solutions for the displacement fields can now be used to form the dynamic stiffness matrix.

The spectrum and dispersion curves obtained from classical and non-classical Timoshenko beam theories are shown in Figs. 6 and 7. Figure 6 shows the variation

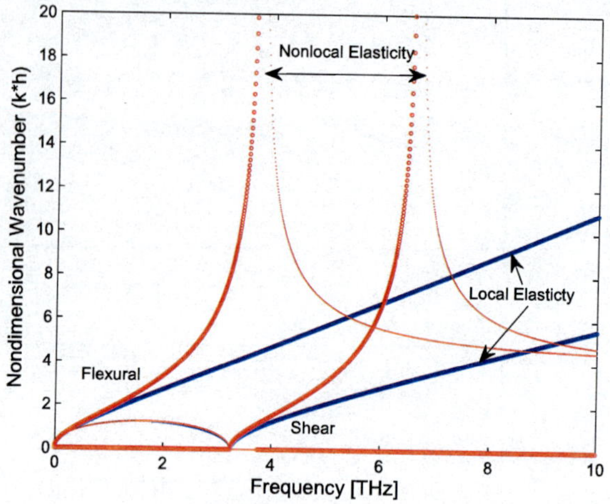

Fig. 6 Spectrum curves (Wavenumber dispersion: Real wavenumbers—thick lines; Imaginary wavenumbers—thin lines), for nonlocal beam obtained from both local and nonlocal Timoshenko beam theories

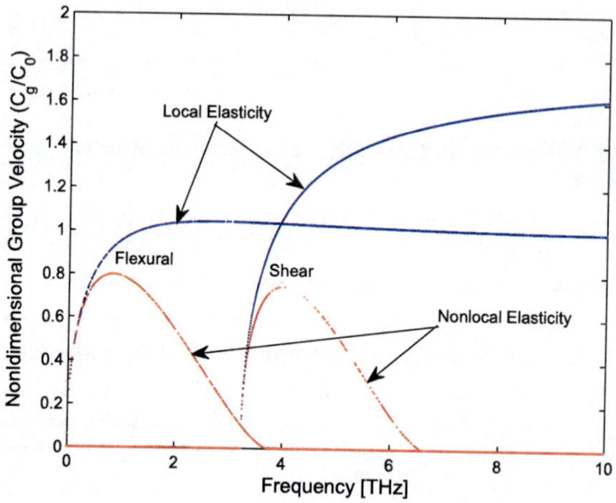

Fig. 7 Dispersion curves (Group speed dispersion), for nonlocal beam obtained from both local and nonlocal Timoshenko beam theories

of the wavenumbers with the wave frequency for both classical and non-classical elasticity. This figure shows only two modes, namely, flexural and shear. Flexural wave mode starts from zero wave frequency and shear wave mode propagates only after shear cut-off frequency, the frequency at which the imaginary part of wavenumber becomes real. For $e_0a = 0$, which is the case of local theory of elasticity solution, wavenumbers increase monotonically with the increase in frequency, which is shown in Fig. 6 and correspondingly, the group speeds, shown in Fig. 7) increases with increase in wave frequency. However, at higher frequencies, they attain a constant value, which is typical of Timoshenko beam solution. However, with the introduction of nonlocal scale effects, the wave behavior is altered drastically. Both the flexural and shear wave modes escapes to infinity at a particular frequency called the "escape frequency", beyond this frequency there is no wave propagation.

3.4 Computation of Dynamic Stiffness Matrix for Timoshenko Beam

The dynamic stiffness matrix for the spectral Timoshenko beam element is formulated following the same direct method as the spectral Ruler-Bernoulli dynamic stiffness. The displacements and forces at each node are written in terms of the unknown coefficients $\overline{A}$, $\overline{B}$, $\overline{C}$ and $\overline{D}$. These relations can then be used to form an expression for the dynamic stiffness matrix, The spectral beam element developed will use two nodes with two DOF, transverse displacement (ω_1, ω_2) and rotation (ψ_1, ψ_2) at each node as shown in Fig. 8.

Evaluating the transverse displacement and rotation, in Eq. (15), at the nodes and writing in matrix form gives

$$\begin{Bmatrix} \omega(0,t) \\ \psi(0,t) \\ \omega(L,t) \\ \psi(L,t) \end{Bmatrix} = \begin{Bmatrix} \omega_1 \\ \psi_1 \\ \omega_2 \\ \psi_2 \end{Bmatrix} e^{i\omega t} = [D] \begin{Bmatrix} \overline{A} \\ \overline{B} \\ \overline{C} \\ \overline{D} \end{Bmatrix} e^{i\omega t} \tag{16}$$

for the case of the Timoshenko beam element, $[D]$ is a 4×4 matrix and has the form

$$[D] = \begin{bmatrix} R_1 & R_2 & R_3 & R_4 \\ 1 & 1 & 1 & 1 \\ R_1 e^{ik_1 L} & R_2 e^{ik_2 L} & R_3 e^{ik_3 L} & R_4 e^{ik_4 L} \\ e^{ik_1 L} & e^{ik_2 L} & e^{ik_3 L} & e^{ik_4 L} \end{bmatrix} \tag{17}$$

writing the nodal shear and moment forces in matrix form gives

$$\begin{Bmatrix} V(0,t) \\ -M(0,t) \\ -V(L,t) \\ M(L,t) \end{Bmatrix} = \begin{Bmatrix} V_1 \\ M_1 \\ V_2 \\ M_2 \end{Bmatrix} e^{i\omega t} = [F] \begin{Bmatrix} \overline{A} \\ \overline{B} \\ \overline{C} \\ \overline{D} \end{Bmatrix} e^{i\omega t} \tag{18}$$

For the case of Timoshenko beam element, $[F]$ has the following form

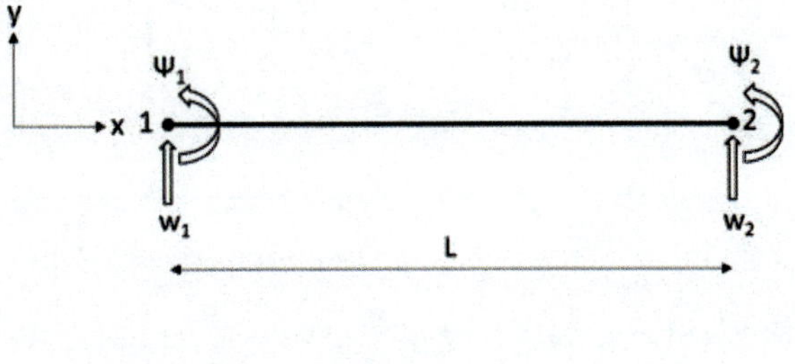

Fig. 8 Schematic of Timoshenko spectral beam element

$$[F]=\begin{bmatrix} -\Gamma(ik_1R_1-1) & -\Gamma(ik_2R_2-1) & -\Gamma(ik_3R_3-1) & -\Gamma(ik_4R_4-1) \\ -EIik_1 & -EIik_2 & -EIik_3 & -EIik_4 \\ \Gamma(ik_1R_1-1)e^{ik_1L} & \Gamma(ik_2R_2-1)e^{ik_2L} & \Gamma(ik_3R_3-1)e^{ik_3L} & \Gamma(ik_4R_4-1)e^{ik_4L} \\ EIik_1e^{ik_1L} & EIik_2e^{ik_2L} & EIik_3e^{ik_3L} & EIik_4e^{ik_4L} \end{bmatrix} \quad (19)$$

where $\Gamma = GA\kappa$. The transverse displacement and rotation evaluated at the nodes in Eq. (19) form the vector of nodal DOF $\{d\} = \left\{ \omega_1\ \psi_1\ \omega_2\ \psi_2 \right\}^T e^{i\omega t}$. The shear force and bending moments are evaluated at the nodes in Eq. (18) from the vector of nodal forces $\{f\} = \{V_1\ M_1\ V_2\ M_2\}^T e^{i\omega t}$. The dynamic stiffness matrix $[K_{dyn}]$ relates the nodal DOF $\{d\}$ to the nodal loads $\{f\}$ as $\{f\} = [K_{dyn}]\{d\}$. The dynamic stiffness matrix can be solved from Eqs. (16) and (18) to be $[K_{dyn}] = [F][D]^{-1}$.

The dynamic stiffness matrix is symmetric dynamic stiffness matrix includes the effects of transverse inertia, rotary inertia, shear deformation, and bending. The dynamic stiffness elements variation with frequency is shown in Figs. 9 and 10. Due to space limitation the dynamic stiffness for diagonal terms of K_{11}, K_{44} and off-

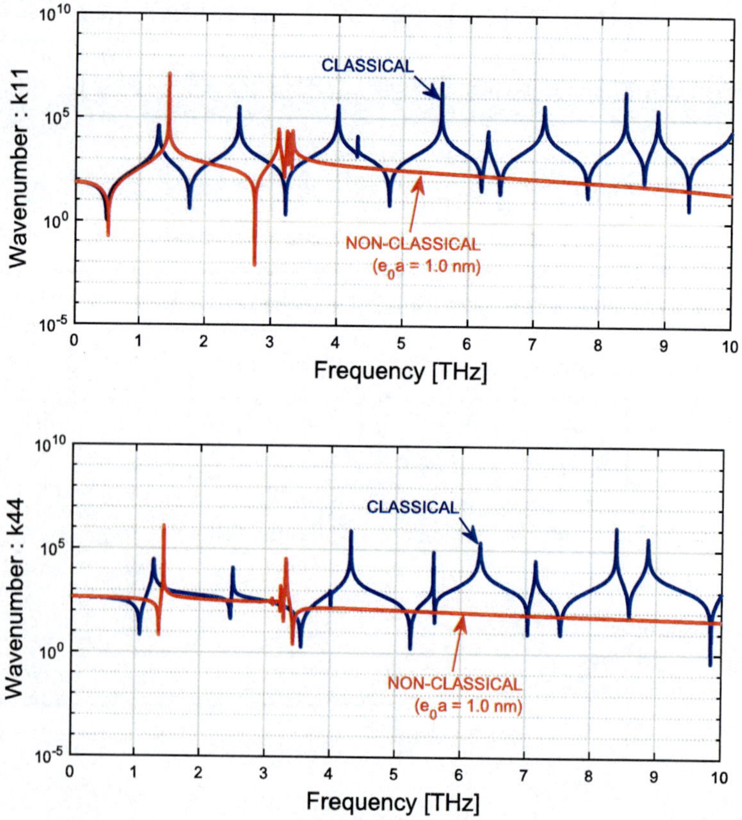

Fig. 9 Dynamic stiffness (K_{11}, K_{44}) behavior of non-classical Timoshenko beam at lower frequencies

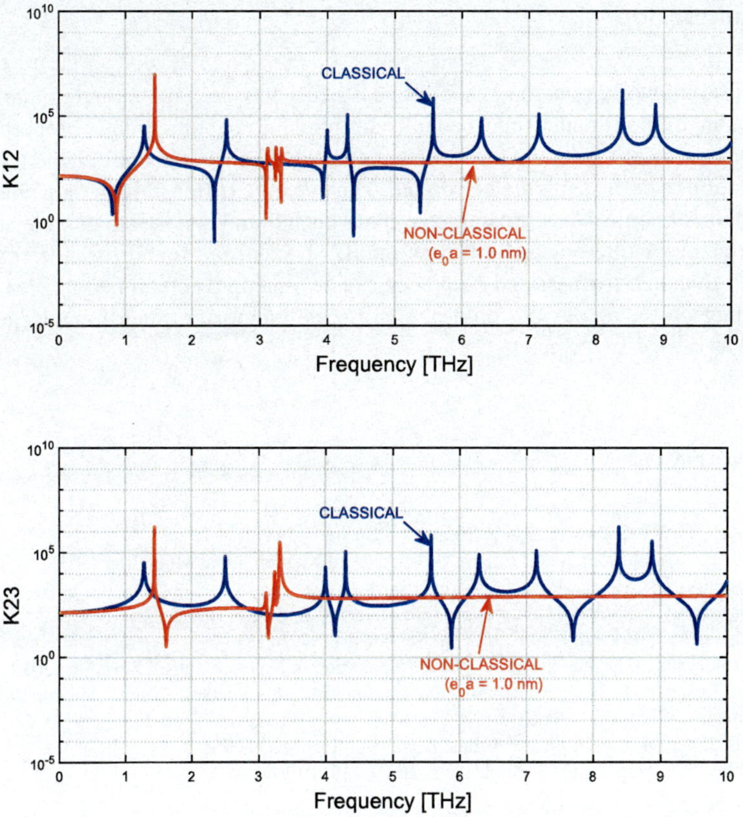

Fig. 10 Dynamic stiffness (K_{12}, K_{23}) behavior of non-classical Timoshenko beam at lower frequencies

diagonal terms of K_{12}, K_{23} are shown in the results. The variation of the diagonal stiffness elements with frequency are shown in Fig. 9 for classical and non-classical solutions. A slight good match in the dynamic stiffness obtained for very low frequencies, i.e., <1 THz between classical and non-classical solutions. A tremendous variation in the dynamic stiffness is seen in the non-classical solutions at higher frequencies. As the non-classical scale coefficient increases the amplitude of the dynamic stiffness response reduces and also this effect introduces extra number of peaks within the given frequency range. This effect cannot be captured in classical solutions. Similar variations are observed in the off-diagonal stiffness elements and are shown in Fig. 10. These variations in dynamic stiffness elements are directly related to the dynamic response of the non-classical beams. It has been shown that the non-classical Timoshenko beams will have maximum of 2 escape frequencies.

4 Conclusion

In the present work, non-classical elasticity theory has been incorporated into classical Euler-Bernoulli and Timoshenko beam model to capture unique properties of these beams under the continuum mechanics theory. The spectral finite element (SFE) formulation of both beams are performed. The dynamic stiffness matrix are obtained as function of nonlocal scale parameter. It has been found that the small scale affects the elements of the dynamic stiffness matrix. It has seen found that the relation between dynamic stiffness and frequency can be seen upto certain frequencies only, depending on the values of nonlocal small scale parameter. Such frequencies are termed as escape frequencies.

References

1. Doyle, J.F.: Wave Propagation in Structures. Springer, NY (1997)
2. Eltaher, M.A., Khater, M.E., Emam, S.A.: Applied Mathematical Modelling (2016)
3. Eringen, A.C.: J. Appl. Phys. **54**, 4703 (1983)
4. Eringen, A.C.: Nonlocal Continuum Field Theories. Springer, NY (2002)
5. Eringen, A.C., Edelen, D.G.B.: J. Eng. Sci. **10**, 1 (1972)
6. Gopalakrishnan, S., Narendar, S.: Wave Propagation in Nanostrucutres. Springer (2013)
7. Gopalkrishnan, S., Roy Mahapatra, D., Chakraborty, A.: Spectral Finite Element Method. Springer London Ltd. (2008)
8. Graff: Wave Motion in Elastic Solids. Dover Publications (1995)
9. Reddy, J.N., Panga, S.D.: J. Appl. Phys. **103**, 023511 (2008)

Fem Simulations for Fatigue Life Estimation of Big Turbo-generator Shaft During Various Fault Disturbances Under Active Control

Tarun Kumar, Rajeev Kumar, and S. C. Jain

1 Introduction

Turbo-generator coupled shaft subjected to torsional vibrations in power generation system due to various short circuits in network, it may be 3-Phase, 2-Phase, 1-Phase to line or mal-synchronization [1]. These faults are very sensitive to turbo generator rotor system because of its bulky size and long length. Various types of fault occurring due to grid interaction cause an impulsive electromagnetic torque on the coupled shaft system which generates torsional vibrations in it with amplitude 2 to 6 times of the nominal torque [2]. Sometimes these vibrations lead to severe damage to coupled shaft system or complete failure of it due to aggregation of stress. Low cycle fatigue wears of 70% of life of rotor and creep accounts for remaining 30% [3, 4]. The synchronous generator draws power from a prime mover (turbine) as shown in Fig. 1. The different researchers simulated generator vibrations during various electrical faults under loaded and unloaded conditions using different models [5–12]. Further, various researchers presented computational methods to measure torsional vibration produced on coupled shaft due to electrical faults on generator under unloaded condition and estimated the fatigue life of rotor under vibrations [13–19]. Although no complete analysis is available in the literature in which turbo-generator shaft life is numerically simulated under loaded conditions for various electrical faults. Further no comprehensive literature is available in which such turbo-generator rotor's vibrations is semi-actively controlled using piezoelectric material and vibrations produced are compared with the uncontrolled system. Therefore, in this paper, a comparative result of vibrations under uncontrolled and controlled rotor

T. Kumar (✉) · R. Kumar · S. C. Jain
School of Engineering, Indian Institute of Technology Mandi, Kamand 175005, HP, India
e-mail: kvtarun85@gmail.com

S. K. Saha and M. Mukherjee (eds.), *Recent Advances in Computational Mechanics and Simulations*, Lecture Notes in Mechanical Engineering,
https://doi.org/10.1007/978-981-15-8315-5_14

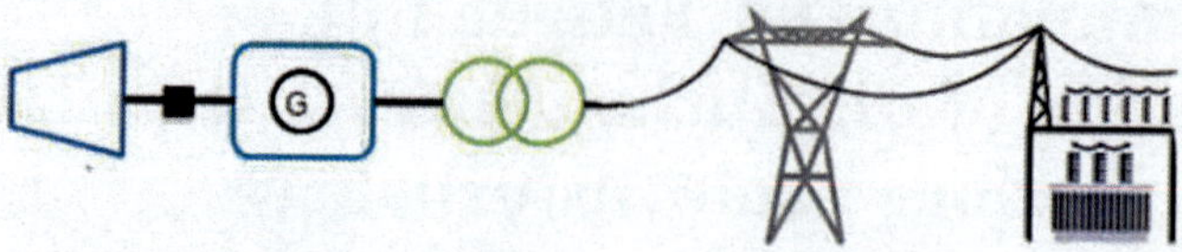

Fig. 1 Schematic of power transmission system

conditions are numerically simulated and fatigue life is compared for different faults using rain-flow counting algorithm.

2 Mathematical Formulation

2.1 Generator Modeling

The generalized *dq0* model is used for modeling synchronous generator [9]. The voltage equations of synchronous generator are given by set of differential Eqs. (1–6) wherein u, i, ψ denote voltages, currents, and flux linkages, respectively. The subscripts $d, q, 0$ are associated with the stator d, q axis components and the zero sequence component of voltages, currents, and flux linkages as given in Eq. (7).

$$[u] = -[R][i] - \frac{d[\psi]}{dt} + \omega[\psi] \tag{1–6}$$

where

$$[x] = \left[x_d\ x_q\ x_0\ x_{fd}\ x_D\ x_Q\right]',\ x \to u, i, \psi$$

$$x_0 = \frac{x_A + x_B + x_C}{3}, x \to u, i, \psi \tag{7}$$

The equation of mechanical motion is given as

$$T_{\text{drive}} = T_{em} + \frac{J}{p}\frac{d\omega}{dt} \tag{8}$$

where

$$T_{em} = \frac{3}{2}p\left(-\psi_d i_q + \psi_q i_d\right) \tag{9}$$

$$\frac{d\gamma}{dt} = -\omega \tag{10}$$

Equations (1–10) are used to drive final dynamical model to find various generator parameters during sudden fault conditions. The same approach as used by [9] is adopted to drive final dynamical model equations for loaded generator. Park-Gorev transformation as given in (11) is used to express $d, q, 0$ parameters (u, i, ψ) in term of a, b, c phase parameters.

T_{drive} = driving turbine torque
J = moment of inertia
p = no of generator pole pairs

$$\begin{bmatrix} x_d \\ x_q \\ x_0 \end{bmatrix} = \frac{2}{3}\begin{bmatrix} \cos\gamma & \cos\left(\gamma + \frac{2\pi}{3}\right) & \cos\left(\gamma + \frac{4\pi}{3}\right) \\ \sin\gamma & \sin\left(\gamma + \frac{2\pi}{3}\right) & \sin\left(\gamma + \frac{4\pi}{3}\right) \\ \frac{1}{2} & \frac{1}{2} & \frac{1}{2} \end{bmatrix}\begin{bmatrix} x_A \\ x_B \\ x_C \end{bmatrix} = [T]\begin{bmatrix} x_A \\ x_B \\ x_C \end{bmatrix} \tag{11}$$

where, $x \rightarrow u, i, \psi$.

2.2 *Fault Modeling of Loaded Generator*

As shown in Fig. 2 generator is in loaded condition, i.e., it is delivering power to the grid system. Terminal voltage at three phases of generator has value u_a, u_b, and u_c. Generator is transferring power through line with inductance L_e and resistance R_e. Line voltage at the point of fault is $u_{\infty a}$, $u_{\infty b}$, $u_{\infty c}$. During occurrence of fault the phase voltages at generator terminals are different in loaded condition in contrary to unloaded case [9] as given by Eqs. (12–13).

$$[u_{abc}] = [u_{\infty abc}] + R_e[I][i_{abc}] + L_e[I]\left[\frac{d}{dt}[i_{abc}]\right] \tag{12}$$

From (11) and (12), we get

$$\left[u_{dq0}\right] = [T][u_{\infty abc}] + R_e[I][T][i_{abc}] + L_e[I]\left[\frac{d}{dt}[T][i_{abc}]\right] \tag{13}$$

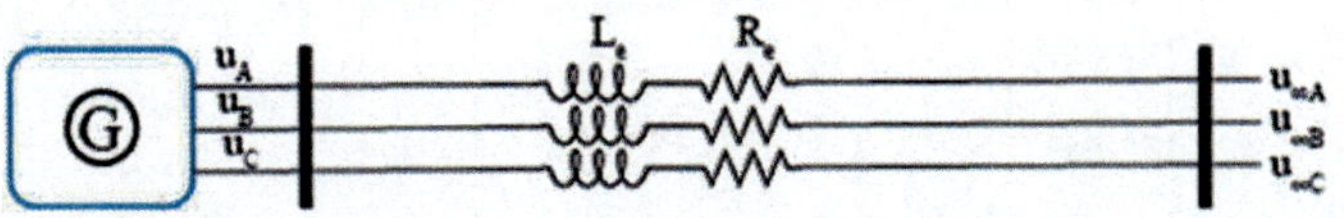

Fig. 2 Loaded generator connected to infinite bus bar

Above equations are further modified based on fault modeling for loaded generator for four different cases of electric faults as below.

Line to ground short circuit. Line to ground short circuit occurred when a phase line come in contact with ground. The mathematical model in case of line to ground short circuit consists of four numbers of differential equations given by Eqs. (14–17). Let us assume that phase A is grounded then boundary conditions will be as

$$u_{\infty A} = 0,\ i_B = i_C = 0$$

Equations (14–16) are directly obtained from [9].

$$L_{fd\sigma}\frac{di_{fd}}{dt} - L_{D\sigma}\frac{di_D}{dt} = -R_{fd}i_{fd} + R_D i_D + u_{fd} \tag{14}$$

$$L_{md}\frac{di_d}{dt} + L_{md}\frac{di_{fd}}{dt} + L_D\frac{di_D}{dt} = -R_D i_D \tag{15}$$

$$L_{mq}\frac{di_q}{dt} + L_Q\frac{di_Q}{dt} = -R_Q i_Q \tag{16}$$

For the loaded generator Eq. (17) is derived using boundary conditions for the specific fault as.

At generator's terminal

$$-u_A = R\,i_A + \frac{d\psi_A}{dt}$$

Also from converse Park-Gorev transformation, we get

$$\psi_A = \psi_d \cos\gamma + \psi_q \sin\gamma + \psi_0$$

Differentiating the above equation w.r.t time and using same terminology as in [9], finally get

$$\frac{d\psi_d}{dt}\cos\gamma + \frac{d\psi_q}{dt}\sin\gamma + \frac{d\psi_0}{dt} = -Ri_A - \omega\psi_d \sin\gamma + \omega\psi_q \cos\gamma - u_A$$

$$\begin{bmatrix} \frac{1}{3}L_0 \\ L_{d1} \\ L_{q2} \\ L_{md1} \\ L_{md1} \\ L_{mq2} \end{bmatrix} \frac{d}{dt} \begin{bmatrix} i_A \\ i_d \\ i_q \\ i_{fd} \\ i_D \\ i_Q \end{bmatrix} = \begin{bmatrix} -R \\ -\omega L_{d2} \\ \omega L_{q1} \\ -\omega L_{md2} \\ -\omega L \\ \omega L_{mq1} \end{bmatrix} \begin{bmatrix} i_A \\ i_d \\ i_q \\ i_{fd} \\ i_D \\ i_Q \end{bmatrix} - u_A \tag{17}$$

u_A can be evaluated using (12) by applying boundary conditions.

The electromagnetic torque of synchronous generator is given by Eq. (18) (Ref. [9].)

$$T_{em} = -P\begin{bmatrix} \frac{2}{3}(L_{d1} - L_{q1})\sin\gamma \\ L_{md2} \\ L_{md2} \\ -L_{mq1} \end{bmatrix}^T \begin{bmatrix} i_A \\ i_{fd} \\ i_D \\ i_Q \end{bmatrix} i_A \tag{18}$$

Line to line short circuit Line to line short circuit occurred when two phases come in contact with each other. The mathematical model in case of line to line short circuit consists of four numbers of differential equations given by Eqs. (14–16) and Eq. (19). Let us assume that phase B and phase C are shorted then boundary conditions will be as

$$u_{\infty B} = u_{\infty C},\ i_A = 0,\ i_B = -i_C$$

For the loaded generator Eq. (19) is derived using boundary conditions for the line to line short circuit fault as.

At generator's terminal

$$-u_B = R\,i_B + \frac{d\psi_B}{dt},\ -u_C = R\,i_C + \frac{d\psi_C}{dt}$$

$$\frac{d}{dt}(\psi_B - \psi_C) + R\,(i_B - i_C) = -u_B + u_C$$

$$\frac{d}{dt}(\psi_B - \psi_C) + 2R\,i_B = -u_B + u_C$$

Also from converse Park-Gorev transformation, we get

$$\psi_B - \psi_C = \sqrt{3}\big(-\psi_d\sin\gamma + \psi_q\cos\gamma\big)$$

Differentiating the above equation w.r.t time and using the same terminology as in [9], finally get

$$\begin{bmatrix} L_{d2} \\ -L_{q1} \\ L_{md2} \\ L_{md2} \\ -L_{mq1} \end{bmatrix}^T \cdot \frac{d}{dt}\begin{bmatrix} i_d \\ i_q \\ i_{fd} \\ i_D \\ i_Q \end{bmatrix} = \begin{bmatrix} \frac{2}{\sqrt{3}}R \\ \omega L_{d1} \\ \omega L_{q2} \\ \omega L_{md1} \\ \omega L_{md1} \\ \omega L_{mq2} \end{bmatrix}^T \begin{bmatrix} i_B \\ i_d \\ i_q \\ i_{fd} \\ i_D \\ i_Q \end{bmatrix} - u_B + u_C \tag{19}$$

u_C and u_B can be evaluated using (12) by applying boundary conditions.

The electromagnetic torque of synchronous generator is given by Eq. (20) (Ref. [9].)

$$T_{em} = -p \begin{bmatrix} 2(L_{d1} - L_{q1}) sin\, \gamma \\ -\sqrt{3} L_{md1} \\ -\sqrt{3} L_{md1} \\ -\sqrt{3} L_{mq2} \end{bmatrix}^T \begin{bmatrix} i_B \\ i_{fd} \\ i_D \\ i_Q \end{bmatrix} i_B \tag{20}$$

Three-phase short circuit Three-phase short circuit occurred when all three phases come in contact with each other. The mathematical model in case of three-phase short circuit consists of five numbers of differential equations given by (1), (2), (4), (5). and (6) modified as in [9]. The boundary conditions in this fault will be as

$$u_{\infty A} = u_{\infty B} = u_{\infty C} = 0$$

The electromagnetic torque of synchronous generator is given by Eq. (21) (Ref. [9].)

$$T_{em} = \frac{3}{2} \mathrm{p} \left(- L_{md}\, i_{md}\, i_q + L_{mq}\, i_d\, i_{mq} \right) \tag{21}$$

Phase synchronization fault In phase synchronization fault the frequency of the generator goes out of phase with frequency of the network. The mathematical model for phase synchronization fault consists of five numbers of differential equations given by (1), (2), (4), (5), and (6). The phase difference over transmission line at the time of fault is defined as δ. The torque equation for this type of fault is same as Eq. (21).

2.3 *Dynamic Modeling*

The complete dynamic model for line to ground and line to line faults are represented by set of coupled Eqs. (22), (24), (25) and for three-phase and mal-synchronization faults are represented by set of coupled Eqs. (23)–(25). The coupled turbine-generator equations are solved numerically using fourth-order Adams predictor–corrector scheme with startup by fourth-order Runge–Kutta method in MATLAB.

$$L \cdot \frac{d}{dt} \begin{bmatrix} i_{phase} \\ i_{fd} \\ i_D \\ i_Q \end{bmatrix} + X \begin{bmatrix} i_{phase} \\ i_{fd} \\ i_D \\ i_Q \end{bmatrix} = \begin{bmatrix} u_{fd} \\ 0 \\ 0 \\ 0 \end{bmatrix} \tag{22}$$

$$L \cdot \frac{d}{dt}\begin{bmatrix} i_d \\ i_q \\ i_{fd} \\ i_D \\ i_Q \end{bmatrix} + X\begin{bmatrix} i_d \\ i_q \\ i_{fd} \\ i_D \\ i_Q \end{bmatrix} = \begin{bmatrix} u_d \\ u_q \\ u_{fd} \\ 0 \\ 0 \end{bmatrix} \tag{23}$$

$$\frac{d\omega}{dt} = -\frac{P}{J}T_{em} + T_{drive} \tag{24}$$

$$\frac{d\gamma}{dt} = -\omega \tag{25}$$

2.4 Active Control of Turbo-generator Shaft

The turbo-generator shaft is modeled using finite element method. The rotor is divided into number of elements with 2 rotational dofs (single dof at each node). The Hamilton's principle is used to drive elemental stiffness and mass matrices. Three types of sections, solid circular, hollow circular, and tapered, are used to model the rotor elements. The elemental matrices are combined to form the global matrices and complete FEM equation is represented by Eq. (26).

$$[I]_g\{\ddot{\theta}\}_g + [C]_g[\dot{\theta}]_g + [K]_g\{\theta\}_g = [T_p]_g + [T_{\text{uniform}}]_g \tag{26}$$

where

$[I]$ = Global moment of inertia matrix
$[C]$ = Global damping matrix
$[K]$ = Global stiffness matrix
$[T_p]$ = Global point torque matrix
$[T_{\text{uniform}}]$ = Global distributed torque matrix.

The piezoelectric material is used as sensor as well as actuator to actively control the rotor vibration. The sensor material is placed at specific elements of rotor and feedback is generated using PD control as shown in Fig. 3. The torque generated by the actuator in closed-loop plant is given by Eq. (27). The specific meaning of various parameters for sensor and actuators are given in Table 3.

$$T_{act} = \frac{4d_{15}}{3h_e}U_A\left[\left(R + t_s + h_{pe}\right)^3 - (R + t_s)^3\right]G_{pe} \tag{27}$$

where

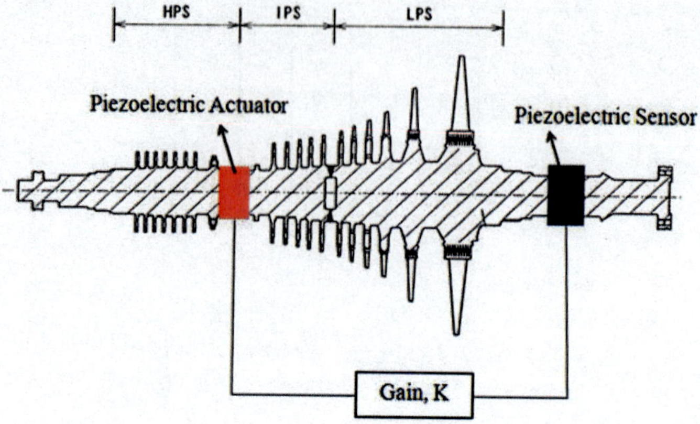

Fig. 3 Piezoelectric sensor and actuator fixed on the rotor system

$$U_A = k_p U_s + k_d \frac{dU_s}{dt} \tag{28}$$

$$U_s = \frac{G_{pe} d_{15} R}{\varepsilon_0 \varepsilon_{pe}} \left[\phi_{pe}\left(x_{s_2}, t\right) - \phi_{pe}\left(x_{s_1}, t\right) \right] \tag{29}$$

Fatigue life calculation begins with determining number of cycles (life L) of component at particular stress amplitude by using Basquin's equation given as

$$A = S_f L^B \tag{30}$$

where, S_f shear is stress level at which component is undergoing L number of cycles, A and B are constants which are determined by using two boundary conditions, at $S_f = S_e$(endurance stress), $L = 10^6$, and $S_f = 0.9S_{ut}$ (S_{ut} = ultimate tensile strength), $L = 10^3$. From A and B life of component at any stress level can be calculated by using Eq. (30). Linear damage accumulation rule or Palmgren–Miner's is given as

$$\frac{N_1}{L_1} + \frac{N_2}{L_2} + \cdots + \frac{N_n}{L_n} = 1 \tag{31}$$

At shear stress level S_{f1} component is subjected to N_1 number of cycles is known. By using Eq. (30) at stress level S_{f1} life of component is calculated as L_1. Therefore, life completed in N_1 cycles is $\frac{N_1}{L_1}$ and if there are $S_{f2}, S_{f3}, \ldots, S_{fn}$ stress levels completing $N_2, N_3, \ldots, N_n$ cycles then the total life/amount of damage is calculated by using Eq. (31).

3 Validation

The dynamic model described in Sect. 2.3 is validated by the results from [9]. The values of inductance L_e and resistance R_e are assumed to be zero for validation purpose only. Since dynamic model is same for line to ground and line to line short circuit fault (six coupled equations), the results are validated with generator torque in line to ground short circuit for the most unfavorable conditions given in [9] as shown in Fig. 4.

Similarly dynamic model is same for three-phase short circuit and phase synchronization fault. So results are validated with generator torque in three-phase short circuit for most unfavorable condition as given in [9] and shown in Fig. 5.

To validate the code in all circumstances the numerical results obtained from MATLAB are validated with peak value of generator torque for line to ground short

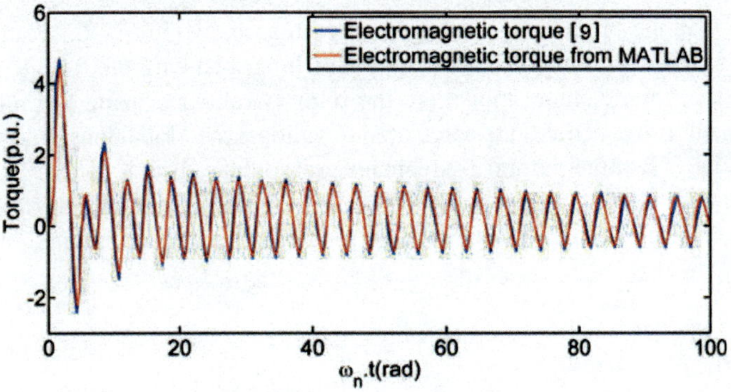

Fig. 4 Comparison of calculated electromagnetic torque during line to ground short circuit with [9] results

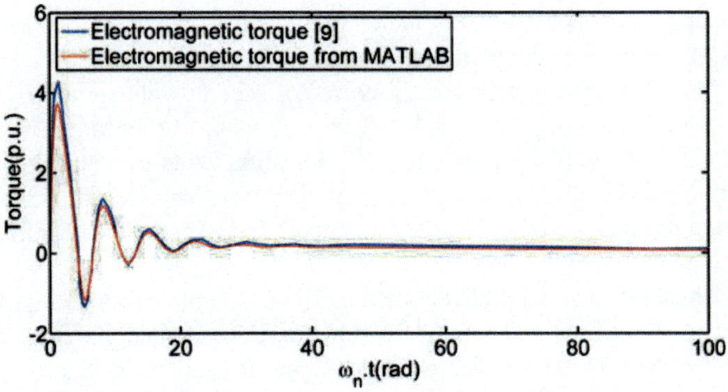

Fig. 5 Comparison of calculated electromagnetic torque during three-phase short circuit with [9] results

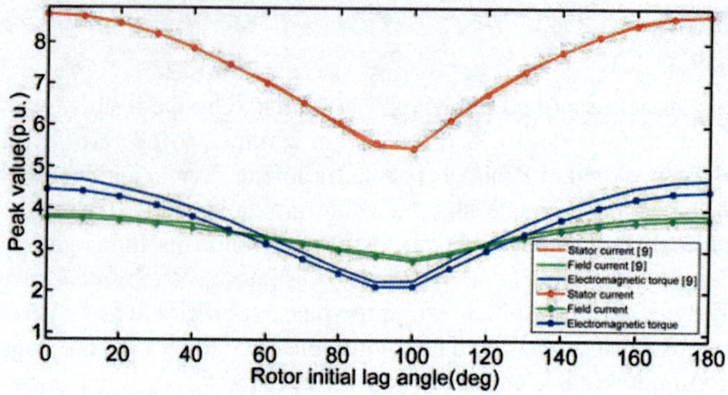

Fig. 6 Comparison of peak value stator current, field current, and electromagnetic torque during line to ground short circuit with [9] results

circuit over a range of rotor lag angle given by [9] as shown in Fig. 6. From Figs. 4, 5 and 6 it is clear that numerical results are matching well with the Ref. [9] results.

Further the finite element model for the rotor is validated using Ref. [19]. The first three natural frequencies are calculated to validate the global mass and stiffness matrices. The first three natural frequencies are compared with the Ref. [19] and convergence of the code is established in Table 1. The first three natural frequencies of the tube as given in Ref. [19] show close resemblance to the simulated results.

4 Results and Discussion

The generator is modeled using dq0 formulation. The complete system is connected with the infinite load through grid with resistance R_e and inductance, L_e. The generator torque vibrations are plotted in Fig. 7 for 1 phase to ground, 2 phase to ground, Three Phase and mal-synchronization faults. Per unit system is used to calculate the peak values of torque. The coupled turbine-generator equations are solved numerically using fourth-order Adams predictor-corrector scheme with startup by fourth-order Runge–Kutta method in MATLAB. The peak values of pu torque for line to ground, line to line, three-phase and mal-synchronization faults are 12 pu, 7.9 pu, 17.1 pu, and 8.1 pu, respectively. The actual value of electromagnetic torque is calculated by multiplying per unit value by torque base $3pU_nI_n\omega_n^{-1}$. The various parameters of generator are given in Table 2.

The rotor is modeled using finite element method. Complete rotor is divided into 187 numbers of elements and 188 numbers of nodes. Each element have two degrees of freedom, i.e., one rotational degree of freedom at each node. Since torsional vibrations are independent from translational and axial vibrations so other DOFs are neglected to keep the formulation simple. The whole rotor is modeled using three types of elements (Solid cylinder, hollow cylinder and tapered) and stiffness

Table 1 Comparison of first three natural frequencies from MATLAB and Ref. [19]

S. No	No of elements considered in MATLAB	First 3 natural freqs. from MATLAB ($x10^3$Hz)	First 3 natural freqs. of Ref. [19]. ($x10^3$Hz)	% error	Avg. % error
1	5	1.682	1.534	9.647	174.46
		7.737	4.911	57.544	
		46.247	8.315	456.187	
2	6	1.576	1.534	2.737	14.294
		5.553	4.911	13.072	
		10.566	8.315	27.071	
3	7	1.557	1.534	1.499	6.047
		5.045	4.911	2.728	
		9.472	8.315	13.91	
4	8	1.551	1.534	1.108	1.904
		4.859	4.911	−1.058	
		8.786	8.315	5.664	
5	9	1.547	1.534	0.847	−0.302
		4.774	4.911	−2.789	
		8.401	8.315	1.034	
6	10	1.547	1.534	0.847	0.0353
		4.773	4.911	−2.810	
		8.487	8.315	2.068	

is calculated accordingly. The generator torque is applied on nodes associated with the generator position on rotor length. The complete dynamic system is solved using ODE15s in MATLAB. The diameter of various sections of coupled rotor along the length is shown in Fig. 8.

Further the piezoelectric material is used as sensor and actuator to control the rotor vibrations. PVDF polymer layer is used due to its low brittleness and high shear constant. The sensor layer is glued on the element just near the generator and actuator is glued near high-pressure turbine. The PD feedback control law is used to give appropriate torque to actuator layer. The various parameters for controlled system are given in Table 3.

The element 39 is selected for the analysis of uncontrolled and controlled torque comparisons. The comparative results for uncontrolled and actively controlled system are plotted in Fig. 9a–c for line to ground, three-phase and mal-synchronization fault. The peak torque in uncontrolled system are 0.34 MNm, 0.48 MNm, and 0.23 MNm and in actively controlled system peak torque is 0.23 MNm, 0.33 MNm, and 0.16 MNm for the mentioned faults, respectively. From Fig. 9 it is clear that there is significant reduction in the peak torque. Further to calculate the fatigue life at element 39, the shear stress is calculated using torsion equation. The numbers of

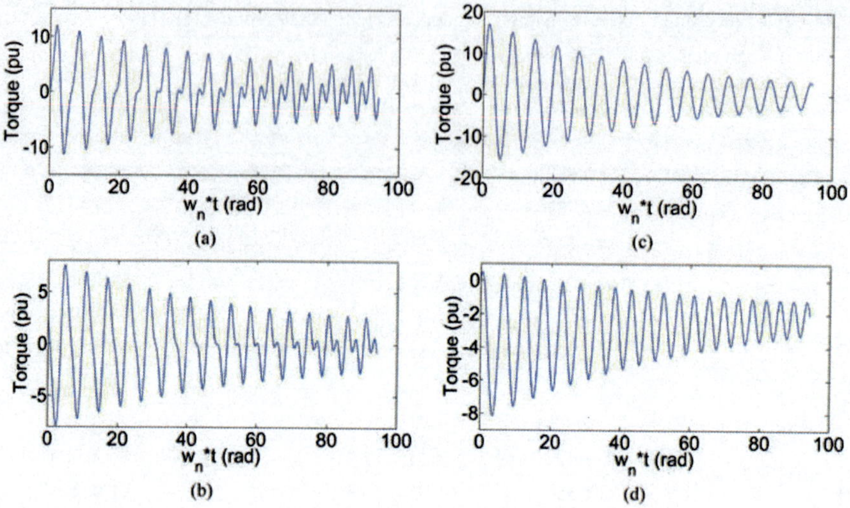

Fig. 7 **a** Per unit generator torque during line to ground fault. **b** Per unit generator torque during line to line fault. **c** Per unit generator torque during three phase fault. **d** Per unit generator torque during mal-synchronization fault

Table 2 Various parameters used in generator modeling

Parameter	Symbol[a]	Value
Parameters related to generator	S_n	588 MVA
	$U_{line,n}$	11,000 V
	$Cos\varphi_n$	0.85
	n_n	3000 rpm
	f	50 Hz
Stator winding parameters	R	0.00193 p.u
	L_σ	0.1256 p.u
Rotor windings parameters	R_{fd}	0.0009 p.u
	R_D	0.0093 p.u
	R_Q	0.008 p.u
	$L_{fd\sigma}$	0.02721p.u
	$L_{D\sigma}$	0.33 p.u
	$L_{Q\sigma}$	0.08 p.u
d-q axis magnetizing inductances	L_{md}	1.9539 p.u
	L_{mq}	1.8504 p.u
Line resistance	R_e	0.1235 p.u
Line inductance	L_e	6.773 p.u
Infinite Voltage	U_∞	8.7639 p.u

[a]Various symbols and related subscripts are given in [9]

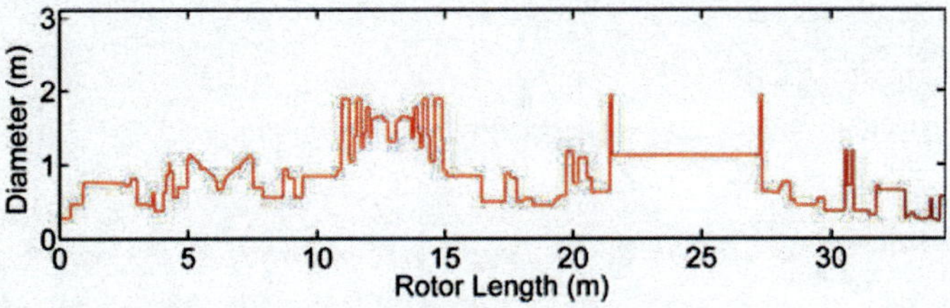

Fig. 8 Diameter of various sections of rotor along the length

Table 3 Various parameters for active vibration control of rotor

Parameter	Symbol	Value
Piezoelectric coefficient	d_{15}	-27 pCN^{-1}
Shear modulus of PVDF	G_{pe}	3.5 Gpa
Relative dielectric permittivity of PVDF	ε	12
Glue layer thickness	t_s	1×10^{-6} m
Piezoelectric layer thickness	h_{pe}	1×10^{-3} m
Amplification coefficients	k_p k_d	10^{-9} 10^{-8}

different amplitude stress cycles are calculated using rain-flow counting algorithm. Relative damage index (D) in uncontrolled system is found to be 0.005011, 0.002209, 0.3991 and for controlled system is 0.004124, 0.002106, 0.005868 for line to ground, three-phase and mal-synchronization faults, respectively. From above results, it is clear that there is increase in fatigue life of a particular section in actively controlled system as compared to the uncontrolled system.

5 Conclusions

The dynamic models of loaded synchronous generator under four types of electric are analyzed in MATLAB. The evolution of electromagnetic torque with time is plotted and compared under the most unfavorable conditions. The rotor is modeled using FEM and complete dynamic system is numerically simulated using ode15s in MATLAB. Further the coupled turbo-generator shaft is controlled using piezoelectric PVDF sensor and actuator and PD feedback control law is used for closed-loop system. The controlled system shows significant decrease in the peak torque which further enhance the fatigue life of rotor under vibrations during electrical faults. Further study can be performed to control the turbo-generator shaft vibrations using intelligent control like neural network.

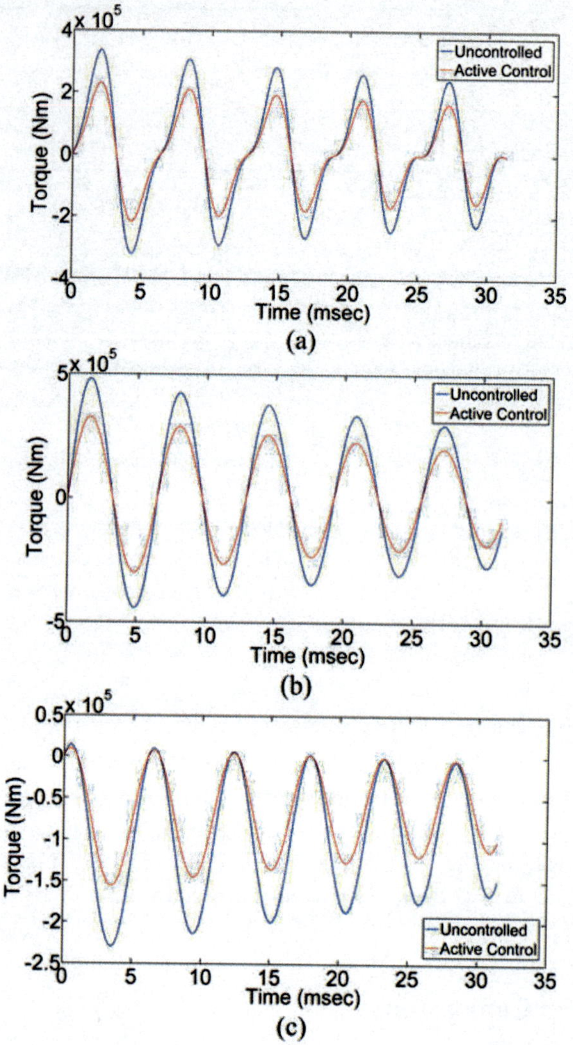

Fig. 9 **a** Torque on element 39 for uncontrolled and actively controlled system in line to ground fault. **b** Torque on element 39 for uncontrolled and actively controlled system in three-phase fault. **c** Torque on element 39 for uncontrolled and actively controlled system in mal-synchronization fault

References

1. Nikander, A., Janne, A.: Connecting induction generators in parallel with main synchronous generators of a hydroelectric power plant in electrical network—simulation study. In: 18th International Conference on Electricity Distribution (CIRED) 2005 Jun 6, pp 1–4
2. Wachel, J.C., Szenasi, F.R.: Analysis of torsional vibrations in rotating machinery. In: Proceedings of the 22nd Turbomachinery Symposium 1993. Texas A&M University, Turbomachinery Laboratories
3. Bordeasu, I., Popoviciu, M.O., Marsavina, L., Voda, M., Negru, R., Pirvulescu, L.D.: Numerical simulation of fatigue cracks initiation and propagation for horizontal axial turbines shafts. Ann. DAAAM Proc. 407–409

4. Loewenthal, S.H.: Factors that affect the fatigue strength of power transmission shafting and their impact on design. J. Mech. Transmiss. Automat. Design **108**(1), 106–114 (1986)
5. Khangar, V.S., Jaju, S.B.: A review of various methodologies used for shaft failure analysis. Int. J. Emerg. Technol. Adv. Eng. ISSN **2**(6), 2250–2459 (2012)
6. Asi, O.: Fatigue failure of a rear axle shaft of an automobile. Eng. Fail. Anal. **13**(8), 1293–1302 (2006)
7. Marín, J.J.C.: Fracture mechanics approach of repaired top roll shafts in sugar cane mills. J. Mech. Behavior Mater. **16**(6), 419–429 (2005)
8. Jackson, M.C., Umans, S.D., Dunlop, R.D., Horowitz, S.H., Parikh, A.C.: Turbine-generator shaft torques and fatigue: part I-simulation methods and fatigue analysis. IEEE Trans. Power Appar. Syst. **6**, 2299–2307 (1979)
9. Lupsa-Tataru, L.: Comparative simulation study on synchronous generators sudden short circuits. Model. Simul. Eng. 8 (2009)
10. Jackson, M.C., Umans, S.D.: Turbine-generator shaft torques and fatigue: part III-refinements to fatigue model and test results. IEEE Trans. Power Appar. Syst. **3**, 1259–1268 (1980)
11. Chan, K.S., Enright, M.P., Golden, P.J., Naboulsi, S., Chandra, R., Pentz, A.C.: Probabilistic high-cycle fretting fatigue assessment of gas turbine engine components. J. Eng. Gas Turbines Power **134**(6), 062502 (2012)
12. Stojanovic, D., Petrovic, D., Mitrovic, N.: Torsional torques of big turbine-generator shafts due to malsynchronization. In: 2000 10th Mediterranean Electrotechnical Conference. Information Technology and Electrotechnology for the Mediterranean Countries. Proceedings. MeleCon 2000 (Cat. No. 00CH37099), vol. 3, pp. 1051–1054. IEEE (2000)
13. Cudworth, C.J., Smith, J.R.: Steam turbine generator shaft torque transients: a comparison of simulated and test results. IEE Proc. C (Generat. Transmiss. Distribut.) **137**(5), 327–334. IET Digital Library (1990)
14. Chyn, C., Wu, R.-C., Tsao, T.-P.: Torsional fatigue of turbine-generator shafts owing to network faults. IEE Proc.-Generat. Transmiss. Distribut. **143** (5), 479–486 (1996)
15. Tsai, C.Y., Huang, S.C.: Transfer matrix method to vibration analysis of rotors with coupler offsets. Shock Vibrat. **20**(1), 97–108 (2013)
16. Chan, D.S.: The transient torsional vibration behaviour of a turbine-generator system under short circuit excitation. In: ASME 1996 International Gas Turbine and Aeroengine Congress and Exhibition, 1996, pp. V003T07A001–V003T07A001 (1996)
17. Tsai, J.-I., Lin, C.-H., Tsao, T.-P.: Assessment of long-term life expenditure for steam turbine shafts due to noncharacteristic subharmonic currents in asynchronous links. IEEE Trans. Power Syst. **19**(1), 507–516 (2004)
18. Oliquino, R., Islam, S., Eren, H.: Effects of types of faults on generator vibration signatures. In: Australasian Universities Power Engineering Conference, 2003, pp. 1–6
19. Sung, C.C., Varadan, V.V., Bao, X.Q., Varadan, V.K.: Active torsional vibration control experiments using shear-type piezoceramic sensors and actuators. J. Intell. Mater. Syst. Struct. **5**(3), 436–442 (1994)

Complete Creep Life Prediction Using Continuum Damage Mechanics and XFEM

V. B. Pandey, I. V. Singh, and B. K. Mishra

1 Introduction

Several components of nuclear power plants are usually subjected to high temperature applied at several hours leading to creep phenomenon. It is noticed that creep plays an important role in the failure of these components. Therefore, the life estimation of such parts under creep loading is quite essential. The creep life of a component can be divided into two parts: the life under crack nucleation (i.e., microvoids and microcrack growth period) and the life during crack growth. In literature, several approaches have been developed to analyze the creep failure. In these approaches, continuum damage mechanics (CDM) is considered most suitable for the estimation of the life during crack nucleation. In fact, CDM was developed for creep analysis. Generally, finite element method (FEM) is used along with CDM to estimate the life. Although, CDM well predicts the crack initiation life. However, the pathological mesh dependency limits the use of CDM for life prediction during crack propagation. Although, several attempts have been suggested in literature to resolve this issue [1]. However, the use of CDM was limited to crack initiation only. Recently, CDM and extended finite element method (XFEM) are combined to compute the life during crack propagation. The application of this combination has been used to replicate the crack growth in different types of problems such as elastic crack growth [2], plastic crack growth [3], fatigue crack growth [4], and creep crack growth [5]. All the aforementioned studies suggest that the crack propagation can be simulated successfully through this combination. Sebra et al. [3] proposed a framework in which crack nucleation and propagation can be done through CDM-XFEM combination. They developed this framework to compute the crack growth for plasticity problems.

V. B. Pandey (✉) · I. V. Singh · B. K. Mishra
Indian Institute of Technology Roorkee, Roorkee 247667, India
e-mail: vibhuti.pandey27@gmail.com

S. K. Saha and M. Mukherjee (eds.), *Recent Advances in Computational Mechanics and Simulations*, Lecture Notes in Mechanical Engineering,
https://doi.org/10.1007/978-981-15-8315-5_15

Similar to [3], we are proposing a framework to determine the complete creep life through CDM and XFEM. In this approach, a non-discontinuous domain is considered in the beginning. In the absence of discontinuity, the XFEM is replaced by the FEM. The life of this non-discontinuous domain is evaluated through CDM. Once the damage reaches its critical value, a developed macrocrack is modeled through the XFEM. In short, CDM estimates the life and XFEM demonstrates crack as well as its growth.

2 Theoretical Formulation

The analytical formulation and essential mathematical equations are detailed in this section for performing creep simulations.

2.1 Elastic–Plastic Creep Analysis

Similar to [5], elasto-plastic creep analysis is performed in this work. Elasto-plastic analysis involves time-independent plasticity whereas creep analysis is time-dependent phenomenon, therefore, both can be solved separately to simplify the life estimation. Hence, the creep simulations would be accomplished after the completion of elasto-plastic analysis. Moreover, the damage will be computed during creep only. Based on these assumptions, the required mathematical equations are presented in the following section.

In the framework of small strain, the total strain $\boldsymbol{\varepsilon}^{\text{total}}$ can be written as elastic $\boldsymbol{\varepsilon}^{\text{e}}$, plastic $\boldsymbol{\varepsilon}^{\text{p}}$ and creep strain $\boldsymbol{\varepsilon}^{\text{c}}$,

$$\boldsymbol{\varepsilon}^{\text{total}} = \boldsymbol{\varepsilon}^{\text{e}} + \boldsymbol{\varepsilon}^{\text{p}} + \boldsymbol{\varepsilon}^{\text{c}} \tag{1}$$

To perform the elasto-plastic analysis, isotropic hardening with J_2 plasticity is considered. For small deformation, the total incremental elasto-plastic strain is

$$\text{d}\boldsymbol{\varepsilon} = \text{d}\boldsymbol{\varepsilon}^{\mathbf{e}} + \text{d}\boldsymbol{\varepsilon}^{\mathbf{p}} \tag{2}$$

The elastic strain increment is calculated through the Hooke's law,

$$\text{d}\boldsymbol{\varepsilon}^{\mathbf{e}} = \mathbf{C}\ \text{d}\boldsymbol{\sigma} \tag{3}$$

where $\mathbf{C}$ is the compliance constitutive matrix.

To compute the plastic strain increment, the associated flow rule of plasticity is

$$\text{d}\varepsilon_{ij}^{\mathbf{p}} = \Delta\lambda \frac{\partial F}{\partial \sigma_{ij}} \tag{4}$$

where $\Delta\lambda$ is a scalar multiplier and F is von-Mises yield function.

2.2 Creep Strain Relations

Liu–Murakami established a creep constitutive model to calculate the creep strain rate in an isotropic homogeneous material. The multiaxial form of Liu–Murakami creep constitutive model is [6]

$$\dot{\varepsilon}_{ij}^{c} = \frac{3}{2} A \sigma_{eq}^{n} \frac{\mathbf{S}_{ij}}{\sigma_{eq}} \exp\left(\frac{2(n+1)}{\pi \sqrt{1+3/n}} \left(\frac{\sigma_1}{\sigma_{eq}} \right)^2 \omega^{3/2} \right) \tag{5}$$

where A and n are material constants. $\dot{\varepsilon}_{ij}^{c}$, $\mathbf{S}_{ij}$, σ_1, σ_{eq} is creep strain rate, deviatoric stress, maximum principle stress and von-Mises stress, respectively. A damage variable (ω) is also incorporated in the above equation to demonstrate the deterioration of material. To evaluate the value of ω, a damage evaluation equation is given as

$$\dot{\omega} = D \frac{\left(1 - e^{-q}\right)}{q} \sigma_r^p e^{q\omega} \tag{6}$$

where D, q, and p are material constants. The constants are acquired from the uniaxial creep experiments. σ_r is the rupture stress which is given as

$$\sigma_r = \alpha \sigma_1 + (1 - \alpha) \sigma_{eq} \tag{7}$$

where α is the multi-axiality parameter which determines the effective stress condition in the component.

2.3 Extended Finite Element Method

Extended finite element method (XFEM) is used to model the crack growth without disturbing the original mesh. In XFEM, the crack is represented through the enrichment functions. The enrichment functions are special functions which introduce the local behavior of the discontinuity in the standard finite element (FE) approximation. In the presence of the enrichment functions, the displacement approximation becomes

$$\begin{aligned} \mathbf{u}(\mathbf{x}) = & \sum_{i=1}^{ne} N_i(\mathbf{x}) \mathbf{u}_i + \sum_{j=1}^{np} N_j(\mathbf{x}) \left[H(\mathbf{x}) - H(\mathbf{x}_j) \right] \mathbf{a}_j \\ & + \sum_{k=1}^{nq} N_k(\mathbf{x}) \left\{ \sum_{\alpha=1}^{4} \left[\gamma_\alpha(\mathbf{x}) - \gamma_\alpha(\mathbf{x}_k) \right] \mathbf{b}_k^\alpha \right\} \end{aligned} \tag{8}$$

where ne, $\mathbf{u}_i$, and $N(\mathbf{x})$ are the number of nodes, nodal displacement vector, and shape function of the standard FE mesh, respectively. np and $\mathbf{a}_j$ are the nodes and nodal displacement vector, respectively, corresponding to split elements. $H(\mathbf{x})$ is the Heaviside jump function. nq and $\mathbf{b}_k$ are the nodes and nodal displacement vector, respectively, corresponding to tip elements. The branch function is designated through $\gamma_\alpha(\mathbf{x})$ in the above equation.

The stress–strain values near the crack tip region are limited due to plasticity. Therefore, HRR (Hutchinson, Rice, and Rosengren) singularity is employed as branch functions [5] to capture this phenomenon.

$$\gamma_\alpha(\mathbf{x}) = r^{1/1+m}\left[\cos\frac{\theta}{2},\ \sin\frac{\theta}{2},\ \cos\frac{\theta}{2}\ \sin\theta,\ \sin\frac{\theta}{2}\ \sin\theta\right] \tag{9}$$

where (r, θ) are the cylindrical coordinates at crack tip and m is the Ramberg–Osgood hardening exponent which is evaluated from tensile data.

3 Crack Initiation and Growth Approach

CDM represents the nucleation of microcracks and voids through a damage variable in the continuum domain. As the damage variable reaches its critical value, it is assumed that the damaged portion is not capable enough to resist the external loading. Therefore, this damage region can completely describe the failure of material through the nucleation and propagation of a macrocrack.

To model the nucleation and propagation of crack, the crack characteristics such as initiation point, direction, and length of the crack must be determined. The crack initiation point lies in an element in which damage reaches initially to its critical value. In this work, a notched single-edge notched tensile (SENT) specimen is selected to study the complete creep life of a component. The notched specimen is subjected to mode I loading. Since, the notch works as a stress riser in this problem. Hence, the maximum value of equivalent stress will be at the tip of the notch. The evolution of damage depends directly on the equivalent stress (as presented in Eq. 6). Hence, the damage will reach its critical value in an element which is at the tip of the notch. Therefore, the tip of the notch is considered as crack initiation point in this problem.

After the identification of crack initiation point, the initial crack length and its direction needs to be determined. To find the crack direction, a semi-circle is drawn taken initiation point as the center. On this semi-circle, a set of equidistance points are created. Then, stress triaxiality is calculated on these points. The point with the maximum value of triaxiality determines the direction. The study is conducted on 316 stainless steel. In metals, the size of the representative volume element (RVE) is given as 0.1 mm [7]. Therefore, initial crack length is considered as 0.1 mm in this work.

The growth of crack is the simpler case of crack initiation. The previous tip of the crack becomes the starting point of the new crack segment. The direction of crack and incremental crack length are two unknowns. The direction of new crack segment is determined similarly through a semi-circle whose center is at the crack tip. Similar to crack initiation study, a set of equidistance points are created on this semi-circle and stress triaxiality is calculated on these points. The point with the maximum triaxiality, determines the crack direction. Similar to [5], the crack growth increment is considered as $\Delta a = 0.2$ mm.

There is no crack at the beginning. Therefore, in the absence of any discontinuity, the XFEM works as standard FEM. Once the crack is initiated in the domain, the tip enrichment is required to capture the crack features and hence, XFEM is used to represent the crack and its propagation.

4 Numerical Results and Discussion

In this section, the complete creep life estimation is illustrated by solving a numerical example. Two studies are performed in this section. First, a crack growth simulation is validated on a pre-cracked single edge notched tensile (SENT) specimen. Further, a notched SENT specimen without pre-crack is simulated to predict the complete life. The results of both studies are also compared.

Kim et al. [8] experimentally studied the creep crack growth on a pre-cracked SENT specimen. The specimen is made of 316 stainless steel and subjected to 35 kN load at 550 °C. The schematic figure with dimensions of pre-cracked SENT is shown in Fig. 1a. The tensile and creep properties of 316 SS at 550 °C is provided in Table 1. The thickness of specimen is taken as 10 mm.

The local approach of fracture is used in this work. In the local approach of fracture, mesh works as a numerical parameter. Therefore, Q4 element mesh with 1.6 mm × 1.54 mm is considered. The value of critical damage parameter is considered as 0.44 for this material. The simulation is performed assuming plane strain condition. The simulated creep crack growth life is compared with the experiment. The crack length versus time graph is reported in Fig. 2, which shows that the crack growth life can be estimated efficiently through the CDM-XFEM coupling.

After the validation, a notched SENT specimen is considered for crack initiation and propagation. To compare the results with pre-cracked SENT specimen, the radius of notched SENT specimen and applied load are taken as 7.75 mm and 35 kN, respectively. The other dimensions of notched specimen are given in Fig. 1b itself. All the other required parameters including mesh size are considered same as the validation study. The simulated creep life is demonstrated in the crack length versus time curve in Fig. 2. As expected, the notched specimen takes more time to start the crack growth as compared to pre-cracked specimen. Moreover, this numerical example proves that the complete creep life can be predicted easily through the proposed methodology.

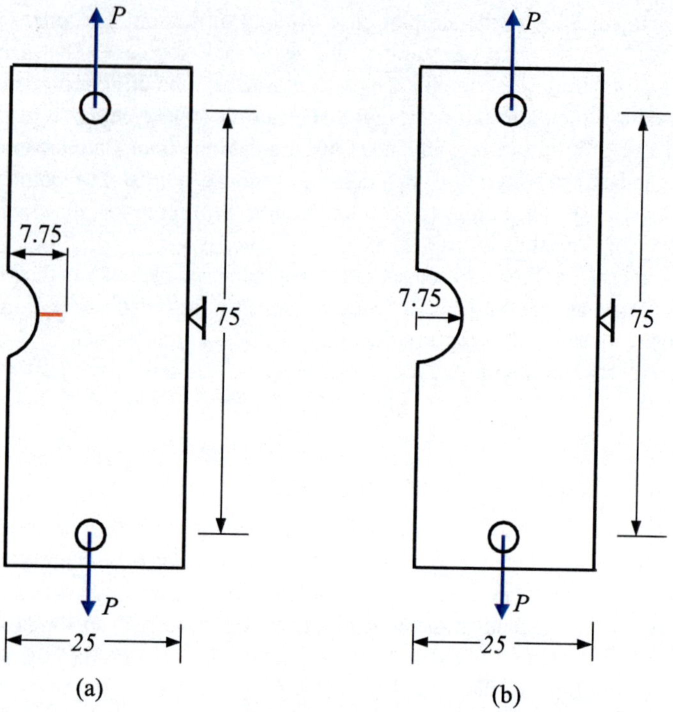

Fig. 1 Schematic figures of **a** pre-cracked and **b** notched SENT specimens

Table 1 Tensile properties and creep constants for 316H steel at 550 °C [5, 9]

E (GPa)	ν	σ_y (MPa)	σ_u (MPa)	m	D	p	q	A	n
140	0.3	170	588	2.3	1.28×10^{-30}	10.594	9.2	2.41×10^{-33}	11.131

5 Conclusion

A methodology is proposed in this paper to simulate the crack initiation and propagation under creep conditions. The complete life is estimated through the coupling of CDM and XFEM. In this approach, CDM determines the creep and damage evolution in the component whereas XFEM models the crack and its propagation. The damage variable and creep strains are estimated through the Liu–Murakami creep damage model. The developed methodology is applied to study the notched SENT specimen subjected to 550 °C. This article concludes that the proposed methodology is capable enough to determine the complete life of the specimen under creep conditions.

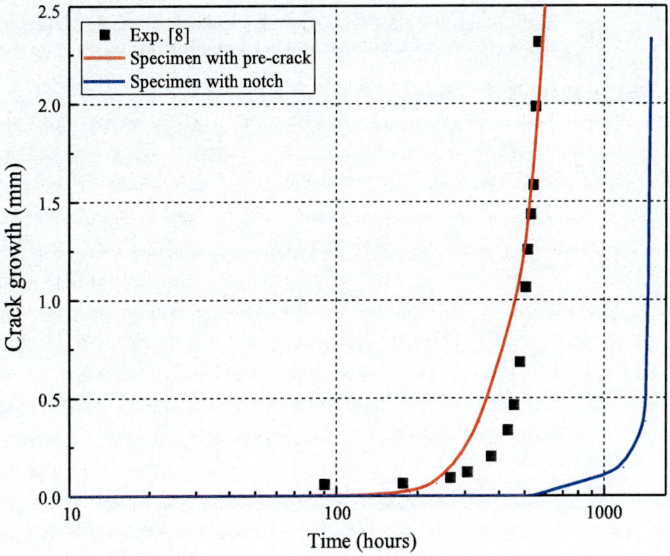

Fig. 2 A comparison of experimental and predicted creep life of pre-cracked and notched SENT specimen

Acknowledgements This work is financially supported by Aeronautics Research and Development Board (AR&DB), DRDO Bhawan, New Delhi (Grant no: AR&DB/01/1051888/M/I).

References

1. Murakami, S., Liu, Y.: Mesh-dependence in local approach to creep fracture. Int. J. Damage Mech. **4**(3), 230–250 (1995)
2. Roth, S.N., Léger, P., Soulaïmani, A.: A combined XFEM–damage mechanics approach for concrete crack propagation. Comput. Methods Appl. Mech. Eng. **283**, 923–955 (2015)
3. Seabra, M.R., Šuštarič, P., de Sa, J.M.C., Rodič, T.: Damage driven crack initiation and propagation in ductile metals using XFEM. Comput. Mech. **52**(1), 161–179 (2013)
4. Pandey, V.B., Singh, I.V., Mishra, B.K., Ahmad, S., Rao, A.V., Kumar, V.: A new framework based on continuum damage mechanics and XFEM for high cycle fatigue crack growth simulations. Eng. Fract. Mech. **206**, 172–200 (2019a)
5. Pandey, V.B., Singh, I.V., Mishra, B.K., Ahmad, S., Rao, A.V., Kumar, V.: Creep crack simulations using continuum damage mechanics and extended finite element method. Int. J. Damage Mech. **28**(1), 3–34 (2019b)
6. Hyde, C.J., Hyde, T.H., Sun, W., Becker, A.A.: Damage mechanics based predictions of creep crack growth in 316 stainless steel. Eng. Fract. Mech. **77**(12), 2385–2402 (2010)
7. Lemaitre, J., Desmorat, R.: Engineering Damage Mechanics: Ductile, Creep, Fatigue and Brittle Failures. Springer Science & Business Media, New York, USA (2005)

8. Kim, N.H., Oh, C.S., Kim, Y.J., Davies, C.M., Nikbin, K., Dean, D.W.: Creep failure simulations of 316H at 550 °C: part II—effects of specimen geometry and loading mode. Eng. Fract. Mech. **105**, 169–181 (2013)
9. Oh, C.S., Kim, N.H., Kim, Y.J., Davies, C., Nikbin, K., Dean, D.: Creep failure simulations of 316H at 550 °C: part I—a method and validation. Eng. Fract. Mech. **78**(17), 2966–2977 (2011)

Numerical Study of Coupled Elasto-Plastic Hydrogen Diffusion at Crack Tip Using XFEM

Anjali Jha, I. V. Singh, B. K. Mishra, Ritu Singh, and R. N. Singh

1 Introduction

The mechanical properties and fracture behavior of material are highly influenced by the hydrogen. Hydrogen diffusion in material can cause reduction in tensile strength, failure time, and fracture toughness of the material. The simultaneous process of hydrogen diffusion with hydride precipitation and mechanical deformation is termed as hydrogen embrittlement. Hydrogen embrittlement can reduce the ductility of the material and change the fracture mode from ductile to brittle. The two main mechanisms widely used for the hydrogen embrittlement are hydrogen-induced cohesion and hydrogen-enhanced local plasticity. The later one is used by Sofronis and Mcmeeking [1] where the flow stress of the material is modeled as a function of total hydrogen concentration. Kim et al. [2] used the hydrogen-enhanced plasticity mechanism to analyze the hydrogen transport phenomenon coupled with elasto-plasticity.

The hydrogen present in the environment diffuses inside the material and then follows by precipitation of hydride when the concentration of hydrogen exceeds the terminal solid solubility limit. These hydrides usually precipitate at the high tensile stress regions such as discontinuities and cracks. Hydrides present at the crack tip will grow up to a critical size followed by hydride fracture and then the process repeats. A numerical model is developed which takes into account the processes of

A. Jha (✉) · I. V. Singh · B. K. Mishra
Indian Institute of Technology Roorkee, Roorkee 247667, India
e-mail: ajha3@me.iitr.ac.in

R. Singh
Atomic Energy Regulatory Board, Mumbai 400085, India

R. N. Singh
Bhabha Atomic Research Centre, Mumbai 400085, India

S. K. Saha and M. Mukherjee (eds.), *Recent Advances in Computational Mechanics and Simulations*, Lecture Notes in Mechanical Engineering,
https://doi.org/10.1007/978-981-15-8315-5_16

hydrogen embrittlement to analyze the process of hydrogen diffusion coupled with the material deformation ahead of the crack tip using XFEM [3, 4].

Many researchers have used various methods such as finite element method and cohesive zone modeling to model the hydride-induced cracking. Very few studies are present in which extended finite element is used for this purpose. Shanati et al. [5] used finite element method (FEM) to model the stress-assisted diffusion process and used a staggered approach to solve the coupled problem. Barrera et al. [6] also used FEM to couple the hydrogen diffusion and mechanical behavior of material using coupled temperature displacement process. Kotake et al. [7] studied the fatigue life of a component for transient hydrogen diffusion near the crack tip by coupling the diffusion and plasticity under cyclic loading. The effect of hydrogen diffusion on the yield stress of the material is also considered. Varias and Feng [8] studied the hydride-induced cracking in metals using the finite element method. Martínez-Pañeda et al. [9] used the phase field method to model the hydrogen-assisted cracking in a cracked square plate and notched cylindrical bars.

In the present work, the effects of stress-assisted diffusion on the hydride precipitation and stress triaxiality are studied in the hydrogen environment. The effective plastic strain in the material depends on the stress triaxiality which directly affects the plastic deformation of the material. The stress triaxiality (λ) is defined as the ratio of the hydrostatic stress and the effective von Mises stress. The higher value of stress triaxiality represents the ductile fracture process [10]. The effect of stress triaxiality on edge crack specimen in the presence or absence of hydrogen ahead of the crack tip should be considered. The variation of stress triaxiality with the change in distance from the crack tip is studied.

2 Governing Equations

2.1 Hydrogen Diffusion

The equation of hydrogen diffusion gives the balance of hydrogen in the material at a certain point under steady-state crack growth conditions. It is assumed that no hydrogen is produced within the material. The equation is given as

$$\frac{\partial C^{HT}}{\partial x_1} = \frac{\partial Q_k^H}{\partial x_k} \tag{1}$$

where C^{HT} is the total concentration of hydrogen and Q_k^H is the hydrogen flux. Some of the hydrogen present gets diffused and precipitates in the form of hydrides while the remaining free hydrogen is present in the solid solution. Therefore, the total hydrogen concentration can be written as

$$C^{HT} = fC^{H,hr} + (1 - f)C^H \tag{2}$$

where $C^{H,hr}$ is the concentration of hydrogen in the hydride, C^H is the concentration of hydrogen in the solid solution, and f is the volume fraction of hydrides. The following relation is used to evaluate the hydrogen flux which gives the diffusion of hydrogen in hydrides,

$$Q_k^H = -(1-f)\frac{D^H C^H}{RT}\frac{\partial u^H}{\partial x_k} \tag{3}$$

where D^H is the diffusivity coefficient and u^H is the chemical potential of hydrogen. The chemical potential of hydrogen depends on the hydrostatic stress in the material,

$$u^H = -\frac{\sigma_{kk}}{3}V^H \tag{4}$$

where V^H is the molar volume of hydrogen and σ_{kk} represents the hydrostatic stress. The evaluated hydrostatic stress is used to solve the diffusion Eq. (1) which requires the gradient of chemical potential of hydrogen and hence the gradient of hydrostatic stress is evaluated as

$$\sigma_{kk,j} = \frac{\partial \sigma_{kk}}{\partial a_p} * J_{ij}^{-1} \tag{5}$$

where J is the jacobian, which maps the natural coordinates with actual coordinates and a_p are interpolation function.

2.2 *Hydride Precipitation*

The hydrides precipitate in the material when the value of hydrogen concentration in the material exceeds the value of terminal solid solubility C^{TS},

$$C^{TS} = C_e^{TS}\exp\left(\frac{\bar{w}_{\text{int}}}{mRT}\right)\exp\left(\frac{\sigma_{kk}\bar{V}^H}{3RT}\right) \tag{6}$$

where C_e^{TS} is the terminal solid solubility under no applied stress, m is the hydrogen mole fraction in hydride, and w_{int} is the interaction energy of precipitating hydride per mol.

2.3 *Material Deformation*

Stress field is required to evaluate the hydrostatic stress used in the hydrogen diffusion equation. Therefore, the extended finite element analysis is done to deal with the hydrogen diffusion crack problem. Elastoplastic analysis is performed to consider the plastic strain around the crack tip. The plastic strain obtained affects the hydrogen diffusion near the crack tip. The constitutive relations used are given as

$$d\sigma_{ij} = D^{ep}_{ijkl} d\varepsilon_{kl} - \frac{E}{3(1-2\nu)} \delta_{ij} d\varepsilon^{H}_{mm} \tag{7}$$

$$D^{ep}_{ijkl} = \left(\frac{E}{1+\nu}\right)\left[\delta_{ik}\delta_{jl} + \frac{\nu}{1-2\nu}\delta_{ij}\delta_{kl} - \frac{3}{2\bar{\sigma}^2} \times \left(\frac{1}{1+\frac{2}{3}\frac{dH}{d\bar{\varepsilon}^p}\frac{(1+\nu)}{E}}\right)\sigma'_{ij}\sigma'_{kl}\right] \tag{8}$$

where E is the elastic modulus, ν is Poisson's ratio, δ_{ij} is Kronecker delta, $\bar{\sigma}$ is effective stress, and $\bar{\varepsilon}_p$ is the effective plastic strain. Isotropic hardening of the material is assumed according to the following relations:

$$\bar{\sigma} = H(\bar{\varepsilon}^p) \tag{9}$$

$$\left(\frac{\bar{\sigma}}{\sigma_0}\right)^{1/n} - \frac{\bar{\sigma}}{\sigma_0} = \frac{3E}{2(1+\nu)}\frac{\bar{\varepsilon}^p}{\sigma_0}, \quad \bar{\sigma} \geq \sigma_0 \tag{10}$$

where n is the hardening exponent and σ_0 is the yield strength of the material. Stress trace in the solid solution of the hydride precipitation zone is determined using the following relation [8]:

$$\sigma^{hz}_{kk} = \left(\frac{3x}{\theta^{hr} V^{hr}}\right)\left[\frac{\sigma_{kk} V^H}{3} + RT \ln\left(\frac{C^{TS}_e}{C^H_b}\right)\right] \tag{11}$$

where C^H_b and σ^{hz}_{kk} are the concentration of hydrogen and the stress trace of a reference particle, respectively.

3 Methodology

Extended finite element method (XFEM) is used to perform the structural analysis of the model. The XFEM helps in eliminating the issue of re-meshing during crack growth. It is more accurate and requires less computational time as compared to FEM. It uses appropriate enrichment functions in the domain of interest using partition of unity. Heaviside and near tip enrichment functions are used for the elements containing the crack discontinuity as shown in Fig. 1. In XFEM, the enriched displacement approximation can be written as [3]

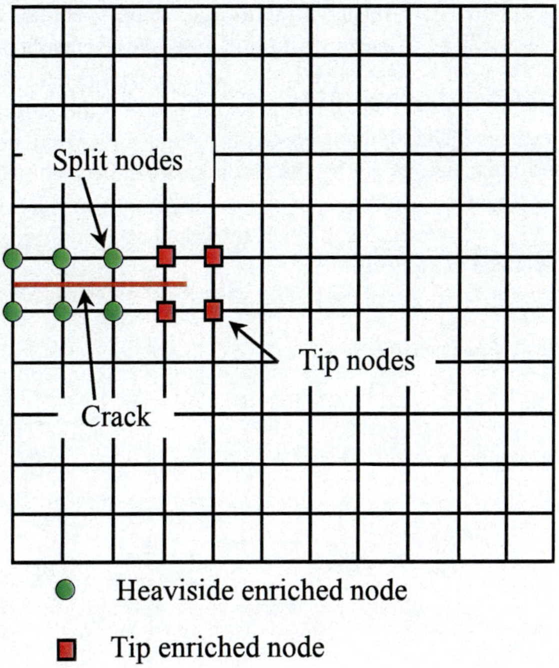

Fig. 1 Elements intersected by crack with enriched nodes

$$\mathbf{u}^h(\mathbf{x}) = \sum_{i=1}^{n} N_i(\mathbf{x})\bar{\mathbf{u}}_i + \sum_{i=1}^{n_A} N_i[\mathrm{H}(\mathbf{x}) - \mathrm{H}(\mathbf{x}_i)]\,\mathbf{a}_i + \sum_{i=1}^{n_B} N_i \sum_{\alpha=1}^{4} [F_\alpha(\mathbf{x}) - F_\alpha(\mathbf{x}_i)]\,\mathbf{b}_i^\alpha \tag{12}$$

where N_i are finite element shape functions, $H(\mathbf{x})$ is a Heaviside enrichment function, and $F_\alpha(\mathbf{x})$ is crack tip enrichment. n is the set of all nodes in the mesh; n_A is the set of nodes associated with those elements which are completely cut by the crack; n_B is the set of nodes associated with those elements which are partially cut by the crack; a_i is the nodal enriched degrees of freedom for $H(\mathbf{x})$; and $\mathbf{b}_i^\alpha$ is the degrees of freedom vector for $F_\alpha(\mathbf{x})$.

The solution strategy of the model is divided into two parts—diffusion analysis and structural analysis. After giving the initial conditions and properties of the material, structural analysis of the specimen is done to evaluate the hydrostatic stresses in the domain. These stresses are then used to evaluate the hydrogen concentration in the diffusion analysis. If the solution converges, it will be updated at the end of time step. This process continues until the end of time iteration loop. The pre-processing part consists of the evaluation of the hydride fraction, terminal solid solubility, and stress trace and stress triaxiality in the hydride precipitation zone.

An edge crack specimen subjected to tensile loading under plane strain conditions is solved by XFEM. A square-shaped specimen of dimension 2 mm × 2 mm (L = 2 mm and H = 2 mm) with an edge crack $a = 0.5$ mm is used for the simulation. The top and bottom edges of the specimen are subjected to a tensile load (σ) of 120 MPa as shown in Fig. 2. x_1 represents the distance from the crack tip in x-direction. A mesh of size (47 × 47 nodes) is taken for the simulations. Higher order nine node quadrilateral elements are used for the evaluation of the hydrostatic stress gradients. The element containing crack tip is enriched with crack tip enrichment and the elements cut by the crack are enriched with Heaviside enrichment. For the hydrogen diffusion part of the model, a uniform distribution of the hydrogen C = 6.33×10^5 mol/m^3 is given as the boundary condition. The material properties used are given in Table 1.

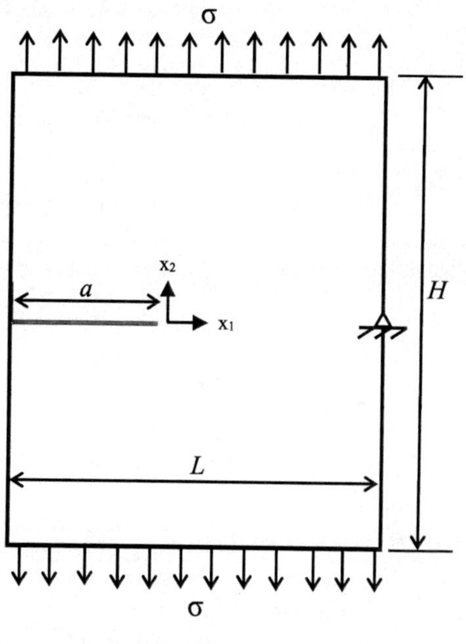

Fig. 2 Specimen of edge crack

Table 1 Value of material properties correspond to Zircaloy-2 and δ hydride [8]

Material properties	Value
Elastic modulus (E), Poisson ratio (υ)	80.4 GPa, 0.369 (573 K)
Yield strength (σ_{-0})	580 MPa (573 K)
Diffusivity coefficient (D^H)	2.17×10^{-7} exp (−35087.06/RT) m^2/s
Terminal solid solubility (C_e^{TS})	6.3741×10^5 exp (−34542.75/RT) mole/m^3
Concentration of hydrogen in hydride ($C^{H,hr}$)	1.02×10^5 mol/m^3
Molar volume of hydrogen in solid solution (V^H)	7×10^{-7} mol/m^3
Molar volume of hydrogen in hydride (V^{hr})	16.3×10^{-6} mol/m^3
Mole fraction of hydrogen in hydride (m)	1.66

4 Results and Discussion

The edge crack specimen is subjected to the tensile loading and the stress–strain behavior of the specimen material predicted is shown in Fig. 3. Stress trace of a reference particle near the crack tip is determined by putting the value of hydrogen concentration in Eq. (11). The value of σ_{kk}^{hz} tends to zero as C_b^H / C_e^{TS} approaches to 1 (hydrogen chemical equilibrium). The results are shown in Fig. 4 and agree well with results available in Varias and Feng [8]. Figure 5 shows the variation of hydride in the edge crack specimen near the crack tip. The hydrogen present in the solid solution will diffuse in the material and hydride precipitates when the hydrostatic stress near the crack tip increases. The volume fraction of hydrogen in hydride is given by hydride fraction (f) whose value varies from 0 to 1. If all the hydrogen is present only in solid solution, then its value is 0 or vice versa. The peak value of hydride fraction is achieved at a small distance from the crack tip where the hydride precipitates and its magnitude increase with the increase in load. Stress triaxiality in the edge crack specimen is also calculated at varying distance from the crack tip (r). As shown in Fig. 6, stress triaxiality increases with increase in distance from crack tip up to a value of 0.08 mm and thereafter it starts decreasing with further increase in r. Higher value of stress triaxiality at the crack tip shows the ductile fracture behavior as suggested by Bao and Wierzbicki [11]. The decrease in stress triaxiality after a distance of 0.08 mm indicates the presence of hydride which reduces the ductility of material ahead of crack tip.

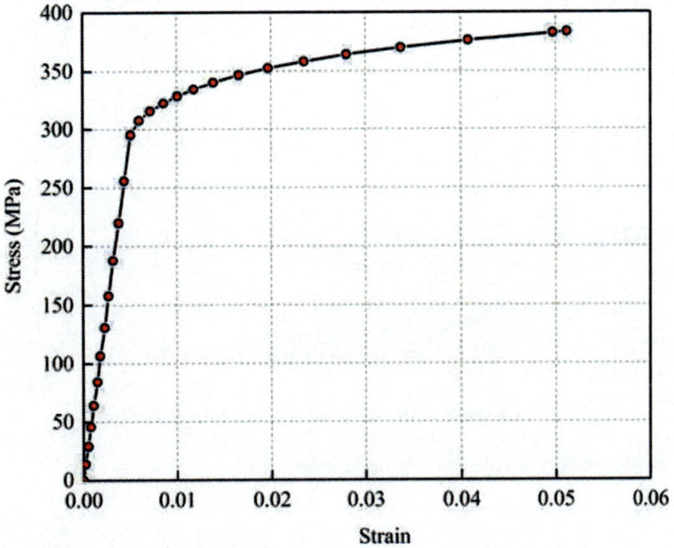

Fig. 3 Stress–strain behavior of the material for Zircaloy-2 at 573 K [8]

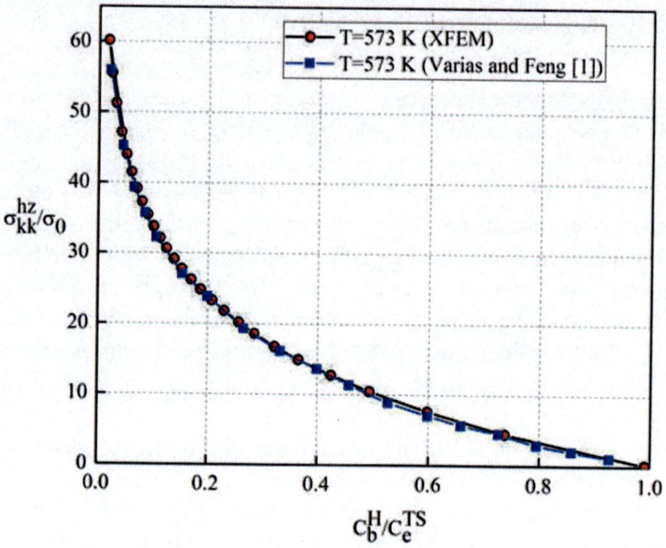

Fig. 4 A comparison of obtained stress trace in the hydride precipitation zone as a function of hydrogen concentration near the crack tip with Varias and Fang [8]

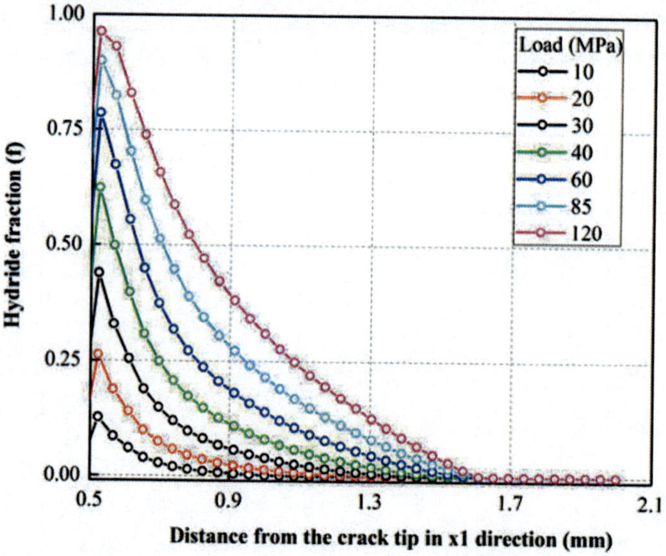

Fig. 5 Hydride fraction with the distance from the crack tip at varying loads

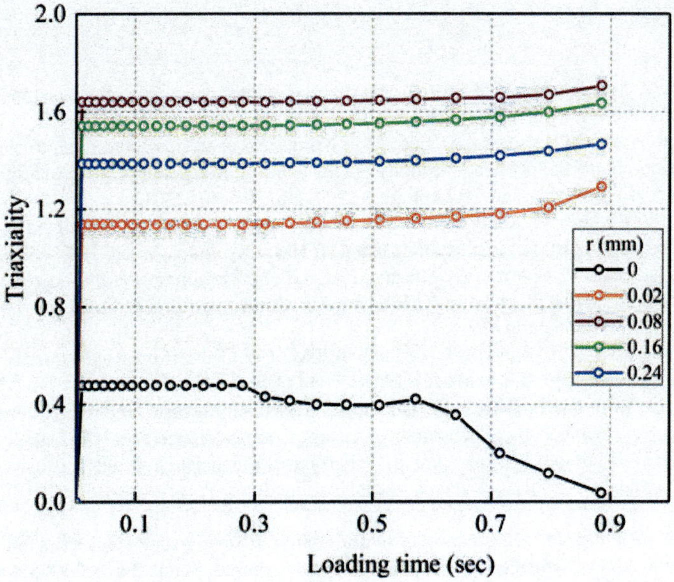

Fig. 6 Variation of stress triaxiality with loading time for different values of distance from the crack tip

5 Conclusion

The hydrogen diffusion coupled with the hydride precipitation and deformation process of the material is modeled for an edge crack specimen using XFEM. Hydrostatic stress is evaluated near the crack tip, and its effect on the hydrogen concentration and hydride precipitation near the crack tip is analyzed. From the present study, the following conclusions can be drawn:

- Hydride fraction increases with increase in load and finally approaches toward its maximum possible value 1.
- With the increase in distance from the crack tip, the hydride fraction first increases to attain a peak value and then decreases.
- The magnitude of peak value of hydride fraction increases with the increase in load.
- Stress triaxiality increases with the increase in distance from the crack tip up to a value of 0.08 and then decreases.

References

1. Sofronis, P., McMeeking, R.M.: Numerical analysis of hydrogen transport near a blunting crack tip. J. Mech. Phys. Solids **37**(3), 317–350 (1989)
2. Kim, S.K., Lee, C.S., Kim, M.H., Lee, J.M.: Numerical analysis of hydrogen transport using a hydrogen-enhanced localized plasticity mechanism. In: Proceedings of the World Academy of Science, Engineering and Technology, vol. 58, pp. 398–401 (2012)
3. Kumar, M., Singh, I.V., Mishra, B.K., Ahmad, S., Rao, A.V., Kumar, V.: Mixed mode crack growth in elasto-plastic-creeping solids using XFEM. Eng. Fract. Mech. **199**, 489–517 (2018)
4. Jha, A., Kukshal, V., Sharma, A., Sharma, K.: Numerical prediction of plastic zone length in straight edge cracked plate using XFEM. In: AIP Conference Proceedings, vol. 1975, no. 1, p. 030003. AIP Publishing (2018)
5. Shanati, S., Ellis, N.S., Randall, T.J., Marshall, J.M.: Coupled diffusion and stress by the finite element method. Appl. Math. Model. **19**(2):87–94 (1995)
6. Barrera, O., Tarleton, E., Tang, H.W., Cocks, A.C.F.: Modelling the coupling between hydrogen diffusion and the mechanical behaviour of metals. Comput. Mater. Sci. **122**, 219–228 (2016)
7. Kotake, H., Matsumoto, R., Taketomi, S., Miyazaki, N.: Transient hydrogen diffusion analyses coupled with crack-tip plasticity under cyclic loading. Int. J. Press. Vessels Pip. **85**(8), 540–549 (2008)
8. Varias, A.G., Feng, J.L.: Simulation of hydride-induced steady-state crack growth in metals–part I: growth near hydrogen chemical equilibrium. Comput. Mech. **34**(5), 339–356 (2004)
9. Martínez-Pañeda, E., Golahmar, A., Niordson, C.F.: A phase field formulation for hydrogen assisted cracking. Comput. Methods Appl. Mech. Eng. **342**, 742–761 (2018)
10. Sung, S.J., Pan, J., Lam, P.S., Scarth, D.A.: Ductile fracture initiation with consideration of strain concentration and stress triaxiality near crack fronts in compact tension specimens of hydrided irradiated Zr-2.5 Nb materials with split circumferential hydrides. Eng. Fract. Mech. **186**, 208–241 (2017)
11. Bao, Y., Wierzbicki, T.: On fracture locus in the equivalent strain and stress triaxiality space. Int. J. Mech. Sci. **46**(1), 81–98 (2004)

Role Played by Grain Boundaries in Plastic Deformation of Polycrystalline Metals: A Discrete Dislocation Dynamics Study

Tawqeer Nasir Tak, Aditya Prakash, Indradev Samajdar, and P. J. Guruprasad

1 Introduction

Plastic deformation in metals and their alloys is a consequence of the collective motion of line defects. These line defects, typically known as dislocations, move by gliding on close-packed planes along close-packed directions [1]. If the dislocations are free to glide on their path, a material undergoes perfect plasticity. However, if the dislocations face obstructions along their path, it may become difficult to move them under a constant applied load. This demands an increase in the amount of applied load in order to have a sustained plastic deformation and is known as strain hardening [1]. Such obstructions may happen due to a variety of reasons, a few of them being precipitates, second-phase particles, and/or other defects.

Dislocations interact with precipitates and second-phase particles and get pinned at the interface as a consequence. Dislocations also interact with other dislocations forming a variety of locks based on the type of interaction, and hence jamming each other. All such interactions, leading to strain hardening of the material, govern the material behavior under applied load. Moreover, dislocations also tend to interact with other crystal defects like point defects, such as interstitials, and two-dimensional defects, such as grain boundaries.

Grain boundaries, being two-dimensional (2D) defects mark the interface between two crystallites that differ from each other due to their orientation. These defects play a critical role in plastic deformation [2]. The interaction between dislocations and grain boundaries is a complex phenomenon and has been a topic of interest to the community for past few decades [3]. This work intends to explore the role

T. N. Tak (✉) · A. Prakash · I. Samajdar · P. J. Guruprasad
Indian Institute of Technology Bombay Powai, Mumbai 400076, India
e-mail: tawqeer.nasir@aero.iitb.ac.in

S. K. Saha and M. Mukherjee (eds.), *Recent Advances in Computational Mechanics and Simulations*, Lecture Notes in Mechanical Engineering,
https://doi.org/10.1007/978-981-15-8315-5_17

played by grain boundaries in deformation of polycrystalline metals using a Discrete Dislocation Dynamics (DDD) framework.

2 Model Formulation

The formulation is based on the original framework laid down by Amodeo and Ghoniem [4]. After further development by Kubin et al. [5] and Lubarda [6], in the next few years, Van der Giessen and Needleman [7] finally adapted the model to solve a boundary value problem by coupling the framework with Finite Element Method (FEM). The procedure was developed by the authors for solving a two-dimensional (2D) plane strain problem, which did not take into account dislocation–dislocation interactions in three dimensions that are primarily responsible for various hardening mechanisms. The key three-dimensional (3D) mechanisms were added to the framework by Benzerga et al. [8]. This included incorporation of dislocation–dislocation interactions at junctions that would ultimately lead to the formation of dynamic sources and obstacles, hence controlling the later stages of strain hardening in the material.

2.1 *Original Framework*

It has been well established that dislocations are the primary carriers of plastic deformation [1]. Collective motion of a plethora of dislocations produces permanent deformation. Dislocations are modeled as line defects embedded in an elastic continuum. Each dislocation is characterized by a Burgers vector, $\mathbf{b}_i$ and a unit normal, $\mathbf{n}_i$ of its slip plane. The problem is solved using the principle of superposition, assuming infinitesimal deformation gradients [7]. The state of the body, at an instant, defined in terms of displacement, strain, and stress is given as

$$u = \tilde{u} + \hat{u}, \epsilon = \tilde{\epsilon} + \hat{\epsilon}, \sigma = \tilde{\sigma} + \hat{\sigma} \tag{1}$$

where ($\tilde{}$) represents the singular fields that are obtained by superposition of the fields $(u^i, \epsilon^i, \sigma^i)$ associated with individual dislocations,

$$\tilde{u} = \sum_{i=1}^{N} u^i, \ \tilde{\epsilon} = \sum_{i=1}^{N} \epsilon^i, \tilde{\sigma} = \sum_{i=1}^{N} \sigma^i \tag{2}$$

N being the total number of the dislocations in the specimen. Here the ($\hat{}$) fields are the image fields used for correcting for the actual boundary conditions and are specified in terms of conventional tractions and displacements applied to portions of the boundary. The governing equations may thus be specified as

$$\nabla \cdot \sigma = 0, \epsilon = \nabla u, \sigma = C : \epsilon \tag{3}$$

where C represents the elasticity matrix. Time-dependent traction, $\mathbf{T} = \mathbf{T_0(t)}$ and displacement, $\mathbf{u} = \mathbf{u_0(t)}$ boundary conditions are applied on the appropriate sections of boundaries to emulate uniaxial tension, compression, or pure shear. Assuming the displacement fields to remain smooth makes the problem a well-posed one and solvable by Finite Element Method (FEM).

In order for the dislocated body to be thermodynamically stable, the dislocations naturally reorganize themselves to minimize the energy of the system. This rearrangement takes place under the action of a gliding force, referred to as Peach–Koehler force, which essentially can be described as the change of the potential energy of the body associated with an infinitesimal variation of the dislocation position in the glide plane. The expression for the Peach–Koehler force is given below:

$$\mathrm{f}^i = \mathbf{m}^i \cdot \{\widehat{\sigma} + \sum_{j \neq i} \sigma^j\} \cdot \boldsymbol{b}^i \tag{4}$$

Peach–Koehler force is the primary driving force behind the dislocation glide. The drag-controlled dislocation glide is represented through the following constitutive rule:

$$Bv^i = \mathrm{f}^i \tag{5}$$

where B is the viscous drag factor and v^{i} is the glide velocity of dislocation i. The position of each dislocation is updated using a first-order time stepping scheme, commonly known as Euler method,

$$x_{i,t+\delta t} = x_{i,t} + v\delta t \tag{6}$$

The problem is solved in an incremental fashion in time, where each time increment involves three main steps: (i) computation of the current dislocation configuration and the associated stress and strain values for the current configuration; (ii) determination of the Peach–Koehler force that drives the dislocations to a stable configuration; and (iii) determination of the instantaneous rate of change of dislocation structure based on a set of constitutive rules, which involves generation, motion, and annihilation of dislocations.

2.2 *Grain Boundaries*

Until recently, grain boundaries have been modeled as obstacles that inhibit the dislocation motion. However, during plastic deformation, a dislocation approaching a grain boundary may interact with the latter in many different ways, which gives rise to a variety of mechanical behaviors. An elastic interaction between a dislocation and a grain boundary is repulsive and hence leads to dislocations piling up at the boundary. However, quite recently [3] several mechanisms based on the plastic interaction between an incoming dislocation and a grain boundary have been distinguished. Few of them being (i) dislocations piling up at the grain boundary, (ii) dislocation absorption into the grain boundary, (iii) dislocation emission from the grain boundary, (iv) residual Burger's vector accumulation at the grain boundary, etc.

Keeping all these things in mind, a basic dislocation transmission mechanism has been implemented in this manuscript. The fundamental principle followed was given by Koning et al. [9]. The transmission can be direct or indirect. The former assumes that the entire dislocation upon reaching the grain boundary is transmitted to the neighboring grain whereas the latter assumes that only a part of a dislocation gets transmitted, and a residue is left on the boundary [10].

When a dislocation in grain with orientation g^1 reaches a grain boundary, a potential plane for transmission is identified in the neighboring grain with orientation g^2. As soon as the plane is identified, a Frank-Read source is placed on the plane at a distance l_n, which is the distance that marks the equilibrium separation between the dislocations when $\tau_R (= \tau_n)$ tries to drive the dislocations apart. The relationship between l_n and τ_n can mathematically be given as

$$l_n = \frac{\mu b}{2\pi(1 - \nu)\tau_n} \tag{7}$$

Although dislocation interaction with grain boundaries is a quite complex phenomenon [3], a simple transmission mechanism [11, 12] has been discussed in this work.

3 Problem Definition

A boundary value problem consisting of a unit cell of dimensions $10 \times 10\ \mu m^2$ was considered. The unit cell was divided into square grains, each side with length 2 μm. Slip planes representative of an FCC slip system in two dimensions (2D) were defined in each grain. Each slip plane was oriented at an angle of ±35.25° with the horizontal. Individual grains were considered elastically isotropic with Young's modulus, E = 70 MPa, and Poisson's ratio, n = 0.33, the values being representative of aluminum. A variable initial source density, ρ_s was used with an average source strength, $\tau_{FR} = 50$ MPa and a standard deviation, $\sigma = 10$ MPa. Critical time for a source to nucleate dislocations was taken as 10 ns. The grains were also supplied with an initial obstacle density, representative of precipitates, etc. with a typical strength, $\tau_{obs} = 150$ MPa. The unit cell did not contain any dislocations before the beginning of the simulation. The dislocations were nucleated on the slip planes as soon as the stress at the location of a Frank-Read source reached a critical value, τ_{FR}. The velocity of a dislocation was given based on Eq. 5, with value of $B = 10^{-4}$ Pa s.

Other short-range interactions between dislocations and obstacles, dislocations and dislocations, etc. were governed by constitutive rules built in the framework.

4 Results and Discussion

Figure 1 shows the schematic of a bicrystal with the grain boundary, separating two crystallites, inclined at an angle ϕ with the horizontal. The slip systems, representative of the crystal orientation in this manuscript, in the two grains were oriented differently with respect to the loading. It can be seen from the schematic that the orientation of one of the grains is g^1 and the orientation of the other grain is g^2.

The bicrystal was subjected to pure shear and two cases were considered (i) when the grain boundary was supposed to be an obstacle and would simply stop dislocations from getting transmitted to the neighboring grain and (ii) in which the grain boundary allowed dislocation transmission from one grain to other.

Figure 2 shows contours of stress development in a bicrystal with dislocation substructure superposed upon them. The grain boundary was considered perpendicular to the horizontal. By looking at the stress contours in both the cases, it can be seen from Fig. 2a that the dislocations piled up at the grain boundary creating stress

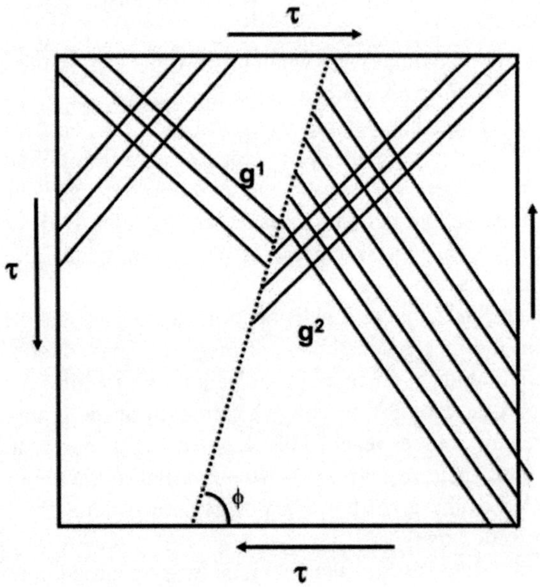

Fig. 1 Schematic of a bicrystal with grains having orientations g^1 and g^2, respectively. The dotted line denotes the grain boundary between two grains oriented at an angle of ∅ with respect to the horizontal

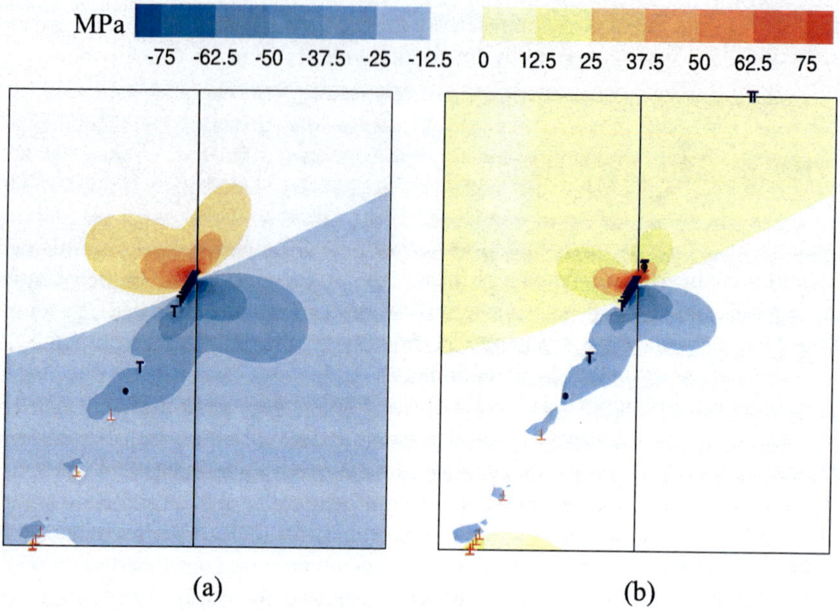

Fig. 2 Stress contours along with dislocation substructure in a bicrystal **a** without and **b** with dislocation transmission across grain boundaries

concentration at the grain boundary and when the transmission is allowed across the grain boundary, the level of stress concentration diminishes (see Fig. 2b).

In a quite similar manner, the same bicrystal with an initial density of Frank-Read sources, $\rho_s = 5 \times 10^{12}$, and no obstacles was subject to pure shear. Figure 3 shows the dislocation substructure superposed upon stress contours post-deformation. Figure 3a represents the case in which grain boundaries act as obstacles to dislocation motion and thus inhibit it; however, Fig. 3b represents the case in which grain boundaries allow dislocation transmission.

It can be seen from Fig. 3 that as the dislocations piled up at the grain boundary, new sources that aid indirect dislocation transmission, represented by red-colored circles near the grain boundary, emerged in the neighboring grain. The strength and location of these sources were governed by the strength of the grain boundary. As the stress near these new sources reached a value that was a representative of grain boundary strength, new dislocation pairs were nucleated in the neighboring grain, one of which traveled to the grain boundary to form a residue and the other one behaved as the transmitted one.

Stress versus strain response of the bicrystal was recorded for both (i) soft boundary, a boundary that allows dislocation transmission and (ii) hard boundary, a boundary that does not let the dislocations transmit to the neighboring grains.

Figure 4 gives a representative stress versus strain response of a bicrystal under pure shear for both the cases. It can be seen from the figure that in case of a grain boundary that doesn't allow slip transmission the response is quite different from

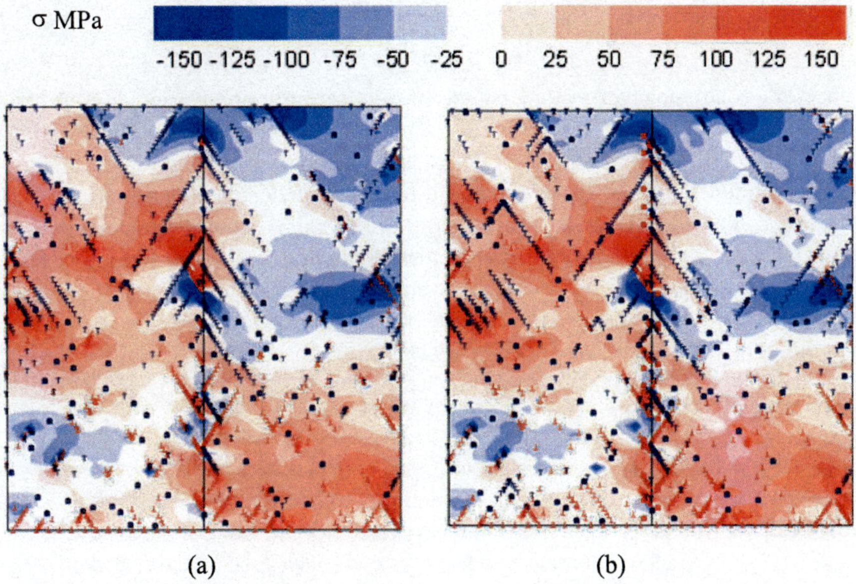

Fig. 3 Stress contours along with dislocation substructure in a bicrystal **a** without and **b** with dislocation transmission across grain boundaries

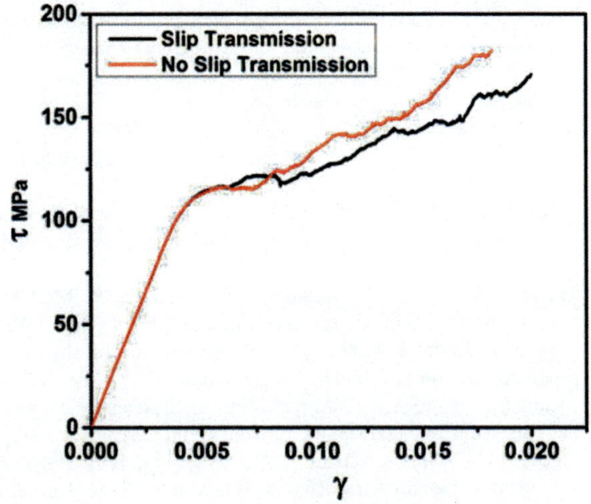

Fig. 4 Representative stress versus strain response of a bicrystal under pure shear for two different cases (i) in which grain boundary allows slip transmission, (ii) in which slip transmission is not allowed

the case in which slip transmission is allowed. One can see that the material hardens more in case of a grain boundary that does not allow slip transmission, whereas shows a softer response in case, dislocation transmission is allowed. This response, hence the role played by grain boundaries in deformation of polycrystalline materials, is worth studying and needs rigorous analysis with modern approaches.

5 Summary

This work is an endeavor to understand the role of grain boundaries in deformation of polycrystalline metallic materials. The work extended Discrete Dislocation Dynamics (DDD) from a single crystal to a polycrystalline setting. Grain boundaries were modeled as (i) hard, which did not allow dislocation transmission; (ii) soft, which allowed dislocation transmission. Deformation of a bicrystal, under pure shear, with the grain boundary perpendicular to the horizontal was studied. Stress contours showed that the hard boundaries were potential sites for stress concentration whereas stress seemed diffused in case of boundaries that allowed dislocations to pass through. Representative stress versus strain response showed that the material hardened more quickly in case of grain boundaries that did not allow slip transmission and remained softer in case slip transmission was allowed.

References

1. Dieter, G.E.: Mechanical Metallurgy, 3rd edn. McGraw-Hill Book Company, New York (1986)
2. Gottstein, G.: Physical Foundations of Materials Science. Springer, Berlin, Heidelberg (2004)
3. Bayerschen, E., McBride, A.T., Reddy, B.D., Bohlke, T.: Review on slip transmission criteria in experiments and crystal plasticity models. J. Mater. Sci. **51**(5), 2243–2258 (2016)
4. Amodeo, R. J., Ghoniem, N. M.: Dislocation dynamics I. A proposed methodology for deformation micromechanics. Phys. Rev. B **41**(10), 6958–6967 (1990)
5. Kubin, L.P., et al.: Dislocation microstructures and plastic flow: a 3D simulation. Solid State Phenom. **23–24**, 455–472 (1992)
6. Lubarda, V.A.: On the elastic energy representation of a dislocated body and dislocation equilibrium conditions. J. Elast. **32**(1), 19–35 (1993)
7. Van der Giessen, E., Needleman, A.: Discrete dislocation plasticity: a simple planar model. Model. Simul. Mater. Sci. Eng. **3**(5), 689–735 (1995)
8. Benzerga, A.A., Brechet, Y., Needleman, A., Van der Giessen, E.: Incorporating three-dimensional mechanisms into two-dimensional dislocation dynamics. Model. Simul. Mater. Sci. Eng. **12**, 159–196 (2004)
9. Koning, M., Miller, R., Bulatov, V.V., Abraham, F.: Modeling grain-boundary resistance in intergranular dislocation slip transmission. Philos. Mag. A **82**(13), 2511–2527 (2002)
10. Zheng, Z., Balint, D.S., Dunne, F.P.E.: Investigation of slip transfer across HCP grain boundaries with application to cold dwell facet fatigue. Acta Mater. **127**, 43–53 (2017)
11. Kumar, R., Nicola, L., Van der Giessen, E.: Density of grain boundaries and plasticity size effects: a discrete dislocation dynamics study. Mater. Sci. Eng. A **527**, 7–15 (2009)
12. Quek, S.S., Wu, Z., Zhang, Y.W., Srolovitz, D.J.: Polycrystal deformation in a discrete dislocation dynamics framework. Acta Mater. **75**, 92–105 (2014)

Effective Properties of Two-Scale Viscoelastic Composites

Vivek Singh, Jayram Desai, and Vikranth Racherla

1 Introduction

Polymers commonly exhibit viscoelasticity. They have dissipation effects (from viscous part) as well as storage effects (from elastic part). Viscoelastic composites have gained interest among researchers because of their applications in energy absorbers, dampers, acoustic insulation, thermal insulation, etc. Many works have been carried out to evaluate the effective properties of viscoelastic composites. Eshelby through his seminal paper [1] introduced a new arena of research in the field of composite materials, Effective Behavior of Composite Materials. Thereafter, the area was pioneered by Hill [2, 3], Walpole [4], Mori and Tanaka [5], Hashin and Shtrikman [6], Christensen [7], Mura and Nemat-Nasser [8], etc. For elastic composites, micromechanical methods are very well used to predict the effective properties if the properties of components and the volume fractions of phases are known. Eshelby's approach for elastic materials has been extended by researchers to viscoelastic materials by Laplace–Carson transform or correspondence principle. In this method, the equations for local and effective properties of linear viscoelastic composite after being transformed to Laplace domain can be treated similar to those for elastic composite with complex moduli. Hashin [9] for the very first time considered the problem of spherical viscoelastic inclusion in elastic and viscoelastic matrices. Laws and McLaughlin [10] extended self-consistent method for the estimation of effective properties of (i) sphere-dispersed viscoelastic composite and (ii) fiber-reinforced viscoelastic composite. Brinson and Lin [11] presented a comparison of Mori-Tanaka and finite element unit cell method for viscoelastic composites. Frequency and volume fraction were found to affect the results for different methods. Yi et al. [12] had applied asymptotic homogenization technique for estimating effective moduli of composite with periodic microstructure. Friebel et al. [13] used the correspondence principle to extend the two-step homogenization procedure to estimate the effective properties of

V. Singh (✉) · J. Desai · V. Racherla
Department of Mechanical Engineering, Indian Institute of Technology Kharagpur, Kharagpur 721302, India
e-mail: mechie.vivek@gmail.com

S. K. Saha and M. Mukherjee (eds.), *Recent Advances in Computational Mechanics and Simulations*, Lecture Notes in Mechanical Engineering,
https://doi.org/10.1007/978-981-15-8315-5_18

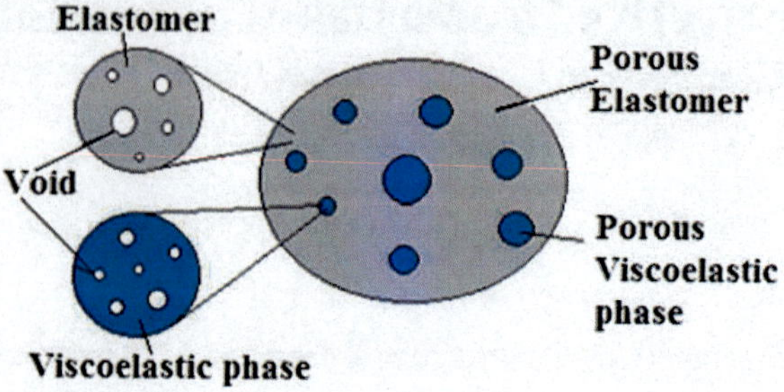

Fig. 1 Schematic of the microstructure considered

multi-phase viscoelastic composites. Nguyen et al. [14] modeled a porous medium as a mixture of viscoelastic matrix with pore inclusion. Both the matrix and inclusion were modeled as Maxwell elements and the results were in agreement with the exact solution. Lahellec and Suquet [15] developed a direct method based on time-integration approach to estimate the effective properties of composite comprising of linear viscoelastic constituents. Daridon et al. [16] estimated the effective properties of a composite which consisted of matrix, inclusion, and the interface. All of the constituents were modeled as Kelvin–Voigt elements but the resulting effective behavior obtained was of Kelvin–Voigt type with an additional memory term.

In the present article, mean-field-homogenization-based effective properties for viscoelastic composite are estimated. The composite consists of elastomeric matrix and inclusions in the form of pores and porous viscoelastic media (Fig. 1). This paper presents a micromechanical approach to estimating the effective viscoelastic properties of the composite that consists of (i) pores and (ii) porous viscoelastic inclusion.

2 Governing Equations

Consider a material with N different phases that are distributed either randomly or periodically, in the specimen occupying a volume Ω_0 with the boundary $\partial\Omega_0$ such that the characteristic length of inclusion is much smaller than the specimen than length scale separation is used to derive macroscopic properties of the composite using the microstuctural properties. In this work, porous viscoelastic composite consisting of pores and porous viscoelastic inclusion is considered. The size of pores in matrix and viscoelastic inclusion is assumed to be much smaller than the size of the viscoelastic inclusion. A representative volume element Ω is considered at the boundary $\partial\Omega$ of which macroscopic strain is related to displacement as $u = \bar{E}x$.

For a linear viscoelastic non-aging material, the stress evolution is related to strain rate as

$$\sigma_{ij}(x,t) = \int_0^t C_{ijkl}\left(x, t-t'\right) \frac{\partial \varepsilon_{kl}\left(x, t'\right)}{\partial t'} dt', \qquad \forall x \in \Omega \tag{1}$$

where C_{ijkl} is the relaxation modulus. In case of small displacement and neglecting body force, the compatibility condition and equilibrium equation are as follows [17]:

$$\varepsilon_{ij} = \frac{1}{2}\left[u_{i,j} + u_{j,i}\right], \qquad \forall x \in \Omega \tag{2}$$

$$\sigma_{ij,j} = 0, \quad \forall x \in \Omega \tag{3}$$

For viscoelastic material, the stress–strain relations are converted to linear form by applying Laplace–Carson (LC) transform. Thereafter, the linear viscoelastic problem becomes similar to a linear elastic problem. The viscoelastic properties of the composite are estimated in the LC space by relating the local behavior

$$\sigma^*(x) = \mathbb{C}^*(x) : \varepsilon^*(x) \tag{4}$$

and macroscopic behavior of an RVE

$$\Sigma^*(x) = \mathbb{C}^*_{hom} : E^* \tag{5}$$

In the above equations, (*) superscript terms represent LC transform of the corresponding quantities as

$$\varphi^*(x,p) = p\int_0^\infty \varphi(x,t)e^{-pt}dt \tag{6}$$

The local and average strain tensors are related by the equation

$$\varepsilon^*(x) = \mathrm{A}^*(x) : E \tag{7}$$

where $A^*(x)$ is the strain localization tensor at point x. Strain localization tensor for an inclusion embedded in the matrix is obtained by solving Eqs. (1)–(3) along with uniform strain boundary condition. In elastic domain, the problem is solved by displacement approach. Perfect interface is considered between the matrix and inclusion and both the phases are considered to be isotropic. The shear moduli (μ) and bulk moduli (K) are estimated by prony series

$$\mu(t) = \mu_0\left[1 - \sum_{i=1}^{m} w_i\left(1 - e^{-\frac{t}{\xi_i}}\right)\right] \tag{8}$$

$$K(t) = K_0\left[1 - \sum_{i=1}^{m} w_i\left(1 - e^{-\frac{t}{\xi_i}}\right)\right] \tag{9}$$

Table 1 Mechanical properties of the idealized viscoelastic materials used by Friebel et al. [13]

Shear		Bulk	
τ_i [s]	$\mu_0 w_i$ [bar]	τ_i [s]	$K_0 w_i$ [bar]
0.032	2.512	100.000	3000
0.100	10.000	316.228	100
0.316	56.234		
1.000	316.228		
3.162	1000.000		
10.000	199.526		
31.623	50.119		
100.000	19.953		
316.228	12.589		
1000.000	2.512		
3162.278	1.698		
10,000.000	1.202		
31,622.777	1.148		
100,000.000	1.096		
	$\mu_0 = 1677.979$		$K_0 = 3300$

where μ_0 and K_0 are instantaneous shear and bulk moduli, and $w_i K_0$ and $w_i \mu_0$ are weights corresponding to relaxation time ξ_i . The representative values for prony series have been taken from Friebel et al. [13] and are listed in Table 1.

The relaxation tensors for the phases are given as

$$\mathbb{C}^{i*} = 3K_i^* \mathbb{J} + 2\mu_i^* \mathbb{K} \tag{10}$$

$$\mathbb{J}_{ijkl} = \frac{1}{3}\delta_{ij}\delta_{kl}, \quad \mathbb{K}_{ijkl} = \frac{1}{2}\left(\delta_{ik}\delta_{jl} + \delta_{il}\delta_{jk} - \frac{2}{3}\delta_{ij}\delta_{kl}\right) \tag{11}$$

where K_i^* is the bulk relaxation and G_i^* is the shear relaxation. For particulate composites, the global strain localization tensor is related to local strain localization tensor as

$$\mathbb{A}^* = \mathrm{A}^{i*}\left(c_0 I + \sum_{r=1}^{N} c_r \mathrm{A}^{r*}\right)^{-1} = \mathrm{A}^{i*}\left(\sum_{r=0}^{N} c_r \mathrm{A}^{r_*}\right)^{-1} \tag{12}$$

where c_0 is the volume fraction of the matrix, c_r is the volume fraction of different inclusions, and A^{i*} is the local strain localization tensor. Also, $A^{0*} = I$ has been used. For any phase(s),

$$\mathrm{A}_s{}^{i*} = \left[I + \mathbb{P}_s^{i*}\left(C^s - C^0\right)\right]^{-1} \tag{13}$$

where the polarization tensor $\mathbb{P}$ for a spherical inclusion in a matrix with shear and bulk moduli μ_0, K_0 is given by

$$\begin{aligned}\mathbb{P}^0 &= \frac{1}{3K_{p0}}\mathbb{J} + \frac{1}{2\mu_{p0}}\mathbb{K}\\ 3K_{p0} &= 3K_0 + 4\mu_0, \quad 2\mu_{p0} = \frac{5\mu_0}{3}\frac{3K_0 + 4\mu_0}{K_o + 2\mu_0}\end{aligned} \tag{14}$$

The Mori–Tanaka scheme when applied to matrix-inclusion composite considers the matrix to be reference medium [19]. The subscript 0 denotes matrix whereas the subscripts 1, 2, ..., N denote different inclusions. Effective modulus of the composite obtained using Mori–Tanaka model is

$$\mathbb{C}^*_{hom} = \sum_{r=0}^{N} c_r \mathbb{C}^r \mathbb{A}^i_r \left(\sum_{s=0}^{N} c_s A^i_s\right)^{-1} \tag{15}$$

The individual behavior of porous elastomer and porous viscoelastic phase is given by

$$\begin{aligned}\bar{K_1}^* &= \frac{4}{3}\frac{1-p_1}{p_1}\mu_1^{0*}, & \bar{\mu_1}^* &= \frac{3\,(1-p_1)}{3+2p_1}\mu_1^{0*}\\ \bar{K_2}^* &= \frac{4\mu_2^{0*}K_2^{0*}(1-p_2)}{4\mu_2^{0*} + 3p_2K_2^{0*}}, & \bar{\mu_2}^* &= \frac{(9K_2^{0*} + 8\mu_2^{0*})(1-p_2)\mu_2^0*}{K_2^{0*}(9+6p_2) + \mu_2^0 * (8+12p_2)}\end{aligned} \tag{16}$$

where p_1 and p_2 represent porosity in elastomer and viscoelastic phase, respectively, also (μ_1^{0*}, K_1^{0*}) and (μ_2^{0*}, K_2^{0*}) are shear and bulk moduli of elastomer and viscoelastic phase in Laplace–Carson space.

Composite Sphere Assemblage (CSA) [20] model states that the composite can be modeled with the assembly of concentric composite spheres of different sizes. The composite sphere consists of an inner sphere—inclusion and an annular sphere matrix. The size of composite sphere is such that in each sphere the inclusion volume fraction is same as that of the composite. The radius ratio of a (internal sphere) and b (outer sphere) is related to inclusion volume fraction in composite as $(a/b)^3 = f$, f being the volume fraction of inclusion in composite. CSA model gives an exact solution only for bulk modulus as

$$\bar{K}/K = \frac{(1-p)(2-4\nu)}{(2-4\nu) + (1+\nu)p} \tag{17}$$

where p is porosity and ν is Poisson's ratio.

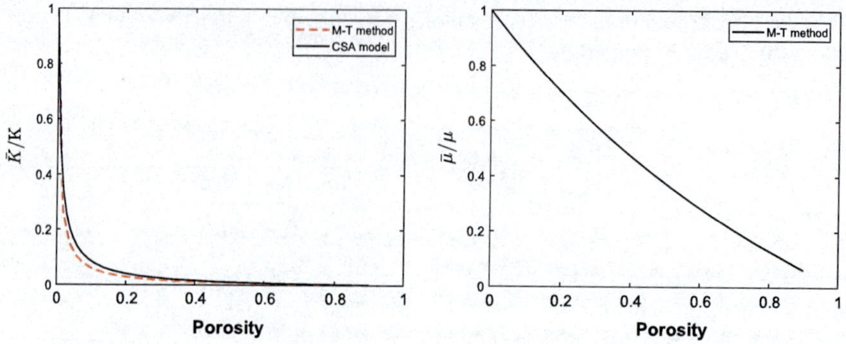

(a) Effective normalized bulk modulus of porous elastomer

(b) Effective normalized shear modulus of porous elastomer

Fig. 2 Effective normalized modulus of porous elastomer

3 Results

In this section, mean-field estimate for macroscopic properties is presented. Silicone rubber is used as matrix material. For silicone rubber, bulk modulus (K) is 1.8 GPa and shear modulus (μ) is 0.01 GPa. By substituting Laplace variable (s) with $i\omega$ we can get frequency-dependent behavior. Here we have taken the frequency to be 1 Hz and the results below are for that. Porosity in the matrix as well in inclusion varies from 0 to 70%. The effective bulk modulus ($\bar{K}$) is observed to decay exponentially with the variation of porosity (Fig. 2a). $\bar{K}$ drops to $\sim$7% at 20% porosity which indicates the sensitivity of $\bar{K}$ (i.e., resistance to uniform compression) to porosity. However, in case of shear modulus, the behavior is different as it decreases gradually with porosity (Fig. 2b).

Mechanical loss factor or dissipation factor (tan δ) is measure of material's ability to dissipate energy in addition to storage of energy. Higher mechanical loss factor ensures better dissipation of energy. For a composite with only porous inclusions and no porosity in matrix, tan δ is observed to increase up to 0.85 volume fraction of viscoelastic inclusion and thereafter it decreases but with the increase in volume fraction of viscoelastic inclusion tan δ keeps increasing (Fig. 3). Viscoelastic phase thus helps in easy energy dissipation. For the second case, when non-porous inclusions are embedded in porous matrix, it is observed that dissipation is better at high volume fraction of inclusion but matrix porosity has adverse effect on its dissipation property (Fig. 4).

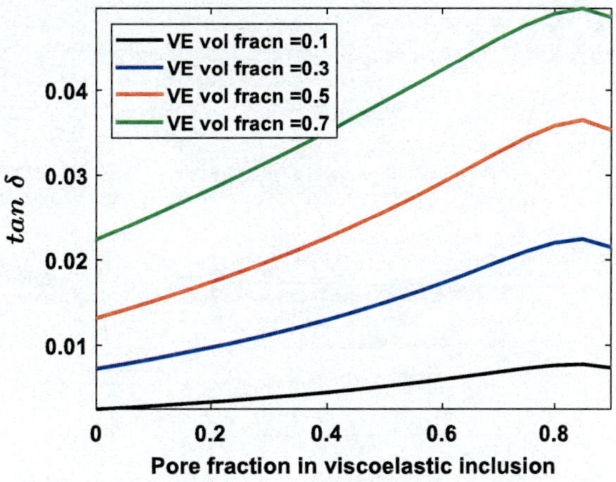

Fig. 3 Effect of inclusion pore fraction on tan δ of the composite

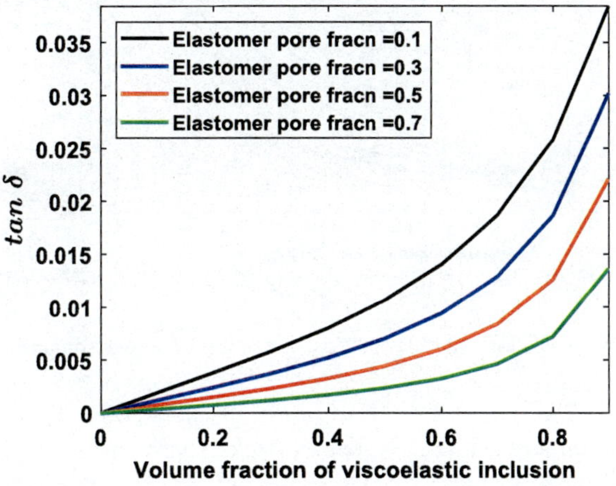

Fig. 4 Effect of viscoelastic phase volume fraction on tan δ of the composite

3.1 Effect of Porosity on Effective Behavior

In this section, we present the results about the effect of matrix and inclusion porosity on effective properties of the composite. Figure 5 shows the results for three cases of porosity in inclusion: (i) 0.3, (ii) 0.5, (iii) 0.7.

Bulk modulus has an exponential decay with the addition of porous inclusion as well as matrix porosity (Fig. 5a). The composite's effective bulk modulus $\bar{K}$ is sensitive to porosity and inclusion. Young's modulus and shear modulus are almost similar in nature with a gradual growth (Fig. 5b, c). The growth of shear modulus and hence Young's modulus is due to the fact that shear modulus for viscoelastic phase is much higher than that of silicone rubber. Also it can be seen that addition of porous inclusion reduces the normalized modulus of composite. Addition of porous inclusions enhances the ability of composite to dissipate energy whereas the addition

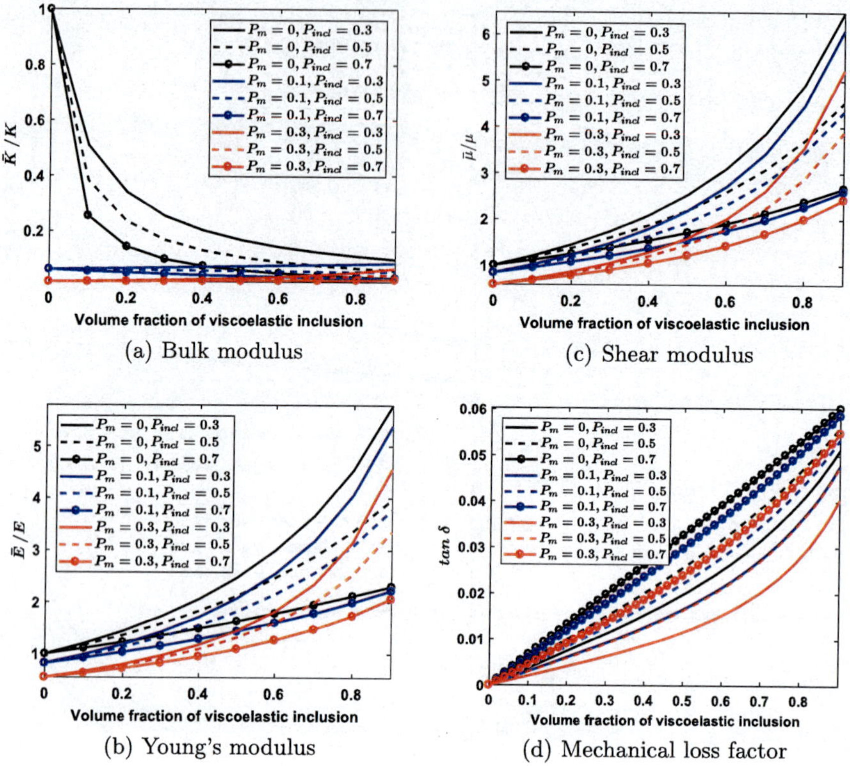

Fig. 5 Effective normalized elastic modulus of porous composite

of porosity in matrix is seen to weaken the energy dissipation behavior of composite. Also the increase of viscoelastic phase also improves the composite's ability to dissipate energy.

4 Conclusion

The present study for two-scale viscoelastic composites was carried out in linear elasticity and viscoelasticity. Linear viscoelastic problem was converted to linear elastic one using Laplace–Carson transform. The effective behavior of heterogeneous material can be determined by mean-field estimation technique (Mori–Tanaka method). Here Mori–Tanaka method was used to estimate effective properties of the composite.

1. The microstructure of the composite is governed by the porosity in the constituent phases and their volume fractions.

2. The dissipation effect in the composite is affected by the microstructure of the composite, i.e., the volume fraction of inclusion and porosity in the inclusion and matrix.
3. Bulk modulus has an exponential decay whereas shear modulus has a gradual decay with an increase in matrix porosity, and addition of viscoelastic inclusion improves Shear and Young's modulus.
4. With the addition of porous inclusions dissipation effect gets enhanced.

References

1. Eshelby, J.D.: The determination of the elastic field of an ellipsoidal inclusion, and related problems. Proc. R. Soc. Lond. A **241**(1226), 376–396 (1957). https://doi.org/10.1098/rspa.1957.0133
2. Hill, R.: Elastic properties of reinforced solids: some theoretical principles. J. Mech. Phys. Solids **11**(5), 357–372 (1963). https://doi.org/10.1016/0022-5096(63)90036-X
3. Hill, R.: A self-consistent mechanics of composite materials. J. Mech. Phys. Solids **13**(4), 213–222 (1965). https://doi.org/10.1016/0022-5096(65)90010-4
4. Walpole, L.J.: On the overall elastic moduli of composite materials. J. Mech. Phys. Solids **17**(4), 235–251 (1969). https://doi.org/10.1016/0022-5096(69)90014-3
5. Mori, T., Tanaka, K.: Average stress in matrix and average elastic energy of materials with misfitting inclusions. Technical report. Acta Mater. **21**(5), 571–574 (1973). https://doi.org/10.1016/0001-6160(73)90064-3
6. Hashin, Z., Shtrikman, S.: A variational approach to the theory of the elastic behaviour of multiphase materials. J. Mech. Phys. Solids **11**(2), 127–140 (1963). https://doi.org/10.1016/0022-5096(63)90060-7
7. Christensen, R.M.: Viscoelastic properties of heterogeneous media. J. Mech. Phys. Solids **17**(1), 23–41 (1969). https://doi.org/10.1016/0022-5096(69)90011-8
8. Nemat-Nasser, S., Hori, M.: Micromechanics: Overall Properties of Heterogeneous Materials, vol. 37. Elsevier (2013)
9. Hashin, Z.: The inelastic inclusion problem. Int. J. Eng. Sci. **7**(1), 11–36 (1969). https://doi.org/10.1016/0020-7225(69)90020-2
10. Laws, N., McLaughlin, R.: Self-consistent estimates for the viscoelastic creep compliances of composite materials. Proc. R. Soc. Lond. A **359**(1697), 251–273 (1978). https://doi.org/10.1098/rspa.1978.0041
11. Brinson, L.C., Lin, W.S.: Comparison of micromechanics methods for effective properties of multiphase viscoelastic composites. Comp. Str. **41**(3–4), 353–367 (1998). https://doi.org/10.1016/S0263-8223(98)00019-1
12. Yi, Y.M., Park, S.H., Youn, S.K.: Asymptotic homogenization of viscoelastic composites with periodic microstructures. Int. J. Solids Struct. **35**(17), 2039–2055 (1998). https://doi.org/10.1016/S0020-7683(97)00166-2
13. Friebel, C., Doghri, I., Legat, V.: General mean-field homogenization schemes for viscoelastic composites containing multiple phases of coated inclusions. Int. J. Solids Struct. **43**(9), 2513–2541 (2006). https://doi.org/10.1016/j.ijsolstr.2005.06.035
14. Nguyen, S.T., Thai, M.Q., Vu, M.N., To, Q.D.: A homogenization approach for effective viscoelastic properties of porous media. Mech. Mat. **100**, 175–185 (2016). https://doi.org/10.1016/j.mechmat.2016.06.015
15. Lahellec, N., Suquet, P.: Effective behavior of linear viscoelastic composites: a time-integration approach. Int. J. Solids Struct. **44**(2), 507–529 (2007). https://doi.org/10.1016/j.ijsolstr.2006.04.038

16. Daridon, L., Licht, C., Orankitjaroen, S., Pagano, S.: Periodic homogenization for Kelvin-Voigt viscoelastic media with a Kelvin-Voigt viscoelastic interphase. Eur. J. Mech. A Solid **58**, 163–171 (2016). https://doi.org/10.1016/j.euromechsol.2015.12.007
17. Gurtin, M. E.: An Introduction to Continuum Mechanics, vol. 158. Academic Press (1982)
18. Christensen, R.: Theory of Viscoelasticity: An Introduction. Elsevier (2012)
19. Mori, T., Tanaka, K.: Average stress in matrix and average elastic energy of materials with misfitting inclusions. Acta Mater. **21**(5), 571–574 (1973). https://doi.org/10.1016/0001-6160(73)90064-3
20. Hashin, Z.: The elastic moduli of heterogeneous materials. J. Appl. Mech. **29**(1), 143–150 (1962). https://doi.org/10.1115/1.3636446

Thermo-Mechanical Analysis of Shape Memory Polymer Composite

L. Bhola, P. M. Mujumdar, and P. J. Guruprasad

1 Introduction

The intent of this paper is to formulate an analytical framework to analyze different shape memory polymer-based composite and sandwich structures. Main goal of this study is to evaluate the applicability of shape memory polymer in self-deployable large structures. Currently existing approaches to deploy large structures in space such as solar arrays, antennas, solar sails, sunshields, etc. all rely upon conventional systems [1–4]. The support structure and mechanisms can sometime comprise 90% weight of the total deployable assembly. In addition, they increase the failure rate, cost of the system, and complexity of overall structure. These disadvantages can be overcome by using smart materials such as shape memory materials, due to their ability to store a shape which can be recovered back with the help of external stimuli such as hygro/thermo-mechanical cycle, electromagnetism, chemical reaction, and photo-activity [3].

Shape memory polymer is used due to its high recovery rate compared to shape memory alloys and shape memory ceramics which is a major requirement for self-deployable application. Due to low stiffness and strength of shape memory polymer, it is used in different topological architectures such as SMP sandwich structure or an SMP composite (SMPC) structure. However, these configurations have to be assessed and a proper analysis and design has to be carried out since combining such a low stiffness material with high stiffness fiber or material layers without any shape memory effect in different topological configurations will significantly affect the structural recovery and stored strain.

L. Bhola (✉) · P. M. Mujumdar · P. J. Guruprasad
Indian Institute of Technology Bombay, Mumbai 400076, India
e-mail: lalit.b@aero.iitb.ac.in

S. K. Saha and M. Mukherjee (eds.), *Recent Advances in Computational Mechanics and Simulations*, Lecture Notes in Mechanical Engineering,
https://doi.org/10.1007/978-981-15-8315-5_19

Shape memory polymer due to its ability to store a shape which can be recovered with the help of an external stimuli has many engineering applications in bio-medical [4], heat shrinkable tubes [5, 6], self-deployable space shields [7], various medical devices, and other intelligent structures. This characteristic of shape memory polymer is due to its differential physical nature above and below its glass transition temperature. Significant work has been done on the shape memory polymer-based composites, including carbon nanotubes composite, chopped fiber configuration [8], fiber-reinforced composite, and nanotube composite [9–12]. Most of the work, however, mainly focuses on the recovery characteristic of SMP composite. There is not much work done focusing on the deformation characteristics of structures made of these material systems. This paper deals with the development of an analytical and numerical framework of a unidirectional shape memory polymer-based composite and sandwich structure which can be extended to other composite architectures.

2 Analytical Formulation

In this section, the analytical framework for SMP-based unidirectional composite is described. First the constitutive model of the material will be extended to calculate micro-mechanical properties of a laminae based on SMP matrix and fiber. This material configuration will be used throughout the paper to investigate the behavior of a one-dimensional SMP-based composite laminate under a transverse loading. The modeling strategy presented can be extended to sandwich structure, other composite architectures, and various boundary conditions.

2.1 SMP Constitutive Model

SMP constitutive model used for analytical formulation is based on phase transformation theory, i.e., at any arbitrary temperature, SMP model is a mixture of two distinct phases: "Active Phase" and "Frozen Phase." The frozen phase (glassy phase) refers to the phase with locked confirmational rotation corresponding to higher temperature entropic deformation. In contrast, active phase (rubbery phase) refers to the phase with free confirmational motion of molecules [13]. According to the model volume, fraction of these phases depends on the change in temperature and by changing the ratio of these phases during a thermo-mechanical cycle shape memory behavior can be captured.

The constitutive equation for the material describing normal stress versus normal strain can be given by

$$\sigma = E(T)(\epsilon_m - \epsilon_s - \int_{T_h}^{T} \alpha dT) \tag{1}$$

where ϵ_s represent stored strain in SMP during cooling, α is the coefficient of thermal expansion, T_h is reference temperature above glass transition temperature, and $E(T)$ is effective elastic modulus which can be given as

$$E(T) = \frac{1}{\frac{\phi_f}{E_i} + \frac{1-\phi_f}{E_e}} \tag{2}$$

where ϕ_f is the volume fraction of frozen phase, and E_e and E_i are the elastic modulus of active and frozen phases, respectively.

Storage or inelastic strain can be determined from the following equation:

$$\frac{d\epsilon_s}{dT} = \frac{\epsilon_m - \epsilon_s - \int_{T_h}^{T} \alpha dT}{E_e[\frac{\phi_f}{E_i} + \frac{1-\phi_f}{E_e}]} \frac{d\phi_f}{dT} \tag{3}$$

The phase transformation ratio ϕ_f can be computed empirically from the following relation:

$$\phi_f = 1 - \frac{1}{1 + c_f(T_h - T)^n} \tag{4}$$

The variation of elastic modulus with respect to temperature can be determined using the above set of equations, which is shown in Fig. 1. A drastic change in elastic modulus can be observed near glass transition temperature. This is due to the transformation of frozen phase into active phase of SMP [14].

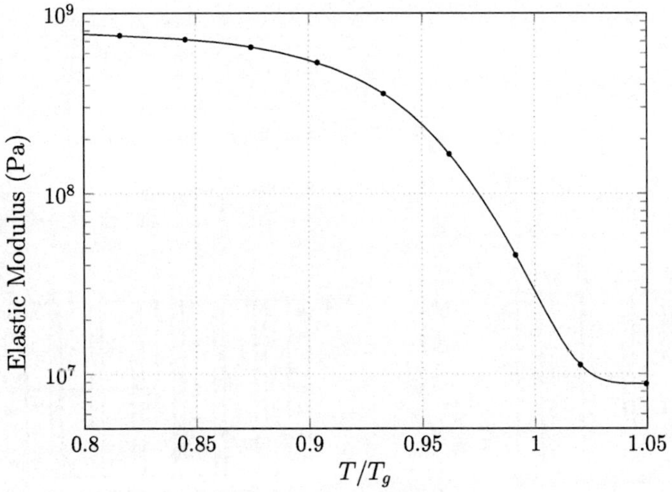

Fig. 1 Variation of elastic modulus with respect to temperature

2.2 SMP Composite Modeling

This formulation will involve only laminate made up of unidirectional laminate for the sake of micro-mechanical simplicity. However, the methodology described here can be extended for other composite architectures using appropriate micro-mechanical analysis to determine their material properties.

Unidirectional Laminate Properties: Mechanical properties of a laminate can be calculated in all the principal directions using basic micro-mechanical approach based on the constitutive model of both fibers and matrix. A pseudo-time history variable will be present in the constitutive relation of SMP. The constitutive relation for fiber and matrix is given as

$$\sigma_f = E_f \epsilon_f$$

$$\sigma_m = E_{mat}(T)(\epsilon_m - \epsilon_s - \int_{T_h}^{T} \alpha dT)$$

where σ_f and σ_m represent stress in fiber and matrix, respectively.

Along fiber direction: Let's assume an RVE of laminate with load in fiber direction which causes equal strain in fibers and matrix, Fig. 2a. The final constitutive equation for the laminate along fiber direction is given as

$$\sigma_1 = E_1\epsilon_1 - E_1^s(\epsilon_s - \epsilon_T) \tag{5}$$

Here, micro-mechanical properties are given as

$$E_1 = (E_f V_f + E_{mat} V_m) \tag{6}$$

$$E_1^s = V_m E_{mat} \tag{7}$$

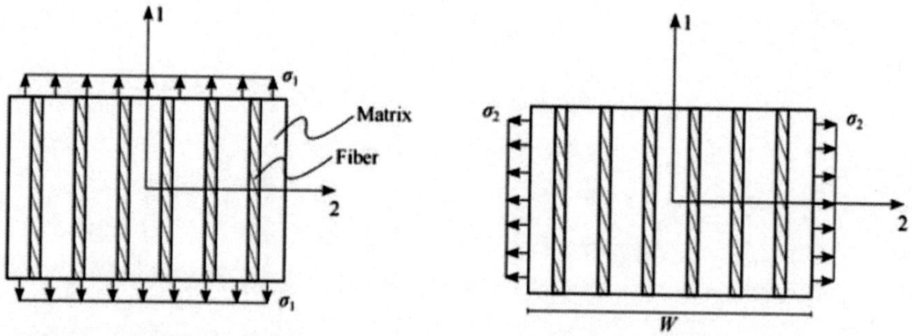

(a) Along fiber direction (b) Across fiber direction

Fig. 2 Loading representation on RVE (representative volume elements)

Across fiber direction: Let's assume an RVE of lamina with load across fiber direction which causes different deflections in both fiber and matrix as shown in Fig. 2b. The final across fiber direction is given as

$$\sigma_2 = E_2\epsilon_2 - E_2^s(\epsilon_s - \epsilon_T) \tag{8}$$

which results in the following expressions:

$$E_2 = \frac{E_{mat}E_f}{E_f V_m + E_{mat}V_f} \tag{9}$$

$$E_2^s = \frac{E_{mat}E_f V_m}{E_f V_m + E_{mat}V_f} \tag{10}$$

Similarly, we can calculate shear modulus and Poisson's ratio of the respective SMP composite.

Using above equivalent properties one can write constitutive equation for a lamina of 0° and 90° layups, which can be used to construct constitutive equation for any layup using the transformation scheme.

$$\begin{Bmatrix} \sigma_x \\ \sigma_z \\ \tau_{xz} \end{Bmatrix} = [\overline{Q}] \begin{Bmatrix} \epsilon_x \\ \epsilon_z \\ \gamma_{xz} \end{Bmatrix} - [\overline{Q}]^s \begin{Bmatrix} \epsilon_x^s + \epsilon_T \\ \epsilon_z^s + \epsilon_T \\ \gamma_{xz}^s + \epsilon_T \end{Bmatrix} \tag{11}$$

where given matrix can be calculated as

$$[\overline{Q}] = [T(\theta)]^{-1}[Q][R][T(\theta)][R]^{-1}$$

$$[\overline{Q}]^s = [T(\theta)]^{-1}[Q]^s[R][T(\theta)][R]^{-1}$$

$$[Q] = \begin{bmatrix} Q_{11} & Q_{12} & 0 \\ Q_{21} & Q_{22} & 0 \\ 0 & 0 & Q_{66} \end{bmatrix} \tag{12}$$

where $Q_{11} = E_1/(1 - \nu_{12}\nu_{21})$, $Q_{22} = E_2/(1 - \nu_{12}\nu_{21})$, $Q_{12} = \nu_{21}E_2/(1 - \nu_{12}\nu_{21})$, and $Q_{66} = G_{12}$.

$$[Q]^s = \begin{bmatrix} Q_{11}^s & Q_{12}^s & 0 \\ Q_{21}^s & Q_{22}^s & 0 \\ 0 & 0 & Q_{66}^s \end{bmatrix} \tag{13}$$

where $Q_{11}^s = E_1^s/(1 - \nu_{12}^s\nu_{21}^s)$, $Q_{22}^s = E_2^s/(1 - \nu_{12}^s\nu_{21}^s)$, $Q_{12}^s = \nu_{21}^s E_2^s/(1 - \nu_{12}^s\nu_{21}^s)$, and $Q_{66} = G_{12}^s$.

The storage strain for a lamina can be given as

$$\frac{d\epsilon^s{}_k}{dT} = [E_e]^{-1}\sigma_{mk}\frac{d\phi_f}{dT}$$

$$\frac{d\{\epsilon^s\}_k}{dT} = [E_e]^{-1}\Big([\overline{Q'}]\{\epsilon\}_k - [\overline{Q'^s}](\{\epsilon^s\}_k + \{\epsilon_T\}_k)\Big)\frac{d\phi_f}{dT} \tag{14}$$

where given matrix can be calculated as

$$[\overline{Q}]' = [T(\theta)]^{-1}[Q]'[R][T(\theta)][R]^{-1}$$

$$[\overline{Q}]'^s = [T(\theta)]^{-1}[Q]'^s[R][T(\theta)][R]^{-1}$$

The prescribed constitutive model for shape memory polymer composite is implemented on UMAT, for further finite element-based analysis of various SMPC-based structures.

3 UMAT Implementation

User material subroutine (UMAT) is a Fortran program written for finite element software ABAQUS to define underlying constitutive equation. In general, finite element software has inbuilt material libraries along with various types of material characteristics like elastic, plastic, viscoelastic, viscoelastoplastic, temperature dependent, etc. But sometimes either observed material behavior is not available or some new material type is absent from the library of FEM softwares, the need to provide material type and behavior from outside becomes important. UMAT is thus one such way to guide ABAQUS toward new material model.

An ideal UMAT requires following definitions for its correct implementation:

- Definition of Cauchy stress or true stress for user element.
- Definition of stress rate for co-rotational framework.
- Dependence of material properties on temperature, time, deformation, or any other variable.
- Solution-dependent variables denoted by SDVs. In each time increment, they get updated and used in next increment.
- Constant material properties are denoted by PROPS.
- Correct definition of variables, use of nomenclature, and strict following of Fortran conventions so that it can easily be implemented in ABAQUS.
- Definition of proper Jacobian matrix (DDSDDE) (Fig. 3).

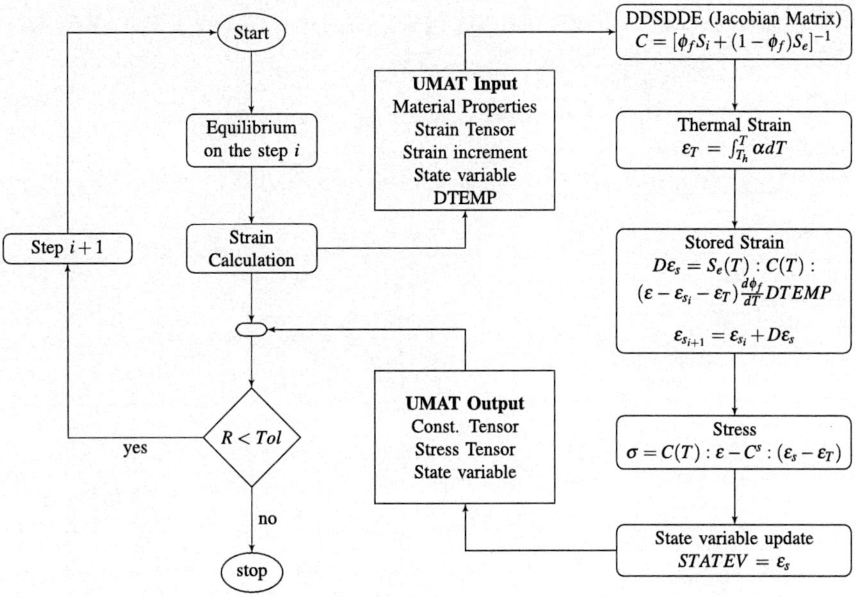

Fig. 3 Shape memory polymer-based UMAT algorithm

3.1 Finite Element Simulation Procedure in ABAQUS

To simulate a shape fixation response of a shape memory polymer-based structure in solver module of ABAQUS, the thermo-mechanical cycle is divided into four steps which include a initial step and three static-thermal steps. In the initial step, temperature is defined constant throughout the geometry using pre-defined field. Boundary conditions are also defined subsequently in the step, which will be propagated throughout all the steps involved. The pre-defined temperature field is propagated to the next step (loading step). However, in forward steps, it is modified as per the thermo-mechanical programming of material. In the second step (loading), loads are created, as per the problem definition. The step time has been taken as one second to reduce the simulation time. In the second step (cooling step), the pre-defined temperature field was modified from T_h to T_l. Another boundary condition was applied to constrain the whole geometry in the deformed position. The loads are propagated in this step. The proccss introduced an inelastic strain also known as stored strain and the updated values of stress are computed with the increase in time. The time increment was taken quite low for higher accuracy, which will be seen in the result section. In the final step (unloading step), loads and second boundary conditions applied in the previous steps are made inactive. However, the pre-defined temperature field was propagated. By the end of this step, the shape fixation simulation will be completed and the results can be checked and validated with the analytical results. In case of a shape recovery analysis, an additional step is required with same inactive boundary

Table 1 Finite element model steps, analysis parameters, and material's elastic and thermal properties

	STEPS definition				
	Initial step	Loading step	Cooling step	Unloading step	
Boundary condition	*Created*	*Propagated*	*Propagated*	*Propagated*	
Loading condition	–	*Created*	*Propagated*	*Inactive*	
Disp. constrain	–	–	*Created*	*Inactive*	
Pre-defined field	*Created* (assigned T_h)	*Propagated*	*Modified* (reduced to T_l)	*Propagated*	
Material's properties					
E_e (MPa)	n	c_f	T_g (K)	k (m^2 kg/s^2 K)	α ($K^{-1} \times 10^{-4}$)
813	4	2.76×10^{-5}	343	1.30×10^{-23}	$-3.16 + 0.0142T$
Analysis parameters					
E_f (MPa)	L (m)	b (mm)	t_{face} (mm)	t_{core} (mm)	P (N)
70	1	50	0.5	49	$0.025(EI_{eff})T_h$

conditions as the unloading step. The pre-defined temperature field is modified from T_l to T_h. Additional things such as heating and cooling rate can be added to the pre-defined field. The whole procedure is provided in tabular form in Table 1.

4 Results and Discussion

In this section, the proposed quasi-static finite element model will be implemented on an SMP sandwich cantilever beam with a concentrated load at the tip of the beam as shown in Fig. 4, to examine the shape fixation and recovery behavior of such SMP structural architecture. Basic assumptions based on Euler–Bernoulli beam theory were made with very small rate of heating and cooling to exclude the dynamic effect of inelastic strains while changing temperature.

4.1 Analysis Results

Results section will provide insight into the thermo-mechanical response of sandwich beam under linearly varying bending moment or concentrated load at the tip of the beam. The important parameters analyzed or simulated are shape fixation and recovery of the sandwich beam due to SMP core. The thermo-mechanical analysis is mainly divided into four steps which are given below:

- Displacement loading of the beam at higher temperature above glass transition temperature.
- Inelastic strain formulation while cooling the beam below glass transition temperature.
- Force unloading of the beam below glass transition temperature.
- Inelastic strain recovery while heating the beam.

The second step of thermo-mechanical cycle can be performed with constant loading condition or constrained displacement condition, a constant loading condition is analyzed in this case as it is much practical compared to the other in physical applications.

As a UMAT is formulated for analysis of SMP-based structure in ABAQUS, a validation study of the solution was required. Results for a pure SMP-based beam with earlier parameters were computed using formulated UMAT in ABAQUS and

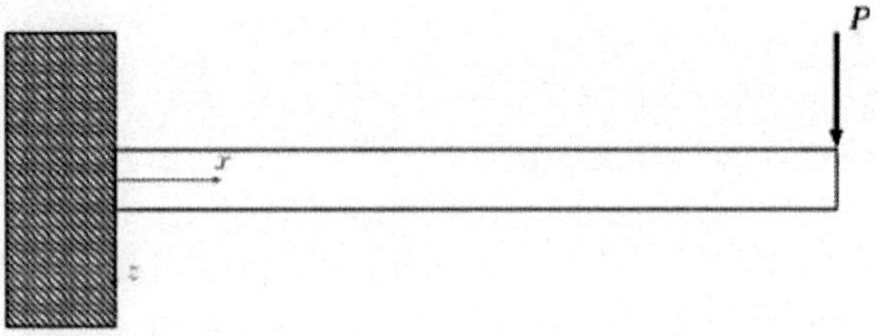

Fig. 4 Cantilever sandwich beam with tip load

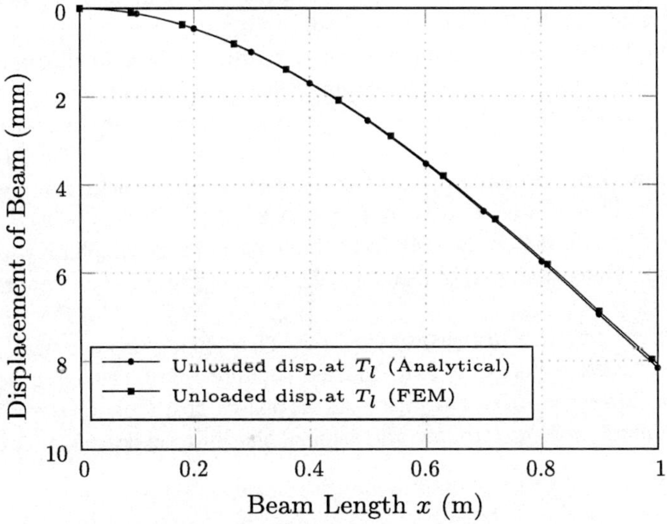

Fig. 5 Validation of shape fixation response of SMP beam computed with analytical and ABAQUS-UMAT formulation

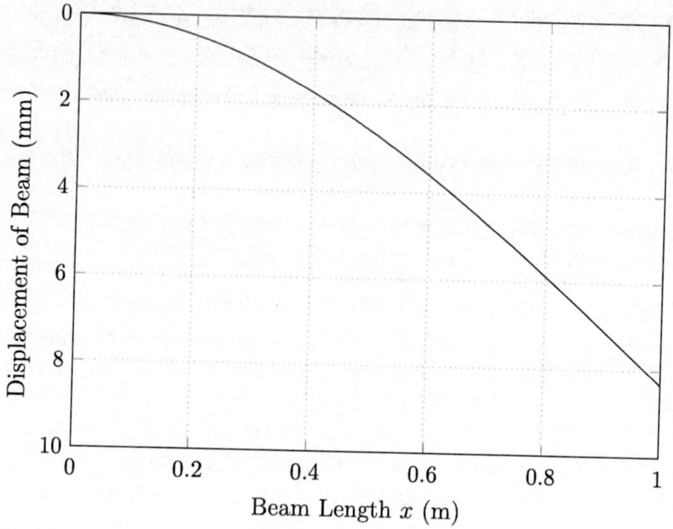

Fig. 6 Displacement response of SMP sandwich beam under tip load at temperature T_h

analytically. It can be seen in Fig. 5 that the UMAT results agree with the analytical results for SMP beam.

To initiate the thermo-mechanical analysis, the sandwich beam is loaded with a concentrated load at the tip above glass transition temperature. Deformation response at T_h is computed first which can be seen in Fig. 6. Since the applied load is a multiple of EI (flexural stiffness) at higher temperature, the shape of deformation does not change regardless of the sandwich parameters.

The beam is then cooled from temperature T_h to T_l with tip load held constant, to compute the shape fixation response under quasi-static condition. As the beam is cooled, a component of inelastic strain (storage strain and thermal strain) will be formed due to which an additional deformation can be seen in Fig. 7a, b. This inelastic strain is caused due to change in mechanical and structural properties of SMP above and below glass transition temperature.

The phase transformation of SMP from active phase to frozen phase will lead to the locking or storage of a part of mechanical or bending strain in the core region of sandwich structure.

Finally, shape recovery response of the sandwich beam is computed while heating the unloaded structure quasi-statically from temperature T_l to T_h, which causes transformation of frozen phase to active phase. As the phase transformation starts, internal modeled moment tends to zero, and the original shape of the beam is recovered as seen in Fig. 8.

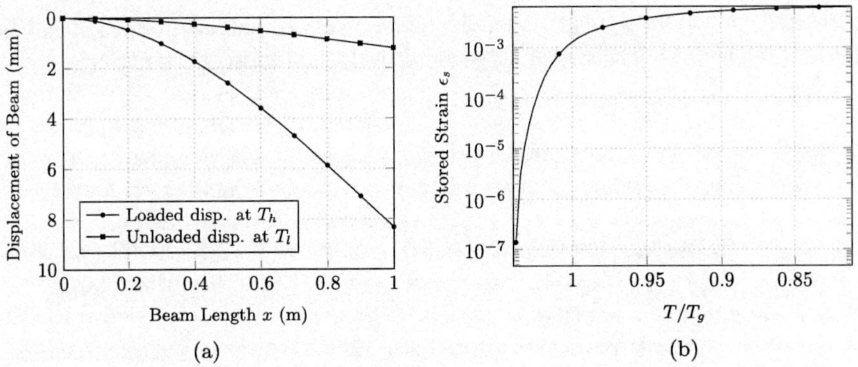

Fig. 7 Shape memory response of beam while cooling: **a** displacement of beam at various temperature; **b** bending strain at various temperatures

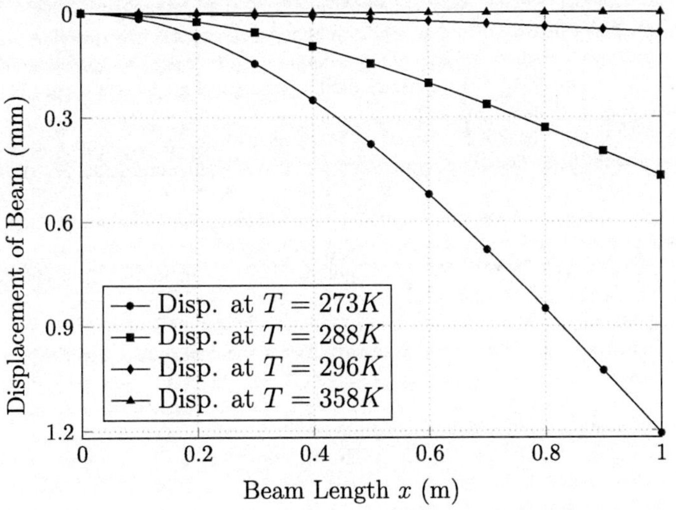

Fig. 8 Unconstrained shape recovery response of sandwich beam while heating

5 Concluding Remarks

The goal of this study was to develop an analytical and numerical framework for composite and sandwich shape memory polymer structure for engineering applications. Analytical framework is divided into three parts: constitutive model for SMP, computation of micro-mechanical properties of SMP laminate, and finally constituting the governing equation for a laminate following the general classical laminated plate theory. The proposed model is capable of analyzing any unidirectional laminate or sandwich structure based on SMP.

Model predictions were capable of highlighting the salient features of SMP composite and sandwich beam shape fixation, recovery, and spring back response.

References

1. Lendlein, A., Kelch, S.: Shape memory polymers. Angew. Chem. Int. Ed. **41**(12), 2034–2057 (2002)
2. Lendlein, A., Langer, R.: Biodegradable, elastic shape-memory polymers for potential biomedical applications. Science **296**(5573), 1673–1676 (2002)
3. Charlesby, A.: Atomic radiation and polymers: international series of monographs on radiation effects in materials. Elsevier (2016)
4. Siskind, R.D.: Model development for shape memory polymers. North Carolina State University (2008)
5. Liu, Y., Gall, K., Dunn, M.L., McCluskey, P.: Thermomechanical recovery couplings of shape memory polymers in flexure. Smart Mater. Struct. **12**(6), 947 (2003)
6. Tobushi, H., Hayashi, S., Hoshio, K., Ejiri, Y.: Shape recovery and irrecoverable strain control in polyurethane shape-memory polymer. Sci. Technol. Adv. Mater. **9**(1), 015009 (2008)
7. Ghosh, P., Srinivasa, A.R.: A two-network thermomechanical model of a shape memory polymer. Int. J. Eng. Sci. **49**(9), 823–838 (2011)
8. Ohki, T., Ni, Q.Q., Ohsako, N., Iwamoto, M.: Mechanical and shape memory behavior of composites with shape memory polymer. Compos. Part A Appl. Sci. Manuf. **35**(9), 1065–1073 (2004)
9. Bergman, D., Yang, B., Fang, H.: A quasi-static nonlinear model of shape memory polymer composite beams for space applications. In: Spacecraft Structures Conference, p. 0672 (2014)
10. Ni, Q.Q., Zhang, C.S., Fu, Y., Dai, G., Kimura, T.: Shape memory effect and mechanical properties of carbon nanotube/shape memory polymer nanocomposites. Compos. Struct. **81**(2), 176–184 (2007)
11. Liu, Y., Gall, K., Dunn, M.L., McCluskey, P.: Thermomechanics of shape memory polymer nanocomposites. Mech. Mater. **36**(10), 929–940 (2004)
12. Wang, Z.D., Li, Z.F.: Theoretical analysis of the deformation of SMP sandwich beam in flexure. Arch. Appl. Mech. **81**(11), 1667–1678 (2011)
13. Butaud, P., Foltête, E., Ouisse, M.: Sandwich structures with tunable damping properties: on the use of shape memory polymer as viscoelastic core. Compos. Struct. **153**, 401–408 (2016)
14. Liu, Y., Gall, K., Dunn, M.L., Greenberg, A.R., Diani, J.: Thermomechanics of shape memory polymers: uniaxial experiments and constitutive modeling. Int. J. Plast. **22**(2), 279–313 (2006)

Energy Absorption Characteristics of Balsa Wood Filled Aluminum Honeycomb Structures Subjected to Axial Quasi-static Loading

Nadeem Ahmad, Devendra K. Dubey, and Rohit Sankrityayan

1 Introduction

Over the last few decades, man-made honeycombs have been widely used in many industries due to their properties such as high strength to weight ratio and excellent energy absorption capabilities. Honeycombs are manufactured from materials such as Aluminum, NOMEX, polymer, and ceramic. Aluminum honeycombs are used as industrial products, as well as core materials in sandwich panels in various fields of engineering such as aerospace, automotive, and naval engineering [1]. Energy absorbers made from honeycombs are thin-walled structures which absorb kinetic energy typically through progressive folding and/or bending collapse [2]. In order to enhance the energy absorption with low value of the peak load while keeping lightweight design, low-density fillers, such as polymeric and metallic foams, have demonstrated considerable potential.

Several studies have been conducted on the out-of-plane compression of Aluminum honeycombs for quasi-static and impact cases [3–6]. Zhou and Mayer studied the effects of strain rate in out-of-plane compression scenario on Aluminum honeycombs and found that the plateau stress, increased with strain rate [6]. Apart from the studies conducted on the crushing characteristics of bare honeycomb, there have been studies on enhancing the energy absorption of honeycomb using other lightweight materials in combination with it.

As a low-density material, Polyurethane (PU) foam was first attempted by Reddy and Wall [7], and it was shown that in very thin cylindrical tubes filling with PU foam prevented irregular overall buckling and forced development of more symmetric stable buckling patterns; thus the stability of the crushing process was improved

N. Ahmad · D. K. Dubey (✉) · R. Sankrityayan
Department of Mechanical Engineering, Indian Institute of Technology, New Delhi, India
e-mail: dkdubey@mech.iitd.ac.in

S. K. Saha and M. Mukherjee (eds.), *Recent Advances in Computational Mechanics and Simulations*, Lecture Notes in Mechanical Engineering,
https://doi.org/10.1007/978-981-15-8315-5_20

by the presence of the filler. Alavi and Sadeghi conducted experiments on foam-filled Aluminum hexagonal honeycombs under the out-of-plane compression loads [8]. They observed that the crushing strength of bare honeycombs and foam-filled honeycombs increased with strain rate and bare honeycombs were more sensitive to strain rate than foam-filled honeycombs. Mozafari et al. employed ABAQUS software and observed that the mean crushing strength and energy absorption of foam-filled honeycomb were higher than the sum of those of bare honeycomb and foam [9]. Wu et al. [10] improved the mechanical properties of the honeycomb core by filling with rigid PU foam. Such a foam-filled honeycomb was used to construct sandwich panels with graphite/epoxy composite face-sheets, which were showed to have a higher impact resistance through the low-velocity impact tests.

Other than the metallic or polymeric foams, there have been studies performed on naturally occurring and biodegradable lightweight energy absorbing materials. Among these are balsa wood and cork. Balsa wood is a naturally occurring porous bio-composite that offers remarkable mechanical and physical properties. As its cellular/porous microstructure allows the application of large deformations, fine composite nano-architecture of wood cell material increases its specific strength and stiffness, giving rise to a high specific energy dissipation capacity. Balsa wood is found in a wide range of densities from 40 to 380 kg/m^3, depending on the average size and the wall thickness of cells, which provides the flexibility in design since the strength is a monotonic function of its density [11]. According to Borrega and Gibson, there are no engineered materials suitable for sandwich panel cores with a similar combination of mechanical properties and low-density as balsa wood [12]. Atas and Sevim investigated low-velocity impact response of sandwich composite panels made of balsa wood cores and glass/epoxy face-sheets [13]. They concluded that balsa wood was able to absorb more energy for less deformation.

Unlike the honeycomb filled with foams, a study on the honeycomb structures combined with balsa wood or cork has not been conducted. In this work, an experimental and numerical study on the energy absorption characteristics of bare honeycomb and honeycomb filled with balsa wood in different configurations have been carried out. The out-of-plane quasi-static compression tests have been carried on bare honeycomb and two different styles of filled honeycombs. Subsequently, the experimental results have been used to validate the finite element analysis of quasi-static simulations.

2 Experimental Setup

The Out-of-plane quasi-static crushing tests were performed on the bare and filled honeycomb panels at room temperature in Shimadzu Universal testing machine with a loading capacity of 10 KN (Fig. 1). The specifications of honeycombs used in the experiments are mention in Table 1. Each honeycomb core sample tested, consists of 25 cells. The specimen was placed in between the support and loading platen

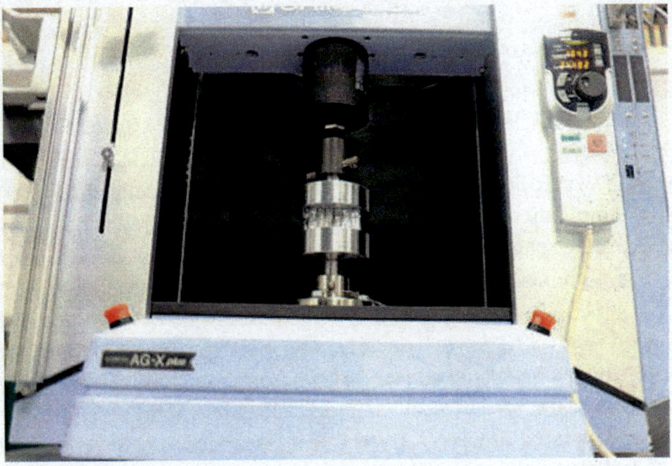

Fig. 1 Shimadzu universal testing machine used for carrying out compression tests

Table 1 Geometrical and physical properties of Aluminum honeycomb

Designation	Material	Cell size (mm)	Foil thickness (μm)	$\rho_{honeycomb}$ (kg/m^3)	Panel height (mm)
H37	Al 3003	12	80	37	30.8

without any fixture and was loaded at a constant rate of 2 mm/min. Load and displacement data were directly recorded by an automatic data acquisition system, and then nominal stress (defined as force over the original cross-sectional area) and nominal strain (defined as displacement over the original thickness of the specimen) were calculated. Stress and strain data were subsequently used to calculate the densification strain (critical strain when the cell walls are squeezed together), peak strength (corresponding to the peak load in force-displacement curve over the cross-sectional area), plateau strength (average taken from the peak stress to the densification strain), absolute energy absorption (AEA), and specific energy absorption (SEA).

3 Materials

Aluminum honeycombs used in the experiments were manufactured with Al 3003 H19 alloy. Due to the unavailability of Al 3003 properties, approximate material properties were acquired from the literature [14] as $\sigma_Y = 195$ MPa and $\sigma_U = 211$ MPa and plastic stress-strain data as shown in Table 2.

Experiments were carried out on bare honeycombs, and two different configurations of balsa wood filled honeycombs. In Type A filled honeycomb, all the cells of

Table 2 True plastic stress-strain data for Al 3003 H19 [14]

Strain	0	0.006	0.011	0.018
Stress (MPa)	195.0	209.8	215.4	216.0

the honeycomb were filled with wood. In type B, only the cells along the periphery of the honeycomb panel were filled. Figure 2 shows the bare, Type A, and Type B filled honeycombs. It should be noted that the wood fillings have fibers perpendicular to the length, which means, the compression is carried out perpendicular to the fiber direction.

Balsa wood with 112 kg/m^3 density was used to fill the honeycomb cells. Although, various wood species differ largely from one another, the cell wall material of the wood is the same for all of them. Density, modulus and yield strength of the cell wall material are 1469 kg/m^3, 38 GPa and 120 MPa, respectively [11]. The mechanical properties of balsa wood have been found to be largely dependent on its density [12]. Young's modulus in the longitudinal direction (EL), compressive strength in the longitudinal direction (σ_L), and tangential direction (σ_T) are 2.5 GPa, 8 MPa, and 0.8 MPa, respectively [15]. Other properties of balsa wood used in simulations are mentioned in Table 3.

Quasi-static compression tests were carried out in the tangential direction on balsa wood of cross-section of 10 mm by 10 mm and height 20 mm to obtain the compression behavior of the wood. The stress-strain graph obtained is shown in Fig. 3.

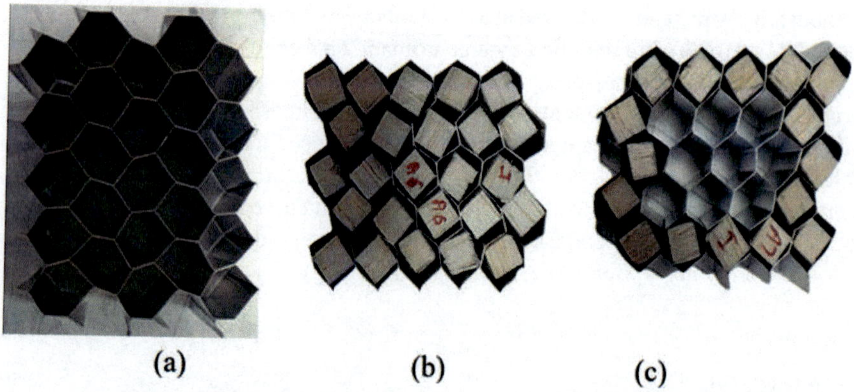

Fig. 2 Honeycomb samples filled with balsa wood, **(a)** bare honeycomb, **(b)** type A filled honeycomb, and **(c)** type B filled honeycomb

Table 3 Elastic ratios for Balsa [15]

E_T/E_L	E_R/E_L	G_{LR}/E_L	G_{LT}/E_L	G_{RT}/E_L	μ_{LT}
0.015	0.046	0.054	0.037	0.005	0.488

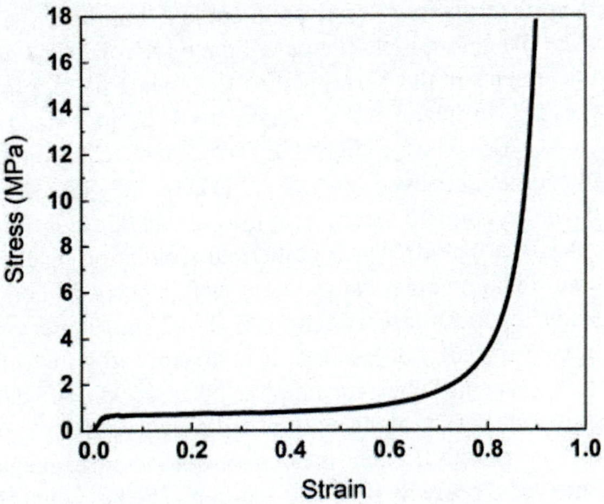

Fig. 3 Experimental result obtained from compression test of balsa wood in tangential direction

4 Finite Element Modeling for Quasi-static Simulation

Finite Element (FE) modeling of Aluminum honeycombs were carried out using LS-DYNA [16] package. Mass scaling and time scaling techniques were used to simulate quasi-static condition. Figure 4 shows representative FE models of the bare, Type A, and Type B filled honeycombs.

Belytschko-Tsay Shell 163 elements with four integration points were employed to simulate the honeycomb cell walls for high computational efficiency [17]. In each honeycomb cell, four oblique walls are modeled with single wall thickness while two vertical walls with double wall thickness. Since, the time-step for shell element is directly proportional to its mass and characteristic length, Ls, therefore, for increasing time-step, mass scaling is employed. For implementing mass scaling in LS-DYNA, CONTROL_TIMESTEP card was used.

Aluminum honeycomb walls were simulated using a PIECEWISE LINEAR PLASTICITY material model, without strain rate sensitivity for foil material. Input

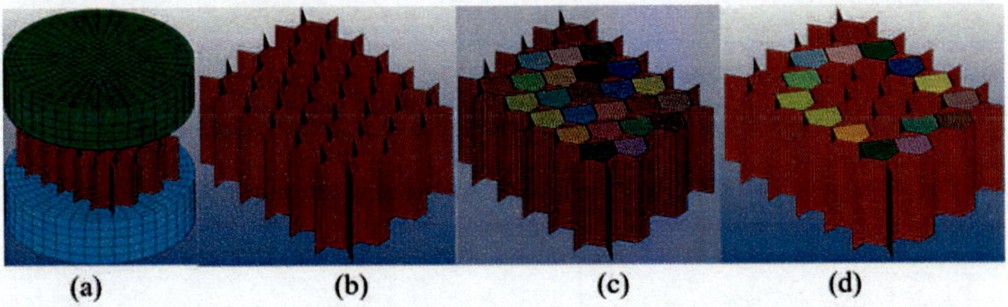

Fig. 4 FE model of **(a)** bare honeycomb with impactor (top), honeycomb (middle), and fixed support (bottom), **(b)** only bare honeycomb **(c)** type A filled honeycomb, and **(d)** type B filled honeycomb

material properties are given in Table 2. For the lower platen, all degrees of freedom were fixed. For upper platen, all the three rotational movements and two transitional movements in the X and Y directions were fixed. The upper platen was kept free to move in the negative Z direction at a constant velocity to compress.

AUTOMATIC_SURFACE_TO_SURFACE contacts were employed between platens and honeycomb, and AUTOMATIC_SINGLE_SURFACE was provided for the honeycomb to incorporate the contact during folding of the cell wall. Mesh characteristics can affect the numerical simulation results as it is important in building up a reliable numerical model. For this purpose, convergence and mesh sensitivity study for mesh size variation of 0.5–2 mm is performed for the meshed models. In case of bare honeycomb, it is discretized with 2, 1.4, 1, 0.73, and 0.5 mm size shell elements. Stress-strain plots corresponding to different element mesh sizes are shown in Fig. 5, along with resulting curve from experiments. It is observed that, FE model with 0.5 mm mesh shows good agreement with the experimental results followed closely by 0.73 and 1 mm meshed models. However, the time required for 0.5 and 0.73 mm meshed models are very large as compared to 1 mm meshed model. Significant fluctuations in stress values are observed in the case of 1.4 mm meshed model and in case of 2 mm meshed model. Hence, mesh size of 1 mm is considered for further studies on Type A and Type B filled honeycombs.

In the FE model, balsa wood was modeled with a constant stress solid element. The material model chosen for balsa wood was MAT_HONEYCOMB. MAT_MODIFIED_HONEYCOMB and MAT_WOOD were the other two material models used for modeling wood in FEA. Among these three, MAT_HONEYCOMB has been noticed to be the fastest and reasonably accurate in modeling compression of balsa in a confined space [18].

The contact between the impactor/support and the wood filling were modeled by AUTOMATIC_SURFACE_TO_SURFACE contacts. The same was used to model contact between wood and honeycomb. The size of the wood filling was

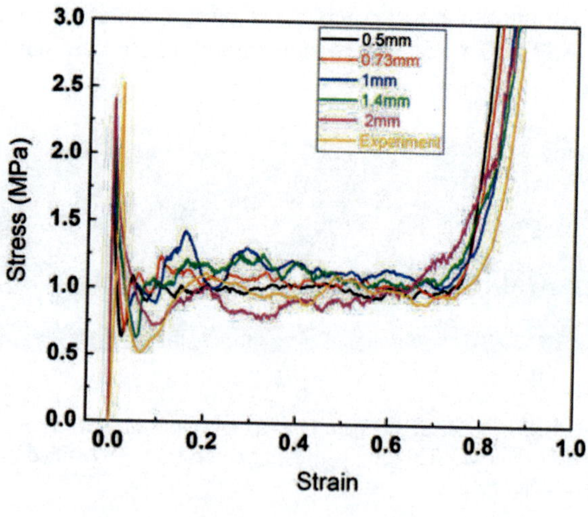

Fig. 5 Comparison between stress-strain graphs of 0.5, 0.73, 1, 1.4, and 2 mm meshed model for bare honeycomb case

kept slightly smaller than the cells of honeycomb to avoid initial penetration. The self-contact of the wood filling during buckling was modeled with AUTOMATIC_SINGLE_SURFACE contact algorithm. Mesh sizes used for compression of Type A honeycomb, for both honeycomb as well as wood, were 0.5, 0.73, 1, and 1.4 mm.

5 Results and Discussion

In this section detailed analysis of the experimental, as well as FE results are presented in terms of compression and energy absorption characteristics of the bare and filled honeycombs.

5.1 Experimental Results

Quasi-static compression of two types of wood filled honeycomb has been investigated along with out-of-plane quasi-static compression of bare honeycomb. Stress-strain behavior of the bare, Type A, and Type B honeycombs is obtained and shown in Fig. 6. The graphs in all the three cases, show three regions of elastic compression (first peak), progressive buckling (plateau region), and densification (final ramping up of stress), as demonstrated by a typical honeycomb in compression (Fig. 7).

It is observed from Fig. 8a, that the SEA (Specific Energy Absorption) in case of both Type A and Type B honeycombs decreases compared to bare honeycomb case. Furthermore, the SEA of Type A is noticed to be lower than that of Type B. The reason for this decrease in SEA could be due to the relatively high density of balsa

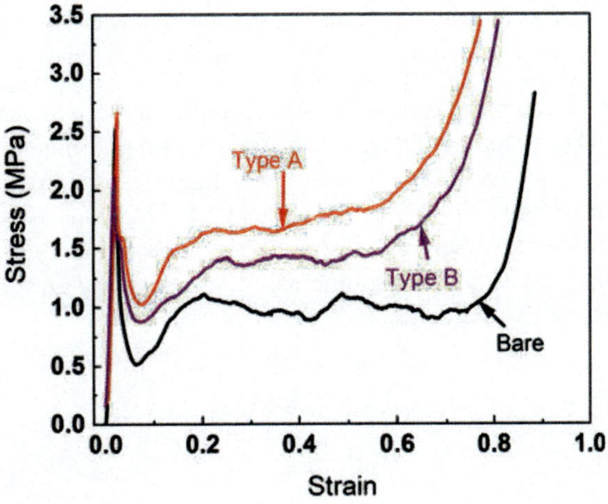

Fig. 6 Comparison between stress-strain graphs of bare, type A filled, and type B filled honeycombs

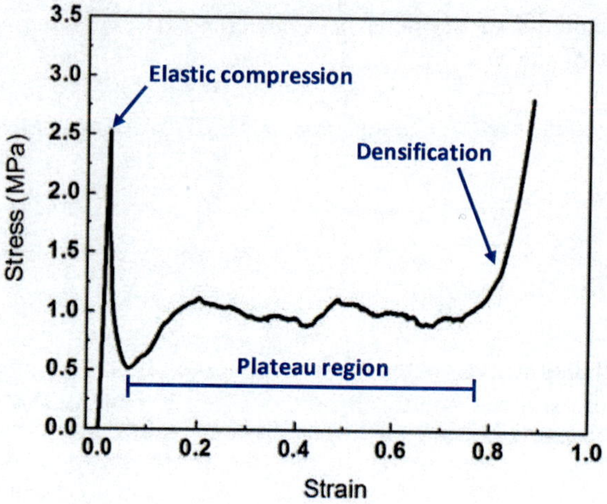

Fig. 7 Stress-strain curve for bare Aluminum honeycomb labeled with different regions of the curve

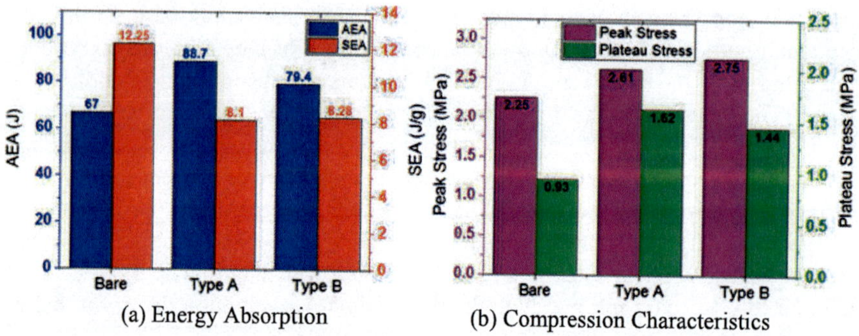

Fig. 8 **(a)** Energy absorption behavior and **(b)** compression characteristic of bare, type A, and type B filled honeycombs (AEA: Absolute Energy Absorption, SEA: Specific Energy Absorption)

wood compared to honeycomb. A combination of lower density of balsa wood with the honeycomb of present density may lead to different results.

As shown in Fig. 8b, there is a sharp increase in the plateau stress of the filled honeycombs (74% for Type A and 55% for Type B) when compared to the bare honeycomb case. Peak stress for both type A and Type B honeycomb is nearly the same and is higher than the bare honeycomb by 16% and 22.22% for type A and type B, respectively.

5.2 FEA Results for Quasi-static Simulation

The experimental results given in the previous section have been used to validate quasi-static simulations for bare, Type A, and Type B honeycombs. Figures 9, 10, and 11 shows the comparison between FE and experimental results for bare, Type A, and Type B honeycombs, respectively.

FE Simulations were performed using LS-DYNA software. As the peak stress is very short-lived, the slight difference in the experimental and the FE peak stress may be due to the difference in the frequency of capturing the output data, which is many times higher in the experimental result than the FE result. The plateau region of FE result for bare honeycomb shows good agreement with the experimental result, see Fig. 9. In the case of Type A and Type B filled honeycombs, FE results slightly

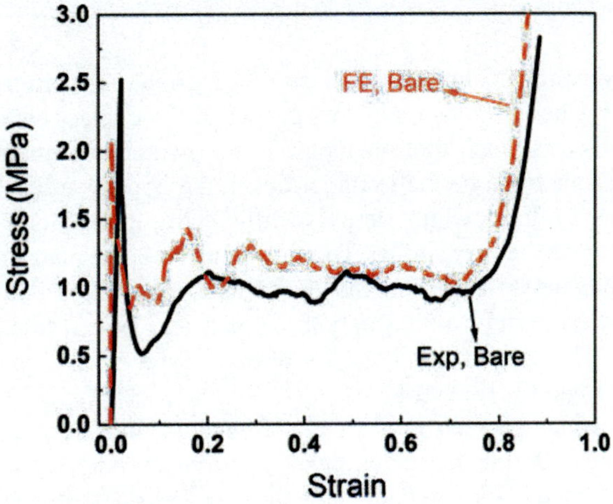

Fig. 9 Stress-strain curve showing comparison between experimental and FE compression result for bare honeycomb

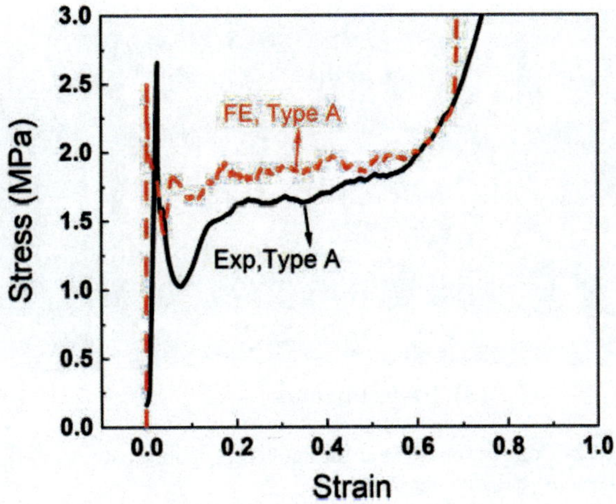

Fig. 10 Stress-strain curve showing comparison between experimental and FE compression result for type A honeycomb

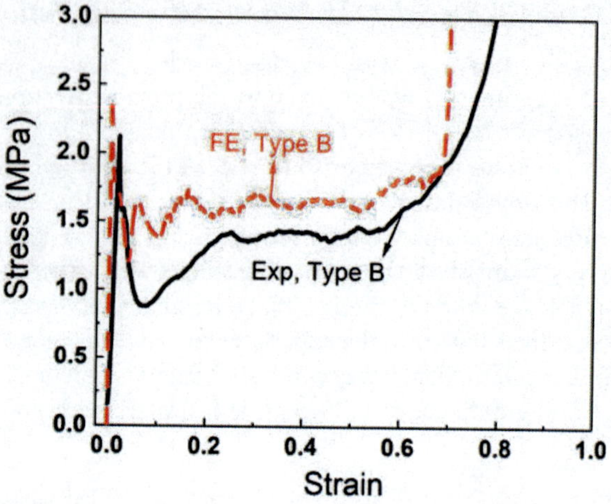

Fig. 11 Stress-strain curve showing comparison between experimental and FE compression result for type B honeycomb

overestimate the plateau stress as compared to experimental results, see Figs. 10 and 11. The difference may be because of the contact between wood and honeycomb, which is much more complex to be accurately simulated in the FE model. Other reason could be the irregularities in the honeycomb cell shapes, and the shapes of wood fillings, which were also difficult to control. However, the trend in the curves is found to be very similar. The slow increase in the plateau stress as the strain increases is captured well in both the FE results. Densification also starts approximately around the same strain values in both, Type A and Type B filled cases.

The deformation behavior of bare, Type A, and Type B honeycombs are shown in Figs. 12, 13, and 14.

It can be seen in Figs. 12, 13, and 14, that honeycomb cell walls deformation pattern during the beginning of experiment is not like the deformation characteristics obtained in FE simulations. In the experiment, plastic deformation starts somewhere

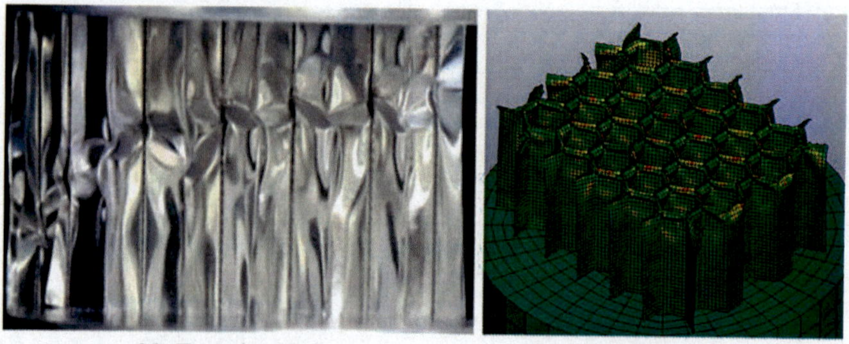

(a) Experimental (b) FEA

Fig. 12 Comparison between the initial deformation pattern of experiment (**a**) and FE compression (**b**) results for bare honeycomb

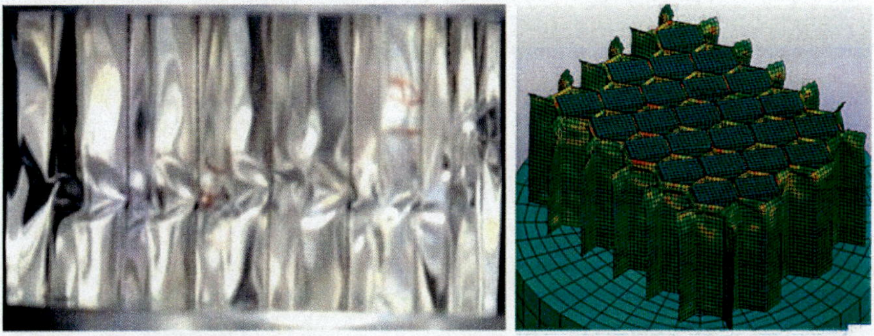

(a) Experimental (b) FEA

Fig. 13 Comparison between the initial deformation pattern of experiment (**a**) and FE compression (**b**) results for type A honeycomb

(a) Experimental (b) FEA

Fig. 14 Comparison between the initial deformation pattern of experiment (**a**) and FE compression (**b**) results for type B honeycomb

in the middle while buckling begins from one of the interfaces (moving or stationary) in the FE model. The difference in the deformation behavior could be due to the adhesive failure in the experimental case while in the FE model, perfect adhesive bonding between the walls was assumed. Hence, in the experimental tests, material starts to fail from the weaker region, i.e., from the failure of the adhesive bond. The unavailability of the actual material properties of the Aluminum honeycombs could be another reason for the difference in the initial deformation pattern.

The final deformed shapes for experimental and FE results of bare, Type A, and Type B honeycombs are shown in Figs. 15, 16, and 17. The final deformed shapes show a good agreement between the experimental result and FE simulation results for all the three cases. In both FE and experimental results, outer cell walls of honeycombs are deformed irregularly, predominantly in bare honeycomb case. In all the three cases, the inner walls show more regular deformation.

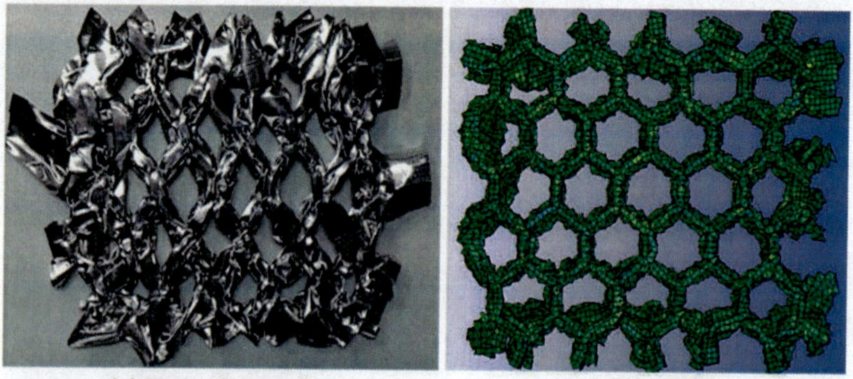

(a) Experimental (b) FEA

Fig. 15 Comparison between the final deformed shape of experimental and FE compression results for bare honeycomb

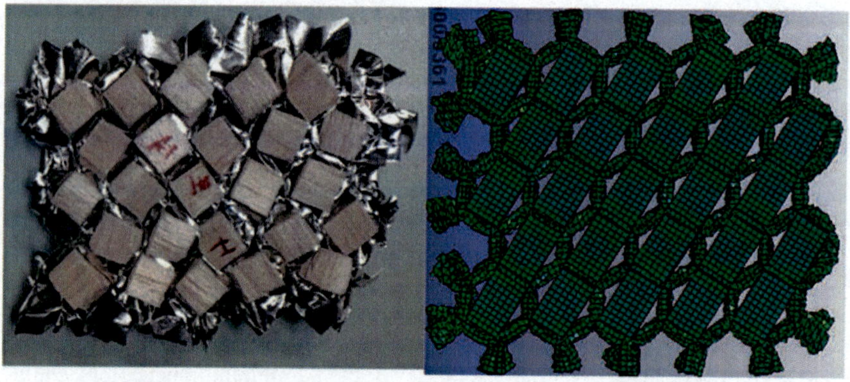

(a) Experimental (b) FEA

Fig. 16 Comparison between the final deformed shape of experimental and FE compression results for type A honeycomb

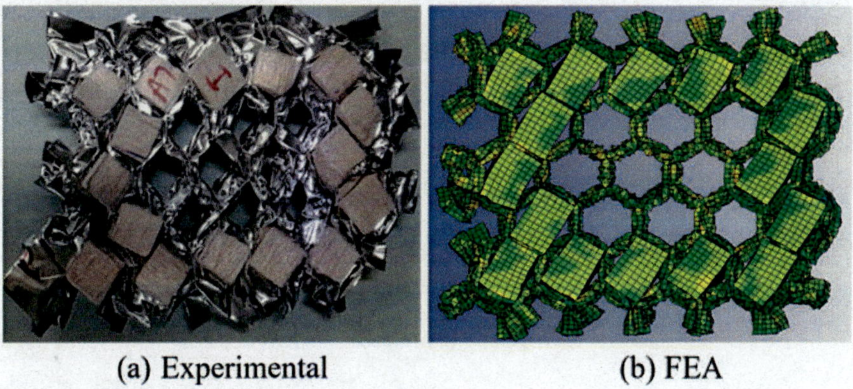

Fig. 17 Comparison between the final deformed shape of experimental and FE compression results for type B honeycomb

6 Conclusion

Present work focuses on experimental and numerical analysis of bare and balsa wood filled honeycombs for comparison in the energy absorption characteristics under compressive loading for two different configurations. In the first configuration, (Type A) balsa wood is filled to each cell of the honeycomb (Fig. 2b), and in the second configuration (Type B) balsa wood is filled in the cells along the periphery of the honeycomb panel (Fig. 2c). The FE simulation under quasi-static condition is validated using experimental results. The FE results show good agreement with experimental results (Figs. 9, 10, and 11). The slight overestimation of the plateau stress in the case of Type A and Type B filled honeycombs may be because of the following two reasons. Firstly, the contact between wood and honeycomb is much more complex than the contact algorithm used in the present FE model. The irregularities in the shapes of honeycomb and wood filling may be the second one.

The energy absorption behavior, in terms of absolute energy absorption and specific energy absorption, is compared for bare and filled honeycombs in Fig. 8a. It is observed that AEA for filled honeycombs is higher than bare honeycomb by 32 and 18.51% for Type A and Type B honeycombs, respectively. This increase shows the possibility of using balsa wood filled honeycomb in areas where significant energy absorption along with larger in-plane stiffness of the sandwich core is required. Specific energy absorption, however, decreases for balsa filled honeycombs as compared to the bare honeycomb. This can be attributed to the relatively higher density of balsa wood, 112 kg/m^3 with respect to the density of present honeycomb 37 kg/m^3. The compression characteristics, in terms of peak stress and plateau stress in Fig. 8b, shows that the balsa filled honeycomb exhibit higher plateau stress as compared to the bare honeycomb by about 74% and 55% for Type A and Type B, respectively.

Experimental and FE deformation behavior of bare and filled honeycomb for all the cases is shown in Figs. 12, 13, and 14. The initial deformation of honeycomb during experiments show failure around the middle, while in the results from FE analysis, it begins from the interface (moving or stationary). The adhesive failure of the honeycomb walls and the unavailability of material properties of the Aluminum honeycomb material are the possible reasons for the difference. The final deformed shapes from the FE results are quite like that of the experiments with outer walls irregularly deformed while the inner ones deformed regularly. Overall, this study contributes towards developing an understanding of energy absorption characteristics of light wood (balsa) filled Aluminum honeycomb materials.

References

1. Gibson, L.J., Ashby, M.F.: Cellular Solids: Structure and Properties, 2nd edn. Cambridge University Press (2014)
2. Liu, Q., Ma, J., He, Z., Hu, Z., Hui, D.: Energy absorption of bio-inspired multi-cell CFRP and aluminum square tubes. Compos. Part B Eng. **121**, 134–144 (2017)
3. Baker, W.E., Togami, T.C., Weydert, J.C.: Static and dynamic properties of high-density metal honeycombs. Int. J. Impact Eng. **21**(3), 149–163 (1998)
4. Liu, Q., et al.: Axial and lateral crushing responses of aluminum honeycombs filled with EPP foam. Compos. Part B Eng. **130**, 236–247 (2017)
5. Wu, E., Jiang, W.-S.: Axial crush of metallic honeycombs. Int. J. Impact Eng **19**(5–6), 439–456 (1997)
6. Zhou, Q., Mayer, R.R.: Characterization of aluminum honeycomb material failure in large deformation compression, shear, and tearing. J. Eng. Mater. Technol. **124**(4), 412–420 (2002)
7. Reddy, T.Y., Wall, R.J.: Axial compression of foam-filled thin-walled circular tubes. Int. J. Impact Eng **7**(2), 151–166 (1988)
8. Alavi Nia, A., Sadeghi, M.Z.: An experimental investigation on the effect of strain rate on the behavior of bare and foam-filled Aluminium honeycombs. Mater. Des. **52**, 748–756 (2013)
9. Mozafari, H., Khatami, S., Molatefi, H.: Out of plane crushing and local stiffness determination of proposed foam filled sandwich panel for Korean Tilting Train express—numerical study. Mater. Des. **66**, 400–411 (2015)
10. Wu, C.L., Weeks, C.A., Sun, C.T.: Improving honeycomb-core sandwich structures for impact resistance. J. Adv. Mater. **26**(4), 41–47 (1995)
11. Borrega, M., Ahvenainen, P., Serimaa, R., Gibson, L.: Composition and structure of balsa (Ochroma pyramidale) wood. Wood Sci. Technol. **49**, 403–420 (2015)
12. Borrega, M., Gibson, L.J.: Mechanics of balsa (Ochroma pyramidale) wood. Mech. Mater. **84**, 75–90 (2015)
13. Atas, C., Sevim, C.: On the impact response of sandwich composites with cores of balsa wood and PVC foam. Compos. Struct. **93**, 40–48 (2010)
14. Gattas, J.M., You, Z.: The behavior of curved-crease fold cores under low-velocity impact loads. Int. J. Solids Struct. **53**, 80–91 (2015)
15. Wood Handbook. Wood as an Engineering Material. General Technical Report FPL: GTR-190 (2010)
16. Livermore Software Technology Corporation: LS DYNA Keywords User's Manual, Version 971. Livermore Software Technology Corporation. Livermore, CA (2016)
17. Ayman Ashab, A.S.M., Ruan, D., Lu, G., Bhuiyan, A.A.: Finite element analysis of aluminum honeycombs subjected to dynamic indentation and compression loads. Materials (Basel) **9**(3), 162 (2016)

18. Maillot, T., Lapoujade, V., Grippon, E., Toson, B., Bardon, N., Pesque, J.-J.: Comparative study of material laws available in LS-DYNA to improve the modeling of Balsa wood. In: 13th International LS-DYNA Users Conference (2014)

Recent Advances in Computational Fluid Dynamics

Simulation of Flow and Transport Process—Scope of Meshless Methods

Aatish Anshuman and T. I. Eldho

1 Introduction

Groundwater is a vital resource of water as it is widely used for drinking, agricultural and industrial purposes. Due to overuse/pumping of groundwater, its availability is decreasing continuously. Apart from that, it is being contaminated due to various sources of pollution such as waste storage facilities, industries and sanitary landfills. Transport of contaminants in groundwater is incredibly complex which includes advection, dispersion, adsorption and single or, multispecies reactions. Therefore, prediction/assessment of quality and quantity of groundwater in an aquifer is essential for efficient management of the resource. Groundwater flow and contaminant transport simulation is an important step in this direction which enables the prediction of groundwater availability fate of contaminants.

Flow and contaminant transport modelling in the aquifer requires solving partial differential equations derived from conservation of mass and other applicable equations. Analytical solutions to these equations can be obtained for simplistic cases. However, field aquifers often encountered with irregular boundaries, complex boundary conditions, heterogeneity, anisotropy, recharge zones and pumping/injection wells. Hence, numerical modelling is done to estimate head and concentration in these problems. Conventionally, these are modelled using grid/mesh-based methods such as Finite Difference (FDM) and Finite Element Method (FEM). Grid/mesh creation necessitates expensive computations. Mesh/grid refinement is required nearby the high-stress areas such as pumps/injection well locations. Further, contaminant transport modelling using these methods is prone to instabilities such as numerical dispersion and artificial oscillation [14]. One of the

A. Anshuman · T. I. Eldho (✉)
Department of Civil Engineering, Indian Institute of Technology Bombay, Mumbai, India
e-mail: eldho@civil.iitb.ac.in

S. K. Saha and M. Mukherjee (eds.), *Recent Advances in Computational Mechanics and Simulations*, Lecture Notes in Mechanical Engineering,
https://doi.org/10.1007/978-981-15-8315-5_21

techniques to stabilize this issue is adaptive meshing or, re-meshing [5] in which the mesh is modified at each time step thus increasing the computational effort further. Apart from that, the introduction of artificial oscillation, crosswind diffusion and upwinding are some of the techniques for improving the stability in mesh-based methods [5].

Meshless methods are alternatives to solving flow and contaminant transport problems. These methods eliminate the use of mesh which reduces the computational effort and adds adaptability to the models. The models using these methods are actively being developed for different field and hypothetical problems. Based on the formulation of governing equations, these methods are either strong or, weak form-based. Further, these methods use some kind of basis functions for interpolation of state variables in the domain. Some examples of weak form-based techniques include Radial Point Interpolation Method (RPIM) [10], Element Free Galerkin Method (EFGM) [13] and Meshless Petrov Galerkin method (MLPG) [16]. An example of a strong form meshless method is Point Collocation Method (RPCM) [1–3, 11, 12, 15]. This method uses Radial Basis Functions (RBF) for interpolation of shape functions and their derivatives. These shape functions are used to discretize the governing equations and boundary conditions in the problem domain. Kansa used this technique for the very first time to solve partial differential equations [8, 9]. This technique is further improved with local support domains which eliminate the ill-conditionality issue associated with it. Meenal and Eldho [11, 12] used rectangular support domain which required dummy nodes around the boundary for uniformity number of nodes in support domains. Singh et al. [15] used circular support domains which does not necessitate creation of dummy nodes. Anshuman et al. [1] extended the application to reactive transport problems and performed sensitivity analysis with respect to the model parameters. Further, reactive simulation of decay chain in porous media is developed and verified against RT3D [2, 3].

2 Methodology

2.1 Governing Equations

2.1.1 Flow in Confined and Unconfined Aquifers

The flow in a confined aquifer is governed by the following equation [4].

$$\frac{\partial}{\partial x_i}\left[T_{x_i}\frac{\partial h}{\partial x_i}\right] = \mathrm{S}\frac{\partial h}{\partial t} + Q_w\delta \tag{1}$$

where h is the head in groundwater, T_{x_i} is the transmissivity, S is the storativity, t is time, Q_w is the pumping rate and δ is the Kronecker delta function. The governing Eq. (1) is subjected to the following initial and boundary conditions:

$$h(x_i, 0) = h_1 \quad x_i \in \Omega \tag{2}$$

$$h(x_i, t) = h_2 \quad x_i \in \partial\Omega_1 \tag{3}$$

$$T\frac{\partial h}{\partial n} = q_1 \quad x_i \in \partial\Omega_2 \tag{4}$$

where h_1 is the known head in the problem domain Ω at $t = 0$. h_2 and q_1 are known head and flux along the boundaries $\partial\Omega_1$ and $\partial\Omega_2$, respectively. Similarly, the flow in unconfined aquifers is governed by Eq. (5) [4].

$$\frac{\partial}{\partial x_i}\left[k_{x_i}\frac{\partial h}{\partial x_i}\right] = \mathrm{S_y}\frac{\partial h}{\partial t} + Q_w\delta - q \tag{5}$$

where k_{x_i} is the hydraulic conductivity and $\mathrm{S_y}$ is the specific yield of the aquifer. For an unconfined aquifer, the initial and boundary conditions are given by [4]

$$h(x_i, 0) = h_1 \quad x_i \in \Omega \tag{6}$$

$$h(x_i, t) = h_2 \quad x_i \in \partial\Omega_1 \tag{7}$$

$$k_{x_i}h\frac{\partial h}{\partial n} = q_1 \quad x_i \in \partial\Omega_2 \tag{8}$$

2.1.2 Contaminant Transport

Due to the presence of hydraulic head gradient, the contaminants in the groundwater are migrated from one location to another. The governing equation for transport of reactive contaminant species in groundwater is given by [4].

$$R\frac{\partial C}{\partial t} = \frac{\partial}{\partial x_i}\left[D_{xx_i}\frac{\partial C}{\partial x_i}\right] - \frac{\partial\left(V_{x_i}C\right)}{\partial x_i} - R\lambda C + \frac{Q_w C}{n_e} \tag{9}$$

where C is the concentration of the contaminant species, n_e is the porosity of the medium, Q_w is the pumping rate, D_{xx_i} is the dispersion coefficient and V_{x_i} is the seepage velocity. The reactions related to the species are denoted by R and λ which are retardation factor and order decay rate, respectively. The initial and boundary conditions associated with the governing Eq. (9) are given by

$$C(x_i, 0) = 0 \quad x_i \in \Omega \tag{10}$$

$$C(x_i, t) = C_1 \quad x_i \in \partial\Omega_1 \tag{11}$$

$$\left[D_{xxi}\frac{\partial C}{\partial x_i}\right]n_{xi} = g_2 \quad x_i \in \partial\Omega_2 \tag{12}$$

In a coupled environment, the velocities and dispersion coefficients in Eq. (9) are obtained with the help of a groundwater flow model as shown in Eqs. (13) and (14).

$$V_{x_i} = -\frac{k_{x_i}}{n_e}\frac{\partial h}{\partial x_i} \tag{13}$$

$$D_{xx_i} = \frac{\sum \alpha_{x_i} V_{x_i}^2}{\bar{V}} \tag{14}$$

where α_{x_i} is the dispersivity and $\bar{V}$ is the resultant velocity.

2.2 *Radial Point Collocation Method*

In RPCM, a set of scattered nodes are used for the representation of the domain. For approximation of shape functions and their derivatives, intersecting local support domains are formed around all nodes. Shape function interpolation is carried out using radial basis functions (RBF) and polynomial basis functions (PBF) [10]. For instance, the state variable C in Eq. (9) can be written as a combination of n number of RBFs (R) and m number of polynomial basis functions (PBFs) as follows [10].

$$C(\mathrm{x_i}) = \sum_{i=1}^{n} R_i a_i + \sum_{j=1}^{m} p_j b_j \tag{15}$$

where a and b are the unknown coefficients computed by enforcing C to pass through the nodes in the local support domain. A total of n number of equations are generated with $n + m$ variables due to the above-mentioned condition. The following m constraint conditions can be used to obtain the remaining m equations.

$$\sum_{i=1}^{m} p_j(x_i)a_i = 0,\ j = 1, \ldots, m \tag{16}$$

Here multi-quadrics radial shape functions (MQ-RBF) are used which are given as [7]

$$R_i = (r_i^2 + c_s^2)^q \tag{17}$$

The terms q and c_s are referred to as MQ-RBF shape parameters. In standard MQ-RBF, q is fixed as 0.5 [10]. However, studies have demonstrated that fixing the value to 0.98 or 1.03 gives good results for fluid- and structure-related problems [10]. The parameter c_s is represented as a factor of nodal distance [10]. The method for calculating d_c for different nodal distributions is presented in Liu and Gu [10]. The term r_i denotes the distance between the point of interest and ith node in the support domain. The procedure for solving the system of Eqs. (15) and (16) is explained in the previous studies [2, 3, 10]. Here, the use of PBF is avoided as satisfactory results have been reported using MQ-RBFs [1, 3, 11, 12, 15]. The shape functions (ϕ) are derived at n nodes in the support domain by solving Eqs. (15) and (16). The state variable C is approximated at a node as follows:

$$C(x, y) = \sum_{i=1}^{n} \phi_i C_i \tag{18}$$

Similarly, the shape function derivatives can be employed for computation of derivative of state variables at all nodes. The point collocation method is then used for discretization of governing. For discretization in time domain in Eq. (9), the forward difference scheme [14] as shown in Eq. (20) is implemented.

$$\frac{\partial C}{\partial t} = \frac{C_i^{t+\Delta t} - C_i^t}{\Delta t} \tag{19}$$

The concentration C in Eq. (7) is replaced by the semi-implicit formulation as follows.

$$C = \theta C^{t+\Delta t} + (1-\theta)C^t \tag{20}$$

Here, θ is a temporal weighting parameter which ranges between 0 and 1 [14]. Here, $\theta = 0.5$ is used which is known as Crank–Nicolson scheme. It is a fairly stable, accurate and widely used method [1–3, 15]. Using Eqs. (19) and (20), Eq. (9) can be expressed as follows:

$$\begin{aligned} &\left[\frac{R\phi}{\Delta t} - \theta\left\{D_{xx_i}\frac{\partial^2\phi}{\partial x^2} - V_{x_i}\frac{\partial\phi}{\partial x} - \lambda\phi\right\}\right]C_i^{t+\Delta t} \\ &= C_i^t\left[\frac{R\phi}{\Delta t} + (1-\theta)\left\{D_{xx_i}\frac{\partial^2\phi}{\partial x^2} - V_{x_i}\frac{\partial\phi}{\partial x} - \lambda\phi\right\}\right] + \frac{q_w C'}{n_e} \end{aligned} \tag{21}$$

Equation (21) is applied at the internal nodes in the domain Ω which yields a system of equations. For the nodes on the Dirichlet boundary $\partial\Omega_1$ (see Eq. (11)), the discretization is implemented as

$$\phi C_i = C_1 \tag{22}$$

Here, the Neumann type boundary ($\partial\Omega_2$) is discretized using direct collocation approach given by

$$D_{xx}\frac{\partial\phi}{\partial n}C_i = g_1 \tag{23}$$

For impervious boundaries, the fluxes (g_1) are set to zero. Combining Eqs. (21), (22) and (23), for N boundary and internal nodes, the following system of equations will be obtained.

$$[K]_{N\times N}\{C\}_{N\times 1}^{t+\Delta t} = [F]_{N\times N}\{C\}_{N\times N}^{t} + \{Q\}_{N\times 1} \tag{24}$$

The system matrix K is sparse as it is formed using local support domains. The sparsity of the matrix decreases with increase in the number of nodes in the support domain. The governing equations for confined (Eq. (1)) and unconfined aquifer (Eq. (5)) and the corresponding boundary can also be discretized in a similar manner using the shape functions and its derivatives. These equations are used to determine the head values at the nodes throughout the aquifer. The system of equations for determination of head can be written as follows.

$$[K]_{N\times N}\{h\}_{N\times 1}^{t+\Delta t} = [F]_{N\times N}\{h\}_{N\times N}^{t} + \{Q\}_{N\times 1} \tag{25}$$

2.3 Model Development

The development of coupled flow and transport model in RPCM consists of the following steps:

1. *Node definition and local support domain*: The problem domain is represented using nodes scattered throughout the domain. At this step, the size of the local support domain is input. The aquifer characteristics such as transmissivity, hydraulic conductivity, pump/recharge well location, porosity and dispersivity are also input at this stage.
2. *Computation of matrices containing shape function information*: The MQ-RBF shape parameters q and c_s are specified. The shape function and derivatives are computed for each node and its support domain using MQ-RBF.
3. *Development of the flow model and solution*: The governing equation and boundary conditions for the concerned aquifer type (confined or, unconfined) (Eq. (5)) are discretized using shape function values and its derivative values. This system of equations is represented in the form of Eq. (25). The unknown heads are obtained by solving the system of equations using matrix inversion or, Gaussian solver.
4. *Computation of seepage velocities and dispersion coefficients*: Based on the difference of the head, the seepage velocities at all nodes are computed using the

corresponding support domain. For instance, the velocity components at node k may be computed using values of head and derivatives of shape function.

$$V_{x_i} = -\frac{k_{x_i}}{n_e} \sum_{i=0}^{n} \frac{\partial \phi_i}{\partial x_i} h_i \tag{26}$$

where n is the number of nodes in the support domain. The dispersion coefficients are computed using the nodal velocities as per Eq. (14).

5. *Development of the transport model and solution*: The governing equations and boundary conditions are assembled in a similar manner as the flow model as shown in Eq. (24). These equations are solved till the maximum simulation time is reached. For multispecies transport problems, the system of Eq. (24) is obtained for each species.

The steps involved in the model development are presented in Fig. 1.

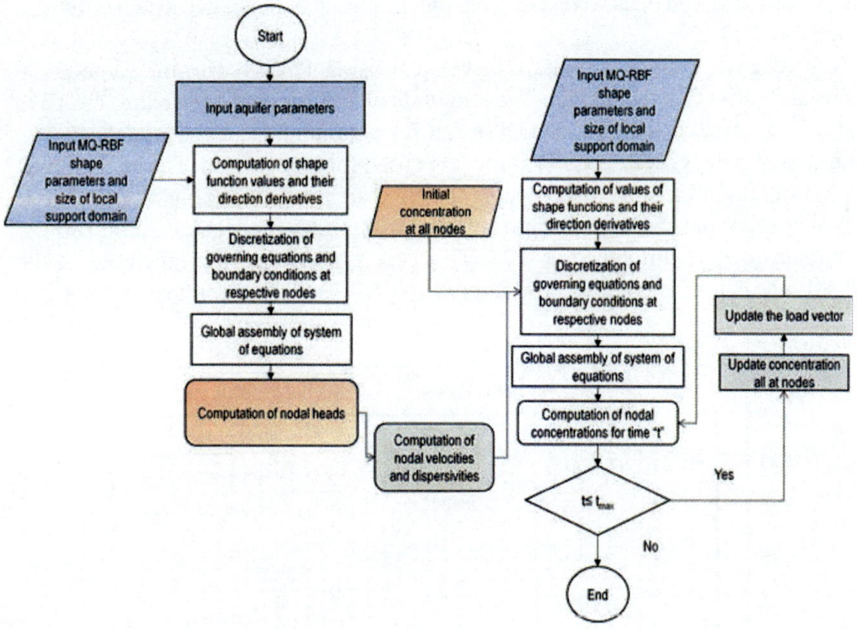

Fig. 1 Flow chart for model development

3 Model Applications

3.1 Case Study 1

Here, a hypothetical homogenous unconfined aquifer of dimensions 200 m × 150 m is considered. The hydraulic conductivity and porosity of the aquifer material are 50 m/d and 0.3, respectively. The north and south boundaries are no flow boundaries whereas the west and east boundaries are Dirichlet boundaries with head values of 10 m and 8 m, respectively. There is a pump located in the aquifer at (120, 110 m) which extracts groundwater at the rate of 150 m^3/d. A contaminant source which is located at (50, 70 m) injects Tritium of 2000 ppm at the volumetric rate of 2 m^3/d. The contaminant undergoes first-order degradation with decay rate of 0.000153 d^{-1}. The steady-state flow in the aquifer is modelled by RPCM-Flow model with 26 × 16 uniformly distributed nodes (see Fig. 2). The size of the local support domain is 4 times the nodal distance. The MQ-RBF shape parameters q and c_s are 0.98 and 5 times the nodal distance. The model results are compared with MODFLOW [6] in Fig. 3. It is observed that the results from the meshless model match excellently with that of MODFLOW.

The reactive transport model is developed using RPCM with the same nodal configuration as the flow model. The longitudinal and lateral dispersivities are 10 m and 0.1 m, respectively. The model is run for a simulation period of 1 year with time step of 5 days. MT3DMS [17] model is used for comparison. The contaminant concentrations at the end of the simulation period are presented in Fig. 4 and Table 1. It is observed that the concentrations obtained by both models are similar to the maximum relative difference of 0.28. This case study shows the efficiency of the RPCM model for the coupled flow and reactive transport in unconfined aquifers.

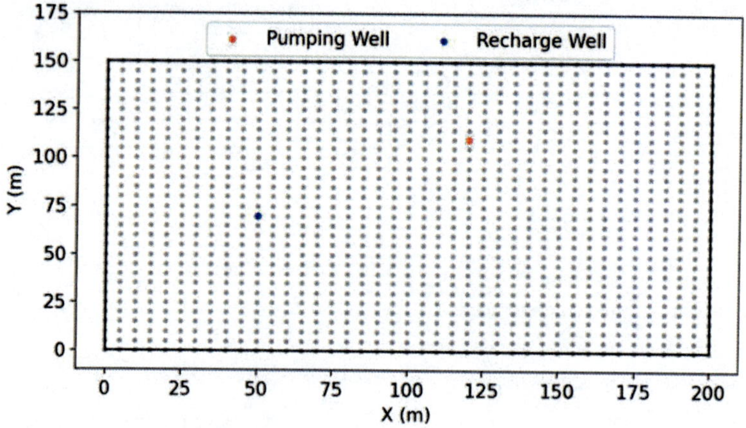

Fig. 2 Representation of nodal distribution for case study 1

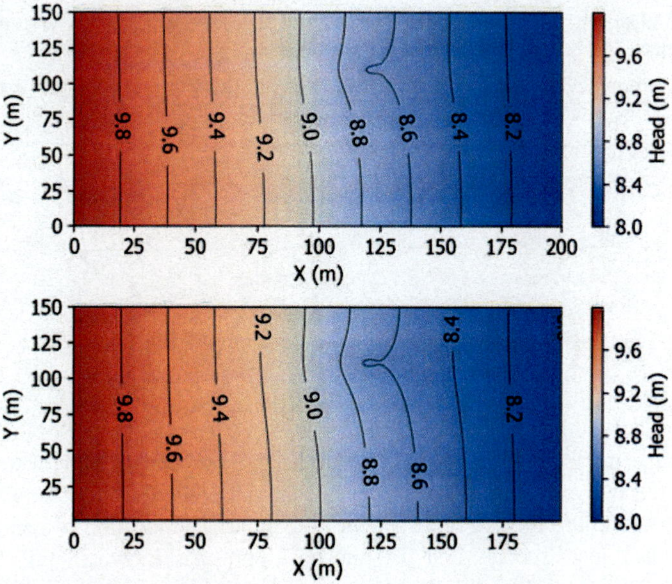

Fig. 3 Distribution of head in the aquifer as simulated by RPCM (left) and MODFLOW (right)

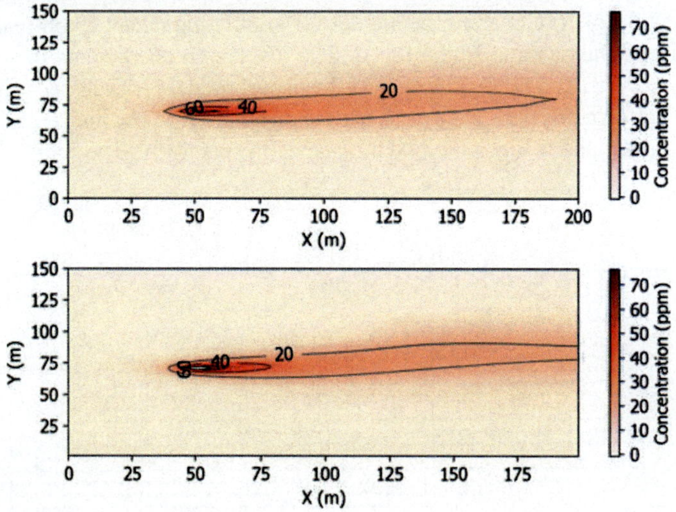

Fig. 4 Distribution of Tritium at the end of 1 year in the aquifer as simulated by RPCM (left) and MT3DMS (right)

Table 1 Comparison of concentrations obtained by RPCM and MODFLOW-MT3DMS at randomly selected locations

Location	Coupled-RPCM (ppm)	MODFLOW-MT3DMS (ppm)	Relative difference
125, 125 m	0.49	0.38	0.28
95, 70 m	31.27	30.30	0.03
70, 80 m	19.57	16.97	0.15
160, 65 m	16.07	14.10	0.13
145, 45 m	3.78	3.03	0.24

3.2 Case Study 2

In this case, a hypothetical field-like aquifer is considered as shown in Fig. 5. The velocities in x- and y-directions are assumed to be 1.25 m/d and 0.2 m/d (in downward direction), respectively. A three-species decay chain is considered for study in which species 1, 2 and 3 are formed in a sequential manner. The location of contaminant source which injects species 1 at the rate of 1 ppm/d is shown in Fig. 5. Here, direct sources for species 2 and 3 are not considered. Species 1 acts as the source for species 2, and species 2 acts as the source for species 3. The first-order decay rates of the species 1, 2 and 3 are 0.002 d^{-1}, 0.003 d^{-1} and 0.004 d^{-1}, respectively. The formulation of RPCM transport model for simulating decay chain reaction is discussed in Anshuman and Eldho [2, 3]. The dispersion coefficients in x- and y-directions are taken as 5 m^2/d. The RPCM-Transport model is developed with 1841 nodes with uniform nodal distance of 25 m as shown in Fig. 5. The model parameters q and c_s are 0.98 and 5 times the nodal distance, respectively. The radius of local

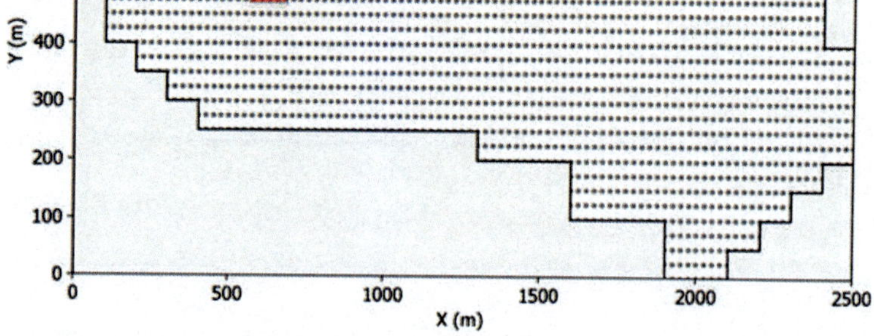

Fig. 5 Schematic diagram of study area and nodal distribution for case study 2

support is taken as 4 times the nodal distance. The model is run for a simulation period of 2 years with a time step of 5 days.

The developed model is compared with a FEM-based model developed in COMSOL [5]. The results from both models are presented in Fig. 6. It is observed that the results from the proposed model matches with the FEM-based model. The concentrations of the contaminants at randomly selected locations are presented in

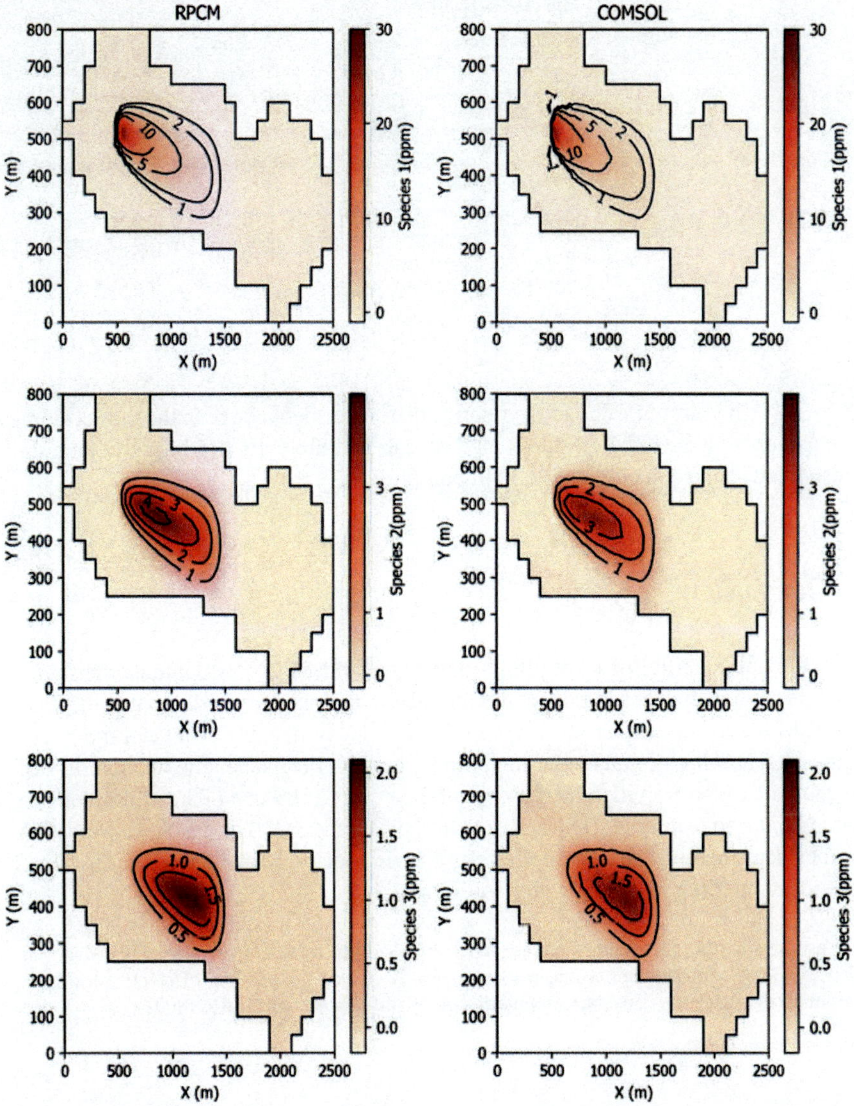

Fig. 6 Distribution of species 1, 2 and 3 at the end of 2 years in the aquifer as simulated by RPCM (left) and COMSOL (right)

Table 2 Comparison of concentrations obtained by RPCM and COMSOL at randomly selected locations

Location	COMSOL	RPCM	Rel. Diff.
Species 1 (ppm)			
600, 525 m	20.34	21.73	0.07
675, 525 m	13.76	14.43	0.05
700, 550 m	8.02	7.42	0.08
850, 550 m	4.26	3.65	0.14
Species 2 (ppm)			
600, 525 m	2.39	2.58	0.08
675, 525 m	3.03	3.23	0.07
700, 550 m	2.19	1.91	0.13
850, 550 m	1.93	1.64	0.15
Species 3(ppm)			
600, 525 m	0.24	0.27	0.10
675, 525 m	0.53	0.58	0.08
700, 550 m	0.47	0.39	0.18
850, 550 m	0.67	0.57	0.15

Table 2. The maximum value of the relative difference between both the model simulations is 0.18. Hence the meshless method can be efficiently used for studying the transport of species in a decay chain in aquifers.

4 Conclusion

Here, the applicability of a meshless method named RPCM is demonstrated with two case studies. It is observed that the model simulations are similar to that of FDM and FEM based models. In meshless methods, the connectivity between the nodes is not present which reduces the modelling steps and enhances the adaptability of the model. It also provides increased stability due to the use of local domains in case of high advective [1, 15] and reactive problems [1]. The study showed that the meshless methods can be applied for a variety of problems related to flow and transport modelling in groundwater aquifers.

Acknowledgements The authors of the paper are thankful to Board of Research in Nuclear Sciences (BRNS) for supporting this work through the project "Modelling of reactive transport in groundwater using meshfree based numerical methods" (Project no. 16BRNS002).

References

1. Anshuman, A., Eldho, T.I., Singh, L.G.: Simulation of reactive transport in porous media using radial point collocation method. Eng. Anal. Bound. Elem. **104**, 8–25 (2019)
2. Anshuman, A., Eldho, T. I.: Meshfree radial point collocation-based coupled flow and transport model for simulation of multispecies linked first order reactions. J. Contam. Hydrol. **229.** 103582 (2020)
3. Anshuman, A., Eldho, T.I.: Modeling of transport of first order reaction networks in porous media using meshfree radial point collocation method. Comput. Geosci. (2019)
4. Bear, J., Cheng, A.H.D.: Modeling groundwater flow and contaminant transport, vol. 23. Springer Science & Business Media (2010)
5. Comsol Multiphysics Reference Manual
6. Harbaugh, A.W., Banta, E.R., Hill, M.C., McDonald, M.G.: MODFLOW-2000, The U. S. Geological Survey Modular Ground-Water Model-User Guide to Modularization Concepts and the Ground-Water Flow Process. Open-File Report. U. S. Geological Survey, vol. 92, p. 134 (2000)
7. Hardy, R.L.: Multiquadric equations of topography and other irregular surfaces. J. Geophys. Res. **76**(8), 1905–1915 (1971)
8. Kansa, E.J.: Multiquadrics—a scattered data approximation scheme with applications to computational fluid-dynamics—I surface approximations and partial derivative estimates. Comput. Math Appl. **19**(8–9), 127–145 (1990)
9. Kansa, E.J.: Multiquadrics—a scattered data approximation scheme with applications to computational fluid-dynamics—II solutions to parabolic, hyperbolic and elliptic partial differential equations. Comput. Math Appl. **19**(8–9), 147–161 (1990)
10. Liu, G.R., Gu, Y.T.: An Introduction to Meshfree Methods and Their Programming. Springer Science & Business Media (2005)
11. Meenal, M., Eldho, T.I.: Simulation of groundwater flow in unconfined aquifer using meshfree point collocation method. Eng. Anal. Bound. Elem. **35**(4), 700–707 (2011)
12. Meenal, M., Eldho, T.I.: Two-dimensional contaminant transport modeling using meshfree point collocation method (PCM). Eng. Anal. Bound. Elem. **36**(4), 551–561 (2012)
13. Pathania, T., Bottacin-Busolin, A., Rastogi, A.K., Eldho, T.I.: Simulation of groundwater flow in an unconfined sloping aquifer using the element-free Galerkin method. Water Resour. Manag. 1–19 (2019)
14. Rastogi, A.K.: Numerical Groundwater Hydrology. Penram International Publication, India (2007)
15. Singh, L.G., Eldho, T.I., Kumar, A.V.: Coupled groundwater flow and contaminant transport simulation in a confined aquifer using meshfree radial point collocation method (RPCM). Eng. Anal. Bound. Elem. **66**, 20–33 (2016)
16. Swathi, B., Eldho, T.I.: Groundwater flow simulation in unconfined aquifers using meshless local Petrov-Galerkin method. Eng. Anal. Bound. Elem. **48**, 43–52 (2014)
17. Zheng, C., Wang, P.P.: MT3DMS: a modular three-dimensional multispecies transport model for simulation of advection, dispersion, and chemical reactions of contaminants in groundwater systems; documentation and user's guide. Alabama University (1999)

Computational Modelling of Stochastic Buffeting of Fin-Like Structures

Rahul Sundar, Avisha Ghorpade, Jithin Jith, Sayan Gupta, and Sunetra Sarkar

1 Introduction

Buffeting is defined as the response of a structure to the aerodynamic excitation provided by separated flows [1]. Fin buffeting refers to the response of an aircraft fin to the separated wake from upstream components like the wing, foreplane, fuselage, etc. Depending on the flight conditions and geometry configurations, the buffeting response of a fin may be very large, and may lead to the accumulation of significant fatigue damage [2], thus, warranting its study in the design of a delta wing aircraft.

A thorough understanding of the sources of fin buffeting enables better modelling and estimation of buffeting loads on the structure. For flow over a delta wing, there exist four major flow phenomena responsible for fin buffeting—(i) vortex wandering [3], (ii) vortex breakdown [4], (iii) fluctuations of the breakdown location, [5], and (iv) vortex shedding [6]. All of these mechanisms are due to the vortices generated by the delta wings themselves, or by leading-edge extensions and forebodies.

The fluctuations in the velocity field due to these sources occur in different frequency regimes as identified by Gursul et al. [7] which can be visualised using a power spectrum of velocity fluctuations. From Gursul et al.'s [7] experimental work, it was evident that the helical mode instability (arising from vortex breakdown) has the highest frequency and occurs at smaller angles of attack of the delta wing. On the other hand, vortex shedding has a lower frequency and occurs at larger angles of attack. Vortex interaction (the major cause of fluctuation of the breakdown location)

R. Sundar (✉) · S. Sarkar
Department of Aerospace Engineering, IIT Madras, Chennai 600036, India
e-mail: rahul.sundar95@gmail.com

A. Ghorpade · J. Jith · S. Gupta
Department of Applied Mechanics, IIT Madras, Chennai 600036, India

S. K. Saha and M. Mukherjee (eds.), *Recent Advances in Computational Mechanics and Simulations*, Lecture Notes in Mechanical Engineering,
https://doi.org/10.1007/978-981-15-8315-5_22

has the lowest frequency and is seen to occur at intermediate angles of attack, and vortex wandering happens to be the weakest source compared to others.

The structural response of the fin to various buffeting sources has been studied in a statistical manner by various authors. The earliest work was done by Liepmann [8], in which the forces acting on a thin plate in a turbulent flow were determined using analytical methods. Once the power spectrum of the forces was obtained, it was applied to the frequency response function of a single degree of freedom model representing the thin plate. In this manner, it was possible to determine the response of the plate. In the following work, Liepmann [9] was able to extend this approach to wings of finite spans. A similar approach has been used by Ribner et al. [10] as well. The main drawback of these approaches is that they require analytical methods to determine the forces acting on the fin. This is not always possible and, when possible, may not be as accurate as compared to a CFD simulation.

In this work, a stochastic approach is proposed to model the unsteady wake of a delta wing to estimate buffeting loads acting on the fin-like structure. The approach is demonstrated by computing the buffeting loads on a 2D symmetric NACA airfoil representing a fin structure located in the wake of an idealised delta wing. The computational studies are carried out using OpenFOAM [11], an open-source CFD software based on the finite volume method.

2 Computational Methodology

2.1 *Statistical Model of the Flow Field and Computational Implementation*

Using the spectral characteristics of the various buffeting sources described in Sect. 1, it is possible to construct a temporally consistent statistical model of the flow downstream of a delta wing. The present study proposes to achieve this by reconstructing the flow field as a collection of random processes derived from the available experimental power spectra. To this end, two tasks are undertaken—(i) fitting the noisy experimental power spectra to a smoother numerical approximation, and (ii) generating the velocity field as random processes based on the fitted spectra.

To obtain a smoother numerical fit to the experimental power spectra, the present study utilises a linear combination of Gaussian distributions and linear interpolation functions as most experimental power spectra consist of multiple peaks [12]. Figure 5 shows an experimental spectrum which has been approximated as mentioned.

2.1.1 Generation of Random Processes

This section deals with the generation of velocity fields from a prescribed power spectral density (obtained by fitting a smoother numerical approximation) using random processes. A random process also requires an associated probability distribution. When prior knowledge of the probability distribution is unavailable, as is the case here, it is customary to assume that the random process follows a Normal distribution [13]. One of the most popular methods of generating stationary random processes with a Normal distribution from a given power spectral density is the Spectral Representation Method (SRM) [14]. Suppose the power spectral density of a random process, in this case, the velocity fluctuation magnitude $V(t)$, is given by $S(\omega)$, a realisation $v(t)$ of $V(t)$ can be generated using SRM by

$$v(t) = \sum_{n=0}^{N-1} \sqrt{2S(\omega_n)\Delta\omega}\cos(\omega_n t + \phi_n), \tag{1}$$

where $\Delta\omega = \omega_u/N, \omega_n = n\Delta\omega$, and ϕ_n is a random phase angle which follows a uniform distribution in the range $[0, 2\pi]$. Here, ω_u is called the cut-off frequency that represents an upper frequency limit for $S(\omega)$, and N represents the number of segments into which the frequency range of $S(\omega)$ needs to be discretised. Shinozuka et al. [14] have shown that the time step Δt used to evaluate Eq. 1 needs to be $\Delta t \leq 2\pi/2\omega_u$. Another point to keep in mind is that $v(t)$ generated from equation (1) will be periodic with a time period $T = 2\pi/\Delta\omega$. Therefore, $\Delta\omega$ should be chosen such that $v(t)$ does not show periodic behaviour in the timespan chosen for generation. Though SRM can also be used to generate stationary vector random processes [15], in this work we limit ourselves to a univariate random process.

2.1.2 Implementation of SRM in a CFD Solver

Existing OpenFOAM boundary conditions suffered from the drawback that the random process had to be generated beforehand which led to the computational process being cumbersome. The new boundary condition ensures that the random process generation is performed on the fly. The method described in the previous subsection (SRM) is implemented in the open-source CFD software OpenFOAM [11] as a custom boundary condition written in C++. It can be used as an inlet boundary condition which generates the velocity field (see Eq. (2)) at the inlet as a random process using SRM. The user simply needs to specify the mean free stream velocity ($\mathbf{U_0}$), an input power spectral density ($S(\omega)$), and a unit direction ($\hat{\mathbf{g}}$) for the velocity fluctuation ($v(t)$) at the inlet. The velocity field at the inlet can be given by

$$\mathbf{U(t)} = \mathbf{U_0} + v(t)\hat{\mathbf{g}}. \tag{2}$$

Here, $v(t)$ is obtained as described in Sect. 2.1.1.

2.2 2D Computational Study of Fin Buffeting

2.2.1 Study Conditions

Because of the symmetric nature of the fin and its zero angle of attack, fluctuations in the streamwise velocity are less significant when it comes to buffeting, and therefore the focus is on fluctuations of velocity in the transverse direction [16]. It is to be noted that the fin is assumed to have a zero dihedral angle. In case of non-zero dihedral angles, other components of the velocity also become significant [16]. In the present study, the airfoil is subjected to a random vertical gust emanating from the inlet. The frequency range of the spectrum is decided from experimental measurements [12] which show that the buffeting phenomena over a delta wing are present in the range $fc/U_\infty \in [0, 10]$ (where f is the frequency of the gust and U_∞ is the free-stream velocity). A flat spectrum is chosen for initial validation of the boundary condition so that the energy contribution of all the frequencies in the range of interest are equal. The final study is performed with a realistic spectrum, the details of which will be discussed in Sect. 3. The time history of the vertical gust is generated using the Spectral Representation Method [14]. The vertical gust is assumed to be perfectly correlated along the vertical axis.

2.2.2 Liepmann's Analytical Model

Liepmann's model [8] provides a way to determine the power spectrum of the lift coefficient of a thin airfoil which encounters a random vertical gust. It is based on Sears' function [17] which can be used to obtain the lift coefficient of a thin airfoil subjected to a sinusoidal vertical gust. Liepmann's model states that the power spectral density of the lift coefficient is given by

$$S_{cl}(\omega) = 4\pi^2 S_v(\omega)|\phi(k)|^2 \tag{3}$$

where $S_v(\omega)$ is the power spectral density of the vertical gust, and

$$\phi(k) = \frac{J_0(k)K_1(ik) + iJ_1(k)K_0(ik)}{K_1(ik) + K_0(ik)}. \tag{4}$$

Here, $k = \omega c/2U_\infty$ and J_n and K_n are Bessel functions.

2.2.3 Computational Domain

The computational domain of the study is shown in Fig. 1. It consists of an airfoil of chord $c = 1$ m and angle of attack $\alpha = 0°$ placed in a rectangular domain of dimensions $16c \times 22c$. The leading edge of the airfoil is placed at a distance of $3c$

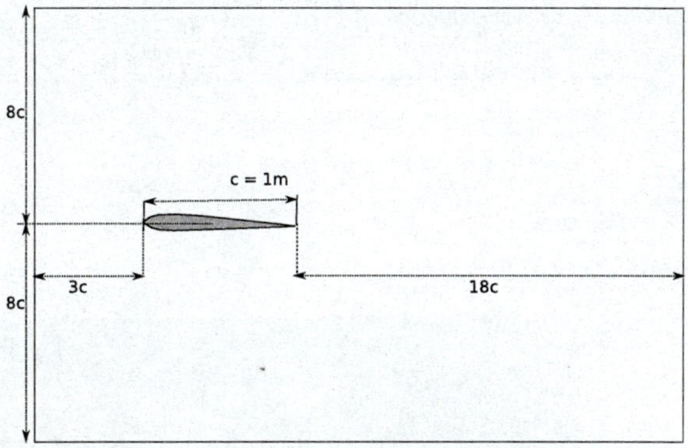

Fig. 1 Computational domain

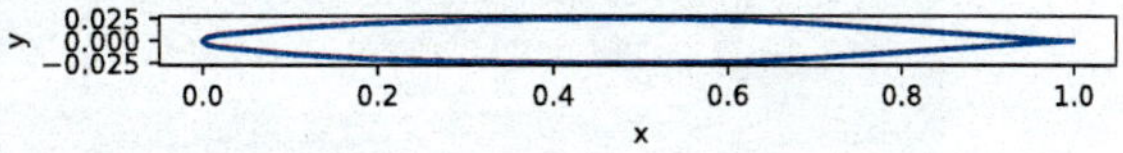

Fig. 2 NACA 66005 airfoil

from the inlet (leftmost edge). NACA 66005 (see Fig. 2) is considered in this study which has a symmetric profile, representative of vertical tail fins on the delta wing aircraft. The details of the study parameters are given in Table 1.

A hybrid mesh, which is structured within an elliptical region around the airfoil and unstructured outside of it, is used to discretise the computational domain. The mesh near the airfoil, where it is much finer, is shown in Fig. 3. The computational mesh is parameterised by the number of grid points (n_a) present on the surface of the airfoil. A mesh and time step convergence study was performed for different mesh resolutions and time step resolutions. In order to maintain brevity, the details and results of the study aren't presented. However, the mesh with $n_a = 600$ and a time step, $\Delta t = 2.5 \times 10^{-5}$ was found to be the suitable choice for further computational studies.

2.2.4 Solver Details

Computational studies have been carried out using OpenFOAM, a finite volume-based computational fluid dynamics framework. The solver used is PisoFOAM, which is an unsteady Navier–Stokes solver with the pressure velocity coupling achieved through the PISO algorithm [18]. The spatial and temporal discretisation schemes are second-order accurate. The pressure equation is solved using Geometric Algebraic Multi Grid (GAMG) solver with Diagonal Incomplete Cholesky (DIC)-Gauss Seidel-based smoother and the velocity is solved using Preconditioned Bi-conjugate Gradient (PBiCG) iterative solver with Diagonal Incomplete LU (DILU) preconditioning.

Table 1 Study parameters

Parameter	Value
Chord, c	1 m
Angle of attack, α	0°
Free-stream velocity, U_∞	1 m/s
Reynolds number, Re	10^4

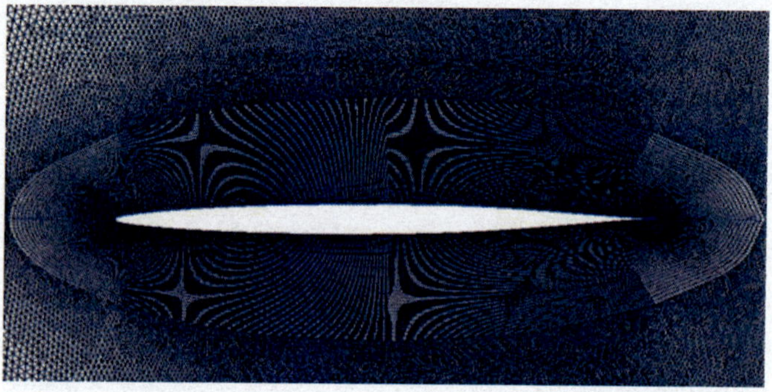

Fig. 3 Closer view of the mesh around the airfoil

3 Results and Discussion

In this section, we discuss the key results obtained from computational studies for a flat spectrum and a realistic input spectrum obtained from experimental data [16]. Figure 4 compares the power spectral density of c_l of NACA 66005 airfoil for the converged mesh with Liepmann's model for the flat spectrum. It is quite evident that the two are quite close to each other. This is therefore a confirmation of the validity of the implemented boundary condition.

The realistic power spectrum is obtained from experimental studies conducted by Breitsamter et al. [16] on the wind-tunnel model of a high-agility aircraft of canard-delta wing type. Figure 5 represents the power spectrum of the non-dimensional vertical gust v/U_∞ for a delta wing angle of attack of $\alpha = 30°$ and a gust intensity of $v_{rms}/U_\infty = 3.41\%$ at a Reynolds number of $\mathrm{Re} = 0.97 \times 10^6$. However, we have chosen to restrict the study in $\mathrm{Re} = 10,000$ regime using the same experimental input power spectrum. The experimentally measured power spectrum (represented by the red curve in Fig. 5) is approximated by a smoothed power spectrum (represented by the blue curve) for numerical convenience (see Sect. 2.1). The time history of the vertical gust is generated using SRM as discussed in Sect. 2.1.

The time series of the lift coefficient for the first 60 s is shown in Fig. 6. The power spectrum of the lift coefficient is shown in Fig. 7 where it is compared with the analytical prediction from the Liepmann model [8]. It can be seen that the peak frequencies of the numerical and analytical spectra coincide, however, the magnitude

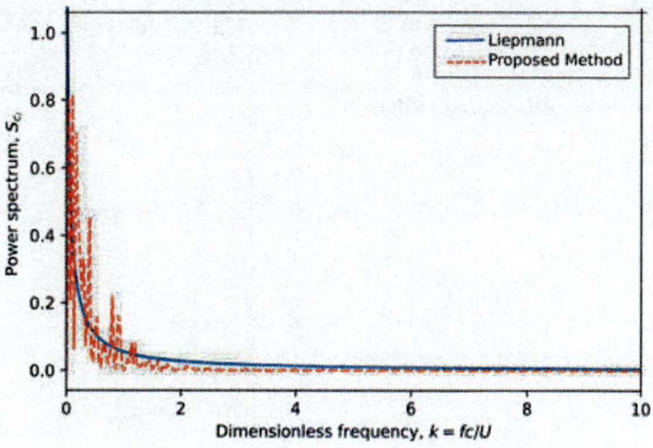

Fig. 4 Power spectral density of c_l for $n_a = 600$ (NACA 66005) compared with Liepmann's model for a flat spectrum

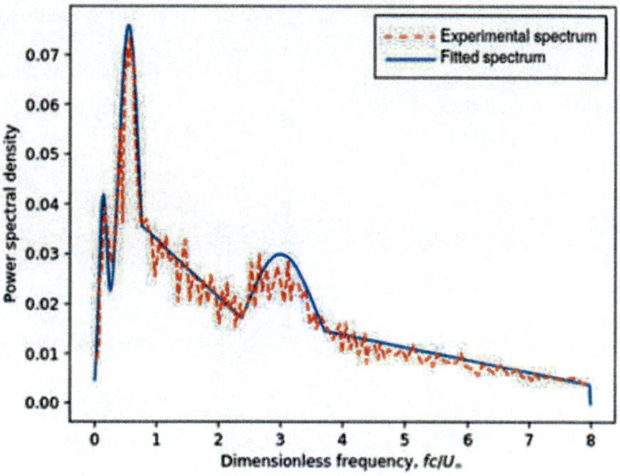

Fig. 5 Realistic vertical gust spectrum. Dashed red lines and smooth blue lines indicate the experimental and fitted spectrum, respectively (see Sect. 2.1)

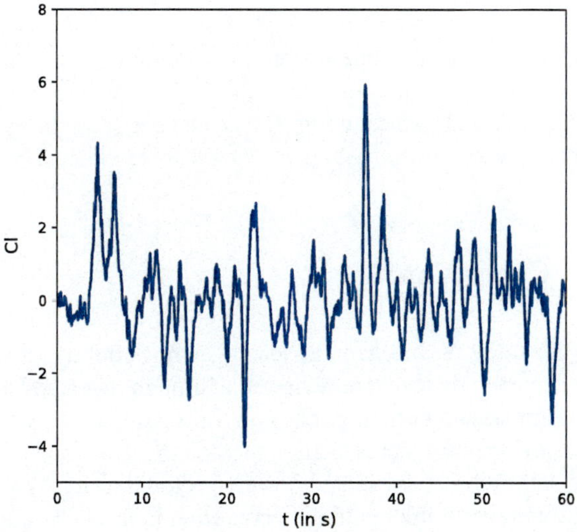

Fig. 6 Time history of lift coefficient (c_l) for the realistic input spectrum

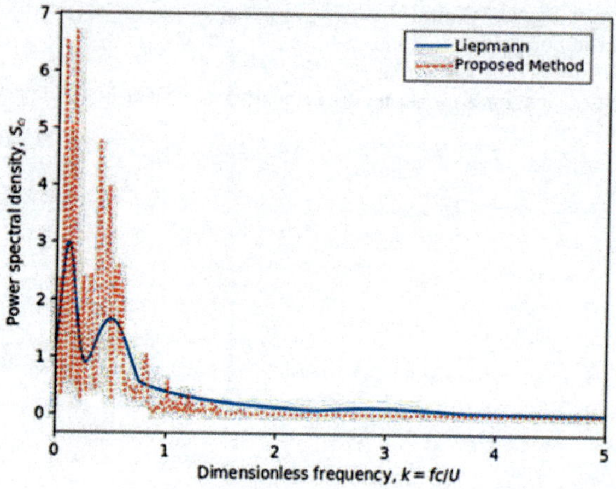

Fig. 7 Comparison of power spectra of lift coefficient with Liepmann model at $Re = 10^4$ for a relistic input spectrum

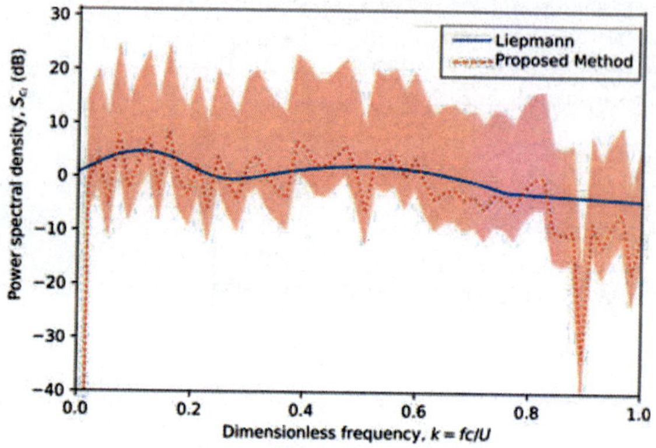

Fig. 8 Power spectra of lift coefficient with 95% confidence interval

of the spectra are quite different. Keeping in mind that the time series of the lift coefficient is random in nature, the 95% confidence interval of the spectrum [19] is computed and plotted in Fig. 8. It can be seen that the upper bound of the confidence interval is significantly higher than what is predicted by the Liepmann model.

4 Conclusion

In this study, the mechanisms leading to fin buffeting were investigated and their spectral characteristics were reviewed. Through this study, a novel stochastic approach to estimate the buffeting loads on a fin-like aircraft structure was presented which uses the spectral representation method to reconstruct a temporally consistent velocity field from its associated power spectral density. The reconstructed velocity field was used as an inlet boundary condition in the CFD studies carried out on a 2D fin section using OpenFOAM. The novel boundary condition implemented in OpenFOAM allowed us to estimate the buffeting loads (c_l) in a computationally efficient manner. The c_l power spectra for a flat as well as a realistic input spectrum were compared with that of Liepmann's analytical model. The studies revealed that for the flat spectrum the numerical results matched with that of Liepmann's. However, for the realistic input spectrum, it was found that the Liepmann's model underpredicted the amplitude of c_l, thus showing the necessity to use a full-fledged CFD simulation to accurately obtain the buffeting loads over analytical models. The authors believe that the novel boundary condition implemented in this study is generalised and can be used to even study other problems involving a random gust input spectra as long as the power spectral density is known and the process is stationary.

Acknowledgements We are grateful to Aeronautical Development Agency, India, for its support along the course of this work. We would also like to thank the P. G. Senapathy Computing Centre, IIT Madras, for providing us with the high-performance computing facility.

References

1. Mabey, D.G.: Some aspects of aircraft dynamic loads due to flow separation. Progress Aerosp. Sci. **26**(2), 115–151 (1989)
2. Chesneau, T.R., Wood, N.J.: Fin buffeting characteristics of a generic single fin aircraft. Proc. Inst. Mech. Eng. G J. Aerosp. Eng. **210**(3), 247–260 (1996)
3. Menke, Matthew, Gursul, Ismet: Unsteady nature of leading edge vortices. Phys. Fluids **9**(10), 2960–2966 (1997)
4. Leibovich, S.: Vortex stability and breakdown: survey and extension. AIAA J. **22**(9), 1192–1206 (1984)
5. Gursul, Ismet, Yang, Houbin: On fluctuations of vortex breakdown location. Phys. Fluids **7**(1), 229–231 (1995)
6. Rediniotis, O.K., Stapountzis, H., Telionis, D.P.: Vortex shedding over delta wings. AIAA J. **28**, 944–946 (1990)
7. Gursul, I., Xie, W.: Physics of buffeting flows over delta wings. In: 36th AIAA Aerospace Sciences Meeting and Exhibit, Reno, NV. American Institute of Aeronautics and Astronautics (1998)
8. Liepmann, H.W.: On the application of statistical concepts to the buffeting problem. J. Aeronaut. Sci. **19**(12), 793–800 (1952)
9. Liepmann, H.W.: Extension of the statistical approach to buffeting and gust response of wings of finite span. J. Aeronaut. Sci. **22**(3), 197–200 (1955)
10. Ribner, H.S.: Spectral theory of buffeting and gust response: unification and extension. J. Aeronaut. Sci. **23**(12), 1075–1077 (1956)
11. Hrvoje, J., Aleksandar, J., Zeljko, T. et al.: Openfoam: A c++ library for complex physics simulations. In: International Workshop on Coupled Methods in Numerical Dynamics, vol. 1000, pp. 1–20. IUC Dubrovnik Croatia (2007)
12. Gursul, I., Xie, W.: Buffeting flows over delta wings. AIAA J. **37**(1), 58–65 (1999)
13. Papoulis, A.: Probability, Random Variables, and Stochastic Processes. McGraw-Hill (1991)
14. Shinozuka, Masanobu, Deodatis, George: Simulation of stochastic processes by spectral representation. Appl. Mech. Rev. **44**(4), 191–204 (1991)
15. Yongxin, Wu., Gao, Yufeng., Li, Dayong., Feng, Tugen, Mahfouz, Ali H.: Comparison of the spectral representation method to simulate spatially variable ground motions. J. Earthquake Eng. **18**(3), 458–475 (2014)

16. Breitsamter, Christian, Laschka, Boris: Fin buffet pressure evaluation based on measured flow-field velocities. J. Aircraft **35**(5), 806–815 (1998)
17. Sears, William R.: Some aspects of non-stationary airfoil theory and its practical application. J. Aeronaut. Sci. **8**(3), 104–108 (1941)
18. Versteeg, H.K., Malalasekera, W.: Computational fluid dynamics. The finite volume method (1995)
19. Wayne, A.F.: Introduction to Statistical Time Series, 2nd edn. Wiley, New York (1995)

Investigating the Dynamical Behaviour of Dipteran Flight-Inspired Flapping Motion Using Immersed Boundary Method-Based FSI Solver

Chhote Lal Shah, Dipanjan Majumdar, and Sunetra Sarkar

1 Introduction

Biologically inspired insect flight mechanisms have recently become a primary research subject, mainly, to use in micro air vehicles (MAVs) for defence surveillance and environmental monitoring [1]. The prime source of inspiration in this regard with the reliance on natural selection comes from insects, which constitutes nature's successful flyers [2]. The insect flight motors have considerably evolved over ages making their fapping flight energetically efficient by storing the kinetic energy as elastic energy. These make use of deformed flexible elements such as flight muscles, flight thorax and wing hinges for storing the elastic energy and recovering it during the wing stroke reversal [3]. The Dipteran flight mechanism is one such mechanism. In particular, it has been discovered that Dipteran flight systems and their muscular control over the resulting flight dynamics are an efficient growth strategy of the mechanical wing-motor systems that are bio-inspired [4–6].

Diptera enhances the aerodynamic forces by significantly modifying the wing stroke dynamics through muscle-induced variations in the flight motor [7]. The bistable "click" mechanism [8, 9] is one of the most common modelling procedures to illustrate the muscle to a wing interface mechanism accounted for in the literature so far. Most of the research on the Diptera flight have focused on the formulation of an accurate mathematical model of the click mechanism. A few investigations have directed on studying the dynamical aspects associated with the click mechanism at a low Reynolds number. Although the majority of the research works [10–12] have focused on investigating the "complex" flight patterns performed by the natural

C. L. Shah (✉) · D. Majumdar · S. Sarkar
Department of Aerospace Enginering, IIT Madras, Chennai 600036, India
e-mail: lal.chhote.shah@gmail.com
URL: https://www.lalchhoteshah.com/

S. K. Saha and M. Mukherjee (eds.), *Recent Advances in Computational Mechanics and Simulations*, Lecture Notes in Mechanical Engineering,
https://doi.org/10.1007/978-981-15-8315-5_23

flyers such as insects and birds, there are significant gaps in the understanding of underlying physics. The Dipteran flight motor-inspired flapping system is inherently non-linear and further encounters external actuation from the muscle forces. If a non-linear dissipative system is stimulated by external forcing, it veers from its state of equilibrium and either oscillates or goes through a complex temporal transformation called chaos.

Modelling the Dipteran flight mechanism as an irrationally non-linear forced oscillator structure, Cao et al. [13] investigated the interesting bifurcation phenomenon involving various flight patterns along with complicated coexisting flight mechanisms. They stated a period of doubling cascade, contributing to a chaotic attractor when the system is introduced to an external excitation force. In all the aforementioned literature, the direct contribution from the surrounding unsteady flow has not been considered. Hence, the system dynamics comprising the coupling between the non-linear framework and the unsteady fluid flow need to be studied as the flight pattern can be remarkably amended in the presence of fluid–structure interaction (FSI) conditions. Recently, Bose et al. [14] have reported chaotic transients, and intermittent transitions to stable chaos in the study of Dipteran flight-inspired flapping motion. Their FSI solver was developed using the Lumped Vortex Method (LVM)-based flow solver and did not consider the effect of viscosity. However, the Reynolds number ($\mathrm{Re} = U_{\infty}c/\nu$) regime in which insects fly, the viscous effects are significant that may alter the dynamic behaviour of the system.

In this paper, we present an in-house FSI solver developed based on the forced Duffing oscillator model as a structure given by Brennan et al. [9]. The structural model [9] is coupled with the IBM solver to capture the non-linear interaction between the wing and unsteady viscous flow. The current study is primarily focused on examining the coupled non-linear FSI behaviour of the Dipteran flight motor system. This paper is arranged in the following way: Sect. 2 gives the details about the problem statement, formulation of the FSI solver and boundary conditions along with convergence study and validation of the solver. Section 3 describes the characterisation of the different system dynamics using non-linear time-series analyses. Section 4 briefs the significant results of the paper.

2 Computational Methodology

2.1 *Structural Model*

A simplified structural model for the Dipteran flight motor system has been proposed by Brennan et al. [9]. A schematic of the flight motor system is presented in Fig. 1a. The two cantilever beams C1 and C2 represent the cuticle (notum (C1), pleural apophysis (C2)). The columns C1 and C2 are considered to have equal stiffness k_s and are one of the simplifications found in this model as in the case of cantilever C2; the pleurosternal muscle adds to its stiffness. The wing is modelled at B and

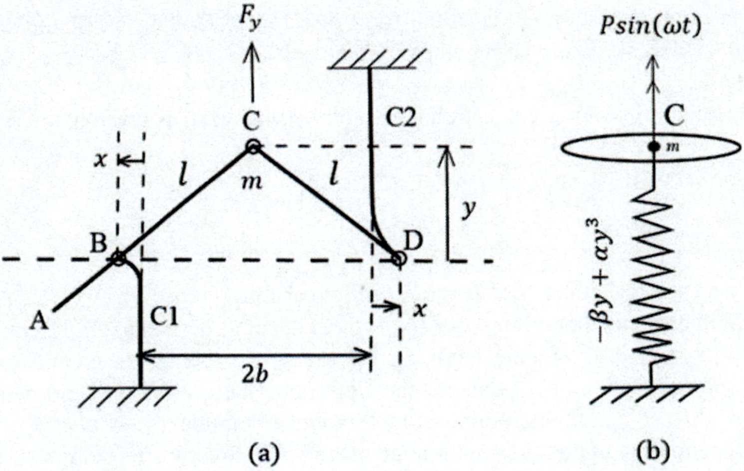

Fig. 1 Schematic of Dipteran flight motor model: **a** Click mechanism [9], and **b** Duffing oscillator model

C as a rigid ABC link. CD is also a rigid connection representing the parascutum, articulating at C and D. The hinge at B is the wing mechanism, and the hinge at C is the first axillary sclerite. The scutellar lever driving force $F_y(t)$ is applied at C. BC = CD = l is presumed for simplicity. It is also assumed that the cantilevers are massless, while at C the equivalent mass m of the rigid connections is lumped. As shown in Fig. 1, y is the instantaneous vertical displacement of C from BD and x is the horizontal displacement of the cantilevers. The initial distance between the columns is $2b$. In the present two-dimensional FSI simulation, the shape of the wing is considered to be an elliptical foil of thickness to chord ratio of 0.12 and the centre of the ellipse is at point C as shown in Fig. 1b.

The structural equation for the insect flight motor governing the heaving motion of the wing can be expressed in the following way as given in [9]:

$$\ddot{y}(t) - \frac{{\omega_0}^2}{2} y(t) \left[1 - \frac{y^2(t)}{D^2} \right] = \frac{P}{m} \sin(\omega t) + \frac{L(t)}{m}, \tag{1}$$

where $y(t)$ denotes the vertical displacement of the centre of ellipse, $D = \sqrt{\frac{2\left(1-\frac{b}{l}\right)}{\left(\frac{b}{l^3}\right)}}$;

$\omega_0 = \sqrt{\frac{4k_s\left(1-\frac{b}{l}\right)}{m}}$ is the structural natural frequency. Here, P and ω are the amplitude and frequency of the sinusoidal driving forcing $F_y(t)$, respectively. The readers are requested to refer to [8, 9] for a deep understanding of the biological connections and detailed formulation of the mathematical model of the insect flight motor. Under the effect of the aerodynamic load $L(t)$, the resulting system forms a single degree-of-freedom two-way-coupled FSI system mimicking the click mechanism in the

Dipteran flight. Equation (1) is in the form of a forced Duffing oscillator model, where the external forcing term comes from muscle actuation and the aerodynamic lift.

The structural governing equation is non-dimensionalized as given below:

$$\frac{\mathrm{d}^2\overline{y}}{\mathrm{d}\tau^2} - \frac{1}{2}k_0^2\overline{y}\left[1 - \left(\frac{\overline{y}}{\overline{D}}\right)^2\right] = \overline{F}\sin(k\tau) + \frac{1}{2\mu}C_L, \tag{2}$$

where $\overline{y} = y/c$ (c is the chord length of elliptical foil), $\overline{D} = D/c$, $\tau = U_\infty t/c$ is the non-dimensional time, $k_0 = \omega_0 c/U_\infty$ is the reduced structural natural frequency and $k = \omega c/U_\infty$ is the reduced forcing frequency, U_∞ being the incoming free stream velocity. $\mu = m/\rho_f c^2$ denotes the equivalent mass (ρ_f is the fluid density) and $\overline{F} = Pc/mU_\infty^2$ is the non-dimensional forcing amplitude.

The $\frac{b}{l}$ ratio is an important structural parameter as it directly affects the structural stiffness enabling the system to undergo complex dynamic behaviours. Here, it has been chosen to be 0.9 for all simulations. The structural damping has been considered to be zero, and the non-dimensional forcing frequency (k) has been taken as $0.5k_0$. Equation (2) is integrated using the explicit fourth-order Runge–Kutta scheme.

2.2 *Flow Solver*

The discrete forcing Immersed Boundary Method proposed by Kim et al. [15] is used in the present work to solve the incompressible Navier–Stokes equations for computing the flow-field around the flapping foil. The flow governing equations are given in the non-dimensional form below:

$$\frac{\partial \mathbf{u}}{\partial \tau} + \nabla.(\mathbf{uu}) = -\nabla p + \frac{1}{\mathrm{Re}}\nabla^2\mathbf{u} + \mathbf{f}, \tag{3}$$

$$\nabla.\mathbf{u} - q = 0, \tag{4}$$

where $\mathbf{u}$ denotes the flow velocity non-dimensionalised by U_∞, and p is the pressure non-dimensionalised by $\rho_f U_\infty^2$. Here, $\mathbf{f}$ is the momentum forcing applied to enforce the no-slip boundary condition at the solid boundary immersed in the fluid. The source/sink term, q, is added to the continuity equation (Eq. 4) to ensure mass conservation across the immersed boundary. A finite volume-based second-order semi-implicit Fractional Step Method (FSM) is used to solve flow equations (Eqs. 3 and 4). The diffusion term is discretised using a second-order Crank–Nicolson method, and Adams-Bashforth discretisation is used to advance for the convection term. The velocity is corrected using a pseudo pressure correction term so that the continuity is satisfied at every time step. The discretised form of governing equations (Eqs. 3 and 4) are as follows:

$$\frac{\hat{\mathbf{u}}^n - \mathbf{u}^{n-1}}{\Delta\tau} = -\nabla p^{n-1} - \tfrac{3}{2}H\left(\mathbf{u}^{n-1}\right) + \tfrac{1}{2}H\left(\mathbf{u}^{n-2}\right) + \tfrac{1}{2\mathrm{Re}}L\left(\mathbf{u}^{n-1}\right) + \tfrac{1}{2Re}L\left(\mathbf{u}^{n-2}\right) + \mathbf{f}^n, \tag{5}$$

$$\nabla^2\phi^n = \frac{1}{\Delta\tau}\left(\nabla.\hat{u}^n - \hat{q}^n\right), \tag{6}$$

$$\mathbf{u}^n = \hat{\mathbf{u}}^n - \Delta\tau\nabla\phi^n, \tag{7}$$

$$p^n = p^{n-1} + \phi^n - \frac{\Delta\tau}{\mathrm{Re}}\nabla^2\phi^n, \tag{8}$$

where $L(*)$ is the discretised form of the Laplace operator and $H(*)$ denotes the discretised form of the convective term, $\hat{\mathbf{u}}^n$ is the intermediate velocity at n^{th} time step, $\hat{q}^n$ is the source term calculated from $\hat{u}^n$, ϕ is the pseudo pressure correction term and $\Delta\tau$ is the time increment. For further details regarding the solution procedure of flow equations the readers are requested to see [15].

At every time step, the flow equations are solved to get the flow-field around the body, and the aerodynamic loads acting on the body are evaluated. Then this lift force calculated by the flow solver is supplied to the structural equation to compute the position of the body in the next time step. Thus, at every time step, the flow solver and the structural solver exchange information in a staggered manner resulting in a weak coupling FSI solver.

2.3 Convergence Study

The flow domain and the mesh grid used in the present study are shown schematically in Fig. 2. At the initial time, the centre of the elliptic foil lies at the origin. A Dirichlet

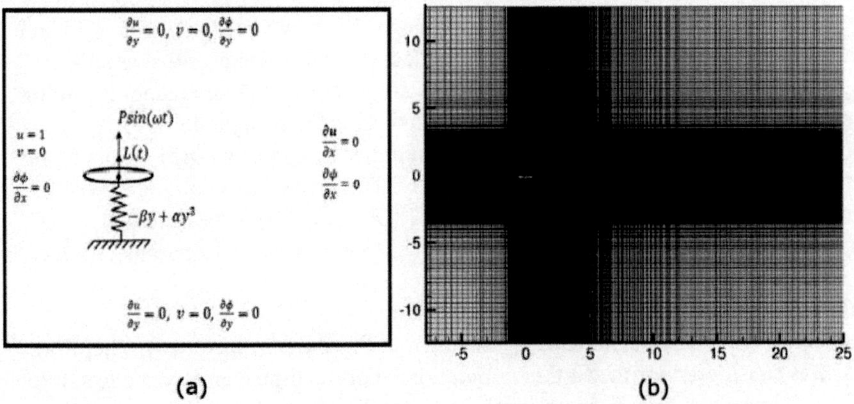

Fig. 2 **a** Flow domain and corresponding boundary conditions, and **b** Cartesian grid

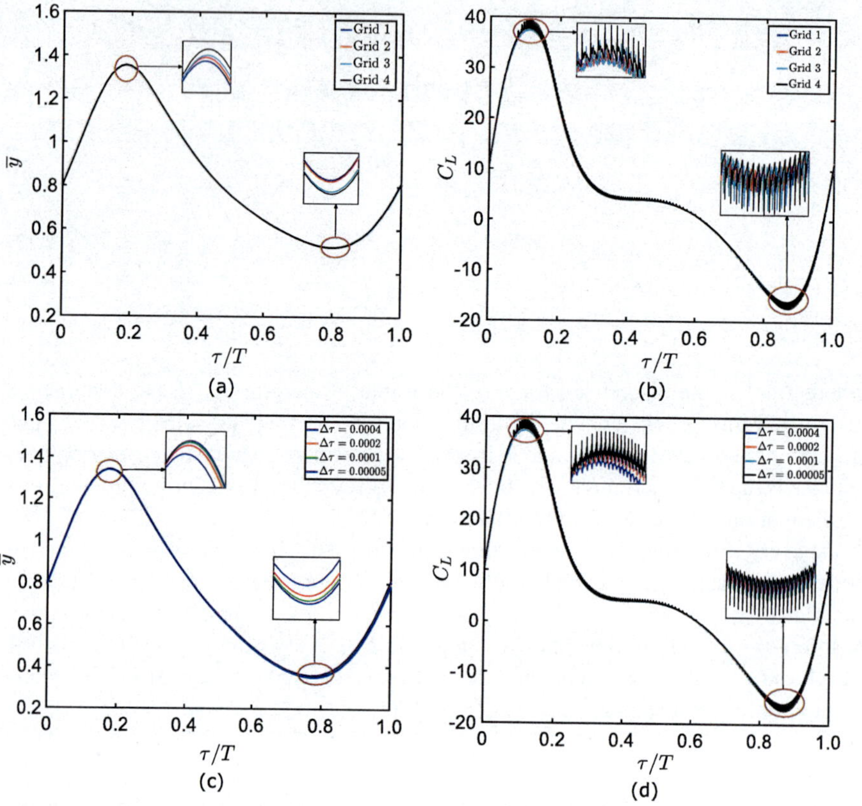

Fig. 3 Convergence: **a** and **c** grid and time convergence for displacement, **b** and **d** grid and time convergence for C_L, respectively

condition is applied at the left-hand side boundary as inflow, and the slip boundary conditions are applied at upper and lower boundaries. A Neumann type condition is employed at the outlet. The boundary conditions are shown in Fig. 2a. The grid and time convergence tests are carried out to select the appropriate grid size and time step Δt, respectively. The convergence tests are performed at an actuating forcing amplitude of $\bar{F} = 12.07$. The flow domain is discretised using a mesh of $N_x \times N_y$ grid points, where N_x and N_y indicate the number of Cartesian grid points along horizontal and vertical directions, respectively. The values of $N_x \times N_y$ for Grid-1, Grid-2, Grid-3 and Grid-4 are 796×1192, 718×1017, 660×892 and 614×799, respectively. The convergence studies have been performed for the parameters $\overline{D} = 0.9428$, $\mu = 1.0$ and Re $= 100$. The steady-state solutions for vertical displacement of the foil ($\bar{y}$) and lift coefficient (C_L) are shown in Fig. 3. Time evolutions of $\bar{y}$ and C_L match closely for Grid-1 and Grid-2 as shown in Fig. 3a, b, respectively. Therefore, Grid-2 is considered for the further simulations. For the time convergence test, four different time steps have been chosen ($\Delta\tau = 0.0004$, 0.0002, 0.0001 and 0.00005)

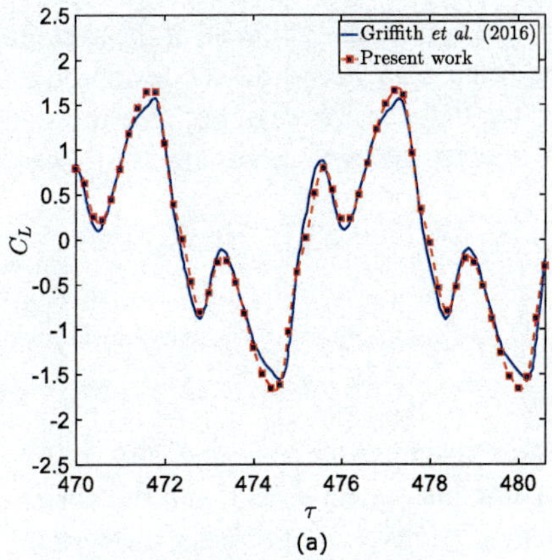

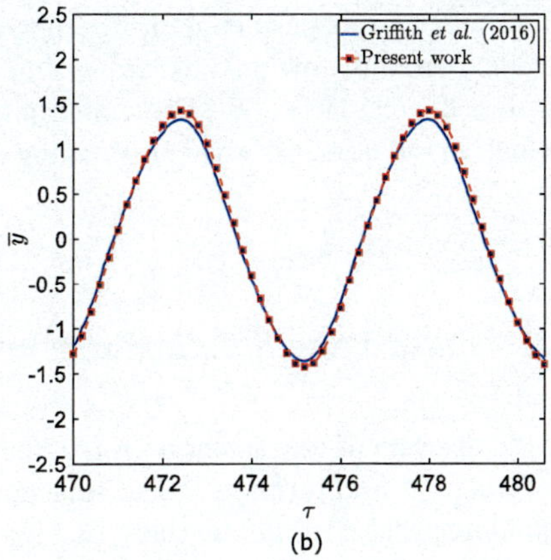

Fig. 4 Comparison of **a** aerodynamic lift coefficients (C_L) and **b** vertical displacement ($\bar{y}$) (present study and Griffith et al. (2016) [16])

to perform the simulations. The time histories of $\bar{y}$ and C_L at these four different $\Delta\tau$ are compared in Figs. 3c, d, respectively. The results for $\Delta\tau = 0.00005$ and 0.0001 are in very good agreement with each other as shown in Figs. 3c, d. Therefore, Grid-2 and time-step size $\Delta\tau = 0.0001$ are selected for the rest of the simulations in this paper.

2.4 Validation

To establish the efficacy of the current FSI solver, it is validated thoroughly with the work of Griffith et al. [16], where the flow past an elastically mounted elliptic cylinder is studied for various aspect ratios (ratio of major to minor axis) of the elliptic section. The harmonic oscillator model governing the heaving motion of the cylinder and the active control implemented via a pitching motion is given in [16]. We present the time histories of lift coefficient (C_L) and vertical displacement ($\bar{y}$) for an aspect ratio of $\Gamma = 6.0$ at Re $= 200$, $\frac{c\zeta}{MU} = 0.503$ (ζ is the damping coefficient), $f_N^* = 0.2$ and $f_D^* = 0.18$ in Fig. 4. Very good agreement is observed between the results from the present FSI solver and that of Griffith et al. [16].

3 Results and Discussion

Due to their small size and low flight speed, the insect flight manifests a low Reynolds number flow around their wing. The present work investigates the dynamical behaviour of the Dipteran flight motor model at Re $= 100$. First, the driving force amplitude ($\bar{F}$) is considered to be zero to examine the natural response of the system inside a viscous flow field. Next, an actuation force with $\bar{F} = 4.82$ is considered to explore the response of the Dipteran flight system in the presence of both the muscle actuation force and aerodynamic loading. The other parametric values are as follows: $\bar{D} = 0.9428$, $\mu = 1.0$.

3.1 Fixed Point Solution at $\bar{F} = 0$

In the absence of any actuation force, the system response damps down and settles to an equilibrium position. The displacement time history for $\bar{F} = 0$ and the corresponding phase portraits are shown in Figs. 5a, b, respectively. In this case, the system has damped down solely due to the flow-induced damping as structural damping has not been considered in the model. The basin of attraction of the system consists of two stable spiral sinks in the two potential wells and a saddle point which is similar to the case of an unforced damped Duffing oscillator. The potential energy of the present system is given by the following equation (Eq. 9) and is plotted as shown in Fig. 6.

$$V = -\frac{1}{4}k_0^2\bar{y}^2 + \frac{1}{8}\frac{k_0^2}{\bar{D}^2}\bar{y}^2. \tag{9}$$

Two potential wells marked with a and c are stable equilibrium points, and the peak point marked with b is an unstable equilibrium point. It is observed that the response settles down at any one of the stable equilibrium points corresponding to one of the potential wells depending on the initial conditions.

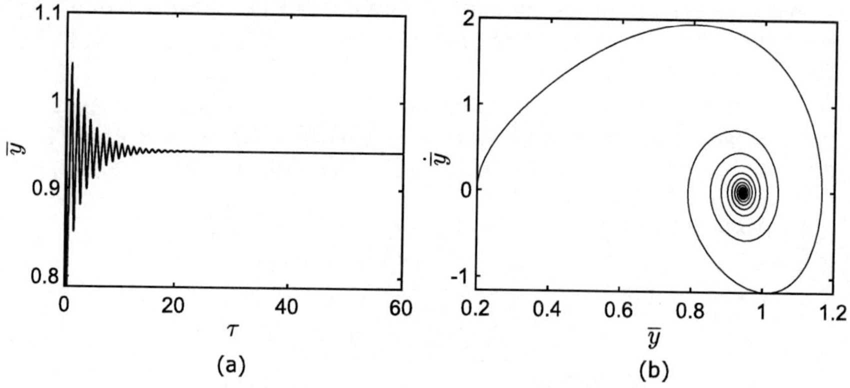

Fig. 5 **a** Displacement time histories and **b** phase portrait for $\bar{F} = 0.0$

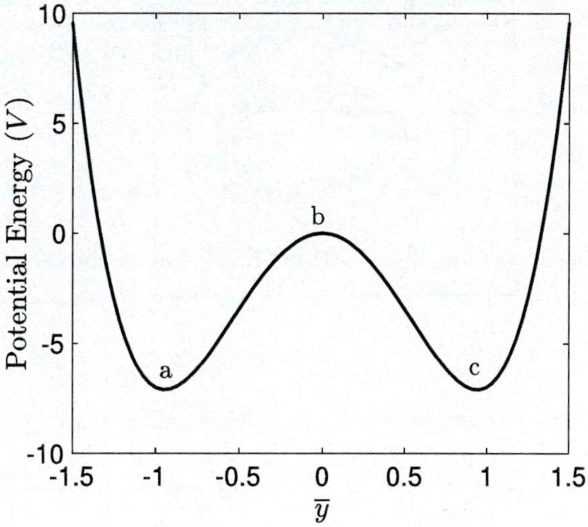

Fig. 6 Schematic of potential well

3.2 Periodic Orbits at Small Amplitudes Forcing

As $\bar{F}$ is increased, the spiral sinks turn into period-1 orbits. One such period-1 response and the corresponding phase portrait and frequency spectrum observed at $\bar{F} = 4.82$ are shown in Figs. 7 and 8, respectively. It is interesting to note from the time history shown in Fig. 7 that the system response jumps into different periodic orbits as the solution progresses in time. These jumps in amplitude and settling into different stable periodic orbits are seen in the corresponding phase portrait (see Fig. 8a). Though the system response jumps into different amplitude orbits, the frequency of oscillation remains the same. The frequency spectrum shown in Fig. 8b

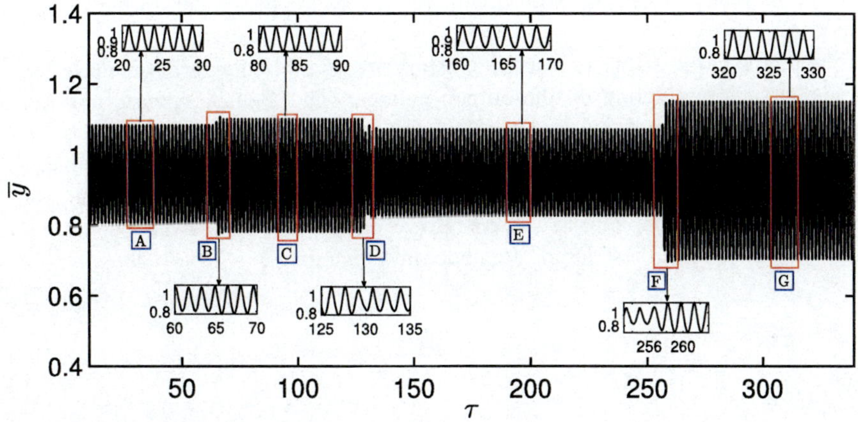

Fig. 7 Displacement time history for $\bar{F} = 4.82$

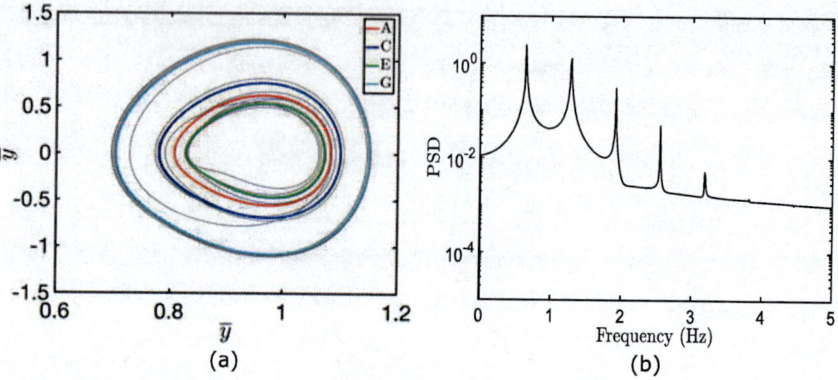

Fig. 8 **a** Phase portrait and **b** frequency spectra for $\bar{F} = 4.82$

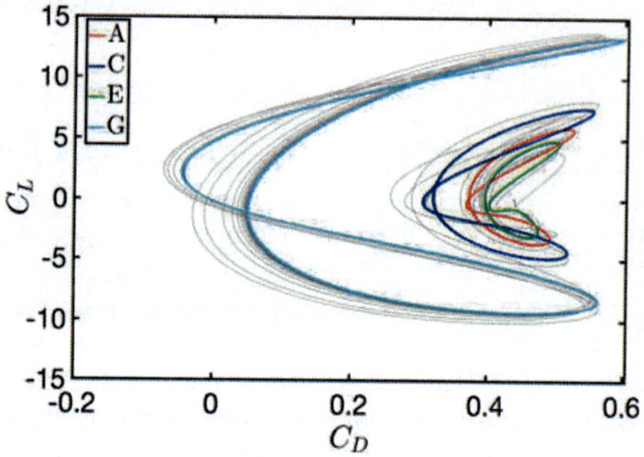

Fig. 9 $C_L - C_D$ phase portraits for $\overline{F} = 4.82$

shows the dominant peak depicting the forcing frequency and its super harmonics with gradually decreasing power, which indicates the periodicity in the system response.

The jumps in the displacement time history are possibly due to the jump in the aerodynamic forces acting on the elliptic cylinder. The $C_L - C_D$ phase portrait, as shown in Fig. 9 is not the same for the complete $C_L - C_D$ time histories, but it gradually changes from having two loops to one single loop and finally to three loops. These changes in the aerodynamic loads seem to be reflected in the displacement time history of the coupled FSI system. However, the specific reasons behind the underlying physics of these jumps are under investigation.

4 Conclusions

In this paper, the dynamic behaviour of a Dipteran flight motor inspired flapping system has been investigated in the low Reynolds number regime. The Dipteran flight motor, coupled with aerodynamic forces resembles a forced Duffing oscillator system. The simulations are performed using an in-house FSI solver developed by combining the Duffing oscillator model with an Immerse Boundary Method-based flow solver. In the absence of muscle actuation, the system response damps down to a fixed point solution due to the fluid added damping. On the other hand, in the presence of a harmonic muscle actuation, the system shows periodic oscillation. However, the response jumps into different periodic orbits as the simulation is run for a long time. Detailed flow-field studies are further being undertaken for understanding the underlying physics associated with this jump phenomenon.

References

1. Ward, T.A., Rezadad, M., Fearday, C.J., Viyapuri, R.: Review of biomimetic air vehicle research: 1984–2014. Int. J. Micro Air Veh. **7**, 375–394 (2015). https://doi.org/10.1260/1756-8293.7.3.375
2. Ellington, C.P.: The novel aerodynamics of insect flight: applications to micro-air vehicles. J. Exp. Biol. **202**, 3439–3448 (1999)
3. Alexander, R.M., Bennet-Clark, H.: Storage of elastic strain energy in muscle and other tissues. Nature **265**, 114–117 (1977). https://doi.org/10.1038/265114a0
4. Harne, R., Wang, K.: Dipteran wing motor-inspired flapping flight versatility and effectiveness enhancement. J. R. Soc. Interface **12**, 20141367 (2015). https://doi.org/10.1098/rsif.2014.1367
5. Lau, G.-K., Chin, Y.-W., Goh, J.T.W., Wood, R.J.: Dipteran-insect-inspired thoracic mechanism with nonlinear stiffness to save inertial power of flapping-wing flight. IEEE Trans. Rob. **30**, 1187–1197 (2014). https://doi.org/10.1109/TRO.2014.2333112
6. Chin, Y.W., Lau, G.K.: 'Clicking' compliant mechanism for flapping-wing micro aerial vehicle. In: IEEE/RSJ International Conference on Intelligent Robots and Systems (IROS), Vilamoura, Portugal, Oct. 7–12, pp. 126–131 (2012). https://doi.org/10.1109/IROS.2012.6385809
7. Dickinson, M. H., and Tu, M. S., "The Function of Dipteran FlightMuscle," Comp. Biochem. Physiol., Part A: Mol. Integr. Physiol.,116, 223-238 (1997)
8. Thomson, A.J., Thompson, W.A.: Dynamics of a Bistable System: The Click Mechanism in Dipteran Flight. Acta Biotheor. **26**, 19–29 (1977). https://doi.org/10.1007/BF00115924
9. Brennan, M., Elliott, S., Bonello, P., Vincent, J.: The 'Click' Mechanism in Dipteran Flight: If It Exists, Then What Effect Does It Have? J. Theor. Biol. **224**, 205–213 (2003). https://doi.org/10.1016/S0022-5193(03)00158-9
10. Liu, H., and Kawachi, K., "A numerical study of insect flight", J. of Comp. Phys., 146, 124-156 (1998). https://doi.org/10.1006/jcph.1998.6019
11. Coen van den, B., and Charles P. E., "The three-dimensional leading-edge vortex of a hovering model hawkmoth", Philosophical Transactions of the Royal Society of London. Series B: Biological Sciences, 352, 329-340 (1997). https://doi.org/10.1098/rstb.1997.0024
12. Jun-duo, Z. and Wei-Xi, H., "On the role of vortical structures in aerodynamic performance of a hovering mosquito", Phys. of Fluids, 31, 051906 (2019). https://doi.org/10.1063/1.5090878
13. Cao, Q., Xiong, Y., Wiercigroch, M.: Novel Model of Dipteran Flight Mechanism. Int. J. Dyn. Control **1**, 1–11 (2013). https://doi.org/10.1007/s40435-013-0001-5

14. Chandan, B., Vikas, R., Sayan, G., Sunetra, S.: Transient and Stable Chaos in Dipteran Flight Inspired Flapping Motion. J. of Comp. and Nonlinear Dyn. **13**, 021014 (2018). https://doi.org/10.1242/jeb.025007
15. J. Kim, D. Kim, H. Choi., "An immersed-boundary finite-volume method for simulations of flow in complex geometries," Journal of Computational Physics ,171, 132-150 (2001). https://doi.org/10.1006/jcph.2001.6778
16. Martin, D., G., David, L. J., Sheridan, J., and Justin, S. L., "Passive heaving of elliptical cylinders with active pitching–from cylinders towards flapping foils", J. of Fluids and Struct., 67, 124-141, (2016). https://doi.org/10.1016/j.jfluidstructs.2016.09.005

Probing into the Efficacy of Discrete Forcing Immersed Boundary Method in Capturing the Aperiodic Transition in the Wake of a Flapping Airfoil

Dipanjan Majumdar, Chandan Bose, and Sunetra Sarkar

1 Introduction

The fundamental understanding of aero/hydro-dynamics of the natural bio-propulsion system has a great significance in efficient design of futuristic Micro Aerial Vehicles (MAV) and Autonomous Underwater Vehicles (AUV). The dynamical signature of the aerodynamic loads generated by a flapping foil inherently depends on the near-field wake topology. The flow topology around the foil, in turn, depends on the kinematic parameters such as flapping amplitude (say, A) and frequency (say, f_e). The combined effect of amplitude and frequency can be better described by the non-dimensional plunge velocity (κh), here κ $(= 2\pi f_e c/U_\infty)$ is the reduced frequency, h $(= A/c)$ is the non-dimensional plunge amplitude, U_∞ is the free stream velocity, and c is the chord length of the airfoil. In a low κh regime, a drag producing Kármán vortex street is observed [1–3] whereas the thrust generation at higher κh is associated with a reverse Kármán vortex street [1–3] in the trailing-wake of a flapping foil. With further increase in κh, a symmetry-breaking bifurcation takes place giving way to a deflected reverse Kármán vortex street [2]. Eventually, the wake becomes aperiodic at high values of κh [6]. With the growth in the computational resources, various low and high fidelity numerical methodologies [4–6] have been employed to study these phenomena. Among these, the Navier-Stokes (N-S) framework can accurately capture the nonlinear wake-structure interactions arising from fluid non-linearity at high κh regime. In recent studies, an interesting chaotic transition has been reported in the flow-field behind a flapping airfoil at significantly high κh values

D. Majumdar (✉) · S. Sarkar
Department of Aerospace Engineering, Indian Institute of Technology Madras, Chennai 600036, India
e-mail: dipanjanmajumdar100@gmail.com

C. Bose
Department of Applied Mechanics, Indian Institute of Technology Madras, Chennai 600036, India

S. K. Saha and M. Mukherjee (eds.), *Recent Advances in Computational Mechanics and Simulations*, Lecture Notes in Mechanical Engineering,
https://doi.org/10.1007/978-981-15-8315-5_24

($\kappa h > 1.5$) [5, 6]. Furthermore, Bose and Sarkar [6] have reported that the leading-edge separation and the interaction of the primary leading-edge vortex (LEV) with the trailing-edge vortex (TEV) structures hold the key to the aperiodic trigger and its subsequent propagation in the far-field.

It is to be noted that most of the aforementioned studies are carried out using body conformal mesh methods (for example, Arbitrary Lagrangian Eulerian (ALE) method). However, in the ALE approach, re-meshing is required to incorporate any movement and/or deformation of the body as the solution is advanced in time, thus increasing the computational overhead and algorithmic complexity. A possible alternative to this approach is the Immersed Boundary Method (IBM). In the IBM framework, the presence of a solid boundary immersed inside a fluid is incorporated by imposing constraints in the governing equations instead of direct application of physical boundary condition at the grid points. Therefore, the complex geometry problems involving moving boundaries can be handled using simple structured orthogonal mesh without regenerating the mesh at every time step. A detailed review of both the approaches can be found in [7, 8]. In IBM, the implementation of boundary conditions is not trivial and suffers from non-physical oscillation of the pressure and velocity fields across the immersed boundary (IB) [12]. These flow-field oscillations near the boundary can affect the formation and growth of the primary LEV structures which are key to the manifestation of the aperiodic trigger. Small discrepancies in capturing these boundary structures may not have significant effect on the overall flow-field in the periodic regime. However, these small perturbations may result in a different dynamical state of the flow-field and lead to an erroneous prediction of the transitional route in the aperiodic regime. Therefore, it is important to thoroughly test the potency of IBM methods in capturing the transitional flow dynamics.

The primary focus of the present work is to investigate if the onset of chaotic transition in the unsteady flow-field behind a plunging foil (at a high κh value: $\kappa h = 2.0$ and $\text{Re} = 300$) can be accurately captured using an in-house IBM solver in comparison to a well-established ALE-based N-S solver. The flow solver has been developed in the C++ environment, broadly following the discrete direct forcing IBM approach proposed by Kim et al. [9]. The simulation results obtained from the IBM solver are compared with the results obtained from a well-established body-fitted ALE solver, in terms of vorticity contours of the flow-field and time history of drag coefficient. The present study aims to compare the accuracy of the IBM solver in capturing the boundary structures such as, the formation, growth, and separation of the primary LEV which provides the trigger to aperiodicity in the near-field. The rest of the paper is organized as follows: The kinematics of body movement and detailed computational methodology of the in-house solver are discussed in Sect. 2. Details of the computational domain, boundary condition, convergence study, and the validation study are presented in Sect. 3. The time evolutions of the aerodynamic forces and the corresponding flow fields around the solid body obtained through the two different numerical approaches are compared in Sect. 4. Finally, the major findings of the present study are concluded in Sect. 5.

2 Computational Methodology

2.1 Governing Equation

A 2D elliptic foil is considered to be plunging harmonically in a uniform free stream and the equation of motion of the center of the foil is given in the non-dimensional form as below,

$$\bar{y}_c(\bar{t}) = h\sin(\kappa\bar{t}), \tag{1}$$

$$\dot{\bar{y}}_c(\bar{t}) = \kappa h\cos(\kappa\bar{t}). \tag{2}$$

Here $\bar{t}$ $(= tU_\infty/c)$ is the non-dimensional time. The present work is focused at low Reynolds number regime where the flow is governed by the incompressible Navier-Stokes equation. The governing equations are solved using the discrete direct forcing Immersed Boundary Method [9]. A forcing term, $\bar{\mathbf{f}}$, is added to the momentum conservation equation to reconstruct the velocity field around the IB and thus the no slip-no penetration boundary conditions are satisfied. Also a source/sink term, $\bar{q}$, is added to the continuity equation to ensure the mass conservation. Therefore, in the framework of the discrete forcing IBM [9], the governing equations are given by,

$$\frac{\partial \bar{\mathbf{u}}}{\partial \bar{t}} + \bar{\nabla}\cdot(\bar{\mathbf{u}}\bar{\mathbf{u}}) = -\bar{\nabla}\bar{p} + \frac{1}{Re}\bar{\nabla}^2\bar{\mathbf{u}} + \bar{\mathbf{f}}, \tag{3}$$

$$\bar{\nabla}\cdot\bar{\mathbf{u}} - \bar{q} = 0. \tag{4}$$

Here, $\bar{\mathbf{u}}$ $(= \mathbf{u}/U_\infty)$ is non-dimensional flow velocity and $\bar{p}$ $\left(= p/(\rho U_\infty^2)\right)$ is non-dimensional pressure; Re $(= U_\infty c/\nu)$ is the Reynolds number. $\bar{\nabla}$ is the non-dimensional form of the operator ∇. Here ρ and ν are the density and kinematic viscosity of the fluid, respectively. Henceforth, all the discussion will be done based on the non-dimensional quantities and for the sake of typographical ease, the over-bar sign $(\bar{\ })$ on the symbols, representing non-dimensional quantities, will not be used.

2.2 Numerical Method

The flow governing equations are solved on a background Cartesian grid and the location of the solid body is tracked by a set of Lagrangian markers. The grid points that are on or inside the solid domain are marked as solid points and the grid points that are outside the solid domain are marked as fluid points. Figure 1 shows a schematic of the mesh grid classification. The grid points, on or inside the solid domain, where the flow velocities are reconstructed by applying the momentum forcing, will be referred as forcing points (Fig. 1). The momentum forcing is applied only to the grid points residing just inside the solid domain and have at least one neighbor in the fluid

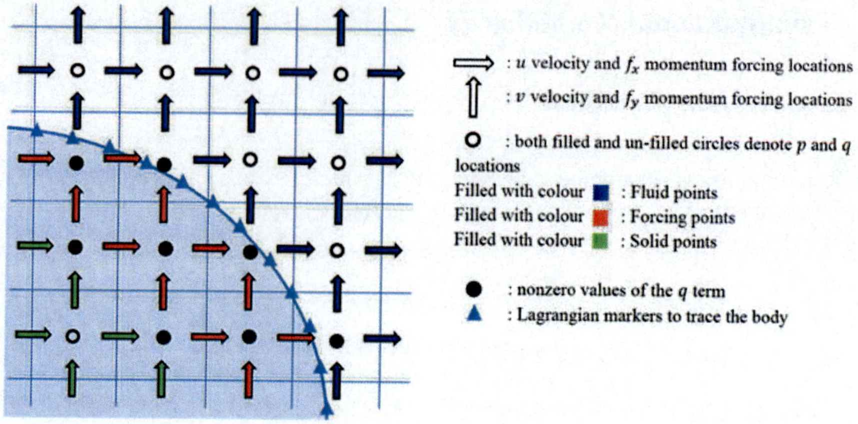

Fig. 1 Schematic of the mesh grid depicting the u, v, f_x, f_y, p, and q locations and the grid points classifications. Shaded area inside the IB denotes the solid domain

domain. Flow governing equations (Eqs. (3) and (4)) are solved using a finite-volume-based second-order semi-implicit Fractional Step Method (FSM) on a staggered grid arrangement as shown in Fig. 1. A second-order Crank-Nicolson method is used to discretize the diffusion term whereas the convection term is advanced in time using Adams-Bashforth discretization. A pseudo pressure correction term is used to correct the velocity to satisfy continuity at every time step. In the present framework of FSM, the discretized form of the governing equations are as follows:

$$\frac{\hat{\mathbf{u}}^k - \mathbf{u}^{k-1}}{\Delta t} = -\nabla p^{k-1} - \frac{3}{2} H\left(\mathbf{u}^{k-1}\right) + \frac{1}{2} H\left(\mathbf{u}^{k-2}\right) + \frac{1}{2\mathrm{Re}} L\left(\hat{\mathbf{u}}^k\right) + \frac{1}{2\mathrm{Re}} L\left(\mathbf{u}^{k-1}\right) + \mathbf{f}^k, \tag{5}$$

$$\nabla^2 \phi^k = \frac{1}{\Delta t}\left(\nabla \cdot \hat{\mathbf{u}}^k - q^k\right), \tag{6}$$

$$\mathbf{u}^k = \hat{\mathbf{u}}^k - \Delta t \nabla \phi^k, \tag{7}$$

$$p^k = p^{k-1} + \phi^k - \frac{\Delta t}{Re} \nabla^2 \phi^k, \tag{8}$$

where $L(*)$ is the discretized form of the Laplace operator and $H(*)$ denotes the discretized form of the convection term; $\hat{\mathbf{u}}^k$ denotes intermediate velocity at the kth time step, ϕ is pseudo pressure correction term and Δt denotes time step size.

To solve the discretized momentum equation Eq. (5) for intermediate velocities $\hat{\mathbf{u}}^k$, the momentum forcing term $\mathbf{f}^k$ must be known a priori. Kim et al. [9] suggested to evaluate $\mathbf{f}^k$ at the forcing points by advancing momentum equation explicitly. Thus an estimation of the intermediate velocity is computed at all the grid points around the IB as follows:

$$\frac{\widetilde{\mathbf{u}}^k - \mathbf{u}^{k-1}}{\Delta t} = -\nabla p^{k-1} - \frac{3}{2} H\left(\mathbf{u}^{k-1}\right) + \frac{1}{2} H\left(\mathbf{u}^{k-2}\right) + \frac{1}{\mathrm{Re}} L\left(\mathbf{u}^{k-1}\right) + \mathbf{f}^k. \quad (9)$$

Equation (9) is solved for $\widetilde{\mathbf{u}}^k$ (an estimate of intermediate velocity $\hat{\mathbf{u}}^k$) considering $\mathbf{f}^k = 0$. Then, $\mathbf{f}^k$ at a forcing point, is evaluated using the following equation,

$$\mathbf{f}^k = \frac{\mathbf{U}_f^k - \widetilde{\mathbf{u}}^k}{\Delta t}, \quad (10)$$

where $\mathbf{U}_f^k$ is the desired velocity at a forcing point that ensures no slip-no penetration boundary conditions on the IB surface. The discrete forcing approach of IBM is essentially setting desired values of velocity ($\mathbf{U}_f^k$) to the forcing points such that appropriate boundary conditions are satisfied exactly on the immersed boundary itself. For the special case of a forcing point lying exactly on the immersed boundary, $\mathbf{U}_f^k = \mathbf{U}_\Gamma^k$, where $\mathbf{U}_\Gamma^k$ is the velocity of the IB at that particular point at the kth time step. However, in general, most of the forcing points do not lie on the IB as the mesh grid does not conform with the IB. For those cases, the desired velocities at the forcing points are to be evaluated using appropriate extrapolation scheme from the velocity at the surrounding fluid nodes and the velocity of the boundary movement.

The mass source/sink term (q^k) in the discretized pressure correction equation (Eq. 6) can be evaluated using the intermediate velocity field ($\hat{\mathbf{u}}^k$) in the following way [10]:

$$q^k = \frac{1}{\Delta A} \sum_{i=1}^{4} \alpha_i \left(\hat{\mathbf{u}}^k - \mathbf{U}_\Gamma^k\right) \cdot \mathbf{n} \Delta l_i, \quad (11)$$

where ΔA is the area of a grid cell, $\mathbf{n}$ is unit outward normal at each cell face, Δl_i is the length of each cell face and α_i is a flag at each cell face and is defined as 1 only at the cell faces where momentum forcing is applied (i.e., where $\mathbf{f}^k \neq 0$) and otherwise zero.

The fluid dynamic loads $\mathbf{F}$ (in non-dimensional form) on the solid body are evaluated using the following equation as given in Lee et al. [11]:

$$\mathbf{F} = -\int_{\Omega_b} \mathbf{f}\, \mathrm{d}V + \int_{\Omega_b} \left(\frac{\partial \mathbf{u}}{\partial t} + \nabla \cdot (\mathbf{uu})\right) \mathrm{d}V, \quad (12)$$

where Ω_b is the control volume bounded by the solid boundary. The lift coefficient $\left(C_L = L/0.5\rho u_\infty^2 c\right)$ and drag coefficient $\left(C_D = D/0.5\rho u_\infty^2 c\right)$ are defined in the usual way and evaluated as $C_L = 2F_y$ and $C_D = 2F_x$, where F_x and F_y are the components of the non-dimensional aerodynamic load $\mathbf{F}$ along the x and y-axis, respectively.

Discrete forcing IBM approaches are known to suffer from spurious oscillations of the pressure and velocity field near the IB due to the temporal and spatial discontinuities [11, 12]. The present study explores the capability of the IBM tools in-hand in capturing the chaotic behavior in near-field vortex structures in the unsteady wake of a plunging foil, where any small discrepancies in resolving the boundary structures will result in incorrect prediction of the aperiodic transition. In this context, the present study investigates the ability of the discrete forcing IBM in capturing the near-field vortex interactions as compared to a well-validated body-fitted ALE solver. The convergence and validation of the ALE solver was presented in Bose and Sarkar [6]. The simulations for the ALE results are performed using a finite-volume-based CFD solver OpenFOAM [13]. One can find an extensive validation of the OpenFOAM tool in Boss Ph.D. Thesis [14].

3 Convergence and Validation of IBM Solvers

A rectangular computational domain of size $[-10c, 25c] \times [-12.5c, 12.5c]$ is considered in the IBM solver. A Dirichlet boundary condition ($u = 1$, $v = 0$) is used at the inlet of the computational domain. Slip boundary condition ($\partial u/\partial y = 0$, $v = 0$) is applied at the top and bottom boundaries and a convective flow boundary condition ($\partial \mathbf{u}/\partial t + u_s \partial \mathbf{u}/\partial x = 0$, u_s is the average stream wise velocity) is used at the outlet. A time step size of $\Delta t = 0.0001$ is chosen after performing a time convergence study. The time convergence results are not shown here for the shake of brevity. A grid independence test is also performed considering four minimum grid sizes of $\Delta x = \Delta y = 0.003$, $\Delta x = \Delta y = 0.004$, $\Delta x = \Delta y = 0.005$, and $\Delta x = \Delta y = 0.006$ in the region of body movement. Figure 2 shows that the time traces of the aerodynamic loads at the grid sizes of $\Delta x = \Delta y = 0.003$ and $\Delta x = \Delta y = 0.004$ have almost no difference and also are in good agreement with the reference results of the ALE solver. Therefore we choose the minimum grid size of $\Delta x = \Delta y = 0.004$ for all the following simulations.

The present IBM solver has been validated thoroughly by comparing the results with that of Lewin and Haj-Hariri [5] at various κh values. Figure 3 shows that the

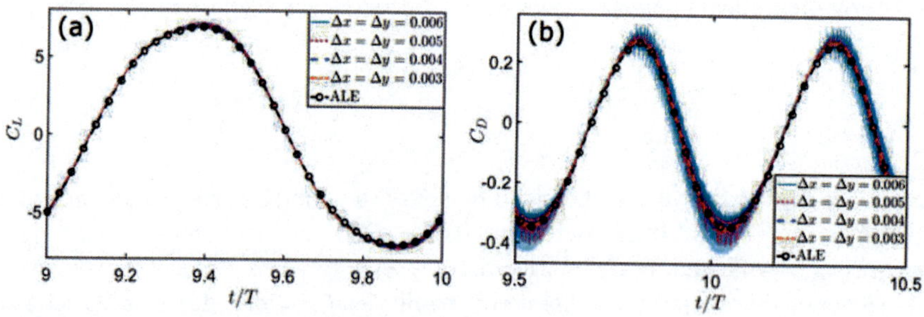

Fig. 2 Grid convergence study (**a** Lift coefficient and **b** Drag coefficient) of IBM solver for the flow over an plunging foil at $h = 0.25$ wtih $\kappa h = 1.0$

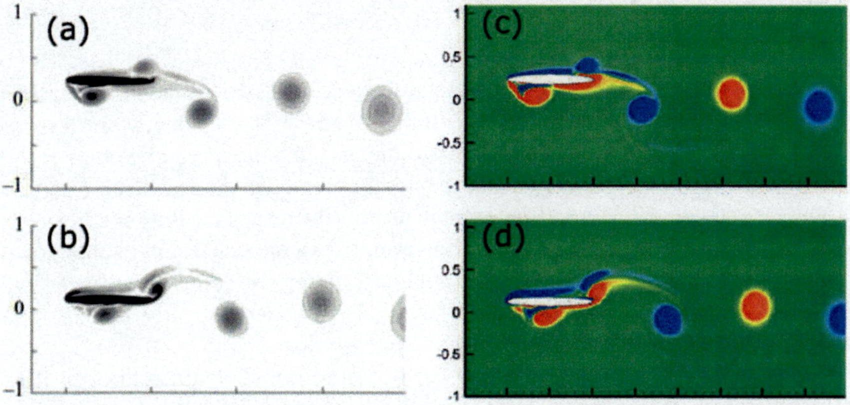

Fig. 3 Vorticity contours at different time instants during the down-stroke of a plunging foil at $\kappa = 3.333$ and $\kappa h = 0.8$ at $\mathrm{Re} = 500$. In the figure **a, b** Lewin and Haj-Hariri [5] and **c, d** present IBM solver

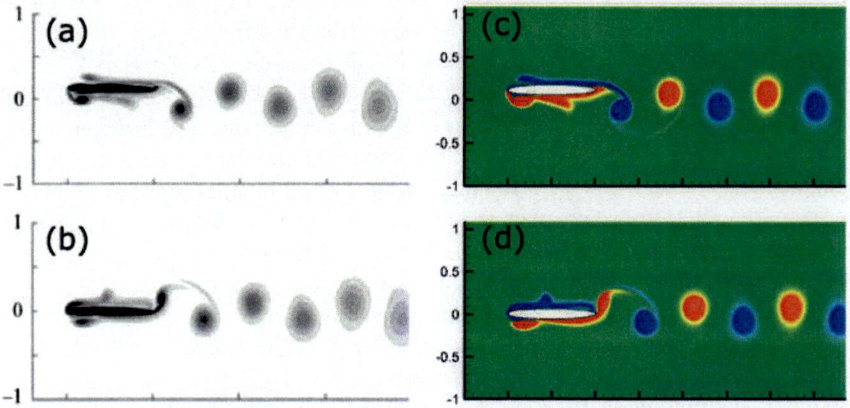

Fig. 4 Vorticity contours at different time instants during the down-stroke of a plunging foil at $\kappa = 6.667$ and $\kappa h = 0.8$ at $\mathrm{Re} = 500$. In the figure **a, b** Lewin and Haj-Hariri [5] and **c, d** present IBM solver

LEV is getting entrained in the TEV and only one vortex is shed into the wake per stroke for $\kappa = 3.333$ and $\kappa h = 0.8$ at $\mathrm{Re} = 500$. Instantaneous vorticity contours during the down-stroke at $\kappa = 6.667$ and $\kappa h = 0.8$ are depicted in Fig. 4 which shows the presence of a well-organized reverse Kármán vortex street in the trailing-wake. These results show very good agreement with the results of Lewin and Haj-Hariri [5].

4 Results and Discussions

Since the present study focuses on investigating the accuracy of discrete forcing IBM approach in capturing the chaotic trigger in the unsteady flow-field, numerical simulations have been performed at a high non-dimensional plunge velocity of $\kappa h = 2.0$ at Re = 300. At this κh value, chaotic behavior is observed in the near-field structures around the foil which, in turn, is reflected in the time history of the drag coefficient as captured by the ALE solver; see Fig. 5. However, IBM approach results in unphysical oscillations in the velocity field near the boundary giving rise to inaccurate near-field structures cascading into incorrect flow-field behavior. Consequently, it ends up in predicting the wrong onset of chaos in the unsteady flow-field.

Time histories of C_D and the $C_L - C_D$ phase portraits, obtained through IBM and ALE, are compared in Fig. 6. It is seen that after some initial transients (beyond $t/T = 3$), the ALE approach results in chaotic time history of C_D (Fig. 6b). On the contrary, IBM approach fails to demarcate the aperiodic trigger from the flow-field correctly and the flow-field remains almost periodic (or quasi-periodic) till $t/T = 10$, beyond which it becomes aperiodic as reflected in the C_D time history as shown in Fig. 6a. The $C_L - C_D$ phase portraits also reflect the similar transition. In order to clearly visualize the delayed onset to aperiodicity captured by IBM, the trajectories in the $C_L - C_D$ phase-space are plotted in two different colors and line types, as can be seen from Fig. 6c, d; the solid-blue color line denotes the trajectories between 4th and 10th cycles and the dashed-ash color line represents the trajectories between 11th and 20th cycles. It is seen that the initial part (solid-blue) of the phase-portrait between 4th and 10th cycles, obtained from IBM approach (Fig. 6c), is indicative of a quasi-periodic attractor whereas the latter part (dashed-ash) reflects a chaotic attractor. Contrastingly, the phase portraits obtained from the body-fitted ALE solver (Fig. 6d) are representative of a chaotic attractor throughout.

It is also important to investigate the underlying vortex interactions which lead to the incorrect prediction of the chaotic onset in the case of IBM. Next, we compare the flow fields captured by the IBM and ALE solvers in relation to predict the accurate flow dynamics behind a flapping foil. The leading-edge separation phenomenon during the 4th flapping cycle captured by these two solvers are compared in Fig. 7. The flow-field captured by the ALE solver shows that the leading-edge separation

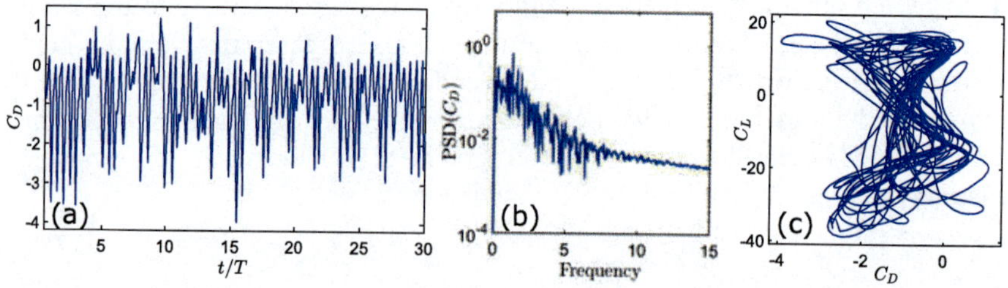

Fig. 5 **a** C_D time history **b** Fourier spectra of C_D and **c** $C_L - C_D$ phase-portrait at $\kappa h = 2.0$

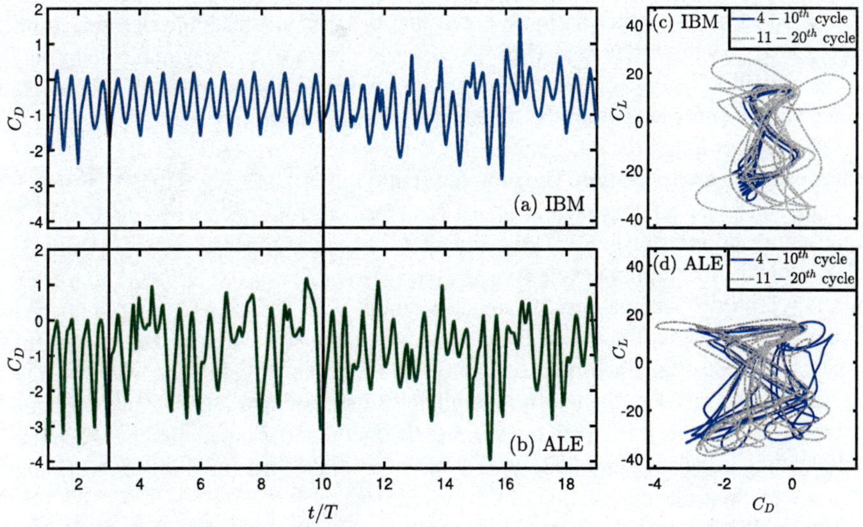

Fig. 6 Comparison of C_D time histories and $C_L - C_D$ phase portraits obtained from the two numerical approaches at $\kappa h = 2.0$

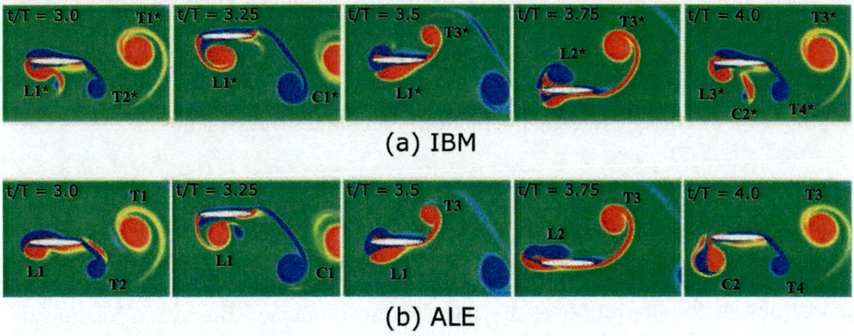

Fig. 7 Comparison of near-field structures during the 4th flapping cycle, obtained from the two numerical solvers

becomes aperiodic at the onset of aperiodicity during the 4th flapping cycle; see Fig. 7b.

However, IBM solver could not capture this aperiodic trigger and predicts a periodic leading-edge separation till $t/T = 10$. In case of IBM, two counter rotating TEVs (T1* and T2*) developed during the 3rd flapping cycle form a vortex couple C1* and become the primary wake structure; see $t/T = 3.0$ to $t/T = 3.25$ of Fig. 7a. However, near the leading-edge a counter-clockwise LEV (L1*) is separated and then is seen to reattach at the mid surface of the foil as the foil moves downward during the 4th cycle ($t/T = 3.0$ to $t/T = 3.5$, Fig. 7a). Same series of events takes place in an opposite sense of rotation during the next half-cycle ($t/T = 3.5$ to

$t/T = 4.0$, Fig. 7a). It is also to be noted that the primary leading-edge structures do not undergo any interaction with the TEVs in this case and have no role to play in the formation of trailing-wake. The wake remains nearly periodic till $t/T = 10$, which can be confirmed from the time history of C_D shown in Fig. 6a.

On the other hand, the ALE solver captures the actual trigger from the flow-field which causes the aperiodicity. The flow-field captured by ALE shows that the primary counter-clockwise leading-edge structure (L1) form a strong vortex couple (C2) with equal contribution from the clockwise shear layer from the upper surface of the elliptic foil during the 4th flapping cycle (see $t/T = 3.0$ to $t/T = 4.0$, Fig. 7b). This vortex couple (C2) traverses in a circular arc and collides with the stronger couple formed by the TEVs T3 and T4, the otherwise primary wake structure, in the subsequent cycles; thus destroying the organized pattern of the wake. Notably, the leading-edge separation patterns are seen to be completely different in the subsequent cycles sustaining the chaos. It is worth mentioning the fact that the interaction between the main leading-edge structure (C2) with the primary wake structures is seen to be the key factor behind the destruction of the periodic wake.

The IBM solver could not capture the actual aperiodic trigger at the 4th flapping cycle due to its inability to capture the formation of the strong vortex couple near the leading-edge. Though the IBM solver captures a weak vortex couple C2* (see $t/T = 4.0$, Fig. 7a), it got diffused and shredded very fast and thus could not interact with the TEVs. The error associated with the unphysical oscillation in the pressure and velocity field near the IB eventually results in a different boundary layer development, thus showing discrepancies in the formation, growth, and evolution of the primary leading-edge structures and their interaction with the trailing-wake.

5 Conclusion

The efficacy of the conventional discrete forcing Immersed Boundary Method in accurately capturing the flow-field transitions behind a plunging airfoil has been investigated in this paper. The flow past a plunging elliptic foil at a high plunge velocity is simulated using an in-house IBM solver. The outcomes of this IBM solver are compared with that of a well-validated body-fitted ALE method-based solver in OpenFOAM. Though the IBM approach is able to capture the flow dynamics and the aerodynamic loads in good agreement in the periodic regime at lower κh values, it fails to accurately predict the correct aperiodic trigger from the flow-field at high κh value as predicted by the well-established ALE solver. The authors are currently working on implementing the needful modification in the IBM solver so that the unphysical oscillation of the velocity field near the immersed boundary can be reduced and the aperiodic transition in the unsteady flow-field can be captured accurately.

References

1. Koochesfahani, M.M.: Vortical patterns in the wake of an oscillating airfoil. AIAA J. **27**, 1200–1205 (1989). https://doi.org/10.2514/3.10246
2. Jones, K.D., Dohring, C.M., Platzer, M.F.: Experimental and computational investigation of the Knoller-Betz effect. AIAA J. **36**, 1240–1246 (1998). https://doi.org/10.2514/2.505
3. Lai, J.S., Platzer, M.: Jet characteristics of a plunging airfoil. AIAA J. **37**, 1529–1537 (1999). https://doi.org/10.2514/2.641
4. Lin, H., Vezza, M., Galbraith, R.M.: Discrete vortex method for simulating unsteady flow. AIAA J. **35**, 494–499 (1997). https://doi.org/10.2514/2.122
5. Lewin, G.C., Haj-Hariri, H.: Modelling thrust generation of a two-dimensional heaving airfoil in a viscous flow. J. Fluid Mech. **492**, 339–362 (2003). https://doi.org/10.1017/S0022112003005743
6. Bose, C., Sarkar, S.: Investigating chaotic wake dynamics past a flapping airfoil and the role of vortex interactions behind the chaotic transition. Phys. Fluids **30**, 047101 (2018). https://doi.org/10.1063/1.5019442
7. Mittal, R., Iaccarino, G.: Immersed boundary methods. Annu. Rev. Fluid Mech. **37**, 239–261 (2005). https://doi.org/10.1146/annurev.fluid.37.061903.175743
8. Hou, G., Wang, J., Layton, A.: Numerical methods for fluid-structure interaction: a review. Commun. Comp. Phys. **12**, 337–377 (2012). https://doi.org/10.4208/cicp.291210.290411s
9. Kim, J., Kim, D., Choi, H.: An immersed-boundary finite-volume method for simulations of flow in complex geometries. J. Comp. Phys. **171**, 132–150 (2001). https://doi.org/10.1006/jcph.2001.6778
10. Lee, I., Choi, H.: A discrete-forcing immersed boundary method for the fluid-structure interaction of an elastic slender body. J. Comp. Phys. **280**, 529–546 (2015). https://doi.org/10.1016/j.jcp.2014.09.028
11. Lee, J., Kim, J., Choi, H., Yang, K.-S.: Sources of spurious force oscillations from an immersed boundary method for moving-body problems. J. Comp. Phys. **230**, 2677–2695 (2011). https://doi.org/10.1016/j.jcp.2011.01.004
12. Nicolaou, L., Jung, S.Y., Zaki, T.A.: A robust direct-forcing immersed boundary method with enhanced stability for moving body problems in curvilinear coordinates. Comp. Fluids. **119**, 101–114 (2015). https://doi.org/10.1016/j.compfluid.2015.06.030
13. Jasak, H., Jemcov, A., Tukovic, Z.: OpenFOAM: A c++ library for complex physics simulations. In: International workshop on coupled methods in numerical dynamics, pp. 1–20. IUC Dubrovnik Croatia (2007). https://foam-extend.fsb.hr/wp-content/uploads/2018/04/Jasak_Jemcov_Tukovic_OpenFOAM_C_library_for_complex_physics_simulations_2007.pdf
14. Bos, F.M.: Numerical simulations of flapping foil and wing aerodynamics: mesh deformation using radial basis functions (2010). https://repository.tudelft.nl/islandora/object/uuid:edecee10-7db2-443e-8cdf-d41aaddaa1b2?collection=research

Numerical Investigation of Switching of a Jet Generated by a Foil Pitching in Still Fluid

Chayanit Nigaltia and Sachin Y. Shinde

1 Introduction

Flying and swimming animals have motivated researchers to study flapping foils. Wings of aircraft and birds, propeller and helicopter rotor blades, tails of fish all have cross-sections of airfoil shape. For this reason, airfoil has remained a topic of immense interest for a long time.

How the natural flyers generate power for propulsion has been a topic of keen interest for researchers. Flapping foils have been studied extensively for the same. The phenomenon of thrust generation by a flapping foil in a fluid flow is known as Knoller–Betz effect. Jones et al. [1] experimentally and numerically investigated the Knoller–Betz effect on sinusoidally plunging airfoil. They reported that these unsteady thrust producing wakes are mainly due to inviscid phenomenon. Deflected wakes were also observed by them, leading to the production of lift and drag. Examination of the mean streamwise velocity field of a plunging NACA0012 airfoil in the absence of freestream velocity by Lai and Platzer [2] revealed that the velocity profiles were independent of the frequency of oscillation, when the jet velocity and the lateral distance were nondimensionalized by the peak plunge velocity and the amplitude of oscillation, respectively. Numerical viscous simulations by Lewin and Haj-Hariri [3] on a sinusoidally heaving symmetric airfoil, over a range of frequencies and amplitudes showed distinct solutions. Aperiodic and asymmetric solutions were also reported. Heathcote and Gursul [4] studied the jet switching phenomenon for a periodic plunging airfoil with varying stiffness in zero freestream velocity. The jet switching phenomenon was observed to be quasi-periodic. It was also observed that the switching frequency increases with increasing plunge frequency and plunge

C. Nigaltia (✉) · S. Y. Shinde
Indian Institute of Technology, Kanpur 175 005, India
e-mail: f1dr.chayanit@gmail.com

S. K. Saha and M. Mukherjee (eds.), *Recent Advances in Computational Mechanics and Simulations*, Lecture Notes in Mechanical Engineering,
https://doi.org/10.1007/978-981-15-8315-5_25

amplitude, and the amplitude of jet switching showed a decrease with decreasing Strouhal number ($St = fA/U_{\infty}$). Godoy-Diana et al. [5] experimentally studied the vortex wakes produced by a high aspect ratio flapping foil in a hydrodynamic tunnel. Transition from Benard-von Karman (BvK) to reverse Benard-von Karman (rBvK) and the symmetry breaking to asymmetric wakes was discussed. It was also observed that the transition from a BvK to rBvK wake precedes the actual drag-thrust transition. Godoy-Diana et al. [6] studied the vortex wakes, and the deflection angle of the mean jet is used to specify the asymmetry in the wake. A quantitative model based on the effective phase velocity ($U_P{}^*$), which is the velocity along the dipole direction induced by the vortex street, excluding the dipole velocity and the free-stream velocity, to characterize the symmetry breaking was derived. Cleaver et al. [7] did an experimental investigation on plunging NACA0012 airfoil undergoing small amplitude and high-frequency oscillations with varying angle of attack (α). Bifurcations were observed at high St for the cases in which $\alpha \leq 10°$. Shinde and Arakeri [8] experimentally studied the flow produced by a rigid pitching airfoil in quiescent water. A weak jet was observed which changes direction continuously. This meandering was random and independent of initial conditions.

To study this switching phenomenon in great detail, the data has been captured for large number of cycles and at highly resolved time-steps. Acquiring this data experimentally is difficult, so numerical simulations have been performed.

The chapter is organized as follows: Sect. 2 presents the numerical formulation of the problem. Section 3 describes the jet switching pattern from a rigid pitching foil in still fluid. Some criteria are laid to classify the jet deflection as upward, downward or aligned along the centerline. The data obtained has been organized in a way to observe the switching process clearly at highly resolved time-steps. Finally, in Sect. 4, the conclusions are collected and the future scope of the research work has been discussed.

2 Numerical Formulation

2.1 Problem Statement and Flapping Kinematics

Simulations have been conducted to study the flow generated by an airfoil pitching in the absence of freestream velocity. We used open source software OpenFOAM 5.0 for simulations. NACA0015 airfoil has been considered for investigating the flow field. The motion profile is given as

$$\theta = \theta_{\max} \sin(2\pi f t) \tag{1}$$

where θ is the instantaneous pitching angle, $\theta_{\max}$ is the amplitude of oscillation, f is the pitching frequency and t is the time.

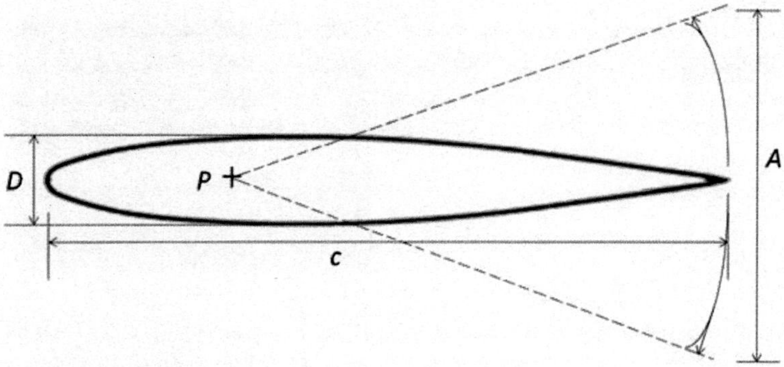

Fig. 1 Schematic diagram of a pitching foil

The peak-to-peak amplitude of the airfoil TE is denoted by A, chord by c, maximum airfoil thickness by D and the airfoil pitching point by P (Fig. 1).

The amplitude of oscillation ($\theta_{\max}$) and the pitching frequency (f) have been varied, keeping the mean angle of attack of the airfoil equal to zero and the flow has been studied. We studied four amplitudes ($\pm 5°$, $\pm 10°$, $\pm 15°$ and $\pm 20°$) and four frequencies (1, 2, 3 and 4 Hz) for each amplitude, thus constituting an all 16 cases.

2.2 *Governing Equations and Discretization*

Flow equations are solved using the arbitrary Lagrangian Eulerian (ALE) formulation on a mesh, which deforms with time. ALE form of incompressible Navier–Stokes equations is written as

$$\frac{\partial \mathrm{U}}{\partial x} + \nabla.[(U - U_m)U] = -\nabla\left(\frac{p}{\rho}\right) + \upsilon\nabla^2 U \tag{2}$$

where ρ is the fluid density, U is the velocity of the fluid, U_m is the mesh velocity, p is the pressure and υ is the kinematic viscosity of the fluid.

Temporal discretization is first-order implicit Euler and a maximum Courant number ($Co_{\max}$) of 0.8 is used to vary the corresponding time-step. The flux values are interpolated from cell centers to face centers linearly. Convective term is discretized using second-order upwinding and the diffusive term using central differencing.

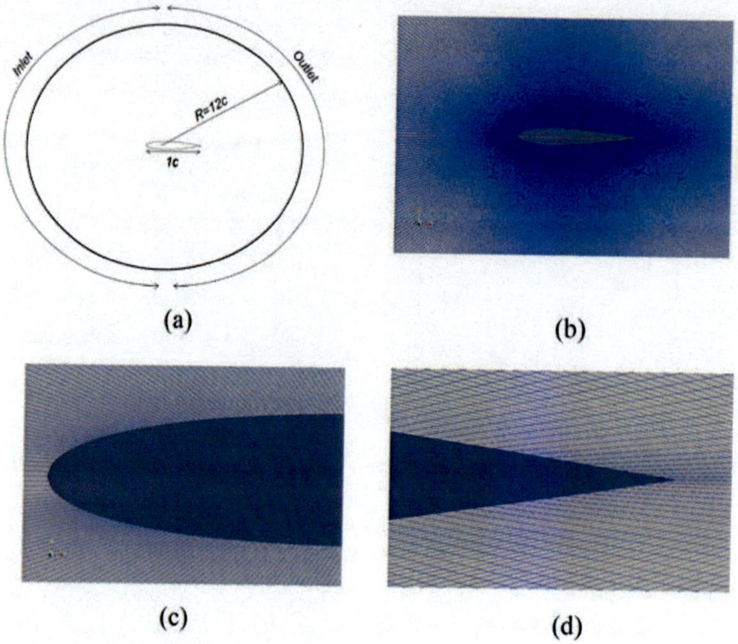

Fig. 2 **a** Computational domain, **b** grid around the airfoil, **c** closer view near LE, **d** closer view near TE

2.3 Computational Domain and Boundary Conditions

A schematic diagram of the computational domain is shown in Fig. 2. The domain is circular in shape with radius 12*c*, where *c* is the chord length of the airfoil with center at the quarter-chord point, which is the pitch point of the airfoil. A circular region of radius 5*c* undergoes a pure pitching motion, while the outside grid remains stationary. The boundaries are far enough to eliminate boundary effects.

For velocity, a constant free stream boundary condition is kept at the inlet and a zero-gradient condition at the outlet boundary. On the airfoil, a moving wall boundary condition is imposed with no flux normal to the wall. A zero pressure gradient is kept at the inlet and on the airfoil, while a constant pressure condition is set at the outlet boundary.

2.4 Solver Details and Validation

Treatment of velocity equation is done using preconditioned bi-conjugate gradient (PBiCG) solver with diagonal incomplete-LU (DILU) preconditioner. The pressure equation has been handled using preconditioned conjugate gradient (PCG)

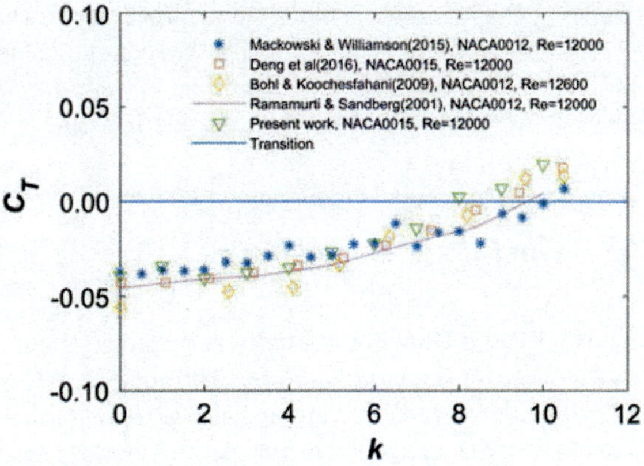

Fig. 3 Mean thrust coefficient (C_T) versus reduced frequency (k) plot of the airfoil with a pitching amplitude of $\theta_{\max} = \pm 2°$. Data from the literature are included for comparison

solver with diagonal incomplete-Cholesky (DIC) preconditioner. The velocity and the pressure equations are coupled using PISO algorithm.

The flow solver is validated quantitatively with the data from the literature. NACA0015 airfoil is used for validation with pitching point at quarter-chord distance from the LE. Figure 3 shows quantitative validation of the solver wherein the thrust coefficient (C_T) is plotted against the reduced frequency k defined as $k = (2\pi fc)/(2U_\infty)$. The plot shows good match with the previous studies. In comparison to the experimental study by Mackowski and Williamson [9] with NACA0012, for low and high k values, our C_D and C_T values are higher, respectively. This is due to the fact that NACA0015 being thicker than NACA0012 will produce larger drag force at low frequencies and at high frequencies will produce higher thrust. This fact is also reported by Jian Deng et al. [10]. The results have also been compared with the viscous simulations on NACA0012 by Ramamurti and Sandberg [11] and experimental results of Bohl and Koochesfahani [12].

A qualitative validation of the solver is also carried out by comparing the vorticity fields of the flow generated with the flow from Schnipper et al. [13] and Bohl and Koochesfahani [12]. We also performed a mesh independence study and observed that a mesh with 4,27,880 nodes gives satisfactory results. More details about the solver and validation can be found in Nigaltia's M.Tech thesis [14].

3 Results and Discussion

The results are discussed in detail for one case, namely $\theta_{\max} = \pm 10°, f = 2$ Hz. The detailed features of the instantaneous and the time-averaged flow are presented, followed by discussion on the criterion used in the literature to detect and classify jet deflection. Later, our own criterion has been defined and described based on location of maximum resultant velocity magnitude. Further, a detailed investigation of the jet switching and meandering patterns is presented, and the jet switching is classified into two patterns—abrupt and gradual. The temporal evolution of abrupt and gradual switching is studied in detail. Finally, the section is concluded by presenting a detailed and thorough investigation of the effect of variation in amplitude and frequency of pitching on the jet switching and meandering patterns.

3.1 The Flow

Figures 4 and 5 show the vorticity and velocity fields for three different instances, respectively, for the case $\theta_{max} = \pm 10°$ and $f = 2$ Hz, when the jet shows different deflection: downward, upward and along the centerline. In the absence of freestream velocity, there is no agency to convect the vortices away from the place of shedding. This is the main reason for the production of an inclined flow as commented by Shinde and Arakeri [8].

The time averaged flow is shown in Fig. 6. The averaged velocity and vorticity fields do not show the irregular patches as seen in the instantaneous flow fields. As will be evident later, the jet deflects and meanders randomly, which result in the

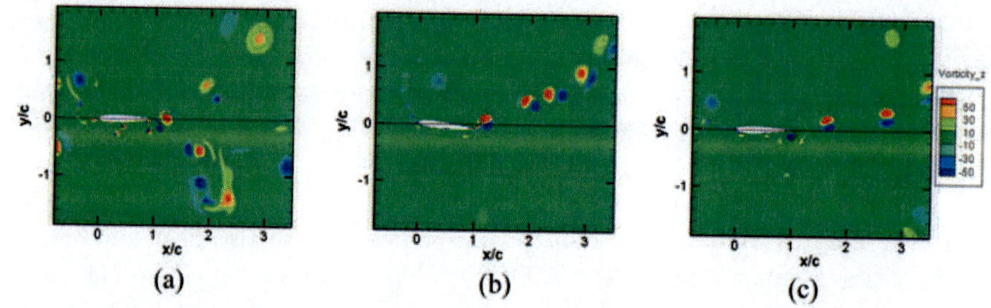

Fig. 4 Instantaneous spanwise vorticity contours at three different instances for the case $\theta_{max} = \pm 10°$ and $f = 2$ Hz, when the jet is **a** inclined in the downward direction, **b** inclined in the upward direction and **c** nearly along the centerline

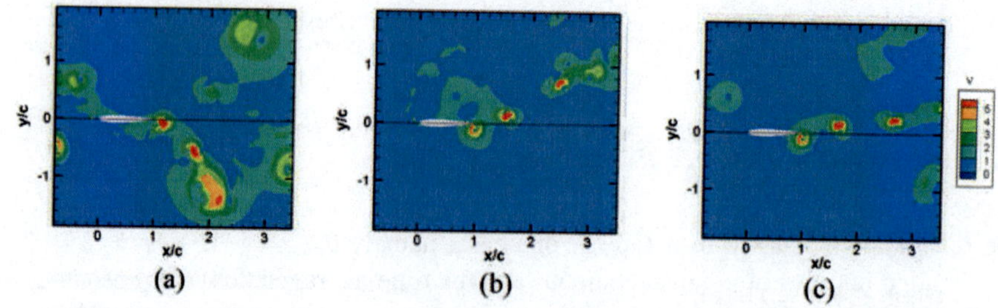

Fig. 5 Instantaneous resultant velocity contours at three different instances for the case $\theta_{max} = \pm 10°$ and $f = 2$ Hz, when the jet is **a** inclined in the downward direction, **b** inclined in the upward direction and **c** nearly along the centerline

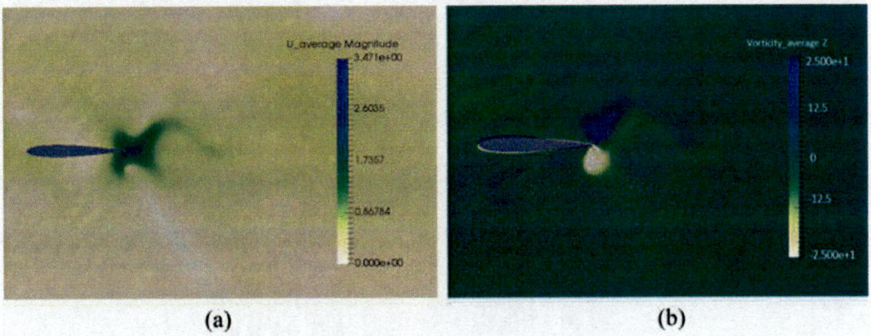

Fig. 6 Time averaged **a** resultant velocity and **b** spanwise vorticity field for the case $\theta_{max} = \pm 10°$ and $f = 2$ Hz

cancellation of the velocity and vorticity in the downstream over long time average. Some flow, however, is seen near the trailing edge region.

3.2 Criterion for Jet Deflection

Researchers have used different criteria to identify jet deflection. Heathcote and Gursul [4] while studying rigid and flexible foils observed the streamwise velocity distribution values at a point $(x/c, y/c) = (1, 0.5)$, where origin is assumed to be located at the airfoil TE. However, they did not present a clear picture of directionality of the jet. They observed quasi-periodicity in the variation of the streamwise velocity with time. Shinde and Arakeri[8] studied this switching phenomenon in pitching airfoil. They manually observed the jet flow field, and classified it into three different categories: upward (U), downward (D) and spread along centerline (S-C). The jet picks up directionality after some cycles and this direction keeps changing with time.

To identify and classify the jet deflection, maximum resultant velocities are plotted for all the time instants at five different sections x/c = 0.5, 0.75, 1.0, 1.25, 1.5 to look at the velocity variation over time, as shown in Fig. 7. It is decided to focus on two x-stations $x/c = 0.5$ and 1 as the deciding criteria for jet deflection. In case, it is difficult to decide based on the two sections, one more x-station at $x/c = 1.25$ is considered.

3.3 Jet Inclination

To get the idea of the switching pattern of the generated jet, interrogation points are set at $y/c = \pm 0.25$, at different x/c sections, after looking at the velocity and the vorticity fields. If the maximum resultant velocity locations are present beyond these

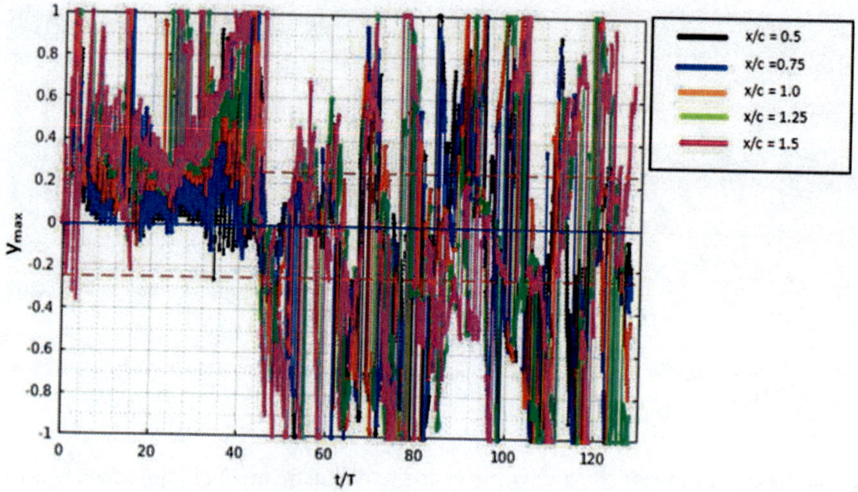

Fig. 7 Plots showing location of maximum resultant velocity at every instant at different streamwise × locations, for $\theta_{max} = \pm 10°$ and $f = 2$ Hz

y-locations, then this may be because of stray vortices present in the flow, and hence we neglect it. Let the numbers 1, 2 and 3 denote the sections at $x/c = 0.5$, 1 and 1.25, respectively, so $y1_{max}$ will give the transverse location of maximum resultant velocity at Sect. 1 at a particular instant.

Using the criteria detailed in Table 1, the jet inclination data is obtained for the case when $\theta_{max} = \pm 10°$ and $f = 2$ Hz, for 128 cycles (Fig. 8). The jet switches direction randomly, as observed from the figure, and the switching is sometimes abrupt and sometimes gradual. It is evident from Fig. 8 that the jet stays in one particular orientation for a time period which ranges from a fraction of a cycle to several cycles, and also it keeps changing the direction continuously and randomly.

Table 1 Criteria used for classifying jet deflection. $y1$, $y2$, $y3$ represent the y-location of maximum resultant velocity at three x-stations, respectively, at $x/c = 0.5$, 1 and 1.25

$y1_{max}$	$y2_{max}$	$y3_{max}$	Deflection
>0.25	>0.25	–	Upward (U)
0< , <0.25	>0.25	–	Upward (U)
>0.25	0< , <0.25	>0.25	Upward (U)
>0.25	0< , <0.25	<0.25	Center (C)
<−0.25	<−0.25	–	Downward (D)
0> , >−0.25	<−0.25	–	Downward (D)
<−0.25	0> , >−0.25	<−0.25	Downward (D)
<−0.25	0> , >−0.25	>−0.25	Center (C)

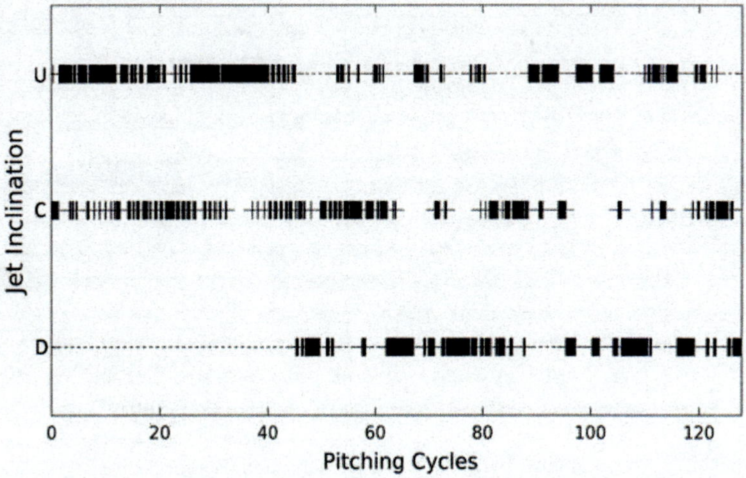

Fig. 8 Jet inclination data for the case when $\theta_{max} = \pm 10°$ and $f = 2$ Hz. The figure shows random switching. U, D, C, respectively, indicate upward, downward deflected jet and jet along the centerline

3.4 Temporal Evolution of the Switching Process

By looking at the jet inclination (Fig. 8) and the maximum velocity transverse location plot (Fig. 9), it is observed that the jet sometimes changes direction abruptly and sometimes gradually. Cycle time at section $x/c = 0.5$ is extracted to observe these switching patterns.

From the maximum velocity transverse location plot, it can be seen that gradual switching is taking place for 3 cycles, from 60 to 63 cycles and the sudden switching happens several times between 35 and 40 cycles. Figure 9 shows the zoomed view of the cycle period of interest and the jet inclination data in case of gradual switching. Similarly, Fig. 10 shows the zoomed view of the cycle period of interest and the jet inclination data in case of sudden switching. Sudden switching indicates that the time scale of switching is even smaller than the smaller time scale used in the simulations, that is, $t/T = 0.01$ and it needs further investigation with a time-step much smaller than that used in present simulations.

3.5 Parametric Study

A similar kind of study has been carried out by varying four amplitudes ($\theta_{max} = \pm 5°, \pm 10°, \pm 15°$ and $\pm 20°$) and four frequencies ($f = 1, 2, 3$ and 4 Hz), for each amplitude. We observed a deflected jet that continually and randomly changes the directions in all the 16 cases studied. We present the data for two representative

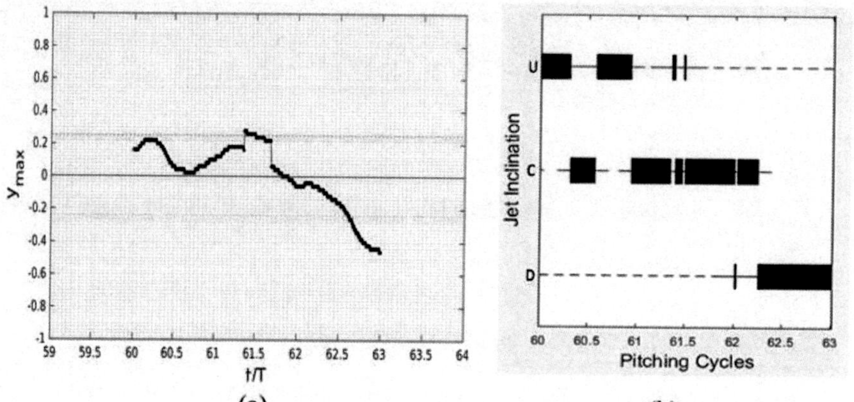

Fig. 9 Figure showing **a** gradual switching and **b** jet inclination data for the case $\theta_{max} = \pm 10°$ and $f = 2$ Hz

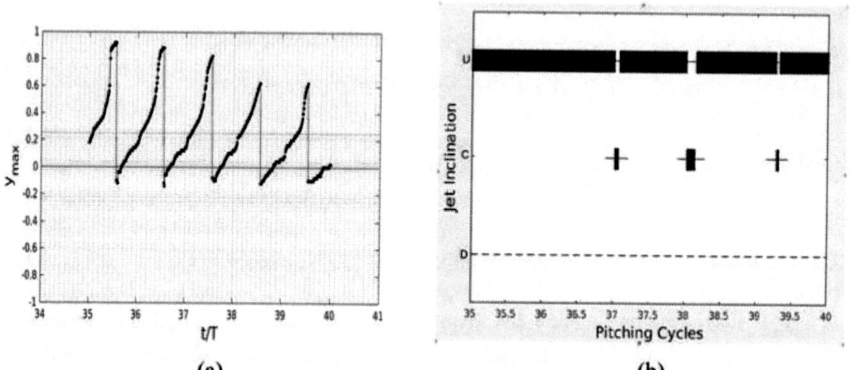

Fig. 10 Figure showing **a** sudden switching and **b** jet inclination data for the case $\theta_{max} = \pm 10°$ and $f = 2$ Hz

cases corresponding to $\theta_{max} = \pm 15°, f = 1$ Hz and $\theta_{max} = \pm 20°, f = 3$ Hz. The jet inclination plot is shown in Fig. 11. It appears that amplitude and frequency does not have any particular effect on the jet switching pattern.

4 Conclusions and Future Scope

The flow generated by a symmetric NACA0015 foil pitching about quarter-chord point from the leading edge in the absence of freestream velocity is studied. The flow is studied by varying amplitude and frequency of pitching. The present work is

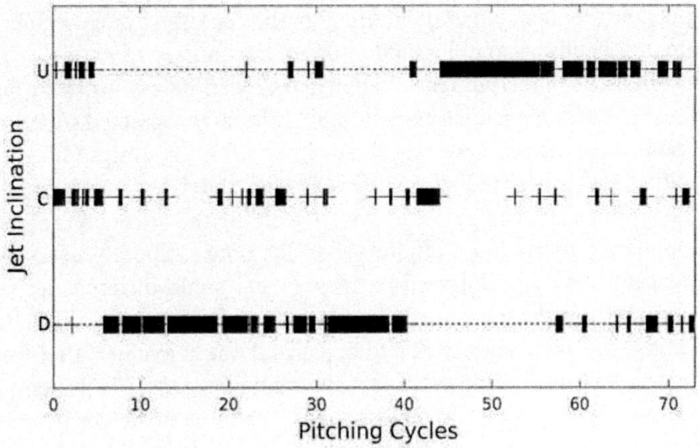

(a)

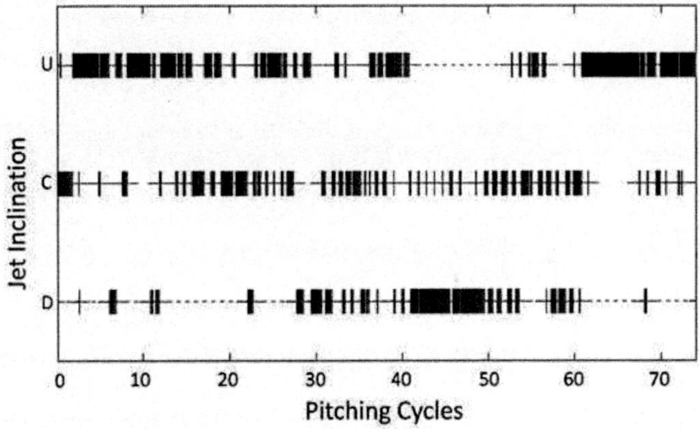

(b)

Fig. 11 Jet inclination data for the case for **a** $\theta_{max} = \pm 15°, f = 1$ Hz and **b** $\theta_{max} = \pm 20°, f = 3$ Hz

aimed to explore the jet deflection and meandering trends generated by a foil pitching in an otherwise still fluid. To identify and classify the jet deflection, a criterion is designed based on the y-location of the maximum resultant velocity at two or three x-stations located at $x/c = 0.5, 1, 1.25$. It was observed that the jet remains in one particular orientation for time periods which range from some fraction of a cycle to several cycles and the jet changes the inclination continuously and randomly. No pattern, quasi-, or periodicity in the jet switching pattern was observed. Two types of switching process—gradual and sudden—were observed. The time resolution used in the simulations was found to be insufficient, to study sudden switching. The parametric study conducted by varying amplitude and frequency of pitching reveals that amplitude and frequency do not have any particular effect on switching pattern, and the switching is observed to be random in all the cases.

The controlled jet deflection from flapping foils can be exploited in man-made devices for turning and maneuvering activities. If by any regulated process, this jet deflection can be eliminated, the straight jet can be used for straight line propulsive motions. Both these things, the straight and the deflected jets can be utilized efficiently by the humans and can pave the way of building the artificial propellers. The artificial propellers may be more efficient than the conventionally used propellers. Flow investigation for even longer time may reveal some different jet switching pattern. Hence, the flow demands further investigation for longer duration. In case of abrupt switching, the simulations need to be carried out at even smaller time-steps, which can reveal very interesting details on switching patterns. The criterion used to classify the jet inclination seems to be less robust for some instances when the stray vortices arrive at the investigation stations.

References

1. Jones, K.D., Dohring, C.M., Platzer, M.F.: Experimental and computational investigation of the Knoller-Betz effect. AIAA J. **36**, 1240 (1998)
2. Lai, J.C.S., Platzer, M.F.: Characteristics of a plunging airfoil at zero freestream velocity. AIAA J. **39**, 531 (2001)
3. Lewin, G.C., Haj-Hariri, H.: Modelling thrust generation of a two-dimensional heaving airfoil in a viscous flow. J. Fluid Mech. **492**, 339 (2003)
4. Heathcote, S., Gursul, I.: Jet switching phenomenon for a periodically plunging airfoil. Phys. Fluids **19**, 027104 (2007)
5. Godoy-Diana, R., Aider, J., Wesfreid, J.E.: Transition in the wake of a flapping foil. Phys. Rev. E **77**, 016308 (2008)
6. Godoy-Diana, R., Marais, C., Aider, J., Wesfreid, J.E.: A model for the symmetry breaking of the reverse B´enard–von K´arm´an vortex street produced by a flapping foil. J. Fluid Mech. **622**, 23 (2009)
7. Cleaver, D.J., Wang, Z., Gursul, I.: Bifurcating flows of plunging aerofoils at high Strouhal numbers. J. Fluid Mech. **708**, 349 (2012)
8. Shinde, S.Y., Arakeri, J.H.: Jet meandering by a foil pitching in quiescent fluid. Phys. Fluids **25**, 041701 (2013)
9. Mackowski, A., Williamson, C.: Direct measurement of thrust and efficiency of an airfoil undergoing pure pitching. J. Fluid Mech. **765**, 524–543 (2015)
10. Deng, J., Sun, L., Teng, L., Pan, D., Shao, X.: The correlation between wake transition and propulsive efficiency of a flapping foil: a numerical study. Phys. Fluids **28**, 094101 (2016)
11. Ramamurti, R., Sandberg, W.: Simulation of flow about flapping airfoils using finite element incompressible flow solver. AIAA J. **39**, 253–260 (2001)
12. Bohl, D.G., Koochesfahani, M.M.: MTV measurements of the vortical field in the wake of an airfoil oscillating at high reduced frequency. J. Fluid Mech. **620**, 63 (2009)
13. Teis, S., Andersen, C.A., Bohr, T.: Vortex wakes of a flapping foil. J. Fluid Mech. **633**, 23 (2009).
14. Nigaltia, C.: Numerical investigation of switching of a jet generated by a foil pitching in still fluid. M.Tech thesis (Indian Institute of Technology, Department of Mechanical Engineering, India, 2019).

Studying the Channel Confluence Hydraulics Using Eddy Viscosity Models and Reynolds Stress Model

Abhishek K. Pandey, Pranab K. Mohapatra, and Vikrant Jain

Notations

h_d	Tailwater depth (m)
U_d	Tailwater average velocity (m/s)
W	Channel width (=0.914 m)
$u*$	Non-dimensional longitudinal velocity
$x*$, $y*$ and $z*$	Non-dimensional Cartesian coordinate axes
W_s^*, L_s^*	Non-dimensional maximum width and length of the separation zone
k	Turbulent kinetic energy (m^2/s^2)
$\in$	Turbulent dissipation rate (m^2/s^3)
ω	Specific dissipation rate (1/s)

1 Introduction

Channel confluence is an essential part of a river system. A better understanding of the confluence hydraulics is required to study the fluvial dynamics, irrigation and drainage network, pollution dispersion and sediment transport. Channel confluence hydraulics is three-dimensional (3D) and is associated with complicated flow features.

Best [1] studied the open channel confluence and identified six different flow zones, viz. Flow stagnation; Flow separation; Maximum velocity zone; Flow diversion; Flow recovery zone; and Shear layer with associated vortex and turbulence

A. K. Pandey · P. K. Mohapatra (✉) · V. Jain
Indian Institute of Technology Gandhinagar, Gandhinagar 382355, India
e-mail: pranabm@iitgn.ac.in

S. K. Saha and M. Mukherjee (eds.), *Recent Advances in Computational Mechanics and Simulations*, Lecture Notes in Mechanical Engineering,
https://doi.org/10.1007/978-981-15-8315-5_26

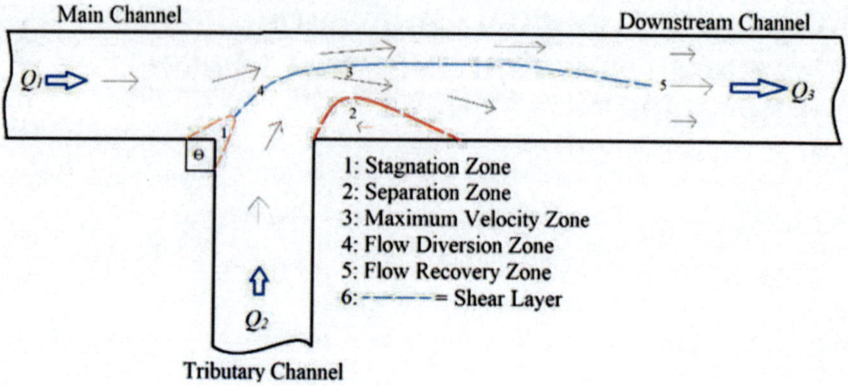

Fig. 1 Flow structures at the channel confluence [1]

(Fig. 1). The separation zone is formed below the downstream of the confluence corner, and it reduces the effective width of the downstream channel. Channel confluence has been studied extensively by the researchers analytically [13, 23], experimentally [1, 2, 12, 14, 17] and numerically [11, 15, 16, 18, 22]. Weber et al. [25] obtained a 3D velocity field in a 90-degree horizontal, equal-width channel confluence, using Acoustic Doppler Velocimeter (ADV). Above studies listed different parameters such as confluence angle, discharge ratio, bed topography and planform geometry as the main controlling parameters of the 3D flow features of the channel confluence.

Most of the previous numerical studies on confluence hydraulics were carried out utilizing RANS equations. RANS equations require additional equations for the turbulence closure. Several turbulence models were reported in the earlier studies. None of the turbulence models is appropriate for all types of flows. Thus, the selection of the turbulence model is critical in a study. For example, standard k-$\in$ and standard $k - \omega$ turbulence models do not perform satisfactorily in case of the adverse pressure gradient [4, 21]. Therefore, the separation zone, which is characterized by an adverse pressure gradient, may not be predicted accurately by such models. These two models are insensitive to high strain, high swirl and rotation [4, 8]. Other variants of the $k - \in$ and $k - \omega$ turbulence model are also not reliable when the turbulence is highly anisotropic [8]. However, Shear Stress Transport (SST) variant of $k - \omega$ model is seen as more sensitive for the prediction of the separation zone (Fluent Theory Guide 2017; [16]. Due to the inherent demerits of the eddy viscosity models (Spalart-Allmras; $k - \in$ and $k - \omega$), RSM is attempted here to have a comparison between the two types of models.

Huang et al. [15] and Luo et al. [16] studied the channel confluence hydraulics using the $k - \omega$ model and SST $k - \omega$ model, respectively. Bradbrook et al. [4–6] and Geberemariam [11] studied the flow structures of the channel confluence for different bed topography, junction angles and flow conditions using $k - \in$ turbulence model. Bradbrook et al. [4–6] used a renormalized group (RNG) $k - \in$ model, while

Geberemariam [11] used a standard variant of $k - \epsilon$ turbulence model to study the separation zone. Biron et al. [3] studied the role of bed discordance in the mixing process of the channel confluence using the RNG variant of the turbulence model. Đorđević and Stojnić [9] studied the flow structures for the discordant bed using a standard variant of $k - \epsilon$ turbulence model. Baranya and Józsa [7] also used a standard variant of $k - \epsilon$ turbulence model to study the flow structure of the scaled-down model of the actual river confluence. Roberts [20] studied the flow structures of the confluent meander bend using standard and the realizable variant of the $k - \epsilon$ turbulence model.

The studies cited above used a wide range of turbulence models to predict the confluence hydraulics. To the authors' knowledge, only different eddy viscosity models have been utilized in these studies. The present study aims to assess the capabilities of two different classes of turbulence models in predicting the channel confluence hydraulics. Three eddy viscosity turbulence models (Spalart-Allmras model; Standard $k - \epsilon$ model; SST $k - \omega$ model) and RSM are adopted for the purpose. Simulated results are compared with the experimental results for velocity fields and WSE by Weber et al. [25].

2 Methodology

2.1 Numerical Modelling

The present study uses the CFD software Fluent, version 17.2, which solves 3D RANS equations. For eddy viscosity turbulence models, (a) Spalart-Allmras (SA) model; (b) Standard $k - \epsilon$ model; (c) SST $k - \omega$ model are chosen. SA model is a one-equation turbulence model that solves the transport equation of kinematic turbulent viscosity. The standard $k - \epsilon$ model and SST $k - \omega$ model are two-equation turbulence models. They solve transport equations for turbulent kinetic energy and its dissipation rate. For eddy viscosity turbulence models, Reynolds stresses are related to mean strain rate as per the Boussinesq eddy viscosity assumption. RSM is a seven-equation model that solves transport equations for the Reynolds stresses and their dissipation rates. Linear pressure-strain model has been chosen to model the pressure-strain term in transport equations of the Reynolds stresses. Standard model constants for all the turbulence models are adopted. A multiphase flow approach is adopted and governing equations are discretized using the finite volume method [24]. The water surface is tracked using the Volume of Fluid (VOF) method. SIMPLE algorithm is adopted for pressure-velocity coupling. Time discretization is second-order accurate. Normalized scaled residuals for all the flow variables are set to 10^{-4} to satisfy convergence.

Hexahedral grids are generated throughout the computational domain. Mesh sizes are chosen based on the grid independence test. Mesh size is finer near the wall region, at the water-air interface and at the channel confluence as the flow variables are expected to change sharply in these areas [11, 16]. Spatial and temporal boundary conditions are specified as the governing equations are elliptic. Uniform velocity and turbulence parameters are specified for water inlets at upstream channels. For air inlets, pressure inlet and turbulence parameters are specified at upstream channels. The pressure outlet boundary condition with appropriate backflow turbulence parameters is specified at the outlet. Water flow depth at the outlet ($h_d = 0.296$ m for all the cases) is imposed. The symmetry boundary condition is set on the top of the computational domain. Stationary and no-slip boundary conditions with zero roughness height are specified on all solid boundaries. The wall function is adopted for wall treatment. A steady-state flow condition is achieved through the false transient approach. More details and other information can be found in Fluent [10].

3 Results and Discussion

The computational domain of the present study is the same as that of the physical model of Weber et al. [25]. The Cartesian coordinate system is shown in Fig. 2 with z^*-direction in vertically upward and origin at the channel bed. Width of the channel (=0.914 m) and tailwater average velocity (=0.628 m/s) are used as length and velocity scales, respectively.

Mesh independency tests are conducted before finalizing the mesh sizes as the turbulent flows are more grid sensitive. Figure 3 shows the value of simulated u^* near the water surface ($z^* = 0.278$) at $x^* = -1.33$ and -2 using standard $k - \epsilon$ model. The results of mesh independency tests are shown for three different numbers of meshes: coarser (504000), medium (672000) and fine (973000). All the numerical simulations are carried out for the medium size of the meshes.

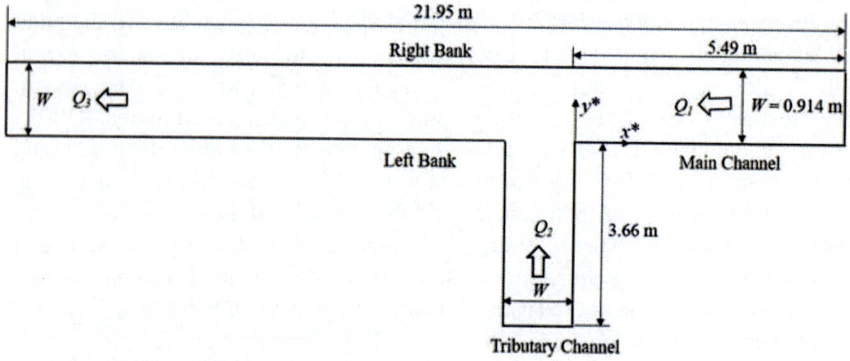

Fig. 2 Computational domain used in the present study

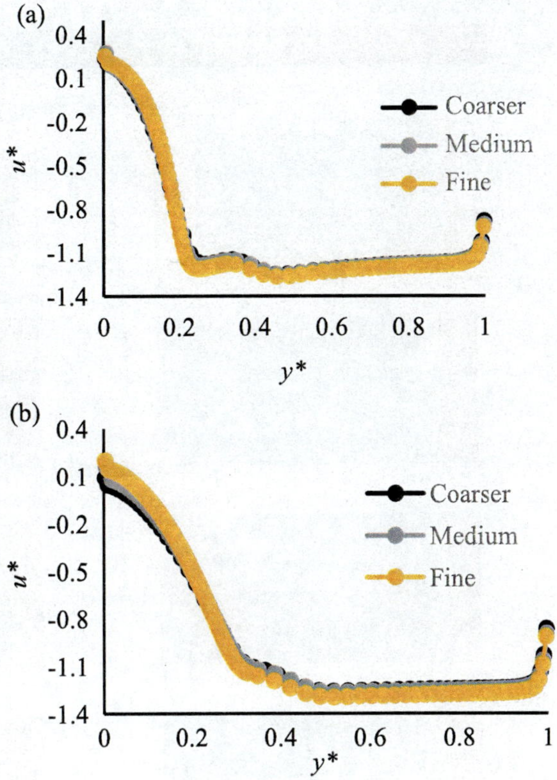

Fig. 3 Simulated u^* along the width of the channel near the water surface for three different mesh sizes: **a** $x^* = -1.33$; **b** $x^* = -2$

3.1 Longitudinal Velocity

Figure 4 compares the simulated and experimental [25] longitudinal velocities. The non-dimensional values at $x^* = -1.33, -2, -3$ and -5 for discharge ratio, $Q_r = 0.75$ (Q_1/Q_3 in Fig. 1) are considered. As shown in Fig. 4, simulated results compare satisfactorily with those obtained experimentally except at two regions: (i) near the bed and (ii) towards the left bank. The mismatch near the bed may be attributed to the limitations of ADV, which was used to measure velocities near the bed. Note that wall function has been imposed near the solid boundaries in the present numerical model. Prediction of u^* is satisfactory in other regions of the cross sections. At $x^* = -1.33$, high values of u^* near the water surface are well predicted by the standard $k - \epsilon$ model and RSM model [(Fig. 4c, e), while SA model and SST $k - \omega$ model slightly over-predict the values (Fig. 4b, d)]. The spatial extension of high values of u^* is more for SA model and SST $k - \omega$ model at this section. Farther at the downstream cross sections, high values of u^* decrease (Fig. 4 corresponding to $x^* = -2, -3$ and -5]. At $x^* = -2$, SA and SST $k - \omega$ model over-predicts the high region of velocity (Fig. 4b, d), while standard $k - \epsilon$ model under-predicts this region (Fig. 4c). RSM model predicts more accurate high region of u^* region at $x^* = -$

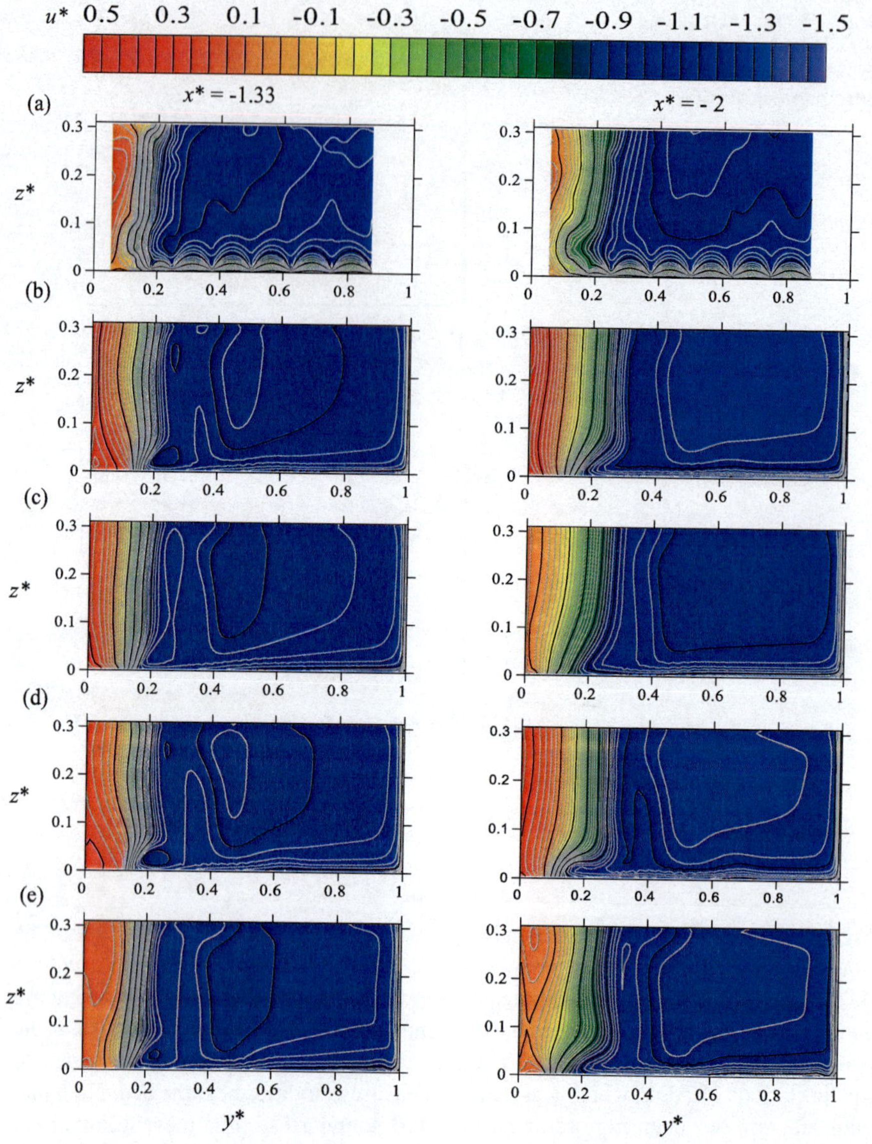

Fig. 4 Isovels of u^*: **a** and **f** Experimental; **b** and **g** Present study (SA); **c** and **h** Present study (Standard $k-\in$); **d** and **i** Present study (SST $k-\omega$); **e** and **j** Present study (RSM)

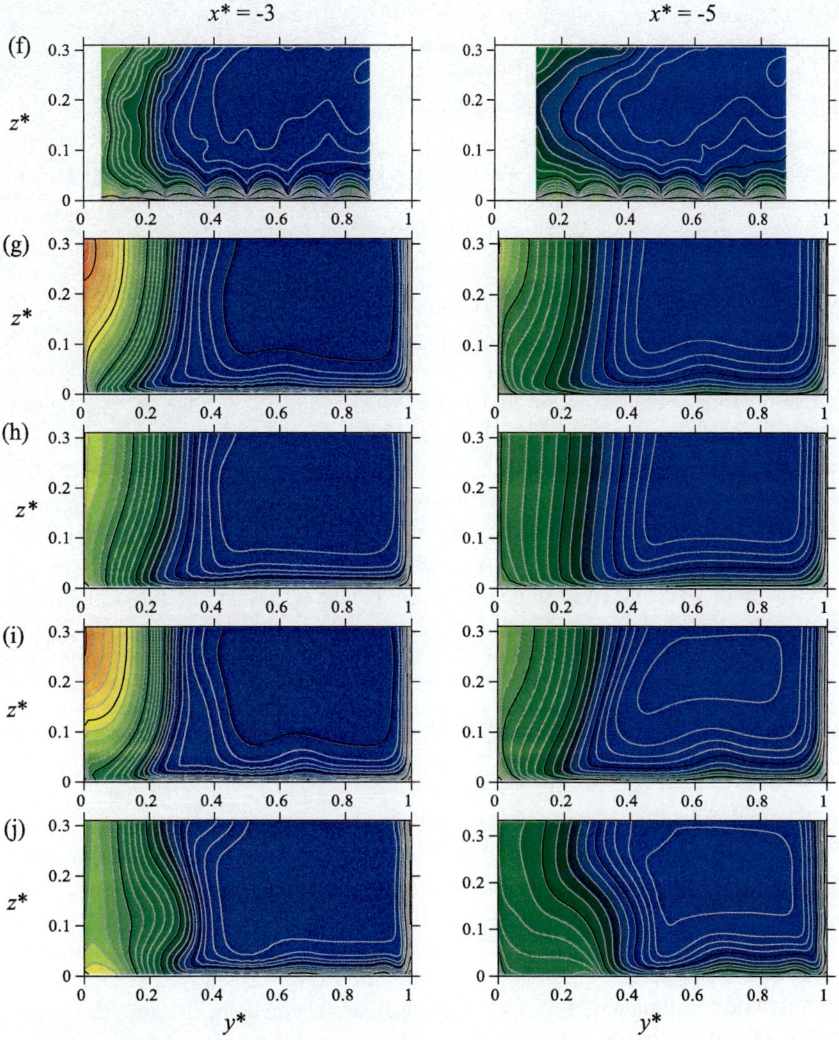

Fig. 4 (continued)

2 (Fig. 4e). Region of high values of simulated u^* over-predicts the experimental values at $x^* = -3$ and -5 for all the turbulence models. In all the cases, simulated u^* under-predicts the corresponding experimental values towards the upper region of the left bank at $x^* = -3$ and -5 (Fig. 4f–j). Under-prediction of u^* is less for the standard $k - \in$ model and RSM model (Fig. 4h, j), while more for the SA and SST $k - \omega$ model (Fig. 4g, i).

Table 1 Simulated flow separation zone

Case	z^*	W_s^*	L_s^*
Experimental	0.278	0.130	1.34
Spalart-Allmras	0.278	0.118	2.13
Standard $k-\in$	0.278	0.090	1.18
SST $k-\omega$	0.278	0.120	2.11
RSM	0.278	0.180	1.98

3.2 Flow Separation Zone

The boundary of the separation zone is obtained using zero isovels of u^* [19]. Flow separation zone is characterized by its maximum width, W_s and length, L_s. Table 1 summarizes the results for the flow separation zone through the non-dimensional parameters, W_s^*, and L_s^*. Values are presented near the water surface ($z^* = 0.278$). Standard $k - \in$ model is weak in predicting the dimensions of the separation zone. Standard $k - \in$ model under-predicts the values of W_s^*, and L_s^*. RSM model over-predicts the values of W_s^*, and L_s^*. SA and SST $k-\omega$ models predict the value of W_s^* more accurately compared to standard $k-\in$. However, SA and SST $k-\omega$ models highly over-predict the value of L_s^*.

3.3 Water Surface Elevation

Simulated WSE along $y^* = 0.167$, 0.5 and 0.833 are presented in Fig. 5. As the flows combine at the junction, the water surface starts falling at the junction area and the depression continues in the downstream direction and then it starts rising again. WSE in the downstream channel is always less than the WSE at upstream of the junction. Minimum WSE is associated with the separation zone region. For all the cases, simulated WSE underestimates the experimental WSE in upstream of the junction. Simulated WSE in the downstream channel shows more deviation as compared to the experimental values. RSM over-predicts water depression. The stretch of the water depression is larger compared to that of the other cases. Simulated water depression by the SA and SST $k-\omega$ models is satisfactory. However, large localized water depression is predicted along $y^* = 0.5$ using the SST $k-\omega$ model (Fig. 5b). Model performance in predicting the WSE using SA model and SST $k-\omega$ model shows a similarity with respect to the experimental WSE for most of the channel reach. Water depression is captured accurately using standard $k-\in$ model. WSE prediction by standard $k-\in$ model is accurate at other places also. WSE prediction appears similar in the farther downstream channel for all eddy viscosity turbulence models considered in the present study. The above discussion shows that free surface tracking by VOF method in the current simulations is more suitable using standard $k-\in$ turbulence model.

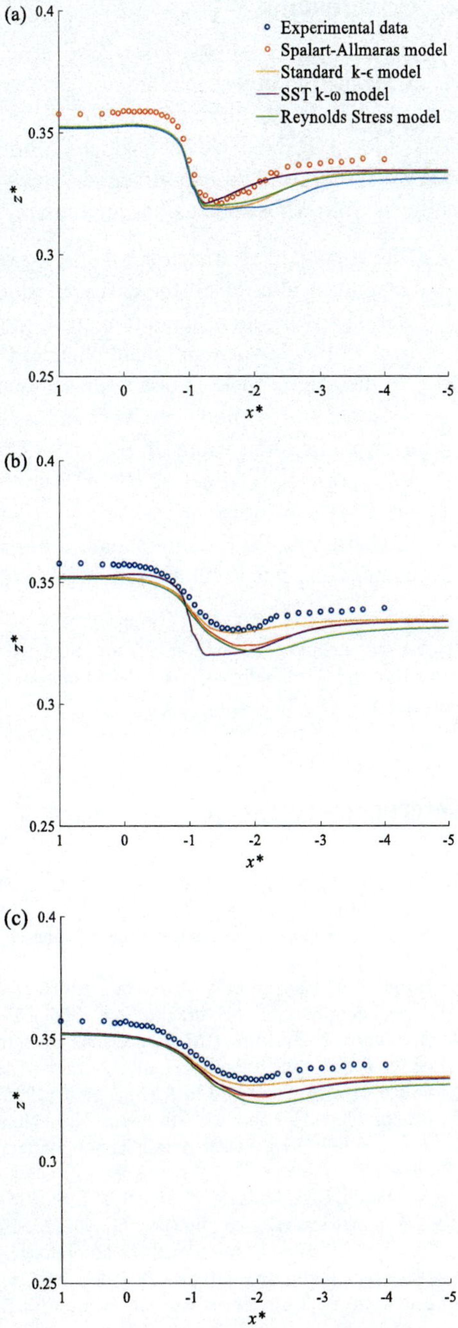

Fig. 5 Comparison of flow depth along **a** $y^* = 0.167$; **b** $y^* = 0.5$; **c** $y^* = 0.833$

4 Conclusions

Open-channel confluence hydraulic has been studied numerically by using CFD software, Fluent. For turbulence modeling, three eddy viscosity models and Reynolds Stress Model (RSM) have been adopted. Simulated $u*$ and WSE has been compared with the corresponding experimental values of Weber et al. [25]. From the present study, the following conclusions are drawn:

(a) The simulated $u*$ match satisfactorily with the corresponding values of experimental results for all the cases considered in the present study. RSM shows comparatively more deviation from the experimental results towards the left bank of the downstream main channel (Fig. 4e, j).
(b) No turbulence model is best suited for the prediction of W_s^* and L_s^* (Table 1). Standard $k-\in$ model under-predicts the values of W_s^* and L_s^*. RSM model over-predicts the value of W_s^* and L_s^*. SA and SST $k-\omega$ models are more appropriate for estimating W_s^*. However, these two models highly over-predict the L_s^*.
(c) Prediction of WSE using standard $k-\in$ model is more accurate than all the other cases. The Prediction of WSE is less accurate using RSM.

Acknowledgements The authors acknowledge the financial help by Prime Minister Research Fellowship and IIT Gandhinagar to conduct this study.

References

1. Best, J.L.: Flow dynamics and sediment transport at river channel confluences. Ph. D. thesis, Univ. of London, London (1985)
2. Best, J.L., Reid, I.: Separation zone at open-channel junctions. J. Hydraulic Eng. **110**(11), 1588–1594 (1984)
3. Biron, P.M., Ramamurthy, A.S., Han, S.: Three-dimensional numerical modelling of mixing at river confluences. J. Hydraulic Eng. **130**(3), 243–253 (2004)
4. Bradbrook, K.F., Biron, P.M., Lane, S.N., Richards, K.S., Roy, A.G.: Investigation of controls on secondary circulation in a simple confluence geometry using a three dimensional numerical model. Hydrol. Process. **12**(8), 1371–1396 (1998)
5. Bradbrook, K.F., Lane, S.N., Richards, K.S.: Numerical simulation of three dimensional, time averaged flow structure at river channel confluences. Water Resour. Res. **36**(9), 2731–2746 (2000)
6. Bradbrook, K.F., Lane, S.N., Richards, K.S., Biron, P.M., Roy, A.G.: Role of bed discordance at asymmetrical river confluences. J. Hydraulic Eng. **127**(5), 351–368 (2001)
7. Baranya, S., Józsa, J.: Numerical and laboratory investigation of the hydrodynamic complexity of a river confluence. Periodica Polytechnica Civil Eng. **51**(1), 3–8 (2007)
8. Davidson, P.: Turbulence: An Introduction for Scientists and Engineers. Oxford University Press (2015)
9. Đorđević, D., Stojnić, I.: Numerical simulation of 3D flow in right-angled confluences with bed elevation discordance in both converging channels. Procedia Eng. **154**, 1026–1033 (2016)
10. Fluent, ANSYS.: Theory Guide 17.2. Ansys Inc. UA, (2016)

11. Geberemariam, T.K.: Numerical analysis of stormwater flow conditions and separation zone at open-channel junctions. J. Irrigation Drainage Eng. **143**(1), 05016009 (2016)
12. Gurram, S.K., Karki, K.S., Hager, W.H.: Subcritical junction flow. J. Hydraulic Eng. **123**(5), 447–455 (1997)
13. Hager, W.H.: Discussion of 'separation zone at open-channel junctions' by James L. Best and Ian Reid (November, 1984). J. Hydraulic Eng. **113**(4), 539–543 (1987)
14. Hsu, C.C., Wu, F.S., Lee, W.J.: Flow at 90 equal-width open-channel junction. J. Hydraulic Eng. **124**(2), 186–191 (1998)
15. Huang, J., Weber, L.J., Lai, Y.G.: Three-dimensional numerical study of flows in open-channel junctions. J. Hydraulic Eng. **128**(3), 268–280 (2002)
16. Luo, H., Fytanidis, D.K., Schmidt, A.R., García, M.H.: Comparative 1D and 3D numerical investigation of open-channel junction flows and energy losses. Adv. Water Resour. **117**, 120–139 (2018)
17. Mosley, M.P.: An experimental study of channel confluences. J. Geol. **84**(5), 535–562 (1976)
18. Pandey, A.K., Mohapatra, P.K.: 3D simulation of flow in a right angled channel junction with a pit. In: World Environmental and Water Resources Congress 2019: Hydraulics, Waterways, and Water Distribution Systems Analysis. American Society of Civil Engineers, Reston, VA (2019)
19. Qing-Yuan, Y., Xian-Ye, W., Wei-Zhen, L., Xie-Kang, W.: Experimental study on characteristics of separation zone in confluence zones in rivers. J. Hydrol. Eng. **14**(2), 166–171 (2007)
20. Roberts, M.V.T.: Flow dynamics at open channel confluent-meander bends. Ph. D. thesis, Univ. of Leeds (2004)
21. Roulund, A., Sumer, B.M., Fredsøe, J., Michelsen, J.: Numerical and experimental investigation of flow and scour around a circular pile. J. Fluid Mech. **534**, 351–401 (2005)
22. Schindfessel, L., Creëlle, S., De Mulder, T.: How Different cross-sectional shapes influence the separation zone of an open-channel confluence. J. Hydraulic Eng. **143**(9), 04017036 (2017)
23. Shabayek, S., Steffler, P., Hicks, F.E.: Dynamic model for subcritical combining flows in channel junctions. J. Hydraulic Eng. **128**(9), 821–828 (2002)
24. Versteeg, H.K., Malalasekera, W.: An Introduction to computational fluid dynamics: the finite volume method. Pearson Education Ltd., Harlow, England (2007)
25. Weber, L.J., Schumate, E.D., Mawer, N.: Experiments on flow at a 90 open channel junction. J. Hydraulic Eng. **127**(5), 340–350 (2001)

Computational Modelling of Turbulent Flows Using an Adaptive Mesh Finite Element Method: A Benchmarking Study

Naman Agarwal and Gaurav Bhutani

1 Introduction

Turbulence is an essential feature of many engineering applications, such as multiphase process flows, due to the mixing it induces. Thus, understanding turbulence is critical for increasing the efficiency of these processes. Flow fields—velocity and pressure—in turbulence are chaotic, which causes diffusion of mass, momentum and energy. Turbulence is characterised using eddies of different length scales, the smallest of which scale as $Re^{-0.75}$, which makes it challenging to perform experiments for turbulent flows. Modelling turbulence using computational fluid dynamics (CFD) provides an economical alternative to experiments in many cases for a better understanding of the physics of these flows. With an advancement of computational infrastructure turbulent flows can now be simulated for industrial applications.

Turbulence can be modelled using different approaches. In the past, engineers have used two-equation Reynolds-averaged Navier–Stokes (RANS) k–ϵ model [1, 3, 7, 10], where k and ϵ represent the turbulent kinetic energy and the rate of dissipation of turbulent kinetic energy in the flow, respectively. This model is computationally economical; however, it lacks accuracy because of the simplifying assumption of modelling all scales of turbulence in the flow. Moreover, it can only be used for modelling fully developed turbulent flows. On the other end of the spectrum, direct numerical simulation (DNS) resolves all scales of turbulence making it extremely expensive computationally; computation time scales as Re^3 in DNS. Large Eddy Simulation (LES) turbulence model provides a bridge between the above two models as it uses spatial filters to filter out smaller universal scales of motion and

N. Agarwal (✉) · G. Bhutani
Multiphase flow research lab (MFRL), School of Engineering, Indian Institute of Technology Mandi, Suran, Himachal Pradesh, India
e-mail: s17001@students.iitmandi.ac.in

S. K. Saha and M. Mukherjee (eds.), *Recent Advances in Computational Mechanics and Simulations*, Lecture Notes in Mechanical Engineering,
https://doi.org/10.1007/978-981-15-8315-5_27

resolves the larger anisotropic scales. It has been shown to provide a better result as compared to the RANS models [8, 9].

The finite element method (FEM) has become popular in recent times in the engineering community for solving flow transport equations. It shows promise as compared to the finite difference method (FDM) and the finite volume method (FVM) due to the ease with which it can handle unstructured meshes and the ease in obtaining higher order accurate schemes (through the use of higher order shape functions), respectively. FEM is increasingly used in computational mechanics nowadays, especially in interdisciplinary problems such as fluid–structure interaction (FSI) applications. This motivated the choice of an open-source, finite element (FE) framework for modelling turbulence in the present work. This framework, known as *Fluidity*, is highly parallelised and utilises adaptive mesh refinement, which makes it a highy tractable for solving industrial-scale turbulent flow problems. A two-dimensional (2D) backward-facing step (BFS) was chosen as a model problem in the present work to model turbulence due to the simplicity and popularity of this benchmarking flow problem. A 2D problem was chosen as opposed to a fully 3D one due to the computation economy of the simulations, which enabled more numerical experiments to be performed.

A robust FEM for LES turbulence models is required with a clarity on the performance of different discretisation schemes and solvers. Although some work has been carried out on LES with FEM [4], however, lack of guidelines leaves the user confused when deciding the schemes and numerical parameters for the stable solution of LES turbulence model. The current work presents a detailed numerical analysis of incompressible turbulent flow over a 2D BFS using a continuous Galerkin FE method in an open-source finite element framework, which allows fully unstructured anisotropic adaptive mesh refinement along with the use of distributed parallelism. The use of adaptive and fixed meshes, two LES models (second-order Smagorinsky and dynamic tensorial LES model), non-linear relaxation parameters and velocity–pressure shape function pairs are investigated and thoroughly analysed in this work. The primary reattachment length was calculated for the different simulation cases and compared against experimental data found in the literature [2], which has been presented and discussed in this work.

2 Model Details

2.1 Governing Equations

Turbulence can be described completely by the incompressible Navier–Stokes equation, which for Newtonian flow is given as:

$$\frac{\partial u_i}{\partial t} + \frac{\partial u_i u_j}{\partial x_j} + \frac{1}{\rho}\frac{\partial p}{\partial x_i} - \nu \frac{\partial^2 u_i}{\partial x_j \partial x_j} = 0, \tag{1}$$

where u_i is the flow velocity, ρ is the flow density, p is the fluid pressure and ν is the kinematic viscosity of the fluid. The continuity equation for incompressible flows is given as

$$\frac{\partial u_i}{\partial x_i} = 0. \tag{2}$$

LES works by filtering the velocity field, which decomposes it into its filtered and sub-filtered components, $\overline{u_i}$ and u_i', respectively, given as

$$u_i = \overline{u_i} + u_i'. \tag{3}$$

The filtered velocity is resolved and the fluctuating velocity, whose behaviour is universal, needs to be modelled. The filtering operation on a field $\phi(x_i, t)$ is achieved using a convolution product of filter kernel $\mathbf{G}$ with ϕ, given as:

$$\overline{\phi} = \mathbf{G} \star \phi(x_i, t) = \int_{-\infty}^{\infty} G(r_i, x_i)\phi(x_i - r_i, t)\mathrm{d}r, \tag{4}$$

where r_i is the radial distance associated with the filter. This helps us write the filtered momentum equation as

$$\frac{\partial \overline{u_i}}{\partial t} + \frac{\partial \overline{u_i u_j}}{\partial x_j} + \frac{1}{\rho}\frac{\partial \overline{p}}{\partial x_i} - \nu \frac{\partial^2 \overline{u_i}}{\partial x_j \partial x_j} = 0. \tag{5}$$

The filtered velocity product $\overline{u_i u_j}$ is expressed in terms of the product of filtered velocities $\overline{u}_i \overline{u}_j$ as

$$\overline{u_i u_j} = \overline{u}_i \overline{u}_j - \tau_{ij}, \tag{6}$$

where τ_{ij} is the sub-filter-scale (SFS) stress tensor, which in the LES method is modelled. The SFS stress tensor contains contribution from the interaction between filtered scales, the filtered and sub-filter scales, and the interaction between the SFS terms. Thus, the final filtered Navier–Stokes equation can be written as

$$\frac{\partial \overline{u}_i}{\partial t} + \overline{u}_j \frac{\partial \overline{u}_i}{\partial x_j} + \frac{1}{\rho}\frac{\partial \overline{p}}{\partial x_i} - \nu \frac{\partial^2 \overline{u}_i}{\partial x_j \partial x_j} - \frac{\partial \tau_{ij}{}^R}{\partial x_j} = 0, \tag{7}$$

where τ_{ij}^R is the deviatoric part of SFS stress, with the isotropic part of the SFS stress is absorbed into the filtered pressure term $\overline{p}$. The SFS stress is generally modelled using the Boussinesq eddy viscosity hypothesis, which assumes a linear relationship between deviatoric SFS stress τ_{ij}^R and the filtered rate of strain $\overline{S}_{ij}$. The two LES models explored in this work are described below.

2.1.1 Second-Order Smagorinsky Model

The second-order Smagorinsky model assumes a balance between energy production and energy dissipation rates. The SFS stress is expressed as

$$\tau_{ij}^{R} = -2\nu_T \overline{S_{ij}}, \tag{8}$$

where ν_T is known as the eddy viscosity. For the second-order Smagorinsky model, it is given as

$$\nu_T = C_s^2 \overline{\Delta}^2 |\overline{S}|, \tag{9}$$

where C_s is the Smagorinsky coefficient (=0.17), $\overline{\Delta}$ is the filter length which is calculated as scalar multiple of the square root of area of an element, and $|\overline{S}| = \sqrt{2\overline{S}_{ij}\overline{S}_{ij}}$.

2.1.2 Dynamic Tensorial LES Model

This model uses a tensorial eddy viscosity along with a dynamic Smagorinsky coefficient. That is, $C_s(x_i, t)$, which helps address the anisotropy and inhomogeneity of the turbulent flow. The tensorial eddy viscosity model is given as

$$\nu_{T,ij} = C_s^2 (\overline{\Delta}^2)_{ij} |\overline{S}|, \tag{10}$$

where $(\overline{\Delta}^2)_{ij}$ is the tensorial squared filter width which is calculated as scalar multiple of the squared mesh size tensor $(\Delta)_{ij}^2$ (see [5] for details). The anisotropic nature of the eddy viscosity helps in the calculation of diffusion independently in each spatial direction. The dynamic model uses a two-step filtering to calculate the Smagorisnky coefficient C_s. A sequential application of the *mesh* (represented using an overbar) and the *test* (represented using a tilde) filters results in an expression for C_s, given as

$$C_s(x_i, t) = -\frac{1}{2} \frac{\mathcal{L}_{ij} \overline{S}_{ij}}{(\tilde{\overline{\Delta}}^2 |\tilde{\overline{S}}| \tilde{\overline{S}}_{ij} \tilde{\overline{S}}_{ij} - \overline{\Delta}^2 \widetilde{|\overline{S}| \overline{S}_{ij}} \, \overline{S}_{ij})}, \tag{11}$$

where $\mathcal{L}_{ij}$ is the Leonard tensor expressed as

$$\mathcal{L}_{ij} = \widetilde{\overline{u}_i \overline{u}_j} - \tilde{\overline{u}}_i \tilde{\overline{u}}_j. \tag{12}$$

Bull [5] can be referred for further details.

2.2 *Finite Element Formulation*

The FEM helps discretise the filtered momentum equation which can then be solved numerically. In the continuous Galerkin (CG) FE method, the momentum equation is multiplied by a test function $\widetilde{w}$ and integrating it over the volume Ω to obtain the weak form of the equation, given as:

$$\int_{\Omega} \widetilde{w}\left(\frac{\partial u_i}{\partial t} + u_j\frac{\partial u_i}{\partial x_j} + \frac{1}{\rho}\frac{\partial p}{\partial x_i} - (\nu + \nu_T)\frac{\partial^2 u_i}{\partial x_j \partial x_j}\right) = 0 \tag{13}$$

The overbar representing filtering has been removed for simplicity. The above equation is discretised using a finite approximation for u_i, given as $u_i^j = \sum_{j=1}^{N_{nodes}} u_i^j \phi^j$, where ϕ^j are basis functions of velocity (which take a value one at the node j and zero at all other nodes) and u^j are the unknown coefficients. For pressure, the discretisation is performed by approximating $p = \sum_{i=1}^{N_{nodes}} p^i \varphi^i$, where φ^i are basis functions of pressure. The final discretised equation, written in matrix form, is given as

$$M\frac{\mathrm{d}\underline{\mathrm{u}}_i}{\mathrm{d}t} + A(u_i)\underline{\mathrm{u}}_i + K\underline{\mathrm{u}}_i + C\underline{\mathrm{p}} = 0, \tag{14}$$

where a mass matrix M, advection matrix A, viscosity matrix K and pressure gradient matrix C. $\underline{\mathrm{u}}_i$ and $\underline{\mathrm{p}}$ represents unknown vectors of velocity and pressure coefficients, respectively.

2.3 *Mesh Adaptivity*

Fixed meshes are seldom suitable for a dynamic problem since the solution fields change continuously, so adaptive meshes are required for capturing anisotropy and inhomogeneity of the flow fields. Mesh adaptivity is implemented in fluidity using an interpolation-based method for the estimation of an a posteriori error metric, which optimises the mesh iteratively until a given tolerance is met. Mesh adaptivity is carried out in three steps in Fluidity. (1) Metric estimation: desired error metric of the field is estimated, (2) Mesh generation: desired mesh (which agrees with the above error metric) is generated using an iterative technique and (3) Field interpolation: all fields are transferred from the old to the new mcsh.

3 Problem Set-up and Model Parameters

As stated previously, turbulent incompressible flow in a 2D BFS with air as the working fluid was simulated using the finite element fluidity framework. The primary recirculation length was calculated for the different cases and the results were compared with experimental data available in the literature [2]. The following analyses have been presented in the present paper:

1. Grid independence test
2. Comparison of different LES models
3. Comparison of velocity–pressure pairs
4. Comparison of non-linear relaxation parameter.

The geometry chosen for the present work is illustrated in Fig. 1. It is exactly the same as used by Armaly et al. [2] in their experiments. The Reynolds number (Re) expression is also same as used by [2], given as

$$\mathrm{Re} = \frac{uD}{\nu},$$

where D is twice the step height, i.e. $2(H - h)$. All simulations are presented for Re = 2000.

The boundary conditions are listed in Table 1. A parabolic x-velocity profile with a mean value of 4.1 m s^{-1} was selected for the inlet. A no-slip boundary condition at the walls and a homogeneous Neumann condition for velocities at the outlet was applied. The physical and numerical parameters chosen in the solver are listed in Tables 2 and 3, respectively. Implicit scheme with a CFL number of 2 was used. All simulations were executed on a multicore machine with 20 threads to save on computation time. The flow velocity was initialised to zero for all simulations.

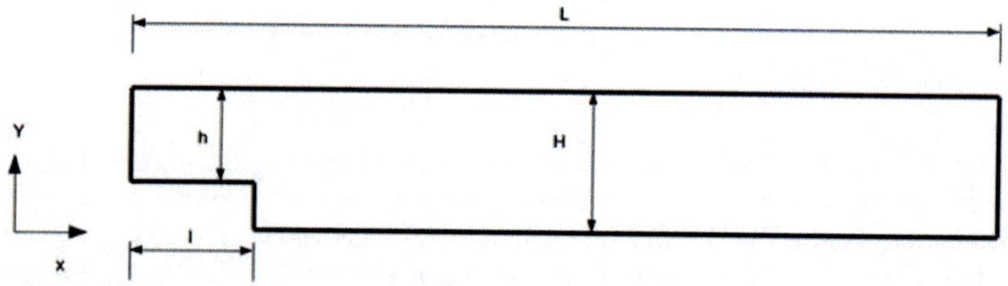

Fig. 1 The 2D BFS geometry used for modelling turbulence in the present work. $L = 460, l = 60, h = 5.2, H = 10.1$. All dimensions in mm

Table 1 Boundary condition for 2D BFS simulation

Boundary	Velocity	Pressure
Inlet	u = parabolic, $v = 0$	$\frac{\partial p}{\partial n} = 0$
Walls	No slip	$\frac{\partial p}{\partial n} = 0$
Outlet	$\frac{\partial u}{\partial n} = \frac{\partial v}{\partial n} = 0$	$p = 0$

Table 2 Physical parameters for 2D BFS simulation

Physical parameter	Value
Air density ($kg\,m^{-3}$)	1.225
Air viscosity ($Pa \cdot s$)	1.7894×10^{-5}

Table 3 Numerical parameters for 2D BFS simulation

Numerical parameter	Value
Overall simulation time (s)	20
Number of Picard iterations	2
Tolerance for Picard iterations (L^2-norm)	10^{-12}

4 Results and Discussion

4.1 Grid Independence Test

In order to study the effect of grid resolution on the flow prediction, the second-order Smagorinsky LES model was chosen. A P2–P1 velocity–pressure pair, which corresponds to quadratic shape functions for velocity and linear shape functions for pressure, was used with adaptive time stepping. The results are presented for 6 different unstructured meshes, which include fixed and adaptive meshes, the details for which are given in Table 4. An initial timestep of 2×10^{-5} s was chosen for each simulation to ensure that the initial CFL number was below 1.0. The mesh was adapted to the velocity field in the present work, the numerical adaptivity parameters for which are shown in Table 5.

The flow pattern for fixed and adaptive meshes is presented in Figs. 2 and 3, respectively. For the fixed-mesh case, the recirculation zone size increased with time and attained a steady value after 2.3 s. However, the flow pattern in the adaptive mesh case

Table 4 Meshes chosen for grid independence test

Nodes	Δx (mm)	Simulation time to steady state (s)
11k	0.75	1.59
38k	0.398	2.26
65k	0.3	2.393
65k adaptive	0.3	–
97k	0.246	2.36
130k	0.212	2.337

Table 5 Adaptivity settings for adaptive mesh simulation

Number of timesteps between two adapts	10
Gradation parameter	2
Minimum edge length (mm)	0.2122
Maximum edge length (mm)	0.75

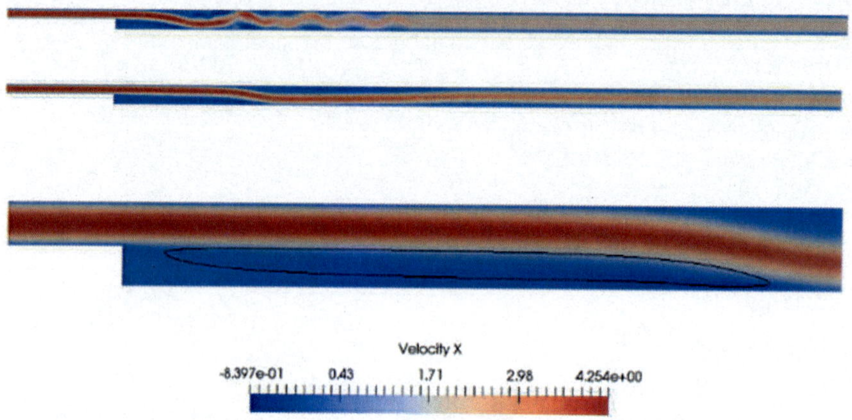

Fig. 2 Flow x-velocity for a fixed mesh with 97 K nodes. Top: 0.1 s simulation time, middle: 2.36 s simulation time (corresponds to steady state) and bottom: magnified view for steady state showing the recirculation zone

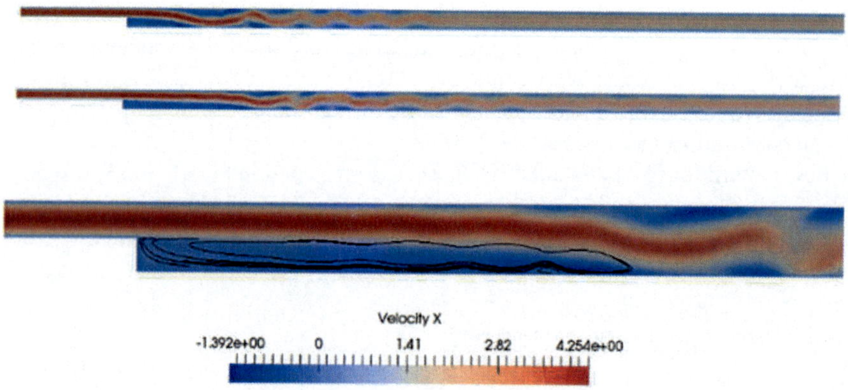

Fig. 3 Flow x-velocity for the adaptive mesh. Top: 0.5-s simulation time, middle: 1.5-s simulation time and bottom: magnified view for 6.64-s simulation time

Fig. 4 Mesh for **a** 97k nodes fixed mesh, **b** 65k nodes adaptive mesh

is different from the fixed-mesh case, as seen in Fig. 3. The solution never achieved a steady state in the adaptive mesh case because of the instability caused by flow near the step for the present Reynolds number, which corresponds to a transition flow regime. The fixed and adaptive meshes near the recirculation zone are shown in Fig. 4. A grid independence (or mesh convergence) behaviour is clearly visible in Fig. 5, which shows results for the fixed meshes. The computation time is seen to scale linearly with the number of nodes. This was the reason for the choice of the 97k mesh as reference for the fixed-mesh results to be compared with the adaptive mesh case, which is discussed in the forthcoming paragraphs.

Demonstrated in Fig. 6, the predicted value of the recirculation length agrees well with the experimental results, with a small discrepancy, which could be attributed to either the experimental uncertainty or due to the three-dimensional nature of the experiments [2]. Since the recirculation zone length fluctuated with time for the adaptive mesh simulation for reasons discussed above, a moving average of the flow fields was analysed instead of instantaneous values same as Bull et al. [4]. Figure 7 shows the recirculation length calculated from moving averaged flow velocities plotted against the simulation time for different averaging time window sizes (Δt). The recirculation length can be seen to stabilise with increasing Δt, as expected. For this reason, the moving average recirculation zone for $\Delta t = 0.5$ s is presented in Fig. 6 corresponding to the adaptive mesh results, which agree well with the experiments. Mesh adaptivity, therefore, provides better results than fixed-mesh simulations, as seen from the validation results in Fig. 6, which makes it preferable to fixed meshes. Recirculation length of 65k adaptive mesh case was 0.069 m as

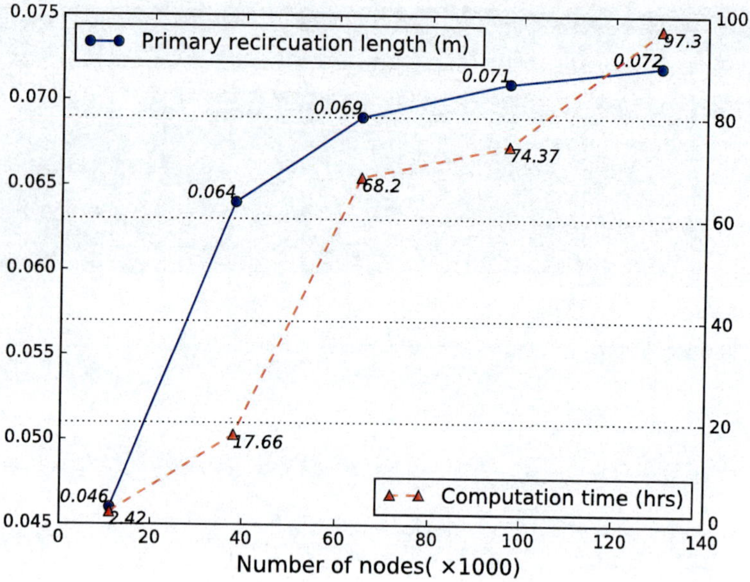

Fig. 5 Recirculation length and computation time on 20 cores plotted for various fixed-mesh simulations

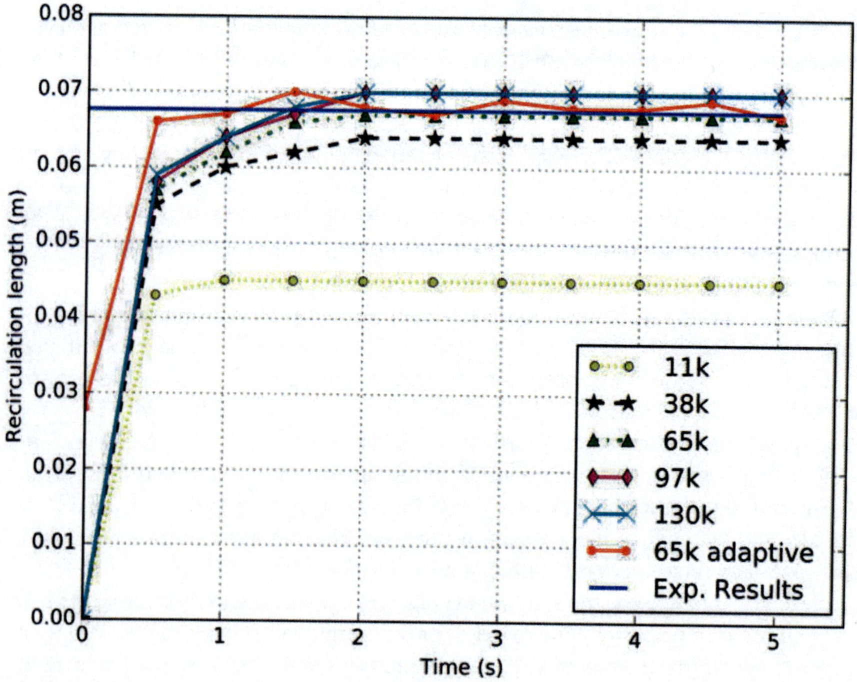

Fig. 6 Recirculation length with simulation time

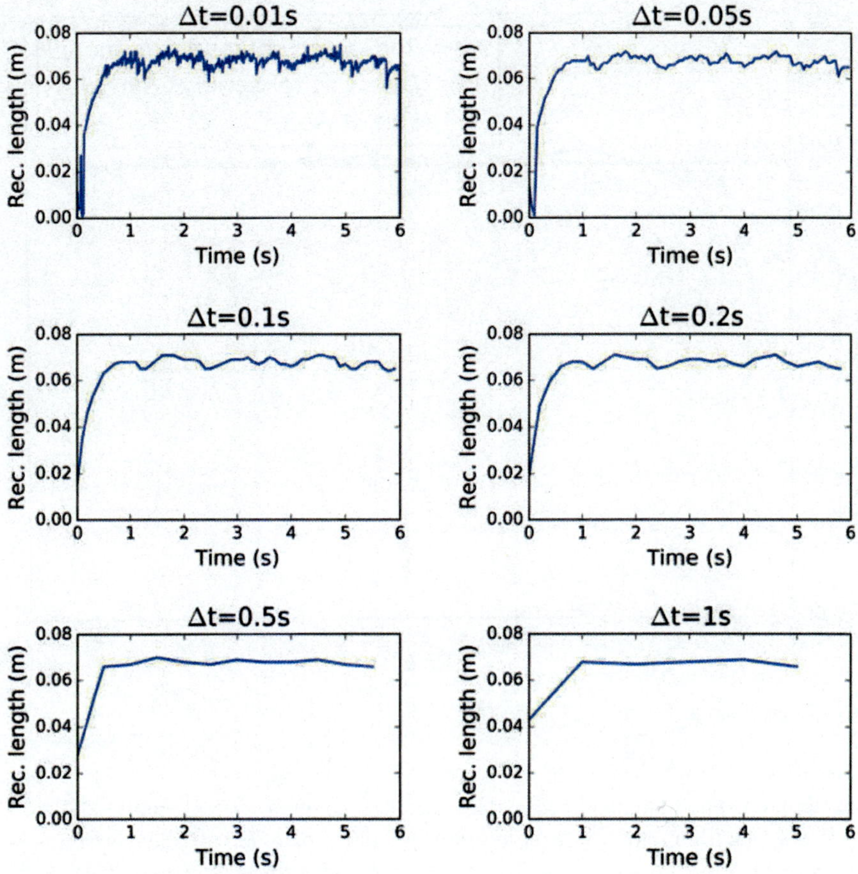

Fig. 7 Recirculation length with simulation time for adaptive meshes—effect of moving average time window Δt

compared to the 97k fixed-mesh case which was 0.071 m, whereas the computation time of the 65k adaptive mesh case was 8 % less as compared to the 97k fixed-mesh case, as seen from Fig. 5. Therefore, for the comparison of different LES models, adaptive- esh simulations were used and the results are presented in the next subsection.

4.2 LES Turbulence Models

Two LES models—second-order Smagorinsky and dynamic tensorial LES model—are compared in this section. For the simulations, the adaptive mesh case (with 65k mesh nodes) was selected, with a P2–P1 velocity–pressure element pair and adaptive

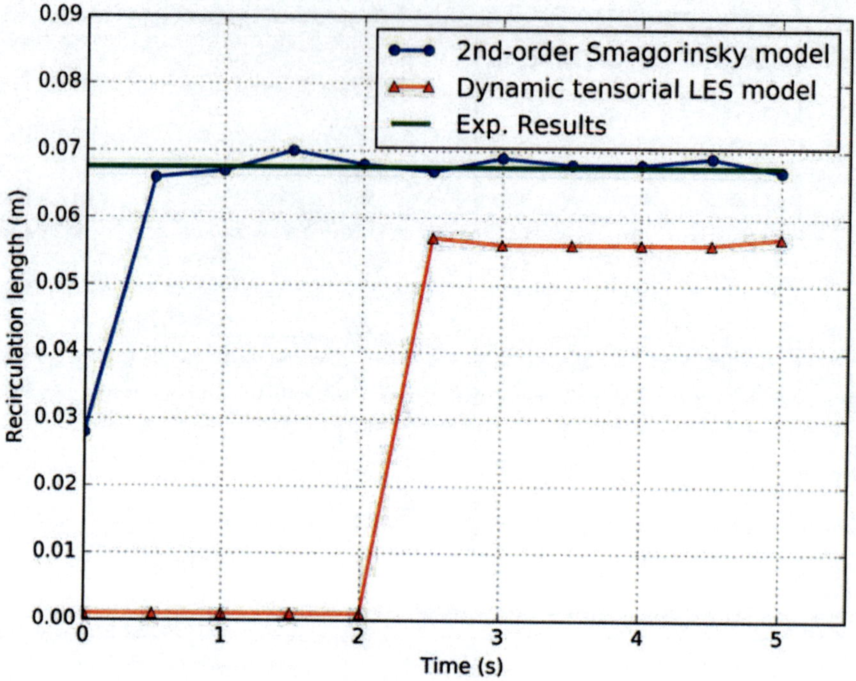

Fig. 8 Comparison of different LES models

time stepping. The numerical parameters used for the dynamic tensorial model are as follows: first filter ratio = 2, test filter ratio = 2 and a scalar length scale. First filter ratio is defined as the ratio of the first filter width to length scale, whereas the test filter ratio is the ratio of the second filter width to the first filter width.

A comparison of the primary recirculation length predicted by the two turbulence models is shown in Fig. 8. The second-order Smagorinsky model clearly shows a result that is much closer to the experimentally measured value. The dynamic tensorial model, however, shows a sub-optimal result due to the present choice of the use of first filter ratio. The prediction is similar to Bull et al. [5] who could not obtain a decent match with the experiments with a filter width ratio of 2. Germano et al. [6], however, reported a superior result with the same choice of filter width ratio. The effect of the filter width ratio on the prediction of turbulence dynamics is not understood well and requires further investigation.

Moreover, the computation cost associated with the dynamic tensorial model was significantly more as compared to the second-order Smagorinsky model, as shown in Fig. 9. Hence, for 2D simulations on unstructured meshes, second-order Smagorinsky model was found to be the optimum choice for modelling turbulence in the transition flow regime.

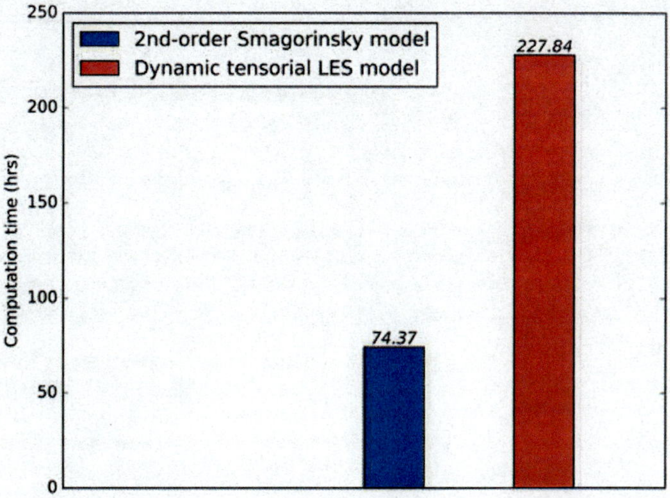

Fig. 9 Comparison of computational time for different LES models (for a simulation time of 2.36 s). All simulations performed on 20 cores

4.3 *Velocity–Pressure Element Pairs*

P1–P1 (linear–linear) and P2–P1 (quadratic–linear) velocity–pressure pairs were compared for a second-order Smagorinsky model. A fixed mesh with 97k nodes was chosen for this comparison, with adaptive time stepping.

Figure 10 presents the recirculation length for the two element pairs, with very similar prediction. The P2–P1 element is expected to be computationally more expensive due to the quadratic shape functions used for velocity and the same was confirmed from simulation results, as shown in Fig. 11.

4.4 *Nonlinear Relaxation Parameter*

The Navier–Stokes equation, which is non-linear in the velocity in the convective term, is linearised before the standard FEM can be applied. A relaxation parameter $\theta \in (0, 1)$ is used to calculate the nonlinear velocity as a weighted sum of the velocity at the nth timestep and the present Picard iteration. Two values of θ (0.5 and 1.0) were tested in the present work for a second-order Smagorinsky model. As seen from Fig. 12, the recirculation length calculated using both these values were very similar. Thus, the results are not too sensitive to the choice of θ.

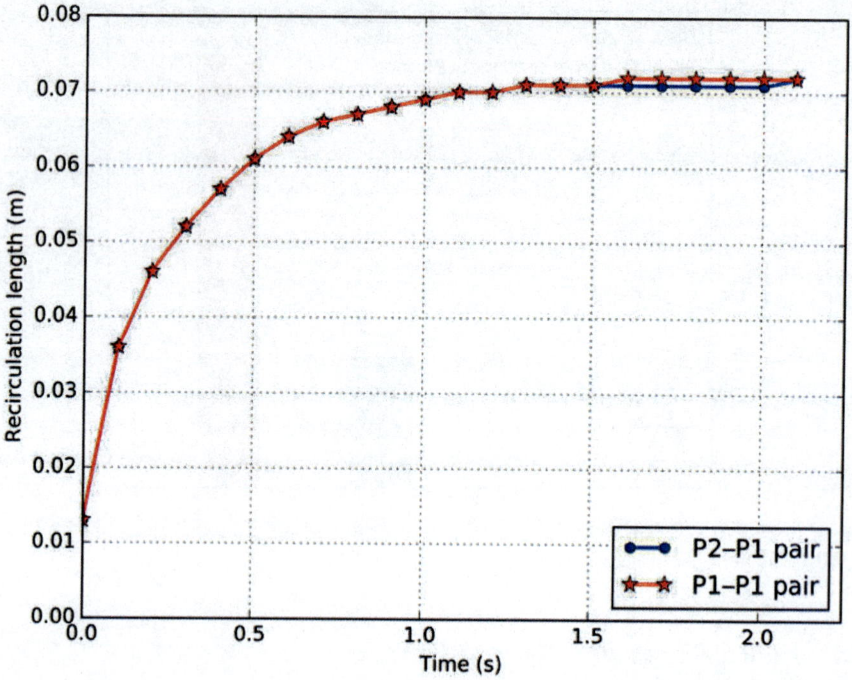

Fig. 10 Comparison of recirculation length for different velocity–pressure pairs. All simulations on 20 cores

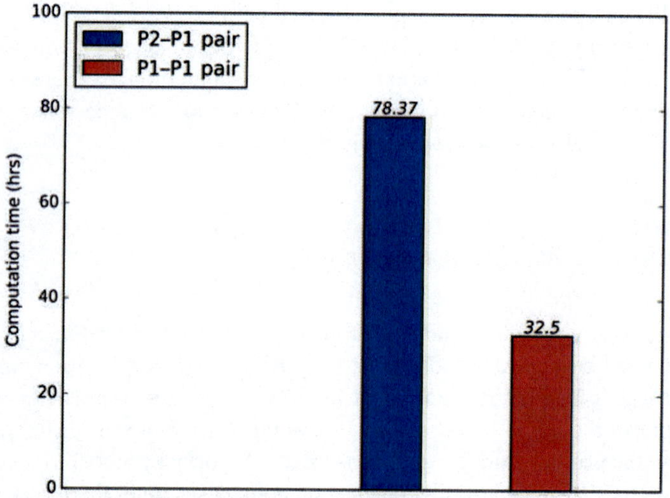

Fig. 11 Comparison of computational time for different velocity–pressure pairs. All simulations on 20 cores

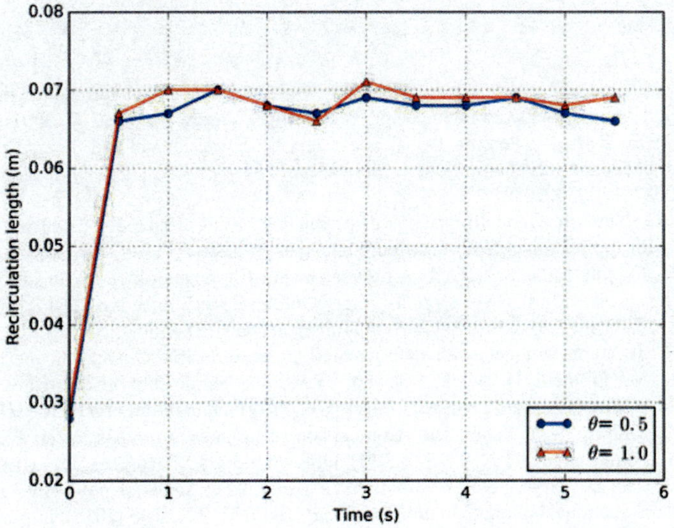

Fig. 12 Comparison of recirculation length for different values of nonlinear relaxation parameter θ

5 Conclusion

The following conclusions can be drawn based on the results presented in this work:

1. The validation error in the recirculation length for the fixed-mesh case (with 97k nodes) was larger than the adaptive mesh case. It was found to be 4.4% as opposed to 1.47% for the latter. At the same time, the adaptive mesh simulations were 8% quicker than the fixed-mesh case. This clearly showed the superiority of using mesh adaptivity in the CFD modelling of turbulence.
2. For fixed meshes, both P1–P1 and P2–P1 element pairs provided similar result for the recirculation length, with a difference of only 0.7% between the two. However, the P1–P1 element pair was 58% faster than the P2–P1 element pair. Thus, P1–P1 element pair is preferred for fixed meshes.
3. The choice of non-linear relaxation parameter was not found to affect the result for the prediction of the validation results.

This work can be extended to study the effect of first filter ratio on the dynamic tensorial LES model. Also, the effect of filter functions with mesh-adaptivity on recirculation length can be investigated in the future.

The results presented in this work will help the user choose LES models and parameters more effectively for a finite element solution method. Although the study was presented for a 2D BFS, the results are expected to hold for 3D simulations too.

References

1. Anwar-ul-Haque, F.A., Yamada, S., Chaudhry, S.R.: Assessment of turbulence models for turbulent flow over backward facing step. Proc. World Congress Eng. **2**, 2–7 (2007)
2. Armaly, B.F., Durst, F., Pereira, J.C.F., Schönung, B.: Experimental and theoretical investigation of backward-facing step flow. J. Fluid Mech. **127**, 473–496 (1983). https://doi.org/10.1017/S0022112083002839
3. Bhutani. G.: Numerical modelling of polydispersed flows using an adaptive-mesh finite element method with application to froth flotation. Ph.D. thesis (2016)
4. Bull, J.R., Piggott, M.D., Pain, C.C.: A finite element LES methodology for anisotropic inhomogeneous meshes. In ICHMT Digital Library Online. Begel House Inc. (2012). https://doi.org/10.1615/ICHMT.2012.ProcSevIntSympTurbHeatTransfPal.1560
5. Bull, J.R.: Turbulence models with adaptive meshing for industrial CFD. Ph.D. thesis (2013)
6. Germano, M., Piomelli, U., Moin, P., Cabot, W.H.: A dynamic subgridscale eddy viscosity model. Phys. Fluids A Fluid Dyn. **3**(7), 1760–1765 (1991). https://doi.org/10.1063/1.857955
7. Khan, Z., Bhusare, V.H., Joshi, J.B.: Comparison of turbulence models for bubble column reactors. Chem. Eng. Sci. **164**, 34–52 (2017). https://doi.org/10.1016/j.ces.2017.01.023
8. Ma, T., Lucas, D., Ziegenhein, T., Fröhlich, J., Deen, N.G.: Scale-adaptive simulation of a square cross-sectional bubble column. Chem. Eng. Sci. **131**, 101–108 (2015). https://doi.org/10.1016/j.ces.2015.03.047
9. Mattson, M.D., Mahesh, K.: A one-way coupled, Euler-Lagrangian simulation of bubble coalescence in a turbulent pipe flow. Int. J. Multiphase Flow **40**, 68–82 (2012). https://doi.org/10.1016/j.ijmultiphaseflow.2011.11.013
10. Schäfer, F., Breuer, M., Durst, F.: The dynamics of the transitional flow over a backward-facing step. J. Fluid Mech. **623**, 85–119 (2009). https://doi.org/10.1017/S0022112008005235

Finite Element Computational Modelling of Non-Newtonian Fluids Using Anisotropic Mesh Adaptivity

Neeraj Kr. Singh and Gaurav Bhutani

1 Introduction

Non-Newtonian fluids are widely encountered in many fields of science and engineering, such as geophysics, hydrogeology, chemistry, food, petroleum, lubricants, to name but a few. Moreover, many modern materials and manufacturing processes also require further understanding of the behaviour of non-Newtonian fluids, due to their wide applications in practice [1]. Compared to Newtonian fluids, non-Newtonian fluids usually have a complex constitutive behaviour, which may complicate their investigation through the use of numerical methods. In the past few years, progress has taken place in the solution of non-Newtonian flows using the following methods: finite element method (FEM) [2], finite volume method [3], lattice Boltzmann method (LBM) [4] and smoothed particle hydrodynamics (SPH) [5]. Using the methods mentioned above, some complex flow features that differ from the Newtonian fluids have been reported.

With the flow systems becoming large, an accurate analysis of non-Newtonian flows through the use of CFD has become computationally expensive. In this paper, we demonstrate the use of a highly parallelised adaptive mesh finite element solver—Fluidity—to study the behaviour of power-law non-Newtonian fluid flows in a two-dimensional (2D) square cavity. The FEM has been proved to be a powerful numerical technique for simulating Newtonian and non-Newtonian fluid flows and partic-

N. Kr. Singh · G. Bhutani (✉)
Multiphase flow research lab (MFRL), School of Engineering, Indian Institute of Technology Mandi, Kamand, India
e-mail: gaurav@iitmandi.ac.in

S. K. Saha and M. Mukherjee (eds.), *Recent Advances in Computational Mechanics and Simulations*, Lecture Notes in Mechanical Engineering,
https://doi.org/10.1007/978-981-15-8315-5_28

ularly successful in dealing with complex boundaries. Non-Newtonian fluids have non-linear dependence on the rate of shear strain, which makes the flow equations non-linear, and therefore more complex. Compared to the finite volume and finite difference methods, FEM allows for accuracy improvement by changing the order of polynomial in test and trial functions for the flow fields, helping in the handling of non-linearity. Hence, the FEM is considered to offer excellent possibilities for simulating non-Newtonian flows.

The flow in a 2D lid-driven square cavity, which is a classic benchmark problem in fluid mechanics, has been widely studied by many researchers in the past decades. Driven cavity flow is of great importance as it can offer an ideal framework in which meaningful and detailed comparisons can be made between results obtained from theory and computation [6]. However, to the best of the knowledge of the authors, most available works are limited to the Newtonian fluids and only a few works associated with the non-Newtonian (power-law) fluid flows in a square cavity have been reported. To address this limitation, in the present work, power-law non-Newtonian fluid flow is simulated in a 2D square cavity using a finite element adaptive mesh solver. Non-Newtonian effects in the lid-driven cavity (LDC) flows are investigated and a detailed comparison of the results with published results have been shown. Along with the LDC benchmark, code verification is also presented for a non-Newtonian Poiseuille flow problem.

2 Model Details

2.1 Governing Equations

Following are the governing equations for an incompressible non-Newtonian fluid flow:

Mass Conservation Equation

$$u_{i,i} = 0 \tag{1}$$

Momentum Conservation Equation

$$\rho \partial_t u_i + u_j u_{i,j} = \sigma_{ij,j} + \rho b_i \tag{2}$$

$$\sigma_{ij} = -p\delta_{ij} + \tau_{ij} \tag{3}$$

$$\tau_{ij} = 2\eta(\dot{\gamma})\dot{\gamma}_{ij} \tag{4}$$

$$\dot{\gamma}_{ij} = \frac{1}{2}(u_{i,j} + u_{j,i}) \tag{5}$$

Here, $\eta(\dot{\gamma})$ denotes the dependence of viscosity on the rate of strain tensor. Other symbols have their usual meaning. It must be noted that Einstein's notation had been used in this text.

Shear Strain Rate

The shear strain rate ($\dot{\gamma}$) is defined by using I_2, the second invariant of shear strain rate tensor, which makes $\dot{\gamma}$ a scalar. It is dependent on the rate of the strain tensor, and independent of the coordinate system:

$$I_2 = \dot{\gamma}_{ij}\dot{\gamma}_{ij}, \tag{6}$$

$$\dot{\gamma} = \sqrt{2\dot{\gamma}_{ij}\dot{\gamma}_{ij}}. \tag{7}$$

Here $\dot{\gamma}_{ij}$ is the rate of strain tensor and $\dot{\gamma}$ is magnitude of strain tensor.

Power-Law Model

$$\eta = K\dot{\gamma}^{n-1} \tag{8}$$

Here, K is consistency index, $\dot{\gamma}_{ij}$ is shear strain rate and n is power-law index.

2.2 *Finite Element Formulation*

The finite element method helps to discretise the momentum equation, which can then be solved numerically. In the Continuous Galerkin (CG) FE method, the momentum equation is multiplied by a test function ψ and integrated over the volume Ω to obtain the weak form of the equation, given as

$$\int_{\Omega} \psi\left(\frac{\partial u_i}{\partial t} + u_j\frac{\partial u_i}{\partial x_j} + \frac{1}{\rho}\frac{\partial p}{\partial x_i} - \frac{\partial \nu}{\partial x_j}\frac{\partial u_i}{\partial x_j}\right) = 0 \tag{9}$$

The above equation is discretised using a finite approximation for u_i, given as $u_i^j = \sum_{j=1}^{N_{nodes}} u_i^j \phi^j$, where ϕ^j are basis functions of velocity (which take a value one at the node j and zero at all other nodes) and u^j are the unknown coefficients. For pressure, the discretisation is performed by approximating $p = \sum_{i=1}^{N_{nodes}} p^i \varphi^i$, where φ^i are basis functions of pressure. The final discretised equation, written in matrix form, is given as

$$M\frac{\mathrm{d}\underline{u}_i}{\mathrm{d}t} + A(u_i)\underline{u}_i + K\underline{u}_i + C\underline{p} = 0, \tag{10}$$

where mass matrix M, advection matrix A, viscosity matrix K and pressure gradient matrix C are known. $\underline{u}_i$ and $\underline{p}$ represent unknown vectors of velocity and pressure coefficients, respectively, which are determined by solving the above system.

2.3 *Mesh Adaptivity*

Fixed meshes are seldom suitable for a dynamic problem since the solution fields change continuously, so adaptive meshes are required for capturing anisotropy and inhomogeneity of the flow fields. An important step is the construction of an error metric that allows the size and shape quality of individual elements to be assessed, so that local operations may be applied to improve overall mesh quality. This approach results in anisotropic meshes, and so a metric that incorporates information on the anisotropy of the solution variables being approximated is advantageous [7].

Mesh adaptivity is implemented in Fluidity using an interpolation-based method for the estimation of an a posteriori error metric, which optimises the mesh iteratively until a given tolerance is met. Mesh adaptivity is carried out in three steps in Fluidity. (1) Metric estimation: desired error metric of the field is estimated, (2) Mesh generation: desired mesh (which agrees with the above error metric) is generated using an iterative technique, and (3) Field interpolation: all fields are transferred from the old to the new mesh.

3 Problem Setup and Model Parameters

3.1 *Poiseuille Flow—Verification*

For the verification of the implementation of non-Newtonian power-law model in Fluidity (as a Python script), flow between two semi-infinite parallel plates, also known as Poiseuille flow, was considered in the fully developed region. The following analyses were performed for this case:

1. Comparison of numerical results with analytical results for three cases: $n = 0.5, 1.0, 1.5$.
2. Relative percentage error in numerical and analytical solution with time.

The Reynolds number in the present work is given by

$$\text{Re} = \frac{U^{2-n}h^{n-1}}{K}$$

where U is velocity of the fluid at the inlet, h is the characteristic length that is equal to distance between two parallel plates, and K is the consistency parameter and n is power-law index.

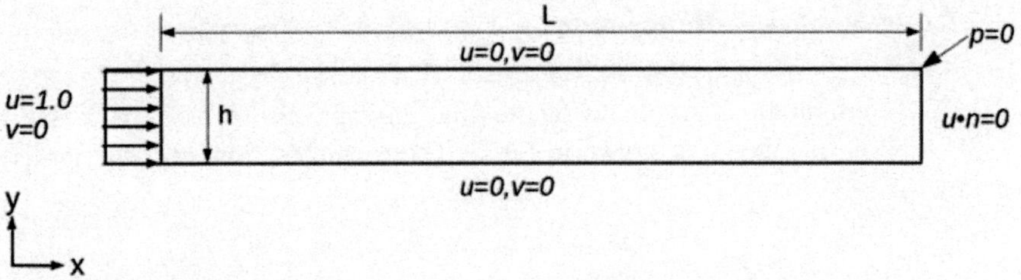

Fig. 1 The 2D parallel plate geometry used for verification in the present work. $L = 20$, $h = 1.0$. All dimensions in *m*

Table 1 Boundary condition for parallel plate simulation

Boundary	Velocity	Pressure
Inlet	$u = 1.0,\ v = 0$	$\frac{\partial p}{\partial n} = 0$
Plates	No slip	$\frac{\partial p}{\partial n} = 0$
Outlet	$\frac{\partial u}{\partial n} = \frac{\partial v}{\partial n} = 0$	$p = 0$

Table 2 Physical parameters for parallel plate simulation

Physical parameter	Value
Consistency index (K)	0.01
Power index (n)	0.5, 1.0, 1.5

Table 3 Numerical parameters for parallel plate simulation

Numerical parameter	Value
Overall simulation time (s)	17
Number of Picard iterations	2
Tolerance for Picard iterations (L^2-norm)	10^{-12}

The boundary conditions are shown in Fig. 1. A uniform x-velocity profile with a magnitude of $1.0\,\mathrm{m\,s^{-1}}$ was selected for the inlet. A no-slip boundary condition at the walls and a homogeneous Neumann condition for velocities at the outlet was applied. Details are specified in Table 1.

The physical and numerical parameters chosen in the solver are listed in Tables 2 and 3, respectively. An implicit scheme with a CFL number of 5.86 was used. All simulations were executed on a multicore machine with 20 threads to save on computation time. The flow velocity was initialised to zero for all simulations.

3.2 Couette Flow—Benchmark

A 2D-LDC configuration, a type of Couette flow, the same as that used by Li et al. [8] was used in this work, as shown in Fig. 2. The Reynolds number definition is given by

$$\mathrm{Re} = \frac{U^{2-n} L^{n-1}}{K},$$

where U is velocity of the fluid at the top wall, L is the characteristic length that is equal to length of the square cavity, K is the consistency parameter, and n is power-law index. The following analyses were carried out for this case:

1. Grid independence test for four fixed meshes,
2. Effectiveness of anisotropic mesh adaptivity,
3. Effect of timestep on accuracy,
4. Effect of Reynolds number,
5. Effect of power-law index,
6. Effectiveness of parallel computation.

The initial condition for velocity was 0.0 ms^{-1}. Boundary conditions are listed in Table 4 and shown in Fig. 2. The x-velocity was constant on the top wall for grid independence test. No-slip boundary condition at the remaining three walls with homogeneous Neumann boundary conditions for pressure was used. The physical and numerical parameters are listed in Tables 5 and 6. A 2D unstructured mesh was used for this exercise.

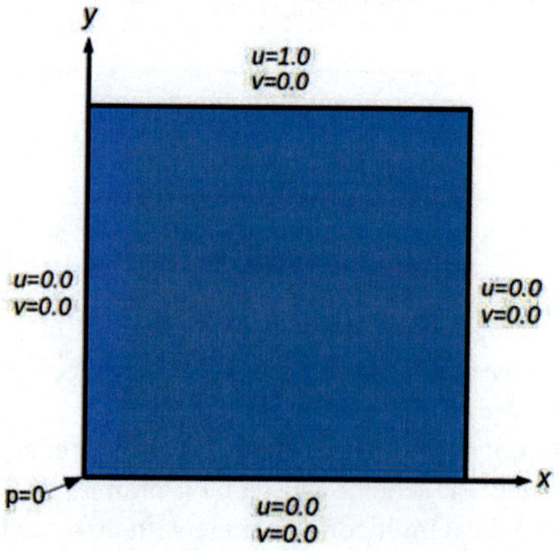

Fig. 2 The 2D-LDC geometry used for modelling non-Newtonian fluid in the present work. $L = 1$, all dimensions in m

Table 4 BCs for 2D-LDC simulation

Boundary	Velocity	Pressure
Top wall	u = constant = 1.0, $v = 0$	$\frac{\partial p}{\partial n} = 0$
Remaining walls	No slip	$\frac{\partial p}{\partial n} = 0$
Origin	$u = 0, v = 0$	$p = 0$

Table 5 Physical parameters for 2D-LDC simulation

Power-law index (n)	0.5, 1.0, 1.5
Re	100

Table 6 Numerical parameters for 2D-LDC simulation

Numerical parameter	Value
Time step Δt (s)	1×10^{-2}
Overall simulation time (s)	200

4 Results and Discussion

4.1 *Verification*

The results for simulation of flow between two parallel plates with power-law fluids for different flow indices ($n = 0.5, 1.0, 1.5$) are compared with analytical solution, as shown in Figs. 3, 4, 5. The agreement was excellent as seen from the figures. Percent relative error (with analytical results) was calculated and plotted in Fig. 6 at $h = 0.5$ m. The convergence can be clearly seen from the plot with a large initial error and decreases to within 0.15% after 6 s.

4.2 *Benchmarking*

Fluidity results for the 2D-LDC were compared with the results obtained using the Lattice Boltzmann method [8].

4.2.1 Grid Independence Test

In order to study the effect of grid resolution on the flow prediction, a P1–P1 velocity–pressure pair, which corresponds to linear shape functions for velocity with no sta-

Fig. 3 Shear thinning ($n = 0.5$)

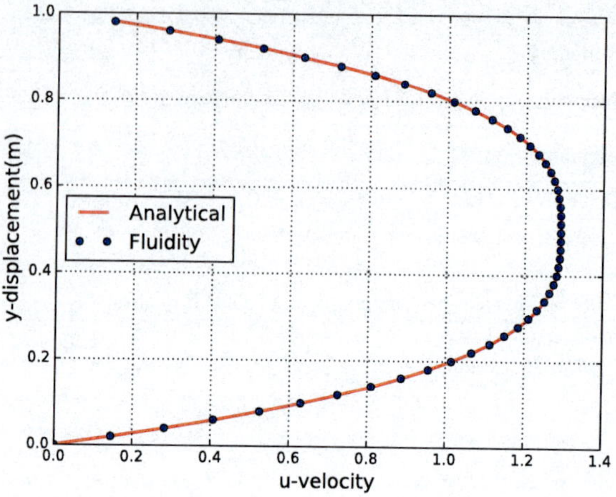

Fig. 4 Newtonian ($n = 1.0$)

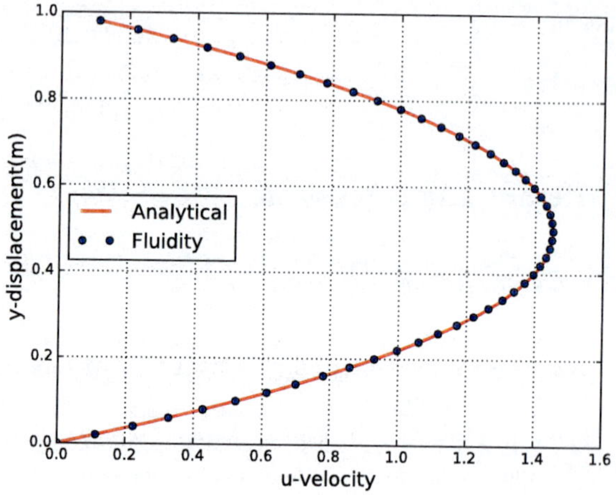

Fig. 5 Shear thickening ($n = 1.5$)

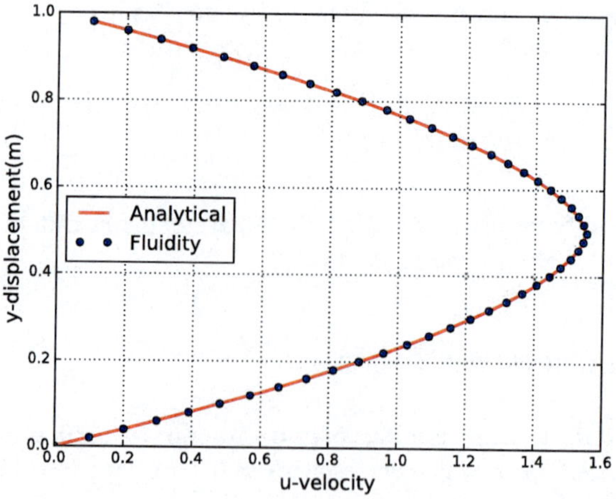

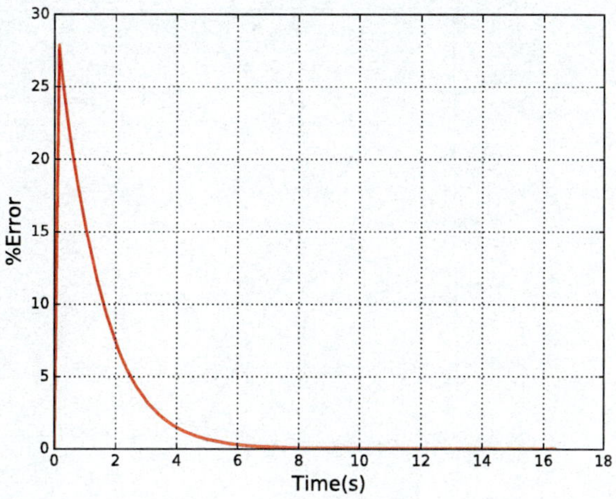

Fig. 6 Relative percentage error

Table 7 Meshes chosen for grid independence test

Nodes	Δx (m)	Simulation time to steady state (s)
336	0.0625	15
1271	0.03125	30
4831	0.015625	55
19324	0.0078125	350
1400 adaptive	–	18

Table 8 Adaptivity settings for adaptive mesh simulation

Number of timesteps between two adapts	10
Gradation parameter	2
Minimum edge length (m)	0.001
Maximum edge length (m)	0.5

bilisation and linear shape functions for pressure, was used with fixed time stepping. The Crank–Nicolson ($\theta = 0.5$) discretisation is used in time. The results are presented for 4 different fixed unstructured meshes. One adaptive mesh simulation is also presented to analyse the effect of mesh adaptivity on simulation run-time and error. The details of meshes are given in Table 7. An initial timestep of 0.01 s was chosen for each simulation to ensure that the initial CFL number was below 1.0. The mesh was adapted to the velocity field in the present work, the numerical adaptivity parameters for which are shown in Table 8.

A grid independence (or mesh convergence) behaviour is clearly visible in Fig. 7, which shows results for the fixed meshes. The computation time scales linearly with the number of nodes, hence the mesh with 4831 nodes is the most optimal.

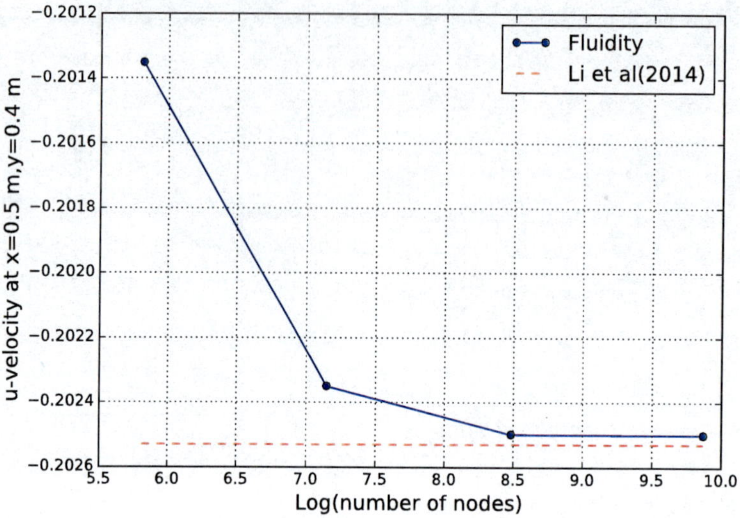

Fig. 7 Grid independence test

4.2.2 Anisotropic Mesh Adaptivity

In anisotropic mesh adaptivity, the mesh resolution depends on the interpolation error for the fields of interest—x-velocity, y-velocity and pressure in this work. The adaptive mesh is shown in Fig. 8. The upper edge of the square cavity mesh is refined due to the velocity shear. The interpolation error was within 10, 6 and 1% for horizontal velocity (u), vertical velocity (v) and pressure fields, respectively. Anisotropy allows to refine the mesh in particular directions which reduces the computational time significantly, as shown in Fig. 9. The characteristic mesh length for fixed mesh is plotted on the x-axis with computation time on y-axis. The accompanying adaptive mesh histograms correspond to similar accuracy as the fixed mesh.

Parallel and serial simulations were also run for the above-mentioned four fixed mesh cases on a 20-core machine. A clear reduction in run-time can be seen through the use of multicore simulations (Fig. 10).

4.2.3 Effect of Reynolds Number

A large number of simulations are carried out for shear-thickening (n = 1.5) and shear-thinning (n = 0.5) fluids with Re =1000, 3000, 5000 and 8000. Figures 11 and 12 show the flow streamlines for shear-thickening and shear-thinning cases, respectively. With an increase in Re, the primary vortex generated by the motion of

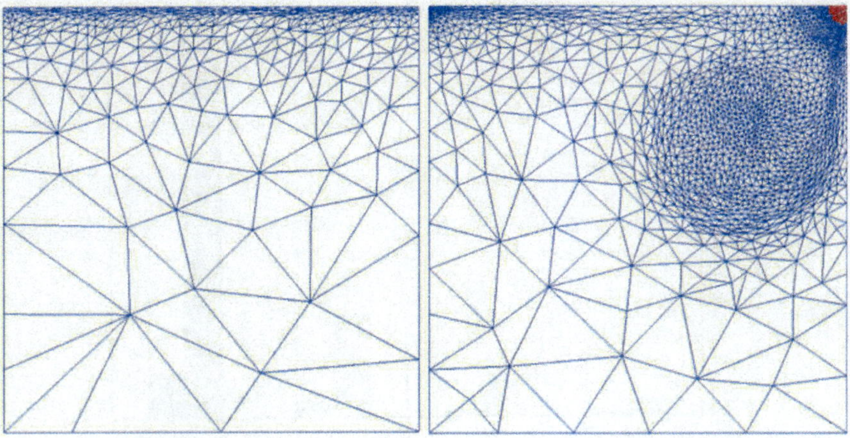

(a) Mesh at t=0 s

(b) Mesh at t=10.0 s

Fig. 8 Adaptive mesh

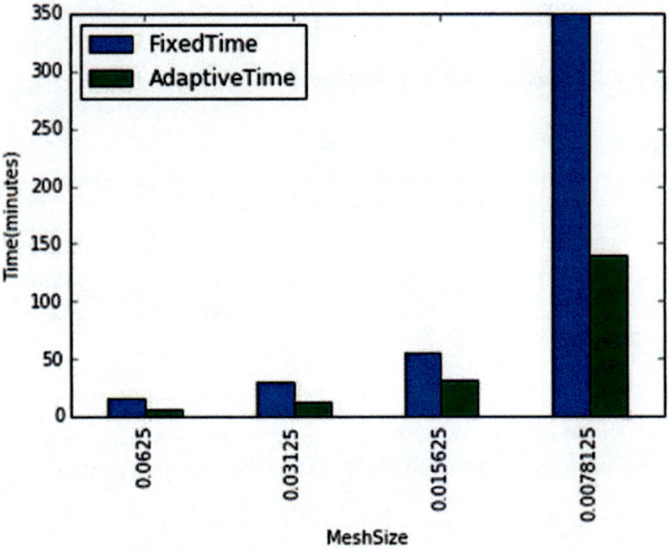

Fig. 9 Comparison between fixed and adaptive mesh

upper lid gradually deviates from the top right corner and moves towards the centre. Also, the number of secondary vortices increases with an increase in Re. In short, increasing Reynolds number not only leads to an increase in the number of vortex structures but also brings more complexity to the flow structure. These effects are more pronounced for shear-thinning fluids.

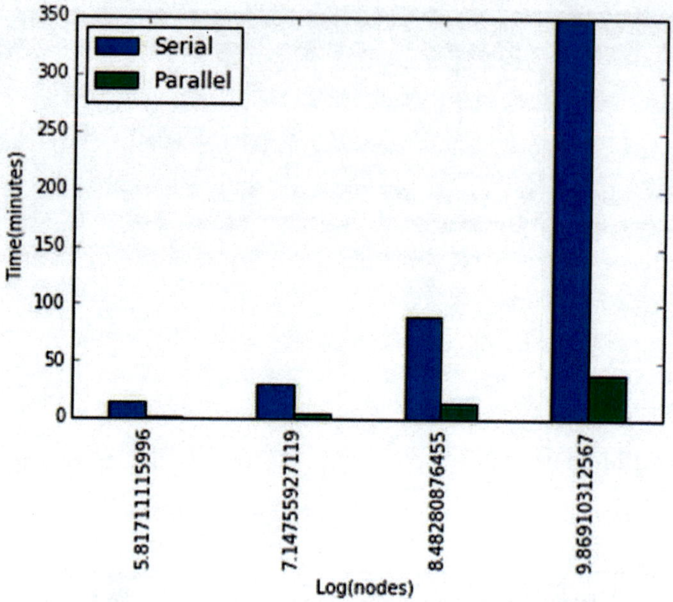

Fig. 10 Serial–Parallel computational time comparison

Table 9 Comparison of location of centre of primary vortex

	Li et al. [8]		Fluidity	
Re	Xc	Yc	Xc	Yc
100	0.5628	0.7282	0.5698	0.7284
500	0.5495	0.6377	0.5477	0.6287
1000	0.5345	0.6097	0.5337	0.6001

The comparison of location of centre of primary vortex can be seen in Table 9. A clear shift of the primary cortex centre to down-left can be seen from the data.

4.3 Detailed Benchmarking of Numerical Results

The horizontal velocity (x-velocity) in the y-direction at $x = 0.5$ m has been compared with the published data of Li et al. [8]), which shows a good agreement for both shear-thinning ($n = 0.5$) and shear-thickening ($n = 1.5$) fluids, as shown in Figs. 13 and 14. The vertical velocity also shows a good agreement with previous work.

Since the implementation of non-linear viscous terms is explicit in the present solver, the effect of the choice of timestep size is presented in Fig. 15. The error drops with timestep size as expected.

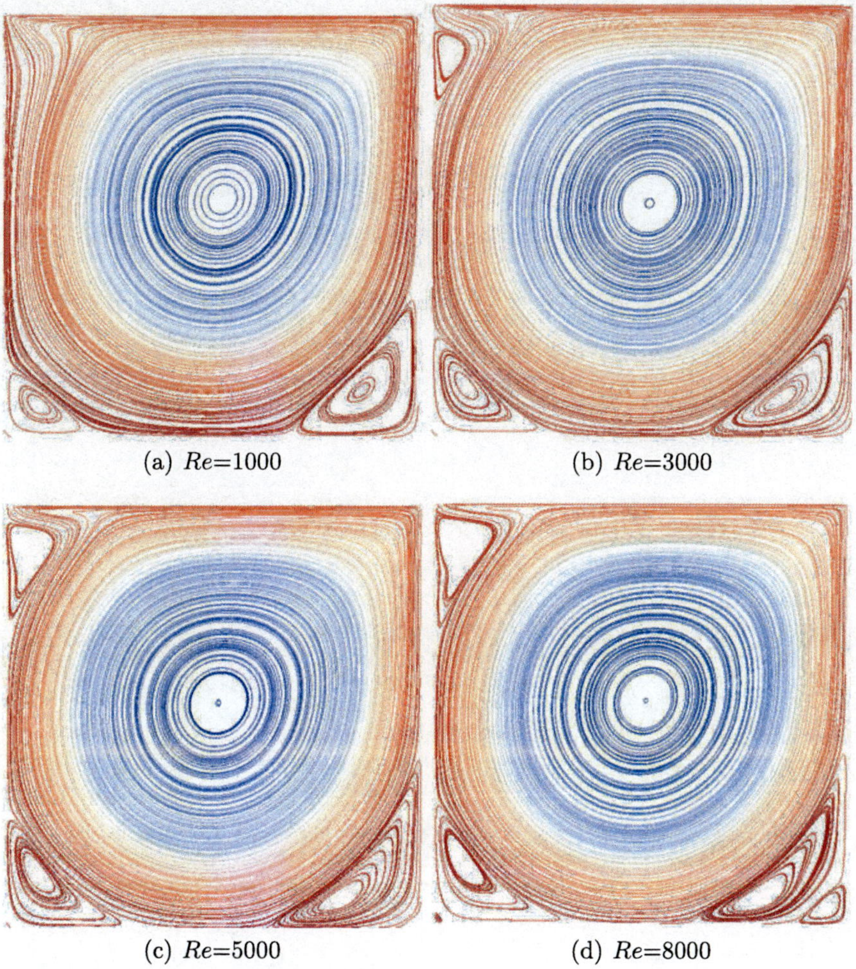

Fig. 11 Effect of Re on Shear-thickening fluids

4.3.1 Effect of Power-Law Index (n)

The increase in the power-law index (n) results in the velocity to increase near the top and bottom of the cavity (Fig. 16). The v-velocity appears sinusoidal and becomes sharper with an increase in n.

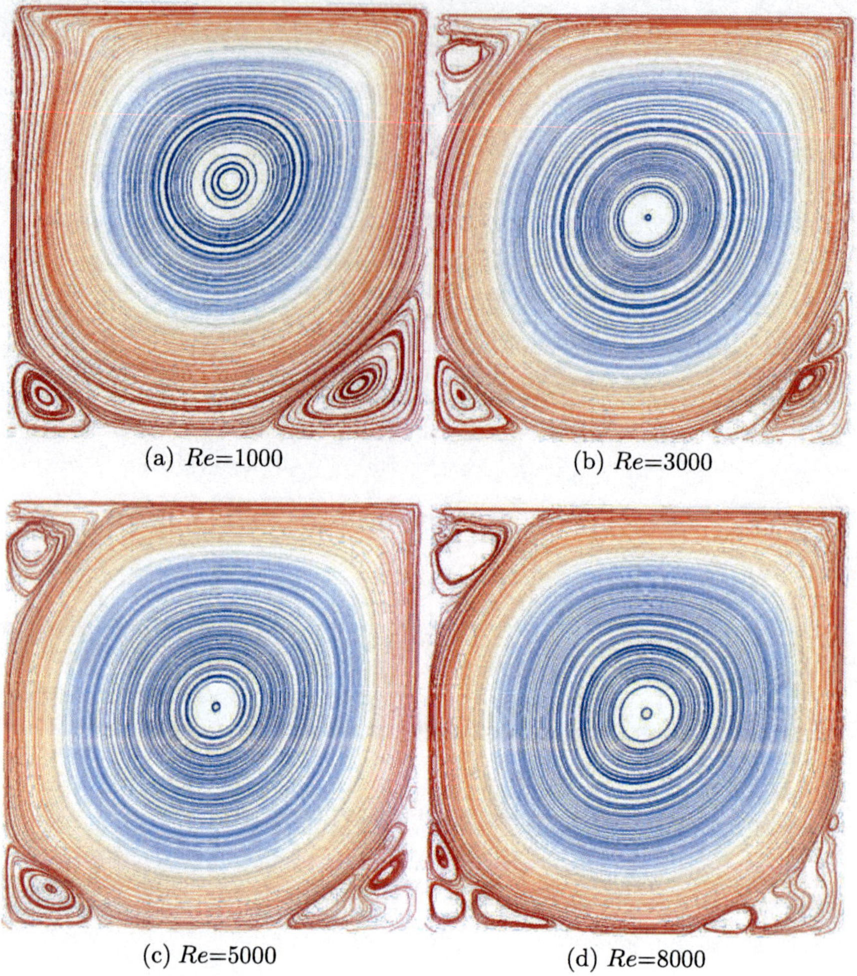

(a) Re=1000 (b) Re=3000 (c) Re=5000 (d) Re=8000

Fig. 12 Effect of Re on Shear-thinning fluids

5 Conclusion

In this paper, power-law (non-Newtonian) fluid flow in a 2D square cavity is studied using the finite element Fluidity code. The effects of mesh adaptivity, change in Reynolds number, and power-law index are presented. A detailed comparison between velocity values with published results is also shown. The following conclusions can be drawn based on the results obtained:

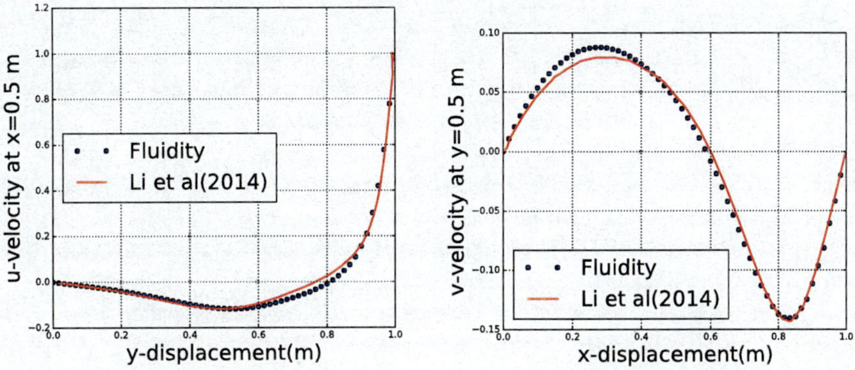

(a) x-velocity (b) y-velocity

Fig. 13 Velocity benchmarking for shear-thinning fluids

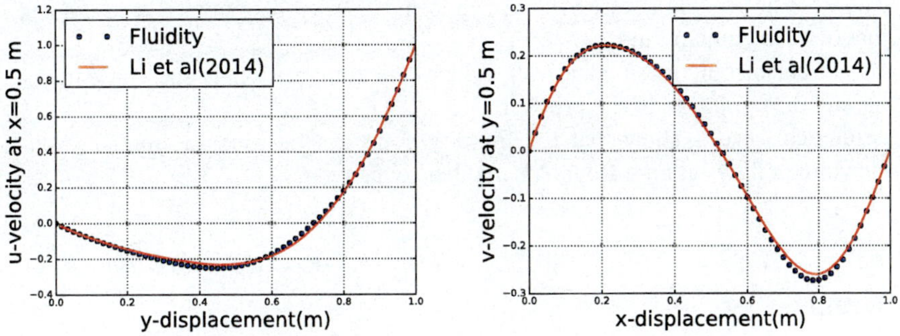

(a) x-velocity (b) y-velocity

Fig. 14 Velocity benchmarking for shear-thickening fluids

Fig. 15 Effect of timestep size

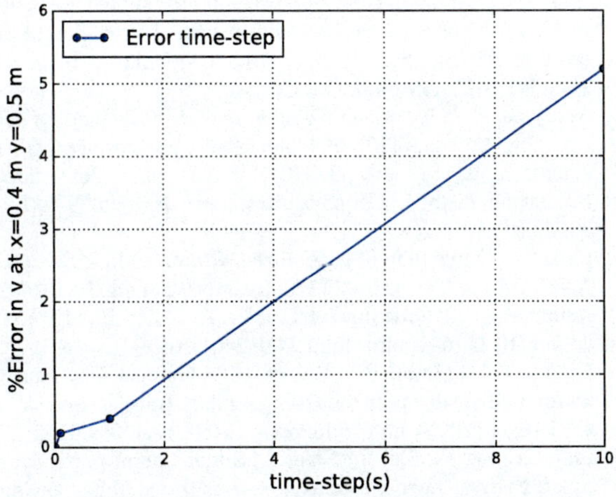

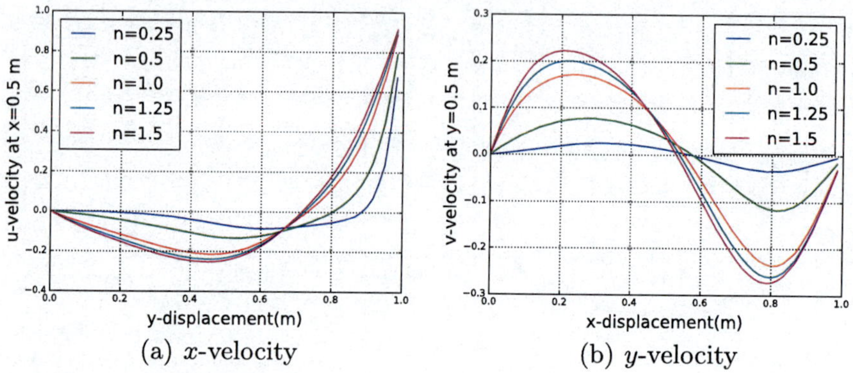

Fig. 16 Effect of power-law index (n)

1. There is a noticeable difference between the flow structure of power-law fluid and that of a Newtonian fluid.
2. Use of anisotropic mesh adaptivity reduces the computational time while maintaining the dynamics of the problem.
3. Numerical results show that the Fluidity code can be used to simulate non-Newtonian flows at high Reynolds number robustly.

References

1. Pearson, J.R.A., Tardy, P.M.J.: Models for flow of non-Newtonian and complex fluids through porous media. J. Non-Newton. Fluid Mech. **102**(2), 447–473 (2002). https://doi.org/10.1016/S0377-0257(01)00191-4
2. Bell, B.C., Surana, K.S.: A space -time coupled p-version least squares finite element formulation for unsteady two-dimensional Navier-Stokes equations. Int. J. Numer. Methods Eng. **39**(15), 2593–2618 (1996). https://doi.org/10.1002/(SICI)1097-0207(19960815)39:15<2593::AID-NME968>3.0.CO;2-2
3. Neofytou, P.: A 3rd order upwind finite volume method for generalised Newtonian fluid flows. Adv. Eng. Softw. **36**(10), 664–680 (2005). https://doi.org/10.1016/j.advengsoft.2005.03.011
4. Aharonov, E., Rothman, D.H.: Non-Newtonian flow (through porous media): A lattice-Boltzmann method. Geophys. Res. Lett. **20**(8), 679–682 (1993). https://doi.org/10.1029/93GL00473
5. Rafiee, A.: Modelling of generalized Newtonian lid-driven cavity flow using an SPH method. ANZIAM J. **49**(3), 411–422 (2008). https://doi.org/10.1017/S1446181108000011
6. Kamm, R.D.: Cellular fluid mechanics. Ann. Rev. Fluid Mech. **34**(1), 211–232 (2002). https://doi.org/10.1146/annurev.fluid.34.082401.165302
7. Piggott, M.D., Farrell, P.E., Wilson, C.R., Gorman, G.J., Pain, C.C.: Anisotropic mesh adaptivity for multi-scale ocean modelling. Philos. Trans. R. Soc. A Math. Phys. Eng. Sci. **367**(1907), 4591–4611 (2009). https://doi.org/10.1098/rsta.2009.0155
8. Li, Q., Hong, N., Shi, B., Chai, Z.: Simulation of power-law fluid flows in two-dimensional square cavity using multi-relaxation-time lattice Boltzmann method. Commun. Comput. Phys. **15**(1), 265–284 (2014). https://doi.org/10.4208/cicp.160212.210513a

Computational Study of Shear Flow Past Square Cylinder with Horizontal Control Plate

Ashwani and Rajendra K. Ray

1 Introduction

The study of uniform flow beyond a cylindrical impediment has fascinated researchers for a long term, and exact data is to be had in the literature. However, many sensible cases exist where the flow coming near a frame is sheared rather than uniform. For example, cylindrical structures submerged in wind, tides or currents are usually exposed to a non-uniform loose circulation. When the oncoming free stream is a linear drift, a consistent vorticity is embedded within the free stream. This might also result in a complicated interplay with the boundary layers that become independent from the body which ends up in a exclusive type of vortex shedding structure. It is essential to recognise the consequences of the imminent shear flow on the mechanism of vortex shedding. Such type of study would cause higher expertise of the vortex-precipitated vibration, its subsequent suppression and control. In our problem, shear flow past a square cylinder with horizontal control plate is studied systematically over more than a few shear rates by making use of a valuable numerical scheme, that is, HOC (Higher order compact) scheme. Much of the prior research has already been done in various fields using uniform flow past bluff bodies till date. Cheng et al. [1] investigated the linear shear flow over a square cylinder at a Reynolds number of 100 using the lattice Boltzmann method. Their results show the suppression effect of the flow on vortex shedding with some basic features of the flow phenomenon with respect to K. Later on, Cheng et al. [2] numerically simulated shear flow around square cylinder, examined vortex shedding frequency and aerodynamic forces exerted on

Ashwani · R. K. Ray (✉)
Indian Institute of Technology Mandi, Mandi 175005, India
e-mail: rajendra@iitmandi.ac.in

Ashwani
e-mail: mr.punia11@gmail.com

S. K. Saha and M. Mukherjee (eds.), *Recent Advances in Computational Mechanics and Simulations*, Lecture Notes in Mechanical Engineering,
https://doi.org/10.1007/978-981-15-8315-5_29

the cylinder for $Re = 100$ and K ranging from 0.0 to 0.4 using lattice Boltzmann method. Their results show that vortex phenomenon is strongly dependent on Re and K. In 1960, Lax and Wendroff [3] use an idea to raise the temporal accuracy from one to two. They use this idea on transient hyperbolic PDEs and use the original PDE to approximate the second-order time derivative in a Taylor series expansion. After this, in 1982 Gartland [4] uses discrete weighted mean approximation to get higher order compactness while Noye and Tan [5] apply weighted modified PDE method and in 1984, Gupta et al. [6] apply series to get higher order compactness expansion to the differential equations.

In 2002, Kalita et al. [7] works on higher order compact schemes for the unsteady two-dimensional convection–diffusion equation with variable convection coefficients and they showed the region where complexity is more, very accurate results can be achieved through HOC schemes with significantly small number of grid points. Later, Ray and Kalita [8, 9] develop HOC scheme for incompressible viscous flows in polar coordinate system (r, θ). Therefore, of late, the Higher Order Compact (HOC) finite difference schemes for the computation of incompressible viscous flows are steadily gaining reputation due to their immoderate accuracy and advantages related to compact difference stencils.

The purpose of the present study is to investigate the effect of shear rate on the unsteady flow separation and the vortex shedding phenomenon using a structural bifurcation. In the process, we aim to know the effect of added control plate with the square cylinder on flow. The schematic diagram of the problem is shown in Fig. 1.

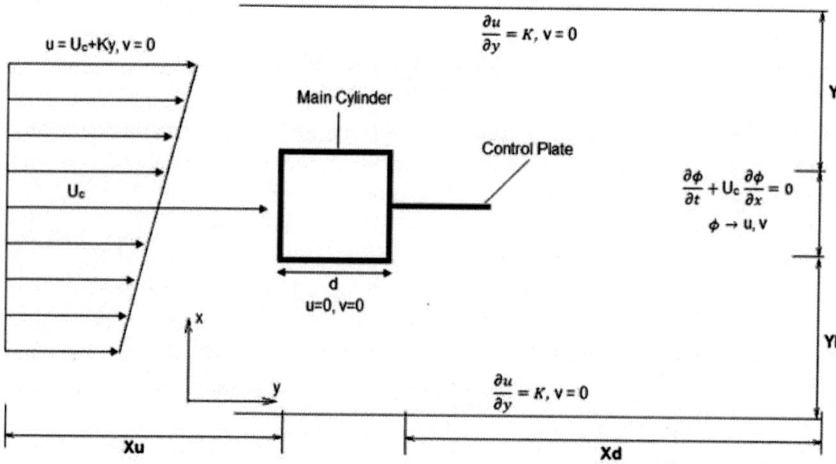

Fig. 1 Schematic diagram of the problem

2 Problem Statement and Governing Equations

In our problem we are considering an unsteady, incompressible shear flow past a square cylinder with a horizontal control plate of unit length and placed at a distance 'd' from the square cylinder, where 'd' is the length of the side of the cylinder. Here centre of the square cylinder is fixed at origin (0, 0) and a horizontal plate is attached to the cylinder as shown in schematic diagram of the problem. The physical domain is infinite but for the computational feasibility of problem, the computational area must be finite. Thus the usage of right boundary conditions can reduce the scale of computational domains. we apply a linear combination at inlet,

$$u = U_c + Ky, \quad v = 0 \tag{1}$$

where U_c is the centre line velocity of flow and K is the shear parameter. The slip boundary situations are applied on the above and lower side of the domain. That is,

$$\frac{\partial u}{\partial y} = K, \quad v = 0 \tag{2}$$

No-slip boundary conditions are considered on the surface of the cylinder [10]. We use convective boundary conditions at the outer boundary of domain which is as follows:

$$\frac{\partial \phi}{\partial t} + U_c \frac{\partial \phi}{\partial x} = 0 \tag{3}$$

The flow is governed by the incompressible Navier–Stokes equation. The stream-function vorticity ($\psi - \phi$) formulation of the Navier–Stokes equation in Cartesian coordinate (x, y), is given as

$$\frac{\partial^2 \phi}{\partial x^2} + \frac{\partial^2 \phi}{\partial y^2} = R_e \left(u \frac{\partial \phi}{\partial x} + v \frac{\partial \phi}{\partial y} + \frac{\partial \phi}{\partial t} \right) \tag{4}$$

$$\frac{\partial^2 \psi_1}{\partial x^2} + \frac{\partial^2 \psi_1}{\partial y^2} = -\phi \tag{5}$$

where ϕ = vorticity, ψ_1 = Stream function, u, v = x-direction and y-direction velocity component, respectively.
Velocity component u, v in terms of stream function (ψ_1) can be written as

$$u = \frac{\partial \psi_1}{\partial y}, v = -\frac{\partial \psi_1}{\partial x} \tag{6}$$

so vorticity,

$$\phi = \frac{\partial v}{\partial x} - \frac{\partial u}{\partial y} \tag{7}$$

2.1 Numerical Discretisation

As flow is governed by Navier–Stokes equations. So, first, we have to discretise the governing equations. Discretisation of governing equations by using Higher Order Compact (HOC) scheme in Cartesian coordinate as follows:

$$\frac{\partial^2 \phi}{\partial x^2} + \frac{\partial^2 \phi}{\partial y^2} = R_e \left(u\frac{\partial \phi}{\partial x} + v\frac{\partial \phi}{\partial y} + \frac{\partial \phi}{\partial t} \right) \tag{8}$$

$$\frac{\partial^2 \psi_1}{\partial x^2} + \frac{\partial^2 \psi_1}{\partial y^2} = -\phi \tag{9}$$

HOC discretisation of (8) at (i, j)th node is given as

$$\begin{aligned}[Re+&A_{i,j}\delta_x^2 + B_{i,j}\delta_y^2 + C_{i,j}\delta_x + D_{i,j}\delta_y + G_{i,j}\delta_x\delta_y + H_{i,j}\delta_x\delta_y^2 + K_{i,j}\delta_x^2\delta_y + \\ &L_{i,j}\delta_x^2\delta_y^2]\phi_{i,j}^{n+1} = [Re + A_{i,j}\delta_x^2 + B_{i,j}\delta_y^2 + C_{i,j}\delta_x + D_{i,j}\delta_y + G_{i,j}\delta_x\delta_y \\ &+H_{i,j}\delta_x\delta_y^2 + K_{i,j}\delta_x^2\delta_y + L_{i,j}\delta_x^2\delta_y^2]\phi_{i,j}^n \end{aligned} \tag{10}$$

Similarly, HOC discretisation of (9) is

$$[\delta_x^2 + \delta_y^2 - (h2 + k2)\delta_x^2\delta_y^2]\psi_{i,j} = [-1 + H2\delta_x^2 + K2\delta_Y^2]\phi_{i,j} \tag{11}$$

where

$$\begin{aligned}
A_{i,j} &= -H12Re - 0.5\Delta t A1_{i,j} \\
B_{i,j} &= -K12Re - 0.5\Delta t A2_{i,j} \\
C_{i,j} &= -H11Re - H12u_{i,j}Re^2 - 0.5\Delta t A3_{i,j} \\
D_{i,j} &= -K11Re - K12v_{i,j}Re^2 - 0.5\Delta t A4_{i,j} \\
G_{i,j} &= -0.5\Delta t A5_{i,j} \\
H_{i,j} &= -0.5\Delta t A6_{i,j} \\
K_{i,j} &= -0.5\Delta t A7_{i,j} \\
L_{i,j} &= -0.5\Delta t A8_{i,j}
\end{aligned}$$

where

$$\begin{aligned}
&A1_{i,j} = 1 + H11Reu_{i,j} + H12Re^2u_{i,j}^2 + 2H12Re(u_x)_{i,j} \\
&A2_{i,j} = 1 + K11Rev_{i,j} + K12Re^2v_{i,j}^2 + 2K12Re(v_y)_{i,j} \\
&A3_{i,j} = -Reu_{i,j} + H11Re(u_x)_{i,j} + K11Re(u_y)_{i,j} + H12Re^2u_{i,j}(u_x)_{i,j} \\
&+H12Re(u_{xx})_{i,j} + K12Re(u_{yy})_{i,j} + K12Re^2v_{i,j}(u_y)_{i,j} \\
&A4_{i,j} = -Rev_{i,j} + H11Re(v_x)_{i,j} + K11Re(v_y)_{i,j} + H12Re^2u_{i,j}(v_x)_{i,j} \\
&+H12Re(v_{xx})_{i,j} + K12Re(v_{yy})_{i,j} + K12Re^2v(v_y)_{i,j}
\end{aligned}$$

$$
\begin{aligned}
A5_{i,j} &= H11Rev_{i,j} + k11Reu_{i,j} + H12Re^2u_{i,j}v_{i,j} \\
&+2H12Re(v_x)_{i,j} + 2k12Re(u_y)_{i,j} + k12Re^2u_{i,j}v_{i,j} \\
A6_{i,j} &= -H11 - H12Reu_{i,j} + K12Reu_{i,j} \\
A7_{i,j} &= -K11 + H12Rev_{i,j} - K12Rev_{i,j} \\
A8_{i,j} &= -H12 - K12 \\
H2 &= -h^2/12, \qquad K2 = -h^2/12, \\
H11 &= K11 = Reu_{i,j}h^2/6, \\
H12 &= K12 = -h^2/12.
\end{aligned}
$$

2.2 Validation with Existing Results

The instantaneous streak-line patterns are shown in Fig. 2, which are in good agreement with the results of Cheng et al. [2] at $Re = 100$ and $K = 0.0$ and $K = 0.4$. The vortices behind the cylinder are in a regular alternating arrangement.

2.3 Results and Discussion

Figures 3, 4 and 5 show the vorticity contours for different K values. The centre of the cylinder is fixed at origin. We can easily see that for $K = 0.0$ the pattern of flow is symmetric. We found that for $K = 0.1$ the vorticity contour is different from $K = 0.0$ because as K changes in positive direction, the speed of the fluid flow decreases at bottom of cylinder. For $K = 0.4$, it is totally suppressed. In this case, vortex formation region does not change with respect to time. We plot the transverse velocity component fluctuation on a centerline in Figs. 6 and 7 for $Re = 100$ at different K values. Figure 7 shows that as K increases, u values increase in the upper part and decrease in the lower part of the cylinder, whereas with increment

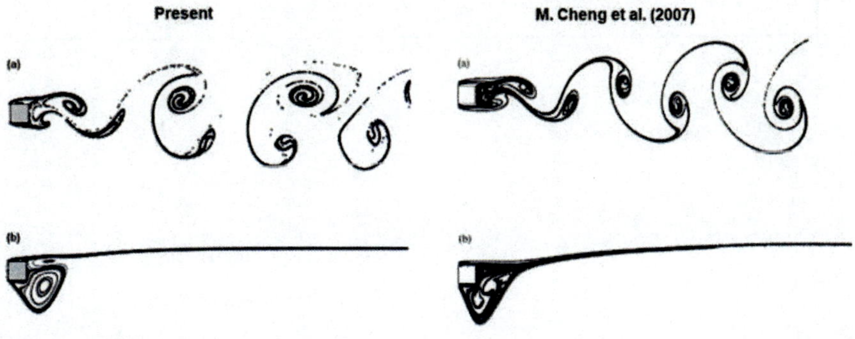

Fig. 2 Comparison of streak-line pattern at $Re = 100$ for **a** $K = 0.0$ and **b** $K = 0.4$ with the results of Cheng et al.

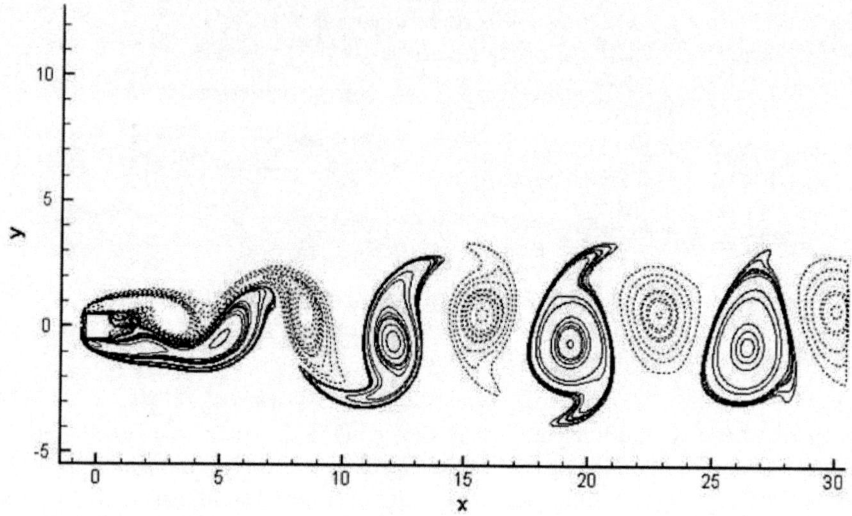

Fig. 3 Vorticity contours for $K = 0.0$

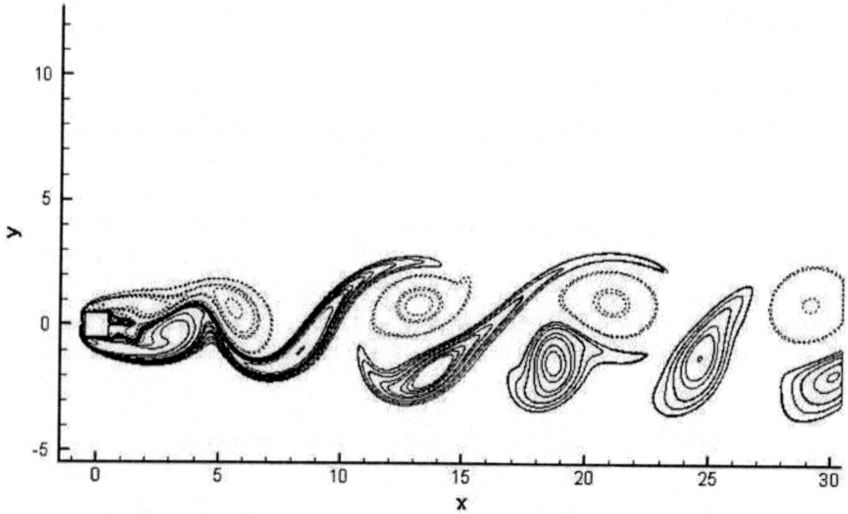

Fig. 4 Vorticity contours for $K = 0.1$

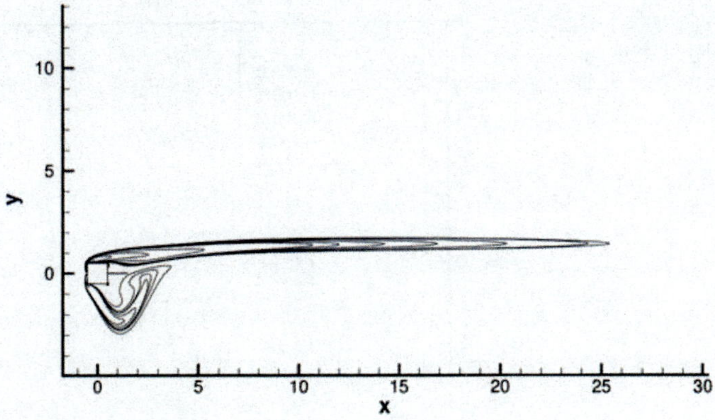

Fig. 5 Vorticity contours for $K = 0.4$

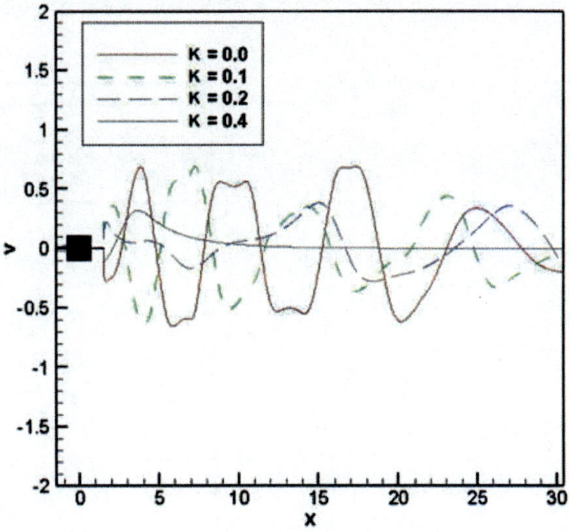

Fig. 6 v along x-axis for different values of K at $Re = 100$

of shear parameter values, v values decrease along the x-axis as shown in Fig. 6. A phase diagram (Fig. 8) is also drawn between u versus v at a monitoring point (0.75, 0.5) behind the cylinder at different K. These plots are the indication of a periodic solution.

2.4 *Study of Structural Bifurcation Analysis*

Bifurcation is the mathematical analysis of qualitative adjustments in the topological shape of a given family. A bifurcation takes place when a small clean shift within the

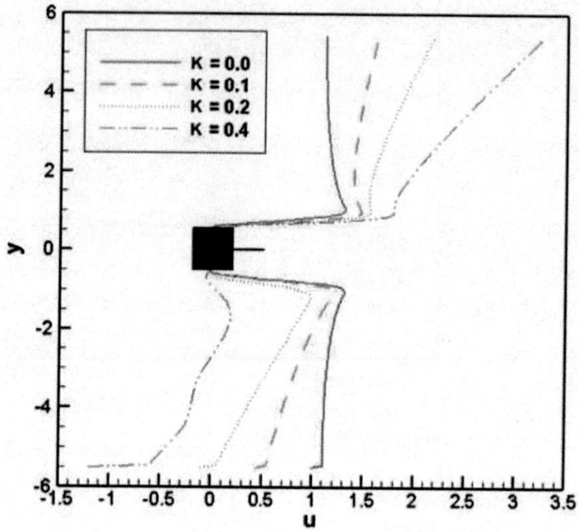

Fig. 7 u along y-axis for different values of K at $Re = 100$

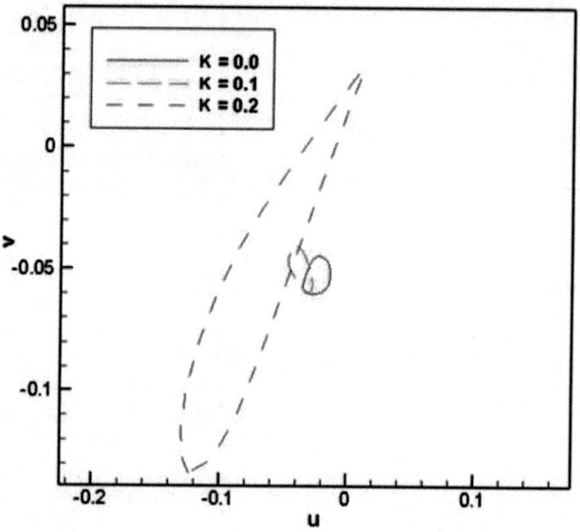

Fig. 8 (u-v) phase diagram for $Re = 100$, at different K values

parameter values of a system leads to a fast shifting in its qualitative or topological behaviour. The conditions for the separation at a point P^* on the floor of the cylinder, at time T^* for a fluid flow governed by N-S equations with no-slip condition on the boundary of solid, are given by

$$\omega = 0, \quad \frac{\partial \omega}{\partial \tau} = 0, \quad \frac{\partial^2 \omega}{\partial \tau^2} > 0, \quad \frac{\partial \omega}{\partial t} < 0 \tag{12}$$

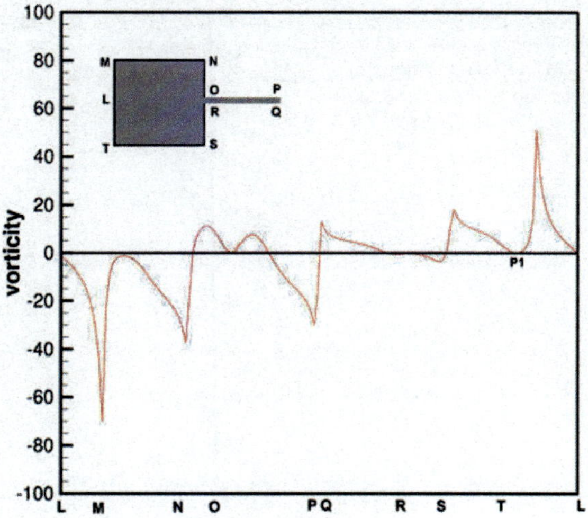

Fig. 9 Vorticity distribution on the surface of the cylinder and horizontal plate during the structural bifurcation at **t = 0.626** for $K = 0.1$

Table 1 Occurrences of first and second structural bifurcation points at $Re = 100$

$Re = 100$	$[P1(x, y), T1]$	$[P2(x, y), T2]$
$K = 0.0$	[(−0.22, −0.5), 1.201]	[(−0.22, 0.5), 1.201]
$K = 0.1$	[(−0.26, −0.5), 0.626]	[(−0.22, 0.5), 2.587]
$K = 0.4$	[(−0.34, −0.5), 1.122]	[(−0.22, 0.5), 4.601]

at (P^*, T^*). It is thought that a structural bifurcation takes place at time T^* and has a degenerate singular factor P^* at the solid floor with a downward flow. In the case of an upward shear flow, remaining two inequalities will change the symptoms.

From Figs. 9 and 10, we can easily see that at point $P1$ and $P2$, $\omega = 0$ and also there exist extremum (relative minimum or maximum) at $P1$ and $P2$ so $\frac{\partial \omega}{\partial \tau} = 0$. At the point $P1$, there is relative minima and at point $P2$ there is relative maxima. This implies that $\frac{\partial^2 \omega}{\partial \tau^2} > 0$ and $\frac{\partial^2 \omega}{\partial \tau^2} < 0$ at $P1$ and $P2$, respectively. Now for fourth condition we plot the vorticity variation with time at point $P1(-0.26, -0.5)$, $P2(-0.22, 0.5)$ as in Figs. 11 and 12. From Fig. 11, we can observe that vorticity is decreasing with respect to time. This implies $\frac{\partial \omega}{\partial t} < 0$. In Fig. 12, vorticity is increasing with respect to time. This implies $\frac{\partial \omega}{\partial t} > 0$, which implies that $P1$ and $P2$ are our bifurcation points for $K = 0.1$. Similarly, we found bifurcation point for different values of K as shown in Table. From Table 1, we found that time of occurrence of second bifurcation is increasing but time of occurrence of first bifurcation it is not monotonic.

Fig. 10 Vorticity distribution on the surface of the cylinder and horizontal plate during the structural bifurcation at **t = 2.587** for $K = 0.1$

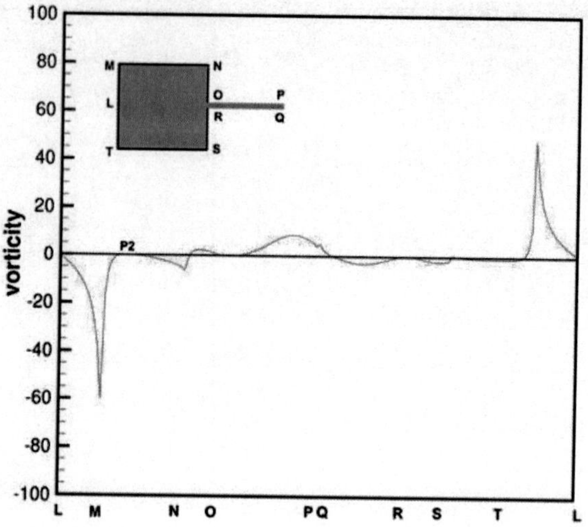

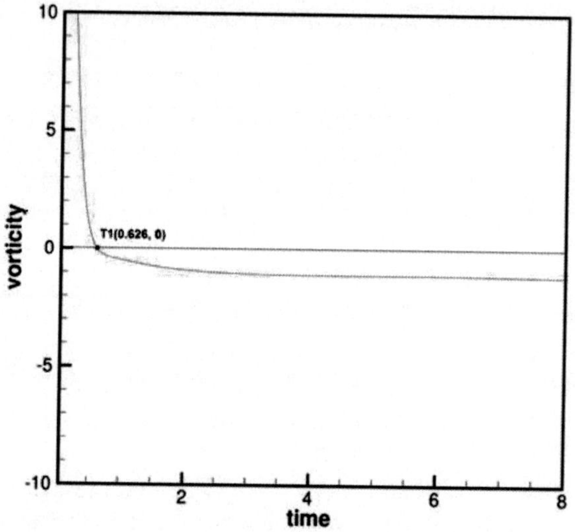

Fig. 11 Vorticity variation with time at point $P1(-0.26, -0.5)$

3 Conclusion

The current work presents the study of shear flow past a square cylinder with horizontal control plate. By using the HOC scheme, we can increase the accuracy of the solution. The study of vorticity contours concludes that these are mainly dependent on the shear parameter (K) and control plate. As K increases, the speed of the flow from lower side of the cylinder decreases and for different value of shear parameter flow behaves differently. The u-v phase diagrams for different K are the indication

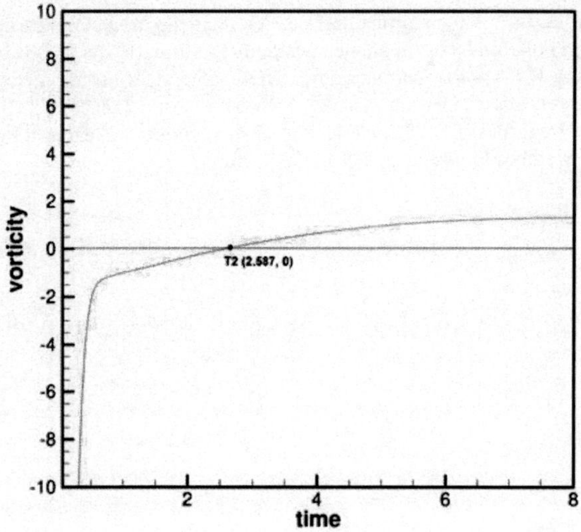

Fig. 12 Vorticity variation with time at point $P2(-0.22, 0.5)$

of a periodic solution. Bifurcation analysis concludes that first bifurcation occurs at the bottom of the cylinder and second bifurcation occurs at the top of the cylinder. We also observed that for $K = 0.0$ first and second bifurcation occur at same time because flow is symmetric in the case of $K = 0.0$.

References

1. Cheng, M., Tan, S.H.N., Hung, K.C.: Linear shear flow over a square cylinder at low Reynolds number. Phys. Fluids **17**(7), 078103 (2005)
2. Cheng, M., Whyte, D.S., Lou, J.: Numerical simulation of flow around a square cylinder in uniform-shear flow. J. Fluids Struct. **23**(2), 207–226 (2007)
3. Lax, P., Wendroff, B.: Systems of conservation laws. Commun. Pure Appl. Math. **13**(2), 217–237 (1960)
4. Gartland Jr., E.C.: Discrete weighted mean approximation of a model convection-diffusion equation. SIAM J. Sci. Stat. Comput. **3**(4), 460–472 (1982)
5. Noye, B.J., Tan, H.H.: A thirdorder semi-implicit finite difference method for solving the one-dimensional convection-diffusion equation. Int. J. Numer. Methods Eng. **26**(7), 1615–1629 (1988)
6. Gupta, M.M., Manohar, R.P., Stephenson, J.W.: A single cell high order scheme for the convection-diffusion equation with variable coefficients. Int. J. Numer. Methods Fluids **4**(7), 641–651 (1984)
7. Kalita, J.C., Dalal, D.C., Dass, A.K.: A class of higher order compact schemes for the unsteady two-dimensional convection-diffusion equation with variable convection coefficients. Int. J. Numer. Methods Fluids **38**(12), 1111–1131 (2002)

8. Ray, R.K., Kalita, J.C.: A transformation-free HOC scheme for incompressible viscous flows on non-uniform polar grids. Int. J. Numer. Methods Fluids **62**(6), 683–708 (2010)
9. Kalita, J.C., Ray, R.K.: A transformation-free HOC scheme for incompressible viscous flows past an impulsively started circular cylinder. J. Comput. Phys. **228**(14), 5207–5236 (2009)
10. Lankadasu, A., Vengadesan, S.: Shear effect on square cylinder wake transition characteristics. Int. J. Numer. Methods Fluids **67**(9):1115–1134 (2011)

Numerical Study of Shear Flow Past an Inclined Square Cylinder with Vertical Control Plate

Rishabh Saxena and Rajendra K. Ray

1 Introduction

One of the most considered problems when flow past through square cylinder in which different types of experimental and numerical studies has been done. This problem is interesting because of its practical importance in different fields like bridges, buildings, vortex flow metre, etc., apart from the experimental studies, this problem also plays an important role for academic importance in which most of the researchers are doing numerical simulation. Kumar and Ray [1] has already done numerical simulation for inclined square cylinder ($\alpha = 45°$) for different shear parameter (K) values at Reynolds number ($Re = 100$). Yoon et al. [2] present the study for the laminar flow past an inclined square cylinder in uniform free stream. Sohankar [3] presents the study for different incidence angles ($\alpha = 0°–45°$) of square cylinder, Sohankar has done computational work with the help of SIMPLEC code with non-staggered grid for computing the flow behaviour.

In this paper, we extend the problem of the inclined square cylinder ($\alpha = 45°$) with the help of vertical control plate. In which we have used higher order compact (HOC) finite difference scheme on uniform grids, which is given by Kalita et al. [4]. We used HOC scheme because it is already tested for the square cylinder and it gives accurate flow behaviour, which was already presented by Kumar and Ray [5]. But in this paper, we extend the use of HOC scheme for the inclined square cylinder with the help of vertical control plate, until no one tested HOC scheme for control

R. Saxena
School of Basic Sciences, Indian Institute of Technology Mandi, Mandi 175001, India
e-mail: rishabh7934@gmail.com

R. K. Ray (✉)
School of Basic Sciences, Indian Institute of Technology Mandi,
VPO. Kamand, Mandi 175005, Himachal pradesh, India
e-mail: rajendra@iitmandi.ac.in

S. K. Saha and M. Mukherjee (eds.), *Recent Advances in Computational Mechanics and Simulations*, Lecture Notes in Mechanical Engineering,
https://doi.org/10.1007/978-981-15-8315-5_30

plate. Also, we analyse the flow behaviour with different K values at $Re = 100$ and apart from the numerical studies we also have done the topological based structural bifurcation analysis which gives us the exact location and exact time when and where first and second bifurcation occurs.

2 Problem Statement and Discretisation

2.1 Governing Equations and Boundary Conditions

We consider the problem of an unsteady, incompressible shear flow past an inclined cylinder with vertical control plate. The square cylinder is inclined at an angle ($\alpha = 45°$). Flow is governed by Navier–Stokes equation. The stream function vorticity ($\psi - \omega$) formulation of the two-dimensional Navier–Stokes equation in Cartesian coordinate (x, y) is

$$\omega_{xx} + \omega_{yy} = R_e(u\omega_x + v\omega_y + \omega_t) \tag{1}$$

$$\psi_{xx} + \psi_{yy} = -\omega \tag{2}$$

here ω = vorticity
ψ = Stream function
R_e = Reynold number
u and v represent the velocity component along x and y, respectively.

Here we define velocity component in terms of stream function ψ

$$u = \psi_y, v = -\psi_x$$

so the vorticity $\omega = v_x - u_y$.

In this problem, We have taken the width and height of the inclined cylinder as 'd' and distance of the vertical control plate from the inclined cylinder as 'd'.

Here we apply a linear combination, for the physical boundary condition at inlet,

$$U = U_c + Ky$$

at $v = 0$, where U_c = centre line velocity of inflow, K = shear parameter. A schematic diagram of the problem is given in Fig. 1. The major issue comes to solve the boundary conditions specially outlet boundary. This happens because our problem is based on physical problem with the infinite domain. So to make the problem computationally feasible we take some finite domain. We choose convective boundary condition at the outlet boundary because it is found as the best fit boundary condition for this problem

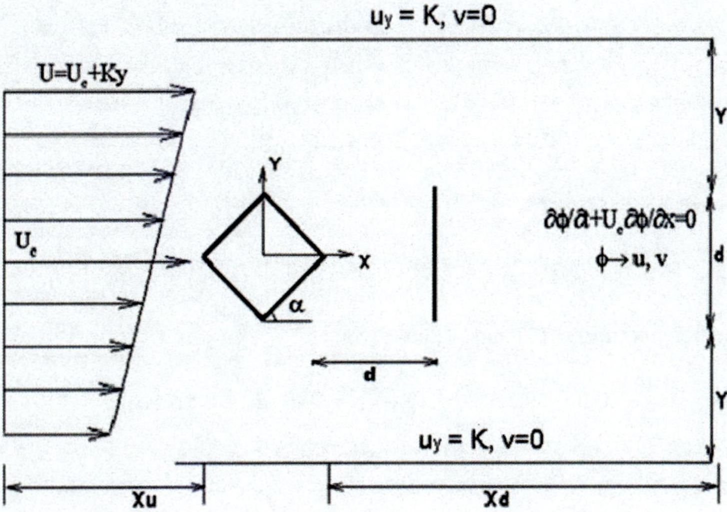

Fig. 1 Schematic diagram of flow control behind the inclined cylinder with an inlet shear using vertical control plate

in the literature [6]. We choose slip boundary condition on the top and bottom of the domain which is shown in Fig. 1.

$$u_y = K, v = 0$$

We have used no-slip boundary condition on the surface of the cylinder and on the control Plate. At the outlet boundary, we have used convective boundary condition

$$\phi_t + U_c \phi_x = 0.$$

2.2 HOC Discretisation

In this section, we discretise the our governing Eqs. (1) and (2) with the help of higher order compact finite difference scheme. To solve the problem, we have used higher order compact finite difference scheme on uniform Cartesian grid. Thus, the higher order compact (HOC) discretisation of the governing Eqs. (1) and (2) at the (i, j) node of the computational domain is given as follows:

$$[Re + A_{11}{\delta_x}^2 + A_{12}{\delta_y}^2 + A_{13}\delta_x + A_{14}\delta_y + A_{15}\delta_x\delta_y + A_{16}\delta_x{\delta_y}^2 + A_{17}{\delta_x}^2\delta_y + A_{18}{\delta_x}^2{\delta_y}^2]\omega_{ij}^{n+1} = [Re + A_{21}{\delta_x}^2 + A_{22}{\delta_y}^2 + A_{23}\delta_x + A_{24}\delta_y + A_{25}\delta_x\delta_y + A_{26}\delta_x{\delta_y}^2 + A_{27}{\delta_x}^2\delta_y + A_{28}{\delta_x}^2{\delta_y}^2]\omega_{ij}^{n} \quad (3)$$

and discretisation for Eq. 2,

$$\left[{\delta_x}^2 + {\delta_y}^2 - (H_2 + K_2){\delta_x}^2{\delta_y}^2\right]\psi_{ij} = \left[-1 + H_2{\delta_x}^2 + K_2{\delta_y}^2\right]\omega_{ij} \quad (4)$$

where coefficients in Eqs. (3) and (4) are given as

$$A_{11(i,j)} = -H_2 R_e - 0.5\Delta t A_{i,j},\ A_{21(i,j)} = -H_2 R_e + 0.5\Delta t A_{i,j},$$
$$A_{12(i,j)} = -K_2 R_e - 0.5\Delta t B_{i,j},\ A_{22(i,j)} = -K_2 R_e - 0.5\Delta t B_{i,j},$$
$$A_{13(i,j)} = -H_1 R_e - H_2 u_{i,j} R_e^2 - 0.5\Delta t C_{i,j},$$
$$A_{23(i,j)} = -H_1 R_e - H_2 u_{i,j} R_e^2 + 0.5\Delta t C_{i,j}$$
$$A_{14(i,j)} = -K_1 R_e - K_2 v_{i,j} R_e^2 - 0.5\Delta t D_{i,j},$$
$$A_{24(i,j)} = -K_1 R_e - K_2 v_{i,j} R_e^2 + 0.5\Delta t D_{i,j}$$
$$A_{15(i,j)} = -0.5\Delta t G_{i,j},\ A_{25(i,j)} = 0.5\Delta t G_{i,j}$$
$$A_{16(i,j)} = -0.5\Delta t H_{i,j},\ A_{26(i,j)} = 0.5\Delta t H_{i,j}$$
$$A_{17(i,j)} = -0.5\Delta t K_{i,j},\ A_{27(i,j)} = 0.5\Delta t K_{i,j}$$
$$A_{18(i,j)} = -0.5\Delta t L_{i,j},\ A_{28(i,j)} = 0.5\Delta t L_{i,j}$$
$$H_1 = Reu_{i,j}\frac{h^2}{6},\ K_1 = Rev_{i,j}\frac{h^2}{6},\ H_2 = \frac{-h^2}{12} - Reu_{i,j}\frac{h^4}{24},$$
$$K_2 = \frac{-h^2}{12} - Rev_{i,j}\frac{h^4}{24},$$
$$A_{i,j} = 1 + H_1 Reu_{i,j} + H_2 Re^2 u_{i,j}^2 + 2H_2 Re(u_x)_{i,j},$$
$$B_{i,j} = 1 + K_1 Rev_{i,j} + K_2 Re^2 v_{i,j}^2 + 2K_2 Re(u_y)_{i,j},$$
$$C_{i,j} = -Reu_{i,j} + H_1 Re(u_x)_{i,j} + K_1 Re(u_y)_{i,j} + H_2 Re^2 u_{i,j}(u_x)_{i,j} + H_2 Re(u_{xx})_{i,j} + K_2 Re(u_{yy})_{i,j} + K2Re^2 v_{i,j}(u_y)_{i,j},$$
$$D_{i,j} = -Rev_{i,j} + H_1 Re(v_x)_{i,j} + K_1 Re(v_y)_{i,j} + H_2 Re^2 u_{i,j}(v_x)_{i,j} + H_2 Re(v_{xx})_{i,j} + K_2 Re(v_{yy})_{i,j} + K2Re^2 v_{i,j}(v_y)_{i,j}$$
$$G_{i,j} = H_1 Rev_{i,j} + K_1 Reui,j + H_2 Re^2 u_{i,j} v_{i,j} + 2H_2 Re(v_x)_{i,j} + 2K_2 Re(u_y)_{i,j} + K2Re^2 u_{i,j} v_{i,j},$$
$$H_{i,j} = -H_1 - H_2 Reu_{i,j} + K_2 Reu_{i,j},\ K_{i,j} = -K_1 + H_2 Rev_{i,j} - K_2 Rev_{i,j},$$
$$L_{i,j} = -H_2 - K_2,\ E_2 = \frac{-h^2}{12},\ F_2 = \frac{-h^2}{12}$$

Equations (3) and (4) represent the discretisation of Eqs. (1) and (2) with the help of HOC scheme. The brief discretisation of the governing equation is explained by Kalita et al. [7].

3 Results and Discussion

3.1 Validation with Existing Result

In this section, we validate our results with the existing results of Kumar and Ray [3], in which we first validated streamline pattern for $Re = 100$ and $K = 0.0$ with different times, then after we validate velocity fluctuation. The qualitative comparisons are presented in Figs. 2 and 3 from the figures we can conclude that our results are almost similar to existing result.

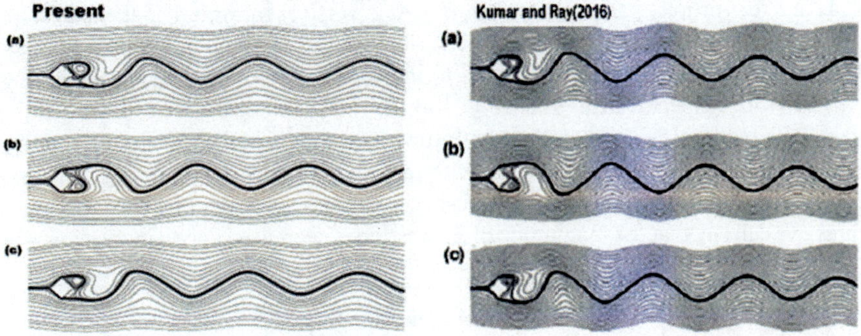

Fig. 2 Comparison of streamline pattern for $Re = 100$, $K = 0.0$ at different times **a** $t = 295$, **b** $t = 298$ and **c** $t = 301$

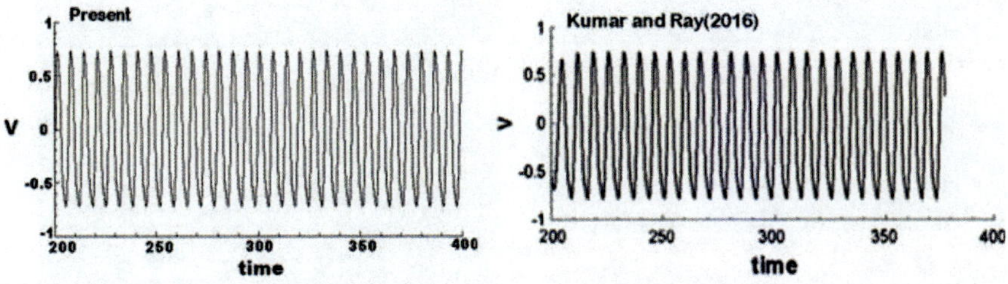

Fig. 3 Comparison of v velocity fluctuation for $Re = 100$ at $K = 0.0$

3.2 Study of Flow Behaviour for $Re = 100$

In our study, we consider three shear parameter values 0.0, 0.1 and 0.2 at $Re = 100$. In Fig. 4, left-hand side we plot the streamlines pattern and corresponding K values on right-hand side is vorticity contour. In vorticity contour, the dotted vortex produces clockwise from the upper surface of the cylinder whereas the solid vortex produces anticlockwise from the lower surface of the cylinder. For $K = 0.0$, the flow pattern is symmetric behind the cylinder which is shown in two ways, first is streamlines pattern and another is vorticity contour, both parts confirms that at $K = 0.0$ the flow is symmetric behind the control plate, which is shown in Fig. 4a. When $K = 0.1$, the flow behaviour is asymmetric behind the control plate, which is clearly shown in Fig. 4b of streamline pattern and vorticity contour. The vorticity contours are completely changed in shape and size behind the control plate for $K = 0.1$ plate is, the dotted vortices are rolling over the solid vortices which are shown in Fig. 4b. When $K = 0.2$, the flow behaviour is again changed and it is completely different from previous K values. In the streamlines, we can see the shape of the contours behind the control plate is completely different from previous K values as shown in Fig. 4c and for vorticity part Fig. 4c when $K = 0.2$, vortices forms only upper surface of the inclined cylinder. So from the observation, we can say that the shape and size of the vortices are changed with respect to the K values. We plot the variation of velocities with respect to time for $Re = 100$, $K = 0.0$ and 0.2. Variation of u and v velocities behind the control plate with time at point (2.58, 0.33), at $Re = 100$. For $K = 0.0$, the velocities fluctuation with respect to time is shown in Fig. 5a. For $K = 0.2$ the velocities fluctuation with respect to time is shown in Fig. 5b. From the figures, we can say that the magnitude of fluctuation is decreased due to increasing K

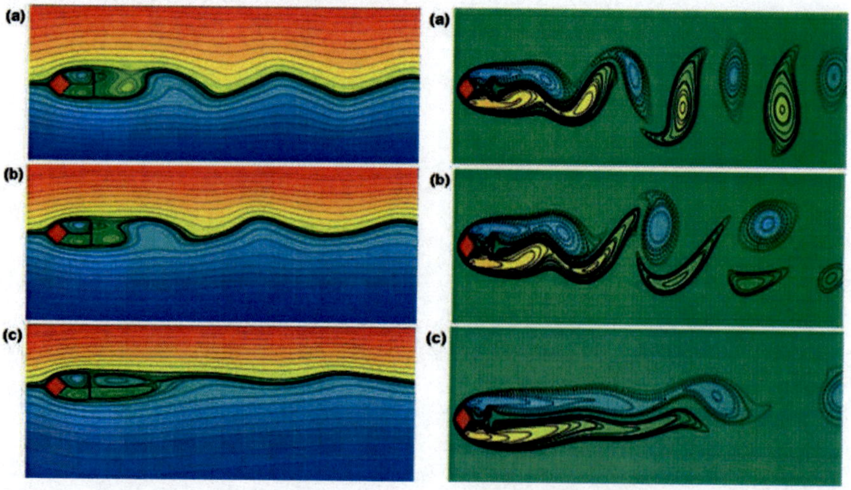

Fig. 4 Streamlines and vorticity for **a** $K = 0.0$, **b** $K = 0.1$ and **c** $K = 0.2$ at $t = 295$

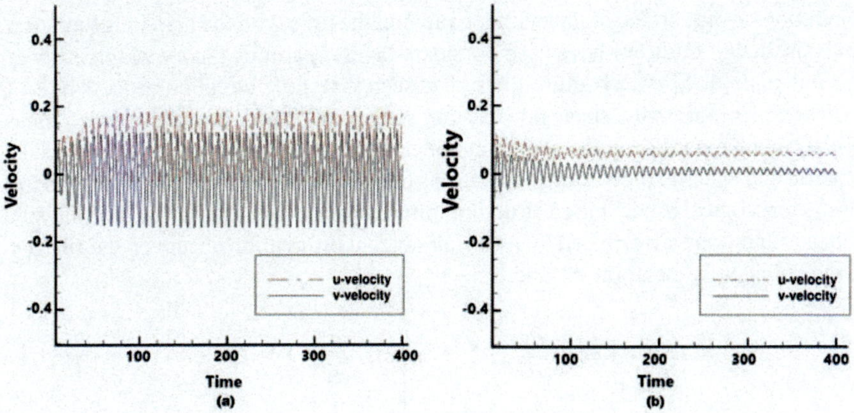

Fig. 5 u and v velocity fluctuation varies with time at **a** $K = 0.0$ and **b** $K = 0.2$

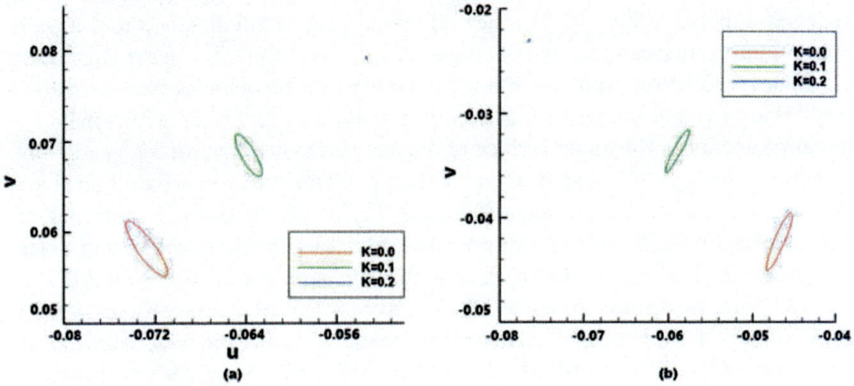

Fig. 6 uv-phase diagram

values. In Fig. 6, we plot the periodicity nature of velocities profile at particular point. In Fig. 6a we plot the velocities at (0.74, 0.3), for different K values at $Re = 100$. In Fig. 6b, we plot the velocities at (0.74, −0.3) for different K values at $Re = 100$. From this figure, we can say that if K increases the periodicity time of velocities will be decreased or in other words, we can say that the fluctuation of velocity profile will be decreased when k increases.

3.3 Study of Bifurcation Analysis

In the qualitative part, we discuss the bifurcation analysis. Structural Bifurcation part plays a major role in mathematics because it describes the structure of any system, if we subject to a change in parameter values. So now, we shall find the first occur separation on the surface of the cylinder and find the time with the help of bifurcation analysis. In this study, we have used boundary layer separation theory which is given by Ghil et al. [4, 5]. Many times it has been seen that the bifurcation process in 2-D unsteady incompressible shear flow past an square cylinder. This study is also done by Ray and Kumar [7] on the square cylinder. From this study, we predict the exact location and specific time of the separations occurs. So these are four conditions (5) for the topological aspect based structural bifurcation analysis that should be satisfy. If these conditions are satisfied then we can say that bifurcation occurs on the surface of the cylinder, at the point P^* and time T^*.

$$\omega(P^*, T^*) = 0, \frac{\partial \omega}{\partial \tau}(P^*, T^*) = 0, \frac{\partial^2 \omega}{\partial \tau^2}(P^*, T^*) > 0$$

$$\frac{\partial \omega}{\partial t}(P^*, T^*) < 0 \tag{5}$$

ω represents the vorticity, t is the time and τ is the tangential direction to the wall. Let us assume a bifurcation occurs at time T^* and has a point P^* on the surface of the cylinder with a flow coming from the lower side of the cylinder then it satisfies the condition (5) and attends a local minima point and decreases with time. If a bifurcation occurs on the upper surface of the inclined cylinder at time T^* and point P^*, then sign will be a change of the condition (5). Our problem is based on shear flow, i.e., first and second bifurcation occurs. We ignore the corner occurrence of the inclined cylinder. So our first motive is to find the bifurcation point and when it happens. We find our results of structural bifurcation for $K = 0.1$, and 0.2 at $Re = 100$. First we discuss for $Re = 100$ and $K = 0.1$, We have already seen in the part of vorticity, from which we can conclude that at $K = 0.1$ the flow behaviour is asymmetric behind the inclined cylinder. So from this analysis, we can say that there exists two bifurcation points which are first and second bifurcation; first bifurcation occurs at the upper surface of the cylinder and second bifurcation occurs at the lower surface of the cylinder. Figure 7a clearly shows us the vorticity reaches to zero at point $P_1^*(x, y) = (0.17, 0.34)$ and also occur local maxima at this point. Figure 7b clearly shows the vorticity reaches to zero at time $T_1^* = 0.234$. So from this analysis, we can say that bifurcation occurs on the upper surface of the inclined cylinder at time T_1^*. Figure 7 satisfies all the condition. $P_1^*(x, y) = (0.17, 0.34)$ occurs at local maxima point and it satisfies $\omega(P^*, T^*) = 0$, $\frac{\partial \omega}{\partial \tau}(P^*, T^*) = 0$, $\frac{\partial^2 \omega}{\partial \tau^2}(P^*, T^*) < 0$ is the condition and increases with time which is clearly shown in Fig. 7b and it satisfies this condition $\frac{\partial \omega}{\partial t}(P^*, T^*) > 0$. So all the conditions are satisfied. The second bifurcation is also occurring because the flow is asymmetric. Figure 8a clearly shows the vorticity reaches to zero at point $P_2^*(x, y) = (0.16, -0.33)$ and also occur at local minima at this point. Figure 8b clearly shows the vorticity reaches to zero at time $T_2^* = 0.427$. So from this analysis, we can say that second bifurcation occurs on the lower surface of the inclined cylinder at time T_1^*. Figure 8 satisfies all the conditions.

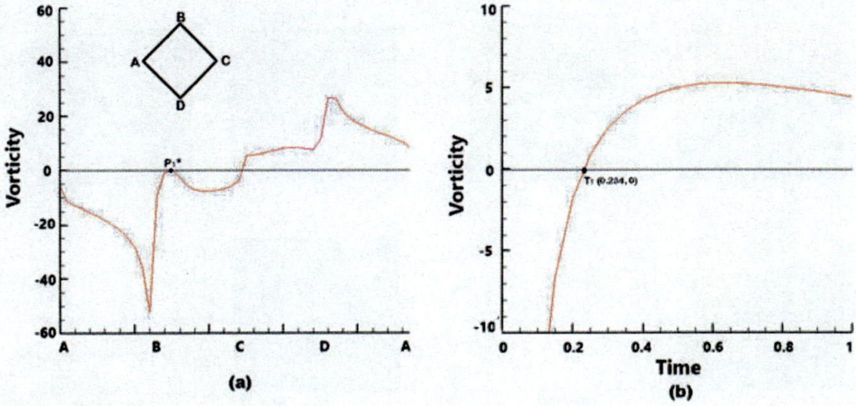

Fig. 7 **a** Surface vorticity at point P_1^* and **b** vorticity variation with time for $Re = 100$, $K = 0.1$

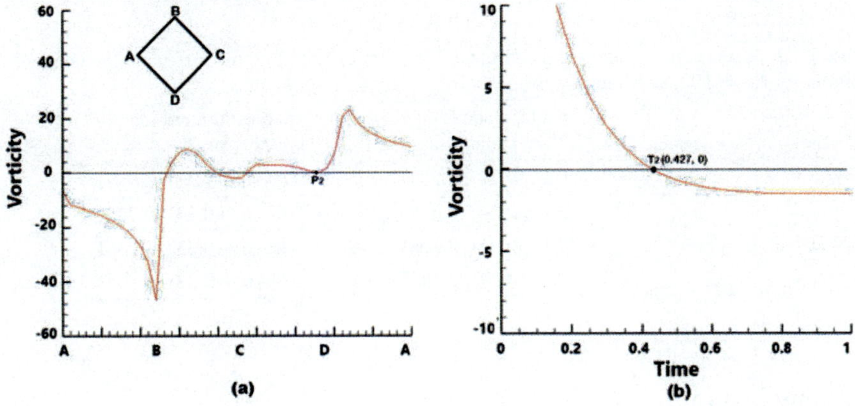

Fig. 8 **a** Surface vorticity at point P_2^* and **b** vorticity variation with time for $Re = 100$, $K = 0.1$

For $P_2^*(x, y) = (0.16, -0.33)$ local minima point occurs and it satisfies $\omega(P^*, T^*) = 0$, $\frac{\partial \omega}{\partial \tau}(P^*, T^*) = 0$, $\frac{\partial^2 \omega}{\partial \tau^2}(P^*, T^*) > 0$ is the condition and decreases with time which is clearly shown in Fig. 8b and it satisfies this condition $\frac{\partial \omega}{\partial t}(P^*, T^*) < 0$. So all the conditions are satisfied. Figure 9 shows a zoomed view of the circulation zones, just before, after and at the time of the occurrence of bifurcation. In Table 1, we present the exact location and the time of occurrence of first and second bifurcations on the surface of the inclined cylinder for different K values at Re = 100. Table shows that there is no fixed pattern of occurrence of bifurcations, but the first bifurcation always occurs earlier as compared to the second bifurcation.

In Table 1, we have done different values of K, from the table we can say that there is no fixed order for bifurcation to occur and also the first bifurcation occurs earlier as compared to second bifurcation (Fig. 9).

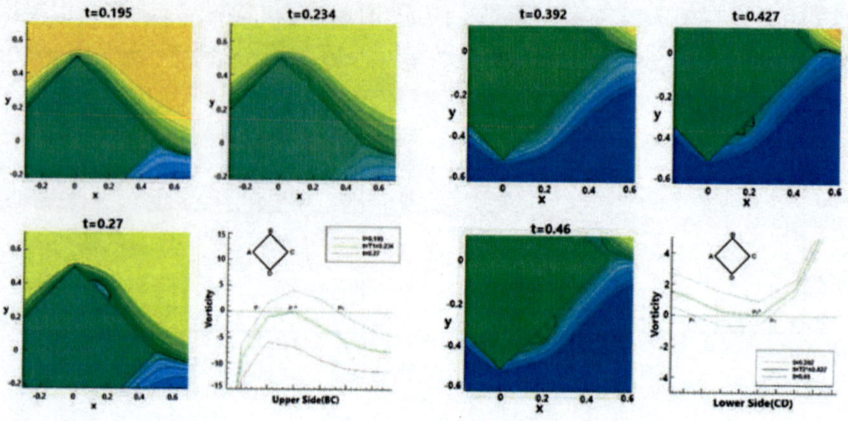

Fig. 9 Zoomed view of the stream function and vorticity profile for the upper and lower portion of the cylinder for $K = 0.1$

Table 1 Occurrence of first and second structural bifurcation points on the surface of the inclined cylinder for $Re = 100$, at different K values

$Re = 100$	First bifuraction $[P_1^*(x, y), T_1^*]$	Second bifurcation $[P_2^*(x, y), T_2^*]$
$K = 0.1$	$[(0.17, 0.34), 0.234]$	$[(0.16, -0.33), 0.427]$
$K = 0.2$	$[(0.17, 0.34), 0.450]$	$[(0.16, -0.33), 0.756]$
$K = 0.3$	$[(0.17, 0.34), 0.168]$	$[(0.16, -0.33), 3.235]$
$K = 0.4$	$[(0.17, 0.34), 0.142]$	$[(0.16, -0.33), 4.505]$

4 Conclusion

In this paper, we have used higher order compact finite difference scheme to discretise the governing equation. Simulations are performed with different K values ($K = 0.0, 0.1$ and 0.2) at $Re = 100$. From the numerical study, we can say that the vortex shedding and velocity fluctuation depend on shear parameter values as well as on the control plate. Apart from the numerical study, we have also done structural bifurcation analysis which gives us exact point and exact time, when and where first and second bifurcation occurs.

References

1. Kumar, A., Ray, R.K.: Higher order compact numerical simulation of shear flow past inclined square cylinder. In: Advanced Computing and Communication Technologies, pp. 305–313. Springer, Singapore (2016)

2. Yoon, D.-H., Yang, K.-S., Choi, C.-B.: Flow past a square cylinder with an angle of incidence. Phys. Fluids **22**(4), 043603 (2010)
3. Sohankar, A., Norberg, C., Davidson, L.: Low-Reynolds-number flow around a square cylinder at incidence: study of blockage, onset of vortex shedding and outlet boundary condition. Int. J. Numer. Methods Fluids **26**(1), 39–56 (1998)
4. Ghil, M., Ma, T., Wang, S.: Structural bifurcation of 2-D nondivergent flows with Dirichlet boundary conditions: applications to boundary-layer separation. SIAM J. Appl. Math. **65**(5), 1576–1596 (2005)
5. Ghil, M., Liu, J.-G., Wang, C., Wang, S.: Boundary-layer separation and adverse pressure gradient for 2-D viscous incompressible flow. Phys. D: Nonlinear Phenom. **197**(1–2), 149–173 (2004)
6. Ray, R.K., Kumar, A.: Numerical study of shear rate effect on unsteady flow separation from the surface of the square cylinder using structural bifurcation analysis. Phys. Fluids **29**(8), 083604 (2017)
7. Kalita, J.C., Dass, A.K., Dalal, D.C.: A transformation free HOC scheme for steady convection diffusion on non-uniform grids. Int. J. Numer. Methods Fluids **44**(1), 33–53 (2004)

Numerical Analysis of the Pressure, Temperature, and Aerodynamic Forces on Hypersonic Blunt Hemispherical Shaped Body

Krishnakumar V. Patel and Prem R. Patel

1 Introduction

Faster than sound speed is a fascinating area in aerospace engineering. To do research in the hypersonic flight has become the vital and extremely interesting topic, because of the complexity of the flow physics during the hypersonic speed. To define the trajectory and altitude of the hypersonic flight, aerodynamic force governed by the shape of the body plays a vital role. Hence, to design the hypersonic vehicle, consolidation of the forces acting on the body is critically important [1]. To accumulate the force history for the flight, it is important to know the equations of motion [2], that define the flight of the vehicle. Generating the equation of motion in the hypersonic regime is a challenging task, since it involves the nature of the high-speed flow complexities [3].

There are many experimental techniques of force balance that could be used for measuring forces. One of the techniques is stress-wave force balance [4], in which the history of the force on the model is determined from the measurement of strain histories. Another method is Impulse force balance [5], in which the history of the force on the model is determined by using the convolution method. However, this technique involved mechanical components and the assumption that the body is rigid. To check the veracity of this technique implemented on the particular component or model is the extremely important subject.

The objective of the present work is to quantify the pressure and temperature variation along the surface of the body and find the trend of the aerodynamic forces

K. V. Patel (✉)
Charotar University of Science and Technology, Gujarat 388421, India
e-mail: krishnaupatel115@gmail.com

P. R. Patel
The University of Texas at Arlington, Arlington, TX 76019, USA

S. K. Saha and M. Mukherjee (eds.), *Recent Advances in Computational Mechanics and Simulations*, Lecture Notes in Mechanical Engineering,
https://doi.org/10.1007/978-981-15-8315-5_31

when the body placed at a different angle of attack in hypersonic flow. The analysis was done on a blunt hemispherical model. Obtained Aerodynamic forces results were validated by comparing with the results from Newtonian and balance theory.

2 Methodology

2.1 Flow Conditions

Hypersonic flow characteristic involves Mach number independence and dissociation phinominan in high enthalpy flow. This complex flow conditions demand high computational power and accurate mashing process. Therefore, the initial condition in the analysis plays an extremely important role. In this research work, we address Mach 8 flow and related flow property at this high-speed. Density-based solver has been implemented in the project so that the compressible flow matrix enabled in the Force analysis. Freestream condition for the model was calculated by using reservoir condition and isentropic relation. Reservoir pressure is calculated from the pressure trace plot, shown in Fig. 1, at the end of the shock tube during the experiment (Table 1).

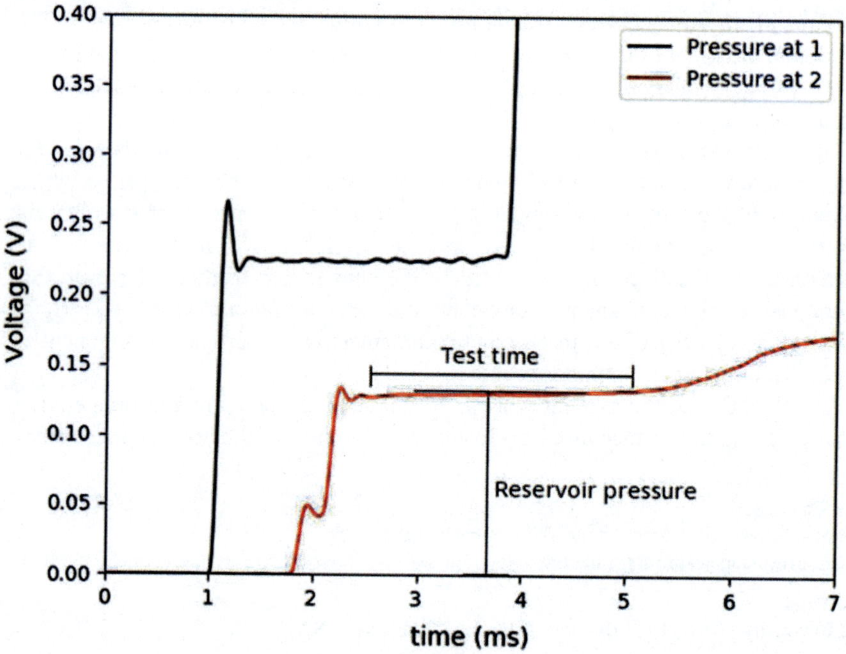

Fig. 1 Pressure traces at the end of the shock tube during the experiment

Table 1 Hypersonic reservoir conditions for the tests in the shock tunnel

Test gas	Air ($\gamma = 1.4$)
P_5 transducer sensitivity	0.1464 V/MPa
P_2 transducer sensitivity	0.7406 V/MPa
Reservoir pressure P_5	10.29 bar
Reservoir temperature T_5	764.65 K

Nozzle is being calibrated for Mach 8 (M_∞) at the exit. Freestream conditions were calculated by the isentropic relations as mentioned in Eqs. 1 and 2

$$\frac{P_5}{P_\infty} = \left(1 + \frac{\gamma - 1}{2} M_\infty^2\right)^{\frac{\gamma}{\gamma - 1}} \tag{1}$$

$$\frac{T_5}{T_\infty} = \left(1 + \frac{\gamma - 1}{2} M_\infty^2\right) \tag{2}$$

Freestream Pressure P_∞ was 105.39 Pa and Temperature $T_\infty = 55.41$ K has been compute as an inlet condition. Form Eq. (3), velocity of the inlet flow $V_\infty = 1193.68$ m/s was define as an boundary condition for inlet surface.

$$V_\infty = M_\infty \sqrt{\gamma\, R\, T_\infty} \tag{3}$$

where, R is the air gas constant and T_∞, M_∞ and V_∞ was the freestream conditions.

During the experiment, the high two-stage suction pump creates nearly 5–3 Pa pressure, therefore, the outlet pressure was set as 3 Pa in the outlet boundary condition. No sleep stationary wall condition has been implemented on the model surface and the boundary of the test section. In the density-based solver, energy or heat transfer plays a significant role, therefore, the energy model was enabled during the setup for analysis.

2.2 Domain and Boundaries

The domain of the test section was rectangular with the height of 45 cm as shown in the Fig. 2, after test section, divergent part was introduced to mitigate the effect of far field conditions. The divergent section of the fluid domain has 1 m length and the 45° divergent angle, this divergent part slow down the supersonic flow and facilitate to capture the wake completely and also to prevent reverse flow to make wake uniform behind the body. The test model has a hemispherical cap from the front and cylindrical further. The position of the model in the test domain shown in the Fig. 2. The 2-D geometry was built in the SolidWorks software and imported into the ANSYS-SpaceClaim module.

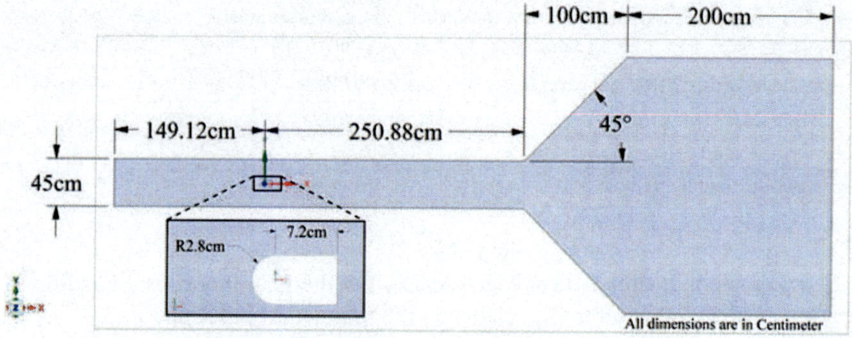

Fig. 2 Geometry of the fluid domain and model position in the test section

2.3 *Mesh Generation*

Meshing of the model, far field, inlet, outlet, and fluid domain has been done in ANSYS workbench. In the hypersonic flow, the boundary layer growth rate and the shock layer is extremely important for the force calculation. Therefore, these paramount parameters must be computed most accurately. In this work, we implemented inflation layer of meshing around the body, which shown in the Fig. 3. In the inflation properties, the maximum layers are 12 and the growth rate is set as 1 so that it can properly measure the boundary layer thickness. Accurately quantify the Bow shock layer and the flow property after the shock wave, body sizing with 0.15 m Sphere of influence and 4 mm element sizing with global origin as Sphere origin had been decided. 0.15 m radios cover the significant shock effect around the blunt body.

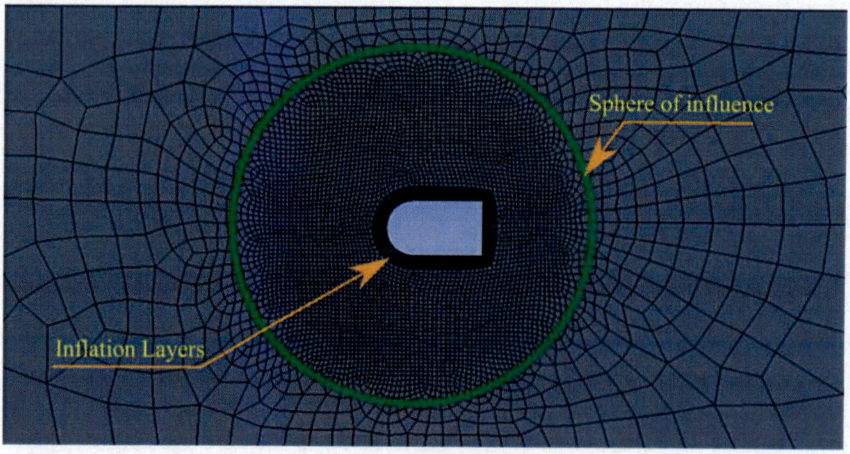

Fig. 3 Meshing of the fluid domain around the model

Table 2 Statistical data for the mesh quality

Mesh aspect ratio (<20)	1.4765
Skewness ($0 < S < 0.25$)	0.135
Jacobian ration: Gauss point ($0.8 < D < 1$)	0.938
Orthogonal quality (close to 1)	0.97

This meshing involves 6928 number of elements and 7057 number of nodes have been generated for the fluid domain. Table 2 shows the criteria which use to check the veracity of the mesh quality.

3 Results and Analysis

The CFD analysis has been done on the hypersonic blunt hemispherical shaped body in ANSYS fluent solver.

3.1 *Shock Structure*

Shock structure depends on the Mach number and the geometry of the body. As shown in the Fig. 4, the bow shock is formed in front of the hemisphere. This detached strong bow shock converts the kinetic energy into the thermal energy; This sudden transition eventually decrease the Mach no form 8 to 0.49 at 0° angle of attack and flow became subsonic after the shock wave. The flow remains subsonic between the shock wave and stagnation point at the center of the hemisphere. However, as flow travel at the downstream, the transition of the subsonic flow into the supersonic flow has been observed. This transition is illustrated in Fig. 4, as the transition lines.

3.2 *Pressure and Temperature Profile*

Pressure and Temperature are the key flow properties in the hypersonic force measurement. Figures 10, 11, and 12, show these property at different angle of attack.

3.2.1 Variation Along Surface

Static pressure has been found out on the polyline created from the intersection of body and XY plane ($z = 0$). This polyline direction has been shown in the Fig. 5. In

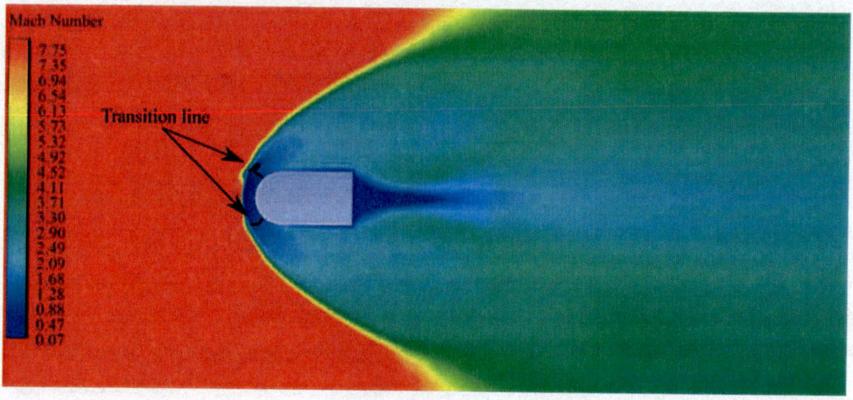

Fig. 4 Shock structure and Mach number at 0° AOA

Fig. 5 Model surface

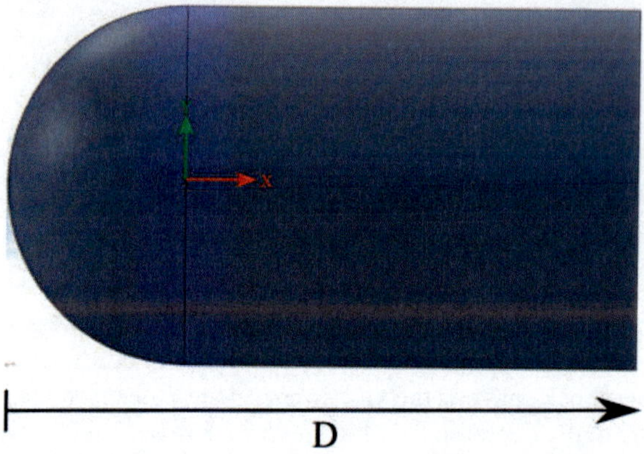

the Fig. 6, lines below the 0° AOA shows the pressure along the upper half of the body. The pressure on the bottom half body illustrated by the lines situated above the 0° AOA. The drop-down of the static pressure was observers till $D = 0$ for 0° AOA. After $D = 0$, the pressure remain constant. As AOA increased, the flatness in the pressure along the bottom half of the body was shifted in negative X direction.

As the angle of attack increased, the static pressure along the upper half of the body decreased. However, the opposite trend noted for the bottom part. The significant improvement in the pressure on the bottom flat surface indicates the deceleration of the supersonic flow. At the 30° AOA, the jump after the decrement indicates the flow septation on the body.

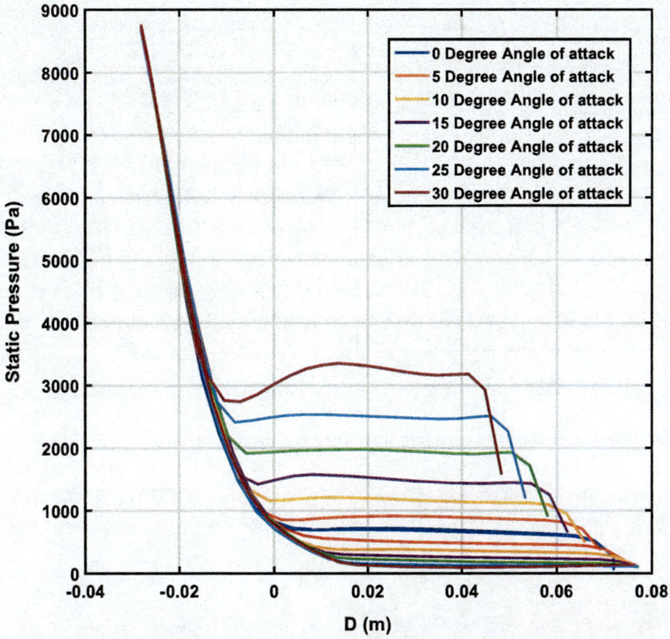

Fig. 6 Pressure along the surface of the model at different AOA

Fig. 7 Distance in front of the model

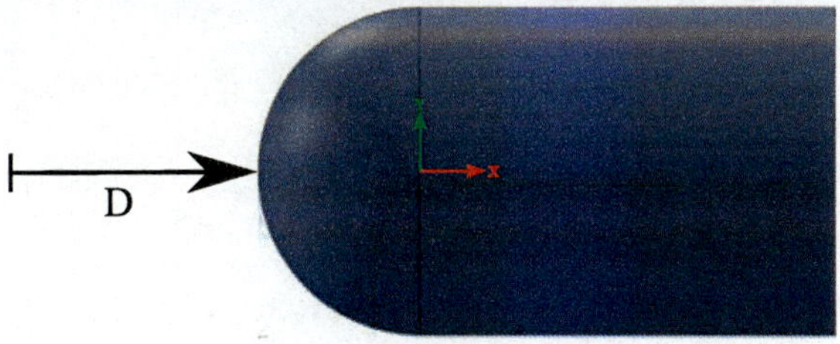

3.2.2 Variation of Flow Properties in Front of the Body

The temperature and pressure change in front of the body (Fig. 7) at different AOA shows in the Fig. 8. In these graphs, the sudden increment in the pressure and temperature between the distance of −36.295 mm and −41.9 mm indicate the effect of the bow shock. As this increment was aline for all AOA, we can say that the strength of the bow shock is independent of the angle of attack. However, the minor change in the pressure and temperature before and after the shock wave for a different angle of attack has been observed.

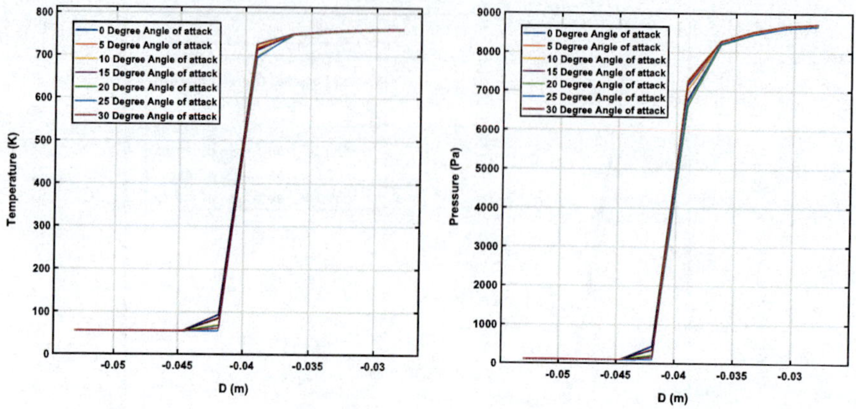

Fig. 8 Temperature (left) and pressure (right) before the model surface at different AOA

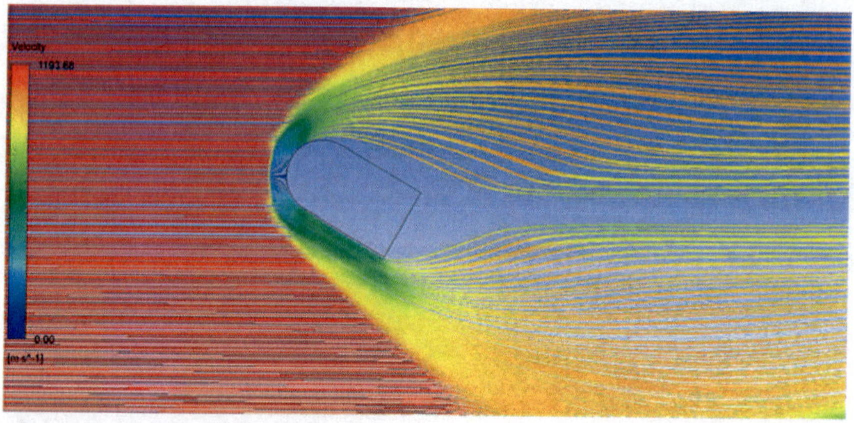

Fig. 9 Streamline at 30° AOA and flow separation visualization

3.2.3 Flow Separation Analysis

As the Fig. 6, indicates that the flow separation occured at the 30° angle of attack. When the pressure starts to increase at $D = 0$ on the bottom half of the body, the flow coming from the upstream direction gets detached. The streamline visualization of this phenomena is shown in the Fig. 9. Because of this flow separation, the hypersonic boundary layer increased significantly and its effect on the temperature is clearly seen at the end of the model in Fig. 12 (Figs. 10 and 11).

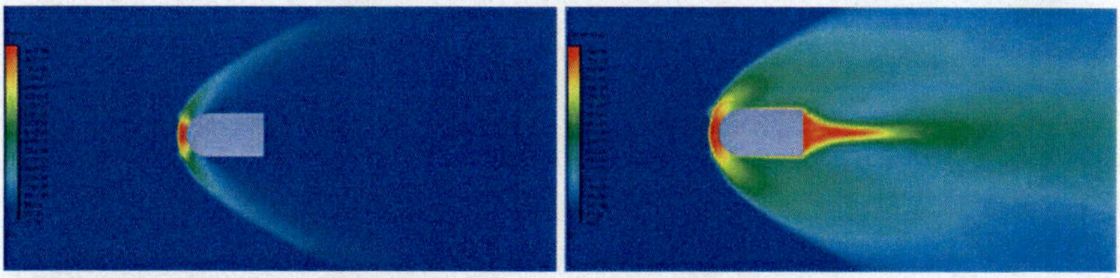

Fig. 10 Static pressure (left) and temperature (right) at 0° AOA

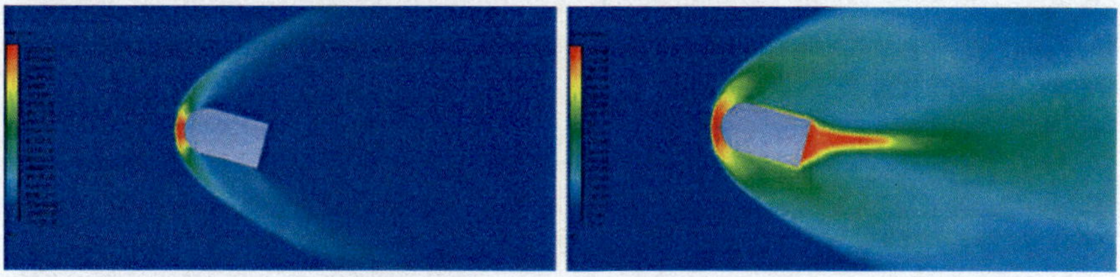

Fig. 11 Static pressure (left) and temperature (right) at 15° AOA

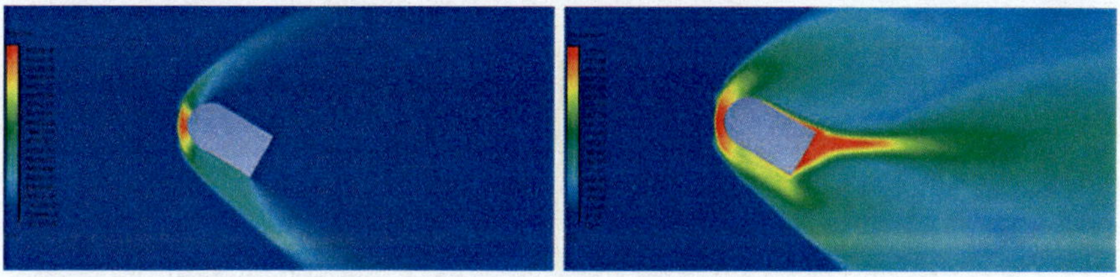

Fig. 12 Static pressure (left) and temperature (right) at 30° AOA

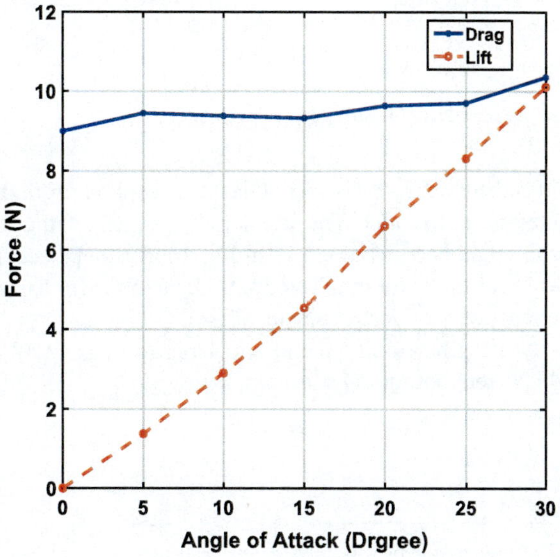

Fig. 13 Drag and lift force at different angle of attack

3.3 Trend of Aerodynamic Forces

The aerodynamic forces on the body were measured by computing the surface integral of the dynamic pressure on the body based on the Eq. (4). Dynamic pressure is the key parameter which is responsible to exert the force on the body. When hypersonic flow was suddenly imparted on the body at a different angle of attack apart from 0°, it induces the pitching moment. However, in this research work, we focus on the fixed model in hypersonic flow, therefore, the Lift and Drag forces were derived in two parts: The force on the hemispherical part and the after body cylindrical part. By combining these separated quantities, the resultant forces on the body were obtained.

$$\textbf{Pressure Force} = \int_{surface} P(x)\,dA \tag{4}$$

$$\textbf{Lift}_{resultant} = \textbf{Lift}_{cylinder} + \textbf{Lift}_{hemisphere} \tag{5}$$

$$\textbf{Drag}_{resultant} = \textbf{Drag}_{cylinder} + \textbf{Drag}_{hemisphere} \tag{6}$$

The Fig. 13, shows the Drag and Lift forces on the body. As AOA will increase the Lift force shows linear increment with the slop of 0.33. At the 30° AOA, this Lift force is nearly similar to the drag force. Therefore, one can say that, at 30° AOA, resultant force act on the body is $\sqrt{2}\times$ **(Lift/Drag)** and the direction is 45° with respect to positive x axis.

The variation in the Drag force is negligible during the increment of the AOA. However, the minor gain in drag can be observed during the transition between 0°–5° and 25°–30° AOA. The total variation in the drag force is only 15% in the span of 0–30 Degree AOA.

3.4 Balance Theory

In this theory, Newton's second law was applied for finding the steady aerodynamic forces on the model. The acceleration signal of the system during aerodynamic loading had been obtained as in Fig. 14. System's acceleration during steady state was 36.13 m s^{-2} for drag force and 12.69 m s^{-2} for Lift force were derived by using the sensitivity of accelerometer sensor.

By using newton's second law, Drag force is **11.52 N**[1] and Lift at 15° AOA is **4.05 N** (see footnote 1) was computed.

[1]**BT** corresponds to balance theory.

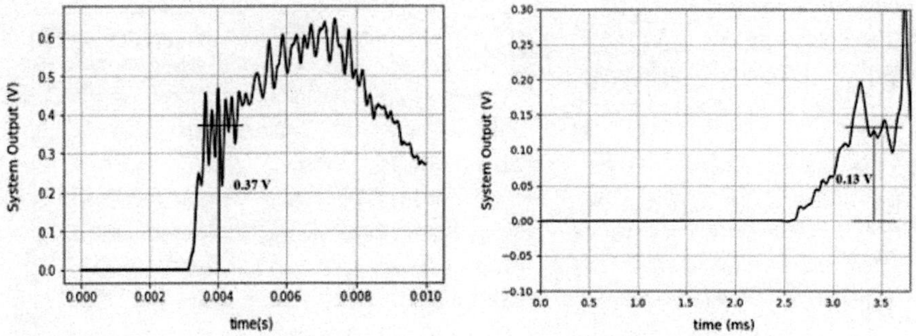

Fig. 14 Drag (left) and lift (right) experimentation output signal obtained by accelerometer

3.5 *Newtonian Theory*

From Newtonian theory, for a very high Mach number, the flowfield becomes independent of Mach number. This phenomenon is commonly known as Mach number independence principle. A direct result from Newtonian theory is that the coefficient of drag and Lift is approximately equal to 1 and 0.3 for a body in hypersonic flow. Aerodynamic force is defined for a body having frontal cross-section area of S and which is in a flow having freestream conditions of M_∞ and P_∞ as:

$$F = \frac{1}{2}\gamma P_\infty M_\infty^2 S \times (C_D \, or \, C_L) \tag{7}$$

by using the Eq. (7), Drag and Lift forces are **11.65 N**[2] and **3.49 N** (see footnote 2) can be quantified, respectively.

Derived aerodynamic forces from numerical analysis are compared with Newtonian and balance theory as shown in the Fig. 15.

The free-flying conditions explained the variation between experimental and computational results. The experimental model was supported by the soft material to create an infinite degree of freedom. Due to this mechanism, the body experience more downstream acceleration cause the increment in drag and the aerodynamic force resultant for Lift was reduced due to pitching moment. On the converse side, in this research work, we analyze the fixed model which experience the hypersonic flow, due to this boundary conditions the surge in the Lift and the decrement in the Drag forces were observed compared to experimental observations.

[2]**NT** corresponds to Newtonian theory.

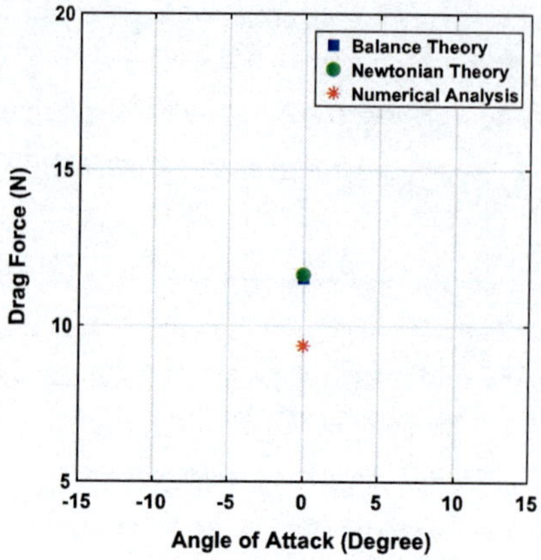

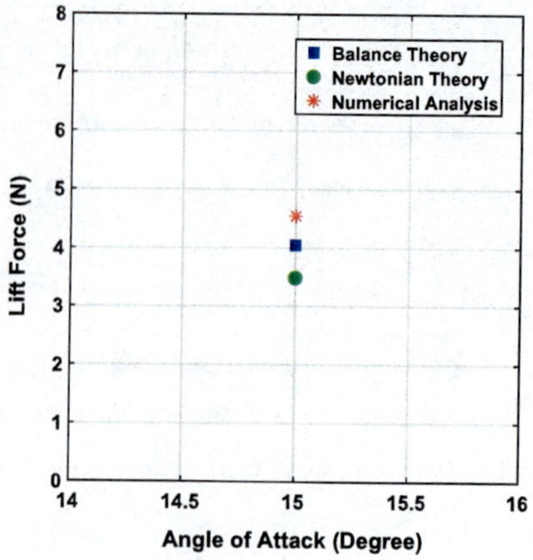

Fig. 15 Steady drag and lift values derived from various methods at AOA = 0° and AOA = 15° respectively.

4 Conclusion

The hypersonic hemispherical model has been analyzed at Mach 8 flow in ANSYS fluent at a different angle of attack. Inflation and spherical body meshing was generated and the appropriate flow conditions which analogues to the experimental condition has been implemented to the fluid domain. Pressure and temperature counters were developed and compare them at different AOA at Mach 8 flow. The trend of the Lift and Drag forces on the body at different AOA were derived. Obtained aerodynamic force components, from numerical analysis, were compared with the results of Newtonian and Balance theory, and the agreement has been found encouraging.

References

1. Anderson, J.D.: Modern Compressible Flow: With Historical Perspective, vol. 12. McGraw-Hill, New York (1990)
2. Anderson Jr., J.D.: Hypersonic and High-Temperature Gas Dynamics. American Institute of Aeronautics and Astronautics (2006)
3. Longo, J.M.A., Hannemann, K., Hannemann, V.: The challenge of modeling high speed flows. In: Proceedings of 6th EUROSIM Congress, vol. 20 (2007)
4. Robinson, M.J., Schramm, J.M., Hannemann, K.: Design and implementation of an internal stress wave force balance in a shock tunnel. CEAS Space J. **1**(1–4), 45–57 (2011)
5. Mee, D.J.: Dynamic calibration of force balances for impulse hypersonic facilities. Shock Waves **12**(6), 443–455 (2003)

Interior Acoustic Analysis of Rectangular-Shaped Rigid Cavity with Opening

Subhankar Pramanik, Sreyashi Das, and Arup Guha Niyogi

1 Introduction

In the present era of enormous industrial development around the globe, the need for acoustic assessment cannot be overruled. The huge emphasis imparted on the research activities in these fields of acoustics is chiefly due to the immense impact of noise on the healthy existence of mankind. Excessive exposure to high sound pressure level (SPL) causes various health hazards along with deafness. It badly affects the level of mental concentration, quality of work, learning and productivity too.

The transport sector is a unique field where the vehicles are most frequently exposed to noise hazards produced by the reciprocating engines. The noise affects the drivers, passengers, as well as the neighbors. Transmission of sound into a cavity due to a sound wave incident on an open window is a very common problem worldwide.

Vehicle cabins and other interior acoustic domains are often provided with windows that allow the acoustic energy to leave the region of interest. At sufficiently low frequencies it can be assumed that the sound pressure is approximately uniform over the opening. Hence, a radiation impedance can be used to represent the opening.

In some of the past researches [1, 2], this condition was simulated by setting the acoustic pressure equal to zero at the leakage surfaces. However, this is not absolutely true in reality since the windows lead only to exterior acoustic fluid and the acoustic pressure is therefore not zero altogether. In 1989, Suzuki [3] solves the cavity noise problem with complicated boundary conditions, including the presence of leakage zone using radiation impedance as described in [4]. Kinsler et al. [5] deduced the expression for the radiation impedance of a circular pipe driven at one end and open at

S. Pramanik (✉) · S. Das · A. G. Niyogi
Jadavpur University, Kolkata 700032, India
e-mail: spsubhankar.ju@gmail.com

S. K. Saha and M. Mukherjee (eds.), *Recent Advances in Computational Mechanics and Simulations*, Lecture Notes in Mechanical Engineering,
https://doi.org/10.1007/978-981-15-8315-5_32

the other end, with wide flanges. In 1990, Seybert et al. [6] adopted general interior and exterior acoustic method and compared the result with radiation impedance method [5] and showed a good agreement of result in low-frequency zone.

In the present study, boundary element method (BEM) has been used to solve the interior acoustic problem of rectangular cavity. Using BEM reduces the computational storage required and hence the time required to solve BEM. This method reduces the order of the problem by one degree. The position of open window is characterized by radiation impedance provided by Kinsler [5]. Variation in SPL at the boundary and at the domain has been shown for different positions and sizes of open window.

2 Theoretical Formulation

The governing equation of a time harmonic acoustic problem is given by the reduced wave (Helmholtz) equation [7],

$$\nabla^2 p + k^2 p = 0 \tag{1}$$

Here, p is the acoustic pressure and k is the wave number.

The equation can be rewritten as a truly boundary integral equation

$$C(P)p(P) + \int_S \left(p\frac{\partial p^*}{\partial n}\right) dS = \int_S \left(p^*\frac{\partial p}{\partial n}\right) dS \tag{2}$$

For the time harmonic cases, the pressure derivative can be replaced by the normal particle velocity, which is a more useful quantity as

$$u_{an} = -\frac{1}{i\rho_o\Omega}\frac{\partial p}{\partial n} \tag{3}$$

Assuming the surface is discretized into M number of eight-noded surface elements, the discretized form of boundary integral equation [8] is given as

$$\begin{aligned} C(p)p(P) + \sum_{m=1}^{M}\sum_{l=1}^{8}\int_{-1}^{+1}\int_{=1}^{+1}\frac{\partial p*}{\partial n}(P,Q)N_1(\xi_1,\xi_2)p_1 J(\xi_1,\xi_2)d\xi_1 d\xi_2 \\ = \sum_{m=1}^{M}\sum_{l=1}^{8}\int_{-1}^{+1}\int_{-1}^{+1}[-i\omega\rho p^*(P,Q)]N_1(\xi_1,\xi_2)v_1 j(\xi_1,\xi_2)d\xi_1 d\xi_2 \end{aligned} \tag{4}$$

Each node of the BE mesh is used once as an observation point, and a boundary element equation is generated. Upon assembly of these equations the system equation for the acoustic enclosure is found in the form of a set of linear algebraic equations.

$$[H]\{p\} = [G]\{v\} \tag{5}$$

Upon assembly of these equations the system equation for the acoustic enclosure is found in the form of a set of linear algebraic equations. The velocity terms $\{v\}$ in this equation are actually fluid particle velocity terms and not the structural velocity terms.

2.1 Boundary Conditions of the Acoustic Cavity

Physically, there may be two types of boundary conditions for this interior rigid acoustic problem. Let a surface, S, enclosing the region of interest, V, may be subdivided into two segments, S_1 and S_2, where $S = S_1 \cup S_2$. The boundary conditions that may practically occur are shown in Fig. 1.

The velocity boundary condition at S_1 is often termed as the Neumann boundary condition, where the fluid particle velocity along the normal to the surface is specified. The S_2 boundary is actually a leakage zone, a window for example.

The radiation impedance function at S_2 zone may be defined as follows:

$$Z_r = \frac{Pressure\ at\ local\ node\ i\ of\ element\ k * Area\ of\ element\ k}{\text{Velocity at local node i of element k}} \tag{6}$$

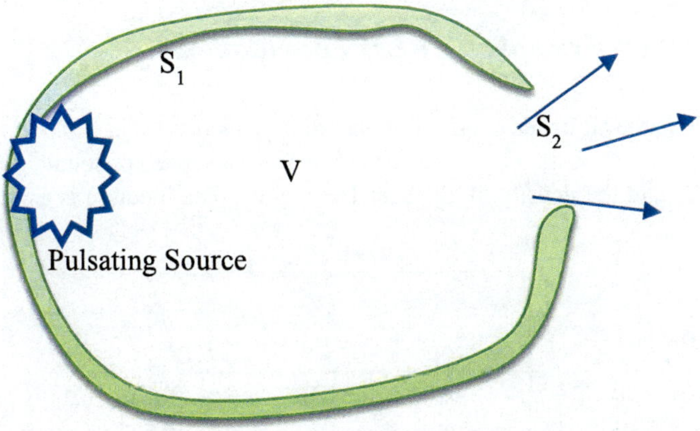

Fig. 1 Different boundary conditions are shown for an interior acoustic domain, V. **a** Normal particle velocity and frequency are specified on S_1 (rigid boundary), **b** S_2 is a window, through which energy escapes into outer space

According to Kinsler et al. [5] the radiation impedance Z_r, for the circular openings is given by

$$Z_r = \rho c A\left\{0.5(ka)^2 + i0.85ka\right\} \tag{7}$$

Here, ρ, c, A, k and a stand for density of air, speed of sound, area of the leakage surface, wavenumber and equivalent radius of the opening, respectively. This expression has been adopted by Seybert et al. [6], wherein for rectangular openings an equivalent radius of the opening has to be used. Since analytical expression for rectangular or square pistons is not available, therefore this simplification has become inevitable. Since the impedance term is now known, a relation between the surface pressure and the normal surface velocity can be established so that the surface velocity terms may be replaced by pressure terms. The pressure quantities are evaluated at first, and then the velocity terms are extracted from the radiation impedance relation.

After the solution is over, the pressure at any internal point inside the acoustic domain is obtained using Eq. (4) while setting C(p) equal to unity.

3 Numerical Results

A MATLAB program has been developed for solving the interior acoustic problem using boundary element method solver. Eight-noded serendipity element has been used in the analysis.

3.1 Validation Example for BEM Formulation

In this problem, the air inside a rigid container of dimensions 1.8 m × 0.6 m × 0.6 m, shown in Fig. 2, acts as an interior acoustic domain. The speed of sound, c, is taken to be 340 m/s and the density of air, ρ, is 1.20 kg/m^3. The medium is excited by a

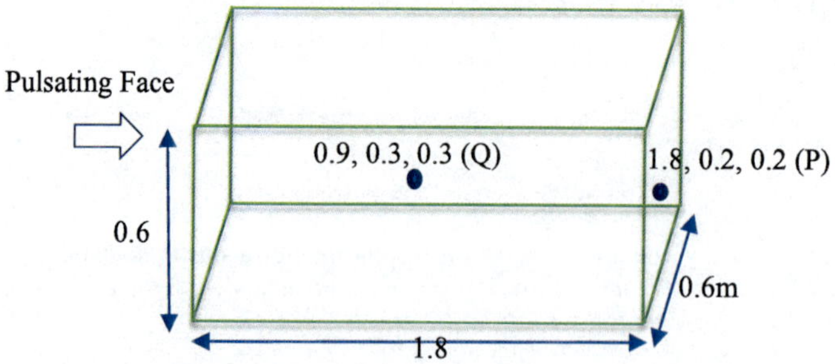

Fig. 2 Geometry of the cavity

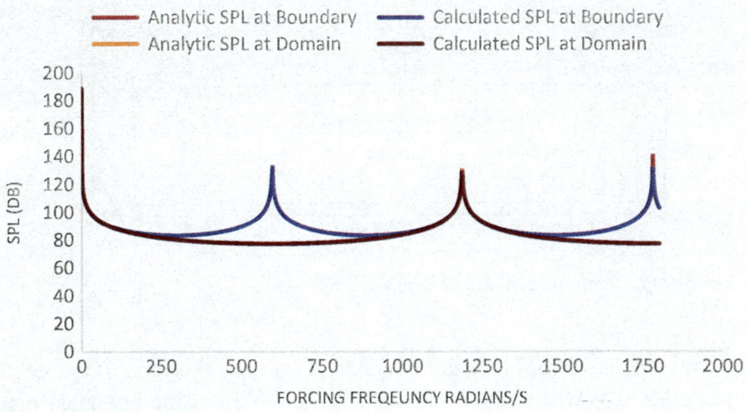

Fig. 3 Analytic and calculated SPL at boundary and domain for closed cavity

sinusoidal motion of the left wall, acting as a rigid piston, with velocity amplitude of 1 mm/s. An element size of 0.2 m × 0.2 m has been taken for the study.

The responses at boundary (P) and at domain (Q) are calculated at a forcing frequency interval of 2 rad/s. Figure 3 shows the sound pressure level (SPL) in decibels (dB) (with a threshold pressure of 2×10^{-5} N/m^2), computed analytically and by using boundary element method at the center of right wall (P), and also at the center of the domain (Q). The analytical solutions and numerical results show very close conformity. Analytically, the acoustic resonance at the right boundary is scheduled to occur at $\Omega = nc\pi/L$, where L is the length of the duct, whereas at the center of the domain, the resonance takes place at $\Omega = 2nc\pi/L$. For the present purpose, analytically, the resonance at the right-side boundary should occur at frequencies of 0, 593.412, 1186.824 rad/s, while that occurs at the center of the domain is of 0, 1186.824, 2373.648 rad/s. This is evidently being observed in the BE solution.

3.2 *Case Study*

3.2.1 Different Opening Position for Rectangular Box with Aspect Ratio (A/R) 1/3

The rectangular cavity as shown in Fig. 4 has been analyzed with different arrangement of openings. The opening size is taken as 0.2 m × 0.2 m. The following four cases have been taken for the study.

i. No opening at the faces of cavity
ii. Central opening provided at the rightmost wall only
iii. Opening provided at the front and back face center of the cavity only
iv. No wall at the right end

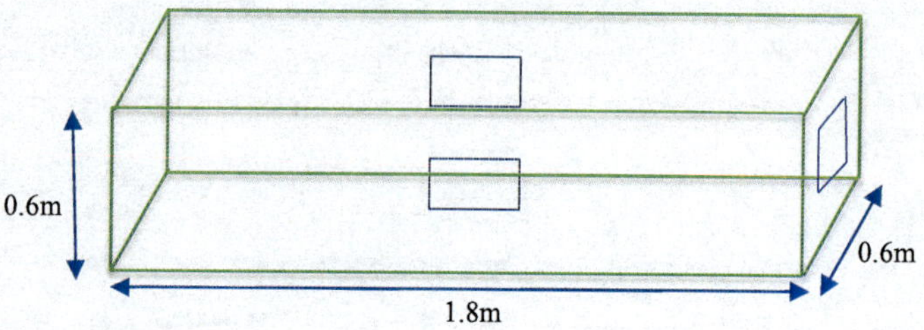

Fig. 4 Opening at different face of size (0.2 m × 0.2 m)

Sound pressure level (SPL) at the boundary (1.8, 0.4, 0.4) and at the center of domain (0.9, 0.3, 0.3) was found out to be 3.2.1 and 3.2.2 using boundary element analysis (BEA). The same boundary conditions as mentioned in the validation study have been used.

From Fig. 5, it is observed that when the opening is present at the rightmost end, there is considerable reduction of SPL. When the opening is placed at both side walls, there is no substantial reduction in SPL. Further, a peak is observed near 200 rad/s for the cases where window is present. If the rightmost wall is fully open then the variation in SPL is prominent compared to other cases. From Fig. 6, at domain center, a dip in the SPL pattern is observed near 1200 rad/s if the opening is 0.2 m × 0.2 m placed at rightmost wall or side walls. But when the right end is fully open the SPL pattern at the center changed totally giving peaks near 300 and 900 rad/s. SPL reduces to 25 dB at 1188 rad/s, whereas there is a peak of 114 dB near 1188 rad/s for no opening condition.

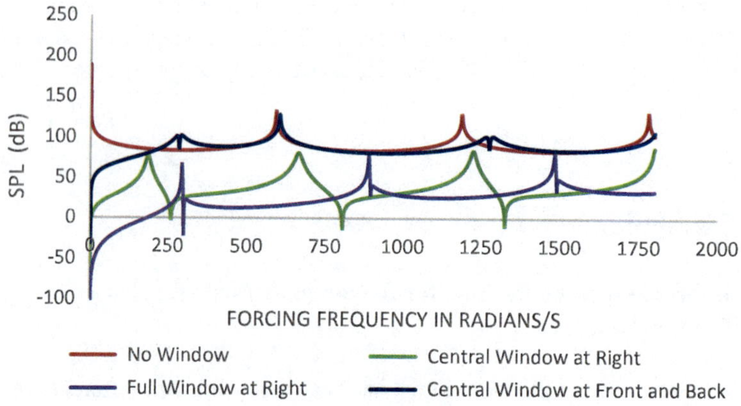

Fig. 5 SPL at boundary point (1.8, 0.4, 0.4) for different opening position

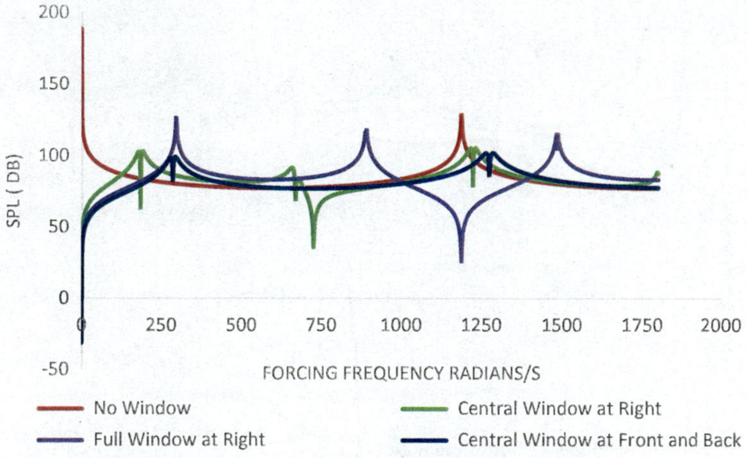

Fig. 6 SPL at domain point (0.9, 0.3, 0.3) for different opening position

3.2.2 Comparison of SPL in dB for Cavities with Different Aspect Ratio

Here two cavities of different aspect ratio have been studied and SPL has been calculated at boundary and domain point with different opening positions. The size of the cavity is taken as 1.8 m × 0.6 m × 0.6 m (A/R = 1/3, rectangular) and 1.8 m × 1.8 m × 1.8 m (A/R = 1, square). The same acoustic property has been used in the study as given above. The following three cases have been taken for the study.

1. For rectangular box (A/R = 1/3) with full opening at right end
2. For square box (A/R = 1) with central opening at right end
3. For square box (A/R = 1) with full opening at right end

The SPL in dB at the boundary and domain points have been plotted in Figs. 7 and 8. At the boundary and domain points, the SPL pattern is very different when a central window is present at the right wall. A kink is visible near 600 rad/s for this case. From both the figures, it is observed that the opening size and its position plays a very important role in variation of SPL. Also, a reduction in overall SPL has been observed for presence of window. The SPL contour at the boundary surface for 600 rad/s has been shown in Fig. 9. It is noticed that the central part of the rectangular box experienced higher SPL. At two ends the SPL value in dB is sufficiently small.

4 Conclusion

In this study the effect of opening on the sound pressure level at different positions at the boundary and inside domain has been shown. Boundary element analysis of the acoustic domain has been performed, thus reducing the dimensionality of the problem

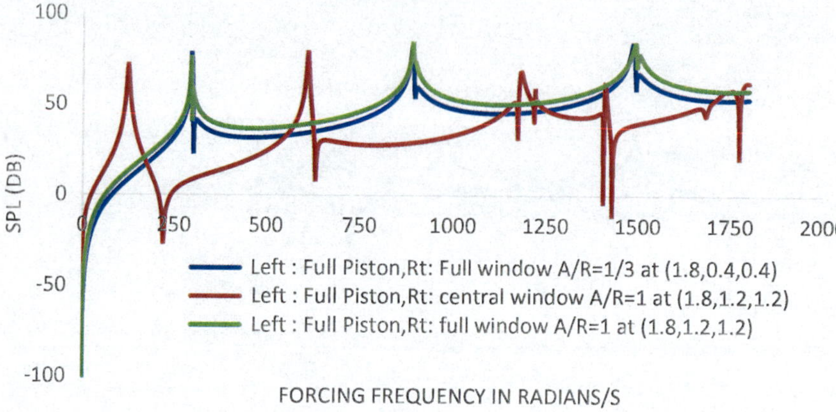

Fig. 7 SPL at window boundary for different aspect ratio

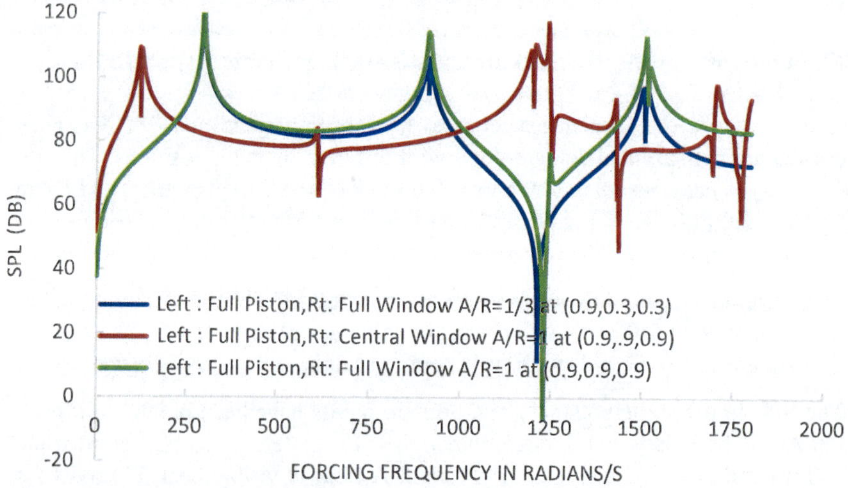

Fig. 8 SPL at center of domain for different aspect ratio

by one. At the opening a radiation impedance has been used in the low-frequency region. It is shown that the position of window and its size is very important in determining the sound pressure level in the cavity. As experimentation with practical models are expensive and produces huge non-biodegradable waste, it is better to use numerical procedure first and then do practical experiments for selected cases.

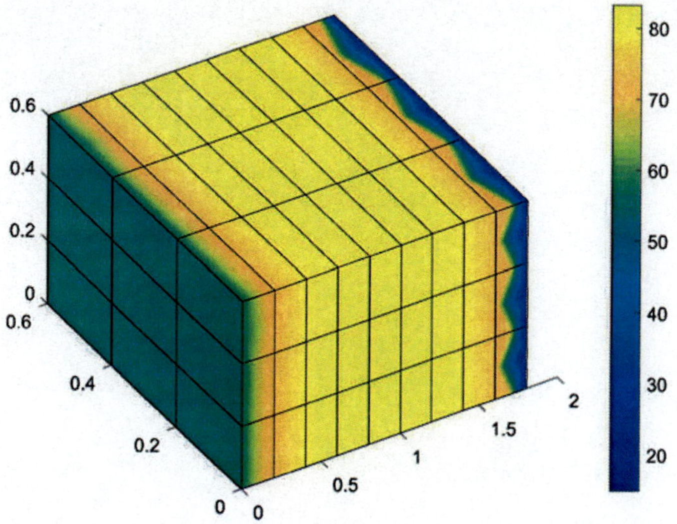

Fig. 9 SPL plot for rectangular box having full opening at the right end at 600 rad/s

References

1. Rajakumar, C., Ali, A.: Boundary element-finite element coupled Eigen analysis of fluid-structure systems. Int. J. Numer. Meth. Eng. **39**, 1625–1634 (1996)
2. Banerjee, P.K., Ahmad, S., Wang, H.C.: A new BEM formulation for the acoustic eigenfrequency analysis. Int. J. Numer. Meth. Eng. **26**, 1299–1309 (1988)
3. Suzuki, S., Maruyama, S., Ido, H.: Boundary element analysis of cavity noise problems with complicated boundary conditions. J. Sound Vib. **130**(1), 79–91 (1989)
4. Takada, K.: On the radiation impedance of a conical shell of pistonic motion in an infinite baffle. J. Acoust. Soc. Jpn. 71–75 (1982)
5. Kinsler, L.E., Frey, A.R., Coppens, A.B., Sanders, J.V.: Fundamentals of Acoustics, 3rd edn. Wiley, Chichester (1982)
6. Seybert, A.F., Cheng, C.Y., Wu, T.W.: The solution of coupled interior/exterior acoustic problems using the boundary element method. J. Acoust. Soc. Am. **88**(3), 1612–1618 (1990)
7. Crocker M.J. (ed.): Encyclopedia of Acoustics. Wiley (1997)
8. Brebbia, C.A.: The Boundary Element Method for Engineers. Pentech Press, London (1978)

Numerical Investigation of Effect of Laser on Natural Convection in Two-Dimensional Geometry

G. Chanakya and Pradeep Kumar

1 Introduction

The study of natural convection in square/rectangular cavities combined with radiation provides a fundamental insight of fluid flow and heat transfer for many engineering applications. From few decades, analysis of these multi-physics problems became interest to the researchers.

Torrance and Rockett [1] had studied the pure natural convection in cylindrical enclosure with the localized heating from bottom numerically for the range of Grashoff numbers ($Ga = 4 \times 10^4$ to 4×10^{10}) for $Pr = 0.71$. Calcagani et al. [2] had numerical and experimentally investigated the fluid flow in a two-dimensional square cavity heated from the bottom with varied heat source lengths, for various Rayleigh numbers 10^3 to 10^6. Researchers like, Ganzorolli and Milanez [3], Aydin and Yang [4], studied the natural convection in cavities heated from below and symmetrically cooled from the sides numerically. The authors observed formation of symmetrical nature of vortices, and this strength increases upon increasing the Rayleigh numbers, and the behaviour of isothermal lines also symmetrical about the centroidal vertical axis of cavity.

Natural convection in vertical and inclined square cavity with uniformly distributed internal energy sources for the inclination angles of 30°, 60°, and 90° had been studied by Acharya and Goldstein [5] for Rayleigh number range 0–10^7.

Webb and Viskanta [6] had studied the radiation-induced buoyancy flow in a rectangular enclosure irradiated from the side experimentally. They observed formation thin thermal boundary layers at the vertical walls and the flow structure loses the centrosymmetry characteristics of natural convection due to the deferentially heated

G. Chanakya (✉) · P. Kumar
Numerical Experiment Laboratory (Radiation & Fluid Flow Physics), School of Engineering, Indian Institute of Technology Mandi, Mandi 175005, Himachal Pradesh, India
e-mail: chandrasekhara_pratap@students.iitmandi.ac.in

S. K. Saha and M. Mukherjee (eds.), *Recent Advances in Computational Mechanics and Simulations*, Lecture Notes in Mechanical Engineering,
https://doi.org/10.1007/978-981-15-8315-5_33

walls. They also developed a theoretical model for the prediction of internal radiative heating of the fluid and motion induced due to buoyancy.

Mezrhab et al. [7] had studied the combined natural convection with radiation with centrally located square body in a cavity, and observed that the radiation homogenized the temperature inside the cavity and increased the Nusselt number. Similarly Sun et al. [8] extended the study for the various Rayleigh numbers, sizes of the inner body and the emissivity of the surfaces. The results showed, that due to the surface radiation exchange, the strengthening of recirculation zones which intern stabilized the flow fields.

Paramanda et al. [9] studied the critical assessment of square enclosure with different geometric configurations for the interaction of thermal radiation with natural convection using three different approaches, like, incompressible (with Boussinesq approximation), purely compressible and quasi-incompressible (i.e., low Mach-number approximation) approach. Their results reveal, radiation effects are sensitive in the bodies with corner edges.

Kumar and Eswaran [10] investigated numerically the combined natural convection with radiation for the two different slanted cavities of angles. Saravanan and Sivaraj [11] studied the natural convection and thermal radiation in cavity with non-uniformly heated thin plate placed horizontal and vertical direction. Fusegi and Farouk [12] investigated the combined natural convection with surface/gas radiation in a cavity both experimentally and numerically.

The natural convection with the volumetric radiation in a square cavity was studied by Mondal and Mishra [13] by using the lattice Boltzmann method for the fluid flow and finite volume method for radiative heat transfer equation.

The performance of DOM, FVM, P1, SP3, and P3 methods for RTE in two-dimensional absorbing, emitting medium were studied by Sun et al. [14]. The results of RTE for above methods have been compared by the solution of Monte Carlo Ray tracing. Xing et al. [15] have studied the natural convection in cylindrical enclosure with heated circular, elliptical, square, and triangular geometry inside the enclosure. The effect of surface radiation was incorporated and concluded that the presence of corners and larger upper space had effect on the heat transfer performance.

The diffuse radiation was considered in all above works, however, little work is available on collimated radiation like work by Andnad et al. [16] and Ben et al. [17]. Both have tried to capture the collimated beam bending phenomenon in a graded refractive index medium by discrete ordinate method for RTE. The numerical study of a collimated beam in the refractive index medium was studied by Ilyushin [18]. Few authors studied the short pulse collimated irradiation [19, 20].

From the above literature, the authors observed there is no much work on the laser (or collimated) beam radiation effects on the combined natural convection with radiation, an attempt made numerically, to implement the collimated beam effects with the natural convection to get the better understanding of the collimated beam application like solar cavity receiver, geometrical optics, etc. In the present work,

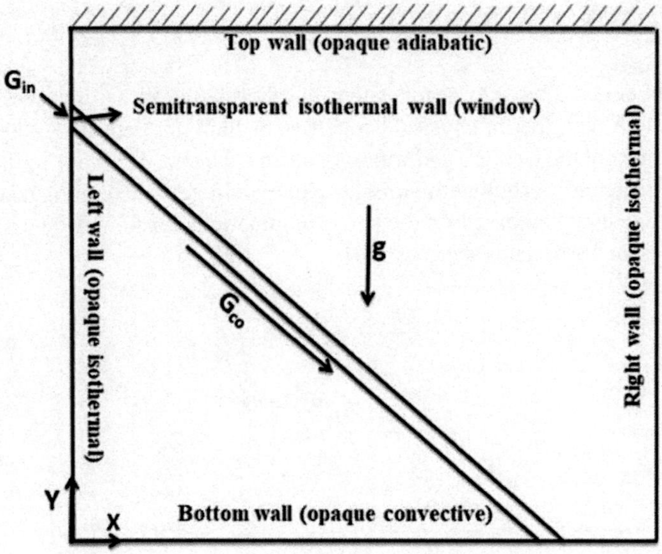

Fig. 1 Schematic diagram of square cavity with laser incidence analysis for present work

the simplified problem was taken for the study, having a square cavity subjected to bottom heating with side cooling, where the collimated beam is entering into the cavity at an angle of 45° from a semitransparent wall, the schematic diagram is shown in Fig. 1.

2 Problem Statement

The combined natural convection with radiation for the various optical thickness ranging from 0, 0.5, 1, 5, 10, and 50 is analysed for the Rayleigh number 10^5 and Prandtl number 0.71 with Boussinesq approximation. The square cavity for the present problem is depicted in Fig. 1. The four walls (left, right, top, and bottom) of the cavity are opaque to the radiation. Whereas a semitransparent wall (i.e., window) is located at a non-dimensional distance of 0.7 from the bottom wall having a non-dimensional width of 0.05 on the left wall. The laser irradiation of 1000 W/m^2 is applied on the semitransparent wall at an angle of 45°. Whereas bottom wall is heated by convective boundary, with free stream temperature of 305K and heat transfer coefficient of 50 $W/m^2 K$. The top wall is subjected to adiabatic boundary, left and right walls are subjected to isothermal boundary conditions.

3 Mathematical Modelling

The present study takes into account that the flow is buoyancy induced laminar, incompressible with thermophysical properties of fluid are constant. Whereas the refractive index of the medium and walls are taken to be one. The walls of the square cavity are treated as opaque with emissivity unity and gray medium is considered. The governing equation based on the above assumptions for flow and temperature, in the Cartesian coordinates given by

$$\frac{\partial u}{\partial x} + \frac{\partial v}{\partial y} = 0 \tag{1}$$

$$u\frac{\partial u}{\partial x} + v\frac{\partial u}{\partial y} = -\frac{1}{\rho}\frac{\partial p}{\partial x} + \nu\left[\frac{\partial^2 u}{\partial x^2} + \frac{\partial^2 u}{\partial y^2}\right] \tag{2}$$

$$u\frac{\partial v}{\partial x} + v\frac{\partial v}{\partial y} = -\frac{1}{\rho}\frac{\partial p}{\partial y} + \nu\left[\frac{\partial^2 v}{\partial x^2} + \frac{\partial^2 v}{\partial y^2}\right] + g\beta(T - T_{ref}) \tag{3}$$

$$u\frac{\partial T}{\partial x} + v\frac{\partial T}{\partial y} = \alpha\left[\frac{\partial^2 T}{\partial x^2} + \frac{\partial^2 T}{\partial y^2}\right] - \frac{1}{\rho C_p}\nabla \cdot q_r \tag{4}$$

where $\nabla\cdot q_R$ is the divergence of the radiative flux, which can be calculated as

$$\nabla\cdot q_R = \kappa(4\pi I_b - G) \tag{5}$$

where κ_a is the absorption coefficient, I_b is the black body intensity and G is the irradiation, evaluated by integrating the radiative intensity, (I), in all directions, i.e.,

$$G = \int_{4\pi} I d\omega$$

The radiative intensity obtained by solving the radiative transfer equation. The radiation in the participating medium is governed by integrodifferential equation known as radiative transfer equation (RTE), is given by

$$\frac{\partial I(r,s)}{\partial s} = \kappa_a I_b(s) - (\kappa_a + \sigma_s)I(r) + \frac{\sigma_s}{4\pi}\int_{4\pi} I(r,\hat{s}_i)\phi(s,s_i)d\Omega_i \tag{6}$$

3.1 Boundary Conditions

The following boundary conditions employed for the present study.
cavity walls: u = v = 0

Left wall at x = 0; T = 296 K
Right wall at x = 1; T = 296 K
Top wall at y = 1; $q_c + q_r = 0$
The bottom wall is heated by convection and given by

$$q_{conv} = h_{conv}(T_{free} - T_w) \tag{7}$$

The Radiative transfer Eq. (6) is subjected to the following boundary condition for a diffusely emitting and reflecting wall

$$I(\mathbf{r_w}, \hat{\mathbf{s}}) = \epsilon_w I_b(\mathbf{r_w}) + \frac{1-\epsilon_w}{\pi}\int_{\hat{\mathbf{n}}\cdot\hat{\mathbf{s}}>0} I(\mathbf{r_w}, \hat{\mathbf{s}})|\hat{\mathbf{n}}\cdot\hat{\mathbf{s}}|d\Omega. \quad \text{for } \hat{\mathbf{n}}\cdot\hat{\mathbf{s}} < 0 \tag{8}$$

The window, which is treated as semitransparent wall subjected to the following boundary condition:

$$I(\mathbf{r_w}, \hat{\mathbf{s}}) = \epsilon_w I_b(\mathbf{r_w}, \hat{\mathbf{s}}) + I_{co} + \frac{1-\epsilon_w}{\pi}\int_{\hat{\mathbf{n}}\cdot\hat{\mathbf{s}}>0} I(\mathbf{r_w}, \hat{\mathbf{s}})|\hat{\mathbf{n}}\cdot\hat{\mathbf{s}}|d\Omega.\text{for } \hat{\mathbf{n}}\cdot\hat{\mathbf{s}} < 0 \tag{9}$$

The laser irradiation feature has been developed in open source computational fluid dynamic (CFD) package OpenFOAM framework and integrated with other heat transfer and fluid flow applications. Subsequently laser beam feature is tested thoroughly. The combined diffuse radiation and natural convection feature has been validated in section [5], where upon the new results for combined laser irradiation and natural convection for geometry in Fig. 1 have been produced.

3.2 Non-dimensional Numbers

The results are presented in the non-dimensional numbers which govern the fluid flow and heat transfer characteristics. The scales for velocity, temperature, length are expressed in dimensionless foam as

$$U = \frac{u}{u_o} \quad V = \frac{v}{v_o} \quad X = \frac{x}{L} \quad Y = \frac{y}{L} \quad \theta = \frac{T - T_c}{T_{free} - T_c}$$

where $u_o = v_o = \sqrt{Lg\beta(T_{free} - T_c)}$

$$Ra = \frac{g\beta(T_{free} - T_c)L^3}{\nu\alpha} \quad Pr = \frac{\nu}{\alpha} \quad \overline{G} = \frac{G}{\sigma T_{ref}^4} \quad \tau = \kappa L$$

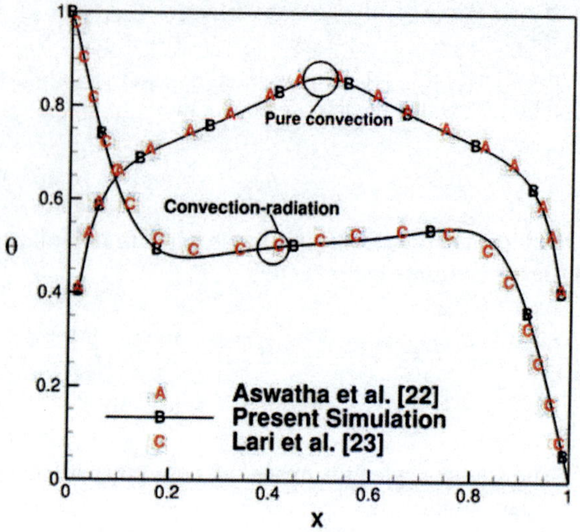

Fig. 2 Validation results for pure convection and combined diffuse radiation and natural convection

4 Validation

In the absence of any benchmark test case for current problem, the validation is performed in three steps. The stand alone feature of collimated irradiation is validated by present authors [21]. In the second step, the current case without collimated and diffuse radiation is validated against Aswatha et al. [22]. The combined diffuse radiation and natural convection in a cavity whose top and bottom walls are adiabatic opaque and vertical walls are opaque isothermal at different temperatures, has been validated against Lari et al. [23] in step three. The validation present for step two and three are presented in Fig. 2. The present simulation results are matching very closely with the published results.

5 Results and Discussion

A simulation is performed in OpenFOAM for laser effect on natural convection in a cavity which is heated from bottom. The laser strikes on the heated bottom wall, the heat transfer and fluid flow characteristics inside the cavity have been studied for different optical thickness of medium and presented below.

The travel of the laser beam inside the cavity can be best represented by irradiation contours; and this non-dimensional irradiation contours for media of optical thickness (τ) 0, 0.5, and 1 are shown in Fig. 3a, b and c, respectively. The optical thickness $\tau = 0$ corresponds to transparent medium, i.e. no absorption by fluid, it is therefore, the laser beam of same irradiation strength reaches to bottom wall, however, strength of laser beam decreases medium having some non-zero optical thickness. This decrement is

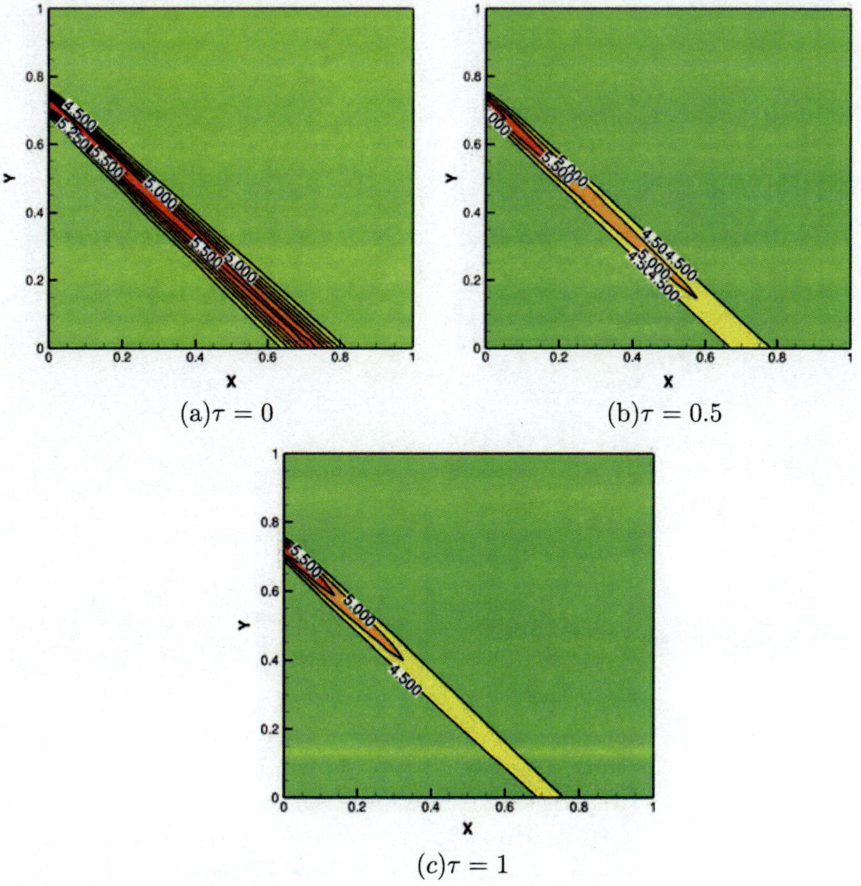

Fig. 3 Contours of non-dimensional irradiation inside the cavity

more for optical thickness 0.5 and 1 is depicted in Fig. 3b and c, respectively, which is also true as medium absorb more radiation energy. This may affect the fluid flow and heat transfer characteristics as can seen in the following paragraphs.

The non-dimensional stream function contours inside the cavity for optical thickness 0, 0.5 and 1 are shown in Fig. 4a, b and c, respectively. This problem without laser irradiation is symmetric about the vertical line at the middle of the cavity ($X = 0.5$) [24] and asymmetricity is induced due to laser incidence. This asymmetricity is due to two reasons (1) a hot spot is created on the bottom wall at the strike point of the laser (2) absorption of laser energy by the fluid. This asymmetricity is clearly visible in contours of the stream function Fig. 4 and non-dimensional temperature Fig. 5. The current problem contains two symmetrical vortices inside the cavity without laser incidence, however, laser incidence causes asymmetricity which vary with behaviour of the medium. As the medium is transparent, all the laser beam energy strikes on

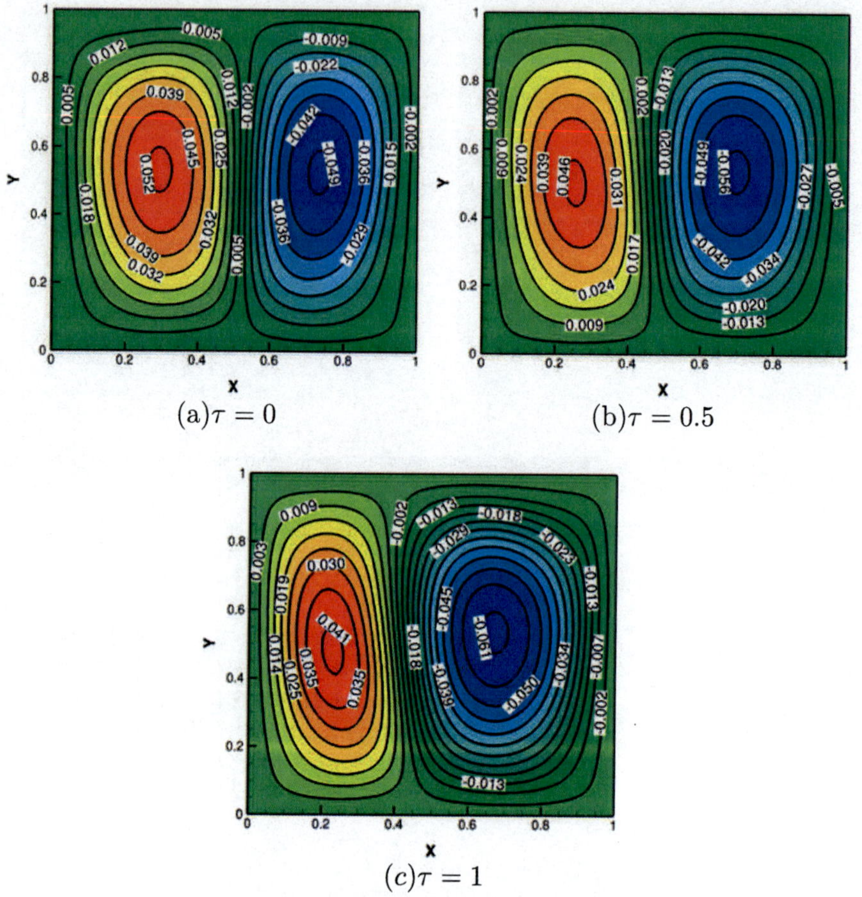

Fig. 4 Contours of non-dimensional stream function inside the cavity

the bottom wall at a non-dimensional distance of 0.7 from origin of the coordinate system for geometry and creates hot spot on wall. In this case more upward buoyancy force to right vortices, these right vortices become thinner and left vortices become thicker, however, opposite behaviour has been observed with non-zero optical thickness in the medium. This is owing to fact that laser beam is travelling through left vortices and transfer the heat to left vortices, this increases in upward buoyancy force. Whereas some heat is also transfered to right vortices through absorption and hot spot: but results in decreases in the size of left vortices and increases in size of right vortices. This phenomenon further enhances with increase in the optical thickness of the medium as shown in Fig. 4c.

The effect of laser beam travel on the temperature characteristics inside the cavity for optical thickness 0, 0.5 and 1 are shown in Fig. 5a, b and c, respectively. A total symmetrical isotherm along the vertical line at $X = 0.5$ in absence of laser beam

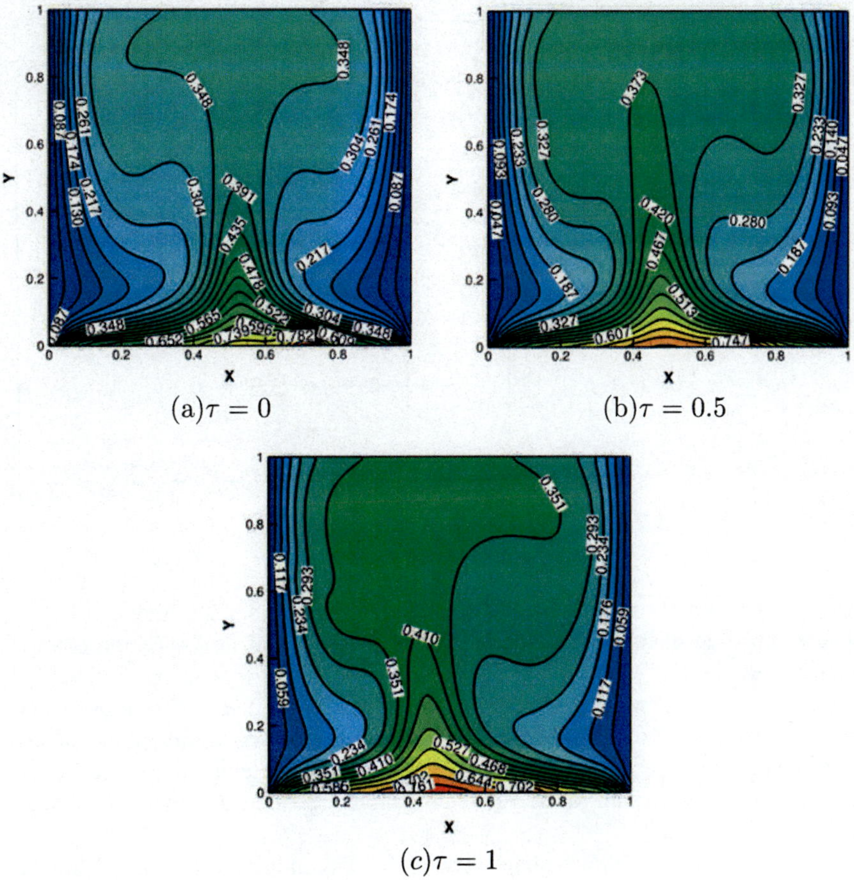

Fig. 5 Contours of non-dimensional temperature inside the cavity

gets tilted either right or left to the vertical line depending upon the behaviour of the medium with radiation energy. The isothermal are tilted towards right for transparent medium, Fig. 5a, this is due to fact that right vortex is smaller than the left vortex and reverse is true for non-zero optical thickness medium, Fig. 5b and c. The more tilt and unsmooth isothermal are formed in core of the cavity for medium having optical thickness, Fig. 5c. A hot zone is created near to bottom wall where two vortices meet. This hot zone shifts, either right or left from the centre depending upon the size of the both vortices.

The non-dimensional temperature variation on the bottom wall is depicted in Fig. 6 for different optical thickness. Each curve has two maximums are corresponds to strike point of the laser beam another for stagnation point developed due to two vortices. Although maxima corresponds to strike point of laser beam remain fixed, but the maximum values are different for different optical thickness medium case.

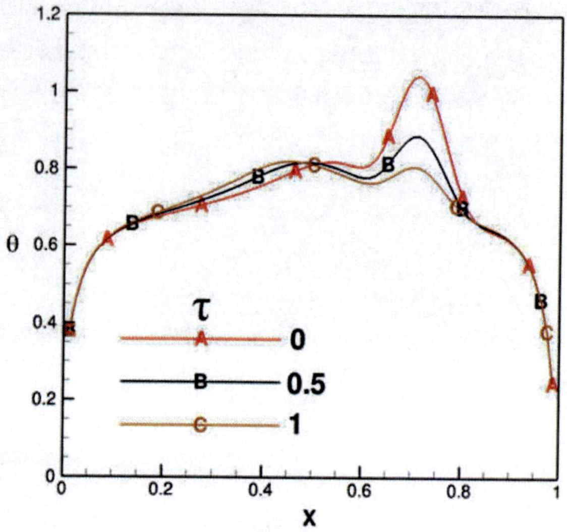

Fig. 6 Variation of non-dimensional temperature on the bottom wall

Truly, it is maximum for transparent case and minimum for optical thickness 1 case because radiation energy get absorbed by the medium and less radiation energy reaches to the bottom wall with further more. The second maxima is decided by the size of two vortices. The right vortex is smaller to the left vortex for transparent case, thus stagnation point shifts right, therefore, the second maximum point is non-dimensional curve whereas, the maximum non-dimensional temperature crosses for same length of bottom wall for transparent medium case indicating reversal of heat transfer phenomenon.

The variation of non-dimensional vertical velocity along the horizontal line at the mid-height of the cavity is shown in Fig. 7. The vertical velocity is non-dimensionalised by convective scale as represented in Sect. 3.2. The vertical velocity is downwards near to the active walls and reaches to minimum some distance away from the wall, then starts going upwards, reaches to maximum some point inside the cavity. Thus maximum velocity is achieved at different locations for optical thickness 0, 0.5 and 1. medium, and this location is at junction of the two vortices inside the medium. It is therefore, the location of maximum velocity is little bit towards right for transparent medium and left for non-zero thickness medium from the centre point of the cavity. The value of maximum velocity also decreases for the medium with optical thickness 0.5 and 1. This is owing to fact that less energy is being transferred to fluid from the bottom wall for case of optical thickness 1 then to case of optical thickness 0.5, than to transparent medium because of increase in bottom wall temperature in ascending order of transparent medium with optical thickness 0.5 and 1.

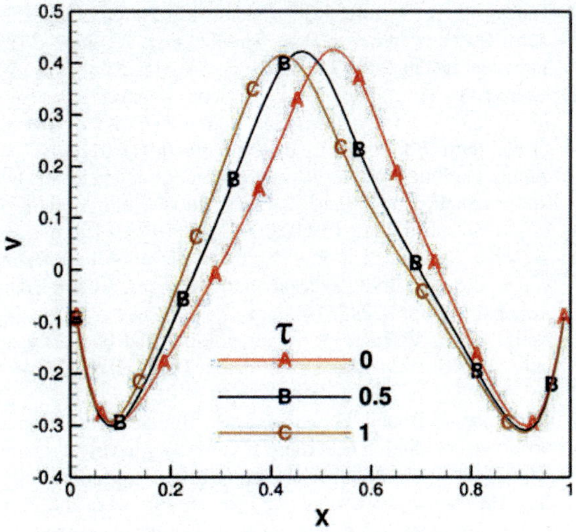

Fig. 7 Variation of non-dimensional vertical velocity along the mid height of the cavity

6 Conclusions

A plain laser of width 0.05 has been passed at an angle of 45° into the natural cavity through semitransparent wall at a height of 0.7. The natural cavity is heated from bottom and effect of radiative behaviour of the medium is analysed on the passage of laser beam in natural cavity. The following conclusions have been drawn out of above results

1. The two vortices problem without laser beam becomes asymmetric due to passage of laser beam.
2. The left vortex become thicker in transparent medium whereas right vortex gets thicker for participating medium and this thickness changes with the optical thickness of the medium.
3. The isotherm is tilted towards right for transparent medium whereas it is tilted towards left for radiation in participating medium. The isotherm becomes irregular inside the cavity for optical thickness case.
4. The maximum vertical velocity is achieved at the junction of two vortices and this maximum velocity decreases with increase of optical thickness of the medium.

References

1. Torrance, K.E., Rockett, J.A.: Numerical study of natural convection in an enclosure with localized heating from below-creeping flow to the onset of laminar instability. J. Fluid Mech. Part 1 **36**, 33–54 (1969). https://doi.org/10.1017/S0022112069001492

2. Calcagani, B., Marsili, F., Paroncini, M.: Natural convective heat transfer in square enclosures heated from below. App. Thermal Eng. **25**, 2522–2531 (2005). https://doi.org/10.1016/j.applthermaleng.2004.11.032
3. Ganzorolli, M.M., Milanez, L.F.: Natural convection in rectangular enclosures heated from below and symmetrically cooled from the sides. Int. J. Heat Mass Transf. **6**(36), 1063–1073 (1995). https://doi.org/10.1016/0017-9310(94)00217-J
4. Aydin, O., Yang, W.-H.: Natural convection in enclosures with localized heating from below and symmetrical cooling from sides. Int. J Num. Methods Heat Fluid Flow **10**(5), 518–552 (2000). https://doi.org/10.1108/09615530010338196
5. Acharya, S., Goldstein, R.J.: Natural convection in an externally heated vertical or inclined square box containing internal energy sources. J. Heat Transf. **107**, 855–866 (1985). https://doi.org/10.1115/1.3247514
6. Webb, B.W., Viskanta, R.: Radiation-induced buoyancy driven flow in rectangular enclosures: experiment and analysis. J. Heat Transf. **109**, 427–433 (1987). https://doi.org/10.1115/1.3248099
7. Mezrhab, A., Bouali, H., Amaoui, H., Bouzidi, M.: Computations of combined natural convection and radiation heat transfer in a cavity having a square body at its centre. Appl. Energy **83**, 627–644 (2006). https://doi.org/10.1016/j.apenergy.2005.09.006
8. Sun, H., Chenier, E., Lauriat, G.: Effect of surface radiation on the breakdown of steady natural convection flows in a square, air-filled cavity containing a centred inner body. Appl. Thermal Eng. **31**, 1252–1262 (2011). https://doi.org/10.1016/j.applthermaleng.2010.12.028
9. Paramanda, M., Khhan, S., Dalal, A., Natarajan, G.: Critical assessment of numerical algorithms for convective-radiative heat transfer in enclosures with different geometries. Int.J. Heat Mass Transf. **108**, 627–644 (2017). https://doi.org/10.1016/j.ijheatmasstransfer.2016.12.033
10. Kumar, P., Eswaran, V.: The effect of radiation on natural convection in slanted cavities of angle 45° and 60°. Int. J. Thermal Sci. **67**, 96–106 (2013). https://doi.org/10.1016/j.ijthermalsci.2012.12.009
11. Saravanan, S., Sivaraj, C.: Coupled thermal radiation and natural convection heat transfer in a cavity with a heated plate inside. Int. J. Heat Fluid Flow **40**, 54–64 (2013). https://doi.org/10.1016/j.ijheatfluidflow.2013.01.007
12. Fusegi, T., Farouk, B.: A computational and experimental study of natural convection and surface/gas radiation interactions in a square cavity. Trans. ASME J. Heat Transf. **112**, 802–804 (1990). https://doi.org/10.1115/1.2910463
13. Mondal, B., Mishra, S.C.: Simulation of natural convection in the presence of volumetric radiation using the lattice Boltzmann method. Num. Heat Transf. Part-A **55**, 18–41 (2009). https://doi.org/10.1080/10407780802603121
14. Sun, Y., Zhang, X., Howell, J.R.: Assessment of different radiative transfer equation solvers for combined natural convection and radiation heat transfer problems. J. Quant. Spectrosc. Radiat. Transf. **194**, 31–46 (2017). https://doi.org/10.1016/j.jqsrt.2017.03.022
15. Yuan, X., Tavakkoli, F., Vafai, K.: Analysis of natural convection in horizontal concentric annuli of varying inner shape. Num. Heat Transf. Part-A **68**, 1155–1174 (2015). https://doi.org/10.1080/10407782.2015.1032016
16. Anand Krishna, N., Mishra, S.C.: Discrete transfer method applied to radiative transfer in variable refractive index. J. Quant. Spectrosc. Radiat. Transf. **102**, 432–440 (2006). https://doi.org/10.1016/j.jqsrt.2006.02.024
17. Ben Abdallah, P., Le Dez, V.: Radiative flux field inside an absorbing-emitting semi transparent slab with a variable refractive index at radiative conductive coupling. J. Quant. Spectrosc. Radiat. Transf. **67**, 125–137 (2000). https://doi.org/10.1016/S0022-4073(99)00200-9
18. Ilyushin Ya, A.: Propagating of collimated beam in the refractive scattering medium. Radiophys. Quantum Electron. **55**, 10–11 (2013). https://doi.org/10.1007/s11141-013-9402-8
19. Verma, A.K., Rath, P., Mahapatra, S.K.: Interaction of short pulse collimated irradiation with inhomogeneity: an accurate model. Int. Com. Heat Mass Transf. **72**, 1–9,(2016). https://doi.org/10.1016/j.icheatmasstransfer.2015.08.025

20. Rath, P., Mahapatra, S.K.: New formulation of radiative flux in ultrashort time scale with its implications. J. Thermophys. Heat Transf. **26**(2), 294–299 (2012). https://doi.org/10.2514/1.58129
21. Garg, A., Chanakya, G., Kumar, P.: Numerical error estimation in finite volume method for Radiative transfer equation for collimated irradiation. In: Proceedings of the 9th International Symposium on Radiative Transfer, RAD-19, June 3–7, 2019, Athens, Greece (2019)
22. Aswatha, C.J., Gowda, G., Sridhara, S.N., Seetharamu, K.N.: Effect of different thermal boundary conditions at bottom wall on natural convection in cavities. J. Eng. Sci. Technol. **6**, 109–130 (2011)
23. Lari, K., Baneshi, M., Nassab, S.G., Komiya, A., Maruyama, S.: Combined heat transfer of radiation and natural convection in a square cavity containing participating gases. Int. J. Heat Mass Transf. **54**, 5087–5099 (2011). https://doi.org/10.1016/j.ijheatmasstransfer.2011.07.026
24. Basak, T., Roy, S., Balakrishnan, A.R.: Effects of thermal boundary conditions on natural convection flows within a square cavity. Int. J. Heat Mass Transf. **49**, 4525–4535 (2006). https://doi.org/10.1016/j.ijheatmasstransfer.2006.05.015

Recent Advances in Multiscale–Multiphysics Problems

Finite Deformation of a Dielectric Cylindrical Actuator: A Continuum Mechanics Approach

Deepak Kumar, Subrat Kumar Behera, and Somnath Sarangi

1 Introduction

In current scenario, engineering materials families are currently expanding with the different types of newly discovered smart materials [13–15]. In line with that, a particular type of smart material whose elastic states and properties change with an application of electric field is known as electro-elastic material [2, 7, 11, 12]. Electro-elastic materials demonstrate a shape change effect with an application of electric field, and this phenomenon is known as electrostriction. This shape change phenomenon with an application of electric field is a strong motivation to develop the different types of smart actuators and sensors made from the dielectric elastomeric materials. Although, the general material properties of smart materials already known through literature [5, 16] and references therein. But, the modeling of an electro-elastic deformation of dielectric actuators restarted the journey due to its diverse applications in the medical equipment, optical devices, and the energy harvesters [17, 27]. Typical examples of the cylindrical dielectric actuator are an elastic nanotube and a pneumatic valve, etc., that may be used to make an artificial muscle and to measure the fluid flow characteristics, respectively. Therefore, the recent success in this area is to develop a new smart material device like DCA that employs in the smart engineering and medical field applications.

In the literature, a well-known tube actuator was developed by Pelrine et al. [20] made of a dielectric elastomer with two compliant electrodes. Therein, the tube wall

D. Kumar (✉)
Department of Mechanical Engineering, Maulana Azad National Institute of Technology Bhopal, Bhopal, Madhya Pradesh, India
e-mail: dkumar.me@iitkgp.ac.in

S. K. Behera · S. Sarangi
Indian Institute of Technology Patna, Bihta 801106, India

S. K. Saha and M. Mukherjee (eds.), *Recent Advances in Computational Mechanics and Simulations*, Lecture Notes in Mechanical Engineering,
https://doi.org/10.1007/978-981-15-8315-5_34

is contracted due to electrostriction phenomenon with an applied electrical voltage or field, causes an axial-elongation. In the parallel work, Wissler and Mazza [26] performed various electrostriction experiments on spring roll type cylindrical actuator made of a VHB 4910 dielectric elastomer. In line with that, they [25] found the field dependent electro-mechanical pressure known as an electrostrictive effect that play a major role in the electro-elastic deformation phenomenon. Further, Carpi and Rossi [3] presented an experimentally validated electro-mechanical model of a cylindrical actuator made of dielectric elastomer. Therein, they [3] formulated the finite deformations expressions by assuming linearly elastic stress-strain theory followed by the Maxwell stress tensor. However, Zhao et al. [29] already pointed out the restriction on the direct use of the Maxwell stress tensor under large deformations. They [29] found that the Maxwell stress may account for voltage-induced deformation only for a very special type of materials, which we call the ideal smart materials, wherein the electric permittivity is deformation independent. In line with that, the direct approach to obtaining the total stress for smart materials is also pointed out by other researchers [4, 21] too on the use of Maxwell stress for the deformation analysis of a continua. Therefore, besides enhancing the material properties and smart material devices, modeling is also a useful step in the development of new smart actuators and sensors, enabling its optimum and reliable engineering design. However, the modeling of smart materials to the external electric field has so far been analyzed mostly through the rigorous continuum electro-mechanical theory given by Toupin [23] since the middle of the last century. But, Dorfmann et al. [7], Bustamante et al. [1] most recently developed the theoretical background for the electro-elastic material based on the general theory of nonlinear electro-elasticity. These developments of nonlinear electro-elasticity were applied to solve the boundary-value problems, which are also currently exemplified in [8] and the recent monograph by Dorfmann and Ogden [6, 18] and references therein.

From the literature, we may conclude that the deformation response of an electro-elastic continua of a DCA may be directly formulated through a continuum mechanics-based approach alongside the second law of thermodynamics in order to the modeling development. In the existing deformation approaches, the Maxwell stress tensor does not provide an effective way for the analysis of smart materials. In one of the many cases [24] wherein the Maxwell stress tensor is utilized to obtain the total stress tensor for the smart materials, may lead to a conceptual inaccuracy. This is because of the stress superposition is physically irrelevant in terms of the definition of stress. In line with that, the present paper aims the development of an unified electro-elastic deformation model through which we may define the deformation response of a DCA analytically within the framework of second law of thermodynamics.

In the present paper, we first develop the general constitutive relationships for an incompressible isotropic electro-elastic continua of a DCA with an amended form of energy density function. This amended energy function successfully overcomes the direct use of Maxwell stress tensor. In addition, we also made an effort to describe the combined electro-elasticity with least material parameters. Further, we analytically formulated an electro-elastic deformation model for a DCA based on the classical continuum mechanics-based approach. Finally, the developed electro-elastic deformation model is then compared with their corresponding experimental data obtained from the different studies presented by Wissler and Mazza [26], Carpi and Rossi [3]. Herein, an alternative electro-elastic deformation approach within the framework of second law of thermodynamics is the major contribution of this work.

The further part of the paper is organized as follows. In Sect. 2, an electro-elastic deformation theory is developed for an isotropic electro-elastic material. In Sect. 3, the formulated electro-elastic deformation theory in the preceding section is applied to a DCA, and an electro-elastic deformation model for the cylindrical actuator is obtained. Next, in Sect. 4, the obtained electro-elastic deformation model for DCA is compared and validated with the corresponding experimental data. Finally, Sect. 5 explains about some concluding remarks.

2 An Electro-elastic Deformation

In current section, a brief overview of the fundamental field equations of physics and the constitutive theories related to electro-elastic materials [9] are developed.

2.1 *Kinematics*

Consider an electro-elastic material continuum in the material space β_0 with an undeformed reference configuration. Herein, the material point in the reference configuration β_0 is labeled by the position vector $\mathbf{X}$ with respect to a fixed origin. Now, the continuum deforms with an electro-elastic force and the material point $\mathbf{X}$ is then takes the position $\mathbf{x} = \kappa(\mathbf{X})$; in the current configuration β. Wherein κ denotes an one-to-one deformation mapping. Therefore, the deformation gradient tensor $\mathbf{F}$ for an incompressible isotropic material and its determinant J may be defined as follows:

$$\mathbf{F} = \mathrm{Grad}\kappa = \frac{\partial \mathbf{x}}{\partial \mathbf{X}}, \quad J = \det\mathbf{F} = 1, \tag{1}$$

wherein Grad is the gradient operator with respect to the position vector $\mathbf{X}$ in the reference configuration β_0.

2.2 *Electro-elastic Field Balance Equations*

2.2.1 Eularian Form

In order to analyse the deformation of an electro-elastic continua, we may consider $\mathbf{E}$, $\mathbf{D}$ and $\mathbf{P}$; the electric field vector, the electric induction or electric displacement vector, and the polarization density, respectively, as the field variables for the current configuration β. Now, for the condensed matter, these field variables are related as follows:

$$\mathbf{D} = \epsilon_0 \mathbf{E} + \mathbf{P}, \tag{2}$$

wherein ϵ_0 is the electric permittivity of free space. In case of isotropic media, the simple and widely used form of the Eq. (2) may be represented as follows:

$$\mathbf{D} = \epsilon_0 \epsilon_r \mathbf{E}, \tag{3}$$

wherein ϵ_r represents the dielectric constant. Now, for the purely static condition in absence of free currents, free electric charge, the electric field $\mathbf{E}$ and the electric displacement $\mathbf{D}$, these field variables satisfy the given Maxwell's equations [22] under the given assumptions

$$\text{curl}\mathbf{E} = 0, \quad \text{div}\mathbf{D} = 0, \tag{4}$$

wherein curl and div represent the curl and divergence operators with respect to the position vector $\mathbf{x}$ in the current configuration β.

2.2.2 Lagrangian Form

In the preceding sub-section, the relations are formulated in Eulerian form with the operators div and curl. However, in the current sub-section, we reformulate the relations in the reference configuration to develop a compact form of the constitutive relations. Herein, the operators Div, Curl are defined with respect to the space variable $\mathbf{X}$. In literature [7] too, the electric field variables in Lagrangian form $\mathbf{E}^l$, $\mathbf{D}^l$ and $\mathbf{P}^l$ were found advantageous to work with it as compared to Eulerian form. Now, the relationships between these field variables may be obtained as follows:

$$\mathbf{E}^l = \mathbf{F}^T\mathbf{E}, \quad \mathbf{D}^l = \mathbf{F}^{-1}\mathbf{D}, \quad \mathbf{P}^l = \mathbf{F}^{-1}\mathbf{P}. \tag{5}$$

The above relationships (5) ensure that Eq. (4) are equivalent to

$$\text{Curl}\mathbf{E}^l = 0, \quad \text{Div}\mathbf{D}^l = 0. \tag{6}$$

2.3 *Constitutive Relations*

In general, the constitutive relations for an incompressible isotropic electro-elastic material may be obtained through the independent field variables $\mathbf{F}$, $\mathbf{E}$. In line with that, in isothermal condition, the free energy density function for an incompressible isotropic electro-elastic material may be defined as a function of these independent field variables in Eularian form given as follows:

$$\varphi = \varphi(\mathbf{F}, \mathbf{E}). \tag{7}$$

Additionally, we may also consider $\mathbf{F}$ and $\mathbf{E}^l$ as independent variables for electro-elastic material in Lagrangian form. Accordingly, the free energy function $\phi(\mathbf{F}, \mathbf{E}^l)$ may also be redefined using relations (5) as follows:

$$\phi(\mathbf{F}, \mathbf{E}^l) = \varphi(\mathbf{F}, \mathbf{F}^{-T}\mathbf{E}^l). \tag{8}$$

Now, following the previous works [7, 10, 19], the total stress tensor $\mathbf{T}$ with the help of Maxwell's concept for electro-elastic material the Cauchy stress tensor $\mathbf{S}$ may be obtained as follows:

$$\mathbf{T} = \mathbf{S} + \mathbf{P} \otimes \mathbf{E} + \epsilon_0 \left[\mathbf{E} \otimes \mathbf{E} - \frac{1}{2}(\mathbf{E}.\mathbf{E})\mathbf{I} \right]. \tag{9}$$

Herein, the total stress tensor $\mathbf{T}$ defined in the above Eq. (9) directly through the superposition of stresses may lead to conceptual inaccuracy, especially for the large deformation of a continua [4, 21]. This is because of the superposition principle applied in the stresses is physically irrelevant in terms of the definition of stress. In line with that, Zhao and Suo [28] and other researchers [4, 21] also raised an issue related to physical interpretation of the Maxwell stress tensor in large deformation. Accordingly, we made an effort to overcome this issue through an amended energy density function $\Omega = \Omega(\mathbf{F}, \mathbf{E}^l)$, which incorporates the Maxwell stress contribution for electro-elastic materials. The expression of the defined amended energy density function is given as follows:

$$\Omega(\mathbf{F}, \mathbf{E}^l) = \rho\phi(\mathbf{F}, \mathbf{E}^l) - \frac{1}{2}\epsilon_0\mathbf{E}^l.(\mathbf{b}^{-1}\mathbf{E}^l). \tag{10}$$

This amended energy function $\Omega(\mathbf{F}, \mathbf{E}^l)$ represents the superposition of the possible form electrical energy and the interaction energy. The superposition of energies from the mechanical and electrical domain is truly accepted unlike the superposition of stress contributions from the mechanical domain and the electrical domain as described in (9). Additionally, the amended energy function $\Omega(\mathbf{F}, \mathbf{E}^l)$ also successfully overcomes the issue related to the physical interpretation of Maxwell stress tensor in smart material for large deformation [29]. For the detailed discussion on the issues related to physical objectivity of Maxwell stress tensor we refer [4, 21] and references therein.

Following the Clausius-Duhem inequality principle based on the second law of thermodynamics, we may formulate the thermodynamically consistent constitutive relations for an incompressible isotropic electro-elastic material. Assuming isothermal conditions, this dissipation inequality may be written in terms of $\Omega(\mathbf{F}, \mathbf{E}^l)$ as follows:

$$\left(\mathbf{T} - \mathbf{F}\frac{\partial \Omega}{\partial \mathbf{F}}\right) : \dot{\mathbf{F}} - \left(\mathbf{D}^l + \frac{\partial \Omega}{\partial \mathbf{E}}\right) : \dot{\mathbf{E}} \geq 0. \tag{11}$$

Now, the set of constitutive relations from the above Eq. (11) in generalized form for an incompressible isotropic electro-elastic material may be written as follows:

$$\mathbf{T} = -p\mathbf{I} + \mathbf{F}\frac{\partial \Omega}{\partial \mathbf{F}}, \quad \mathbf{D}^l = -\frac{\partial \Omega}{\partial \mathbf{E}^l}, \tag{12}$$

wherein p is the indeterminate hydrostatic pressure arising from the incompressibility constraint associated with Cauchy stress tensor.

Additionally, the amended energy density function Ω for an incompressible isotropic electro-elastic material may also be formulated in the invariant form $\Omega(I_1, I_2....I_6)$. These invariants may be derived from the two isotropic tensors left Cauchy green deformation tensor $\mathbf{b} = \mathbf{F}\mathbf{F}^T$ and $\mathbf{E}^l \otimes \mathbf{E}^l$ as follows:

$$\begin{aligned} I_1 = \mathrm{tr}\mathbf{b}, \quad I_2 = \frac{1}{2}[(\mathrm{tr}\mathbf{b})^2 - \mathrm{tr}(\mathbf{b}^2)], \quad I_3 = \det\mathbf{b} = 1, \\ I_4 = [\mathbf{E}^l \otimes \mathbf{E}^l] : \mathbf{I}, \quad I_5 = [\mathbf{E}^l \otimes \mathbf{E}^l] : \mathbf{b}^{-1}, \quad I_6 = [\mathbf{E}^l \otimes \mathbf{E}^l] : \mathbf{b}^{-2}. \end{aligned} \tag{13}$$

From the relations (10), (12) and the definitions of the invariants in (13) the explicit form of $\mathbf{T}, \mathbf{D}$ may be written as follows:

$$\begin{aligned} \mathbf{T} = -p\mathbf{I} + 2\Omega_1\mathbf{b} + 2\Omega_2[I_1\mathbf{b} - \mathbf{b}^2] - 2\Omega_5\mathbf{E} \otimes \mathbf{E} \\ -2\Omega_6[\mathbf{b}^{-1}\mathbf{E} \otimes \mathbf{E} + \mathbf{E} \otimes \mathbf{b}^{-1}\mathbf{E}], \\ \mathbf{D} = -2[\Omega_4\mathbf{b} + \Omega_5\mathbf{I} + \Omega_6\mathbf{b}^{-1}]\mathbf{E}, \end{aligned} \tag{14}$$

wherein the notation Ω_i represents $\Omega_i = \dfrac{\partial \Omega}{\partial I_i}$ for $i = 1, 2, 3.., 6$. The above relations (14) represent the standard constitutive relations based on the second law of thermodynamics for a class of an incompressible isotropic electro-elastic material.

Finally, in this section, a theoretical framework for the analysis of an electro-elastic continua is confined herein with a more physically relevant constitutive relationships (14) within the framework of second law of thermodynamics. These general constitutive relations (14) have sufficient potential to apply directly for the solution of a number of representative boundary-value problems existing in the engineering and medical field applications. Specifically, we are focusing a DCA-based boundary-value problem which is the most generalized configuration used as an actuation and sensing processes in the field of industrial robotics applications.

3 Application to a Dielectric Cylindrical Actuator

In current section, a general electro-elastic deformation phenomenon for a dielectric cylindrical actuator (DCA) is modeled analytically following the developed constitutive relationships (14).

3.1 Electro-elastic Deformation of a Cylindrical Actuator

Consider a dielectric cylindrical actuator (DCA) with (R, Θ, Z) coordinate in the reference configuration β_0. Herein, with an application of radial electric field, the same point is now at (r, θ, z) in the current configuration β. The cross section of the cylindrical actuator in reference configuration is defined as follows:

$$A \le R \le B, \quad 0 \le \Theta \le 2\pi, \quad 0 \le Z \le H, \tag{15}$$

wherein A and B are the inner and outer radii and H is the height of the DCA. The deformed cross section of the cylindrical actuator after deformation in current configuration is defined as follows:

$$a \le r \le b, \quad 0 \le \theta \le 2\pi, \quad 0 \le z \le h, \tag{16}$$

wherein a and b are the inner and outer radii and h is the height of the DCA after deformation. The deformation mapping for an incompressible isotropic cylindrical actuator with the assumed circular symmetry under extension and inflation may be expressed as follows:

$$r^2 = a^2 + \lambda_z^{-1}(R^2 - A^2), \quad \theta = \Theta, \quad z = \lambda_z Z, \tag{17}$$

wherein $\lambda_z = h/H$. Herein, an electrical voltage is applied to the flexible electrodes on the surfaces $r = a$ and $r = b$. This electrical voltage is associated with the equal charges of opposite signs on the two surfaces, which generates a radial electric field E_r within the material. Now, the deformation gradient tensor $\mathbf{F}$ and the electric field vector $\mathbf{E}$ for the above deformation mapping may be obtained as follows:

$$\mathbf{F} = \frac{1}{\lambda\lambda_z}\mathbf{e}_{rr} + \lambda\mathbf{e}_{\theta\theta} + \lambda_z\mathbf{e}_{zz}, \quad \mathbf{E} = E_r\mathbf{e}_r, \tag{18}$$

wherein $\lambda = r/R$ is the stretch in radial direction. The stress components in the corresponding directions may be obtained from the relation $(14)_1$ as follows:

$$T_{rr} = -p + 2\Omega_1 \lambda^{-2}\lambda_z^{-2} + 2\Omega_2(\lambda_z^{-2} + \lambda^{-2}) - 2\Omega_5 E_r^2 - 4\Omega_6 \lambda^2 \lambda_z^2 E_r^2,$$
$$T_{\theta\theta} = -p + 2\Omega_1 \lambda^2 + 2\Omega_2(\lambda_z^{-2} + \lambda^2\lambda_z^2), \quad (19)$$
$$T_{zz} = -p + 2\Omega_1 \lambda_z^2 + 2\Omega_2(\lambda^2\lambda_z^2 + \lambda^{-2}).$$

We may also obtain the associated radial component of electric polarization from the relations (2) and $(14)_2$ given as follows:

$$P_r = -2(\Omega_4 \lambda^{-2}\lambda_z^{-2} - \Omega_5 + \Omega_6 \lambda^2\lambda_z^2)E_r - \epsilon_0 E_r. \quad (20)$$

The equilibrium equation div$\mathbf{T} = 0$ in radial direction is given as follows:

$$\frac{\partial T_{rr}}{\partial r} = \frac{T_{\theta\theta} - T_{rr}}{r}, \quad (21)$$

and the integration of this equilibrium equation (21) provides the relation as

$$\int_{T_{rr}(a)}^{T_{rr}(b)} dT_{rr} = \int_a^b [2\Omega_1(\lambda^2 - \lambda^{-2}\lambda_z^{-2}) + 2\Omega_2(\lambda^2\lambda_z^2 - \lambda^{-2}) + 2(\Omega_5 + 2\Omega_6\lambda^2\lambda_z^2)E_r{}^2]\frac{dr}{r}. \quad (22)$$

By applying the constant volume condition, we may obtain the relation given as follows:

$$\frac{dr}{r} = \frac{d\lambda}{\lambda(1 - \lambda^2\lambda_z)}. \quad (23)$$

Using the above relation (23), the relation (22) may be rewritten as follows:

$$T_{rr}(b) - T_{rr}(a) = \int_{\lambda_a}^{\lambda_b} [2\Omega_1(\lambda^2 - \lambda^{-2}\lambda_z^{-2}) + 2\Omega_2(\lambda^2\lambda_z^2 - \lambda^{-2}) + 2(\Omega_5 + 2\Omega_6\lambda^2\lambda_z^2)E_r{}^2]\frac{d\lambda}{\lambda(1 - \lambda^2\lambda_z)}. \quad (24)$$

Herein, the above relation (24) represents an electro-elastic deformation model for a DCA which is modeled analytically following the second law of thermodynamics.

3.2 An Electro-elastic Material Model

The theory of the preceding sub-section is completely general for an isotropic electro-elastic material and admits many possible specializations. Herein, we recall the general expression of six invariants (13) related to amended energy density function, and we may discard $I_3 = 1$ (incompressibility constraint) essentially. Now, these invariants become as the five independent invariants. Therefore, following the concept of amended energy density function (10), we may generalize a Mooney-Rivlin type electro-elastic material model for an incompressible, isotropic electro-elastic material. Our effort is focussed to obtain an amended energy function with least material parameters for an isotropic electro-elastic material. Therefore, the energy density function that includes elastic and electrical energy contributions is defined as follows:

$$\Omega = C_1(I_1 - 3) + C_2(I_2 - 3) + \frac{\epsilon_0}{2}(C_3 I_4 + C_4 I_5), \tag{25}$$

wherein C_1, C_2, C_3, and C_4 are the material constant parameters. Specifically, we use the following values of the material constants, which were obtained by fitting data for a VHB 4910 dielectric elastomer stress-stretch and electric permittivity-stretch plot available in the literature [25, 26] with the use of relation (19) and (20) as follows:

$$\begin{aligned} C_1(MPa) = 0.0450, \quad C_2(MPa) = 0.0412, \\ C_3 = -1.34, \quad C_4 = -3.36. \end{aligned} \tag{26}$$

Using the above energy function (25) in the relation (24), we may rewrite the relation (24) as follows:

$$\begin{aligned} T_{rr}(b) - T_{rr}(a) = \int_{\lambda_a}^{\lambda_b} & [2C_1(\lambda^2 - \lambda^{-2}\lambda_z^{-2}) + 2C_2(\lambda^2\lambda_z^2 - \lambda^{-2}) \\ & + C_4\epsilon_0 {E_r}^2] \frac{d\lambda}{\lambda(1 - \lambda^2\lambda_z)}, \end{aligned} \tag{27}$$

wherein $\lambda_a = a/A$ and $\lambda_b = b/B$.

3.3 *Electrostatic Pressure Generated in Dielectric Cylindrical Actuator*

In absence of any external force (boundary traction and body force) the body will deform with the application of electric field alone. This phenomenon of electrostriction was modeled by Pelrine et al. [20] through an electrostatic pressure p_{el}. They [20] proposed that the electrostatic pressure p_{el} is acted perpendicular to the applied electric field, and this is the exact reason for transverse expansion of the VHB 4910 film. They [20] also developed a well-known Pelrine equation for free boundary condition in which the applied electric field is related to electrostatic pressure given as follows:

$$p_{el} = \epsilon_0 \epsilon_r E^2. \tag{28}$$

The above Pelrine equation (28) is an equilibrium state equation obtained followed by the first law of thermodynamics. This Eq. (28) represents an effective pressure generated due to electro-mechanical coupling in an dielectric elastomeric actuator with the application of an electric field.

In line with that, we present an alternative method within the framework of second law of thermodynamics, which may illustrate an electrostatic pressure p_{el} dependency on applied electric field analogous to Pelrine et al. [20]. The analytical expression obtained following the second law of thermodynamics may help to understand the electro-mechanical deformation in a more deeper manner as compareed to first law of thermodynamics. In order to obtain the analytical expression for the electrostatic pressure p_{el}, we may use the obtained electro-elastic deformation relation (27) for the different values of applied electric field E_r and axial-strain λ_z. Now, the electro-elastic deformation relation (27) with $T_{rr}(b) = 0$ and $T_{rr}(a) = -p_{el}$ may be rewritten as follows:

$$p_{el} = \int_{\lambda_a}^{\lambda_b} [2C_1(\lambda^2 - \lambda^{-2}\lambda_z^{-2}) + 2C_2(\lambda^2\lambda_z^2 - \lambda^{-2}) + C_4\epsilon_0 {E_r}^2] \frac{d\lambda}{\lambda(1 - \lambda^2\lambda_z)}. \tag{29}$$

Substituting the material parameters values from (26) for an electro-elastic material model (25), the above relation (29) may be solved numerically for the different values of applied electric field and axial-strain. Now, we may plot the numerical values of the electrostatic pressure p_{el} against λ for the different values of electric field $E_r = 20, 40, 60\ \mathrm{V}/\mu\mathrm{m}$ and the fixed axial-strain $\lambda_z = 1.2$ as shown in the Fig. 1. In addition, similar results were also obtained by Melnikov and Ogden [18] for an electro-elastic circular cylindrical tube.

Finally, in this section we have formulated an electro-elastic deformation of a cylindrical actuator with a new electro-elastic energy function (25). An equilibrium state equation (29) is also interpreted following the second law of thermodynamics which is analogous to Pelrine equation (28) obtained by Pelrine et al. [20] within the framework of first law of thermodynamics. Additionally, the numerical values of an effective electro-static pressure are also obtained from the numerical integration of the relation (29).

4 Experimental Validation and Model Comparison

In current section, the analytical model (29) formulated in the preceding section is compared and validated with the experimental results obtained by Wissler and Mazza [26], Carpi and Rossi [3].

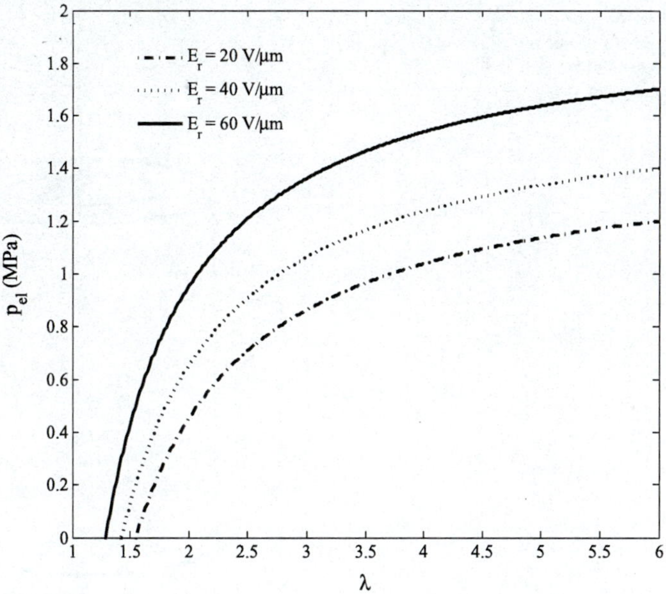

Fig. 1 Plot of the electrostatic pressure p_{el} versus λ for $E_r = 20, 40, 60\,\mathrm{V}/\mu\mathrm{m}$ and $\lambda_z = 1.2$

4.1 Comparison with Wissler and Mazza [26] Experimental Data

Wissler and Mazza [26] designed a spring roll actuator that was cylindrical in shape and VHB 4910 membrane was coated with a mixture of graphite powder (Superior Graphite, ABG1005, 20 g) and silicone oil (Dow Corning, DC 200/50 cs, 45 g) as electrodes. The schematic diagram of the experimental spring roll actuator is shown in the Fig. 2. The experiment was performed at room temperature, and the electric field was applied in radial direction with the help of a linear incremental voltage amplifier (Trek, Model 5/80). Additionally, an electrostatic force due to electrostrictive effect was measured in axial-direction with a force transducer (HBM type S2) by keeping the length of actuator unchanged. This means that the electrostatic forces together with the variation of axial-stresses correspond to a hydrostatic stress state. Therefore, the axial-force (or the corresponding stress component) could be measured directly with a direct measurement of the electrostatic stress. The Pelrine equation (28) was used for the experimental investigation of the electrostrictive effect. Due to electrostrictive effect, a restrained VHB 4910 film will experience an electrostatic force in restraint direction with an application of electric field.

In line with that, we may also analyze the spring roll actuator as shown in the Fig. 2 through the classical continuum mechanics-based approach. Herein, the cylindrical actuator is not subjected to any external force except an application of electric field in radial direction only. Next, the deformation induced by electric field in the cylindrical

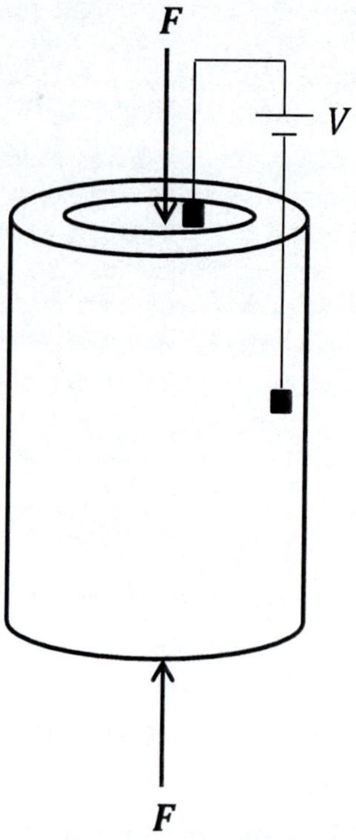

Fig. 2 Schematic experimental set-up of spring roll actuator designed by Wissler and Mazza [26] for electrostriction

actuator may be obtained from the deformation relation (22) through the electro-elastic material model (25). Further, for a constraint axial-direction similar to Wissler and Mazza [26] spring roll actuator, we may introduce $\lambda_z = 1$ in electro-elastic deformation relation (22). Therefore, the electro-elastic deformation relation (22) with $\lambda_z = 1$ may be rewritten for Wissler and Mazza [26] spring roll actuator given as follows:

$$T_{rr}(b) - T_{rr}(a) = \int_a^b [2(C_1 + C_2)(\lambda^2 - \lambda^{-2}) + C_4\epsilon_0 E_r^2]\frac{dr}{r}. \quad (30)$$

The above Eq. (30) is analogous to the Pelrine's equation (28) through which the load F_{th} required to maintain the reference configuration $\lambda = 1$ is obtained as follows:

$$F_{th} = \int_a^b (\pi\epsilon_0 C_4)E_r^2 r dr. \quad (31)$$

By solving the above equation with $b^2 - a^2 = B^2 - A^2$ for $\lambda_z = 1$ and using $A =$ 70.850 mm, $B = 70.725$ mm from the Wissler and Mazza [26] experimental set-up dimensions of spring roll actuator, the above Eq. (31) may be rewritten as follows:

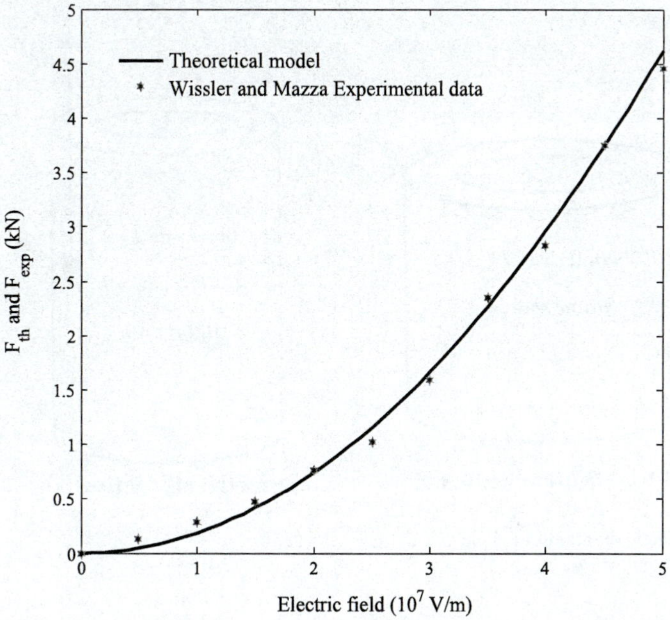

Fig. 3 Comparison of the theoretical load (32) with the Wissler and Mazza [26] experimental data

$$F_{th} = kC_4 E_r^2, \tag{32}$$

wherein $k = \dfrac{\pi\epsilon_0}{2}(B^2 - A^2)$ is a constant. Now, we may plot the load F_{th} (32) required to maintain the original geometry versus electric field plot as shown in the Fig. 3. Additionally, we may also compare this theoretical load versus electric field plot with the Wissler and Mazza [26] experimental load F_{exp} data. Herein, the experimental load F_{exp} is obtained from the electrostrictive pressure versus applied electric field Wissler and Mazza [26] experimental data. In Fig. 3, we may obtain the best fit value of the constant parameter $C_4 = 3.41$ by using the least square method. On the other hand, Wissler and Mazza [26] have considered the electrostatic pressure to calculate the axial-force difference required to keep the length of actuator constant with the help of force transducer (HBM type S2). In line with that, we also obtain the same axial-force through the classical continuum mechanics-based approach, which shows the identical results with more deeper understanding.

4.2 Comparison with Carpi and Rossi [3] Experimental Data

In general, an axial-strain of the dielectric cylindrical actuator (DCA) may be observed with an application of electric field alone. This axial-strain induced by

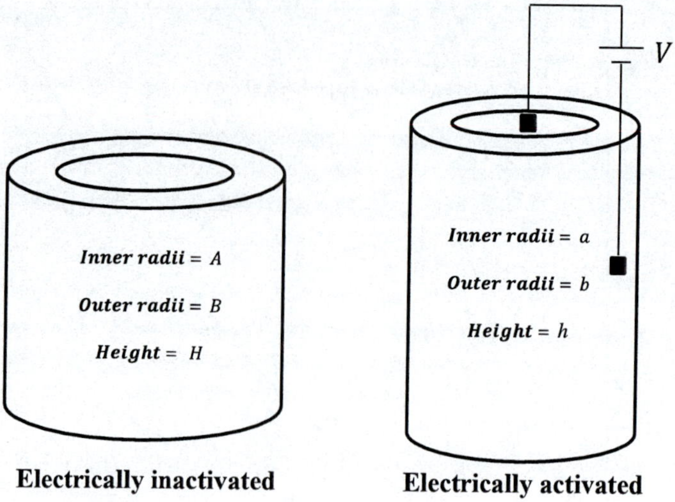

Fig. 4 Schematic diagram of a dielectric cylindrical actuator

electrical loading plays an important role in the actuation mechanism of the actuators and sensors. In line with that, the similar phenomenon was experimentally studied by Carpi and Rossi [3]. Therein, they [3] have measured the axial-strain in the DCA by placing it in the vertical position with its lower end constrained and its upper end connected to an isotonic displacement transducer (Ugo Basile, Italy, 7006). Next, the electric field was applied in radial direction with the help of applied potential difference as shown in the Fig. 4. Further, the step-wise voltages of different value, generated by a DC high-voltage supply, were separately applied to the actuator. Now, the signals of the isotonic axial-displacement were successively recorded for each voltage value between the 10-s time duration. This simple procedure also considers the property of incompressibility constraint, which holds true for most of dielectric elastomeric materials. At last, they [3] have reported an increase in axial-strain with an applied electric field (Voltage/thickness).

In line with that, we may also represent an electrically induced axial-strain for a DCA through the classical continuum mechanics-based approach. Herein, an electrically induced axial-strain expression is formulated through the obtained electro-elastic deformation relation (27) for a DCA. Next, by applying the boundary conditions $T_{rr}(b) = T_{rr}(a) = 0$ similar to Carpi and Rossi [3], we may rewrite electro-elastic deformation relation (27) given as follows:

$$\int_{\lambda_a}^{\lambda_b} \frac{C_4 \epsilon_0 {E_r}^2 d\lambda}{\lambda(1-\lambda^2\lambda_z)} = \int_{\lambda_a}^{\lambda_b} [C_1(\lambda^{-2}\lambda_z^{-2} - \lambda^2) + C_2(\lambda^{-2} - \lambda^2\lambda_z^2)] \cdot \frac{2d\lambda}{\lambda(1-\lambda^2\lambda_z)}. \tag{33}$$

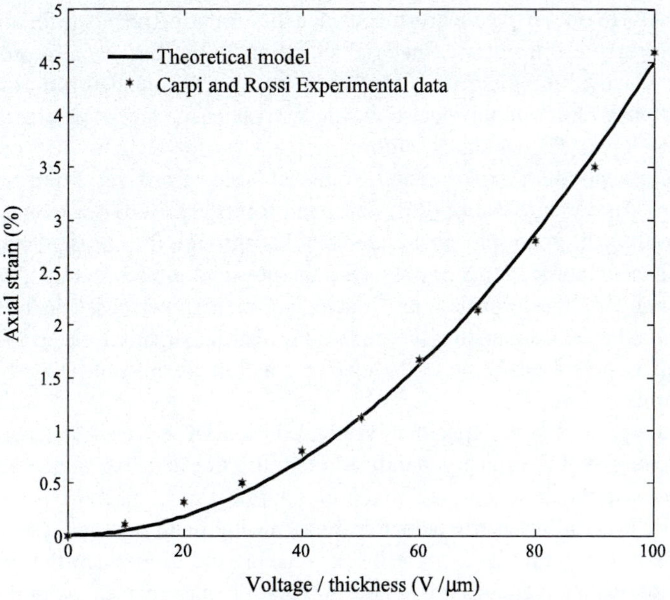

Fig. 5 Comparison of theoretical model (33) with the Carpi and Rossi [3] experimental data

Now, the above electro-elastic deformation model (33) may be numerically solved for the axial-strain λ_z of cylindrical actuator with the different values of applied electric field E_r. Further, the inner and outer radii $A = 0.781$ mm and $B = 0.976$ mm have been considered for the numerical solution of the above electro-elastic deformation relation (33) similar to Carpi and Rossi [3]. At last, we plot the numerical solution of theoretical electro-elastic deformation model (33) shown in Fig. 5, and also compared with the Carpi and Rossi [3] experimental data. Herein, we derive the obtained theoretical model (33) followed by the classical continuum mechanics-based approach, which shows a good agreement with the experimental results obtained by Carpi and Rossi [3].

Finally, in this section we compared and validated the analytical models (22), (27) with the experimental results obtained by Wissler and Mazza [26], Carpi and Rossi [3]. In addition, a good agreement is also achieved between the analytical models (22), (27) and the experimental data [3, 26].

5 Concluding Remarks

In the present paper, we presented a continuum mechanics-based electro-elastic deformation theory through an amended form of energy density function. Next, we applied the formulated electro-elastic deformation theory to a dielectric cylindrical actuator (DCA) to obtain the electro-mechanical deformation response for an incompressible isotropic electro-elastic material analytically. Further, we also proposed a new electro-elastic energy density function based on the amended form of energy function that may represent the electro-elasticity with least material parameters. Further, the developed electro-elastic deformation model for DCA is then compared with their corresponding experimental results obtained from the different studies presented by Wissler and Mazza [26], Carpi and Rossi [3]. Herein, we obtained the identical results with more physical understanding through the classical continuum mechanics-based approach that has resolved the physical interpretation of Maxwell stress tensor under large deformations. Conclusively, it is shown based on the theoretical model that besides enhancing the material properties of smart material devices, modeling also plays a vital role in the electro-elastic deformation phenomenon in smart materials.

Herein, the present study specifically focused on a DCA-based boundary-value problem because it is the most generalized configuration used as an actuation and sensing process in the field of industrial robotics applications. The most useful examples of the cylindrical dielectric actuator are an elastic nano-tube and a pneumatic valve that may be used to make an artificial muscles and to measure the fluid flow characteristics, respectively. An immediate application of the present work lies in the new field of soft robotics which imparts new functionalities as compared to the existing technology. In addition to extending the applicability of the presented work, fibre structures involvement can be incorporated within the proposed constitutive laws for a transversely isotropic electro-elastic deformation that requires an additional set of invariants.

References

1. Bustamante, R., Ogden, R.: Universal relations for nonlinear electroelastic solids. Acta Mechanica **182**(1), 125–140 (2006)
2. Bustamante, R., Dorfmann, A., Ogden, R.: On electric body forces and maxwell stresses in nonlinearly electroelastic solids. Int. J. Eng. Sci. **47**(11), 1131–1141 (2009)
3. Carpi, F., De Rossi, D.: Dielectric elastomer cylindrical actuators: electromechanical modelling and experimental evaluation. Mater. Sci. Eng. C **24**(4), 555–562 (2004)
4. Choi, H.S., Park, I.H., Moon, W.K.: On the physical meaning of maxwell stress tensor. Trans. Korean Inst. Electr. Eng. **58**(4), 725–734 (2009)
5. Damjanovic, D., Newnham, R.: Electrostrictive and piezoelectric materials for actuator applications. J. Intell. Mater. Syst. Struct. **3**(2), 190–208 (1992)
6. Dorfmann, L., Ogden, R.W.: Nonlinear Theory of Electroelastic and Magnetoelastic Interactions. Springer (2016)
7. Dorfmann, A., Ogden, R.: Nonlinear electroelasticity. Acta Mechanica **174**(3–4), 167–183 (2005)
8. Dorfmann, A., Ogden, R.: Nonlinear electroelastic deformations. J. Elast. **82**(2), 99–127 (2006)
9. Eringen, A.C., Maugin, G.A.: Electrodynamics of Continua I: Foundations and Solid Media. Springer Science & Business Media (2012)
10. Kovetz, A.: Electromagnetic Theory. Oxford University Press, Oxford (2000)
11. Kumar, D., Sarangi, S.: Dynamic modeling of a dielectric elastomeric spherical actuator: an energy-based approach. Soft Mater. (2019). https://doi.org/10.1080/1539445X.2019.1616557

12. Kumar, D., Sarangi, S.: Electro-mechanical instability modelling in elastomeric actuators: a second law of thermodynamics-based approach. Soft Mater. 1–13 (2019)
13. Kumar, D., Sarangi, S.: Instability analysis of an electro-magneto-elastic actuator: a continuum mechanics approach. AIP Adv. **8**(11), 115314 (2018)
14. Kumar, D., Sarangi, S.: Data on the viscoelastic behavior of neoprene rubber. Data Brief **21**, 943–947 (2018)
15. Kumar, D., Sarangi, S.: Electro-magnetostriction under large deformation: modeling with experimental validation. Mech. Mater. **128**(1), 1–10 (2019)
16. Lateefi, M.M., Kumar, D., Sarangi, S.: Stability analysis of a hyperelastic tube under large deformation. In: 2018 International Conference on Automation and Computational Engineering (ICACE), pp. 234–239. IEEE (2018)
17. Liu, Y., Ren, K.L., Hofmann, H.F., Zhang, Q.: Investigation of electrostrictive polymers for energy harvesting. IEEE Trans. Ultrason. Ferroelectr. Frequency Control **52**(12), 2411–2417 (2005)
18. Melnikov, A., Ogden, R.W.: Finite deformations of an electroelastic circular cylindrical tube. Zeitschrift für angewandte Mathematik und Physik **67**(6), 140 (2016)
19. Pao, Y.H.: Electromagnetic forces in deformable continua. In: Mechanics Today, vol. 4, pp. 209–305. NSF-supported research (A78-35706 14-70). Pergamon Press, Inc., New York (1978)
20. Pelrine, R.E., Kornbluh, R.D., Joseph, J.P.: Electrostriction of polymer dielectrics with compliant electrodes as a means of actuation. Sens. Actuators A: Phys. **64**(1), 77–85 (1998)
21. Rinaldi, C., Brenner, H.: Body versus surface forces in continuum mechanics: is the Maxwell stress tensor a physically objective cauchy stress? Phys. Rev. E **65**(3), 036615 (2002)
22. Stratton, J.A.: Electromagnetic Theory. Wiley (2007)
23. Toupin, R.A.: The elastic dielectric. J. Ration. Mech. Anal. **5**(6), 849–915 (1956)
24. Volokh, K.: On electromechanical coupling in elastomers. J. Appl. Mech. **79**(4), 044507 (2012)
25. Wissler, M., Mazza, E.: Modeling of a pre-strained circular actuator made of dielectric elastomers. Sens. Actuators A: Phys. **120**(1), 184–192 (2005)
26. Wissler, M., Mazza, E.: Electromechanical coupling in dielectric elastomer actuators. Sens. Actuators A: Phys. **138**(2), 384–393 (2007)
27. Xia, J., Ying, Y., Foulger, S.H.: Electric-field-induced rejection-wavelength tuning of photonic-bandgap composites. Adv. Mater. **17**(20), 2463–2467 (2005)
28. Zhao, X., Suo, Z.: Electrostriction in elastic dielectrics undergoing large deformation. J. Appl. Phys. **104**(12), 123530 (2008)
29. Zhao, X., Hong, W., Suo, Z.: Electromechanical hysteresis and coexistent states in dielectric elastomers. Phys. Rev. B **76**(13), 134113 (2007)

Numerical Studies for Generalized Modified Polarization Saturation (PS) Model in Piezoelectric Media

Kuldeep Sharma and Sandeep Singh

Nomenclature

PS model	Polarization Saturation Model
DDT	Distributed Dislocation Technique
LIF	Local Intensity Factor
COD	Crack Opening Displacement
COP	Crack Opening Potential
DB	Dielectric Breakdown Model

1 Introduction

Smart materials such as piezoelectric materials have unambiguously proved their utility as a key component of sensors, actuators, transducers, etc., in high-tech apparatus of smart devices and structures. Due to the fatigue, cracks develop in them. Therefore, it is crucial and essential to thoroughly investigate the fracture behavior of such materials.

Gao and his co-researchers [7] were the first who considered the role of electric displacement saturation in studying the fracture mechanics of piezoelectric materials and proposed a polarization saturation model. They justified this model by considering piezoelectric materials as mechanically brittle and electrically ductile and proposed the same in analogous to Dugdale's model [4]. In this model, they

K. Sharma (✉) · S. Singh
Department of Mathematics, National Institute of Technology Uttarakhand, Srinagar, Garhwal 246174, India
e-mail: kuldeeppurc@gmail.com

S. K. Saha and M. Mukherjee (eds.), *Recent Advances in Computational Mechanics and Simulations*, Lecture Notes in Mechanical Engineering,
https://doi.org/10.1007/978-981-15-8315-5_35

considered an electrical yielding zone in front of the crack tip in the form of a strip which is bounded by the normal saturated electric displacement value. Bhargava and Jangid [2] extended PS model to study the semipermeable two collinear cracks in an infinite 2-D piezoelectric media. They applied the complex variable technique to obtain the analytical expressions for saturated zone length, COD, COP, and local intensity factor. Apart from the extension of nonlinear (PS/DB and Dugdale) models in 2-D to 3-D or multiple cracks, etc., Dugdale model was also modified by various researchers [1, 8] after varying the condition of normal cohesive stress. Bhargava and Hasan [1] modified the Dugdale model for multiple collinear cracks by quadratically varying normal cohesive stress distribution over the rims of yield zones. Recently, Singh et al. [11–13] applied the polynomial varying polarization saturation models in center crack as well as two collinear cracks using complex variable approach.

Distributed dislocation technique (DDT) is based on the concept of expressing the crack in a material as a continuous distribution of dislocations. By doing this, the problem of crack(s) in the specimen is reduced into simultaneous singular integral equations in terms of dislocation density variable(s) at the crack. After solving these integral equations, one can obtain the dislocation density variable(s) at the crack tip, and hence the fracture parameters which are in terms of dislocation densities at the crack tip. Bilbey et al. [3] analyzed the problem of strip plastic yielding model using DDT where both the crack and the strip plastic zones are simulated by an array of dislocations. Sharma et al. [10] applied the DDT to study the array of equidistant semipermeable inclined cracks in 2-D piezoelectric media. Zhang et al. [14] and Fan et al. [6] applied the DDT to study the impermeable dielectric breakdown and semipermeable polarization saturation models, respectively. In their approach, they simplified the problem of simultaneous integral equations into separate mechanical and electrical dislocation density parameters based singular integral equations defined in a domain, i.e., $b_i(x_0), i = 1, 2, 3$ for $-a \leq x_0 \leq a$ and $b_4(x_0)$ for $-c_1 \leq x_0 \leq c_1$. However, they obtained only the analytical solution of the field variables from these simplified integral equations.

But obtaining the analytical solution for generalized saturated electric displacement conditions defined for a generalized modified PS model is difficult either by the Fan et al. [6] approach or the complex variable approach followed in [11]. Additionally, Fan et al. approach is applied for simplifying the developed simultaneous singular integral equations into individual integral equations with respect to these mechanical and electrical dislocation density parameters. Therefore, in this paper, the DDT with Gauss–Chebychev quadrature (numerical scheme) is applied for solving the reduced Cauchy-type singular integral equations.

Once the generalized dislocation densities b have been obtained, the intensity factors (IFs) at the crack-tips (a) and effective crack-tips (c_1) can be evaluated and further using iterative approach saturated zone lengths are obtained under a different set of conditions.

2 Statement of the Problem

An infinite piezoelectric domain (PZT-4) weakened by a center crack of length $2a$ lying along the x-axis with center at origin is considered. Similar to Gao et al. [7] approach, a saturated zone of length $c_1 - a$ is taken on both sides of the crack tip in the form of a strip or a line segment, i.e., along the x-axis satisfying $a < |x| < c_1$. The problem is studied under electromechanical loading and arbitrary poling direction considering impermeable and semipermeable crack-face conditions. The schematic representation of the problem is presented in Fig. 1.

3 Numerical Solutions for Generalized Modified PS Models Using DDT

The stress and electric displacement can be evaluated by using the following relations:

$$\begin{bmatrix} t_1 \\ t_2 \end{bmatrix} = \begin{bmatrix} -\phi_1 \\ -\phi_2 \end{bmatrix} = \begin{bmatrix} \sigma_{11}, \sigma_{12}, \sigma_{13}, D_1 \\ \sigma_{21}, \sigma_{22}, \sigma_{23}, D_2 \end{bmatrix} \tag{1}$$

After employing the continuous distribution of dislocation and applying surface traction charge conditions as defined in Eq. (1), a system of singular integral equations for dislocation density, b_i are obtained.

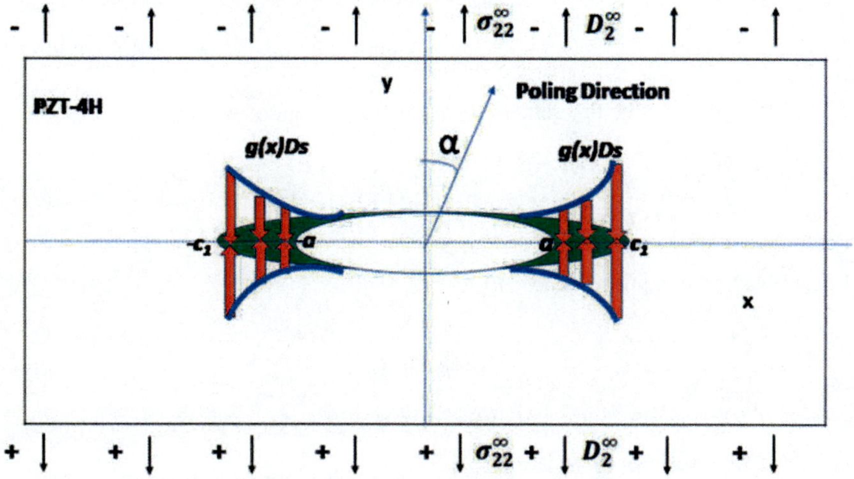

Fig. 1 Schematic representation of the problem

$$\int_{-a}^{a} L_1 \frac{1}{\pi(x-x_0)} \begin{bmatrix} b_1 \\ b_2 \\ b_3 \end{bmatrix} dx_0 + \int_{-c_1}^{c_1} \frac{1}{\pi(x-x_0)} L_2^T b_4 dx_0 + \begin{bmatrix} \sigma_{21}^{\infty} \\ \sigma_{22}^{\infty} \\ \sigma_{23}^{\infty} \end{bmatrix} = 0, \ |x| \le a, \tag{2}$$

$$\int_{-a}^{a} L_2 \frac{1}{\pi(x-x_0)} \begin{bmatrix} b_1 \\ b_2 \\ b_3 \end{bmatrix} dx_0 + \int_{-c_1}^{c_1} \frac{1}{\pi(x-x_0)} L_4 b_4 dx_0 + D_2^{\infty} - D_2^{c} = 0, \ |x| \le a, \tag{3}$$

$$\int_{-a}^{a} L_2 \frac{1}{\pi(x-x_0)} \begin{bmatrix} b_1 \\ b_2 \\ b_3 \end{bmatrix} dx_0 + \int_{-c_1}^{c_1} \frac{1}{\pi(x-x_0)} L_4 b_4 dx_0 + D_2^{\infty} - D_2^{c} = g(x) D_s - D_2^{c}, \ a \le |x| \le c_1, \tag{4}$$

where $L = \begin{bmatrix} L_1 & L_2^T \\ L_2 & L_4 \end{bmatrix} = 2iBB^T$.

Solving Eqs. (2) and (3), we have

$$\int_{-a}^{a} (L_1 L_4 - L_2 L_2^T) \frac{1}{\pi(x-x_0)} \begin{bmatrix} b_1 \\ b_2 \\ b_3 \end{bmatrix} dx_0 = - \begin{bmatrix} \sigma_{21}^{\infty} \\ \sigma_{22}^{\infty} \\ \sigma_{23}^{\infty} \end{bmatrix} L_4 + (D_2^{\infty} - D_2^{c}) L_2^T, \ |x| \le a, \tag{5}$$

Substituting $x = a\xi$ and $x_0 = a\eta$, Eq. (5) becomes

$$\int_{-1}^{1} L^* \frac{v^*(\eta)}{\pi(\xi - \eta)} d\eta = -t^*(\xi), \ |\xi| \le 1, \tag{6}$$

where $L^* = (L_1 L_4 - L_2 L_2^T)$, $v^* = \begin{bmatrix} b_1 \\ b_2 \\ b_3 \end{bmatrix}$ and $t^* = - \begin{bmatrix} \sigma_{21}^{\infty} \\ \sigma_{22}^{\infty} \\ \sigma_{23}^{\infty} \end{bmatrix} L_4 + (D_2^{\infty} - D_2^{c}) L_2^T$.

To obtain the solution of Eq. (6), Gauss–Chebychev quadrature [10] method is applied.

As per this scheme, the discretized form for Eq. (6) can be written as

$$\sum_{k=1}^{m} \frac{1}{m} [\frac{L^*}{(s_{r0} - s_k)}] \nu^*(s_k) = -t_{ni}(s_{r0}), \ \ for \ \ i = 1 \ to \ 5 \tag{7}$$

and single-valued condition reduces to

$$\sum_{k=1}^{m} \nu^*(s_k) = 0 \tag{8}$$

where $\nu(\xi) = \frac{\sum_{k=1}^{m} w_k \nu^*(s_k)}{\sqrt{1-s_k^2}}$, $s_k = cos(\frac{(2k-1)\pi}{2m})$, $k = 1, 2, ..., m$ and $s_{r0} = cos(\frac{r\pi}{m})$, $r = 1, 2, ..., m-1$. Solving Eqs. (7) and (8), the $\nu^*(s_k)$ is evaluated at all the m quadrature points. To evaluate electric dislocation density parameter $b_4(x_0)$, defined in $-c_1 \le x_0 \le c_1$, Eq. (4) is written into the following form.

$$\int_{-c_1}^{c_1} \frac{1}{\pi(x-x_0)} L_4 b_4 dx_0 + D_2^{\infty} - D_2^{c} = g(x)D_s - D_2^{c} - \int_{-a}^{a} L_2 \frac{1}{\pi(x-x_0)} \begin{bmatrix} b_1 \\ b_2 \\ b_3 \end{bmatrix} dx_0, \; a \le |x| \le c_1, \quad (9)$$

or

$$\int_{-c_1}^{c_1} \frac{1}{\pi(x-x_0)} L_4 b_4 dx_0 = g(x)D_s - D_2^{\infty} - \int_{-a}^{a} L_2 \frac{1}{\pi(x-x_0)} \begin{bmatrix} b_1 \\ b_2 \\ b_3 \end{bmatrix} dx_0, \; a \le |x| \le c_1, \quad (10)$$

Further, Eq. (10) can be converted into the domain of Gauss–Chebychev quadrature scheme as

$$\int_{-1}^{1} \frac{1}{\pi(\xi-\eta)} L_4 b_4 d\eta = g(x)D_s - D_2^{\infty} - \int_{-1}^{1} L_2 \frac{1}{\pi(\xi-\eta)} \begin{bmatrix} b_1(\eta) \\ b_2(\eta) \\ b_3(\eta) \end{bmatrix} d\eta, \; \frac{a}{c_1} \le |x| \le 1, \quad (11)$$

Hence, the discretized form for Eq. (11) is

$$\sum_{k=1}^{m} \frac{1}{m}[\frac{L_4}{(s_{r0}-s_k)}] b_4(s_k) = -t_4^*(s_{r0}) \quad (12)$$

and single-valued condition reduces to

$$\sum_{k=1}^{m} b_4(s_k) = 0 \quad (13)$$

where $b_4(\xi) = \frac{\sum_{k=1}^{m} w_k b_4(s_k)}{\sqrt{1-s_k^2}}$, $s_k = cos(\frac{(2k-1)\pi}{2m})$, $k = 1, 2, ..., m$ and $s_{r0} = cos(\frac{r\pi}{m})$, $r = 1, 2, ..., m-1$ and $t_4^*(s_{r0}) = g(c_1 s_{r0})D_s - D_2^{\infty} - \sum_{k=1}^{m} \frac{1}{m}[\frac{L_2}{(s_{r0}-s_k)}]$ $\nu^*(s_k)$. Solving Eqs. (12) and (13), $b_4(s_k)$ is evaluated at all the m quadrature points. Hence, the local intensity factors (LIFs) at the crack-tips a and $-a$ and electric displacement intensity factor at the effective crack-tips c_1 and $-c_1$ are evaluated using the following results:

$$[K_{II}^L \; K_I^L \; K_{III}^L]^T = \sqrt{\frac{\pi}{4a}} G_s[\nu_1(\pm a/c) \; \nu_2(\pm a/c) \; \nu_3(\pm a/c)]^T \quad (14)$$

and

$$[K_{II}^{c_1} \; K_I^{c_1} \; K_{III}^{c_1} \; K_{IV}^{c_1}]^T = \sqrt{\frac{\pi}{4c_1}} G_4 \, \nu_4(\pm 1), \quad (15)$$

respectively; where $L = \begin{bmatrix} G_1 & G_2^T \\ G_2 & G_4 \end{bmatrix}$ and $G_s = G_1 - (G_2^T G_2)/G_4$.

4 Evaluation of Unknown Saturated Zone Length

Since, the approach proposed by the authors is a numerical one so evaluation of saturated zone length is one of the important tasks in this paper, fixing other parameters such as crack length, loadings, etc. Moreover, it is an unknown quantity essential to know prior to getting any numerical solution. Therefore, the authors implemented the iterative approach by varying the zone length and at the same time imposing a supplementary condition of finite electric displacement at the tips of the zone or the zero electric displacement intensity factors at the outer tips of the zone. Further, the saturated zone length is varying arbitrarily from zero to a suitable finite value and correspondingly obtained the normalized electric displacement intensity factor (K^*_{IV}) at each varying tip of the zone. Hence, an interval is obtained where the K^*_{IV} changes its sign. To find the solution accurate up to two decimal places, the authors vary the arbitrary zone length by considering a small step size, i.e., 0.01 units and further applied the bisection method to get the accuracy up to three decimal places. Doing so, approximate outer saturated zone tip is evaluated where K^*_{IV} is zero and hence the saturated zone length is determined. Figure 2 validates the aforementioned approach with established results of PS model considering both impermeable and semipermeable crack conditions and subjected to $D_2^\infty/D_s = 0.5$ and $\sigma_{22}^\infty = 10\,\text{MPa}$ loading.

5 Numerical Solution for Generalized Modified PS Model

The aforementioned approach discussed in Sects. 3 and 4 is a numerical scheme and can be applied very easily to any generalized varying saturated conditions depending upon x/c_1. For example in the polynomial varying modified PS models proposed in [9], the $g(x) = (|x/c_1|)^n$ and for the same $t_4^*(s_{ro}) = g(c_1 s_{ro})D_S - D_2^\infty - \Sigma_{k=1}^m \frac{1}{m}\frac{L_2}{((s_{r0}-s_k))}v^*(s_k)$ defined in Eq. (12) can be reduced into $t_4^*(s_{ro}) = (|s_{ro}|)^n D_S - D_2^\infty - \Sigma_{k=1}^m \frac{1}{m}\frac{L_2}{((s_{r0}-s_k))}v^*(s_k)$. This is the only change required in the numerical codes for evaluation of saturated zone lengths whereas remaining procedure will be the same.

Similarly, the changes required in Eq. (12) can be made very easily for any generalized varying saturated conditions depending upon x/c_1. Hence, this numerical approach could be considered as the numerical solution for generalized PS models with variable saturation condition of the form $g(x) = f(|x/c_1|)$, where $f(|x/c_1|)$ is any arbitrary function of x/c_1 and c_1 is the extended crack length. The absolute value of x/c_1 is taken due to symmetric saturated conditions. To validate this approach, numerical solutions are obtained for all the modified PS models presented in [9] and compared with the analytical results as established. This numerical study is presented in the next section. Moreover, numerical solutions are also presented for biquadratic varying polynomial saturated conditions and compared with the other modified PS models.

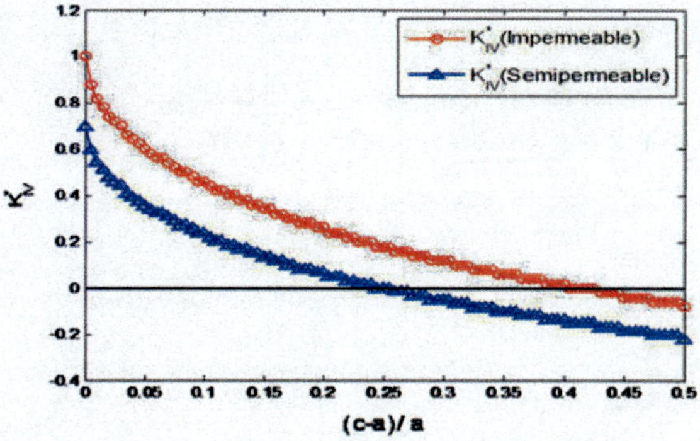

Fig. 2 Variations in K_{IV}^{*} w.r.t $(c_1 - a)/a$ for evaluation of zone length

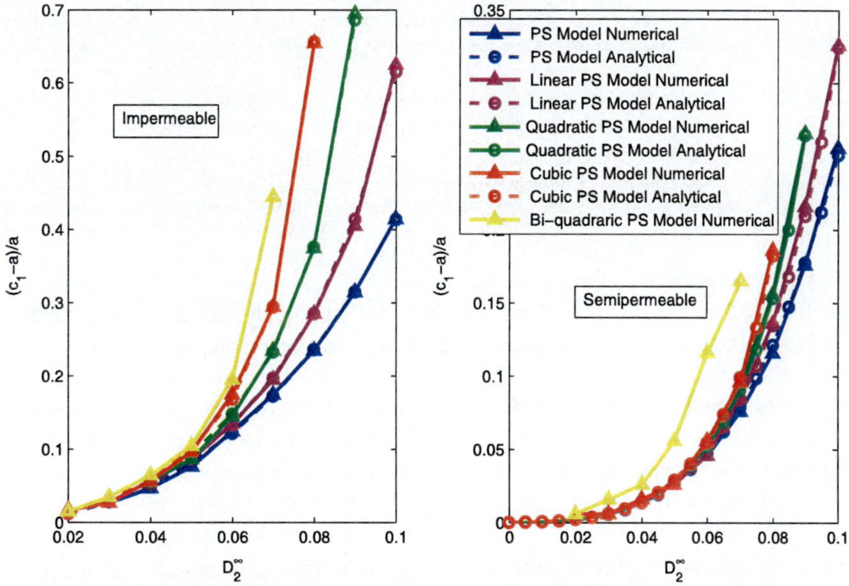

Fig. 3 Variations in normalized saturated zone length with respect to electric displacement loading for all the modified PS models

6 Numerical Studies

In this section numerical studies of PS and generalized modified PS model are presented with respect to electrical loading, polarization angle, and crack-face conditions by considering center crack problem in 2-D piezoelectric media. The study is presented for PZT-4 material with electromechanical loadings $\sigma_{22}^{\infty} = 10\,\text{MPa}$ and electric displacement loading $D_2^{\infty} = 0.08\,\text{C/m}^2$. Throughout the analysis, saturated electric displacement loading is considered equal to $D_s = 0.2\,\text{C/m}^2$.

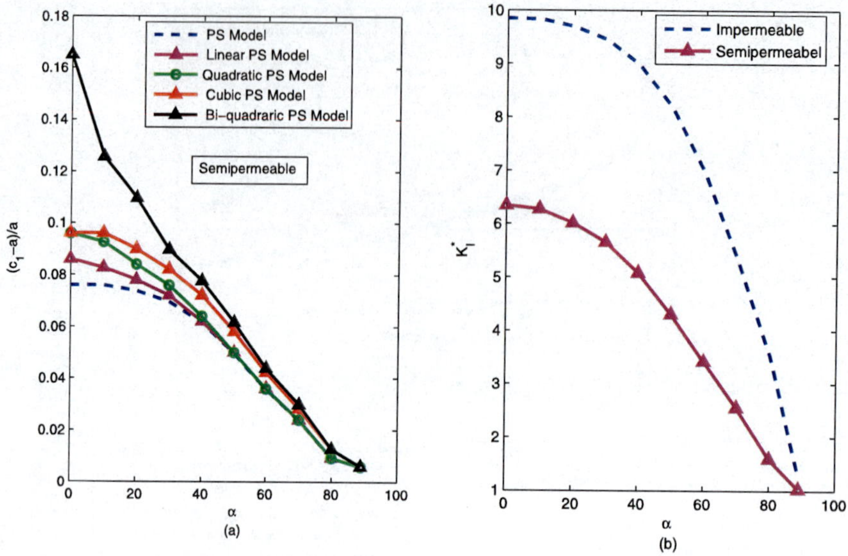

Fig. 4 **a** Variations in normalized saturated zone length with increasing polarization angle **b** variations in K_I^* with increasing polarization angle

To verify the accuracy of the numerical solution and proposed modeling, results of normalized saturated zone length are obtained for PS and polynomial varying modified PS models (with degree 2, 3, and 4) with increasing electric displacement loading under both crack-face conditions (impermeable and semipermeable) are plotted in Fig. 3. These obtained results are then compared with analytical solutions for both PS and modified PS models. A very good agreement of numerical results obtained by this approach is found with analytical results. Also, the behavior of saturated zone length validates the developed numerical approach for finding the solutions for generalized PS model with variable saturation condition of the form $g(x) = f(|x/c_1|)$, where $f(|x/c_1|)$ is any arbitrary function of x/c_1 and c_1 is the extended crack length.

Moreover, the behaviors of saturated zone length and modified polynomial varying saturated condition have been analyzed with respect to poling direction. Results of $(c_1 - a)/a$ are obtained for different polarization angles varying from 0° to 90° and are shown in Fig. 4a. Similar behaviors of $(c_1 - a)/a$ are found for all the modified PS models with increasing polarization angle, i.e., saturated zone length decreases with increasing polarization angle. Results of normalized local intensity factor $K_I^* = (K_I^L)/(\sigma_{22}\sqrt{\pi a})$ are also plotted in Fig. 4b for PS model (as K_I^L defined at the actual crack tip and is independent of the saturated condition) with increasing polarization angle and subjected to both impermeable and semipermeable crack-face conditions. The behavior of K_I^* with respect to increasing polarization angle and crack-face conditions are found to be the same as concluded in [11] and [9].

7 Summary of the Work Done

The DDT along with Fan et al. [5] approach (for separating the integral equations as per mechanical and electrical dislocation density parameters) are successfully implemented to find the numerical solution for the generalized PS model with variable saturation condition of the form $f(|x/c_1|)D_s$, where $f(|x/c_1|)$ is any arbitrary function and c_1 is the extended crack length. Numerical studies are presented for constant, linear, quadratic, cubic, and biquadratic by varying modified PS models with varying electrical loading and polarization angle. Excellent agreement of the numerical results of saturated zone length and local intensity factor is found with the analytical results established in [9]. Saturation zone length and local intensity factor show same behavior as concluded in [9] with respect to variation in electrical loading, crack-face conditions, polarization, angle and saturation condition.

8 Conclusions

The study presents the following conclusions:

1. DDT technique and the proposed approach are successfully implemented to study the generalized PS model and can also be applied to study other kinds of modified PS models.
2. The saturated zone length increases with respect to electric displacement loading for all the PS/modified models. But for a particular electrical loading, saturated zone has higher values as saturated condition varies from constant to biquadratic and this difference increases with the increase in electrical loading.
3. The effect of crack-face conditions is confirmed here as numerical values for normalized saturated zone lengths are found less in semipermeable conditions than the impermeable case.
4. Effect of saturated condition is also observed here on saturated zone length as for a particular polarization angle other than 90°, the numerical values are higher for a higher degree of polynomial varying saturated condition.
5. Both saturated zone length and LIF depend on the poling direction as they decrease with increasing polarization angle.

References

1. Bhargava, R.R., Hasan, S.: Crack opening displacement for two unequal straight cracks with coalesced plastic zones-a modified Dugdale model. Appl. Math. Model. **35**, 3788–3796 (2011)
2. Bhargava, R.R., Jangid, K.: Strip-saturation model for piezoelectric plane weakened by two collinear cracks with coalesced interior zones. Appl. Math. Model. **37**, 4093–4102 (2013)
3. Bilby, B.A., Cottrell, A.H., Swinden, K.H.: The spread of plastic yield from a notch. Proc. R. Soc. A **272**, 304–314 (1963)

4. Dugdale, D.S.: Yielding of steel sheets containing slits. J. Mech. Phys. Solids **8**, 100–104 (1960)
5. Fan, C.Y., Zhao, M.H., Zhao, Y.H.: Numerical solution of polarization saturation/dielectric breakdown model in 2D finite piezoelectric media. J. Mech. Phys. Solids **57**, 1527–1544 (2009)
6. Fan, C.Y., Zhao, Y.F., Zhao, M.H., Pan, E.: Analytical solution of a semi-permeable crack in a 2D piezoelectric medium based on the PS model. Mech. Res. Comm. **40**, 34–40 (2012)
7. Gao, H., Zhang, T.Y., Tong, P.: Local and global energy release rates for an electrically yielded crack in a piezoelectric ceramic. J. Mech. Phys. Solids **45**, 491–510 (1997)
8. Hasan, S., Akhtar, N.: Dugdale model for three equal collinear straight cracks: an analytical approach. Theor. Appl. Fract. Mech. **78**, 40–50 (2015)
9. Sharma, K., Singh, S.: Numerical studies of some modified polarization saturation models in 2-d semipermeable piezoelectric media. In: Proceedings of the International Conference on ACOME 2017, vol. 2017, pp. 79–94 (2018)
10. Sharma, K., Bui, T. Q., Bhargava, R.R., Tiantang, Y., Lei, J., Hirose, S.: Numerical studies of an array of equidistant semi-permeable inclined cracks in 2-D piezoelectric strip using distributed dislocation method. Int. J. Solids Struct. **80**, 137–145 (2016)
11. Singh, S., Sharma, K., Bhargava, R.R.: Complex variable approach in studying modified polarization saturation model in two-dimensional semipermeable piezoelectric media. Appl. Math. Mech. **38**, 1517–1532 (2017)
12. Singh, S., Sharma, K., Bhargava, R.R.: Analytical solution for two equal collinear modified strip saturated cracks in 2-D semipermeable piezoelectric media. Z. Angew. Math. Mech. **99**(9), e201800244 (2019)
13. Singh, S., Sharma, K., Bhargava, R.R.: Modified strip saturated models for two equal collinear cracks with coalesced zones in piezoelectric media. Appl. Maths. Mech. **40**(8), 1097–1118 (2019)
14. Zhong, H.C., Lee, K.Y.: Electroelastic fields induced by two collinear and energetically consistent cracks in a piezoelectric layer. J. Mech. **30**, 361–372 (2014)

Thermal Conduction in One-Dimensional Φ^4 Chains with Colliding Particles

Sankhadeep Bhattacharyya and Puneet Kumar Patra

1 Introduction

Recent advancement in science and technology has enabled the researchers to develop new-age materials such as graphene, carbon nanotubes (CNTs), nanowires, etc. In many of these materials, the motion of the particles (atoms) is severely restricted in one or more dimensions, making them examples of low-dimensional systems. Because of their interesting thermal transport properties, low-dimensional systems have generated significant interest among the researchers. Of specific importance is to identify the link between the macroscopic Fourier's law of heat conduction with its low-dimensional counterpart. For a one-dimensional system, Fourier's law may be stated as follows:

$$J = -\kappa \frac{\partial T}{\partial x}, \tag{1}$$

i.e., the heat current, J, is proportional to the temperature gradient, $\partial T/\partial x$, with the constant of proportionality given by the thermal conductivity of the system, κ. This macroscopic statement of Fourier's law is independent of the system size and length. But, situation becomes complicated when dealing with atomistic-scale low-dimensional systems. For example, CNTs exhibit length-dependent (as well as temperature) thermal conductivity [1]. Similarly, graphene has an exceptionally high thermal conductivity [2].

In a bid to explain these anomalous deviations away from normal thermal transport properties, simplified one-dimensional models have been studied extensively. Two of the most studied pedagogical models are the Fermi–Pasta–Ulam (FPU) chain [3] and the Φ^4 chain [4]. While the FPU chain exhibits anomalous thermal conduction similar to CNTs with κ following a power law with system size, $\kappa \sim L^\alpha$ with $\alpha \neq 0$, Φ^4 chains obey Fourier's law [3]. The generalized Hamiltonian, H, governing the two chains comprising of N particles can be expressed as

S. Bhattacharyya · P. Kumar Patra (✉)
Indian Institute of Technology Kharagpur, Kharagpur 721302, India
e-mail: puneet.patra@civil.iitkgp.ac.in

S. K. Saha and M. Mukherjee (eds.), *Recent Advances in Computational Mechanics and Simulations*, Lecture Notes in Mechanical Engineering,
https://doi.org/10.1007/978-981-15-8315-5_36

$$H = \sum_{i=1}^{N} \left[\frac{p_i^2}{2m} + V_H(\Delta x_{i-1,i}) + V_A(\Delta x_{i-1,i}) + U(x_i) \right], \tag{2}$$

where p_i is the momentum of the ith particle having a mass of m, V_H is the harmonic part of the potential that depends on the distance between the two nearby particles $\Delta x_{i-1,i}$, V_A is the anharmonic part of the potential that also depends on $\Delta x_{i-1,i}$, and $U(x_i)$ is the anharmonic tethering part of potential. FPU chain is obtained when $V_A(\Delta x_{i-1,i}) = \frac{1}{4} c_1 \Delta x_{i-1,i}^4$ and $U(x_i) = 0$, while the Φ^4 chain is obtained when $V_A = 0$ and $U(x_i) = \frac{1}{4} c x_i^4$.

The difference in the nature of the Hamiltonian results in FPU chain being momentum conserving and Φ^4 chain being momentum non-conserving. Across several scientific works, it has been argued in the literature that momentum conservation is a key reason for the FPU chain to display anomalous thermal conduction [5]. The argument, however, does not hold true for all one-dimensional modes as has been identified by Prosen and Campbell [6]. Over the years, several momentum non-conserving models have been proposed that show normal thermal transport properties. For example, working with a chain of coupled rotators, Giardina et al. [7] have shown that the system obeys Fourier's law. Similarly, Xiong et al. [8] have shown through non-equilibrium molecular dynamics that their one-dimensional chain with asymmetric interparticle interactions has a convergent thermal conductivity in thermodynamic limit.

Another possible origin of anomalous thermal transport in momentum-conserving one-dimensional chains has been attributed to the slow diffusion of energy carried by the long-wavelength modes [9]. These long-wavelength modes act as undamped energy transport channels, and result in long distance as well as time correlations in the system. As a result, such momentum-conserving systems possess anomalous thermal transport. Further, it has been suggested that in one-dimensional systems having tethering potential, the energy transported by the long-wavelength modes gets diffused quickly due to the tethering potential. Particularly for the Φ^4 chain, it has been argued that the localized modes having frequency greater than the linear phonon regime are responsible for normal thermal conduction. However, explanation of ballistic thermal conduction in the momentum non-conserving Φ^4 chain, under weakly nonlinear tethering potential, still remains elusive [10].

In the traditional models of FPU and Φ^4 chains, two nearby particles may cross each other. This is in contrast with realistic systems, where any two particles experience large repulsive forces when they come near each other. The presence of such repulsive forces may have a significant bearing on the thermal conductivity properties owing to their ability to create new phonon modes. Keeping this in mind, in this manuscript, we generalize the traditional Φ^4 chain so that no two individual particles can cross each other. This is achieved by addition of a soft-sphere type potential to the Hamiltonian associated with Φ^4 chain. The resulting Hamiltonian is solved numerically to answer the following questions: (i) how does the thermal conductivity and heat flux change when particles are *not allowed* to cross each other, (ii) does enabling collision has any bearing on the temperature gradient being created across the chain, and (iii) does collision alter the density profile within the chain and the diffusive characteristics of the chain.

The manuscript is organized as follows: the next section details the model proposed in this manuscript that prevents two particles from crossing over along with a brief description of the traditional Φ^4 chain, and subsequently we highlight the numerical simulation strategy adopted in this study and the results obtained from our simulations. Lastly, concluding remarks are presented that provides direction for future research.

2 Φ^4 and Φ^{4C} one-dimensional chains

Consider N particles, each of mass m, lying on a one-dimensional line. Let the equilibrium distance between any two particles be l_{eq}. As in Eq. 2, let $V_H(\Delta x_{i,j})$ and $V_A(\Delta x_{i,j})$ be the harmonic and anharmonic interparticle interactions between two particles i and j. These interactions solely depend on the relative displacement of the two particles from their equilibrium positions, $\Delta x_{i,j} = x_i - x_j - (j-i)l_{eq}$. Further, let each particle interact only with its nearest neighbor, i.e., $V(\Delta x_{i,j}) \rightarrow V(\Delta x_{i-1,i})$. In presence of an onsite tethering potential, $U(x_i)$, the Hamiltonian is given by

$$H = \sum_{i=1}^{N} \left[\frac{p_i^2}{2m} + \sum_{i=1}^{N-1} \frac{k}{2} \left(x_i - x_{i-1} - l_{eq}\right)^2 + V_A(\Delta x_{i-1,i}) + U(x_i) \right]. \tag{3}$$

Choosing $k = 1.0$, the traditional Φ^4 chain is obtained from Eq. 3 by substituting $V_A(.) = 0$ and $U(x_i) = c\frac{(x_i - x_{i,0})^4}{4}$. Thus, the Hamiltonian of Φ^4 chain, with $c = 0.1$, is given by

$$\begin{aligned} H_{\Phi^4} = &\sum_{i=1}^{N} \left[\frac{p_i^2}{2m} \right] + \sum_{i=1}^{N-1} \left[\frac{1}{2} \left(x_i - x_{i-1} - l_{eq}\right)^2 \right] \\ &+ \sum_{i=1}^{N} \left[\frac{0.1}{4} \left(x_i - x_{i,0}\right)^4 \right]. \end{aligned} \tag{4}$$

The reason for choosing $c = 0.1$ in Eq. 4 is to ensure that the anharmonic energy contribution is a fraction of harmonic energy for the majority of simulation time. In the limit of large anharmonic contributions ($c = 1$), thermal transport characteristics of ϕ^4 chain have been extensively studied by Hu et al. [5], and Aoki and Kusnezov [11] using deterministic thermostats. Hu et. al argued that as the momentum conservation breaks down due to the tethering potential, thermal conduction follows Fourier's Law where $J \sim \frac{1}{N}$ and dissipation of momentum decays exponentially in time. Using large-scale simulations, it was found that the thermal conductivity depends on temperature according to $\kappa = 2.724/T^{1.382}$ [12]. Patra and Bhattacharya [13] employed a Φ^4 chain to study thermal rectification and differential thermal conduction. Although researchers have extended the Φ^4 model to more than one dimensions [12], we focus only on one-dimensional chains in the present study. However, the existing Φ^4 model does not constrain two particles from crossing each other. In the present work, we modify H_{Φ^4} so that two particles are prevented from crossing each other.

The proposed Φ^{4C} Hamiltonian contains an extra soft-sphere collision term $V_C(.) = a\frac{1}{(x_i - x_{i-1})^6}$ which ensures that two particles, upon coming very close to each other, experience a large repulsive force. The choice of the constant a is governed by two factors: (i) the effective radius of the particles, r, so that when the distance between the two particles is less than $2r$, a large repulsive force is experienced by the particles and (ii) the contribution of $V_C \to 0$ when the distance between the two particles is greater than $2r$. In the present work, we choose $a = 5 \times 10^{-10}$ corresponding to an effective radius of $r = 0.025$. The resulting Hamiltonian is

$$
\begin{aligned}
H_{\Phi^{4C}} = & \sum_{i=1}^{N}\left[\frac{p_i^2}{2m}\right] + \sum_{i=1}^{N-1}\left[\frac{1}{2}\left(x_i - x_{i-1} - l_{eq}\right)^2\right] \\
& + \sum_{i=1}^{N-1}\left[\frac{5\times 10^{-10}}{(x_i - x_{i-1})^6}\right] \\
& + \sum_{i=1}^{N}\left[\frac{0.1}{4}\left(x_i - x_{i,0}\right)^4\right].
\end{aligned} \tag{5}
$$

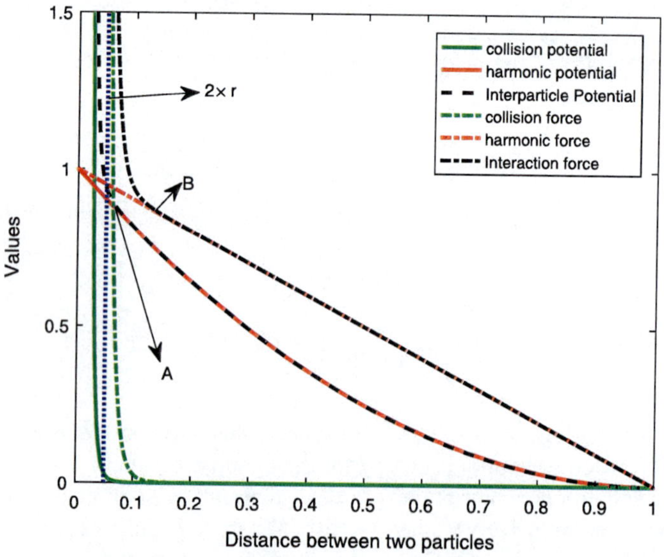

Fig. 1 Different potential and force functions for a particle located at origin assuming a neighbor particle at +1 units. Dotted line (parallel to y axis) shows $2r$ beyond which the soft-sphere potential becomes insignificant. A:(0.065,0.88) is where net potential deviates from harmonic, B:(0.135,0.868) is where net force deviates from the harmonic force

Figure 1 depicts the soft-sphere potential and its corresponding force vis-á-vis the harmonic potential and harmonic force. Beyond $2r$, the interparticle potential and force is dominated by the harmonic potential and the corresponding force, respectively.

3 Simulation Methodology

We now describe the simulation methodology adopted in the study. The particles of both Φ^4 and Φ^{4C} chains are initialized at their equilibrium positions such that $l_{eq} = 1$. Therefore, the initial position of ith particle is $x_{i,o} = i - 1$. So, in a chain comprising of N particles, the initial coordinates range from $0, 1, \ldots .N - 1$. Initial velocities of the particles are randomly sampled from a uniform distribution between ± 0.5. Fixed-fixed boundary conditions have been implemented in the chains by means of fictitious fixed particles present at the coordinates -1 and N. The first (last) boundary particle interacts with only the first (last) particle of the chain through $V_H(.) + V_A(.)$ whose mathematical forms are given in Eqs. 4 and 5. A pictorial representation of the chain is depicted in Fig. 2.

We create a thermal gradient in the chains by controlling the temperature of the first and the last particles. While the first particle is kept in contact with a hot heat reservoir (at temperature T_H), the last particle is kept in contact with a cold heat reservoir (at temperature T_C). Among the different deterministic thermostats [14–17], we choose Nosé–Hoover thermostat [18–20] for temperature control due to its simplicity and wide adoption. The Nosé–Hoover thermostat's mass for both the hot and the cold thermostats is taken as unity. Particles present in between the first and the last particles are governed by the standard Hamiltonian evolution equations. Thus, the resulting equations of motions for all the particles are given by

$$\begin{aligned}
\dot{q}_i &= \frac{\partial H}{\partial p_i} \\
\dot{p}_i &= -\frac{\partial H}{\partial q_i} - \delta_1 \zeta_H p_1 - \delta_N \zeta_C p_N \\
\dot{\zeta}_H &= \frac{p_1^2}{T_H} - 1 \\
\dot{\zeta}_C &= \frac{p_N^2}{T_C} - 1 \\
\text{where} &: \\
\delta_1 &= 1\{i = 1\} \\
\delta_1 &= 0\{i \neq 1\} \\
\delta_N &= 1\{i = N\} \\
\delta_N &= 0\{i \neq N\}.
\end{aligned} \tag{6}$$

These equations of motion are solved using fourth-order Runge–Kutta method with incremental time step equaling $h = 0.0005$. Each simulation run comprises two parts: (i) first 250 million steady-state runs: equations of motion are solved for 250 million time steps under the prescribed boundary conditions so that chains reach steady-state conditions and (ii) last 750 million output runs: in order to compute all time averages

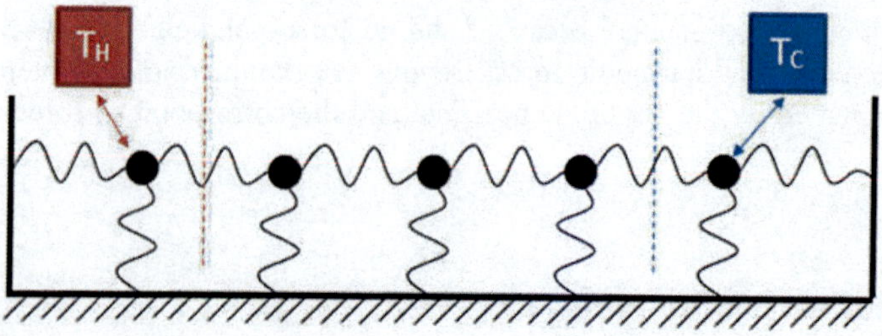

Fig. 2 Pictorial depiction of Φ^4 and Φ^{4C} chains with boundary conditions

equations of motion are solved for a further 750 million time steps. To understand the effect of chain length, simulations are run with different values of N: 16, 32, 64, 128, 256, 384, and 512. For each case, three different temperature values are imposed on the boundary particles: $(T_H, T_C) = (1.1, 0.9)$; $(0.55, 0.45)$ and $(0.11, 0.09)$ such that the mean temperatures, $T_M = 0.5(T_H + T_C)$, are 1.0, 0.50, and 0.10, respectively.

3.1 *Heat Flux Computation*

Let us begin with a hypothetical case where a chain comprising of $N \to \infty$ particles is brought away from equilibrium through a thermal gradient. In such a case, the chain becomes a continuous entity, and the different thermodynamic quantities may be defined on a spatial point x. With local thermodynamic equilibrium [15] conditions prevailing within the system, it is possible to define an instantaneous local temperature field $T(x, t)$ that varies slowly in x and time, t. In a similar way, the local heat current density $J(x, t)$ may be defined. For the chains under consideration, where N is finite, we make an assumption that the temperature gradient is small enough so that the chain is near-equilibrium and local thermodynamic equilibrium hypothesis holds true. In such a scenario, we can now define locally both temperature and heat currents for a particle. From among the different definitions of temperature [21, 22], we restrict ourselves to kinetic temperature, which can locally be expressed as

$$k_B T_i = m v_i^2, \tag{7}$$

where k_B is the Boltzmann constant and taken as unity for the remainder of this study, and T_i is the kinetic temperature of the ith particle. Local heat current may be obtained by taking the time derivative of the local energy density associated with the ith particle [23]:

$$\begin{aligned} \dot{\epsilon}_i = & \tfrac{p_i}{m_i}\dot{p}_i - f_i v_i \\ & + \tfrac{1}{2}\left[f_{i,i-1}(v_i - 1 + v_i) - f_{i,i+1}(v_i + v_i + 1)\right], \end{aligned} \tag{8}$$

which can be written as

$$\dot{\epsilon}_i = \frac{\partial \epsilon_i}{\partial t} + \left[j_{i,i-1} - j_{i+1,i}\right], \tag{9}$$

where $j_{i,j}$ is the energy current flowing from particle j to particle i. When steady state is reached [23], $\langle \dot{\epsilon}_i \rangle = 0$, $\langle \frac{\partial \epsilon_i}{\partial t} \rangle = 0$ along with $\langle \frac{dV(x_i-1-x_i)}{dt} \rangle = 0$, so that

$$\langle j_{i,i-1} \rangle = \langle \tfrac{1}{2}(v_i + v_{i-1}) f_{i,i-1} \rangle = \langle v_i f_{i,i-1} \rangle. \tag{10}$$

This gives $\langle j_{i,i-1} \rangle = \langle j_{i+1,i} \rangle = \langle j(x,t) \rangle = J$. So, the net heat flux, J, and its time-averaged value, $\langle J \rangle$, may be computed as

$$J = \frac{\sum_{i=1}^{N} j_{i,i-1}}{N} \Longrightarrow \langle J \rangle = \left\langle \frac{\sum_{i=1}^{N} j_{i,i-1}}{N} \right\rangle. \tag{11}$$

Thermal conductivity may now be calculated through

$$\kappa = \frac{\langle J \rangle N}{\Delta T}. \tag{12}$$

For large N, systems with normal thermal conductivity gives finite κ, while systems with abnormal thermal conductivity gives

$$\kappa \sim N^{\alpha}, \alpha \neq 0. \tag{13}$$

4 Results and Discussions

4.1 Verification and Check for Steady-State Conditions

Before proceeding further, the simulation code is verified. For this purpose, a Φ^4 chain, having $k = c = 1$ in Eq. 4 and $N = 512$ particles, has been considered. The simulation methodology remains the same as discussed previously. Three values of T_M have been considered: $T_M = 1.0, 0.5$, and 0.1. The higher temperature, $T_H = T_M + 0.1 \times T_M$, while the lower temperature, $T_C = T_M - 0.1 \times T_M$, so that temperature difference, $\Delta T = 0.2 \times T_M$. Thermal conductivity values, κ, obtained from the simulation are compared with $\kappa_{th} = 2.724/T^{1.382}$ [12]. The values are compared in Table 1. Simulation results are comparable with those reported earlier at higher temperatures. At lower temperature, a difference may be observed and can be attributed to the different boundary conditions used for computing κ_{th}.

We now check if steady-state conditions prevail within the simulated chains. The check is based on the argument that $\langle j_{1,2} \rangle = \cdots = \langle j_{i,i-1} \rangle = \langle j_{i+1,i} \rangle = \cdots = \langle j_{N-1,N} \rangle$. In steady-state conditions, the net heat current, J, must equal local heat

Table 1 A Φ^4 chain with $k = 1$ and $c = 1$ is subjected to thermal conduction as per the methodology highlighted before. Our simulation results for thermal conductivity (κ) are compared with those obtained by Aoki and Kusnezov [12]: $\kappa_{th} = 2.724/T^{1.382}$. The difference in the results arises because of different boundary conditions adopted for finding κ_{th}

T	$\langle J \rangle$	$J \times N$	κ	κ_{th}
1	0.0011	0.570	2.849	2.724
0.5	0.0013	0.647	6.470	7.100
0.1	0.0023	1.574	57.870	65.646

current flowing between any two adjacent particles. A significant deviation of local heat current between any two adjacent particle pair indicates that the system has not yet attained steady state. The check is performed for Φ^4 chains with $N = 512$ particles and stiffness parameters as described in the previous section. Results in the format $(J, |J - \min\langle j_{i-1,i}\rangle|, |J - \max\langle j_{i-1,i}\rangle|)$ for $T_M = 1.0, 0.5$, and 0.1 are $(0.0221, 4.71 \times 10^{-5}, 5.88 \times 10^{-5})$, $(0.0237, 3.383 \times 10^{-5}, 2.366 \times 10^{-5})$, $(0.009955, 1.4206 \times 10^{-5}, 7.7032 \times 10^{-6})$, respectively. The small deviation of local heat currents from the total heat current indicates that the chain has steady-state conditions after 750 million time steps, and meaningful time averages may be computed.

4.2 *Temperature Profile*

Under an imposed temperature gradient, assuming local thermodynamic equilibrium, each particle of a chain has a well-defined kinetic temperature. Since three different T_M values have been considered, for each particle i, a scaled kinetic temperature, $T_{S,i}$, is defined:

$$T_{S,i} = 1.0 + 0.2 \times \frac{T_i - T_M}{T_H - T_C}, \tag{14}$$

so that the temperature of a particle always varies between 1.1 and 0.9 and a meaningful comparison can be made. The scaled temperature profile of Φ^4 and Φ^{4C} chains with $N = 512$ is shown in Fig. 3. A typical Φ^4 chain exhibits boundary jumps in temperature profile [24] which becomes more prominent with decreasing T_M. Introduction of soft-sphere collisions drastically reduces the boundary temperature jumps in the Φ^{4C} chain. A stark contrast can be seen for the two chains especially at lower values of T_M. It must be noted that the exact reason for boundary jumps in Φ^4 chains is still open to research.

Researchers have argued that in between the temperature jumps, the temperature profile of a Φ^4 chain varies linearly. Neglecting the temperature jumps, we now calculate the deviation of the profiles from a linear behavior (the green straight line of Fig. 3), using the following distance measure:

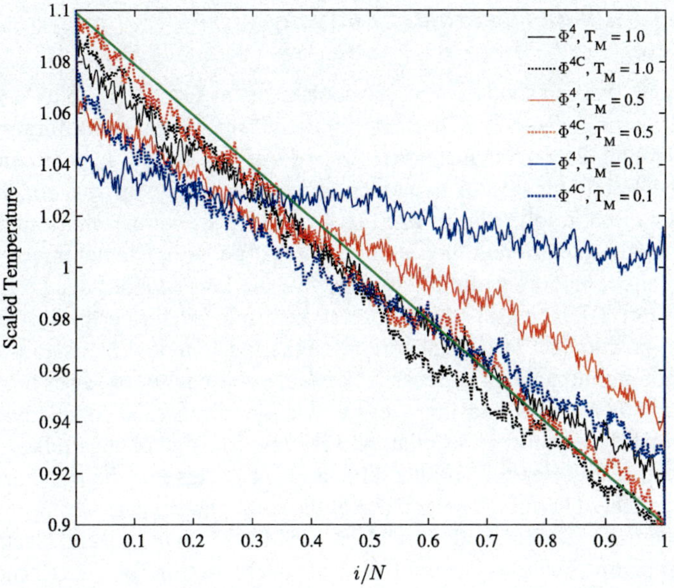

Fig. 3 Scaled temperature profile of Φ^4 and Φ^{4C} chains with $N = 512$ subjected to three different T_M. Scaling is done so that $T_H = 1.1$ and $T_C = 0.9$. T_i values of the remaining particles are interpolated. The green straight line indicates a perfectly linear plot. As can be seen from the figure, deviation from a linear profile decreases for Φ^4 chain as T_M increases. For Φ^{4C} chains, the deviation from linearity is comparatively lesser. Boundary jumps in temperature are more predominant for Φ^4 chains than in Φ^{4C} chains

Table 2 A comparison of deviation from linearity, d_L, for $N = 512$ particles and three values of T_M. As is evident, the deviation from linearity is minimum for Φ^{4C} chain

T_M	Φ^4	Φ^{4C}
1	0.279	0.2568
0.5	0.603	0.1585
0.1	1.175	0.4876

$$d_L = \sum_{i=1}^{N} \sqrt{\left(T_{S,i} - Y_i\right)^2}. \tag{15}$$

Table 2 shows the results of d_L. The results indicate that the deviation from linearity decreases when soft-sphere collisions are introduced in the chain. The reduction in boundary jumps and deviation from linearity suggest that the Φ^{4C} chain allows for quicker thermalization and mimics macroscopic behavior better than the standard Φ^4 chain at lower temperatures.

4.3 Heat Flux and Thermal Conductivity

Figure 4 shows thermal conductivity, κ, calculated using Eq. 12, for both Φ^4 and Φ^{4C} chains with varying N and T_M. Thermal conductivity of Φ^4 chains conforms with the existing literature [25]—with increasing T_M, κ decreases. At low T_M, harmonic interparticle potential dominates the dynamics, and consequently, thermal conduction is near ballistic (see the solid blue line of Fig. 4). In such scenarios, where anharmonic effects are small, the different modes of vibrations (phonons) do not interact significantly, and the temporal evolution of energy of the lower modes occurs relatively unimpeded. As T_M increases, the particles get displaced further away from their equilibrium positions, and the anharmonic tethering potential starts to dominate. Phonon–phonon interactions increase, as a result, and energy of lower modes gets channeled to other modes, causing a severe reduction in thermal conductivity.

The case of Φ^{4C} chains is peculiar. At very low T_M, one obtains relatively high thermal conductivity (see blue dashed line in Fig. 4). This may be attributed to the dominating effect of harmonic potential over the soft-sphere and tethering potentials. However, there is a sudden trend reversal, as is evident from the red and black dashed lines corresponding to $T_M = 0.5$ and 1.0, respectively—κ for $T_M = 1.0$ is more than that of $T_M = 0.5$. Thermal conductivity further increases as T_M is increased from 1.0 to 2.0. This trend reversal—higher thermal conductivity at high temperatures—is typically absent in most one-dimensional chains, and signifies the importance of *not* allowing the particles to cross each other. The exact reasoning behind such trend reversal requires further research, and at this moment we can only offer a

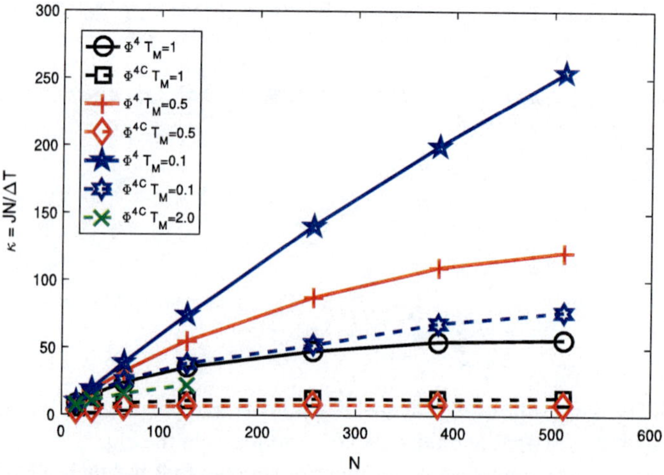

Fig. 4 Variation of thermal conductivity, κ, with N and T_M for Φ^4 and Φ^{4C} chains. While in the Φ^4 chain, κ increases with decreasing temperature owing to harmonic effects being predominant, the behavior of Φ^{4C} chain is not straightforward. It first decreases with temperature and then suddenly increases

conjecture—at higher temperatures, more particles collide, and as a result, energy transfer between two particles increases, causing increased thermal conductivity. It is evident from Fig. 4 that thermal conductivity converges much quickly with N for Φ^{4C} chains vis-á-vis Φ^4 chain, so only a fraction of Φ^{4C} particles is necessary to study the limiting behavior.

5 Conclusions

This manuscript generalizes the Φ^4 model such that no two adjacent particles cross each other. In simpler terms, the Φ^4 particles can now "collide" and repel each other. This is achieved by adding an anharmonic soft-sphere potential to the standard Φ^4 Hamiltonian. The effect of *not* allowing the particles to cross each other is investigated through molecular dynamics. Specifically, we focus on temperature profile and thermal transport properties of the chain.

It is observed that the addition of soft-sphere potential significantly alters the temperature profile while satisfying Fourier's law without significantly altering the diffusion characteristics. The boundary jumps present in Φ^4 chains become negligible in Φ^{4C} chains, with the difference being more pronounced at low temperatures. The results suggest that the Φ^{4C} model may be used to study multiscaling behavior with ease at lower temperatures. Being closer to a realistic system, where two particles *cannot* cross each other, an exhaustive study is in order to understand the behavior of Φ^{4C} model in depth.

Perhaps, the most interesting results arise for thermal conductivity. Like Φ^4 chain, at very low temperatures, one observes large thermal conductivity for Φ^{4C} chain as well. However, unlike Φ^4 chain, thermal conductivity does not *always* increase with decreasing temperature. We suspect that heat flux due to the interplay between soft-sphere and harmonic potentials determines the magnitude of thermal conductivity—with increasing temperature there is a marginal rise in thermal conductivity—possibly because of increased heat flux due to more frequent "collisions" between the particles at higher temperature. Again, an in-depth study is required to ascertain the veracity of our explanation. One can split the contributions of heat flux due to the different potentials along with monitoring the frequency of collisions to understand the interplay. Alternatively, one can look at the Fourier space, and continuously monitor the modal energies to identify if there is more interaction between the phonons (phonon–phonon interaction) resulting in peculiar behavior of thermal conductivity. As of now, the problem remains wide open to research community.

Lastly, we would like to highlight that the numerical solution to the equations of motion relies upon Taylor series expansion. With the inclusion of soft-sphere potential, the nonlinear behavior starts to dominate when two particles come very close, and this requires a very small time step for solution. It is preferable to use adaptive time integration methods for this problem.

Acknowledgements Support for the research provided in part by Indian Institute of Technology Kharagpur under grant DNI is gratefully acknowledged. The expanded version of this manuscript has been submitted for publication in a journal.

References

1. Fujii, M., Zhang, X., Xie, H., Ago, H., Takahashi, K., Ikuta, T., et al.: Measuring the thermal conductivity of a single carbon nanotube. Physical review letters. **95**(6), 065502 (2005)
2. Balandin, A.A., Ghosh, S., Bao, W., Calizo, I., Teweldebrhan, D., Miao, F., et al.: Superior thermal conductivity of single-layer graphene. Nano letters. **8**(3), 902–907 (2008)
3. Lepri, S., Livi, R., Politi, A.: Heat conduction in chains of nonlinear oscillators. Physical review letters. **78**(10), 1896 (1997)
4. Chen, D., Aubry, S., Tsironis, G.: Breather mobility in discrete φ 4 nonlinear lattices. Physical review letters. **77**(23), 4776 (1996)
5. Hu, B., Li, B., Zhao, H.: Heat conduction in one-dimensional nonintegrable systems. Physical Review E. **61**(4), 3828 (2000)
6. Prosen, T., Campbell, D.K.: Normal and anomalous heat transport in one-dimensional classical lattices. Chaos: An Interdisc. J. Nonlinear Sci. **15**(1), 015117 (2005)
7. Benettin, G., Galgani, L., Giorgilli, A.: Numerical investigations on a chain of weakly coupled rotators in the light of classical perturbation theory. Il Nuovo Cimento B (1971–1996). **89**(2), 103–119 (1985)
8. Wang, L., Hu, B., Li, B., et al.: Validity of Fourier's law in one-dimensional momentum-conserving lattices with asymmetric interparticle interactions. Physical Review E. **88**(5), 052112 (2013)
9. Lepri S, Livi R, Politi A. On the anomalous thermal conductivity of one-dimensional lattices. EPL (Europhysics Letters). 1998;43(3):271
10. Tuo, M., Wang, L., Amer, M.R., Yu, X., Cronin, S.B., Xin, H., Microwave properties of suspended single-walled carbon nanotubes with a field-effect transistor configuration. In: IEEE MTT-S International Microwave Symposium. IEEE, pp 1–4 (2011)
11. Aoki, K., Kusnezov, D.: Bulk properties of anharmonic chains in strong thermal gradients: non-equilibrium φ4 theory. Phys. Lett. A. **265**(4), 250–256 (2000)
12. Aoki, K., Kusnezov, D.: Non-equilibrium steady states and transport in the classical lattice φ4 theory. Physics Letters B. **477**(1–3), 348–354 (2000)
13. Patra, P.K., Bhattacharya, B.: Heat pump without particle transport or external work on the medium achieved by differential thermostatting of the phase space. Phys. Rev. E. **93**, 033308 (2016). https://link.aps.org/doi/10.1103/PhysRevE.93.033308
14. Hoover, W.G., Hoover, C.G., Sprott, J.C.: Nonequilibrium systems: Hard disks and harmonic oscillators near and far from equilibrium. Molecular Simulation. **42**(16), 1300–1316 (2016)
15. Patra, P., Bhattacharya, B.: A deterministic thermostat for controlling temperature using all degrees of freedom. The Journal of chemical physics. **140**(6), 064106 (2014)
16. Patra, P.K., Sprott, J.C., Hoover, W.G., Hoover, C.G.: Deterministic time-reversible thermostats: chaos, ergodicity, and the zeroth law of thermodynamics. Molecular Physics. **113**(17–18), 2863–2872 (2015)
17. Braga, C., Travis, K.P.: A configurational temperature Nosé-Hoover thermostat. The Journal of chemical physics. **123**(13), 134101 (2005)
18. Nosé, S.: A unified formulation of the constant temperature molecular dynamics methods. The Journal of chemical physics. **81**(1), 511–519 (1984)
19. Hoover, W.G.: Canonical dynamics: Equilibrium phase-space distributions. Physical review A. **31**(3), 1695 (1985)
20. Hoover, W.G., Aoki, K., Hoover, C.G., De Groot, S.V.: Time-reversible deterministic thermostats. Physica D: Nonlinear Phenomena. **187**(1–4), 253–267 (2004)

21. Casas-Vázquez, J., Jou, D.: Temperature in non-equilibrium states: a review of open problems and current proposals. Reports on Progress in Physics. **66**(11), 1937 (2003)
22. Patra, P.K., Batra, R.C.: Nonequilibrium temperature measurement in a thermal conduction process. Physical Review E. **95**(1), 013302 (2017)
23. Dhar, A.: Heat transport in low-dimensional systems. Advances in Physics. **57**(5), 457–537 (2008)
24. Aoki, K., Kusnezov, D.: Fermi-Pasta-Ulam β model: boundary jumps, fourier's law, and scaling. Phys. Rev Lett. **86**(18), 4029 (2001)
25. Hu, B., Li, B., Zhao, H.: Heat conduction in one-dimensional nonintegrable systems. Phys. Rev. E. **61**, 3828–3831. https://link.aps.org/doi/10.1103/PhysRevE.61.3828

Sequence of Hydrodynamic Phenomena During the Interactions of Drop and Bubble in Vertical Conduit

Subhav Chauhan and Parmod Kumar

1 Introduction

Bubbles and drops comprise the heart of multi-phase flows. They have fundamental importance in many natural processes as well as in industrial processes [1] which involve multi-fluid interactions. Natural phenomena like rainfall, fermentation, oxygen dissolution in lakes due to rain, electrification of atmosphere by sea bubbles, clouds floating in air, bubbles formed by active galactic nuclei [2], and industrial processes like spray drying, liquid–liquid extraction, boiling molten glass globules in air at a glass marble manufacturing industry, air in molten glass in glass-blowing factories, fuel droplets from a fuel injector, and so on [1] involve particle, drop, and bubble motions/interactions.

Commonly, bubbles and drops are defined as blobs of fluid in another continuous phase. However, mathematically, as mentioned by Tripathi [2], they can be defined based on relative density ratio (ρ_r), which is defined as the ratio of the density of the fluid blob to the outer fluid. If $\rho_r < 1$ then the blob is defined as a bubble; otherwise, if $\rho_r > 1$ then it is termed as a drop. Another important term is 'particle', which can be defined as a self-contained body with dimensions between 0.5 μm and 10 cm [2] and is separated from the surrounding medium by a perceivable interface [1]. The material of the fluid surrounding the particle is called 'continuous phase' and the material that forms the particle is called 'dispersed phase' [1]. The particle is termed as 'solid particle' if the dispersed phase is a solid matter. Similarly, if the dispersed phase is fluid, the particle is termed as 'fluid particle'.

S. Chauhan · P. Kumar (✉)
School of Engineering, Indian Institute of Technology Mandi, Kamand 175075, H.P., India
e-mail: parmod@iitmandi.ac.in

S. K. Saha and M. Mukherjee (eds.), *Recent Advances in Computational Mechanics and Simulations*, Lecture Notes in Mechanical Engineering,
https://doi.org/10.1007/978-981-15-8315-5_37

The study of particles' behavior in a fluid is active in research for a long time. There are several studies on kinematics [3–5] and dynamics [2, 6] of these particles in the fluid. There are also frequent interactions among these particles which significantly affect the flow dynamics. Their behavior is governed by fluid properties like viscosity, surface tension and density, hence, is often characterized by the dimensionless numbers like Galilei number($Ga = \rho R\sqrt{gR}/\mu$), Weber number $\left(We = \rho U^2 R/\sigma\right)$, Eotvos number $\left(Eo = gR^2\Delta\rho/\sigma\right)$, and so on, where ρ, μ, σ, g, U, and R denote density, dynamic viscosity, surface tension, acceleration due to gravity, velocity, and radius, respectively, of the fluid particles. Many studies are available in the literature on homogeneous interactions like bubble–bubble [7] and drop–drop [8] dynamics as well as of heterogeneous interactions such as bubble–solid particle [9] and drop–solid particle [10] associated phenomenon. However, interactions of drop and bubble are rarely studied in the literature. In recent years, researchers [11, 12] have studied the bubble–drop interactions for the application of water treatment in the petroleum industry. The small oil particles suspended in the water are removed with the swarm of bubbles. The rising bubble coalesces with the suspended oil such that the oil spread on the periphery of the bubble. The combined bubble and oil then rise to the surface of the water where the air escapes and the oil get deposited in the form of film over the water surface. This process is known as air flotation process [11]. Chakibi et al. [11] have studied the effect of water salinity and bubble size on the efficiency of the flotation process. Dudek and Oye [12] have also shown the spreading of oil droplet across the periphery of the air bubble in micro-fluidic arrangements. It is to be mentioned that instead of its enormous occurrences, the interactions of bubble and drop have not been studied much. Only the recent studies [11, 12] have reported such interactions at micro-scale. Therefore, considering the lacuna in the literature, efforts have been made in the present study to numerically investigate the interactions between drop and bubble in a vertical conduit using the volume of fluid framework.

2 Numerical Methodology

2.1 VOF Approach Using OpenFOAM

The three-dimensional simulations are carried out using the Eulerian approach-based volume of fluid (VOF) framework. The fraction of volume occupied by a particular fluid (primary fluid) in a discretized cell is denoted by volume fraction, α. In the cells at the interface of fluids, this can take values between 0 and 1. Based upon these values the interface between fluids is constructed.

The fluid flow is governed by the continuity and momentum equations, as stated below [13]:

$$\nabla.\boldsymbol{U} = 0, \tag{1}$$

$$\rho_b\left[\frac{\partial \boldsymbol{U}}{\partial t} + \nabla.(\boldsymbol{U}\boldsymbol{U})\right] = -\nabla p + \nabla.\boldsymbol{T} + \boldsymbol{F}_b + \boldsymbol{F}_s, \tag{2}$$

where $\boldsymbol{T} = \mu_b(\nabla \boldsymbol{U} + (\nabla \boldsymbol{U})\prime)$, $\boldsymbol{U}$ and p denote the velocity and pressure fields of the fluid; $\boldsymbol{F}_b$ and $\boldsymbol{F}_s$ denote the body force and surface force, respectively. The fluid properties, viz., bulk density (ρ_b) and bulk viscosity (μ_b), for an n-phase system, are calculated as follows [14]:

$$\mu_b = \sum \alpha_i \mu_i \tag{3}$$

$$\rho_b = \sum \alpha_i \rho_i \tag{4}$$

where subscript i denoted ith fluid. Further volume of fluid framework requires an additional advection equation for spatio-temporal evolution of volume fraction ($\alpha_{i\text{-}}$) as [14]:

$$\frac{\partial \alpha_i}{\partial t} + \nabla.(\alpha_i \boldsymbol{U}) = 0 \tag{5}$$

where

$$\sum \alpha_i = 1 \tag{6}$$

The aforementioned Eqs. (1–6) are discretized using the finite-volume method [13]. The discretized sets of equations are solved in the 'multiphaseInterFoam' solver of the 'OpenFOAM' package. The PIMPLE algorithm is employed for the pressure–velocity coupling along with adaptive time-stepping based on Courant Friedrichs Lewy (CFL) condition.

2.2 Simulation Domain

A 100 mm long cylindrical domain (Fig. 1) having a diameter of 20 mm is used for all the numerical simulations. The structured Cartesian mesh is used to discretize the region around the axis of the domain, while the peripheral region is discretized using the structured polar mesh as shown in Fig. 2. The no-slip condition has been specified for all the faces of the domain. The water, air, and kerosene are used as working fluids. A Taylor water drop is initialized at the top and a similar air bubble is initialized at the bottom of the domain with kerosene as surrounding fluid. The initial topology of the phases is depicted in Fig. 1. An initial velocity of 0.05 m/s is also assigned to both drop and bubble. This is provided to make the fluid particles in motion, which helps in achieving the terminal velocity of drop and bubble in less

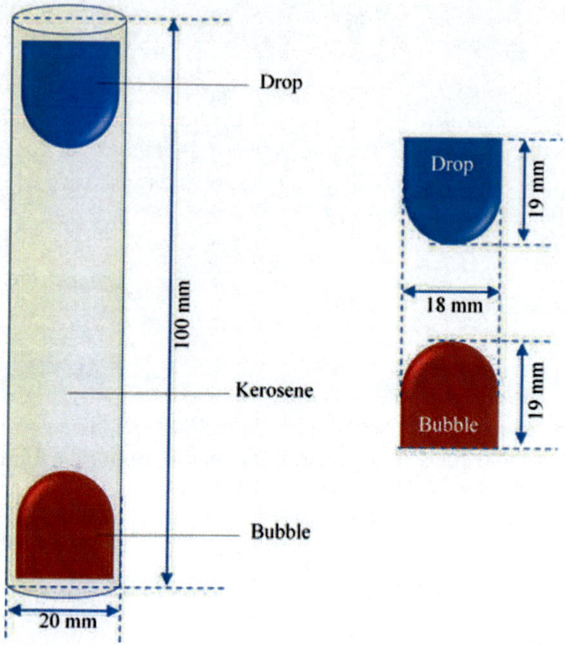

Fig. 1 Schematic representation of the simulation domain, depicting the initial phase configurations

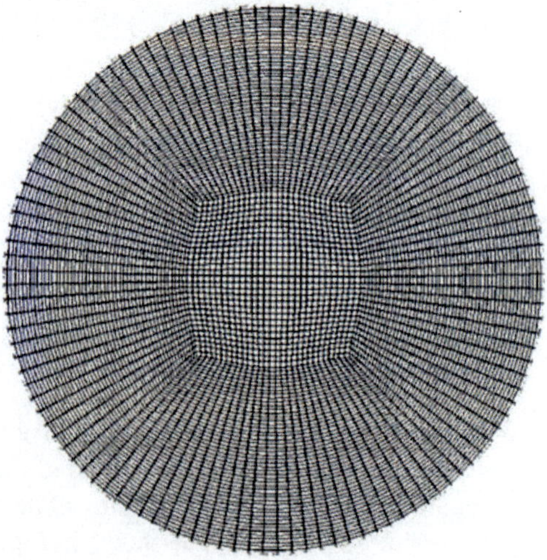

Fig. 2 A cross-sectional view of mesh, perpendicular to the longitudinal axis of the conduit

time. It further results in minimizing the required length of the domain to investigate the interactions in the developed flow region and hence the computational time.

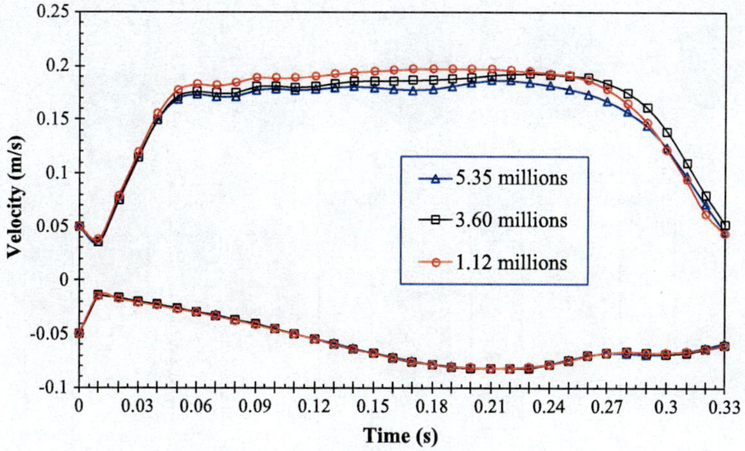

Fig. 3 Velocity vs time plot for bubble and drop at different grid sizes. Curves with negative velocities show the velocity of the drop while those with positive velocities are for the bubble. The legends show the number of hexahedra used to discretize the domain

2.3 Grid Sensitivity Analysis

The grid sensitivity test is done using three grids of the same type. The velocity of the center of mass of each fluid particle is plotted as a function of time to demonstrate the grid sensitivity, as shown in Fig. 3. The complete domain is discretized using hexahedron elements only. The triangular markers are used for a mesh with 5.35 million hexahedra. However, the square and circular markers are used for comparatively coarse meshes. In the case of a drop, all three curves superimpose, indicating that even the coarse mesh with 1.12 million hexahedra can be used to demonstrate the behavior of the drop. In the case of bubble, there is a slight variation in the velocities among all the meshes. However, error in the velocity for the two finer meshes is not much as evident from the figure. Therefore, in the present study, the simulations are performed using the mesh with 3.6 million hexahedra.

3 Results and Discussions

3.1 Pre-collision Motion

The motion of both the bubble and the drop is symmetrical before the collision. There is a formation of a thin film of kerosene oil between the cylinder wall and periphery of both the fluid particles. As the bubble moves upward, the thin film of oil glides downward, maintaining the conservation of mass. Similarly, the oil film between the drop and wall glides upward as the drop moves downward. The breakup of small

Fig. 4 The motion of bubble and drop prior to the collision. The front cross-sectional view along the axis of the tube is shown with a three-dimensional view of the phenomenon at each time step. The water drop is shown in blue while the bubble is shown in red

satellite bubbles from the tail of the Taylor bubble can also be seen from Fig. 4. However, no such breakup is observed in case of a drop. The tail of the bubble takes concave shape as it rises, which is evident from the cross-sectional view given in Fig. 4. At 260 ms, the nose of both the fluid particles started to flatten, which implies that the interaction between the drop and the bubble started slightly before this point in time. The detailed sequence of phenomena taken place during the interactions has been depicted in the next section.

3.2 *Drop–Bubble Interactions*

The sequence of hydrodynamic phenomena during the interaction of a Taylor bubble with a similar-sized drop is shown in Fig. 5. Somewhere between 0.26 and 0.27 s marks the onset of physical interaction of the drop with the bubble, following which the leading portions of both the drop and bubble flatten out. At 290 ms small outward protrusion of the leading surface of the water can be seen in the cross-sectional view. This protrusion has grown around the periphery and the water droplet starts to engulf the bubble. This results in the formation of a thin film of water between the air bubble and the sidewall as evident from the contour at 360 ms. The central cavity formed in the drop continues to expand with the rise of the bubble through its core. The subsequent contours show the complete engulfment of the Taylor bubble within the drop cavity. At 460 ms the bubble started to rise out of drop cavity which results in the formation of a complete annular drop, as can be seen from the cross-sectional view. Another thing to notice from the snapshots following 400 ms is the bending of

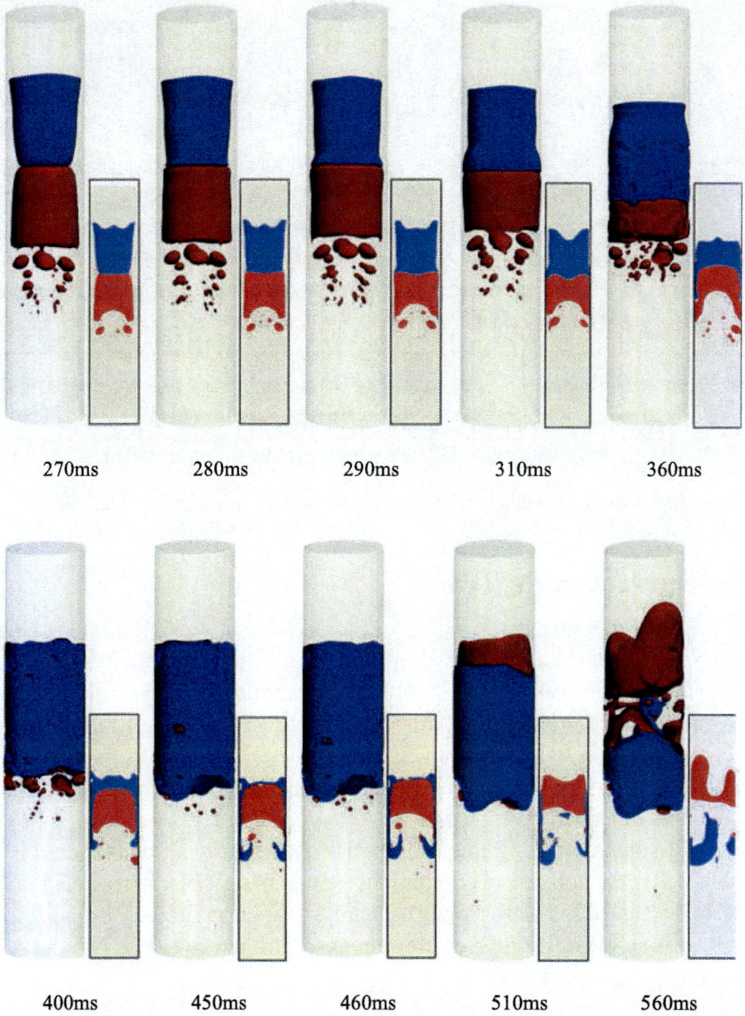

Fig. 5 Three-dimensional view depicting the interactions between a drop and a bubble at different times. The front cross-sectional view along the length of the conduit is shown in insets

downward moving water toward the center of the bubble tail. This is caused due to the circulations in the wake region of the Taylor bubble. Snapshot at 560 ms shows that the physical interaction between bubble and drop is over but their behavior is still affected by one another due to the surrounding flow field. The subsequent series of hydrodynamic phenomena post interactions are presented in the following section.

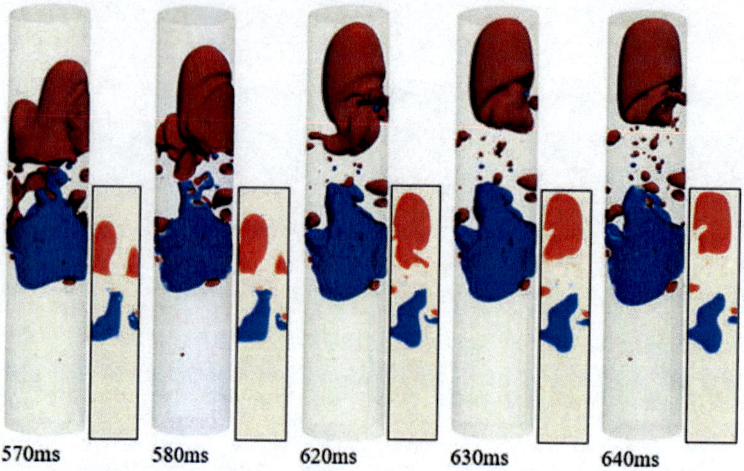

Fig. 6 The sequence of hydrodynamic phenomena after the bypass of drop and bubble

3.3 *Post-interaction Behavior*

After the bypass of the bubble, the drop coalesces filling the cavity. Due to the motion of bulk fluid in wake of the bubble, the central part of the bubble became stagnant, while the circumferential part continues to rise, leading to the formation of two asymmetric lobes as shown in Fig. 6 at time 570 ms. With the further rise of the bubble, the two lobes break and then recombine after some time to form a Taylor bubble as illustrated at time 580–630 ms, respectively. Meanwhile, the drop became almost stagnant during this time period as the gravitational force on it is nearly balanced by the pull due to circulations behind the bubble. After this, the drop has started to descend downward slowly. A detailed variation for the vertical velocity of the drop and bubble during the entire sequence of hydrodynamic phenomena is elucidated in the subsequent section.

3.4 *Variation of Drop and Bubble Velocity*

The velocity analysis of the complete phenomena is depicted in Fig. 7. The graph has been divided into three parts according to previous sub-sections. The first section shows the pre-collision motion of the bubble and drop. The bubble is initialized with 0.5 m/s velocity in an upward direction, while the drop is initialized with the same velocity in the opposite direction. However, there is a sudden fall in magnitude of velocities of both the fluid particles which can be explained to occur because of no gradient in the surrounding pressure field at the initial stage. Then the bubble velocity has risen rapidly till 0.05 s and reached terminal value at around 0.16 s. On the other

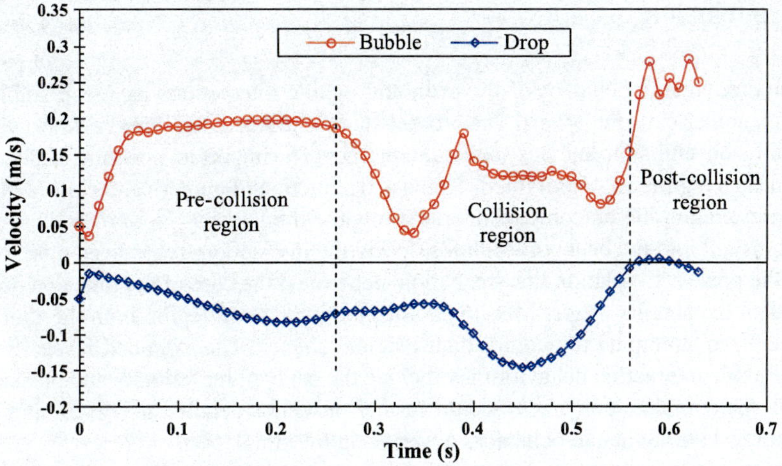

Fig. 7 Variation in the vertical velocity of drop and bubble with time for the entire phenomena

hand, the speed of drop has grown slowly due to comparatively less difference in densities of drop (water) and kerosene as compared to density difference in air and kerosene. The drop reached the terminal velocity at around 0.22 s. The bubble and the drop have come in physical contact after 0.26 s. However, both have started to retard before 0.26 s due to the flow field ahead of these fluid particles.

In the collision regime, the rapid fall in bubble velocity can be observed from 0.27 to 0.34 s. The rise in the droplet speed during the same time is gradual and small. After 0.34 s there has been again a rise in the velocity of the bubble which corresponds to its motion through the drop cavity. Its velocity has continued to rise till 0.39 s. After this time, its velocity again decreased due to additional drag because of the water entrapped in the wake region of the bubble. While the speed of drop continued to rise till 0.45 s, in this time period the magnitude of the drop's velocity has gone even above the terminal speed of Taylor drop. This rise in speed has occurred due to the motion of water through the narrow gap between the bubble and the wall. After 0.45 s, the speed of water again decreased due to the lift applied by the flow field behind the bubble.

The region of the graph after 0.56 s marks the onset of the post-collision regime. From 0.57 to 0.60 s, the velocity of the drop has become positive with maximum velocity achieved at 0.59 s. After 0.59 s, the gravity again became dominant and the drop has started to fall. In the same time period, the velocity of the bubble has picked up again. The oscillations in the velocity of the bubble can be observed which has occurred due to the merging of different lobes of the bubble. During this coalescence process, the bubble also had some velocity in the transverse direction which has caused sudden variations in the shape of the bubble. The further continuation of the phenomenon will result in the dissipation of surrounding flow fluctuations. Hence, this will lead to the formation of a steady interfacial profile and corresponding constant velocity fields.

4 Conclusion

The hydrodynamic behavior of the drop and bubble interactions has been studied through numerical simulations. The present study is divided into three regimes, viz., pre-collision, collision, and post-collision regimes. The important phenomena occurring in each regime are highlighted. The visual depiction of numerical results helped in understanding the outcome of interaction while the velocity plot provided additional insight into the behavior of fluids during the interaction. It has been observed from the present simulation that the bubble penetrated the core of the tube and drop has taken the annular shape. Moreover, after the interaction region both the fluidic entities have attempted to acquire their original shapes. The extended research of drop–bubble interaction behavior may include the study of the velocity and pressure field in the complete domain. It would further provide a detailed understanding of the various hydrodynamic behaviors observed in the simulations.

References

1. Clift, R., Grace, J.R., Weber, M.E.: Bubbles, Drops, and Particles. Courier Corporation (2005)
2. Tripathi, M.K.: Rising bubbles and falling drops. Ph.D. dissertation, Department of Chemical Engineering, IIT Madras (2015)
3. Stokes, G.G.: On the Effect of the Internal Friction of Fluids on the Motion of Pendulums, vol. 9, p. 8. Pitt Press, Cambridge (1851)
4. Allen, H.S.: XXXI. The motion of a sphere in a viscous fluid. The London, Edinburgh, and Dublin Philos. Mag. J. Sci. **50**, 323–338 (1900)
5. Taylor, T.D., Acrivos, A.: On the deformation and drag of a falling viscous drop at low Reynolds number. J. Fluid Mech. **18**(3), 466–476 (1964)
6. Bhaga, D., Weber, M.E.: Bubbles in viscous liquids: shapes, wakes, and velocities. J. Fluid Mech. **105**, 61–85 (1981)
7. Otake, T., Tone, S., Nakao, K. And Mitsuhashi, Y.: Coalescence and breakup of bubbles in liquids. Chem. Eng. Sci. **32**(4), 377–383 (1977)
8. Qian, J., and Chung King Law: Regimes of coalescence and separation in droplet collision. J. Fluid Mech. **331**, 59–80 (1997)
9. Li, S., Zhang, A.M., Wang, S., Han, R.: Transient interaction between a particle and an attached bubble with an application to cavitation in silt-laden flow. Phys. Fluids **30**(8), 082111 (2018)
10. Snyder, B.A., Aston, D.E., Berg, J.C.: Particle—drop interactions examined with an atomic force microscope. Langmuir **13**(3), 590–593 (1997)
11. Chakibi, H., Hénaut, I., Salonen, A., Langevin, D., Argillier, J.F.: Role of bubble-drop interactions and salt addition in flotation performance. Energy Fuels **32**(3), 4049–4056 (2018)
12. Dudek, M., Øye, G.: Microfluidic study on the attachment of crude oil droplets to gas bubbles. Energy Fuels **32**(10), 10513–10521 (2018)
13. Hu, Z.Z., Greaves, D., Raby, A.: Numerical wave tank study of extreme waves and wave-structure interaction using OpenFOAM. Ocean Eng. **126**, 329–342 (2016)
14. Fluent 6.0 User's Guide Volume 1. Fluent Inc., pp. 20_9–20_10 (2001)

Investigating Fluid-Structure Interaction Behavior of a Chord-Wise Flexible Filament in the Wake of a Bluff Body Using Partitioned Strong Coupling Approach

Chandan Bose, Rajanya Chatterjee, Sayan Gupta, and Sunetra Sarkar

1 Introduction

Large-scale flow-induced deformation of thin flexible filament-like structures can be potentially utilized in various engineering applications, such as flow energy harvesting. The idea of producing electrical energy for powering sensors in self-powered devices by exploiting different kinds of flow-induced phenomena, such as, vortex-induced vibrations, flutter, buffeting, and galloping *etc.* has received considerable attention in the recent past [1]. A common configuration of these flow energy harvesters require placing a flexible flapper with piezoelectric converters embedded in it in the wake of a bluff body to harvest energy from its flow-induced deformations [2–5]. Allen and Smits [2] were one of the first to experimentally show that a flexible piezoelectric membrane undergoes resonance when the oncoming vortex frequency of a von Kármán wake has the same value as the natural frequency of the structural first mode which can be exploited for energy harvesting. Akaydin et al. [3, 4] designed a flow energy harvester consisting of a flexible cantilever beam structure placed at the downstream side of a rigid cylinder. In the presence of uniform inflow, this device was able to harvest electrical energy from the periodic motion of the flexible PVDF (polyvinylidene fluoride) beam. In this regard, the flexible filament is often kept at an arbitrary gap from the cylinder in most of the existing literature. However, the fluid-body interactions will be significantly altered if the finite gap is changed between the rigid cylinder and the flexible flapper. Akaydin et al. [3] studied the effect of different locations of the piezoelectric beam on the output electric

C. Bose (✉) · R. Chatterjee · S. Gupta
Department of Applied Mechanics, Indian Institute of Technology Madras, Chennai 600036, India
e-mail: cb.ju.1991@gmail.com

S. Sarkar
Department of Aerospace Engineering, Indian Institute of Technology Madras, Chennai 600036, India

S. K. Saha and M. Mukherjee (eds.), *Recent Advances in Computational Mechanics and Simulations*, Lecture Notes in Mechanical Engineering,
https://doi.org/10.1007/978-981-15-8315-5_38

power and observed that the maximum power is not generated while the beam is very close to the cylinder since the large vortical structures are not fully formed at these locations. They obtained the peak value of power at a gap of two times the cylinder diameter. In designing the placement of these piezoelectric patches, it is typically assumed that these flappers oscillate only in the first mode [2]. However, recent studies by Kumar et al. [6], have shown that more energy can potentially be harvested if the flapper oscillates in higher modes. As has been shown in [6], this can be achieved by introducing an optimum gap between the cylinder and the flapper, which in turn, changes the vortex interactions with the flapper resulting in a change in the dominant mode of the flapper oscillations.

Apart from flow-induced energy harvesting applications, a proper understanding of the FSI behavior of thin elastic structures can directly benefit the efficient design of futuristic biomimetic devices which are inspired by the natural bio-propulsion systems of insect flight or swimming of fishes. Both insect wings and fish fins deform significantly while interacting with the surrounding viscous flow during their locomotion. Generation of desired propulsive forces is greatly dependent on the optimum flexibility of the wing/fin structure [7, 8]. Moreover, swimming efficiency is seen to be benefited in the wake of another surrounding structure. It has been reported that a swimming fish can exploit vortex structures formed by surrounding physical structures to reduce their locomotion efforts [9, 11]. Investigations that study the effects of these perturbations in the flow due to the presence of neighboring structures typically model these perturbations as the wake of a rigid cylinder in a uniform flow [9, 10]. Liao et al. [9] investigated the swimming of a trout in the wake of a D-section cylinder and reported a notable amount of energy saving through the modulation of their flapping behavior depending on the vortex shedding wavelength and frequency. The swimming dynamics get considerably altered in this case as only the anterior axial muscles of fishes are only activated [11]. Even a passive flexible body can generate the desired thrust to propel upstream exploiting the energy of the oncoming vortex street overcoming its own drag [12, 13]. Park et al. [10] also observed that a self-propelled flexible fin could overcome the net drag in the wake of a circular cylinder by resonating its body in synchrony with the incoming vortices. Therefore, it is important to clearly understand the fluid-body interactions of a flexible structure that emerged in an oncoming vortex street.

The high fidelity simulations of FSI problems can be based on either monolithic or the partitioned approach [14]. In the monolithic approach, both the structural and fluid governing equations are solved in a single mathematical and computational framework with implicit interfacial conditions. On the other hand, two different existing solvers are coupled in the partitioned approach to solve the two separated sub-systems of fluid and solid models. Again in the partitioned approach, there exist two different coupling strategies [14]: weak and strong coupling. The weak coupling scheme requires solving each sub-system once per time step and no convergence check is carried out. Although this could be an affordable choice in terms of computational cost, it can lead to a time-lagged solution and numerical instabilities especially when the fluid and solid densities are comparable for a very flexible structure with large deformation leading to unphysical instabilities. The strong coupling scheme demands solving the sub-systems multiple times until the convergence criterion at the interface is satisfied.

Thus, this present paper employs a strongly coupled FSI solver to investigate the dynamics of a flexible filament immersed in the von Kármán wake of a stationary rigid cylinder. This study focuses on understanding the interaction mechanism of the oncoming vortices with the flexible body resulting in changes in its mode-shapes and the consequent effect on the surrounding flow dynamics. In this context, the effect of the finite gap between the cylinder and the filament, length, and thickness of the filament on the dynamics is studied in this paper. The remainder of the paper consists of the following sections. First, the governing equations are presented. Next, the numerical algorithms employed in this paper are described along with a validation study. Subsequently, results and discussions have been presented. The salient features arising from this study are reported in the concluding section.

2 Computational Methodology

2.1 Governing Equations

The high fidelity FSI framework [15], is composed of an incompressible Navier-Stokes (N-S) solver, strongly coupled by a partitioned approach with a nonlinear elastic structural model. The viscous flow-field, governed by an incompressible Navier-Stokes equation, can be cast into arbitrary Lagrangian–Eulerian (ALE) formulation [16] as

$$\nabla.\mathbf{u} = 0, \tag{1}$$

$$\frac{\partial \mathbf{u}}{\partial t} + [(\mathbf{u} - \mathbf{u}^{\mathbf{m}}).\nabla]\mathbf{u} = -\nabla p/\rho + \nu\nabla^2\mathbf{u}. \tag{2}$$

Here, $\mathbf{u}$ is the velocity of the flow, $\mathbf{u}^{\mathbf{m}}$ is the grid point velocity, p is the pressure, ρ is the fluid density, and ν is the kinematic viscosity.

The flexible flapper is modeled as an elastic continuum with its leading edge restrained as a fixed support. The trailing edge is referred to as tip of the flapper. Mathematically, conservation of linear momentum is given by

$$\frac{\partial}{\partial t}\int \rho_s \mathbf{v}\, \mathrm{d}V = \oint \mathbf{n}.\boldsymbol{\sigma}\, \mathrm{d}\Gamma + \int \rho_s \mathbf{b}\, \mathrm{d}V, \tag{3}$$

where, V is the volume of the flapper bounded by the surface Γ with unit normal $\mathbf{n}$, $\mathbf{v}$ is the velocity vector, $\boldsymbol{\sigma}$ is the Cauchy stress tensor, ρ_s is the density of the material of the structure, and $\mathbf{b}$ is the body force per unit mass. Assuming large strain, the material behavior of the flapper has been modeled using St. Venant-Kirchhof hyperelasticity theory. The linear momentum equation can, therefore, be written in terms of the second Piola-Kirchhoff stress tensor (**S**) as: $\sigma = \frac{1}{J}\mathbf{F}.\mathbf{S}.\mathbf{F}^{\mathbf{T}}$; $\mathbf{F} = \mathbf{I} + \Delta\mathbf{w}$ is the deformation gradient, **I** is the identity tensor, and **w** is the deformation vector. According to the constitutive model for the St. Venant-Kirchhoff material, **S** can be defined as: $\mathbf{S} = \lambda(\mathrm{tr}\,\mathbf{E})I + 2\mu\mathbf{E}$, where $\mathbf{E} = \frac{1}{2}(\mathbf{F}^{\mathbf{T}}\mathbf{F} - \mathbf{I})$ and λ, μ are Lame constants ('tr' denotes the trace of a matrix).

2.2 Numerical Algorithms

The incompressible N-S equation has been discretized in ALE approach [16], on a moving grid with a Laplacian mesh motion strategy [15]. The flow solver uses a second order accurate spatial discretization and the temporal discretization is performed using a second order implicit backward differencing scheme. A variable time stepping technique based on a maximal Courant number has been adopted. The pressure velocity coupling is implemented through PISO (Pressure Implicit with Splitting of Operator) algorithm [16]. The absolute error tolerance criteria for pressure and velocity are set to 10^{-6}. The structural part has been solved by a large strain elastic stress analysis solver based on Lagrangian displacement formulation [17]. The absolute tolerance criterion of the structural solver is also taken to be 10^{-6}. A quasi-Newton coupling algorithm with an approximation for the inverse of the residual's Jacobian matrix from a least-square model (IQN-ILS) [14], has been adopted in the strong coupling method.

2.3 Computational Domain and Mesh

A two-dimensional rectangular computational domain, as shown in Fig. 1a, has been considered. The dimensions of the domain are shown in Fig. 1b, in terms of the diameter of the cylinder D. Simulations are performed for varying gap (G), length (L), and thickness (T) of the structure. Standard boundary conditions have been applied: a zero pressure gradient and a constant free-stream at the inlet; a zero velocity gradient and atmospheric pressure condition at the outlet; no slip and zero normal pressure gradient condition on the horizontal walls and traction boundary condition on the flexible flapper. The computational domain is discretized using structured grids as shown in Fig. 1b. The mesh containing 89052 grid points has been finalized through a grid independence test at $Re = 500$.

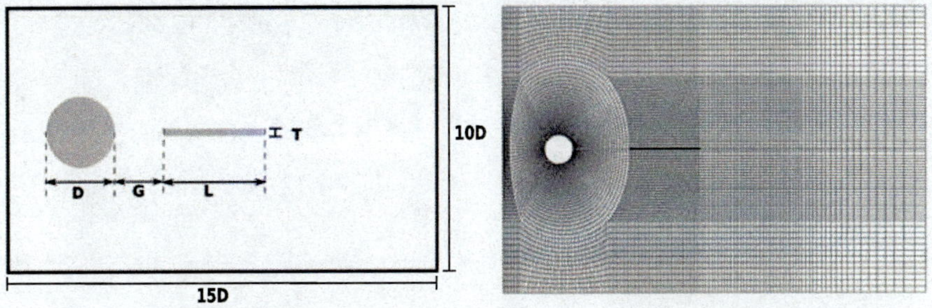

(a) Domain (b) Structured mesh

Fig. 1 Computational domain (not to scale) and mesh

3 Validation of the FSI Solver

The present FSI solver has been quantitatively validated with the benchmark case of a flexible splitter plate attached with a rigid cylinder given by Turek and Hron [18]. The parameters corresponding to 'FSI2' case presented in [18], have been considered for the benchmark validation. It can be clearly seen from Fig. 2, that vertical tip displacement time history of the flexible plate shows an excellent match with the results presented by Turek and Hron [18], in the self-sustained periodic oscillatory state. The mesh, chosen after grid independence study at $Re = 500$, has been used for the validation case at a lower Re. Although the validation study has been performed at $Re = 100$, the present FSI solver reflects accurate physics at $Re = 500$ since the coherent vortex structures are seen to remain intact behind the cylinder at $Re = 500$.

4 Results and Discussions

The parameters considered are as follows: Reynolds number(Re_D) = 500; structure to fluid density ratio ($\mu = \rho_s/\rho_f$) = 1; non-dimensional Young's modulus ($\bar{E} =$

Fig. 2 FSI solver validation with the benchmark case of Turek and Hron [18] at $Re = 100$

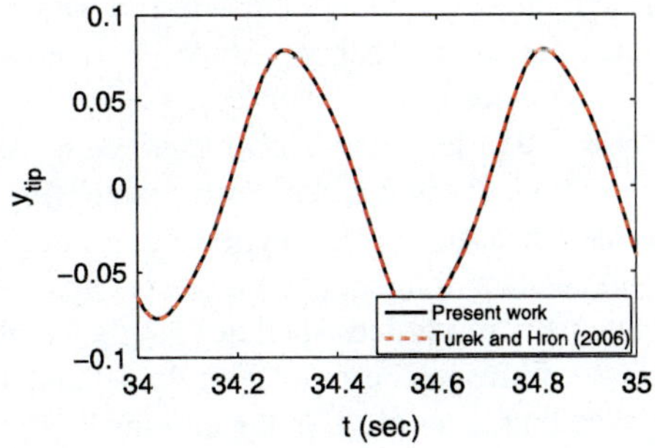

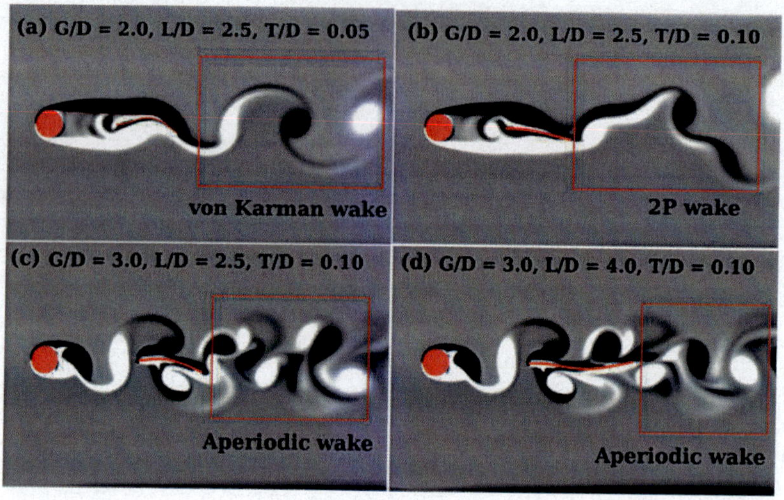

Fig. 3 Various wake patterns behind the flexible filament

$E/(\rho_f * u_m^2)) = 5600$; non-dimensional length $(L/D) = 2.5$ and 4; non-dimensional gap between the cylinder and the filament $(G/D) = 2.0$ and 3.0; the non-dimensional thickness of the filament $(T/D) = 0.05$ and 0.10. Here, ρ_s and ρ_f are the solid and fluid density, respectively; E is the dimensional Young's modulus of the structure; L, T, D, and G are the length and thickness of the filament, diameter of the cylinder, gap between the cylinder and the filament, respectively. u_m is the mean uniform velocity at the inlet.

Simulations have been carried out considering different finite gaps and length scales of the elastic structure to compare its FSI behavior while interacting with oncoming wake vortices. Different wake patterns generated at the trailing edge of the flexible filament is compared in Fig. 3. The variation of flow topology and bending mode-shapes of the filament depends on the Strouhal frequency and the location of vortex impingement. In the case of the lower gap $(G/D = 2.5)$, the shear layers behind the rigid cylinder are seen to roll up and form the vortex cores which evolve with time, growing in strength. Subsequently, the vortex cores convect over the flexible structure triggering vibrations and are shed in the downstream of the flapper. The cylinder-filament system acts as a whole-body together in this case, and vortex formation does not happen in the intermediate space as the gap was less than the formation length of the vortices; see Fig. 3a, b. In case of a thinner filament with $T/D = 0.05$, the wake is characterized by an array of alternate shedding vortices representing a von Kármán vortex street (Fig. 3a). However, when the beam thickness is increased to 0.1D, the bending mode-shape is changed from a mixed-mode oscillation to a dominant first mode oscillation engendering a 2P wake characterized by two pairs of vortices shed in a single flapping cycle (Fig. 3b). On the other hand, when the finite gap between the cylinder and the filament is increased to 3D, the vortex shedding takes place in the intermediate region itself as the gap becomes longer

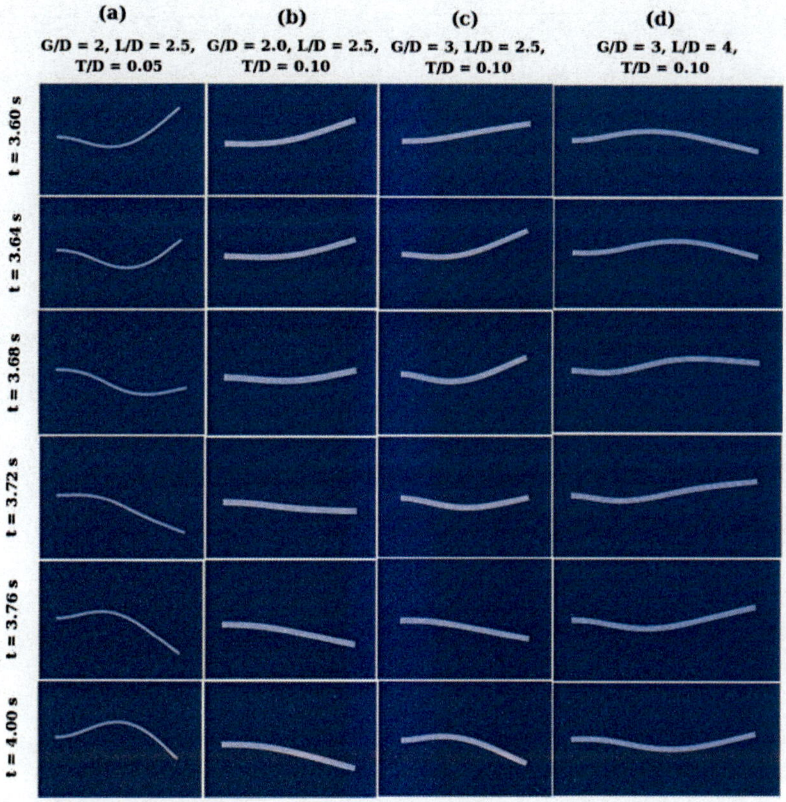

Fig. 4 Different bending mode-shapes of the flexible filament

than the vortex formation length. Interestingly, while convecting downstream, the vortex cores are seen to be intersected by the leading edge of the filament giving rise to complex flow interactions which will be discussed in detail in the latter part of the paper. The organized pattern of the wake is seen to be lost due to these aperiodic interactions; see Fig. 3c, d.

The interaction of the vortex cores on the flexible flapper results in passive pitching of the elastic structure in terms of its different natural bending modes. Being a spatiotemporal problem, the nature of the vibrating mode of the structure depends on the frequency of vortex shedding, as well as the location of vortex interaction. Moreover, the formation length of the wake vortices behind the cylinder plays a key role in exciting different modes in the flapper vibration. In the case of a shorter and thinner filament ($L/D = 1$, $T/D = 0.05$), it appears to oscillate predominantly in its first natural mode-shape (The results are not shown here for the sake of brevity). This can be attributed to the fact that the wake vortices form at the tip of the structure and makes it bend in the upward and downward direction periodically in its first natural mode. However, it attains a mixed-mode vibration of first and second mode-shape

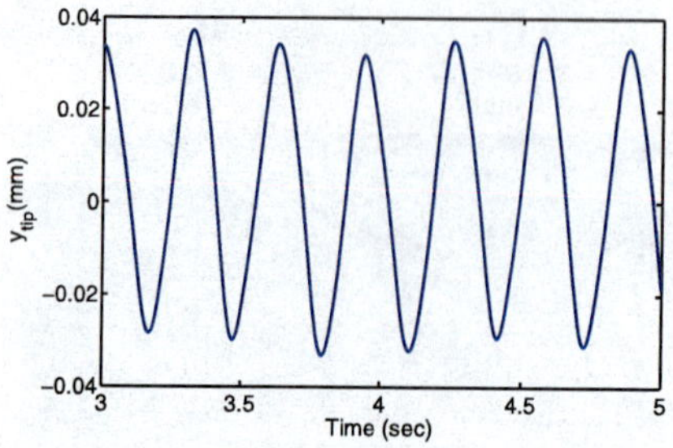

(a) G/D = 3, L/D = 2.5, T/D = 0.10

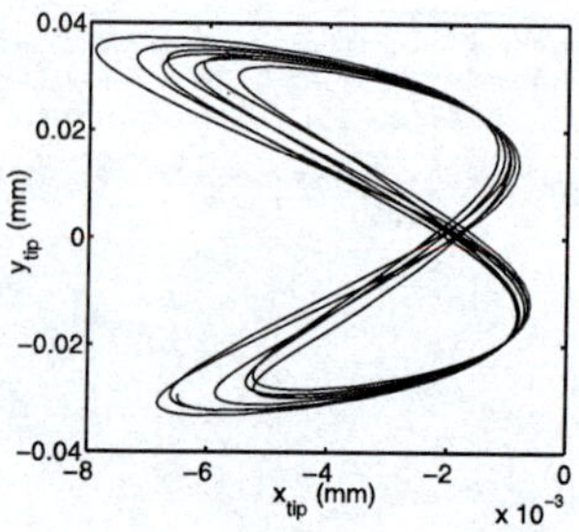

(b) G/D = 3, L/D = 2.5, T/D = 0.10

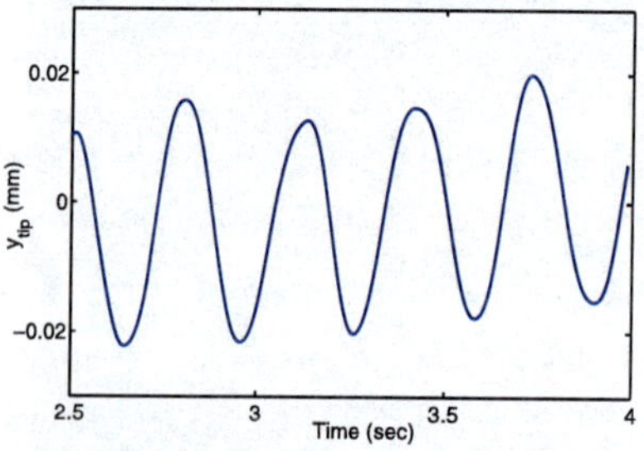

(c) G/D = 3, L/D = 4.0, T/D = 0.10

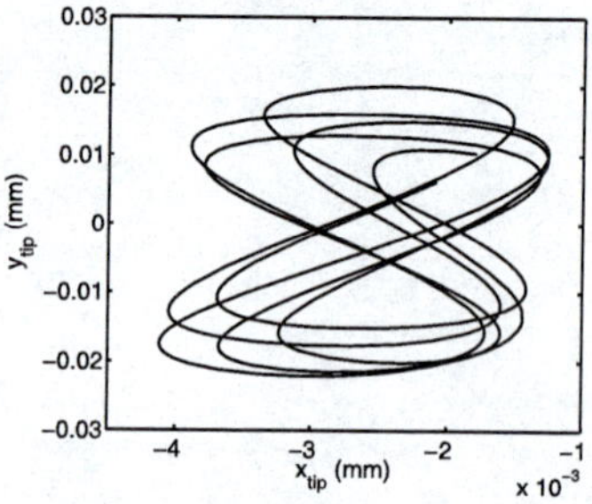

(d) G/D = 3, L/D = 4.0, T/D = 0.10

Fig. 5 Time histories and phase portraits of tip displacement of the flexible filament

for $L = 2.5D$ as the length of the flapper becomes greater than the formation length of the wake vortices. As a result, the vortices impinge in the middle of the longer elastic structure for $L/D = 2.5$, $T/D = 2.5$. Figure 4a shows the typical bending mode-shapes of the self-sustained periodic mixed-mode oscillation involving first and second mode-shapes. Interestingly, as the thickness of the filament is increased from $0.05D$ to $0.10D$, resulting in higher stiffness, a dominant first mode oscillation is observed; see Fig. 4b. On the contrary, the mixed-mode oscillation is again observed as the gap between the cylinder and the filament is increased to $3D$, keeping the same thickness (Fig. 4c). Finally, the filament is seen to exhibit a second type of mixed-mode oscillation involving second and third modes as the length of the filament is increased to $4D$ keeping the gap and the thickness unchanged as can be seen in Fig. 4d. Note that the periodicity of the tip oscillation is seen to be lost for $G/D = 3$ as presented through the time histories and the phase portraits of the y-directional tip oscillations of the filament in Fig. 5. It brings out an interesting regime of aperiodic body-wake interactions which is discussed next.

To understand the aperiodic fluid-body interactions in more detail, the flow physics is analyzed through a sequence of vortex contour snapshots at different time instances for $G/D = 3$, $L/D = 2.5$, $T/D = 0.10$; see Fig. 6. At $t = 2.20$ s, two opposite sense vortices ('1' and '2') are seen to shed alternatively behind the cylinder. At the

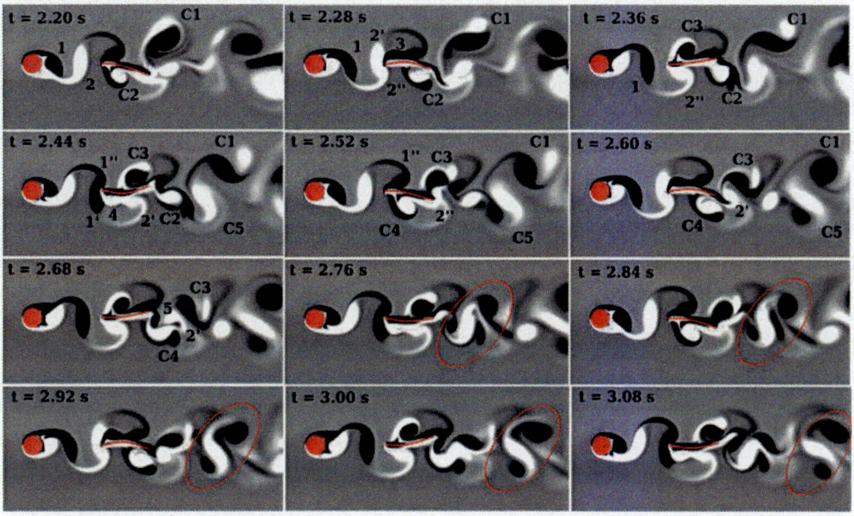

Fig. 6 Aperiodic wake: vortex interactions with the flexible plate ($G/D = 3$, $L/D = 2.5$, $T/D = 0.10$)

same time, one vortex couple 'C2' is seen to interact with the filament and another couple 'C1' is seen to covect downstream behind the filament. The spontaneous vortex couple 'C1' traverses with its self-induced velocity and interacts with the existing downstream flow structures. While convecting downstream, counter-clockwise vortex '2' gets intersected by the leading edge of the filament ($t = 2.28$ s). Consequently, one portion of it ($2'$) forms a vortex couple 'C3' with the nascent leading edge vortex (LEV) '3' on the upper surface of the structure, whereas, the other portion ($2''$) rolls over the lower surface of the structure ($t = 2.36$ s). In the subsequent stages, clockwise vortex '1' also gets intersected by the leading edge and gets divided into two parts similar to '2'. In this case, the lower counterpart ($1'$) forms a couple 'C4' in the lower surface along with the LEV and the upper counterpart ($1''$) rolls over the upper surface ($t = 2.52$ s). Therefore, vortex couples are seen to convect over the filament on the alternative surfaces almost in a periodic manner, causing the filament to bend at its tip in the first mode-shape while they shed from the trailing edge. However, the flow-field becomes complex due to the spontaneous interactions between the vortex couples in the presence of trailing edge vortex downstream. Such an interaction is highlighted next. At $t = 2.60$, 'C3' and 'C4' are seen to collide in a opposite trajectory by sandwiching counter-clockwise vortex $2'$. During this collision, the counter-clockwise component of 'C4' and $2'$ undergoes a partial merging and the resultant stronger vortex attempts to form a new couple with the clockwise component of 'C3' with upward self-induced velocity. At the same time, a trailing edge vortex '5' emerges and tries to capture a part of the counter-clockwise component to form another couple with downward self-induced velocity ($t = 2.68$–3.08 s).

As a result, the clockwise counterpart of 'C4' gets shredded by the opposite sense vortices. These interactions take place spontaneously and get altered in every cycle destroying the periodicity of the wake pattern.

5 Conclusions

The flow-induced dynamics of a chord-wise flexible flapper has been simulated in the wake of a rigid cylinder situated upstream using a strongly coupled FSI solver at a low Reynolds number. It has been observed that the dynamics get altered significantly with the change in G/D, L/D , and T/D. The shorter, as well as stiffer filaments, appear to oscillate dominantly in its first natural mode; whereas the longer and more flexible filaments exhibit a mixed-mode oscillation involving first and second or second and third natural mode-shapes. A steady periodic high amplitude oscillation was observed due to the periodic interaction of the wake vortices into the flexible flapper at lower G/D values. Contrastingly, the flow-field becomes aperiodic through complex vortex interactions in case of higher G/D values. Further studies are being carried out to understand the effect of the Strouhal frequency of the input flow fluctuation and the flexural rigidity of the structure on the FSI dynamics.

References

1. Li, D., Wu, Y., Da Ronch, A., Xiang, J.: Energy harvesting by means of flow-induced vibrations on aerospace vehicles. Prog. Aerosp. Sci. **86**, 28–62 (2016)
2. Allen, J.J., Smits, A.J.: Energy harvesting eel. J. Fluids Struct. **15**, 629–640 (2001)
3. Akaydin, H.D., Elvin, N., Andreopoulos, Y.: Wake of a cylinder: a paradigm for energy harvesting with piezoelectric materials. Exp. Fluids **49**(1), 291–304 (2010)
4. Akaydin, H.D., Elvin, N., Andreopoulos, Y.: The performance of a self-excited fluidic energy harvester. Smart Mater. Struct. **21**(2), 025007 (2012)
5. Gao, X., Shih, W.H., Shih, W.Y.: Flow energy harvesting using piezoelectric cantilevers with cylindrical extension. IEEE Trans. Ind. Electron. **60**(3), 1116–1118 (2013)
6. Kumar, S.K., Bose, C., Ali, S.F., Sarkar, S., Gupta, S.: Investigations on a vortex induced vibration based energy harvester. Appl. Phys. Lett. **111**(24), 243903 (2017)
7. Shelley, M.J., Zhang, J.: Flapping and bending bodies interacting with fluid flows. Ann. Rev. Fluid Mech. **43**, 449–465 (2011)
8. Cleaver, D.J., Calderon, D.E., Wang, Z., Gursul, I.: Lift enhancement through flexibility of plunging wings at low Reynolds numbers. J. Fluids Struct. **64**, 27–45 (2016)
9. Liao, J.C., Beal, D.N., Lauder, G.V., Triantafyllou, M.S.: The Kármán gait: novel body kinematics of rainbow trout swimming in a vortex street. J. Exp. Biol. **206**(6), 1059–1073 (2003)
10. Park, S.G., Kim, B., Sung, H.J.: Self-propelled flexible fin in the wake of a circular cylinder. Phys. Fluids **28**(11), 111902 (2016)
11. Liao, J.C., Beal, D.N., Lauder, G.V., Triantafyllou, M.S.: Fish exploiting vortices decrease muscle activity. Science **302**(5650), 1566–1569 (2003)
12. Beal, D.N., Hover, F.S., Triantafyllou, M.S., Liao, J.C., Lauder, G.V.: Passive propulsion in vortex wakes. J. Fluid Mech. **549**, 385–402 (2006)

13. Eldredge, J.D., Pisani, D.: Passive locomotion of a simple articulated fish-like system in the wake of an obstacle. J. Fluid Mech. **607**, 279–288 (2008)
14. Degroote, J., Haelterman, R., Annerel, S., Bruggeman, P., Vierendeels, J.: Performance of partitioned procedures in fluid structure interaction. Comput. & Struct. **88**(7–8), 446–457 (2010)
15. Tukovic, Z., Cardiff, P., Karac, A., Jasak, H., Ivankovic, A.: OpenFOAM library for fluid structure interaction. In: 9th OpenFOAM Workshop, vol. 2014 (2014)
16. Ferziger, J.H., Peric, M.: Computational Methods for Fluid Dynamics. Springer Science & Business Media (2012)
17. Cardiff, P., Tukovié, Jasak, H., Ivankovié, A.: A block-coupled Finite Volume methodology for linear elasticity and unstructured meshes. Comput. Struct. **175**, 100–122 (2016)
18. Turek, S., Hron, J.: Proposal for numerical benchmarking of fluid-structure interaction between an elastic object and laminar incompressible flow. Fluid-Structure Interaction, pp. 371–385. Springer, Berlin, Heidelberg (2006)

Optimal Design of Rotary Magneto-Rheological Drum Brake for Transfemoral Prosthesis

Radhe Shyam Tak Saini, Hemantha Kumar, Sujatha Chandramohan, and Sujatha Srinivasan

1 Introduction

Smart materials, particularly MR fluids have gained a lot of popularity in the past few decades and have been applied in many areas where controllability, adaptability, and fast response are desirable. Of these applications, vehicular suspension seems to be one area which has seen some major research contribution regarding the implementation of MR fluids in dampers, brakes, etc. The quest for variable walking speed control in prosthetic knee attracted the use of these smart fluids in this domain too. RD-1036 MR damper, originally designed for truck seat applications by Lord Corporation, USA, was the first MR damper to be used by Biedermann Motech GmbH for prosthetic applications [1]. Further advancements in this field suggest that the use of knee-specific damper results in much better performance of prosthetic devices [2]. Studies combining semi-active MR dampers with active elements such as DC motor also has been performed to realize an active knee. Gao et al. [3] used an MR damper in the stance phase of the gait cycle and a DC motor to control swing phase actuation. Their study reported an optimal MR design and no experimental trails of the active prosthesis were conducted.

Knee damping has also been achieved by using rotary MR brakes. Herr and Wilkenfield [4] designed a multi-plate rotary MR brake which can generate an off-state damping of 0.5 Nm and an on-state damping torque of 40 Nm. Gudmundsson et al. [5] designed a multi-plate MR brake with a total of 71 plates. They have

R. S. T. Saini (✉) · H. Kumar
National Institute of Technology Karnataka, Surathkal, Mangalore, India
e-mail: radhe.chandu@gmail.com

S. Chandramohan · S. Srinivasan
Indian Institute of Technology Madras, Chennai, India

S. K. Saha and M. Mukherjee (eds.), *Recent Advances in Computational Mechanics and Simulations*, Lecture Notes in Mechanical Engineering,
https://doi.org/10.1007/978-981-15-8315-5_39

formulated a multi-objective problem with the field induced braking torque, off-state rotary stiffness, and weight of brake as the objectives. The design produced a total of 60 Nm braking torque but the complex design with 71 plates separated by a micron level gap requires high precision manufacturing and thus increased costs.

Broadly classified, continuous rotation MR brakes are of two types: disc and drum types. In disc type MR brakes, the radial faces influence the braking torque produced while in drum type MR brakes, the annular gap coinciding axially with the rotational axis produces the braking torque [6]. The geometric optimization performed by Nguyen et al. [7], resulted in an optimum outer casing diameter of 100 mm with a braking torque of 10 Nm. To produce a braking torque sufficient to mimic the normal gait, the geometric dimensions of the single disc brake will exceed the physiological knee cross-section areas. Further, a drum brake design will generate a nonuniform magnetic field over the activated length and thus results in a lesser braking torque. The shear rates involved in knee braking are very less compared to vehicular applications since the knee has a maximum speed of 60 rpm in an average normal human walking [5]. Therefore, to obtain a high braking torque in a constrained volume, it is necessary to have large shear areas, which justifies the use of multi-disc brakes. In the present study the brake outer dimensions are constrained and thus limiting the shear area, the requirement of high braking torque depends on other parameters like magnetic flux in the gap. Although the braking torque depends on other geometric dimensions along with the fluid gap, the magnetic flux in gap affects the yield strength of MR fluid which is directly proportional to on-state braking torque. Other geometric dimensions affect the off-state braking torque too, which is undesirable. Considering the above limitations, in the present study an inverted drum type MR brake with multi-coil is designed which can generate the torque sufficient for swing phase knee motion.

The following section describes the geometry of the multi-coil brake design and the geometric variables. Later, the design parameters are defined in terms of geometric variables. Four influencing variables are selected and a design of experiments method is used to formulate a set of combinations of these variables. Magnetostatic analysis at each design point is performed to calculate the average magnetic field in the annular and radial gaps which are then used to calculate design parameters. Regression analysis is used later to obtain braking torque as a function of geometric variables. A genetic algorithm optimization is performed to obtain the optimum geometric dimensions of the brake.

2 Geometry

The multi-coil inverted drum brake consists of more than one coil areas wound around the rotor. In the present study, a total of 11 coil areas are considered to get an even distribution of the magnetic field on the active pole length around the axial face of the rotor. Further, the whole casing is considered to be made of magnetic material, which helps in achieving the magnetic field on both the radial and axial faces of the rotor. MR fluid is occupied in the narrow gap of 0.5 mm in between the rotor and

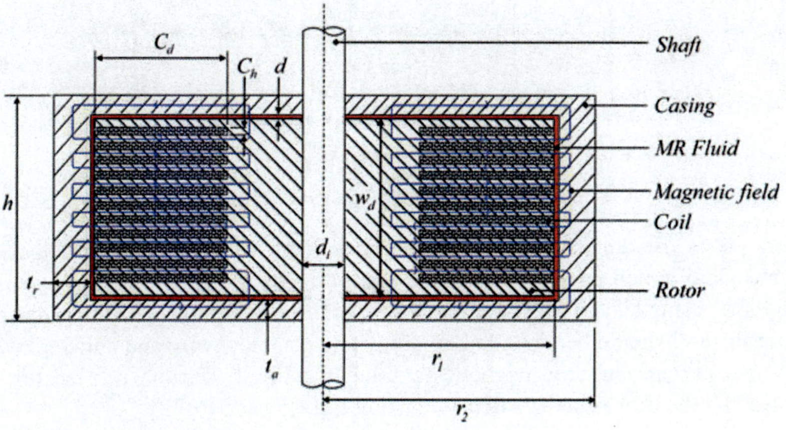

Fig. 1 Cross-section of multi-coil inverted drum brake

the casing. The shaft is assumed to be made of aluminum material to reduce any flux leakages through it. The cross-section of the drum brake along with the geometric variables are shown in Fig. 1.

3 Design Parameters

In this study, MRF-140CG fluid is considered for analysis. The fluid is modeled using a Hershel-Bulkley model given by Eq. (1) [8].

$$\tau = \tau_y + K\dot{\gamma}^n \tag{1}$$

Here, τ_y is field dependent yield stress, K is the fluid consistency index and n is the flow behavior index. All the variables in the model are considered to be field dependent and can be calculated using Eq. (2).

$$Y = Y_\infty + (Y_0 - Y_\infty)\left(2e^{-B\alpha_{sy}} - e^{-2B\alpha_{sy}}\right) \tag{2}$$

Here, Y can represent the variables τ_y, K and n. The value of Y is considered to vary from zero applied value, Y_0 to saturation value, Y_∞. Also, B represents the magnetic flux at the respective location and α_{sy} is the saturation moment index. The parameters for MRF 140CG are: $K_0 = 0.65\text{Pa.s}^n$, $K_\infty = 5400\text{Pa.s}^n$, $\alpha_{sk} = 5\text{T}^{-1}$, $\alpha_{sn} = 35\text{T}^{-1}$, $\tau_{y0} = 25\text{Pa}$, $\tau_{y\infty} = 39000\text{Pa}$, $\alpha_{s\tau_y} = 2\text{T}^{-1}$, $n_0 = 0.915$ and $n_\infty = 0.24$ [8].

When the brake is not supplied with any current, the off-state torque generated by the brake can be calculated using Eqs. (3) and (4) [9].

$$T_0 = \left\{ \begin{array}{c} \frac{4\pi \mu_{eq0} r_1^4}{(n_0+3)d} \left[1 - \left(\frac{d_i}{2r_1} \right)^{n_0+3} \right] \Omega + \frac{4\pi \tau_0}{3} (r_1^3 - r_0^3) \\ +2\pi r_1^2 w_d \left[\tau_0 + K_0 (\frac{\Omega r_1}{d})^{n_0} \right] \end{array} \right\} \tag{3}$$

$$\mu_{eq0} = K_0 \left(\frac{r_1 \Omega}{d} \right)^{n_0 - 1} \tag{4}$$

Here, Ω is the angular velocity at the knee joint. Current supply produces a magnetic field which passes through the core, crosses the MR fluid gap expands through the casing and returns back into the core. It should be noted that the adjacent areas of the coil should be wound alternating between clockwise and anticlockwise direction so as to produce the magnetic field shown in Fig. 1. The total on-state torque generated by the brake can be calculated using Eqs. (5) and (6) [8].

$$T_B = \left\{ \begin{array}{c} \frac{4\pi \mu_{eq} r_1^4}{(n_e+3)d} \left[1 - \left(\frac{d_i}{2r_1} \right)^{n_e+3} \right] \Omega + \frac{4\pi \tau_{ye}}{3} (r_1^3 - r_0^3) \\ +2\pi r_1^2 L_a \left[\tau_{ya} + K_a (\frac{\Omega r_1}{d})^{n_a} \right] \end{array} \right\} \tag{5}$$

$$\mu_{eq} = K_e \left(\frac{r_1 \Omega}{d} \right)^{n_e - 1} \tag{6}$$

Here, τ_{ye} is the torque in the end (or radial) surface of brake and τ_{ya} is the torque in the annular surface of brake. The total torque generated is the sum of both on-state and off-state torques.

$$T = T_0 + T_B \tag{7}$$

The magnetic flux in the gap varies with distance and thus in Eqs. (3)–(7), an average magnetic flux in annular and radial gaps given by Eqs. (8) and (9), is considered.

$$B_a = \frac{1}{L_a} \int_0^{L_a} B dl \tag{8}$$

$$B_r = \frac{1}{(r_1 - 0.5d_i)} \int_{0.5d_i}^{r_1} B dr \tag{9}$$

4 Design of Experiments

The outermost dimensions of the rotary brake namely maximum height, h and the radius of the casing, r_2 are constrained such that the whole knee joint can be contained within the geometrical limits of a normal adult human knee. Therefore, the maximum

Table 1 Parameters and their levels

	Coil depth, C_d (mm)	Coil height, C_h (mm)	Casing radial thickness, t_r (mm)	Casing axial thickness, t_a (mm)
Low level	10	2	3	3
Medium level	15	3	5	5
High level	20	4	7	7

height is restricted to 90 mm while the maximum radius of casing is limited to 35 mm. Further, a fluid gap size of 0.5 mm and a shaft diameter of 8 mm is considered in the analysis. The remaining parameters show a nonlinear variation in the total torque generated by the brake. Increasing the coil height may help accommodate more number of turns of the coil but reduces the pole length, thus effecting the torque. The axial and radial casing thicknesses also show a major effect on the braking torque. Therefore, the brake is designed considering coil height, coil width, casing radial thickness, and casing axial thickness as design variables and three levels are considered for each parameter as listed in Table 1.

All the combinations of design parameters will result in 81 combinations and performing magnetostatic analysis for each will take up a lot of CPU time. Thus, an L27 orthogonal array method is used to determine the various combinations of design variables. For each combination of design variables, an axisymmetric quarter model is used to perform magnetostatic analysis. Low carbon SAE1020 steel is chosen for both casing and rotor because of its high relative permeability and a magnetic saturation flux of 2.3 T [5]. The number of turns in each coil cross-section is determined using Eq. (10).

$$N = \frac{p_f(C_d C_h)}{\left(\frac{\pi}{4}d_c^2\right)} \tag{10}$$

Here, p_f is the packing factor and d_c is the coil diameter. Copper coil of 24 AWG gage is chosen with a packing factor of 0.7. The analysis is performed at 12 V and a current supply of 2 A. Fluid properties of MRF140CG referred from the datasheet [10], were given as inputs to ANSYS, while the rest of material properties were obtained from its own library. A one fourth of brake model is analyzed for each combination of dimensions and average magnetic fields are determined in both annular and radial gaps using Eqs. (8) and (9). Further, the total torque generated by the brake is calculated using Eq. (7). A design speed of 8.4 rpm was selected for the analysis [5]. The combinations of geometric parameters along with their output torque calculated using the above mentioned procedure are listed in Table 2.

Table 2 Response parameters for different geometric variable combinations

Sl. No.	Coil depth, C_d (mm)	Coil height, C_h (mm)	Casing radial thickness, t_r (mm)	Casing axial thickness, t_a (mm)	Total braking torque (Nm)
1	10	4	5	5	8.638019
2	10	3	3	5	9.045624
3	15	2	3	5	12.0034
4	15	3	5	5	9.852269
5	15	4	3	5	9.888702
6	15	3	3	7	8.577636
7	20	2	5	5	7.914665
8	10	3	5	3	8.398466
9	10	3	5	7	7.493893
10	20	3	7	5	5.689815
11	20	3	3	5	8.773237
12	15	2	5	7	8.710436
13	15	3	3	3	10.35281
14	15	3	5	5	8.332232
15	15	4	5	7	5.897144
16	15	3	7	7	6.239161
17	15	2	7	5	7.747744
18	15	4	7	5	6.787517
19	20	3	5	7	6.261766
20	15	4	5	3	9.01323
21	10	3	7	5	7.123916
22	15	2	5	3	10.20791
23	20	3	5	3	7.612614
24	10	2	5	5	9.912466
25	20	4	5	5	5.826426
26	15	3	5	5	8.333187
27	15	3	7	3	7.40763

5 Optimal Design

The calculated weight of the brake for all design points varied in the range of 2.7–2.9 kg. This small variation is because of the restrictions in the outer most dimensions. Owing to this small variation and to simplify the optimization process, weight was not considered as an objective function. Therefore, the optimal design is the one which will yield the maximum total braking torque for the given geometric dimensions.

Since magnetostatic analysis was performed for only discrete values of the dimensions, there may exist an optimum design value in the considered range. To determine this optimum, a regression analysis is performed to obtain total braking torque as a function of the four design variables. The variation of torque with design variables is given by Eq. (11).

$$T = \left\{ \begin{array}{c} 3.88 + 1.357C_d - 1.48C_h - 0.64t_r + 1.54t_a - 0.0392C_d^2 \\ +0.252C_h^2 - 0.028t_r^2 - 0.1362t_a^2 - 0.0407C_dC_h - 0.029C_dt_r \\ -0.0112C_dt_a + 0.144C_ht_r - 0.202C_ht_a + 0.0379t_rt_a \end{array} \right\} \quad (11)$$

A genetic algorithm optimization is performed with an objective function of maximizing braking torque for the given dimensional constraints using MATLAB. The optimum geometric dimensions obtained are coil depth of 14.59 mm, coil height of 2 mm, casing radial thickness of 3 mm, and casing axial thickness 3.98 mm. The optimum braking torque obtained for this design is 11.78 Nm. It can be observed that this value is less than the maximum braking torque shown in Table 2, which is 12 Nm. This difference in value can be attributed to the accuracy with which the regression equation was fit. Magnetostatic analysis using the optimal designs will predict more accurate braking torque value.

6 Results and Discussion

The advantage of multi-coil design over a single coil design is that the former gives a more uniform magnetic field over the activated length. To prove this, a magnetostatic analysis of a 2D axisymmetric model is performed considering only a single coil of the total area of cross-section similar to the multi-coil design. The individual coil area in multi-coil design is $(C_h \times C_d)$ which implies a coil cross-section of $(11 \times C_h \times C_d)$ for the single coil design. Therefore, the activated lengths of both the designs are similar. Models of the multi-coil and single coil designs considered in Finite Elements Magnetics Methods (FEMM) are shown in Fig. 2.

It can be seen from Fig. 2b, that the coil cross-sections are supplied with 2 A and—2 A current alternatively. The number of turns in the coil cross-section is calculated using Eq. (10), and a constant current of 2 A is used in both the analysis. Magnetostatic analysis is performed on both the models. Magnetic flux distribution of both models is shown in Fig. 3. It can be clearly seen that the multi-coil design has much uniform flux density distribution compared to the single coil design. Further, in both the designs, the maximum flux density is less than the saturation field strength of 2.3 T [5].

The magnitudes of the normal magnetic flux density plotted for both the configurations is shown in Fig. 4. As seen in the figure, the multi-coil design resulted in a much uniform magnetic field in the activated pole length region, while the single coil drum brake resulted in a varying one.

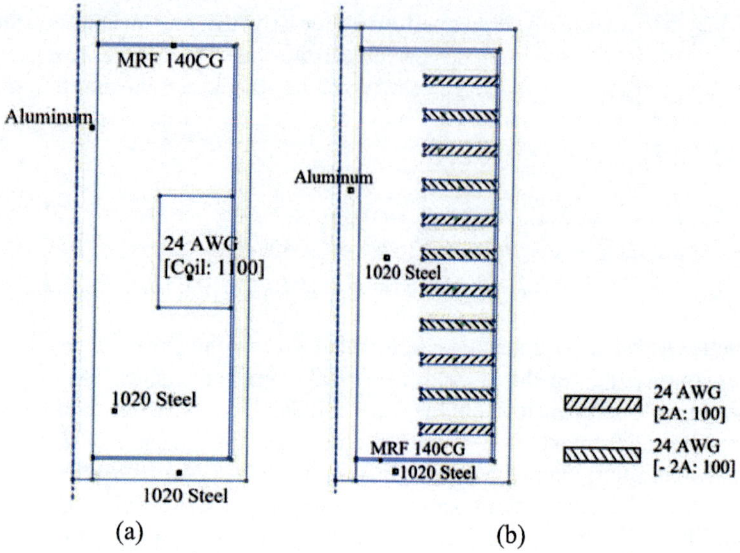

Fig. 2 **a** 2D axisymmetric model of single coil brake and **b** 2D axisymmetric model of multi-coil brake

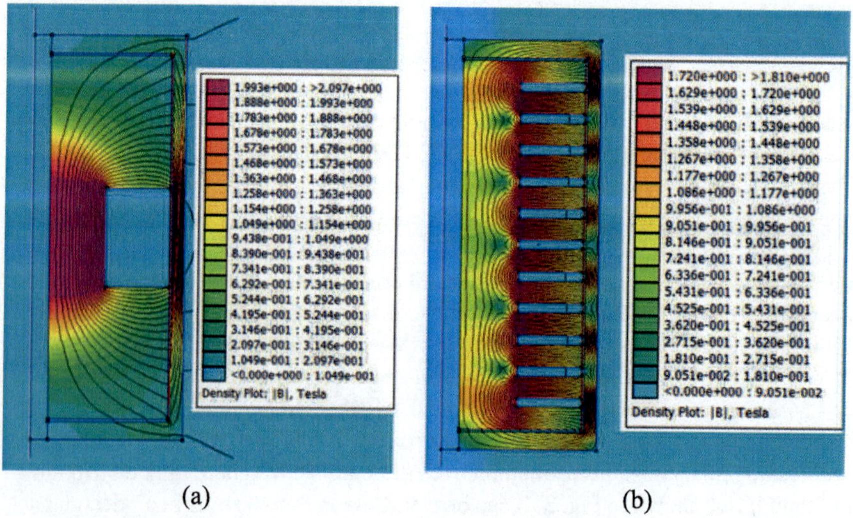

Fig. 3 Magnetostatic analysis of **a** Single coil rotary brake and **b** Multi-coil rotary brake

The magnetic field distribution of multi-coil brake in both the annular, as well as radial faces results in a total braking torque of 13.4 Nm. In a normal gait, the maximum swing phase braking torque required is around 14 Nm [11]. The analysis

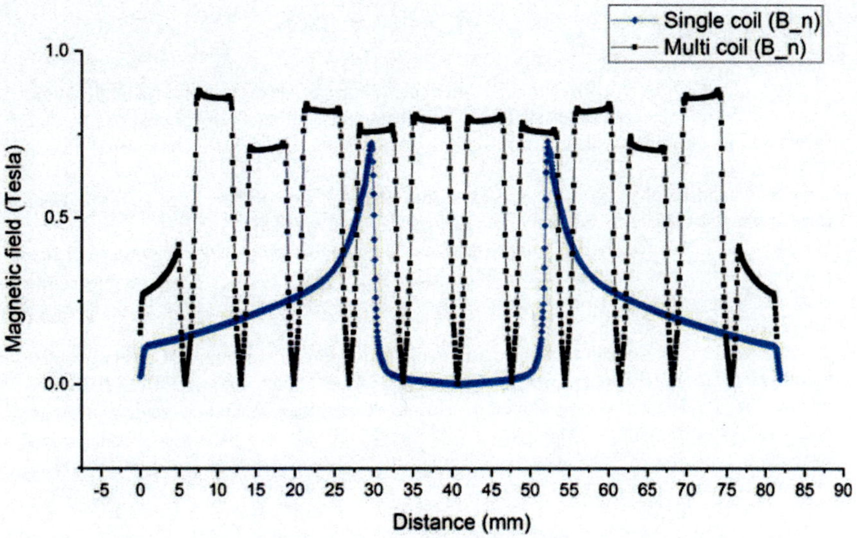

Fig. 4 Normal magnetic flux distribution in single coil and multi-coil brake

was performed at a design current of 2 A, an increase of current produces a much higher field satisfying the torque requirements of a normal swing phase motion.

7 Conclusion

In this paper, a multi-coil inverted drum brake was designed for prosthetic applications. The single coil and the multi-coil designs were compared keeping the total pole length, outer casing dimensions, coil area, design current as constant. The multi-coil drum brake produced a uniform distribution of magnetic flux in the activated pole length. The optimized inverted drum brake is capable of generating a braking torque up to 14 Nm. Although, normal human gait requires a torque of up to 30 Nm, the designed brake can satisfy the swing phase knee requirements of an average human. Moreover, the multi-coil brakes have a much reduced off-state braking torque when compared to the multi-disc brakes because of their lower shearing area.

Acknowledgements The Research was jointly supported by IMPRINT Project (Project No. IMPRINT/2016/7330) which is jointly funded by Ministry of Human Resource Development and Ministry of Road Transport and Highways, Govt. of India and SPARC Project (Project No. 785) funded by Ministry of Human Resource and Development.

References

1. Carlson, J.D., Matthis, W., Toscano, J.R.: Smart prosthetics based on magnetorheological fluids. In: Smart Structures and Materials 2001: Industrial and Commercial Applications of Smart Structures Technologies, vol. 4332, pp. 308–317. International Society for Optics and Photonics (2001)
2. Seid, S., Chandramohan, S., Sujatha, S.: Optimal design of an MR damper valve for prosthetic knee application. J. Mech. Sci. Technol. **32**(6), 2959–2965 (2018)
3. Gao, F., Liu, Y.N., Liao, W.H.: Optimal design of a magnetorheological damper used in smart prosthetic knees. Smart Mater. Struct. **26**(3), 035034 (2017)
4. Herr, H., Wilkenfeld, A.: User-adaptive control of a magnetorheological prosthetic knee. Ind. Robot: Int. J. **30**(1), 42–55 (2003)
5. Gudmundsson, K.H., Jonsdottir, F., Thorsteinsson, F.: A geometrical optimization of a magneto-rheological rotary brake in a prosthetic knee. Smart Mater. Struct. **19**(3), 035023 (2010)
6. Imaduddin, F., Mazlan, S.A., Zamzuri, H.: A design and modelling review of rotary magnetorheological damper. Mater. Des. **51**, 575–591 (2013)
7. Nguyen, Q.H., Lang, V.T., Nguyen, N.D., Choi, S.B.: Geometric optimal design of a magneto-rheological brake considering different shapes for the brake envelope. Smart Mater. Struct. **23**(1), 015020 (2013)
8. Nguyen, Q.H., Choi, S.B.: Optimal design of an automotive magnetorheological brake considering geometric dimensions and zero-field friction heat. Smart Mater. Struct. **19**(11), 115024 (2010)
9. Nguyen, Q.H., Lang, V.T., Choi, S.B.: Optimal design and selection of magneto-rheological brake types based on braking torque and mass. Smart Mater. Struct. **24**(6), 067001 (2015)
10. Lord technical data, https://www.lordmrstore.com/_.../Data_Sheet_-_MRF140CG_Magneto-Rheological_Fluid. Accessed 20 May 2019
11. Winter, D.A.: Biomechanics and Motor Control of Human Movement. Wiley (2009)

Recent Advances in Numerical Techniques

Single Variable New First-Order Shear Deformation Plate Theory: Numerical Solutions of Lévy-Type Plates Using Fourth-Order Runge-Kutta Technique

Himanshu Sawhney, Kedar S. Pakhare, Rameshchandra P. Shimpi, P. J. Guruprasad, and Yogesh M. Desai

1 Introduction

Classical plate theory (*CPT*) is one of the oldest and widely accepted plate theories for thin plates. Since it does not take into account effects of transverse shear in plate deformation, it gives reasonably good results only for thin plates. Whereas it underestimates deflections for shear-deformable plates (Wang et al. [1]).

To overcome drawbacks of *CPT*, Mindlin (Mindlin [2]) proposed displacement-based first-order shear deformation plate theory (*FSDT*). Mindlin's plate theory assumes constant transverse shear strains through the plate thickness. Hence, in order to approximately satisfy constitutive relations between transverse shear stresses and strains, *FSDT* requires shear correction factor.

Recently, Shimpi et al. [3] have developed a variationally inconsistent, single variable new first-order shear deformation plate theory (*SVNFSDT*). It is based on new first-order shear deformation plate theory by Shimpi et al. [3]. The authors have presented analytical solutions for shear-deformable isotropic square plates for various values of plate thickness-to-length ratio and having different boundary conditions. Unlike Mindlin's plate theory which consists of three coupled partial governing differential equations, *SVNFSDT* involves only one partial governing differential equation of fourth-order having one unknown function.

In this paper, numerical results pertaining to flexure of shear-deformable isotropic square plates with two opposite edges simply supported and remaining two edges having simply supported, clamped or free boundary conditions for different values

H. Sawhney (✉) · Y. M. Desai
Department of Civil Engineering, IIT Bombay, Mumbai 400076, India
e-mail: himanshusawhney20591@gmail.com

K. S. Pakhare · R. P. Shimpi · P. J. Guruprasad
Department of Aerospace Engineering, IIT Bombay, Mumbai 400076, India

S. K. Saha and M. Mukherjee (eds.), *Recent Advances in Computational Mechanics and Simulations*, Lecture Notes in Mechanical Engineering,
https://doi.org/10.1007/978-981-15-8315-5_40

of plate thickness-to-length ratio by utilizing fourth-order Runge-Kutta technique for *SVNFSDT* are presented. These results are compared with corresponding results reported in the literature so as to demonstrate the efficacy of the presented numerical solution technique.

2 Plate Under Consideration

The plate analyzed (as shown in Fig. 1) has the following properties:

1. The plate is of uniform thickness h.
2. The material of the plate is linear elastic, isotropic, and homogeneous. Hence, the relationship $G = E/[2(1 + \mu)]$ holds true where G is shear modulus, E is Young's modulus, and μ is Poisson's ratio.
3. The Cartesian coordinate system (right handed) $O - x - y - z$ is used in the paper. With regard to this coordinate system, following points should be noted:
   (a) The mid-surface of undeformed plate coincides with xy-plane.
   (b) The origin O can be placed at a convenient point on the mid-surface of the undeformed plate.
4. The loading $q(x, y)$ acts on the surface $z = -h/2$ of the plate. The sign convention for the loading is assumed to be positive when it acts in the positive z-direction.

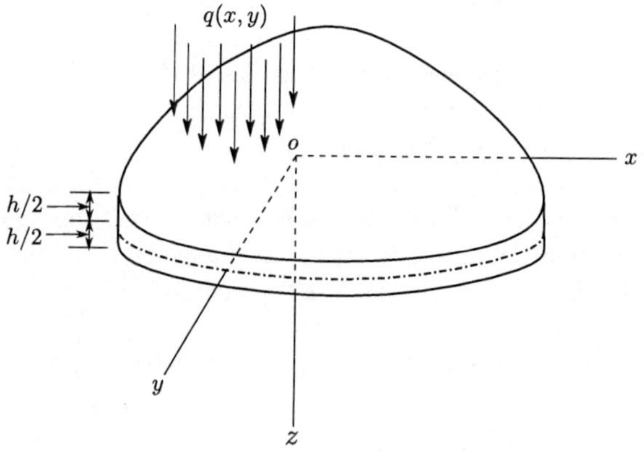

Fig. 1 Geometry of the plate

3 Governing Differential Equation and Boundary Conditions

3.1 Governing Differential Equation

As per Shimpi et al. [3], the transverse displacement w has bending component w_b and shear component w_s, such that

$$w(x, y) = w_b(x, y) + w_s(x, y) \tag{1}$$

Bending component w_b is related to shear component w_s through the following expression:

$$w_s = -\frac{h^2}{6\,k\,(1-\mu)}\left(\frac{\partial^2 w_b}{\partial x^2} + \frac{\partial^2 w_b}{\partial y^2}\right) \tag{2}$$

Here, "k" is a shear correction factor which is same as that of Mindlin [2].
Hence, transverse displacement w can be expressed only in terms of bending component w_b as follows:

$$w = w_b - \frac{h^2}{6\,k\,(1-\mu)}\left(\frac{\partial^2 w_b}{\partial x^2} + \frac{\partial^2 w_b}{\partial y^2}\right) \tag{3}$$

Fourth-order partial governing differential equation of *SVNFSDT* can be given as follows:

$$\frac{\partial^4 w_b}{\partial x^4} + 2\,\frac{\partial^4 w_b}{\partial x^2\,\partial y^2} + \frac{\partial^4 w_b}{\partial y^4} = \frac{q}{D} \tag{4}$$

where D is the plate rigidity. Equation (4) has a striking resemblance with the governing differential equation for *CPT* with the only difference of bending component w_b appearing instead of transverse displacement w.

3.2 Boundary Conditions

Shimpi et al. [3] have also stated physically meaningful plate edge conditions. For illustrative purpose, these conditions for the plate edge $y = b/2$ are as follows:

3.2.1 Simply Supported Edge Conditions

For simply supported edge, transverse displacement w and bending moment M_y are zero.

$$[w]_{y=b/2} = \left[w_b - \frac{h^2}{6\,k\,(1-\mu)} \left(\frac{\partial^2 w_b}{\partial x^2} + \frac{\partial^2 w_b}{\partial y^2} \right) \right]_{y=b/2} = 0 \tag{5}$$

$$\left[M_y\right]_{y=b/2} = -D \left[\frac{\partial^2 w_b}{\partial y^2} + \mu\, \frac{\partial^2 w_b}{\partial x^2} \right]_{y=b/2} = 0 \tag{6}$$

Eventually expressions (5) and (6) lead to following boundary conditions for simply supported edge:

$$[w_b]_{y=b/2} = 0 \tag{7}$$

$$\left[\frac{\partial^2 w_b}{\partial y^2} \right]_{y=b/2} = 0 \tag{8}$$

3.2.2 Free Edge Conditions

For free edge, following boundary conditions are valid:

$$\left[M_y\right]_{y=b/2} = -D \left[\frac{\partial^2 w_b}{\partial y^2} + \mu\, \frac{\partial^2 w_b}{\partial x^2} \right]_{y=b/2} = 0 \tag{9}$$

$$\left[Q_y + \frac{\partial M_{xy}}{\partial x} \right]_{y=b/2} = -D \left[\frac{\partial^3 w_b}{\partial y^3} + (2-\mu)\, \frac{\partial^3 w_b}{\partial x^2\, \partial y} \right]_{y=b/2} = 0 \tag{10}$$

3.2.3 Clamped Edge Conditions

Shimpi et al. [3] have proposed two types of plate clamped edge boundary conditions which are as follows:

1. Clamped edge type 1:

$$[w]_{y=b/2} = \left[w_b - \frac{h^2}{6\,k\,(1-\mu)} \left(\frac{\partial^2 w_b}{\partial x^2} + \frac{\partial^2 w_b}{\partial y^2} \right) \right]_{y=b/2} = 0 \tag{11}$$

$$\left[\frac{\partial w_b}{\partial y} \right]_{y=b/2} = 0 \tag{12}$$

2. Clamped edge type 2:

$$[w]_{y=b/2} = \left[w_b - \frac{h^2}{6\,k\,(1-\mu)} \left(\frac{\partial^2 w_b}{\partial x^2} + \frac{\partial^2 w_b}{\partial y^2} \right) \right]_{y=b/2} = 0 \tag{13}$$

$$\left[\frac{\partial w}{\partial y} \right]_{y=b/2} = \left[\frac{\partial w_b}{\partial y} - \frac{h^2}{6\,k\,(1-\mu)} \frac{\partial}{\partial y} \left(\frac{\partial^2 w_b}{\partial x^2} + \frac{\partial^2 w_b}{\partial y^2} \right) \right]_{y=b/2} = 0 \tag{14}$$

4 Illustrative Examples, Solution Procedure, Results and Discussion

4.1 Illustrative Examples

To show the efficacy of the numerical solution technique for *SVNFSDT*, following examples are solved:

1. Plate with all the edges simply supported (*SSSS*).
2. Plate with two opposite edges simply supported and remaining two edges clamped (*SCSC*). Results are presented for two types of clamped edge boundary conditions (Eqs. 12–14).
3. Plate with two opposite edges simply supported and remaining two edges free (*SFSF*). Results are presented at center of the plate as well as at center of one of the free edges of the plate.

These numerical results are compared with corresponding results reported in the literature. Percentage difference is also presented in Tables 1, 2, and 3 which is obtained using following expression:

$$\%\ \text{difference} = \left| 1 - \frac{\text{Result reported by the particular theory}}{\text{Corresponding result by numerical technique}} \right| \times 100 \tag{15}$$

4.2 Solution Procedure

Fourth-order partial differential equation (Eq. 4) is converted into an ordinary differential equation (*ODE*) using Lévy's method (Reddy et al. [5]). This fourth-order *ODE* is then reduced into a set of first-order *ODE*s which will consist of four unknowns

Table 1 Comparison of non-dimensional parameter for *SSSS* plate (example 1) under the action of uniformly distributed transverse load (Poisson's ratio μ = 0.3)

	Non-dimensional plate transverse displacement $\overline{w} = \frac{w\,D}{q_o\,a^4}$ at $x = a/2,\ y = 0$ with $a/b = 1$ (Values in parentheses indicate % difference †)			
Theory	$h/a = 0.01$	$h/a = 0.05$	$h/a = 0.10$	$h/a = 0.20$
Present$	0.00406	0.00412	0.00427	0.00490
SVNFSDT$ [3]	0.00407 (0.25 %)	0.00412 (0.00 %)	0.00427 (0.00 %)	0.00490 (0.00 %)
CPT$ [3]	0.00406 (0.00 %)	0.00406 (1.46 %)	0.00406 (4.92 %)	0.00406 (17.14 %)
Mindlin plate theory$ [4]	0.00406 (0.00 %)	0.00411 (0.24 %)	0.00427 (0.00 %)	0.00490 (0.00 %)
	Non-dimensional bending moment $\overline{M^x} = \frac{M_x}{q_o\,a^2}$ at $x = a/2,\ y = 0$ with $a/b = 1$ (Values in parentheses indicate % difference †)			
Theory	$h/a = 0.01$	$h/a = 0.05$	$h/a = 0.10$	$h/a = 0.20$
Present$	0.0479	0.0479	0.0479	0.0479
SVNFSDT$ [3]	0.0479 (0.00 %)	0.0479 (0.00 %)	0.0479 (0.00 %)	0.0479 (0.00 %)
CPT$ [3]	0.0479 (0.00 %)	0.0479 (0.00 %)	0.0479 (0.00 %)	0.0479 (0.00 %)
Mindlin plate theory$ [4]	0.0479 (0.00 %)	0.0479 (0.00 %)	0.0479 (0.00 %)	0.0479 (0.00 %)

† % difference is calculated by utilizing Eq. (15)

$ Shear correction factor of 5/6 is utilized

and hence will require four boundary conditions. For a particular case, two boundary conditions are available at each edge. Such type of problem falls under the category of boundary value problem (*BVP*). *BVP* is converted into an initial value problem (*IVP*). This *IVP* is then solved using fourth-order Runge-Kutta method which is one of the most prominently used numerical analysis techniques. The transformation of *BVP* to *IVP* is done by transferring the boundary conditions at the final edge to the initial edge such that there are now four known boundary conditions at the initial edge. This procedure is explained in detail by LomtePatil et al. [6].

Table 2 Comparison of non-dimensional plate transverse displacement for *SCSC* plate (example 2) under the action of uniformly distributed transverse load (Poisson's ratio $\mu = 0.3$)

	Non-dimensional plate transverse displacement $\overline{w} = \frac{w\,D}{q_o\,a^4}$ at $x = a/2$, $y = 0$ with $a/b = 1$ (Values in parentheses indicate % difference †)			
Theory	$h/a = 0.01$	$h/a = 0.05$	$h/a = 0.10$	$h/a = 0.20$
Present clamp type 1$	0.00192	0.00199	0.00222	0.00308
Present clamp type 2$	0.00192	0.00193	0.00196	0.00210
SVNFSDT clamp type 1$ [3]	0.00192 (0.00 %)	0.00199 (0.00 %)	0.00222 (0.00 %)	0.00308 (0.00 %)
SVNFSDT clamp type 2$ [3]	0.00192 (0.00 %)	0.00193 (0.00 %)	0.00196 (0.00 %)	0.00210 (0.00 %)
CPT$ [3]	0.00192 (0.00 %)	0.00192 (3.52 %)	0.00192 (13.51 %)	0.00192 (37.66 %)
Mindlin plate theory$ [4]	0.00192 (0.00 %)	0.00199 (0.00 %)	0.00221 (0.45 %)	0.00302 (1.95 %)

† % difference is calculated by utilizing Eq. (15).
$ Shear correction factor of 5/6 is utilized

4.3 *Results and Discussion*

Following observations can be made from Tables 1, 2, and 3:

1. For *SSSS* plate (example 1, Table 1), it is evident that $\overline{w}$ and $\overline{M}_x$ obtained using numerical technique exactly match with corresponding analytical results of *SVNFSDT* and Mindlin plate theory.
2. For *SCSC* plate (example 2, Table 2), it is evident that $\overline{w}$ obtained using numerical technique exactly match with corresponding analytical results of *SVNFSDT* (for both types of plate clamp edge conditions). These results are also in good agreement with corresponding results of Mindlin plate theory (maximum percentage difference of 1.95 % for $h/a = 0.20$).
3. For *SFSF* plate (example 3, Table 3), it is evident that $\overline{w}$ obtained using numerical technique exactly match with corresponding analytical results of *SVNFSDT*. These results are also in good agreement with corresponding results of Mindlin plate theory (maximum percentage difference of 0.90 % for $h/a = 0.20$ at the center of the plate, maximum percentage difference of 4.58 % for $h/a = 0.20$ at the center of one of the edges of the plate).

Table 3 Comparison of non-dimensional plate transverse displacement for *SFSF* plate (example 3) under the action of uniformly distributed transverse load (Poisson's ratio $\mu = 0.3$)

	Non-dimensional plate transverse displacement $\overline{w} = \frac{w\,D}{q_o\,a^4}$ at $x = a/2$, $y = 0$ with $a/b = 1$ (Values in parentheses indicate % difference †)			
Theory	$h/a = 0.01$	$h/a = 0.05$	$h/a = 0.10$	$h/a = 0.20$
Present$	0.01310	0.01318	0.01342	0.01441
SVNFSDT$ [3]	0.01310 (0.00 %)	0.01318 (0.00 %)	0.01342 (0.00 %)	0.01441 (0.00 %)
CPT$ [3]	0.01309 (0.08 %)	0.01309 (0.68 %)	0.01309 (2.46 %)	0.01309 (9.16 %)
Mindlin plate theory$ [4]	0.01310 (0.00 %)	0.01319 (0.08 %)	0.01346 (0.30 %)	0.01454 (0.90 %)
	Non-dimensional plate transverse displacement $\overline{w} = \frac{w\,D}{q_o\,a^4}$ at $x = a/2$, $y = b/2$ with $a/b = 1$ (Values in parentheses indicate % difference †)			
Theory	$h/a = 0.01$	$h/a = 0.05$	$h/a = 0.10$	$h/a = 0.20$
Present$	0.01501	0.01508	0.01530	0.01616
SVNFSDT$ [3]	0.01501 (0.00 %)	0.01508 (0.00 %)	0.01530 (0.00 %)	0.01616 (0.00 %)
CPT$ [3]	0.01501 (0.00 %)	0.01501 (0.46 %)	0.01501 (1.90 %)	0.01501 (7.12 %)
Mindlin plate theory$ [4]	0.01504 (0.20 %)	0.01522 (0.93 %)	0.01560 (1.96 %)	0.01690 (4.58 %)

† % difference is calculated by utilizing Eq. (15)

$ Shear correction factor of 5/6 is utilized

5 Conclusion

Concluding remarks with regard to the work presented in this paper are as follows:

1. Effectiveness of the fourth-order Runge-Kutta numerical technique used for the flexure of shear-deformable isotropic plates is presented through illustrative examples. The results obtained using this technique are in good agreement with corresponding results reported in the literature.
2. The successful use of numerical solution technique for this theory opens the possibility of using it for the cases for which analytical solutions are difficult to obtain.

References

1. Wang, C.M., Reddy, J.N., Lee, K.H.: Shear Deformable Beams and Plates: Relationships with Classical Solutions. Elsevier Science Ltd., New York (2000)
2. Mindlin, R.D.: Influence of rotary inertia and shear on flexural motions of isotropic, elastic plates. ASME J. Appl. Mech. **18**, 31–38 (1951)
3. Shimpi, R.P., Guruprasad, P.J., Pakhare, K.S.: Single variable new first-order shear deformation theory for isotropic plates. Lat. Am. J. Solids Strut. **15**, 1–25 (2018). https://doi.org/10.1590/1679-78254842
4. Lee, K.H., Lim, G.T., Wang, C.M.: Thick Lévy plates re-visited. Int. J. Solids Struct. **39**, 127–144 (2018). https://doi.org/10.1016/S0020-7683(01)00205-0
5. Reddy, J.N., Khdeir, A.A., Librescu, L.: Lévy type solutions for symmetrically laminated rectangular plates using first-order shear deformation theory. ASME J. Appl. Mech. **54**, 740–742 (1987). https://doi.org/10.1115/1.3173104
6. LomtePatil, Y.T., Kant, T., Desai, Y.M.: Comparison of three dimensional elasticity solutions for functionally graded plates. Compos. Struct. **202**, 424–435 (2018). https://doi.org/10.1016/j.compstruct.2018.02.051

Investigation of Some Recently Proposed Explicit Time Integration Schemes for Nonlinear Problems

Abhijeet Singh, Rishiraj K. Thakur, Vishal Agrawal, and Sachin S. Gautam

1 Introduction

Since the last couple of decades, numerical solutions of partial differential equations (PDEs) that arise from the governing equations of motion have attracted considerable attention. Also, the increase in the computational power of computers in the past few years has enabled the researchers to deal with more and more complex problems. Transient problems like impact, crash, bird strike, etc. constitute some of the most challenging problems. For solving the transient problems, an accurate and efficient solution of corresponding PDEs is required. To solve the PDEs up to a desired accuracy and appropriate computational cost, a suitable time integration scheme is needed [1]. Special attention and care should be given while selecting an analysis suitable time integration scheme for a particular problem. Selection of time integration scheme becomes more difficult in the presence of nonlinearities like geometrical, material, contact, etc [2].

Usually, time integration schemes are classified as (1) direct time integration and (2) mode superposition [3]. Further direct time integration schemes are categorized as (1) explicit schemes and (2) implicit schemes [3]. Explicit schemes are computationally efficient but are limited by time step size for stability [3], while the implicit schemes are unconditionally stable (at least for linear problems) but are computationally expensive due to the repeated computation of the tangent matrix. Therefore, for the multiple degree-of-freedom systems having various nonlinearities, explicit algorithms are used due to the inherent requirement of small time steps to effectively capture the phenomenon. For certain kind of problems like contact-impact problems where the solution is required for a very short span of time, explicit

A. Singh · R. K. Thakur · V. Agrawal · S. S. Gautam (✉)
Indian Institute of Technology, Guwahati 781039, India
e-mail: ssg@iitg.ac.in
URL: https://iitg.ac.in/ssg/

S. K. Saha and M. Mukherjee (eds.), *Recent Advances in Computational Mechanics and Simulations*, Lecture Notes in Mechanical Engineering,
https://doi.org/10.1007/978-981-15-8315-5_41

algorithms are used. Inaccuracies are also very intrinsic to time integration schemes which makes certain schemes more suitable over others. In recent years, a number of explicit time integration schemes have been proposed with improved/desirable stability/accuracy [4–8]. When a problem is numerically solved over a discretized domain, various nonphysical modes get excited which gives rise to spurious oscillations. Traditional schemes like the central difference scheme [3] are unable to resolve these oscillations due to their non-dissipative nature. To cater to such problems, various schemes have been proposed to induce numerical dissipation and hence dampen the nonphysical modes. However, the performance of most of these schemes has not been compared with each other for nonlinear problems. Hence, the objective of the present work is to investigate the performance of some recently proposed explicit time integration schemes for transient problems with the focus on nonlinear problems. The remainder of the paper is structured as follows. Section 2 briefly presents the different time integration schemes used for investigation in this paper. Section 3 presents the results for the schemes for two nonlinear problems. Section 4 concludes the paper.

2 Time Integration Schemes

A general system is said to be in dynamic equilibrium if it satisfies Newton's second law of motion which, in the finite element discretized form [3], can be written for a nonlinear system as

$$\boldsymbol{M}\boldsymbol{A} + \boldsymbol{P} = \boldsymbol{F}, \tag{1}$$

where $\boldsymbol{M}$ is the mass matrix, $\boldsymbol{A}$ is the acceleration vector, $\boldsymbol{P}$ is the internal force vector which has contribution from internal stresses, and $\boldsymbol{F}$ is the external force vector which has contributions from externally applied traction and contact forces. The internal force vector is generally a function of velocity vector $\boldsymbol{V}$ as well as displacement vector $\boldsymbol{U}$. In some special cases, the internal force independently depends on $\boldsymbol{V}$ and $\boldsymbol{U}$ which then proceeds to establish the relationship given by Eq. (1) for linear systems as

$$\boldsymbol{M}\boldsymbol{A} + \boldsymbol{C}\boldsymbol{V} + \boldsymbol{K}\boldsymbol{U} = \boldsymbol{F}, \tag{2}$$

where $\boldsymbol{C}$ and $\boldsymbol{K}$ denote the damping matrix and the stiffness matrix of the system, respectively.

To solve the system of set of equations given by Eqs. (1) or (2) over a time $t \in (0, T_{\text{end}})$ for displacement vector $\boldsymbol{U}$, velocity vector $\boldsymbol{V}$, and acceleration vector $\boldsymbol{A}$, an appropriate time integration scheme is needed which essentially finds the solution of the equations after every Δt. In the present work, explicit time integration schemes are selected. The advantage of an explicit scheme is that the computation of material tangent matrix is generally not needed in every time step Δt. This results in significant saving of the computational time and thus faster per time step calculation as compared to the implicit schemes. However, the choice of the time step is now restricted by the famous Courant–Friedrichs–Lewy (CFL) condition [1, 3]. Next, the details of the explicit time integration schemes used in the present work are presented.

2.1 *Standard Central Difference Scheme (SCDS)*

In the central difference scheme [3], the approximations for the displacement and velocity vectors at time $t + \Delta t$ are given by

$$\boldsymbol{U}_{t+\Delta t} = \boldsymbol{U}_t + \Delta t \boldsymbol{V}_t + 0.5\,(\Delta t)^2\,\boldsymbol{A}_t\,, \tag{3}$$

$$\boldsymbol{V}_{t+\Delta t} = \boldsymbol{V}_t + \frac{1}{2}\Delta t\,(\boldsymbol{A}_t + \boldsymbol{A}_{t+\Delta t})\,, \tag{4}$$

where the quantities at time t are written with subscript t, i.e., $(\cdot)_t$. The advantage of the central difference scheme is that it is very simple to implement and is computationally very efficient.

2.2 *Corrected Explicit Method of Double Time Steps (CEMDTS)*

Yang et al. [4] proposed an explicit scheme which is obtained after correcting the explicit method of double time steps (EMDTS) [4] which is called corrected explicit method of double time steps (CEMDTS) [4]. Its dissipative and stability properties are same as SCDS. The approximations for displacement and velocity at time $t + \Delta t$ are given as

$$\boldsymbol{U}_{t+\Delta t} = \boldsymbol{U}_t + \Delta t \boldsymbol{V}_{t+\Delta t} - \eta\,(\Delta t)^2\,\boldsymbol{A}_t\,, \tag{5}$$

$$\boldsymbol{V}_{t+\Delta t} = \boldsymbol{V}_t + (1+\eta)\,\Delta t \boldsymbol{A}_t - \eta \Delta t \boldsymbol{A}_{t-\Delta t}\,, \tag{6}$$

where η is a variable parameter which varies from 0 to 1. The value of η is suggested as 0.25 for obtaining good results.

2.3 *Stabilized Central Difference Scheme (SzCDS)*

Groβeholz et al. [5] proposed an explicit scheme that aims to stabilize the SCDS by slightly perturbing Eq. (1) as

$$\left[\boldsymbol{M} + a\,(\Delta t)^2\,\boldsymbol{DP}|_{\boldsymbol{U}}\right]\boldsymbol{A} + \boldsymbol{P}(\boldsymbol{U}) = \boldsymbol{F}. \tag{7}$$

The displacement and velocity discretization procedure is carried out similar to SCDS. The term $\boldsymbol{DP}|_{\boldsymbol{U}}$ denotes the Jacobian matrix of $\boldsymbol{P}(\boldsymbol{U})$, evaluated at $\boldsymbol{U}$, and $a = 0.25 \tanh\left(\alpha \Delta t \omega_f\right)$ is a positive scalar function that serves as a limiter. Here, ω_f can be taken as the highest natural frequency of the differential equation. As suggested by Groβeholz et al. [5], the value of α is taken as 0.25, which is determined from the stability analysis of the scheme.

2.4 Explicit Scheme by Chang et al. (CTES)

Chang et al. [6] proposed a one-parameter explicit scheme which possesses unconditional stability under certain conditions. The approximations for displacement and velocity at time $t + \Delta t$ are given as

$$\boldsymbol{U}_{t+\Delta t} = \left[\boldsymbol{I} - (\Delta t)^2 \boldsymbol{B}_1 \left(\boldsymbol{M}^{-1}\boldsymbol{K}\right)\right] \boldsymbol{U}_t + \Delta t \boldsymbol{B}_2 \boldsymbol{V}_t + (\Delta t)^2 \boldsymbol{B}_3 \boldsymbol{A}_t\,, \tag{8}$$

$$\boldsymbol{V}_{t+\Delta t} = \boldsymbol{V}_t + \Delta t \left[\frac{3p-1}{2(p+1)}\boldsymbol{A}_t + \frac{3-p}{2(p+1)}\boldsymbol{A}_{t+\Delta t}\right], \tag{9}$$

where $\boldsymbol{D}$, $\boldsymbol{B}_1$, $\boldsymbol{B}_2$, and $\boldsymbol{B}_3$ are given as

$$\boldsymbol{D} = \frac{2}{p+1}\boldsymbol{M} + \frac{3-p}{2(p+1)}\Delta t \boldsymbol{C}_0 + \frac{1}{(p+1)^2}(\Delta t)^2 \boldsymbol{K}_0\,, \tag{10}$$

$$\boldsymbol{B}_1 = \frac{1}{(p+1)^2}\boldsymbol{D}^{-1}\boldsymbol{M}\,, \tag{11}$$

$$\boldsymbol{B}_2 = \boldsymbol{D}^{-1}\left[\frac{2}{p+1}\boldsymbol{M} - \frac{p^2-2p-1}{2(p+1)^2}\Delta t \boldsymbol{C}_0\right], \tag{12}$$

$$\boldsymbol{B}_3 = \boldsymbol{D}^{-1}\left[\frac{2}{(p+1)^2}\boldsymbol{M} - \frac{(p-1)^2}{4(p+1)^2}\Delta t \boldsymbol{C}_0\right]. \tag{13}$$

Here, $\boldsymbol{C}_0$ and $\boldsymbol{K}_0$ denote initial damping matrix and initial stiffness matrix, respectively. Acceleration at time $t + \Delta t$ can be obtained by modifying Eq. (2) as

$$\left(\frac{2}{p+1}\right)\boldsymbol{M}\boldsymbol{A}_{t+\Delta t} + \left(\frac{p-1}{p+1}\right)\boldsymbol{M}\boldsymbol{A}_t + \boldsymbol{C}\boldsymbol{V}_{t+\Delta t} + \boldsymbol{K}\boldsymbol{U}_{t+\Delta t} = \boldsymbol{F}_{t+\Delta t}. \tag{14}$$

For values of parameter $|p| \leq 1$, the scheme is unconditionally stable. The value of $p = 1$ gives least amplitude decay and as the value of p is decreased, amplitude decay increases.

2.5 Pajand and Rad Predictor–Corrector Explicit Scheme (PRPCES)

Pajand and Rad [7] proposed an explicit scheme having two sub-steps in which first step is the predictor step while the second step is the corrector step. For the first sub-step (predictor step), the approximations for displacement and velocity at time $t + \Delta t$ are given by

$$\tilde{\boldsymbol{U}}_{t+\Delta t} = \boldsymbol{U}_t + \frac{\Delta t}{40}(33\boldsymbol{V}_t + 7\boldsymbol{V}_{t-\Delta t}) + \frac{(\Delta t)^2}{80}(61\boldsymbol{A}_t - 7\boldsymbol{A}_{t-\Delta t}) \tag{15}$$

$$\tilde{\boldsymbol{V}}_{t+\Delta t} = \boldsymbol{V}_t + \frac{\Delta t}{20}\left(11\tilde{\boldsymbol{A}}_{t+\Delta t} + 8\boldsymbol{A}_t + \boldsymbol{A}_{t-\Delta t}\right). \tag{16}$$

The approximation for predicted acceleration at time $t + \dfrac{\Delta t}{2}$ is written as

$$\tilde{\boldsymbol{A}}_{t+\frac{\Delta t}{2}} = \frac{\tilde{\boldsymbol{V}}_{t+\Delta t} - \boldsymbol{V}_t}{\Delta t}. \tag{17}$$

For the second sub-step (corrector step), the approximations for displacement and velocity, respectively, at time $t + \Delta t$ can be written as

$$\boldsymbol{U}_{t+\Delta t} = \boldsymbol{U}_t + \frac{\Delta t}{40}(33\boldsymbol{V}_{t+\Delta t} + 7\boldsymbol{V}_t) + \frac{(\Delta t)^2}{40}\left(7\boldsymbol{A}_t - 25\tilde{\boldsymbol{A}}_{t+\frac{\Delta t}{2}}\right) \tag{18}$$

$$\boldsymbol{V}_{t+\Delta t} = \boldsymbol{V}_t + \frac{\Delta t}{40}\left(33\tilde{\boldsymbol{A}}_{t+\Delta t} - 26\tilde{\boldsymbol{A}}_{t+\frac{\Delta t}{2}} + 33\boldsymbol{A}_t\right). \tag{19}$$

2.6 Pajand and Rad Explicit Scheme (PRES)

Pajand and Rad [8] proposed an explicit scheme having two sub-steps. It is numerically dissipative in nature that helps to suppress unwanted high-frequency modes. For the first sub-step, the approximations for displacement and velocity at time $t + \frac{\Delta t}{2}$ are given by

$$\boldsymbol{U}_{t+\frac{\Delta t}{2}} = \boldsymbol{U}_t + \frac{\Delta t}{2}\boldsymbol{V}_t \tag{20}$$

$$\boldsymbol{V}_{t+\frac{\Delta t}{2}} = \boldsymbol{V}_t + \frac{\Delta t}{2}\boldsymbol{A}_t. \tag{21}$$

Acceleration at time $t + \frac{\Delta t}{2}$ can be obtained by modifying Eq. (2) as

$$\boldsymbol{M}\boldsymbol{A}_{t+\frac{\Delta t}{2}} + \boldsymbol{C}\boldsymbol{V}_{t+\frac{\Delta t}{2}} + \boldsymbol{K}\boldsymbol{U}_{t+\frac{\Delta t}{2}} = \frac{1}{2}(\boldsymbol{F}_{t+\Delta t} + \boldsymbol{F}_t). \tag{22}$$

For the second sub-step, the approximations for displacement and velocity, respectively, at time $t + \Delta t$ can be written as

$$U_{t+\Delta t} = \frac{4}{3}\hat{U}_t - \frac{1}{3}U_t + \frac{\Delta t}{3}V_{t+\Delta t} \tag{23}$$

$$V_{t+\Delta t} = V_t + \Delta t A_{t+\frac{\Delta t}{2}}, \tag{24}$$

where $\hat{U}_t$ is given as

$$\hat{U}_t = U_t + \frac{\Delta t}{2}V_t + \frac{(\Delta t)^2}{4\alpha}A_{t+\frac{\Delta t}{2}} + \frac{(\alpha - 2)(\Delta t)^2}{8\alpha}A_t. \tag{25}$$

In this scheme, least relative period error occurs in the range of $0.8 \leq \alpha \leq 1$. For current work, $\alpha = 0.85$ is considered.

3 Results and Discussion

In the present section, the performance of the schemes discussed in Sect. 2 is presented for two nonlinear problems.

3.1 *Multiple Degree-of-Freedom Nonlinear System Subjected to Harmonic Ground Acceleration*

In the first problem, see Fig. 1, an n-degree-of-freedom spring–mass system [6] is subjected to a harmonic ground acceleration = $100\,[\sin(2t) + \sin(10t)]\ \mathrm{m/s^2}$ as shown in Fig. 1. The problem has been solved for displacement of the nth mass (u_n) against time. For obtaining the solution, $n = 500$ is considered. The plot of displacement of 500th mass m_{500} with time, obtained using SCDS with a very small time step of $\Delta t = 1 \times 10^{-8}$ seconds, is shown in Fig. 2. Since all the schemes would give

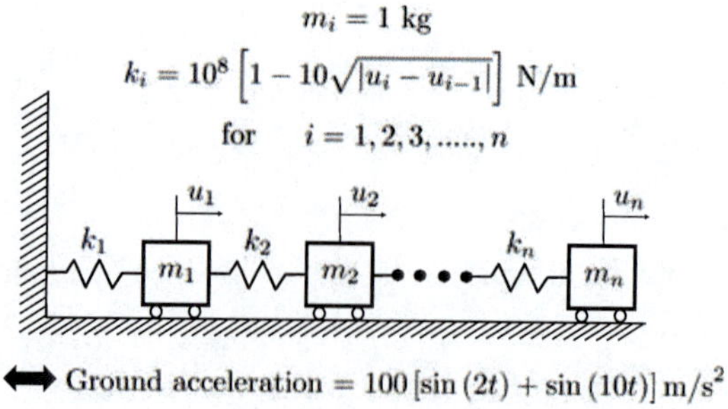

Fig. 1 Multiple degree-of-freedom nonlinear system subjected to harmonic ground acceleration

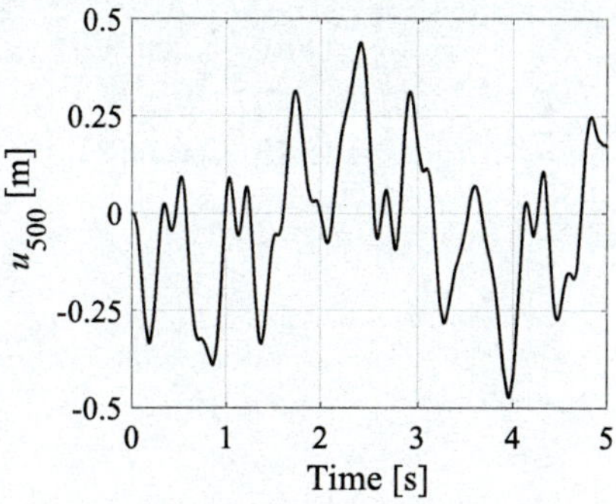

Fig. 2 Displacement (u_{500}) as a function of time for first problem

the similar results for this small time step they are not shown. However, since the system is conservative, a prime concern is the conservation of energy of the system in absence of any dissipative forces. To show the performance of various schemes for energy conservation, a quality called the "Mean absolute error (MAE)" in energy is defined. Mean absolute error (MAE) of any quantity z (z can be any quantity but in the present work it is chosen to be energy of the system) having exact solution $z_{ex}(i)$ and a numerical solution $z_{num}(i)$ at ith data point and having a total of N data points can be defined as

$$\mathrm{MAE}(z) = \frac{1}{N}\sum_{i=1}^{N}\left|\frac{z_{num}(i) - z_{ex}(i)}{z_{ex}(i)}\right|. \tag{26}$$

The exact solution corresponds to the solution obtained using a very small time step which here is taken as one obtained using $\Delta t = 1 \times 10^{-8}$ s. Essentially this means that the errors at each time step are added and a cumulative effect is obtained. Figure 3 shows the plot of MAE in energy for all the schemes for three different time steps.

It can be seen that the SCDS scheme has the lowest error among all the schemes except for two cases where PRPCES has the minimum error. However, it is clear that SCDS outperforms all the other schemes. The PRES is least accurate among all the schemes.

3.2 *One Degree-of-Freedom Adhesive Contact Problem*

Next, a highly nonlinear single degree-of-freedom adhesive contact problem is solved with different explicit time integration schemes [9]. The problem is conservative which means that there is no energy dissipation. Hence, the total energy of the system remains constant. Figure 4 shows the problem setup. The problem consists of one fixed rigid mass and another rigid free mass. Both the masses are acted upon by van der Waals adhesive force which is derived from the Lennard–Jones potential. The

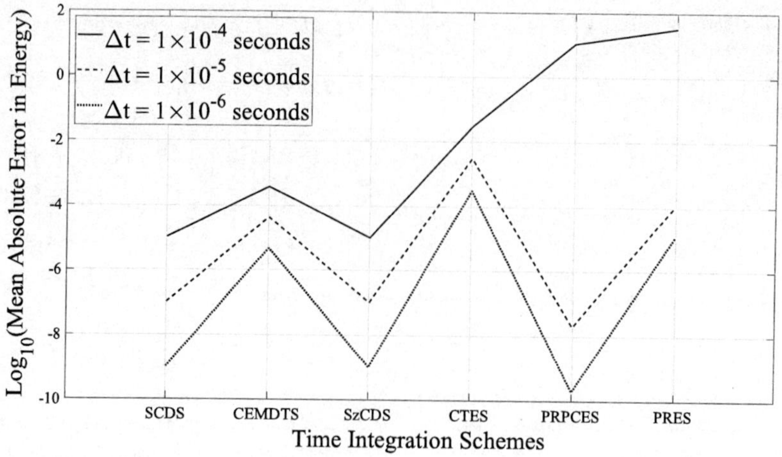

Fig. 3 Mean absolute error (MAE) of energy at different values of Δt for all six time integration schemes under study for first problem

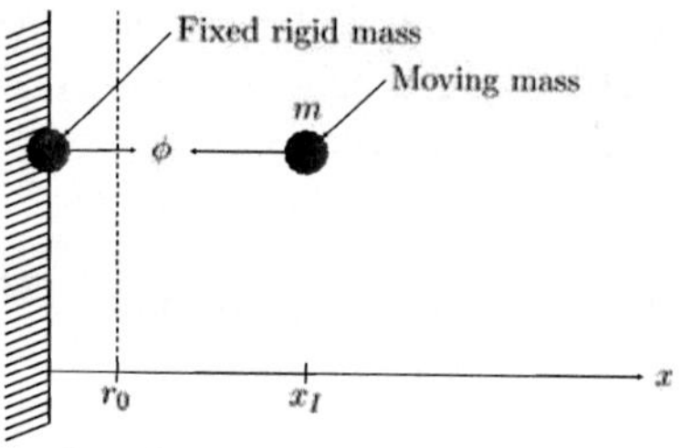

Fig. 4 One degree-of-freedom adhesive contact problem

equilibrium distance for the Lennard–Jones potential $\phi(x) = \epsilon\left(\frac{r_0}{x}\right)^{12} - 2\epsilon\left(\frac{r_0}{x}\right)^{6}$ is given by r_0. Here, ϵ is the material constant related to energy scale. The free mass is left from initial position $\boldsymbol{x}_I$. The free mass is attracted toward the rigid mass due to attractive force component of the van der Waals force till it crosses the equilibrium distance and is repelled by the repulsive part of the van der Waals force. In this system, there are two force contributions, namely, inertia and contact forces which mutually balance out each other to attain dynamic equilibrium. For complete details, the reader is referred to Gautam and Sauer [9]. The complete formulation of the problem including the non-dimensionalization of the governing equation of the system is given in Gautam and Sauer [9]. The resulting non-dimensionalized equation is given by

$$X'' - \frac{\alpha_0}{\gamma_L^8}\left(\gamma_L^{-6}X^{-13} - X^{-7}\right) = 0\,, \tag{27}$$

where X is the non-dimensionalized distance and $(\cdot)''$ denotes the double derivative with respect to non-dimensionalized time ω. The value of the constant α_0 is given by $\alpha_0 = \left(\frac{1}{3}\right)\left(\frac{13}{4}\right)^{\frac{4}{3}}$. For analyzing the system, γ_L is chosen as 1 [9].

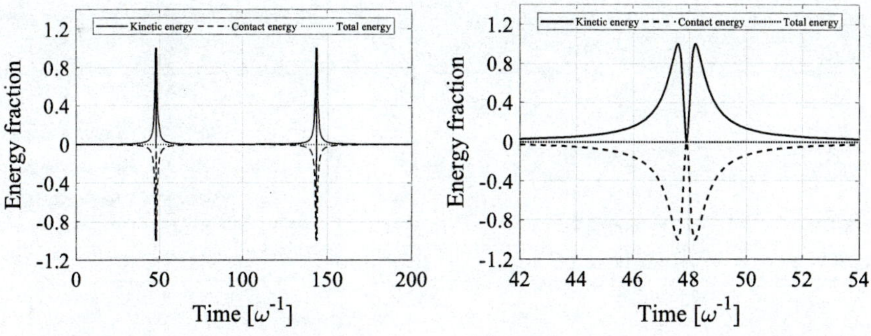

(a) Energy fraction during whole time (b) Zoomed view of energy fraction plot

Fig. 5 Plot of energy fraction versus time for second problem

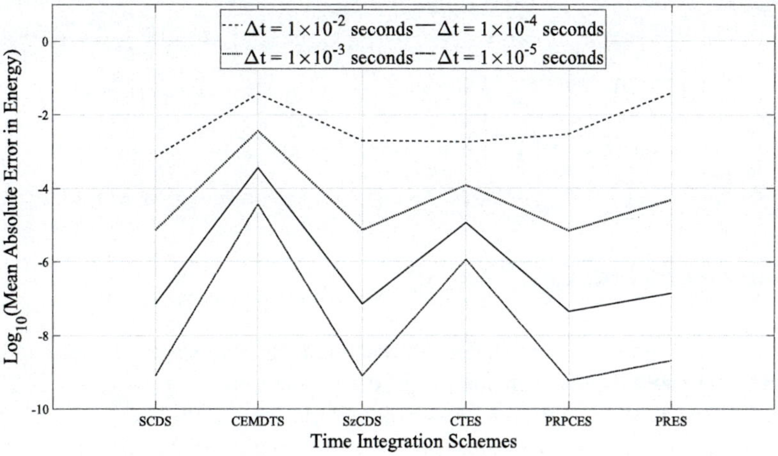

Fig. 6 Mean absolute error (MAE) of energy at different values of Δt for all six time integration schemes under study for second problem

Figure 5 shows the plot of energy fraction with non-dimensionalized time ω. The total time for the analysis is kept at $\omega = 200$ which corresponds to two impacts. The peaks denote the time instance when the moving mass crosses the equilibrium distance and is repelled from the fixed mass and crosses back the equilibrium distance. The SCDS scheme is used as the time integration scheme with $\Delta t = 1 \times 10^{-7}$ seconds and as the kinetic energy increases the contact energy decreases so that the total energy remain constant. The plot of mean absolute error (MAE) of total energy at different values of Δt for all six time integration schemes till $\omega = 200$ is shown in Fig. 6. It can be seen that the SCDS, the SzCDS, and the PRPCES schemes give almost the same error in energy while the other schemes give much higher error. For all the schemes, the error in energy decreases with time step which is as expected.

Table 1 Computational cost for first problem

Scheme	1×10^{-4} s	1×10^{-5} s	1×10^{-6} s
SCDS	0.3280×10^{2} s	0.3246×10^{3} s	0.3264×10^{4} s
CEMDTS	0.3402×10^{2} s	0.3400×10^{3} s	0.3334×10^{4} s
SzCDS	7.9889×10^{2} s	8.0145×10^{3} s	7.9983×10^{4} s
CTES	2.0604×10^{2} s	1.9566×10^{3} s	2.0253×10^{4} s
PRPCES	1.8660×10^{2} s	1.7566×10^{3} s	1.8366×10^{4} s
PRES	0.6700×10^{2} s	0.6673×10^{3} s	0.6819×10^{4} s

Table 2 Computational cost for second problem

Scheme	1×10^{-2} s	1×10^{-3} s	1×10^{-4} s	1×10^{-5} s
SCDS	1.2290×10^{-2} s	1.2512×10^{-1} s	1.2482 s	12.370 s
CEMDTS	2.6865×10^{-2} s	1.6527×10^{-1} s	1.6172 s	16.399 s
SzCDS	44.6880×10^{-2} s	47.1470×10^{-1} s	46.2370 s	463.240 s
CTES	1.2318×10^{-2} s	1.2367×10^{-1} s	1.2860 s	12.480 s
PRPCES	2.2540×10^{-2} s	1.8778×10^{-1} s	1.8779 s	18.713 s
PRES	3.0020×10^{-2} s	1.9170×10^{-1} s	1.8761 s	18.484 s

3.3 *Computational Cost Analysis*

Next, the computational cost associated with each scheme is presented. Tables 1 and 2 show the time taken by each scheme for different time steps for problems 1 and 2, respectively. In each table, the upper row indicates the value of the time step Δt used for evaluating the solution. The remaining rows show the time taken by a particular time integration scheme to obtain the solution for a given value of Δt for a particular problem. It can be seen that the central difference scheme (SCDS) takes least amount of time whereas the stabilized central difference scheme (SzCDS) takes maximum time among all the schemes. This is due to the fact that the Jacobian matrix (see Eq. 7) needs to be evaluated for each time step in the stabilized central difference scheme (SzCDS).

4 Conclusion

In the present work, the performance of a number of explicit time integration schemes is compared for nonlinear problems. It is shown that the central difference scheme gives least error combined with least computational time for any time step among all the schemes. Hence, it can be concluded that although a number of explicit schemes have been proposed in the literature, the central difference scheme performs best for nonlinear problems.

Acknowledgements The authors are grateful to the SERB, DST for supporting this research under project SR/FTP/ETA-0008/2014.

References

1. Paultre, P.: Dynamic of Structures. Wiley, New Delhi (2010)
2. Har, J., Tamma, K.K.: Advances In Computational Dynamics of Particles—Materials and Structures. Wiley (2012)
3. Bathe, K.-J.: Finite Element Procedures. Prentice-Hall India, New Delhi (1996)
4. Yang, C., Xiao, S., Lu, L., Zhu, T.: Two dynamic explicit methods based on double time steps, Proc. IMechE., Part K:: J. Multi-Body Dyn. **228**(3), 330–337 (2014). https://doi.org/10.1177/1464419314530111
5. Groβeholz, G., Soares, D.Jr., and Estorff,O.: A stabilized central difference scheme for dynamic analysis. Int. J. Numer. Methods Eng. **102**(11), 1750–1760 (2015). https://doi.org/10.1002/nme.4869
6. Chang, S., Tran, N., Wu, T., Yang, Y.: A one-parameter controlled dissipative unconditionally stable explicit algorithm for time history analysis. Sci. Iran. **24**(5), 2307–2319 (2017). http://orcid.org/10.24200/SCI.2017.4158
7. Pajand, M.R., Rad., M.K.: An accurate predictor-corrector time integration method for structural dynamics. Int. J. Steel Struct. **17**(3), 1033–1047 (2017). https://doi.org/10.1007/s13296-017-9014-9
8. Pajand, M.R., Rad, M.K.: A family of second-order fully explicit time integration schemes. Comput. Appl. Math. **37**(3) 3431–3454 (2015). https://doi.org/10.1007/s40314-017-0520-3
9. Gautam, S.S., Sauer, R.A.: An energy-momentum-conserving temporal discretization scheme for adhesive contact problems. Int. J. Numer. Methods Eng. **93**, 1057–1081 (2013). http://orcid.org/10.1002/nme.4422

Performance Evaluation of Some Novel Composite Time Integration Schemes for Dynamic Problems

Jasti Mahesh Kumar, Vishal Agrawal, and Sachin Singh Gautam

1 Introduction

The behavior of dynamic systems is often described by partial differential equations (PDEs). It is not always possible to obtain a closed-form analytical solution of the PDEs. In those cases, one has to resort to numerical methods to obtain the solution. Usually, for this, the PDEs are to be first converted into a system of ordinary differential equation (ODEs) and the ODEs are then transformed into a system of algebraic equations, mostly nonlinear, using a suitable finite difference scheme in time often referred to as time integration schemes. Time integration schemes can be classified into implicit and explicit schemes. Explicit schemes are computationally less expensive but are only conditionally stable whereas the implicit schemes are unconditionally stable, at least for linear problems, allow large times but are computationally expensive per iteration due to the cost of computing the tangent matrix at every time step. Explicit schemes are used for wave propagation problems and implicit methods are mostly used for the analysis of short duration structural vibration problems. The implicit schemes are further classified into single substep and multi-substep schemes. In multi-substep schemes, to obtain the solution at the next time step, the current time step is further divided into smaller substeps and a combination of different schemes are used in each substep. These schemes are referred to as composite time integration schemes. Although composite schemes are more computationally expensive as compared to the single substep schemes, they give accurate and stable results even when the single substep implicit schemes fail to do so. After the seminal paper by Bathe and Baig [1], some more composite time integration schemes have been proposed in the literature. Often the performance of

J. M. Kumar · V. Agrawal · S. S. Gautam (✉)
Indian Institute of Technology, Guwahati 781039, India
e-mail: ssg@iitg.ac.in
URL: https://iitg.ac.in/ssg/

S. K. Saha and M. Mukherjee (eds.), *Recent Advances in Computational Mechanics and Simulations*, Lecture Notes in Mechanical Engineering,
https://doi.org/10.1007/978-981-15-8315-5_42

a new composite time integration scheme is done with Bathe and Baig scheme [1]. However, no work exists to compare the performance of most of the recently proposed composite time integration schemes. Hence, the objective of the present work is to carry out a comparative study of these schemes.

In the present work, a number of composite schemes by Bathe and Baig [1], TTBDF (Trapezoidal-Trapezoidal-Backward-Difference-Formula) method by Chandra et al. [2], Wen method by Wen et al. [3], and NTTBDF[1] method by Huang and Fu [4] are investigated for their performance for a number of nonlinear dynamic problems. The remainder of the paper is structured as follows. Section 2 presents the formulation of various composite time integration algorithms. Section 3 presents the performance of the composite schemes for two sets of problems. Conclusions are drawn in Sect. 4.

2 Time Integration Schemes

2.1 *The Bathe Method [1]*

In this algorithm, the time step $\Delta t = t_{n+1} - t_n$ is divided into two substeps of time intervals $\gamma\Delta t$ and $(1 - \gamma\Delta t)$, where $\gamma \in (0, 1)$. Here, $U,\ V,\ A$ denote the displacement, velocity, and acceleration, respectively. The subscripts n, $n + 1$, and $n + \frac{1}{2}$ indicate that the quantity is evaluated at time t_n, t_{n+1}, and $t_{n+\frac{1}{2}}$. The approximations for the first substep are

$$V_{n+\frac{1}{2}} = V_n + \frac{\gamma\Delta t}{2}(A_n + A_{n+\frac{1}{2}}) \tag{1}$$

$$U_{n+\frac{1}{2}} = U_n + \frac{\gamma\Delta t}{2}(V_n + V_{n+\frac{1}{2}}) \tag{2}$$

The approximations for second substep are

$$V_{n+1} = c_1 U_n + c_2 U_{n+\frac{1}{2}} + c_3 U_{n+1} \tag{3}$$

$$A_{n+1} = c_1 A_n + c_2 A_{n+\frac{1}{2}} + c_3 A_{n+1} \tag{4}$$

where $c_1 = \dfrac{1-\gamma}{\gamma\Delta t}$, $c_2 = \dfrac{-1}{(1-\gamma)\gamma\Delta t}$, and $c_3 = \dfrac{2-\gamma}{(1-\gamma)\Delta t}$. Generally, $\gamma = \dfrac{1}{2}$ is used and the optimum value based on least period error and maximum amplitude decay is $\gamma = 2 - \sqrt{2}$. In the current work, we use $\gamma = \dfrac{1}{2}$.

[1] Huang and Fu [4] do not mention in their paper the full name for the scheme but from the present author's viewpoint it stands for *New (N) Trapezoidal-Trapezoidal-Backward Difference-Formula.*

2.2 The TTBDF Method [2]

In this algorithm, the time step Δt is divided into three equal substeps. The approximations for the first substep are

$$U_{n+\frac{1}{3}} = U_n + \frac{\Delta t}{6}\left(V_n + V_{n+\frac{1}{3}}\right) \tag{5}$$

$$V_{n+\frac{1}{3}} = V_n + \frac{\Delta t}{6}\left(A_n + A_{n+\frac{1}{3}}\right) \tag{6}$$

The approximations for second substep are

$$U_{n+\frac{2}{3}} = U_{n+\frac{1}{3}} + \frac{\Delta t}{6}\left(V_{n+\frac{1}{3}} + V_{n+\frac{2}{3}}\right) \tag{7}$$

$$V_{n+\frac{2}{3}} = V_{n+\frac{1}{3}} + \frac{\Delta t}{6}\left(A_{n+\frac{1}{3}} + A_{n+\frac{2}{3}}\right) \tag{8}$$

The approximations for third substep are

$$\frac{\Delta t}{3} V_{n+1} = A\left(\theta_1\right) U_{n+1} + B\left(\theta_1\right) U_{n+\frac{2}{3}} + C\left(\theta_1\right) U_{n+\frac{1}{3}} + D\left(\theta_1\right) U_n \tag{9}$$

$$\frac{\Delta t}{3} A_{n+1} = A\left(\theta_2\right) V_{n+1} + B\left(\theta_2\right) V_{n+\frac{2}{3}} + C\left(\theta_2\right) V_{n+\frac{1}{3}} + D\left(\theta_2\right) V_n \tag{10}$$

where the A, B, C, and D are functions of variable θ and are given as

$$A(\theta) = \frac{11}{6} - \frac{\theta}{3},$$
$$B(\theta) = \theta - 3,$$
$$C(\theta) = \frac{3}{2} - \theta,$$
$$D(\theta) = \frac{-1}{3} + \frac{\theta}{3}.$$

where $\theta = \theta_1$ and $\theta = \theta_s$ in Eqs. (7) and (8), respectively. Here, θ_1 and θ_2 are the algorithmic parameters introduced by the Backward Difference Method and can be varied to control the dissipation characteristics of the scheme. The optimal values based on stability and minimization of numerical dissipation are $\theta_1 = \theta_2 = \frac{3}{4}$ and will be used for further calculations.

2.3 The Wen Method [3]

In this algorithm, the time step is divided into three substeps of time intervals $\gamma \Delta t$, $(1-\gamma)\,\Delta t$ and $\gamma \Delta t$ where the parameter $\gamma \in \left(0, \frac{1}{2}\right)$. The subscripts $n+\frac{1}{3}$ and $n+\frac{2}{3}$ indicate that the quantity is evaluated at time $t_{n+\frac{1}{3}}$ and $t_{n+\frac{2}{3}}$, respectively. The approximations for the first substep are

$$V_{n+\frac{1}{3}} = V_n + \frac{\gamma \Delta t}{2}(A_n + A_{n+\frac{1}{3}}) \tag{11}$$

$$U_{n+\frac{1}{3}} = U_n + \frac{\gamma \Delta t}{2}(V_n + V_{n+\frac{1}{3}}) \tag{12}$$

The approximations for second substep are

$$V_{n+\frac{2}{3}} = \frac{1}{\Delta t}\left[b_1 U_n + b_2 U_{n+\frac{1}{3}} + b_3 U_{n+\frac{2}{3}}\right] \tag{13}$$

$$A_{n+\frac{2}{3}} = \frac{1}{\Delta t}\left[b_1 V_n + b_2 V_{n+\frac{1}{3}} + b_3 V_{n+\frac{2}{3}}\right] \tag{14}$$

where $b_1 = \dfrac{2\gamma - 1}{\gamma(\gamma - 1)}$, $b_2 = \dfrac{1-\gamma}{\gamma(2\gamma - 1)}$, and $b_3 = \dfrac{2-3\gamma}{(\gamma-1)(2\gamma-1)}$. The approximations for third substep are

$$V_{n+1} = \frac{1}{\Delta t}\left[a_1 U_n + a_2 U_{n+\frac{1}{3}} + a_3 U_{n+\frac{2}{3}} + a_4 U_{n+1}\right] \tag{15}$$

$$A_{n+1} = \frac{1}{\Delta t}\left[a_1 V_n + a_2 V_{n+\frac{1}{3}} + a_3 V_{n+\frac{2}{3}} + a_4 V_{n+1}\right] \tag{16}$$

where $a_1 = -1,\ a_2 = \dfrac{1}{(\gamma-1)(2\gamma-1)}$, $a_3 = \dfrac{1}{\gamma(2\gamma-1)}$, and $a_4 = \dfrac{\gamma^2 - \gamma - 1}{\gamma(\gamma-1)}$.

2.4 The NTTBDF Method [4]

In this algorithm, the time step is divided into three substeps of time intervals $\gamma \Delta t, \gamma \Delta t$, and $(1 - 2\gamma \Delta t)$. The approximations for the first substep are

$$U_{n+\frac{1}{3}} = U_n + \frac{\gamma \Delta t}{2}\left(V_n + V_{n+\frac{1}{3}}\right) \tag{17}$$

$$V_{n+\frac{1}{3}} = V_n + \frac{\gamma \Delta t}{2}\left(A_n + A_{n+\frac{1}{3}}\right) \tag{18}$$

The approximations for second substep are

$$U_{n+\frac{2}{3}} = U_{n+\frac{1}{3}} + \frac{\gamma \Delta t}{2}\left(V_{n+\frac{1}{3}} + V_{n+\frac{2}{3}}\right) \tag{19}$$

$$V_{n+\frac{2}{3}} = V_{n+\frac{1}{3}} + \frac{\gamma \Delta t}{2}\left(A_{n+\frac{1}{3}} + A_{n+\frac{2}{3}}\right) \tag{20}$$

The approximations for third substep are

$$U_{n+1} = U_n + \Delta t\left(c_1 V_n + c_2 V_{n+\frac{1}{3}} + c_3 V_{n+\frac{2}{3}} + c_4 V_{n+1}\right) \tag{21}$$

$$V_{n+1} = V_n + \Delta t\left(c_1 A_n + c_2 A_{n+\frac{1}{3}} + c_3 A_{n+\frac{2}{3}} + c_4 A_{n+1}\right) \tag{22}$$

where

$$c_1 = \frac{-(a+1)\gamma^2 + 4\gamma - 1}{4\gamma},$$

$$c_2 = \frac{1 + (a-1)\gamma}{2},$$

$$c_3 = \frac{(1-a)\gamma^2 - 2\gamma + 1}{4\gamma}, \text{ and}$$

$$c_4 = \frac{\gamma}{4} \quad \text{where} \quad a \in [0, 1].$$

The stability conditions and dissipation characteristics for different values of a and γ can be found in the work presented by Huang and Fu [4]. In the current work, we choose $a = 0.5$ and $\gamma = 0.8$. These values are chosen so to have a three substep scheme with higher numerical dissipation than Bathe method, unlike TTBDF [2] and Wen [3] that have very low numerical dissipation.

3 Numerical Examples

In this section, three different numerical examples are considered to investigate the effects of numerical dissipation as well as to compare the relative performance of the four different composite time integrations schemes.

3.1 *One-Dimensional Wave Propagation*

In the first example, the performance of the composite time integration algorithms is evaluated for a linear one-dimensional wave propagation problem that is borrowed

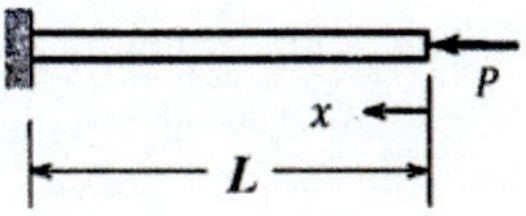

Fig. 1 One-dimensional bar subjected to time dependent end load

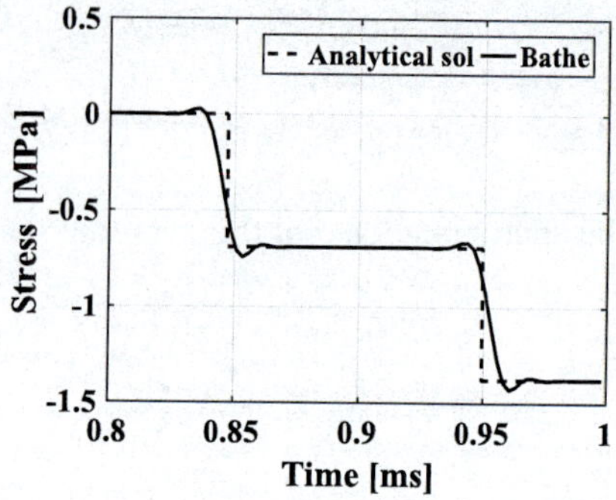

Fig. 2 Stress plot for t_{cr} at the middle of the bar

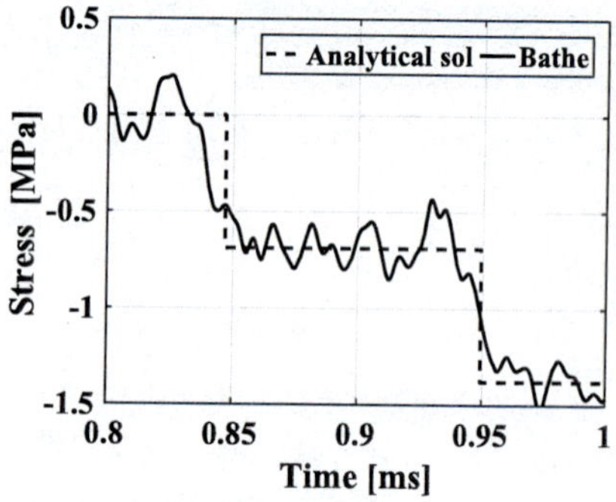

Fig. 3 Stress plot for t_{cr} at the middle of the bar

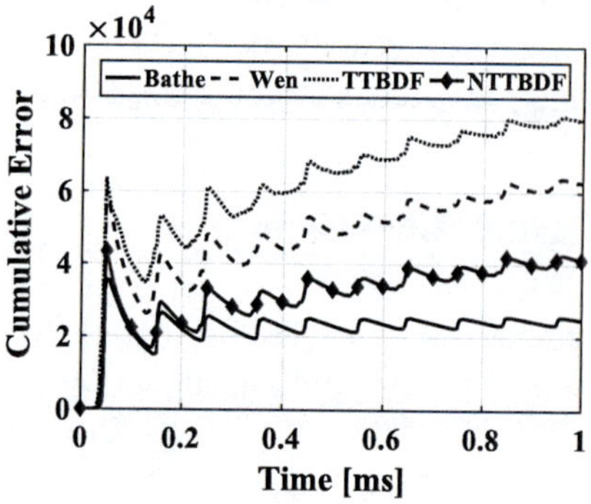

Fig. 4 Cumulative error in stress at midpoint of bar for different schemes

from the monograph by Cook et al. [5]. The length, area, density and Youngs modulus are $L = 0.508$ m, $A = 6.45 \times 10^{-4}$ m^2, $\rho = 7984$ kg/m^3, $E = 206.84$ GPa, respectively. A constant load $P = 444.82$ N is applied as shown in Fig. 1. The critical time step for the given material is given by $\Delta t_{cr} = 2.495 \times 10^{-6}$ s. The stress is evaluated at distance of $x = 0.248$ m from the free end. Figures 2 and 3 represent the analytical and numerical value of stress after a few oscillations to better capture the accuracy of algorithms. The time step $\Delta t = \Delta t_{cr}$ is used in Fig. 2, and $\Delta t = \dfrac{\Delta t_{cr}}{4}$ in Fig. 3. In general, smaller time step should give better solution. However, here, larger time step gives better solution. The reason being the spurious high frequency modes generated due to spacial discretization of domain. These modes have to be dissipated for better accuracy and decreasing the time step decreases the numerical dissipation. To check the relative performance of schemes we define a cumulative error as

$$E_{cumu} = \frac{\sum_{i=1}^{N} \text{abs}\,(X_{act} - X_{num})}{N} \tag{23}$$

where N is the number of data points until time t. It can clearly be seen from Fig. 4 that the high dissipative schemes Bathe and NTTBDF have better performance and Bathe, although a two substep scheme, has better performance as compared to NTTBDF which is a three substep scheme.

3.2 Stiff Pendulum

Many time integration algorithms that are stable for linear systems often fail when applied to nonlinear systems. The pendulum presented by Kuhl and Crisfield [6] is a paradigm to test the stability of algorithms for nonlinear systems. As shown in Fig. 5, the top end of the pendulum is pinned and the free end is given an initial velocity of 7.72 m/s and radial acceleration of 19.577 m/s^2 to balance the centrifugal force. The length, mass per unit length, and axial stiffness are $L = 3.0443$ m, $\rho A = 6.57$ kg/m, and $EA = 10^{10}$ N, respectively.

Fig. 5 The pendulum model

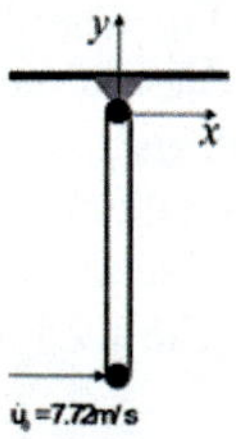

Fig. 6 Plot of energy with time for stiff pendulum ($\Delta t = 0.1$ s)

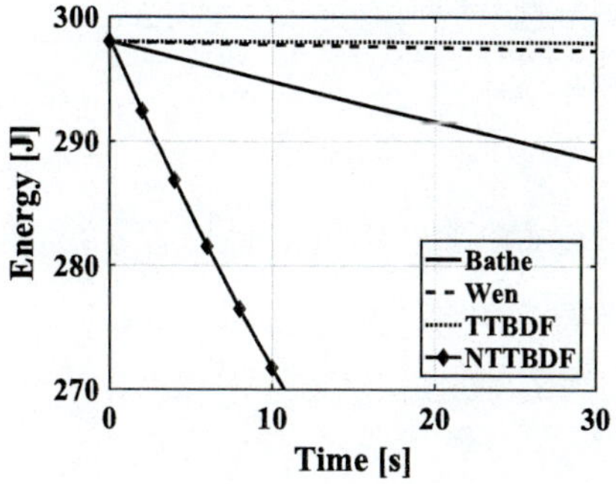

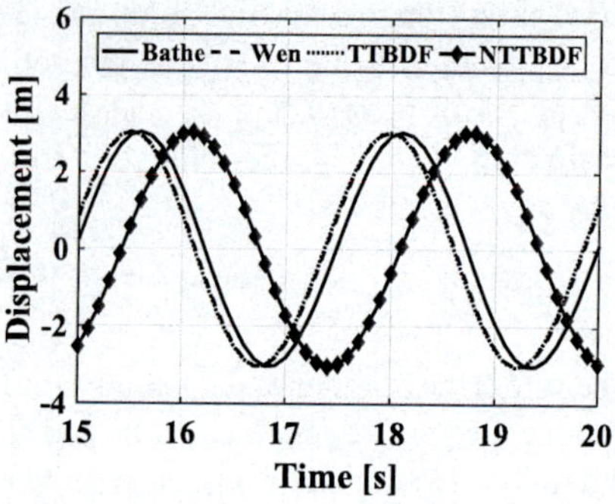

Fig. 7 Plot of displacement with time for $\Delta t = 0.1$ s

Considering $\Delta t = 0.1$ s, the numerical results for energy and the $x-$coordinate of free end are shown in Figs. 6 and 7. The dissipation and period elongation of schemes can be seen clearly.

3.3 Elastic Pendulum

The same problem as in Sect. 3.2 is considered, but with axial stiffness $EA = 10^4$ N and the initial radial acceleration is assumed to be 0 m/s.

Here, we have two different frequency modes, the rotation of pendulum (low-frequency mode) and the vibration of pendulum along the length (high-frequency mode). For a time step of $\Delta t = 0.1$ s, we have the high-frequency mode as shown in Fig. 8 by Bathe method and Fig. 9 by TTBDF. We can see that the high-frequency mode is almost dissipated after 12 s, whereas TTBDF preserves the high frequency and gives better solution for longer periods. This problem can be avoided by choosing a smaller time step as it reduces numerical dissipation.

The plot of x-coordinate of the pendulum with time is shown in Fig. 10. Initially, all the methods give the same results but as time passes we can clearly observe

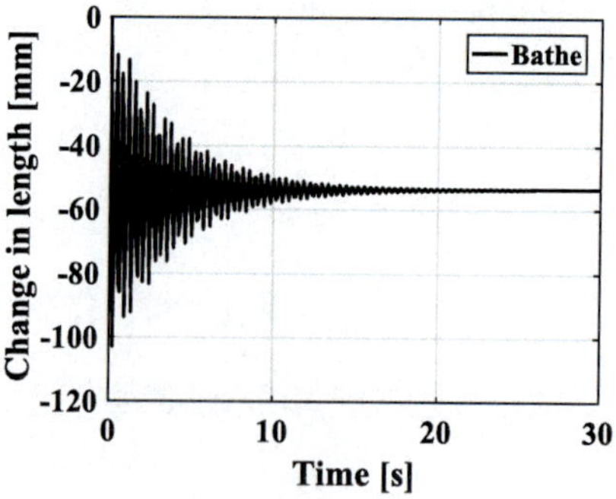

Fig. 8 Plot of change in length with time for $\Delta t = 0.1$ s

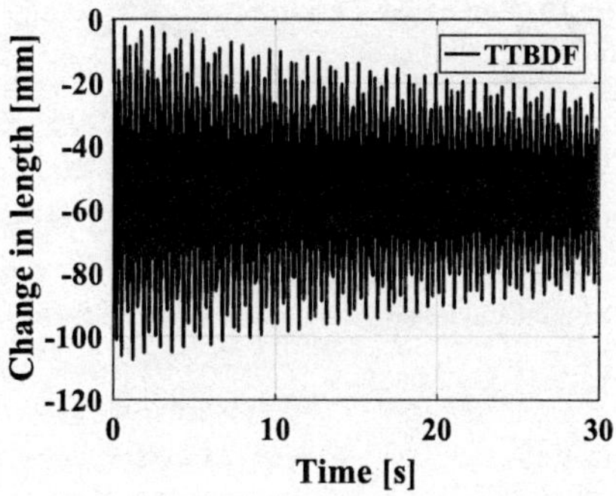

Fig. 9 Plot of change in length with time for $\Delta t = 0.1$ s

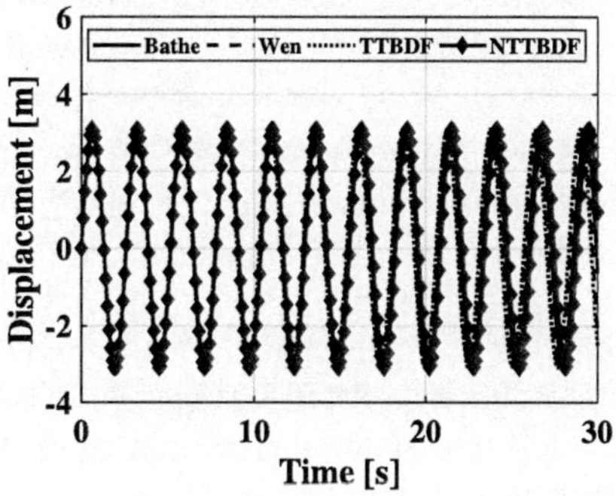

Fig. 10 Plot of displacement with time (0 to 30 s) for $\Delta t = 0.1$ s

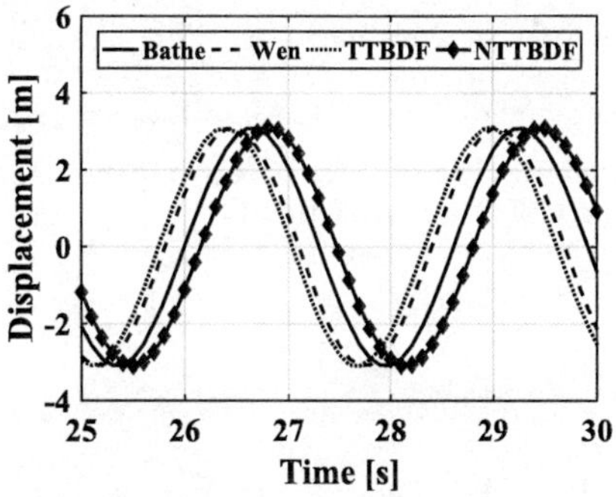

Fig. 11 Plot of displacement with time (25 to 30 s) for $\Delta t = 0.1$ s

the shift in the phase of the solution by different methods as shown in the Fig. 11. TTBDF and Wen schemes have a lower shift and give a more accurate solution. For the relative performance of schemes, we look at the conservation of energy of the system as shown in Fig. 12. TTBDF and Wen can better conserve the energy of the system giving a more accurate solution.

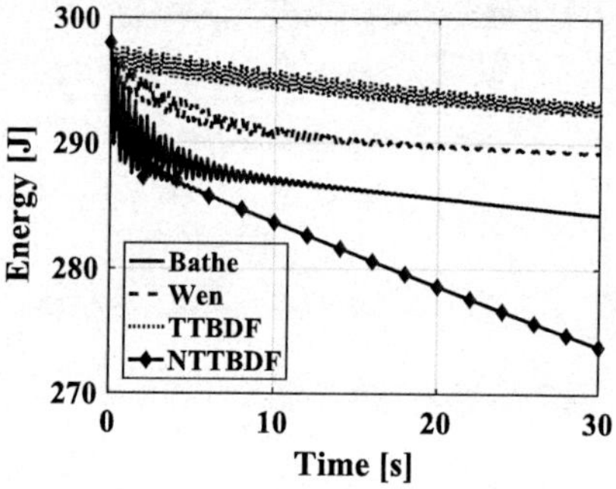

Fig. 12 Plot of energy with time for elastic pendulum ($\Delta t = 0.1$ s)

4 Conclusion

For a given class of problems, it is crucial to choose the appropriate time integration scheme with the desired characteristics for the problem to get a highly accurate solution for the same computational cost involved. For problems involving spurious high-frequency modes as in Sect. 3.1, dissipative schemes like Bathe, NTTBDF with high dissipative parameters should be used. For problems where dissipation is undesirable as in Sects. 3.2 and 3.3, TTBDF and Wen would be a better choice. The superiority of Bathe scheme can be seen in almost all the problems. Even with a larger substep size compared to NTTBDF, Bathe gave better results in comparison to NTTBDF. The high flexibility of damping characteristics of NTTBDF scheme can also be very advantageous. With the right parameter tuning much better results may be obtained.

Acknowledgements The authors are grateful to the SERB, DST for supporting this research under project SR/FTP/ETA-0008/2014.

References

1. Bathe, K.J., Baig, M.M.I.: On a composite implicit time integration procedure for nonlinear dynamics. Comp. Struc. **83**, 2513–2524 (2005)
2. Chandra, Y., Zhou, Y., Stanciulescu, I., Eason, T., Spottswood, S.: A robust composite time integration scheme for snap-through problems. Comput. Mech. **55**(5), 1041–1056 (2015)
3. Wen, W.B., Wei, K., Lei, H.S., Duan, S.Y., Fang, D.N.: A novel sub-step composite implicit time integration scheme for structural dynamics. Comput. Struct. **182**, 176–186 (2017)
4. Huang, C., Fu, M.: A composite collocation method with low-period elongation for structural dynamics problems. Comp. Struc. **195**, 74–84 (2018)
5. Cook, R.D., Malkus, D.S., Plesha, M.E., Witt, R.J.: Concepts And Applications of Finite Element Analysis. Wiley, New Delhi (2003)
6. Kuhl, D., Crisfield, M.A.: Energy-conserving and decaying algorithms in nonlinear structural dynamics. Int. J. Num. Meth. Engrg. **45**, 569–599 (1999)

Development of Efficient Stress Return Algorithms for Simulating Geomaterial Response

Divyanshu Kumar Lal and Arghya Das

1 Introduction

Significant efforts are required to reproduce the stress-strain behavior of soil under different conditions of loading. On the basis of continuum mechanics framework, a great number of constitutive models have been developed to describe the behavior of soil. These mathematical models provide a feasible tool to analyze and simulate the behavior of the materials under different conditions and external stimulus, without actually subjecting the system to those stimuli. Implementation of those models require complex algorithms. One has to make compromise between accuracy and complexity of algorithm. To develop simpler algorithm with high precision is a challenging task.

The solution of elastoplastic boundary value problems typically proceeds by computing a displacement increment on the basis of the current state of the system. The stress increments are computed by solving the local constitutive relations and an improved estimate of the elastoplastic stiffness modulus at the selected points, known as Gauss points. The possible deviation between the initially assumed stiffness and the actual one may yield a residual in the form of a global out-of-balance force which is subsequently sought minimized by a Newton type procedure that alternates between the global equilibrium iterations and the local constitutive updates. The main difference between the various methods lies in how the local stress update is performed and how the required parameters are updated. Generally, methods of solving local constitutive relations can be classified as explicit and implicit and sometimes combination of both is termed as semi-implicit.

D. K. Lal · A. Das (✉)
Indian Institute of Technology Kanpur, Kanpur 208016, India
e-mail: arghya@iitk.ac.in

S. K. Saha and M. Mukherjee (eds.), *Recent Advances in Computational Mechanics and Simulations*, Lecture Notes in Mechanical Engineering,
https://doi.org/10.1007/978-981-15-8315-5_43

Implicit stress integration algorithms have been demonstrated to provide a robust formulation for finite element analyses in computational mechanics but are difficult to formulate for complex nonlinear constitutive laws. Implicit methods find the state of system using both the current state of the system and the later one. The implicit methods, which have received by far the most attention, usually employ the Backward-Euler method for local integration. Using Newton-Raphson iteration, implicit solutions can be converged to get solution. Generally, for this type of integration return mapping schemes are used to find more accurately the plastic corrector. The calculations are performed essentially in two steps: (1) calculating the increment in variables; (2) updating it as predictors. If the trial state lies outside the yield surface, then the trial variables will be relaxed on the yield surface and referred to as the plastic corrector. Methods like closest point projection method (CPPM) developed by Simo and Taylor [1] and Cutting plane method (CPM) suggested by Simo and Taylor [2], based on return mapping are used quite frequently in constitutive modeling, including geotechnical analysis. The concept of numerical stability and accuracy plays a central role in numerical integration for controlling the convergence and consistency of the equation. CPPM is accurate up to second order. However, it is implicit in nature and hence difficult for complex and nonlinear models. So, there is a need of developing simpler and stable algorithm which predicts results close to the one predicted by CPPM or to reduce truncation error.

Explicit methods calculate the next state of a system using the value from the previous state. Compared to implicit, it is simple and easy to implement. However, if some numerical technique is not applied to control the error, the error will propagate during integration. Methods like numerical sub-stepping, line search technique, and error control are proposed by Sloan [3]. Also, one can use numerical techniques like Romberg integration method and Runge-Kutta method to obtain more precise result.

The Present study aims to develop a semi-implicit type Midpoint algorithm by modifying CPM in order to obtain accuracy and stability similar to CPPM. The efficiency of the algorithm is tested stress point analysis and solving boundary value problem using finite elements.

2 Proposed Midpoint Algorithm

Conventionally, Midpoint algorithm is formulated by enhancing explicit stress integration to achieve second order accuracy [4]. However, this approach does not guarantee unconditional stability, especially with large increments. Therefore, an attempt is made to formulate Midpoint algorithm starting from cutting plane method (CPM). Following section briefly discusses the steps involved in CPM and other associated procedure for Midpoint algorithm.

2.1 Cutting Plane Method (CPM)

CPM is a predictor-corrector type scheme which initially predicts an elastic trial stress state and then estimates a plastic corrector to reach the final stress state if the predictor state lies outside the yield stress. In CPM, the total incremental strain tensor $\Delta\varepsilon_{ij}^{t}$ at the nth step is initially assumed to be fully elastic. The trial state is estimated in the following manner:

$$\Delta\sigma_{ij} = D_{ijkl}^{e}\Delta\varepsilon_{kl}^{t} \tag{1}$$

$$\sigma_{ij}^{trial} = \sigma_{ij,n} + \Delta\sigma_{ij} \tag{2}$$

Here, $\sigma_{ij,n}$ is the stress tensor at nth step, σ_{ij}^{trial} is the elastic trial stress tensor, and D_{ijkl}^{e} is the forth order elastic stiffness tensor. If the stress state lies within the yield surface (f), then one can write,

$$f\left(\sigma_{ij}^{trial}, H\right) \leq 0 \Rightarrow \sigma_{ij,n+1} = \sigma_{ij}^{trial} \tag{3}$$

were H represents the hardening variable. However, if the stress state lies outside the yield surface, the elastic increment derived from the total strain is taken as a predictor and a correction is applied using the consistency condition, relaxing on the updated yield surface by correcting the plastic strain increments iteratively. Following the consistency condition and linearizing the yield function f at kth iteration in the nth incremental step to evaluate the value of the plastic multiplier we can get,

$$f^{k+1} = f^{k} + \left(\frac{\partial f}{\partial\sigma_{ij}^{trial}}\right)^{k}\Delta\sigma_{ij} + \left(\frac{\partial f}{\partial H}\right)^{k}\frac{\partial H}{\partial\alpha}\Delta\alpha = 0 \tag{4}$$

In the above expression α in the internal variable that dictates the hardening evolution. The plastic corrector is given by the following expressions using flow rules:

$$\Delta\sigma_{ij} = D_{ijkl}^{e}\left(\Delta\varepsilon_{ij}^{t} - \Delta\varepsilon_{ij}^{p}\right) = D_{ijkl}^{e}\left(-\Delta\lambda\frac{\partial g}{\partial\sigma_{kl}^{trial}}\right) \tag{5}$$

$$\Delta\alpha = \Delta\lambda\xi \tag{6}$$

In Eq. (6), ξ denotes the derivative of the stress conjugate of the internal variable α that controls the strain hardening. Note that in Eq. (5) $\Delta\varepsilon_{ij}^{t}$ taken as zero (0) since it is fully utilized in the predictor stage at Eq. (1). Substituting back the value from Eqs. (5)–(6), into Eq. (4), we get

$$\Delta\lambda = \frac{f^k}{\frac{\partial f}{\partial \sigma_{ij}} D^e_{ijkl} \frac{\partial g}{\partial \sigma_{kl}} - \frac{\partial f}{\partial H}\frac{\partial H}{\partial \alpha}\xi} \tag{7}$$

Calculate the stress correction and hardening parameter for kth step

$$\delta\sigma_{ij} = -\Delta\lambda D^e_{ijkl} \frac{\partial g}{\partial \sigma^{trial}_{kl}} \tag{8}$$

$$\delta H^k = \Delta\lambda \frac{\partial H}{\partial \alpha}\xi \tag{9}$$

Update the stress and hardening parameter as follows:

$$\sigma_{ij,n+1} = \sigma^{trial}_{ij} + \delta\sigma_{ij} \tag{10}$$

$$H^{k+1}_{n+1} = H^k_n + \delta H^k_n \tag{11}$$

The iteration continues until $f\left(\sigma_{ij,n+1}, H^{k+1}_{n+1}\right) \le 0$.

2.2 *Contact Stress*

Formulation of Midpoint algorithm requires to estimate a stress state, namely contact stress σ^c_{ij}, that exists in-between the initial and trial state and also lies on the yield surface, such that

$$f\left(\sigma^c_{ij}, H\right) = 0 \tag{12}$$

Sloan [3] proposed an approach in which by obtaining the σ^c_{ij} on the initial hardening surface, followed by modified Euler integration to finally get correct stress state for a given strain increment. At the integration point, the behavior is either elastic or elastoplastic during loading. If the predicted stress state crosses the yield surface, we need to evaluate the elastic portion of stress increment that is required to assess contact stress. Let β be a scalar such that $\sigma^c_{ij} = \sigma_{ij,n} + \beta\Delta\sigma_{ij}$, with $0 \le \beta \le 1$. For linear yield function the value of β can be obtained by linear interpolation which is

$$\beta = \frac{f_i}{f_i - f_f} \tag{13}$$

where $f_i = f\left(\sigma_{ij,n}, H_{1,n}\right)$ and $f_f = f\left(\sigma^{trial}_{ij}, H_{2,n}\right)$.

2.3 Midpoint Algorithm

A more general class of stress integration method, known as generalized Midpoint [4] was developed, in which new stress return was expressed in terms of both the old and the new trial states. However, in this paper the stress update equation uses contact stress σ_{ij}^{c} instead of initial stress $\sigma_{ij,n}$. The final stress increment after plastic correction and updated hardening term as per the midpoint algorithm is shown here

$$\delta\sigma_{ij} = D\Big((1-\theta)\Delta\lambda_c \frac{\partial g}{\partial \sigma_{ij}^{c}} + \theta\Delta\lambda_2 \frac{\partial g}{\partial \sigma_{ij}^{trial}}\Big) \tag{14}$$

where $\Delta\lambda_c$ and $\Delta\lambda_2$ are obtained by substituting $\sigma_{ij} = \sigma_{ij}^{c}$ and $\sigma_{ij} = \sigma_{ij}^{trial}$ in the consistency condition Eq. (4). Ortiz and Popov [4] proposed that for $\theta = 0.5$, it is second order accurate and for $\theta \geq 0.5$, it is unconditionally stable. Since the proposed midpoint algorithm is formulated from semi-implicit type CPM, unlike conventional midpoint rule, $\theta = 0.0$ does not lead to an explicit solution. However, stress update with $\theta = 1.0$ converges to semi-implicit (CPM) solution. Corrected stress at $n+1$th step follows Eq. (10). Finally, the hardening variable are obtained in the following manner:

$$H_{1,n+1} = H_{1,n} + \Delta\lambda_c \frac{\partial H}{\partial \alpha} \xi_c \tag{15}$$

$$H_{2,n+1} = H_{2,n} + \Delta\lambda_2 \frac{\partial H}{\partial \alpha} \xi_2 \tag{16}$$

$$H_{n+1} = (1-\theta) H_{1,n+1} + \theta H_{2,n+1} \tag{17}$$

3 Finite Element Simulation

3.1 Extended Mohr-Coulomb Model

The elastic-perfectly plastic Mohr-Coulomb model is widely used in Geotechnical constitutive models. However, it provides only a very crude match to the actual shearing behavior of soils and fails to capture the realistic nonlinear stress-strain response and volumetric response. This is because the incremental stiffness of soil changes significantly even for small strain level. Muir Wood [5] introduced a hyperbolic hardening law which accounts for the nonlinearity in the elastoplastic stiffness of the soil. In this constitutive model, hardening parameter evolves with shear plastic strain. This type of hardening is found to be useful for modeling sands, where the rearrangement of particles dominates the response at typical engineering stress levels,

Table 1 Overview of extended Mohr–Coulomb model

Constitutive relation	Formulation
Yield function	$f = q - \eta_y p$
Plastic potential function	$g = q - Mp\ln(p/pr)$
Hardening parameter (H)	$\eta_y = \frac{\epsilon_s^p}{a+\epsilon_s^p}\eta_p$

and irrecoverable volume strains are essentially linked to this rearrangement. The governing constitutive equations of the model is specified in Table 1.

3.2 *Modeling Detail*

The efficacy of the Midpoint algorithm is tested via solving boundary value problem using finite elements. Extended Mohr-Coulomb model given in Table 1 is implemented in ABAQUS (ver. 6.16, [6]), as a user defined material model. In addition, linear elastic soil response is assumed within the yield condition. For the purpose of comparison stress integration algorithm is written using both Midpoint algorithm and CPPM. Deformation response of a rigid strip footing on loose sand is analyzed here. The constitutive parameters of loose sand are given in Table 2.

A rectangular geometry, representing loose sand, as modeled assuming plane strain conditions in finite elements (Fig. 1). Bottom boundary of the model is restrained from the vertical movement, while the side boundaries are restrained from the lateral movement. To simulate rigid strip footing settlement, a displacement-controlled compression is incrementally applied on a 1 m wide loading zone, which is located centrally at the top boundary of the model. In order to avoid any numerical instability at the free surface, a surcharge pressure of 30 kPa is applied at the top boundary and the initial value of hardening parameter.

Table 2 Constitutive parameters

Parameters	Value
K	2000 kPa
G	923.07 kPa
η_p	1.0
M	1.2
a	0.001

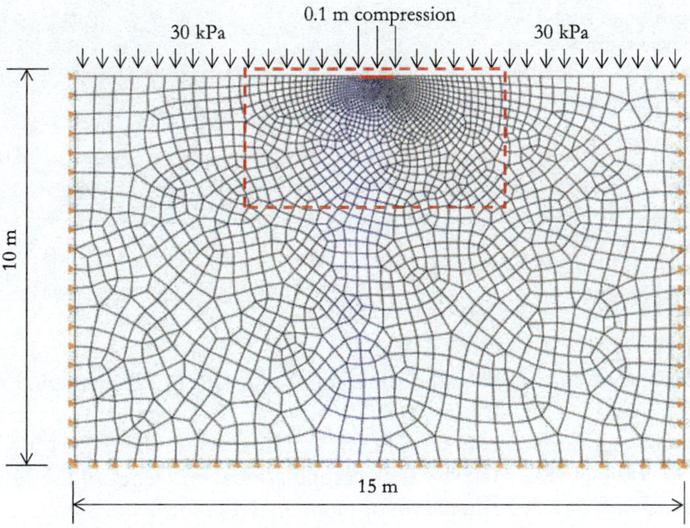

Fig. 1 Finite element model showing mesh and boundary conditions

3.3 *Results and Discussions*

The accumulated plastic (deviatoric) strain at the end of simulation is plotted in Fig. 2, for two different algorithmic treatments. It is noticed from the contour that spatial distribution of plastic strain determined using Midpoint algorithm is near identical to that of CPPM. Note that in Fig. 2, only a part of the entire model is plotted.

The settlement plot shown in Fig. 3, indicates that the evolution of vertical load estimated via Midpoint algorithm is in close agreement to that of CPPM. Two different displacement increments are considered for Midpoint analysis, 0.6 mm and 1.2 mm. The algorithm converges for both the cases. Besides accuracy, the proposed Midpoint algorithm is computationally efficient since the simulation takes lesser

Fig. 2 Deviatoric plastic strain contour during footing settlement

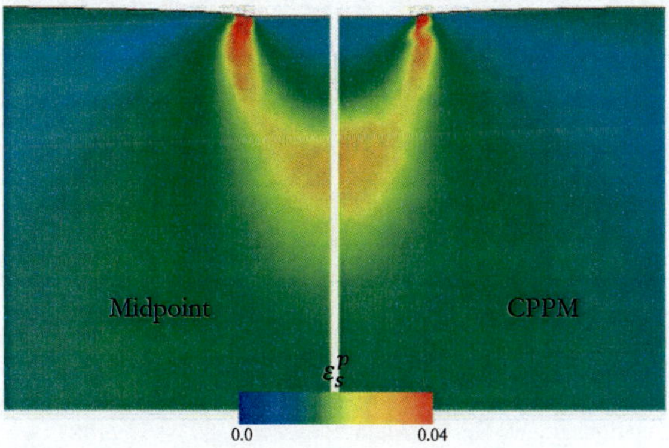

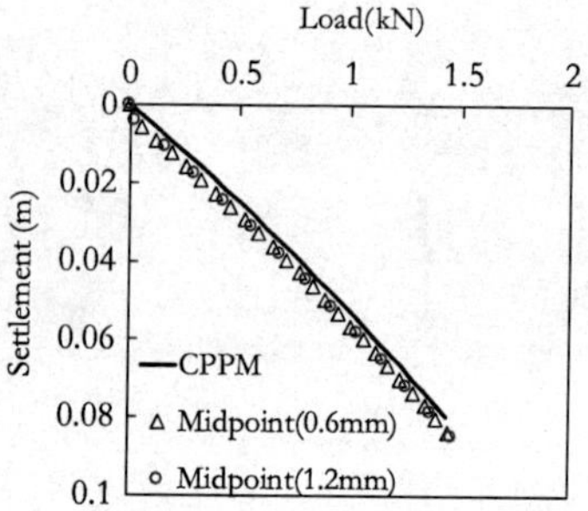

Fig. 3 Comparison of strip footing settlement estimated with different stress integration algorithms

time to complete as compared to the simulation with CPPM. The simulations are performed on a single CPU of a 3.40 GHz Intel Core i7 computer. The simulation using Midpoint algorithm consumes CPU time 215 s, while the CPPM algorithm takes 365 s to complete the analysis. Such computational efficiency of the Midpoint algorithm is a result of its non-iterative formulation at the local stress point.

4 Conclusions

In the present study, we develop a stress integration algorithm that is simple yet accurate enough to that of robust fully implicit methods like CPPM. For this purpose, cutting plane method, which is a simple but conditionally stable scheme, is enhanced to develop a Midpoint type algorithm. The proposed Midpoint algorithm is of semi-implicit in nature and does not involve any formulation beyond the first order differential of constitutive equations. The algorithm is tested for soil constitutive model (extended Mohr-Coulomb) on a finite element platform for simulating the settlement of strip footing. The results indicate that the proposed algorithm responses at par to CPPM even with large step sizes. In addition, the algorithm is faster than CPPM from computational aspect since it does not have any iterative steps.

Acknowledgements Second author wishes to thank SERB for financial support through grant number ECR/2016/001002.

References

1. Simo, J.C., Taylor, R.L.: Consistent tangent operators for rate-independent elastoplasticity. Comput. Methods Appl. Mech. Eng. **48**(1), 101–118 (1985)
2. Simo, J.C., Taylor, R.L.: A return mapping algorithm for plane stress elastoplasticity. Int. J. Numer. Meth. Eng. **22**(3), 649–670 (1986)
3. Sloan, S.W.: Substepping schemes for the numerical integration of elastoplastic stress–strain relations. Int. J. Numer. Meth. Eng. **24**(5), 893–911 (1987)

4. Ortiz, M., Popov, E.P.: Accuracy and stability of integration algorithms for elastoplastic constitutive relations. Int. J. Numer. Meth. Eng. **21**(9), 1561–1576 (1985)
5. Muir Wood, D.: Geotechnical Modelling. CRC Press (2004)
6. ABAQUS, Theory manual, version 6.16. Dassault Systemes (2016)

Kernel Gradient Free Smoothed Particle Hydrodynamics for Transient Boundary Value Problems

K. C. Charan and Siva Prasad AVS

1 Introduction

Smoothed Particle Hydrodynamics (SPH) is the oldest meshfree method developed and is completely meshless when it comes to evaluation of the integrals. The method has been applied to fluid flows, and solid deformation and fragmentation extensively because of its ability to handle the material flow in a more natural manner [1].

One of the major drawbacks of the method is tension instability, which results in clustering of material points or nodes under the tensile state of stress [2]. The onset of tension instability is related to the second derivative of the smoothing function. Several remedies have been suggested to avoid the tension instability [1, 3, 4]. The SPH formulation developed by [5] (Symmetric SPH) and [6] (KGF−SPH), do not require the gradient of the smoothing function. These formulations are exactly the same, although developed independently, and provide better accuracy compared to the previous formulations.

In this work, a FORTRAN code based on KGF−SPH has been developed and validated against benchmark transient problems in heat conduction and elastodynamics. The KGF−SPH formulation is derived in Sect. 2. Following this, the KGF−SPH solutions of heat transfer and elastodynamics problems are discussed in Sects. 3 and 4. The results show good accuracy even with a coarser discretization of the domain and a simple time-stepping scheme is adopted in this work. It has been identified that the size of the neighbourhood of a node (or stress point) and the time-stepping scheme may improve the accuracy of the solution.

K. C. Charan · S. P. AVS (✉)
Department of Mechanical Engineering, Indian Institute of Information Technology, Design and Manufacturing, Kancheepuram, Chennai 600127, India
e-mail: avs@iiitdm.ac.in

S. K. Saha and M. Mukherjee (eds.), *Recent Advances in Computational Mechanics and Simulations*, Lecture Notes in Mechanical Engineering,
https://doi.org/10.1007/978-981-15-8315-5_44

2 Derivation of KGF–SPH Formulation

The KGF–SPH formulation is presented in this section for a 2–D case. To derive the approximation for spatial derivatives of a field $f(\boldsymbol{x})$ at $\boldsymbol{x}$, a Taylor series expansion of the function about $\boldsymbol{x}$ is written as,

$$f(\boldsymbol{x}+\boldsymbol{\xi}) - f(\boldsymbol{x}) = \frac{\partial f}{\partial x_1}\xi_1 + \frac{\partial f}{\partial x_2}\xi_2 + \frac{1}{2}\frac{\partial^2 f}{\partial x_1^2}\xi_1^2 + \frac{1}{2}\frac{\partial^2 f}{\partial x_2^2}\xi_2^2 + \frac{\partial^2 f}{\partial x_1 x_2}\xi_1\xi_2, \quad (1)$$

where, $\boldsymbol{\xi}$ is the position vector relative to $\boldsymbol{x}$, of any neighbouring point within a distance of δ from $\boldsymbol{x}$. Multiplying Eq. 1, with $w(|\boldsymbol{\xi}|)\,\boldsymbol{\xi}$ and $w(|\boldsymbol{\xi}|)\,(\boldsymbol{\xi}\otimes\boldsymbol{\xi})$ and integrating over the circular domain of size δ, gives a system of 5 equations which can be written in the matrix form $\boldsymbol{AF}=\boldsymbol{T}$. Here w is a smooth weight function that depends on the distance, $|\boldsymbol{\xi}|$, of a neighbouring point from $\boldsymbol{x}$. The matrix $\boldsymbol{A}$ can be written as $\boldsymbol{A}=\int w(|\boldsymbol{\xi}|)\,\boldsymbol{\Phi}\boldsymbol{\Phi}^T \mathrm{dV}$, where $\{\boldsymbol{\Phi}\}=\{\xi_1\ \xi_2\ \xi_1^2\ \xi_2^2\ \xi_1\xi_2\}^T$. Further, $\boldsymbol{T}=\int w(|\boldsymbol{\xi}|)\,\boldsymbol{\Phi}\mathrm{dV}$ and $\boldsymbol{F}$ is a column matrix of unknown variables, i.e. $\{\boldsymbol{F}\}=\{\frac{\partial f}{\partial x_1}\ \frac{\partial f}{\partial x_2}\ \frac{1}{2}\frac{\partial^2 f}{\partial x_1^2}\ \frac{1}{2}\frac{\partial^2 f}{\partial x_2^2}\ \frac{\partial^2 f}{\partial x_1 x_2}\}^T$. Solving this matrix equation simultaneously gives the approximation for first and second order spatial derivatives of a field $f(\boldsymbol{x})$. This form was separately derived by [5] and [6], and differs from [4], in that the matrix $\boldsymbol{A}$ is symmetric in the former works. Thus, the approximation for $\nabla^2 f$ in two dimensions is obtained as

$$\nabla^2 f = 2\left(\sum_{I=3}^{4}\sum_{J=1}^{5} B_{IJ}T_J\right), \quad (2)$$

where, $\boldsymbol{B}=\boldsymbol{A}^{-1}$.

In this work, discretization for heat transfer problems involves only nodes, while for elastodynamics, the domain is discretized into stress points and nodes, following the approach in [3]. The discretized forms of the integrals involved in computation of matrices $\boldsymbol{A}$ and $\boldsymbol{T}$ at a node or stress point P can be written as

$$\boldsymbol{A}_P \approx \sum_{Q=1}^{N_P} w\left(|\boldsymbol{\xi}_Q|\right)\boldsymbol{\Phi}_Q\boldsymbol{\Phi}_Q^T \mathrm{V}_Q \text{ and} \quad (3)$$

$$\boldsymbol{T}_P \approx \sum_{Q=1}^{N_P} w\left(|\boldsymbol{\xi}_Q|\right)\boldsymbol{\Phi}_Q \mathrm{V}_Q, \quad (4)$$

where, $Q=1,2,\ldots,N_P$ are the neighbouring points of the point P, N_P the number of neighbours of P within a distance of δ, $|\boldsymbol{\xi}_Q|=|\boldsymbol{x}_Q-\boldsymbol{x}_P|$ and V_Q is the volume associated with the point Q. In this work, the form of the weight function is chosen to be Gaussian as given below.

$$w(\xi) = \begin{cases} \frac{A_n}{(\sqrt{\pi}\delta)^D}\left(e^{-4|\xi|^2/\delta^2} - e^{-4}\right), & \text{for } 0 \le |\xi| \le \delta \\ 0, & \text{for } |\xi| > \delta, \end{cases} \tag{5}$$

where, the factor A_n is determined such that $\int w\mathrm{d}V = 1$ and D is the dimension of the problem. The values of A_n for 1–D and 2–D problems are 2.09646 and 4.40324, respectively.

3 Transient Heat Transfer Simulations

The above KGF–SPH formulation has been applied to three benchmark transient heat transfer problems and their results are discussed in this section.

3.1 1–D Problems

The first two problems are one-dimensional, involving temperature boundary condition and flux boundary condition, respectively, at the left end ($x = 0$) of the problem domain ($0 \le x \le 1$) in cm. In both the 1–D problems the boundary condition at $x = 0.1$ is $T(x = 1) \to T_0$, where, T_0 is the initial temperature, taken to be $T_0 = 273$ K. Assuming Fourier law of heat conduction for heat flux, the governing equation for one-dimensional transient heat transfer is

$$\frac{\partial^2 T}{\partial x^2} = \frac{\rho c_p}{k}\frac{\partial T}{\partial t}, \tag{6}$$

where, k is the thermal conductivity of the material, c_p, the specific heat and ρ, the density. The values of these parameters are chosen such that the thermal diffusivity $\alpha\left(= \frac{k}{\rho c_p}\right) = 1$.

In the first 1–D problem, the boundary condition is $T(x = 0) = T_L = 500$ K. The analytical solution of Eq. 6, with this boundary condition can be determined as

$$T(x, t) = T_L + (T_0 - T_L)\mathrm{erf}\left(\frac{x}{2\sqrt{\alpha t}}\right), \tag{7}$$

where, $\mathrm{erf}\left(\frac{x}{2\sqrt{\alpha t}}\right) = \frac{2}{\sqrt{\pi}}\int_0^{\frac{x}{2\sqrt{\alpha t}}} e^{-s^2}\mathrm{d}s$, is called the error function.

In the second 1–D problem, a heat flux of $q_0 = 1000\ \mathrm{W/cm^2}$ is imposed at $x = 0$. The boundary condition is $\frac{\partial T}{\partial x}(x = 0) = -\frac{q_0}{k}$. The analytical solution of Eq. 6, with this boundary condition can be determined as

$$T(x,t) = T_0 + \frac{2q_0\sqrt{\alpha t}}{\sqrt{\pi}k} exp\left(-\frac{x^2}{4\alpha t}\right) - \frac{q_0 x}{k}\left(1 - \text{erf}\left(\frac{x}{2\sqrt{\alpha t}}\right)\right). \quad (8)$$

For KGF−SPH simulations of these two problems, the 1−D domain is dicretized into 41 points with a spacing of $\Delta x = 0.025$ cm. A simple, explicit forward-difference time marching scheme is chosen with a conditionally stable time step of $\Delta t = 156.25$ μs. Figures 1 and 2 show the distribution of temperature along x at $t = 0.025$ s for both the above 1−D problems. It can be seen that when the number of neighbours is increased by increasing the δ to $4\Delta x$, the accuracy of the solution improves. This observation mandates a more rigorous study on the influence of the parameter δ on the solutions to a partial differential equation.

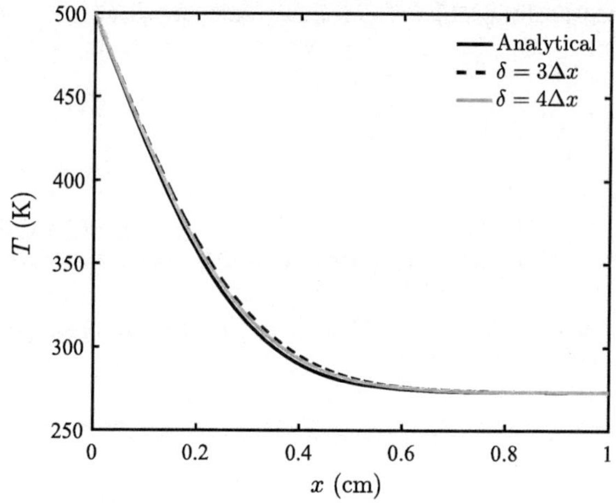

Fig. 1 Temperature distribution for the 1−D problem with $T(x = 0) = 500$ K

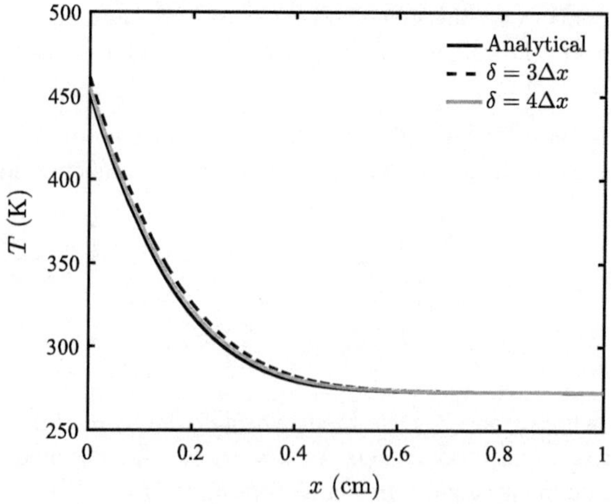

Fig. 2 Temperature distribution for the 1−D problem with $\frac{\partial T}{\partial x}(x = 0) = -\frac{q_0}{k}$

3.2 2−D Problem

For two-dimensional transient heat conduction problem, a square domain defined by $(x, y) \in (-0.05\text{ m}, 0.05\text{ m}) \times (-0.05\text{ m}, 0.05\text{ m})$ is considered with $T = 1.0$ on all the boundaries for $t > 0$ and an initial condition, $T(x, y, t = 0) = 0$. Due to the symmetry of the problem about $x-$ and $y-$ axes, only quarter region of the domain $((x, y) \in (0, 0.05\text{ m}) \times (0, 0.05\text{ m}))$ is considered for both analytical and numerical solutions. The governing equations and boundary conditions for this quarter domain are

$$\frac{\partial T}{\partial t} = \frac{k}{\rho c_p}\left(\frac{\partial^2 T}{\partial x^2} + \frac{\partial^2 T}{\partial y^2}\right), \tag{9a}$$

$$\frac{\partial T}{\partial x}(0, y, t) = \frac{\partial T}{\partial y}(x, 0, t) = 0, \tag{9b}$$

$$T(0.05, y, t) = T(x, 0.05, t) = 1.0. \tag{9c}$$

The analytical solution for Eq. 3.2, can be derived as

$$T(x, y, t) = \sum_{n=1}^{\infty}\sum_{m=1}^{\infty} \frac{(-1)^{m+n}\, 16\,(T_0 - T_L)}{(2m-1)(2n-1)\,\pi^2} \cos(Nx)\cos(My)\, e^{(-C\alpha t)}, \tag{10}$$

where, $M = \frac{(2m-1)\pi}{2L}$, $N = \frac{(2n-1)\pi}{2L}$, $C = \left(M^2 + N^2\right)$, $L = 0.05$ m, T_0 is the initial temperature and T_L is the temperature specified on the boundaries $x = 0.05$ m and $y = 0.05$ m.

The approximation for the laplacian of temperature is obtained using Eq. 2. Temperature boundary conditions are imposed explicitly at the end of every timestep. This requires a sufficiently small timestep to be used in order to get more accurate

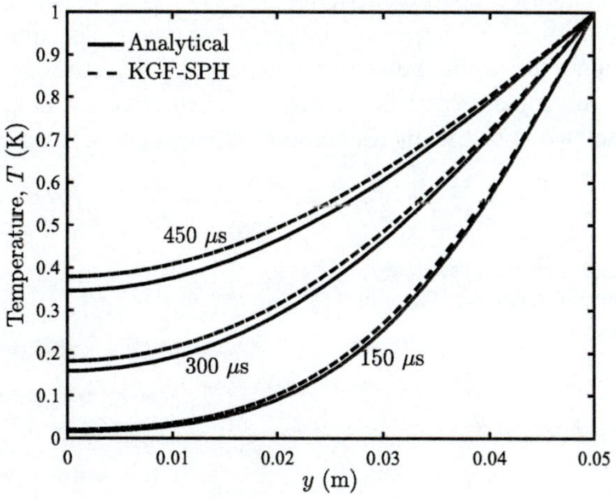

Fig. 3 Variation of temperature along y on $x = 0$ at times 150 μs, 300 μs and 450 μs

results. The timestep used in these simulations is 0.1 μs. The domain is discretized uniformly into 51×51 particles, in $x-$ and $y-$ directions, and the thermal diffusivity $\alpha = 1.0$. Figure 3 shows the temperature variation along y on $x = 0$ obtained from analytical and numerical solutions at times 150 μs, 300 μs and 450 μs. The plot seems to suggest that the deviation of the numerical results from analytical solution keeps increasing with time. A detailed and systematic study on the effects of time-stepping algorithm and the value of δ, the size of the neighbourhood of a SPH point, is necessary to analyze the accuracy of these results.

4 Elastic Wave Propagation in a Long Bar

In this section, the application of KGF–SPH to elastic wave propagation in a long bar is discussed. Figure 4 shows a *long* bar ($l/d \sim 20$) which is given a velocity pulse of the form

$$v\,(x = 0, t) = \psi\,(t) = V_0 \left[\mathcal{H}\,(t) - \mathcal{H}\left(t - \frac{l}{2C} \right) \right], \tag{11}$$

at the left end ($x = 0$) and fixed at the right end, i.e. $(v\,(x = l) = 0)$. Here, C is the 1–D elastic wave speed in the material, related to the Young's modulus E and density of the material, ρ, by $C = \sqrt{E/\rho}$. The problem is of one-dimensional in nature and its solution can be derived from 1–D wave equation as

$$v\,(x, t) = \begin{cases} \psi\left(t - \frac{x}{C}\right), & \text{for } 0 \leq t \leq l/C \\ \psi\left(t - \frac{x}{C}\right) - \psi\left(t + \frac{x}{C} - \frac{2l}{C}\right), & \text{for } l/C \leq t < 2l/C. \end{cases} \tag{12}$$

The displacement and stresses can be determined from the above velocity solution.

We solved this 1–D problem on a 2–D domain, with the Poisson's ratio $\nu = 0$. As mentioned earlier, the dual-point approach [3] is adopted to discretize the domain. An initial uniform particle spacing of 0.1 mm is chosen so that there are 50000 velocity points and 51051 stress points in the domain. The stress points carry stresses and strain rates, while the velocity points carry velocities and accelerations.

We implemented linear elastic constitutive relations within finite deformation framework, so that the Cauchy stress tensor, $\boldsymbol{\sigma}$, is related to the strain rate tensor, $\boldsymbol{d}$, as

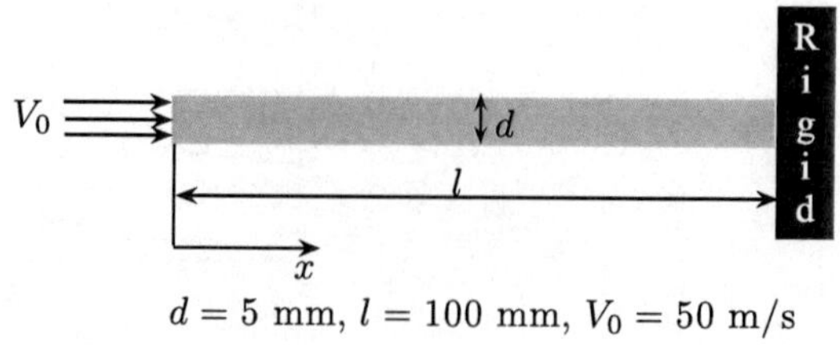

Fig. 4 Schematic of elastic wave propagation problem in a long bar

$$\overset{\nabla}{\sigma} = \mathcal{L} : d, \tag{13}$$

where, for an isotropic solid, $\mathcal{L}^{\alpha\beta\gamma\zeta} = (K_0 - 2\mu/3)\, \delta^{\alpha\beta}\delta^{\gamma\zeta} + \mu\left(\delta^{\alpha\gamma}\delta^{\beta\zeta} + \delta^{\alpha\zeta}\delta^{\beta\gamma}\right)$ in the Cartesian space, $\delta^{\alpha\beta}$ is the Kronecker delta, while $\overset{\nabla}{\sigma}$ $(= \dot{\sigma} - w\sigma + \sigma w)$ is the Jaumann rate of σ and w, the spin tensor. The constants, K_0 and μ are the bulk and the shear moduli, related to the Young's modulus and Poisson ratio of the material by $K_0 = E/\left(3\left(1 - 2\nu\right)\right)$ and $\mu = E/\left(2\left(1 + \nu\right)\right)$, respectively. The material properties are taken to be those of mild steel, so that, Young's modulus, $E = 211$ GPa and $\rho = 7800$ kg/m^3.

Figures 5 and 6 shows the KGF−SPH solutions (dashed line) of velocity and stress, plotted with time, at $x = l/2$. It can be seen that the KGF−SPH solutions from our simulations compare very well with the analytical solutions.

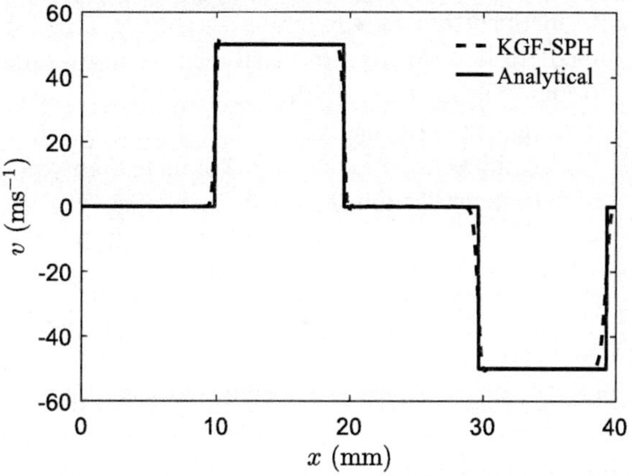

Fig. 5 Comparison of analytical and KGF−SPH solutions of velocity-time history at the midpoint of the bar

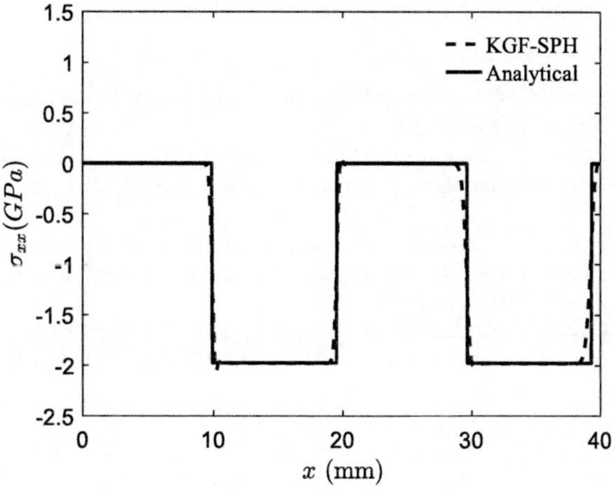

Fig. 6 Comparison of analytical and KGF−SPH solutions of stress-time history at the midpoint of the bar

The need for artificial viscosity term has been avoided by the use of a dual-point approach. Since velocity points are neighbours of stress points and vice-versa, the quantities not available on one type have to be interpolated from those computed on other types. This interpolation, probably, has the same effect as the conservative smoothing filter proposed by [1].

5 Summary and Conclusions

1. The Kernel Gradient Free—Smoothed Particle Hydrodynaimcs (KGF−SPH) formulation was implemented in FORTRAN to simulate transient heat transfer and elasticity problems.
2. The numerical solutions of elastic wave propagation problem obtained using KGF−SPH are quite accurate and match with the analytical solutions.
3. In the heat transfer simulations, it has been found that the numerical solution tends to deviate from the analytical solution with time. The effects of time-stepping scheme and the influence of the δ value on the solutions have to be investigated systematically.
4. The method can be extended to simulate highly nonlinear transient problems in both heat conduction and solid mechanics.

References

1. Randles, P.W., Libersky, L.D.: Smoothed particle hydrodynamics: some recent improvements and applications. Comput. Methods Appl. Mech Eng. **139**, 375–408 (1996). https://doi.org/10.1016/S0045-7825(96)01090-0
2. Swegle, J.W., Hicks, D.L., Attaway, S.W.: Smoothed particle hydrodynamics stability analysis. J. Comput. Phys. **116**, 123–134 (1995). https://doi.org/10.1006/jcph.1995.1010
3. Randles, P.W., Libersky, L.D.: Normalized SPH with stress points. Int. J. Numer. Methods Eng. **48**, 1445–1462 (2000). https://doi.org/10.1002/1097-0207(20000810)48:10<1445::AID-NME831>3.0.CO;2-9
4. Zhang, G.M., Batra, R.C., Modified smoothed particle hydrodynamics method and its application to transient problems. Comput. Mech. **34**, 137–146 (2004). https://doi.org/10.1007/s00466-004-0561-5
5. Batra, R.C., Zhang, G.M.: SSPH basis functions for meshless methods, and comparison of solutions with strong and weak formulations. Comput. Mech. **41**, 527–545 (2008). https://doi.org/10.1007/s00466-007-0209-3
6. Huang, C., Lei, J.M., Liu, M.B., Peng, X.Y.: A kernel gradient free (KGF) SPH method. Int. J. Numer. Methods Fluids **78**, 691–707 (2015). https://doi.org/10.1002/fld.4037

A New Edge-Based Meshless Scheme for High-Speed Inviscid and Viscous Flows

H. S. Krishna

1 Introduction

Many important flow simulations exclusively involve internal flows in which drastic changes in flow properties occur as in wind tunnel blow down test section, gas turbine blade inlet and exit flow passages, combustion chambers of aerospace vehicles, intake ducts, nozzle flows and so on. Complex internal flow analyses involve strong multiple shocks interaction and reflection from wall, shear layers, singular point, stagnation flows, wake region, suction, entrainment and so on. In the past, some well-known test problems devised to simulate these phenomena included carbuncle phenomenon of blunt body problem, NACA airfoil flow characteristics from wind tunnel-derived test data, forward-facing stepped duct flow and the double Mach reflection problem. Some of these highly diffusive flows are amenable to inviscid flow solutions, and the development of their shock structure in the close proximity of the solid wall may be analyzed for gross or averaged effects of flow parameters through appropriate wall-reflecting boundary conditions. But still, errors due to numerical discretization of flow may become the source for spurious oscillations or secondary weak shocks. Conventional CFD solvers employing single time-step marching schemes have inherent limitations of achieving rapid flow stability. For instance, Euler codes require rather time-consuming implicit residual smoothing and enthalpy damping for achieving full flow development and also induced artificial dissipation for obtaining stable Euler flow solution. Furthermore, accurate computation of convective fluxes requires either splitting of fluxes or use of higher order schemes for acceleration of convergence of flow solution. Higher resolution characteristics of shear flows are desirable for greater computational efficiency. This is realized through the flux vector

H. S. Krishna (✉)
Aeronautical Development Agency, P.B. 1718, Bangalore 560017, India
e-mail: hskgm99@gmail.com

S. K. Saha and M. Mukherjee (eds.), *Recent Advances in Computational Mechanics and Simulations*, Lecture Notes in Mechanical Engineering,
https://doi.org/10.1007/978-981-15-8315-5_45

splitting schemes that decompose flux into convective and pressure or acoustic part. Diffusion schemes derived from gas dynamics relations [1] were developed to induce artificial viscosity in regions of discrete shock structure in order to accurately capture shock waves and contact discontinuities. Early researchers included dissipation terms in the finite volume Euler equations formulation to suppress odd–even point oscillations due to pressure–velocity coupling. It was found that second difference diffusion terms with second and fourth difference pressure coefficients rendered wiggle-free computation. The higher-order schemes produce accurate results in smooth flow regions but oscillate at shocks or discontinuities. This problem is partially overcome through Roe linearization, with or without characteristic decomposition and by observing Jameson's local extremum diminishing (LED) criterion for positivity preserving schemes: local extremum cannot change its extreme value. Based on positivity, shock structure model and the number of coefficients in the diffusive flux, different classes of schemes such as CUSP, AUSM and Upwind schemes were developed for inviscid CFD code solution. In this paper, the influence of diffusion on shock structure including the effects of regular shock reflection and meshless code development are illustrated by two classic internal flows at supersonic speeds: a forward-facing step ductflow and oblique shock-reflecting surfaces.

Work on meshless method began in early 1990s with John Batina reporting his pioneering work on meshless Euler flow computation [2] over complete aircraft through least square gradient (LSQ) reconstruction procedure for computing the meshless point coordinates. The time saved by way of less stringent connectivity information, near random distribution of grid points and the ease of modification of code to reconstruct the least square technique make meshless solvers a viable alternative to conventional grid-based solvers. Besides, if stable schemes could be constructed then greater accuracy could be attained through the application of higher order schemes. The diffusion scheme grid-based Euler codes are easily modifiable into meshless or hybrid codes by merely supplementing a routine to existing grid-based solver for construction of least square coefficients. The connectivity between neighboring points is determined by the directional derivatives or gradients of flux vector. Generally, a second-order space discretization is adopted for flow field solution which stipulates C_2 continuity between grid points. The flux gradient is expressed as a linear combination of least square coefficients of edge vector to obtain the direction of flux. The LSQ reconstruction procedure for meshless method is accomplished either by Taylor series expansion of flux function about center of point cloud or by curve fitting of discrete data using polynomial basis function. The edge-based meshless method ensures reciprocity of nodes for proper connectivity and permits the use of approximate Riemann solvers for single midpoint flux computation. The hybrid codes as a by-product of meshless solver are gaining popularity as industry standard codes since they mitigate the problem of generating high-resolution geometry.

2 Modifications for Meshless Solver

Broadly speaking, two categories of meshless method may be identified depending on the method of point distribution: random point cloud and Cartesian grid-free point distribution as depicted in Figs. 1 and 2, respectively. For a Cartesian grid-free solver discussed in this paper, a five-point (due North, South, East and West) stencil in Fig. 2 is used for LSQ reconstruction procedure. The Cartesian meshless solver is ideally suited for geometries having no irregular boundaries. Further, computational accuracy can be improved if the stencil is chosen to lie along the coordinate directions.

2.1 Meshless Convective Fluxes

Table 1 gives the comparison between governing equations, convective meshless fluxes and complete boundary conditions used in inviscid and viscous codes [3, 4]. The procedure for modifying an existing grid-based Euler code to meshless [5, 6] or hybrid code involves evaluating LSQ coefficients, a_{ij} and $b_{ij,}$ along coordinate directions, finding the sum of spatial derivative of linearized convective fluxes, F_N or F_E $(af + bg)$ and then integrating the discretized Euler or Navier–Stokes(N-S) equations to steady-state solution by four-stage Runge–Kutta time integration scheme.

The linearized convective flux is obtained as follows:

$$\sum F_{ij} = \sum\left(\frac{\partial f}{\partial x} + \frac{\partial g}{\partial y}\right) \quad (1)$$

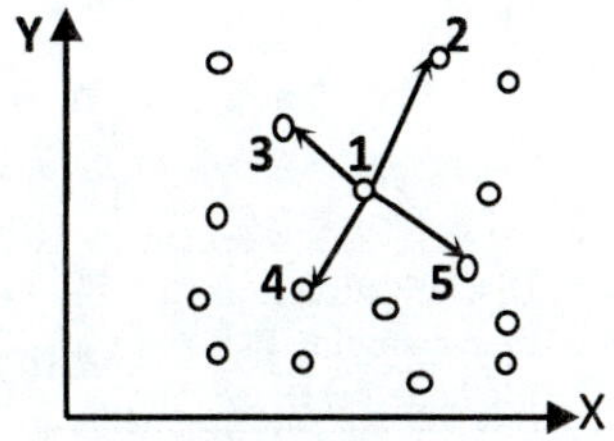

Fig. 1 Arbitrary 5-point stencil at node i

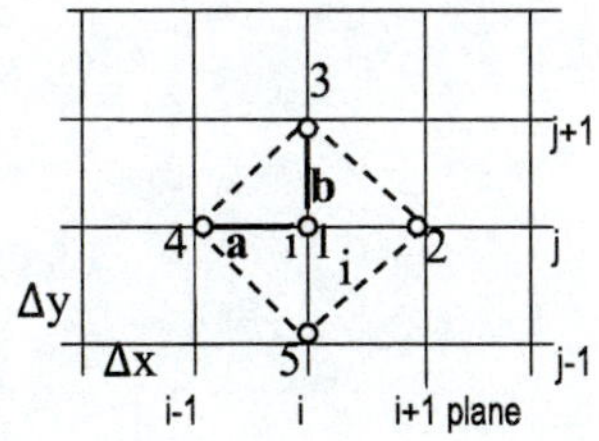

Fig. 2 Cartesian 5-point stencil at node **i**

Table 1 Euler and Navier–Stokes code formulation

Sl. no.	Description	Euler solver	N-S solver
1	Governing equations in conservation law form	$\frac{\partial U}{\partial t} + \nabla F = Q = 0$ assuming adiabatic flow and body force, $f_b = 0$	$\frac{\partial U}{\partial t} + \nabla F_N = Q = 0$ $F_N = F - F_v$
2	State vector	$U = [\rho, \rho u, \rho v, \rho E]^T$	$U = [\rho \rho u \rho v \rho E]^T$
3.1	Convective fluxes: f and g and viscous fluxes, f_V and g_V. Flow variables are pressure p, velocities u and v, density ρ and enthalpy H	$\nabla F = \frac{\partial f}{\partial x} + \frac{\partial g}{\partial y}$ $f = \left[\rho u, p + \rho u^2, \rho uv, \rho uH\right]^T$ $g = \left[\rho u, p + \rho u^2, \rho uv, \rho uH\right]^T$ $f_v = 0, g_v = 0$	$\nabla F_N = \nabla F - \nabla F_V$ $\nabla F_V = \frac{\partial f_V}{\partial x} + \frac{\partial g_V}{\partial y}$ $fv = \left[0, \tau_{xx}, \tau_{xy}, u\tau_{xx} + v\tau_{xy}\right]^T$ $g_v = \left[0, \tau_{yx}, \tau_{yy}, u\tau_{yx} + v\tau_{yy}\right]^T$
3.2	Meshless linearized convective fluxes **a** and **b** are least square coefficients along coordinate directions x and y	$\boldsymbol{F_E} = \boldsymbol{a}\boldsymbol{f} + \boldsymbol{b}\boldsymbol{g}$ $a = \frac{\omega\Delta x \sum \omega\Delta y^2 - \omega\Delta y \sum \omega\Delta y^2}{\sum \omega\Delta x^2 \sum \omega\Delta y^2 - \left(\sum \omega\Delta x \sum \omega\Delta y\right)^2}$ $\omega = \frac{1}{\sqrt{\Delta x^2 + \Delta y^2}}$, inverse distance weighing. Interchange x and y in **a** to obtain **b**	$\boldsymbol{F_N} = \boldsymbol{a}(\boldsymbol{f} - \boldsymbol{f_V}) + \boldsymbol{b}(\boldsymbol{g} - \boldsymbol{g_V})$
4	Shear stresses	$\tau_{xx} = \tau_{yy} = \tau_{xy} = \tau_{yx} = 0$	$\tau_{xx} = 2\mu\frac{\partial u}{\partial x} + \lambda\left(\frac{\partial u}{\partial x} + \frac{\partial v}{\partial y}\right)$ $\tau_{yy} = 2\mu\frac{\partial v}{\partial y} + \lambda\left(\frac{\partial v}{\partial x} + \frac{\partial v}{\partial y}\right)$ $\tau_{xy} = \tau_{yx} = \mu\left(\frac{\partial v}{\partial x} + \frac{\partial u}{\partial y}\right)$
5	Boundary conditions: wall outlet inlet	$v = 0$ Supersonic flow constant $u = 3, v = 0, p = 1, \rho = 1.2$	u = v = 0 (no-slip) Supersonic flow constant $u = 3, v = 0, p = 1, \rho = 1.2$
6	Time step	$\Delta t = \frac{CFL}{\sum_{j=1}^{\infty}\left[\lVert a_i u + b_j v\rVert + \sqrt{(a_i^2 + b_i^2)c}\right]}$	$\Delta t = \frac{CFL}{\sum_{j=1}^{\infty}\left[\lVert a_j u + b_j v\rVert + \sqrt{(a_i^2 + b_i^2)c}\right]}$
7	Discretization explicit time marching R-K Method	$\frac{dU}{dt} = -R_i(U)$ $\boldsymbol{R_i(U)} = \sum \boldsymbol{F_i \Delta S}$	$\frac{dU}{dt} = -R_{Ni}(U)$ $\boldsymbol{R_{Ni}(U)} = \sum \boldsymbol{F_i \Delta S}$

$$\frac{\partial f}{\partial x} = \sum a_{ij} \Delta f_{ij} \tag{2i}$$

$$\frac{\partial g}{\partial y} = \sum b_{ij} \Delta g_{ij} \tag{2ii}$$

for cloud $j = 1, 5$ and i is the node under consideration. The point flux F_E at any node is given by

$$F_E = af + bg \tag{3}$$

where a and b are LSQ coefficients along x and y directions, respectively. The convective fluxes should be computed by complete LSQ reconstruction technique [6] while diffusive flux may be computed as a product of point flux and LSQ coefficients or space coordinates, Δx and Δy.

2.2 Two-Step Time Marching Scheme

In order to ensure proper connectivity between point cloud stencils and flow development, it is essential to smoothen the state vector in a two-step scheme [7] at every node n. The first step differencing of the governing fluid flow equations in conservation law form is:

$$q_n^{k+1} = q_n^k - CFL\sum_i(\Delta F_{ij} + D)\Delta t \tag{4}$$

where linearized convective fluxes for N-S solver, $\Sigma\Delta F_{ij} = F_N$ is given in Table 1, Sl. No. 3.2:

$$\Delta(.)_{ij} = (.)_j - (.)_i \tag{5}$$

and D is the diffusive flux. For Euler solution, viscous fluxes are zero, that is, $F_N = F_E$ when $f_V = g_V = 0$.

In the second step, discretization, smoothing and averaging with previous iteration values of state vector are introduced along with an artificial diffusion term to control expansion shock and to eliminate its distortion as under:

$$Q_n = \left(\frac{q_n + q_p + \varsigma q_t}{2}\right) - CFL \times \Delta t \times R_S \tag{6}$$

$$q_I = \Psi\frac{\frac{q_{i+1,j}+q_{i,j+1}}{2} - q_{i,j}}{3} \tag{7i}$$

$$q_t = \frac{q_I + q_{II} + q_{III} + q_{IV}}{4} \tag{7ii}$$

Here, Q_n represents the smoothened state vector at node n, q_n is the current value of state vector, q_p is its previous iteration value, q_t is the average value of the state vector of four triangles of a quadrant (following the usual convention of coordinate geometry) denoted as q_{I}, q_{II}, q_{III} and q_{IV} that make up the stencil at point (*i, j*) as shown by dashed line in Fig. 2, ζ is a diffusion constant which is assigned a value of unity for highly diffusive flows such as expansion shock and a small fractional value (say 0.1) for all other flows, R_s is the residual, ψ is the van Albada limiter and LSQ coefficient, b_{ij}, is obtained from a_{ij} [6] by interchanging *x* and *y*. Note that q_t tends to integrate the first step numerical solution of Eq. 6 to steady-state condition, thereby improving stability. The reason for splitting point cloud stencil into four triangular elements in Fig. 2 is that triangular mesh produces best results for two-dimensional flow problems. A subtractive corrective term, $\kappa \cdot q_n$ with constant coefficient κ of the order 10^{-5} in the smoothing formula (6) controls the development of expansion shock in a step ducted flow by introducing artificial compressibility into the solution. The stopping criterion for highly diffusive flows is realized by bypassing the second step computation beyond the time step at which the flow fully develops. But for low diffusion flows, this scheme is self-regulating and stable under steady-state condition and as such no controller or regulator is required.

Wall Boundary Conditions Two sets of reflecting wall boundary conditions are found in the literature for viscous and inviscid flows. With usual notations, the wall reflecting boundary conditions [8, 9] for flow velocity along tangential ($_{\|}$) and normal ($_{\perp}$) directions to wall are:

$$V\| = n_y u_i - n_x v_i \tag{8i}$$

$$V\perp = n_x u_i - n_y v_i \tag{8ii}$$

$$V\|_{ghost} = n_y V\| - n_x V\perp \tag{9i}$$

$$V\perp_{ghost} = -n_x V\| - n_y V\perp \tag{9ii}$$

Also, $U = U\times\ isx$, $isx = -1$ for outflow on right plane otherwise 1; $U = U \times isy$, $isy = -1$ for outflow on bottom plane otherwise 1, where U represents velocity component, either streamwise u or normal, v. The remaining boundary conditions of pressure and density are specified as zero normal gradients. For solid wall, the no-slip wall boundary condition replaces zero relative velocity condition normal to the wall in the viscous code.

The new modified four-point boundary conditions for pressure or density that may be applied at the reflecting wall to eliminate Mach reflection is given below:

$$W_{i,j}^{n+1} = W_{i,j}^{n} + \left(3W_{i,j}^{n} - 4W_{i,j}^{n-1} + W_{i-1,j}^{n-1}\right) \tag{10}$$

$$W_{i,j} = \frac{\left(3W_{i,j+1} - W_{i,j+3}\right)}{2} \tag{11}$$

$$W_{i,j} = \left(\frac{\lambda}{2}\right)\frac{\left(W_{i-1,j} + W_{i,j+1} + W_{i+1,j} + W_{i,j-1} + W_{i,j}\right)}{4} \tag{12}$$

where $W(i, j)$ stands either for pressure $p(i, j)$ or density $\rho(i, j)$ and diffusive constant, $\lambda = 0.8$–0.9. Here, Eq. (10) contributes to connectivity of the developing flow by mixing the state vector of the previous iteration and also improves stability. Equation (11) applies four-point formula normal to the wall while Eq. (12) does stencil averaging of pressure to reduce steep pressure gradients in flow.

3 Validation of Case Study

A concrete model for shockwave development does not exist. Among various models [10] put forth by researchers during the last 100 years, the three-shock theory (3ST) by von Neumann describes reasonably well the commonly encountered shock wave phenomena. According to 3ST, non-regular shock reflection called Mach reflection (MR) is produced by three intersecting shock waves: incident wave, reflected wave and Mach stem. Their point of intersection called triple point (T) may move away or parallel to the reflecting surface, giving rise to direct, stationary or inverse Mach reflection as shown in Fig. 3. If local subsonic pockets of flow exist at the wall, then a Mach stem [11] may be formed at the point of reflection O. Two standard test cases described in the ensuing paragraphs validate the two-step scheme for high- and low-diffusion flow phenomena.

3.1 *High Diffusion Problem*

The shock propagation in a wind tunnel with a forward-facing step at high Mach numbers represents a high diffusion problem. The corner at the step represents the

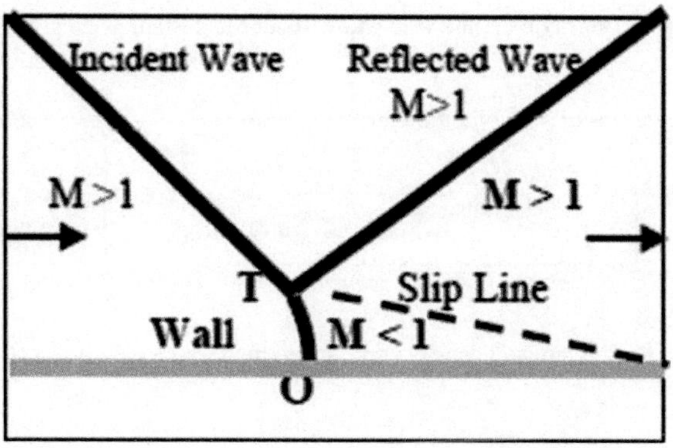

Fig. 3 Inverse reflection at wall O

singular point in the flow because it is the center of rarefaction fan. This point is the origin for all kinds of instabilities, arising from numerical and physical shock interaction. The solution to this problem was first provided by Woodward and Collela [12] in 1984 and is popularly known as expansion shock wave problem. The step in a wind tunnel represents a two-dimensional flow problem due to slab symmetry in direct analogy with symmetric airfoil of an airplane wing. Figure 4 shows the schematics of the forward-facing step. Initial and inlet flow conditions of pressure, $P = 1$, density, $\rho = 1.4$ and velocities, $u = 3.0$ and $v = 0.0$ are enforced as shown in Fig. 4. The outlet boundary condition is taken to be supersonic since free expansion of flow occurs at exit plane. A 300×100 uniform point cloud generated by a Cartesian grid generator forms the computational domain of test problem with uniform point spacing of 0.01. The effectiveness of modified boundary conditions for flow solvers given by Eqs. 10, 11 and 12 can be seen clearly in the density contour plots obtained from generic Euler and N-S codes in Fig. 5, 6 and 7 where the occurrence Mach stem has been successfully eliminated [12].

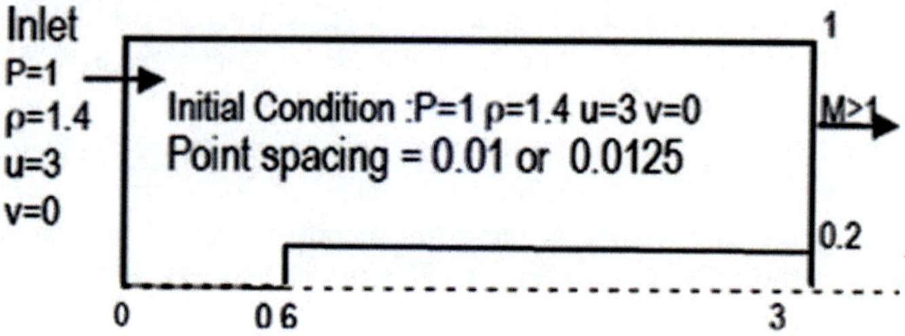

Fig. 4 Schematics of forward-facing step

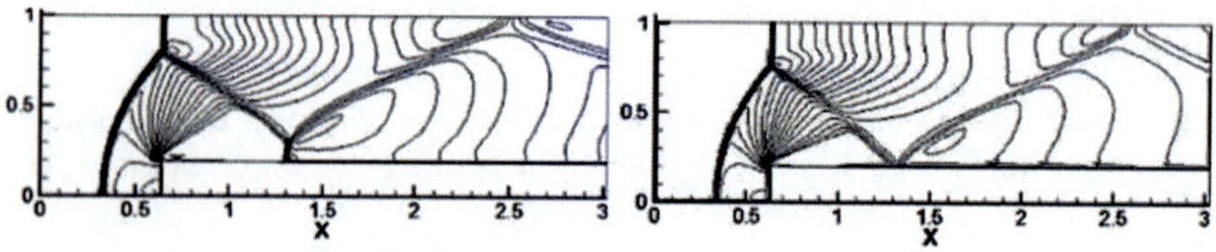

Fig. 5 Density contours with and without Mach stem using generic Euler code

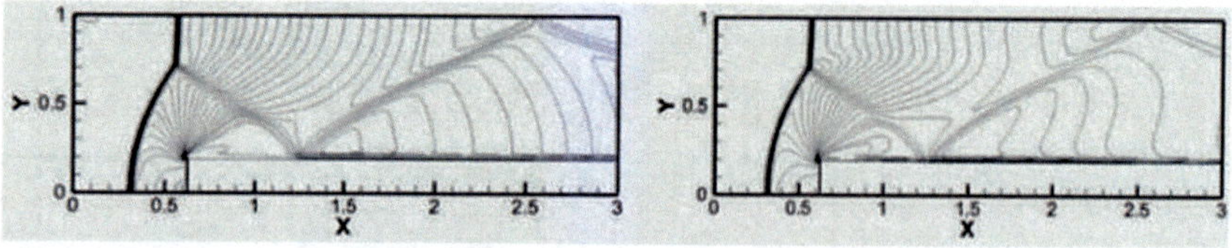

Fig. 6 Pressure and density contours without Mach stem using N-S code

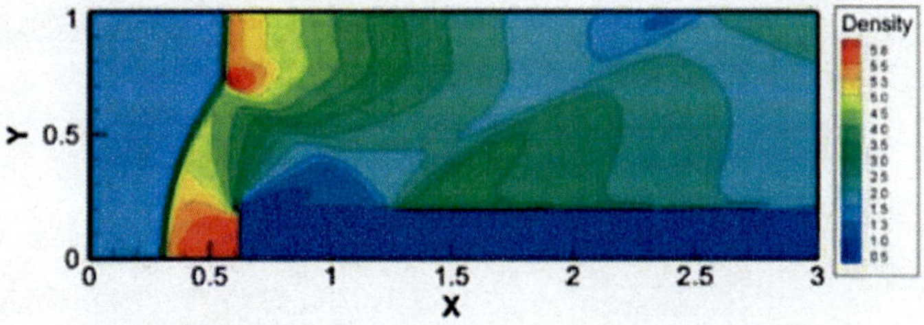

Fig. 7 High diffusion problem: Stepped duct density contour levels

3.2 *Low Diffusion Problem*

The implementation of the newly developed boundary condition for low-diffusion flow is now illustrated through classic examples of oblique shock reflection between reflecting surfaces [13]. A shock wave incident at 29° to the left top corner undergoes regular reflection between parallel reflecting surfaces of length 4.1 and height 1.0 at Mach 2.9 and emerges as reflected wave at nearly 24° at the right end as shown in Fig. 8. The rigid surface is maintained at constant inlet conditions of $P = 1/1.4$, $\rho = 1$, $u = 2.9$, $v = 0$ and the flow M_2, behind the shock is rendered parallel to the reflecting surfaces by deflecting the flow through angle $\theta_1 = \theta_2 = 10.94° = \tan^{-1}(v/u)$ by imposing $(u,\ v,\ \rho,\ p)_{(x,1,t)} = (2.61934, -0.50632, 1.6999, 1.5281)$ on the upper surface with prescribed supersonic outflow condition. The excellent agreement between oblique shock wave theory [13] and results from generic Euler meshless code for pressure and density in the plane $y = 0.5$ is illustrated in Fig. 9. In Fig. 10, the location (1.8 mm) and angles subtended at the reflecting surface (29° and 24°) by the density shock wave match exactly with the theoretical computations obtained from the θ–β–M relationship [14] for oblique shock wave reflection.

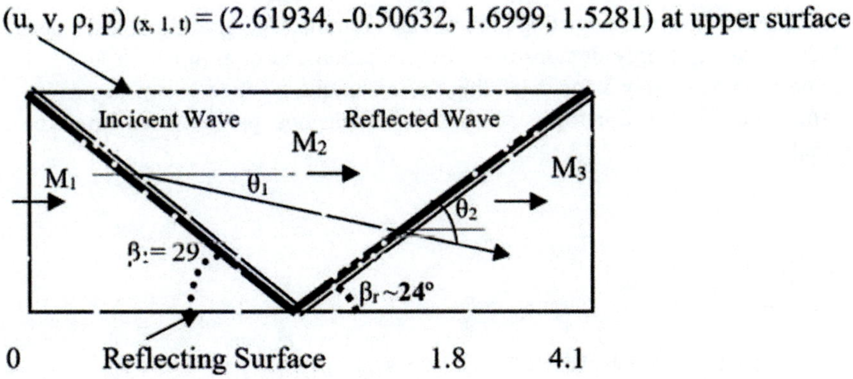

Fig. 8 Schematics of oblique shock reflection

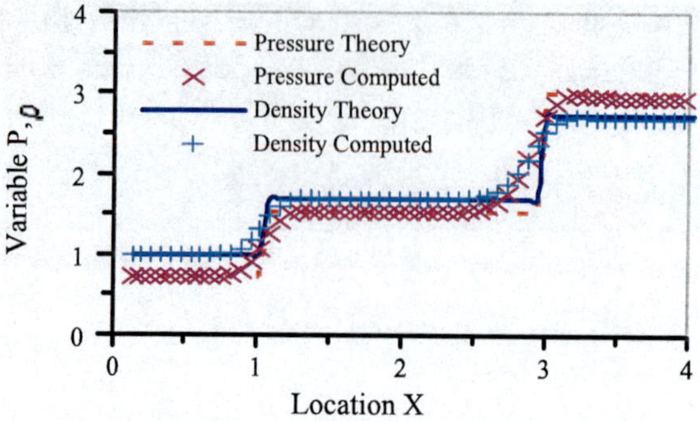

Fig. 9 Comparison of pressure and density between theory and computed value for oblique shock reflection in the plane at $y = 0.5$

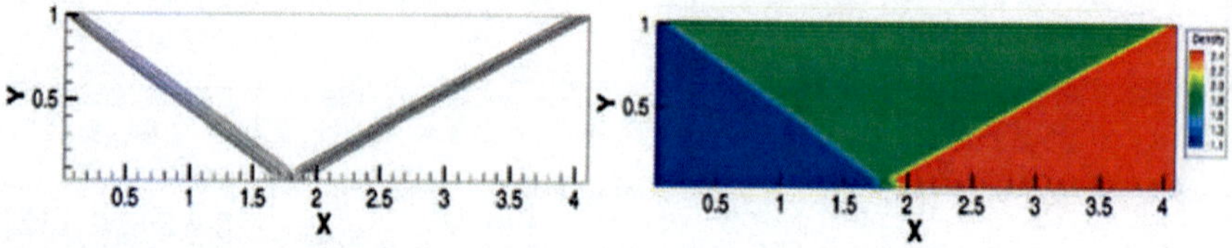

Fig. 10 Low diffusion problem: Reflecting shock density contour levels

4 Conclusion

A new edge-based two-step meshless scheme has been developed and validated for both inviscid and viscous two-dimensional flows. The distinct advantages of this solver are its ease of implementation in coding, applicability to hybrid codes and modular computation of linearized meshless convective fluxes which can be easily supplemented to the existing program as a routine. The versatility of the new meshless solver is amply demonstrated by validation exercises for high and low diffusion problems. Scope for extending the technique of state vector-controlled flow simulation described in this paper to three-dimensional problems remains to be explored.

References

1. Jameson, A.: Analysis and design of numerical schemes of gas dynamics-2: artificial diffusion and discrete shock structure. Int. J. CFD **5**, 1–38 (1995)
2. Batina, J.T.: A Gridless Euler/N-S Solver Algorithm for Complex Aircraft Applications, p. 0333 (1993)
3. Blazek, J.: Computational Fluid Dynamics: Principles and Applications. Elsevier (2001), ISBN: 0 08 043009 0

4. Hirsch, C.: Computational Methods for Inviscid and Viscous Flows, vol. 2. Wiley (1990), ISBN 047192351 6
5. Aaron, K., Jameson, A.: A meshless volume scheme. In: 19th AIAA Computational Fluid Dynamics, p. 3534. San Antonio, Texas, 22–25 June 2009
6. Aaron, K., Jameson, A.:Edge-based meshless methods for compressible flow simulations. In: 46th AIAA Aerospace Sciences Meeting and Exhibit, p. 699 (2008)
7. Krishna, H.S.: A two-step time marching technique for classical CFD codes. In: 14th Annual Symposium of Aeronautical Society of India, CP-39, p. 120. Bangalore (2012)
8. Krishna, H.S.: Development of meshless method boundary conditions for high speed flows. Int. J. Latest Trends Eng. Technol. **9**(1) (2017)
9. Kirshman, D.J., Liu, F.: A gridless boundary condition method for the solution of Euler equations on embedded cartesian meshes with multigrid. J. Comput. Phys. 119–147 (2004)
10. Ben-Dor, G.: Shock-Wave Reflection Phenomena, pp. 12–17. Springer (2007)
11. Jaisankar, S., Rao, R.S.V.: A central Rankine-Hugonoit solver for hyperbolic conservation laws. J. Comput. Phys. **228**, 781 (2009)
12. Woodward, P., Colella, P.: The numerical simulation of two-dimensional fluid flow with strong shocks. J. Comput. Phys. **54**, 115–173 (1984)
13. Hong, L., Joseph, D., Lohner, B.R.: A hybrid cartesian grid and gridless method for compressible flows. In: 43rd AIAA Aerospace Sciences Meeting and Exhibit, 10–13, p. 0492 (2005)
14. Anderson, J.: Aerodynamics for engineers. Mc-Graw Hill Int. Ed., Chap. 9 (1991), ISBN 0–07–100767–9

A Numerical Framework for the Solution of Bivariate Population Balance Equation—Model Implementation and Verification

Deepak Kumar Singh and Gaurav Bhutani

1 Introduction

Population balance equation (PBE) describes the temporal behaviour of particulate systems, which include phenomena such as aggregation, breakage, nucleation and growth. The common application of the PBE is limited to the monovariate systems, which assumes that the evolution of the particle number density function (NDF) depends only on a single internal variable, which is the particle size in most applications. However, there are a number of industrial problems whose description requires the formulation of a multivariate PBE, where two or more internal coordinates of the particles are required to investigate the system. Aerosol systems, powders granulation, synthesis of nanoparticles, mass transfer and chemical reactions in multiphase systems, and extraction columns are common examples of such multivariate systems.

Considering the importance of multivariate PBE, in recent years, PBE solution methods have been extended to solve multivariate problems; the solution methods include Monte Carlo technique, classes and sectional method, and the method of moments (MOM). The bivariate population balance model, which includes two internal variables, has been simulated in commercial and non-commercial codes in the past [5–7], however, the implementation is not available as a standard feature in any CFD code. In general, moment-based methods have been found efficient in implementation and computational cost. Quadrature-based moment methods (QBMM) have been developed as a particular class of very successful moment methods. QBMM is able to overcome the closure problem of source term by using quadrature approximations. Direct quadrature method of moments (DQMOM) is a popular variation of QBMM that sustains its traits of simplicity and computational efficiency when applied to bivariate (and multivariate) PBE, and it can estimate the dependency of the particle velocities on the internal variables [4]. Considering these qualities, DQMOM appears as the most effective choice for the solution of bivariate PBE.

D. K. Singh (✉) · G. Bhutani
Multiphase Flow Research Lab (MFRL), School of Engineering Indian Institute of Technology Mandi, Mandi, India
e-mail: s17012@students.iitmandi.ac.in

S. K. Saha and M. Mukherjee (eds.), *Recent Advances in Computational Mechanics and Simulations*, Lecture Notes in Mechanical Engineering,
https://doi.org/10.1007/978-981-15-8315-5_46

Although a MATLAB implementation of DQMOM exists in the literature [2], there is no open-source implementation of DQMOM in any CFD code as a standard feature; an implementation of DQMOM through the use of user-defined functions in Ansys Fluent was demonstrated though [3, 8].

Fluidity is an open-source finite element code for modelling fluid flows, which has significant advantages over other software products. It allows the use of anisotropic adaptive unstructured meshes resulting in the reduced computational cost of a simulation compared to fixed-mesh simulations. The code is highly parallelised allowing for multinode simulations on high-performance computing (HPC) clusters along with the use of mesh adaptivity. To the best of the authors' knowledge, no comprehensive study on the implementation of the DQMOM for the solution of bivariate PBE problems in a finite element framework has been reported to date. In the present paper, implementation of the DQMOM is presented in Fluidity for solving bivariate PBE. The implementation is verified against analytical solutions, a few of which have been developed as a part of this work.

2 Model Details

2.1 *Bivariate PBE and DQMOM*

The general form of bivariate PBE is given as

$$\frac{\partial n(\boldsymbol{\xi},\boldsymbol{x},t)}{\partial t}+\frac{\partial}{\partial x_i}\langle u_i|\boldsymbol{\xi}\rangle n(\boldsymbol{\xi},\boldsymbol{x},t)-\frac{\partial}{\partial x_i}\left(D_x(\boldsymbol{x},t)\frac{\partial n(\boldsymbol{\xi},\boldsymbol{x},t)}{\partial x_i}\right)=S_{\boldsymbol{\xi}}(\boldsymbol{\xi},\boldsymbol{x},t), \tag{1}$$

where $\boldsymbol{x}$ is the position vector and $\boldsymbol{\xi}=(\xi_1,\xi_2)$ represents the internal coordinate vector for the bivariate PBE, which models the evolution of the number density function $n(\boldsymbol{\xi},\boldsymbol{x},t)$. $\langle u_i|\boldsymbol{\xi}\rangle$ represents mean velocity field of dispersed phase conditional to internal coordinates; particles move in the physical space with this velocity field. D_x is the spatial diffusion tensor, and $S_{\boldsymbol{\xi}}(\boldsymbol{\xi},\boldsymbol{x},t)$ is the source term containing derivatives and integral with respect to $\boldsymbol{\xi}$. These include all possible transformations in the dispersed phase such as aggregation, breakage, nucleation and growth.

The DQMOM approximation for the bivariate NDF can be written as [4]

$$n(\boldsymbol{\xi},\boldsymbol{x},t)\approx\sum_{\alpha=1}^{N}w_\alpha(\boldsymbol{x},t)\delta(\xi_1-\langle\xi_1\rangle_\alpha(\boldsymbol{x},t))\delta(\xi_2-\langle\xi_2\rangle_\alpha(\boldsymbol{x},t)), \tag{2}$$

where α denotes the DQMOM quadrature node, with the weights as w_α and $\langle\xi_1\rangle_\alpha$ and $\langle\xi_2\rangle_\alpha$ as the quadrature abscissas. Weights and quadrature abscissas are scalar fields in the external coordinate space. Using Eq. (2), the moments of the NDF can be estimated as follows:

$$m_{kl} = \int_{-\infty}^{+\infty}\int_{-\infty}^{+\infty} \xi_1^k \xi_2^l n(\boldsymbol{\xi}, \boldsymbol{x}, t) d\boldsymbol{\xi} \approx \sum_{\alpha=1}^{N} w_\alpha \langle\xi_1\rangle_\alpha^k \langle\xi_2\rangle_\alpha^l \tag{3}$$

The quadrature approximation of NDF, as shown in Eq. (2), when substituted in the bivariate PBE (Eq. (1)), followed by taking its (k, l)th moment gives

$$\begin{aligned}&\sum_{\alpha=1}^{N}\left[(1-k-l)\langle\xi_1\rangle_\alpha^k\langle\xi_2\rangle_\alpha^l a_\alpha + k\langle\xi_1\rangle_\alpha^{k-1}\langle\xi_2\rangle_\alpha^l b_{1\alpha} + l\langle\xi_1\rangle_\alpha^k\langle\xi_2\rangle_\alpha^{l-1} b_{2\alpha}\right]\\ &= \sum_{\alpha=1}^{N}\left[k(k-1)\langle\xi_1\rangle_\alpha^{k-2}\langle\xi_2\rangle_\alpha^{l-1} C_{12\alpha} + l(l-1)\langle\xi_1\rangle_\alpha^k\langle\xi_2\rangle_\alpha^{l-2} C_{22\alpha}\right] + \bar{S}_{kl}^{(N)},\end{aligned} \tag{4}$$

where $C_{\alpha\beta\gamma} = D_x \frac{\partial\langle\xi_\beta\rangle_\alpha}{\partial x_i}\frac{\partial\langle\xi_\gamma\rangle_\alpha}{\partial x_i} w_\alpha$ and $\bar{S}_{kl}^{(N)} = \int_{-\infty}^{+\infty}\int_{-\infty}^{+\infty} \xi_1^k \xi_2^l s_\xi(\boldsymbol{\xi}, \boldsymbol{x}, t) d\xi_1 d\xi_2$.

The linear system shown in Eq. (4) is solved for a_α, $b_{1\alpha}$ and $b_{2\alpha}$, which are then used as source terms to evaluate the weights (w_α) and weighted abscissas $(\zeta_{1\alpha} = w_\alpha \xi_{1\alpha}$ and $\zeta_{2\alpha} = w_\alpha \xi_{2\alpha})$ using the following transport equations:

$$\frac{\partial w_\alpha}{\partial t} + \frac{\partial}{\partial x_i}(\langle u_i\rangle_\alpha w_\alpha) - \frac{\partial}{\partial x_i}\left(D_x \frac{\partial w_\alpha}{\partial x_i}\right) = a_\alpha, \tag{5}$$

$$\frac{\partial \zeta_{1\alpha}}{\partial t} + \frac{\partial}{\partial x_i}(\langle u_i\rangle_\alpha \zeta_{1\alpha}) - \frac{\partial}{\partial x_i}\left(D_x \frac{\partial \zeta_{1\alpha}}{\partial x_i}\right) = b_{1\alpha}, \tag{6}$$

and

$$\frac{\partial \zeta_{2\alpha}}{\partial t} + \frac{\partial}{\partial x_i}(\langle u_i\rangle_\alpha \zeta_{2\alpha}) - \frac{\partial}{\partial x_i}\left(D_x \frac{\partial \zeta_{2\alpha}}{\partial x_i}\right) = b_{2\alpha}. \tag{7}$$

Details on the above can be found in Marchisio and Fox [4]. Since the number of unknowns in the vector $\{\ a_\alpha, b_{1\alpha}, b_{2\alpha}\ \}$ is $3N$, hence an equal number of equations are required in the linear system in Eq. (4). This is achieved by taking $3N$ moment sets in Eq. (4) as different (k, l) sets.

2.2 *Finite Element Formulation*

The strong form of the partial differential equation is transformed into a weak form by multiplying it with a test function and integrating over the complete volume Ω. For instance, the weak form of the transport equation for weights is obtained using a test function φ and integrating over the domain Ω:

$$\int_\Omega \varphi\left(\frac{\partial w_\alpha}{\partial t} + \frac{\partial}{\partial x_i}(\langle u_i\rangle_\alpha w_\alpha) - \frac{\partial}{\partial x_i}\left(D_x \frac{\partial w_\alpha}{\partial x_i}\right) - a_\alpha\right) = 0. \tag{8}$$

The above equation is discretised using a finite approximation for w_α, given as $w_\alpha = \sum_{j=1}^{N_{\text{nodes}}} w_\alpha^j \varphi^j$, where φ^j are basis functions for weight and w_α^j are the coefficients, which need to be determined. The known source term is also projected on the FE space as $a_\alpha = \sum_{j=1}^{N_{\text{nodes}}} a_\alpha^j \varphi^j$. The final discretised equation, written in matrix form, is given as

$$M\frac{d\underline{\mathrm{w}}_\alpha}{dt} + A(u_i)\underline{\mathrm{w}}_\alpha + K\underline{\mathrm{w}}_\alpha + S\underline{\mathrm{a}}_\alpha = 0, \tag{9}$$

where M is the mass matrix, A as the advection matrix, K as the diffusion matrix, and S as the source term matrix. The unknown weight vector $\underline{\mathrm{w}}_\alpha$ is determined through the solution of this matrix equation. A similar finite element formulation is applied to the solution of transport equations for the weighted abscissas $\zeta_{1\alpha}$ and $\zeta_{2\alpha}$.

3 Results and Discussion

3.1 Homogeneous Aggregation

Homogeneous bivariate PBE can be written as

$$\frac{\partial n(\xi_1, \xi_2, t)}{\partial t} = S_\xi(\xi_1, \xi_2). \tag{10}$$

The source term for homogeneous aggregation of bivariate PBE is

$$\begin{aligned} S_\xi(\xi_1, \xi_2) = \frac{\beta_0}{2}\int_0^\infty \int_0^\infty n(\xi_1', \xi_2')n(\xi_1 - \xi_1', \xi_2 - \xi_2')d\xi_1' d\xi_2' \\ -\beta_0 n(\xi_1, \xi_2)\int_0^\infty \int_0^\infty n(\xi_1', \xi_2')d\xi_1' d\xi_2', \end{aligned} \tag{11}$$

where a constant aggregation kernel β_0 has been considered.

Applying (k, l)th moment transform on $S_\xi(\xi_1, \xi_2)$, gives

$$\begin{aligned} \bar{S}_{kl} = \frac{\beta_0}{2}\iiiint_0^\infty [(\xi_1 + \xi_1')^k(\xi_2 + \xi_2')^l - \xi_1^k \xi_2^l - (\xi_1')^k(\xi_2')^l] \\ \times n(\xi_1', \xi_2')n(\xi_1, \xi_2)d\xi_1' d\xi_2' d\xi_1 d\xi_2. \end{aligned} \tag{12}$$

Substituting the DQMOM approximation for the NDF, as shown in Eq. (2), in Eq. (12) results in

$$\bar{S}_{kl}^{(N)} = \frac{\beta_0}{2}\sum_{\alpha=1}^{N}\sum_{\gamma=1}^{N} w_\alpha w_\gamma [(\langle\xi_1\rangle_\alpha + \langle\xi_1\rangle_\gamma)^k(\langle\xi_2\rangle_\alpha + \langle\xi_2\rangle_\gamma)^l - \langle\xi_1\rangle_\alpha^k \langle\xi_2\rangle_\alpha^l - \langle\xi_1\rangle_\gamma^k \langle\xi_2\rangle_\gamma^l]. \tag{13}$$

Two cases, polydisperse and monodisperse initial NDF, were used to verify the implementation of this homogeneous aggregation case, which are presented next.

3.1.1 Case 1—Polydisperse Initial NDF

In this case, the bivariate DQMOM implementation in Fluidity, performed as a part of this work, is verified for a homogeneous aggregation PBE with polydisperse initial NDF. The initial NDF in the present case was represented as a DQMOM approximation with $N = 2$ with $w_1(0) = 0.75$ and $w_2(0) = 0.25$, and $\langle \xi_1 \rangle_1 = 3.0$, $\langle \xi_1 \rangle_2 = 2.0$, $\langle \xi_2 \rangle_1 = 4.0$, and $\langle \xi_2 \rangle_2 = 1.0$, which resulted in a non-singular linear system. The constant aggregation kernel $\beta_0 = 1$ was assumed in this case. Figure 1a and b show the evolution of moments m_{00} and m_{10}. A very good agreement can be seen between the Fluidity results and the analytical solution, which is detailed in Appendix 1. The prediction of moment m_{20} and error in the numerical solution are shown in Figs. 2a and 2b, respectively. The accuracy of the numerical solutions is excellent, which is confirmed by the low percentage relative error in the solution.

The evolution of a moment of global order three—m_{30}—is presented in Fig. 3. During the construction of the linear system in Eq. (4) with 6 unknowns, the following (k, l) set was used: $\{(0, 0), (1, 0), (0, 1), (2, 0), (0, 2)\}$, which correspond to pure moments. Since 6 equations are required, three options for the sixth member of the (k, l) set were tried, which included $(2, 1)$, $(2, 2)$ and $(3, 2)$, corresponding to moments of global order three, four and five, respectively. A comparison of the numerical solution for the evolution of m_{30} is presented for these three cases. Numerical solution with $(2, 1)$, i.e. third order, predicts the solution most accurately. In comparison, the predicted numerical solution with fourth and fifth order (k, l) sets over-predict the value of m_{30}. In general, it was noticed that the prediction of a moment (m_{30} in the present case) is excellent if the order of the sixth member of the (k, l) set is equal to the order of the moment of interest; the result is over-predicted if the order of the sixth member is higher than the order of the moment of interest.

3.1.2 Case 2—Monodisperse Initial NDF

To show the effectiveness of Fluidity in handling singular initial conditions, a monodisperse initial NDF $n(\xi_1, \xi_2, 0) = \delta(\xi_1 - 1)\delta(\xi_2 - 1)$, was considered [4]. To surpass singularity, arbitrary perturbations were applied to abscissa values. Figure 4 shows an excellent agreement between Fluidity results and the analytical solutions for the moments m_{11}, m_{00} and m_{10}.

To demonstrate the effect of perturbation, three sets of perturbations to provide initial abscissa values were used, as shown in Fig. 5. For the Fluidity solution best matching the analytical solution, perturbation set was chosen in such a way that initial values of moments m_{10} and m_{01} do not change after perturbing the abscissa values. Initial values of moments m_{10} and m_{01} were close to 1, but not equal to 1, for other sets of perturbations.

Fig. 1 Plots for the time evolution of m_{00} and m_{10} for homogeneous aggregation (Case 1) with $N = 2$

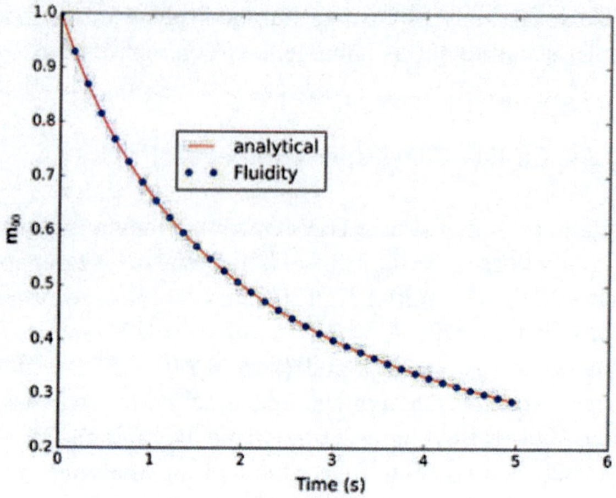

(a) Plot of m_{00}

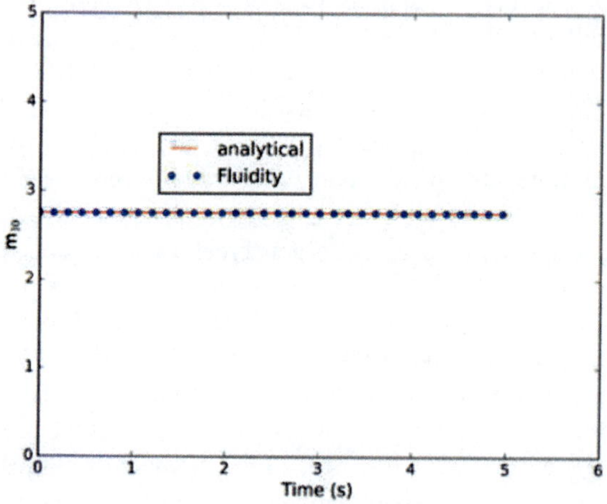

(b) Plot of m_{10}

3.2 *Homogeneous Breakage*

The source term for homogeneous breakage with constant breakage kernel (a_0) is

$$S_\xi(\xi_1, \xi_2) = a_0 \iint_0^\infty n(\xi_1', \xi_2') b(\boldsymbol{\xi} | \boldsymbol{\xi}') d\xi_1' d\xi_2' - a_0 n(\xi_1, \xi_2), \tag{14}$$

where $b(\boldsymbol{\xi} | \boldsymbol{\xi}')$ is daughter distribution function. Applying the moment transform on the $S_\xi(\xi_1, \xi_2)$ gives

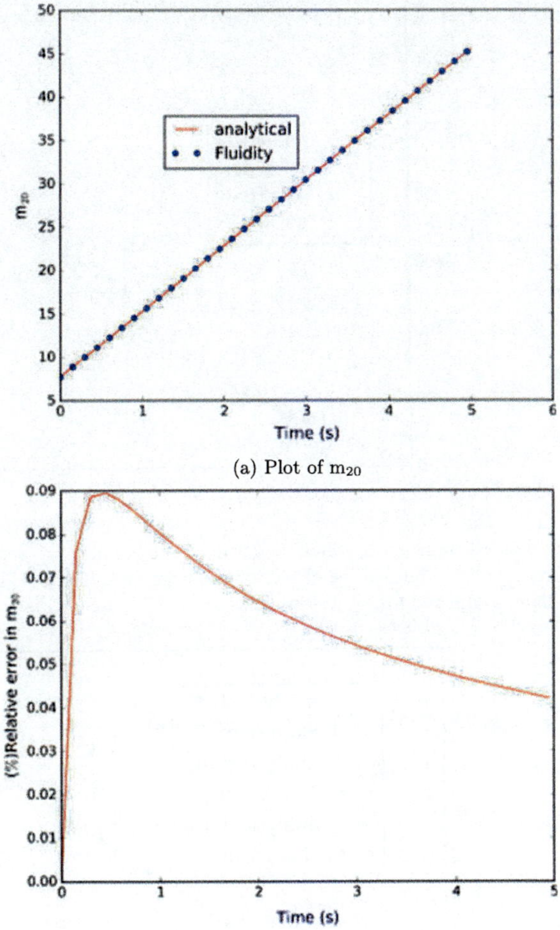

(a) Plot of m_{20}

(b) Errors in m_{20}

Fig. 2 Plots for the time evolution of m_{20} and error in m_{20} for homogeneous aggregation (Case 1) with $N = 2$

$$\bar{S}_{kl}^{(N)} = a_0 \iiiint_0^\infty \xi_1^k \xi_2^l b(\boldsymbol{\xi} | \boldsymbol{\xi}') n(\xi_1', \xi_2') n(\xi_1, \xi_2) d\xi_1' d\xi_2' d\xi_1 d\xi_2 \\ - a_0 \iint_0^\infty \xi_1^k \xi_2^l n(\xi_1, \xi_2) d\xi_1 d\xi_2. \tag{15}$$

Using DQMOM approximation in the above:

$$\bar{S}_{kl}^{(N)} = a_0 \sum_{\alpha=1}^{N} \left[\bar{b}_\alpha^{kl} - \langle \xi_1 \rangle_\alpha^k \langle \xi_2 \rangle_\alpha^l \right] w_\alpha, \tag{16}$$

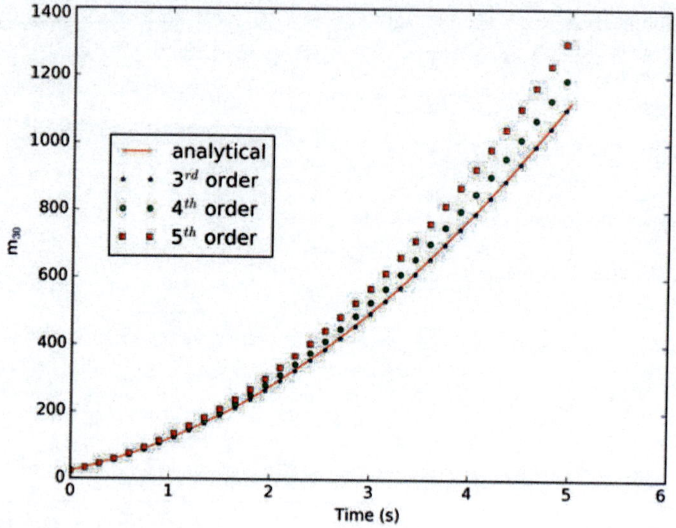

Fig. 3 Prediction of m_{30} with $N = 2$ using the following (k, l) set: {(0, 0), (1, 0), (0, 1), (2, 0), (0, 2)}. Three options of the sixth member of this set were tried: (2, 1), (2, 2) and (3, 2), which corresponded to third, fourth and fifth-order moments

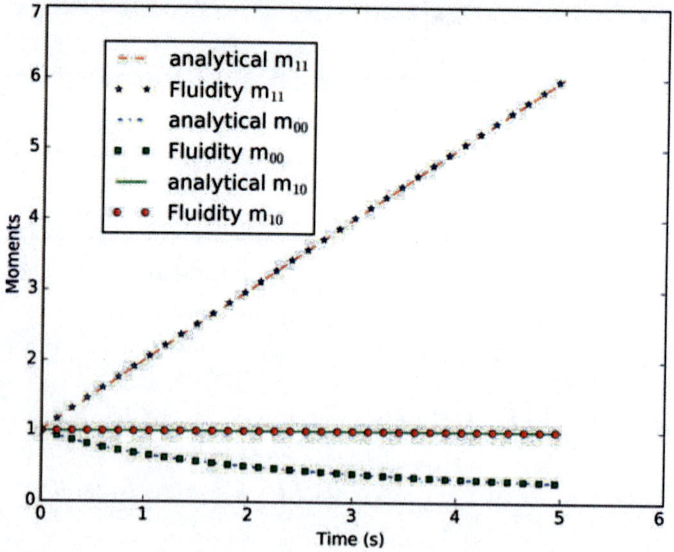

Fig. 4 Moment plot for homogeneous aggregation with $N = 2$ for monodisperse initial NDF (Case 2). Initial conditions are $w_1(0) = 0.65$ and $w_2(0) = 0.35$, and $\langle \xi_1 \rangle_1(0) = \langle \xi_1 \rangle_2(0) = \langle \xi_2 \rangle_1(0) = \langle \xi_2 \rangle_2(0) = 1$. Fluidity results are compared against analytical solution

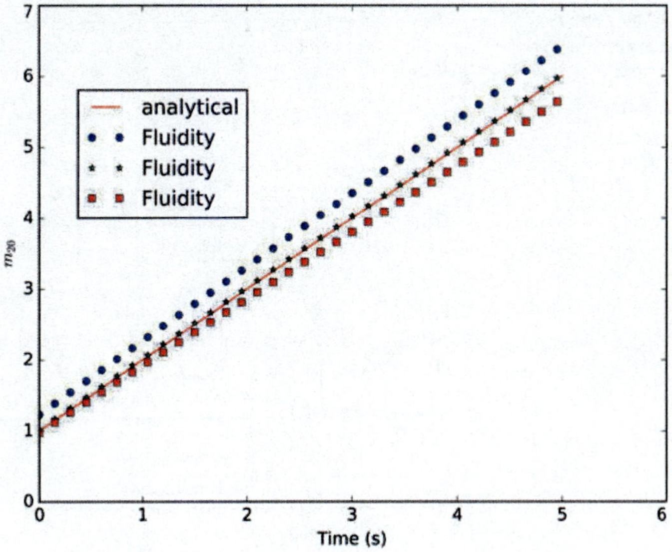

Fig. 5 Effect of perturbation on the prediction of m_{20} for a monodispersed NDF. The moment is estimated accurately for the perturbation set in which the initial values of moments m_{10} and m_{01} remain same after perturbing the initial values of abscissas

where $\bar{b}_\alpha^{kl} = \iint_0^\infty \xi_1^k \xi_2^l b(\xi|\xi')d\xi_1 d\xi_2$.

To verify the implementation of homogeneous breakage case in Fluidity, constant breakage kernel (a_0), and symmetric fragmentation with daughter distribution $b(\xi|\xi') = 2\delta(\xi_1 - \xi_1'/2)\delta(\xi_2 - \xi_2'/2)$ were used.

Example 1 Figure 6a and b demonstrate the agreement between the analytical solutions and numerical solutions of m_{00} and m_{10} for N = 3 with nine moments (m_{00}, m_{10}, m_{01}, m_{20}, m_{11}, m_{02}, m_{03}, m_{30}, m_{22}). Constant breakage kernel, (a_0) = 1.0 was used for this case. Initial values were $w_1 = 0.5$, $w_2 = 0.1$, $w_3 = 0.4$, and $\langle\xi_1\rangle_1(0) = 2.5$, $\langle\xi_1\rangle_2(0) = 3.5$, $\langle\xi_1\rangle_3(0) = 3.5$, and $\langle\xi_2\rangle_1(0) = 2.0$, $\langle\xi_2\rangle_2(0) = 10$ and $\langle\xi_2\rangle_3(0) = 2.70$.

Example 2 Figure 7 shows the variation of accuracy in numerical solutions of m_{11} using Fluidity with N = 2 and N = 3. Six moments (m_{00}, m_{10}, m_{01}, m_{20}, m_{02}, m_{21}) with initial values $w_1(0) = 0.6$, $w_2(0) = 0.4$, and $\langle\xi_1\rangle_1(0) = 1.5$, $\langle\xi_1\rangle_2(0) = 2$, and

Fig. 6 Plots of m_{00} and m_{10} for homogeneous breakage for Example 1 with N = 3

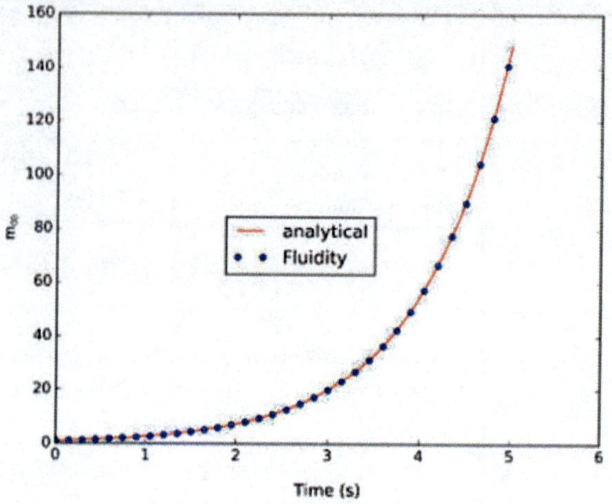

(a) Plot of m_{00}

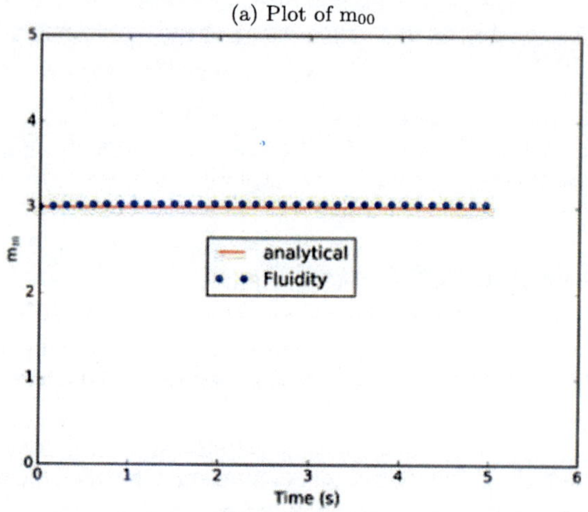

(b) Plot of m_{10}

$\langle\xi_2\rangle_1(0) = 1.0$, $\langle\xi_2\rangle_2(0) = 2$ for N = 2 were used. Moments and initial conditions were the same as described in Example 1 for N = 3. $a_0 = 1.0$ was used for both N = 2 and N = 3. An improvement in accuracy can be seen clearly by using DQMOM approximation with more number of quadrature points.

Example 3 To demonstrate the relation between choice of moment sets and accuracy of numerical solutions, in Fig. 8 three numerical solutions of moment m_{22} are compared to the analytical solution. The first five pure moments (m_{00}, m_{10} , m_{01}, m_{20}, m_{02}) were used in all three cases and the last four moments were of global order three (m_{21}, m_{12} , m_{30}, m_{03}), global order four (m_{22}, m_{31} , m_{13}, m_{40}) and global order five (m_{23}, m_{32} , m_{41}, m_{14}) for cases 1, 2 and 3, respectively. $N = 3$ was used in this analysis. Initial conditions were $w_1 = 0.5$, $w_2 = 0.1$, $w_3 = 0.4$ and $\langle\xi_1\rangle_1(0) = 2.5$,

Fig. 7 Plots of m_{11} for Example 1 with N = 2 and N = 3 for homogeneous breakage. Accuracy of m_{11} increases with N = 3 for a proper set of nine moments

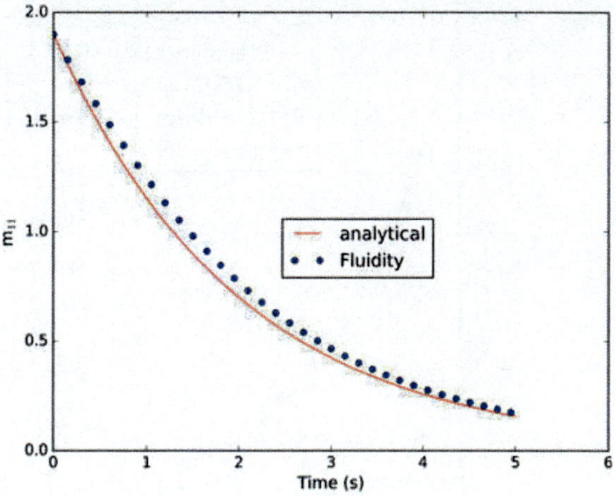

(a) Plot of m_{11} for N = 2

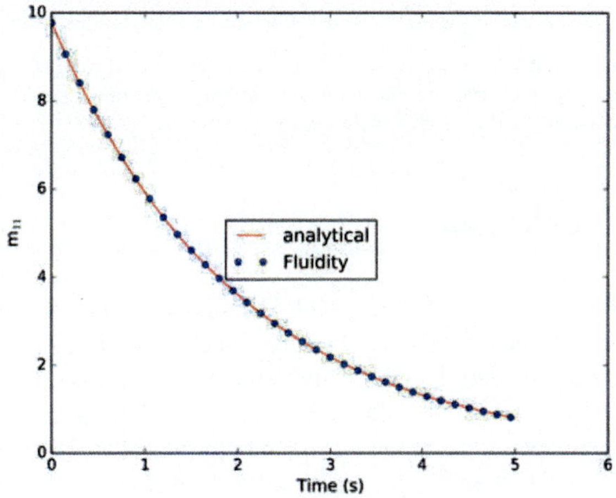

(b) Plot of m_{11} for N = 3

$\langle\xi_1\rangle_2(0) = 3.5$, $\langle\xi_1\rangle_3(0) = 3.5$ and $\langle\xi_2\rangle_1(0) = 2.0$, $\langle\xi_2\rangle_2(0) = 10$ and $\langle\xi_2\rangle_3(0) = 2.70$. $a_0 = 1.0$ was used for this case. Figure 8 shows that the prediction of m_{22} is quite accurate if moments of global order four are used for last four moments, while m_{22} is over-predicted and under-predicted for using moments of global order five and moments of global order three for last four moments, respectively.

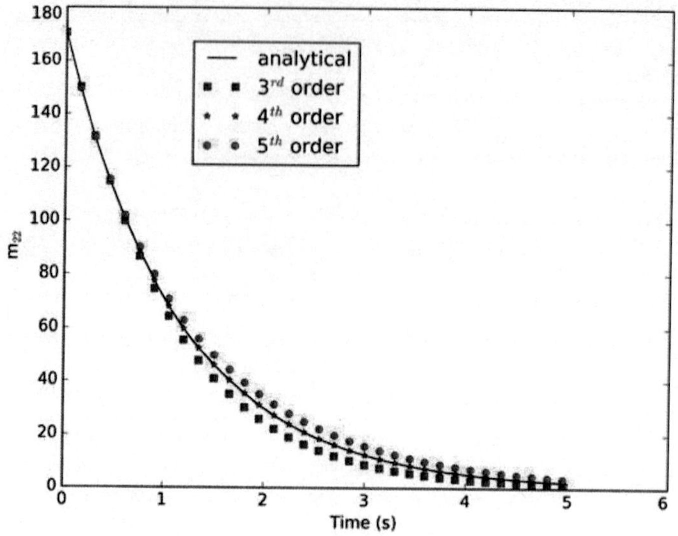

Fig. 8 Predictions of m22 using the first five pure moments ($m_{00}, m_{10}, m_{01}, m_{20}, m_{02}$) and different higher global order moments for the last four moments—global order three, global order four and global order five. $N = 3$ was used in this case

4 Conclusion

Implementation of the DQMOM to solve bivariate PBE in a finite element framework is presented in this work. The analytical solutions to moments were used for verification of bivariate DQMOM implementation in this open-source code—Fluidity. The verification was performed for a bivariate PBE with homogeneous aggregation and homogeneous breakage terms. In both these cases, the Fluidity results converged to the analytical solutions. An important advantage of using Fluidity is that, even in the implementation of DQMOM for bivariate applications, it is capable of maintaining the accuracy of numerical solutions and effectiveness of handling singularity. The effect of the choice of moment set and the number of DQMOM nodes is demonstrated clearly in the present implementation of the DQMOM solution to bivariate PBE. Based on the results obtained in the present work, it can be concluded that the open-source Fluidity code provides an efficient alternative to the other available CFD packages for solving population balance equation, which is not available as a standard feature in any other code.

Appendix 1: Analytical Solutions

Homogeneous Aggregation Problem

Bivariate PBE for homogeneous aggregation with constant aggregation kernel is

$$\begin{aligned}\frac{\partial n(\xi_1, \xi_2, t)}{\partial t} = \frac{\beta_0}{2} \iint_0^\infty & n(\xi_1', \xi_2')n(\xi_1 - \xi_1', \xi_2 - \xi_2')d\xi_1' d\xi_2' \\ & -\beta_0 n(\xi_1, \xi_2) \iint_0^\infty n(\xi_1', \xi_2')d\xi_1' d\xi_2',\end{aligned} \tag{17}$$

and thus the transpo*rt equation for the moment of order $\{k, l\}$ is

$$\begin{aligned}\frac{\mathrm{d}m_{kl}}{\mathrm{d}t} = \frac{\beta_0}{2} \iiiint_0^\infty & [(\xi_1 + \xi_1')^k(\xi_2 + \xi_2')^l - \xi_1^k \xi_2^l - (\xi_1')^k(\xi_2')^l] \\ & \times n(\xi_1', \xi_2')n(\xi_1, \xi_2)d\xi_1' d\xi_2' d\xi_1 d\xi_2.\end{aligned} \tag{18}$$

For the moment of order $\{k, l\} = \{0, 0\}, \{1, 0\}, \{2, 0\}$, the following equations are obtained:

$$\begin{aligned}\frac{\mathrm{d}m_{00}}{\mathrm{d}t} &= -\frac{\beta_0}{2} m_{00}^2, \\ \frac{\mathrm{d}m_{10}}{\mathrm{d}t} &= 0, \\ \frac{\mathrm{d}m_{20}}{\mathrm{d}t} &= \beta_0 m_{10}^2,\end{aligned} \tag{19}$$

and similar equations can be obtained for the other moments. Analytical solutions for the moments can be obtained by integrating the above equations.

Analytical Solution to the Homogeneous Breakage Problem for the Verification of Bivariate DQMOM Implementation

Bivariate PBE for homogeneous aggregation with constant aggregation kernel is

$$\frac{\partial n(\xi_1, \xi_2, t)}{\partial t} = a_0 \iint_0^\infty n(\xi_1', \xi_2')b(\boldsymbol{\xi} | \boldsymbol{\xi}')d\xi_1' d\xi_2' - a_0 n(\xi_1, \xi_2). \tag{20}$$

and thus the transport equation for the moment of order $\{k, l\}$ is

$$\frac{\mathrm{d}m_{kl}}{\mathrm{d}t} = a_0 \iiiint_0^\infty \xi_1^k \xi_2^l b(\boldsymbol{\xi} | \boldsymbol{\xi}')n(\xi_1', \xi_2')n(\xi_1, \xi_2)d\xi_1' d\xi_2' d\xi_1 d\xi_2 - a_0 m_{kl}. \tag{21}$$

Using $b(\boldsymbol{\xi}|\boldsymbol{\xi}') = 2\delta(\xi_1 - \xi_1'/2)\delta(\xi_2 - \xi_2'/2)$, for the moment of order $\{k, l\}$ we can write

$$\frac{\mathrm{d}m_{kl}}{\mathrm{d}t} = a_0(2^{1-k-l} - 1)m_{kl}. \tag{22}$$

Analytical solution for the moment of order $\{k, l\}$ can be obtained by integrating the above equation.

References

1. Bhutani, G., Brito-Parada, P.R., Cilliers, J.J.: Polydispersed flow modelling using population balances in an adaptive mesh finite element framework. Comput. Chem. Eng. **87**, 208-225 (2016). https://doi.org/10.1016/j.compchemeng.2016.01.011
2. Lin, A., Frances, C.: Discussion on DQMOM to solve a bivariate population balance equation applied to a grinding process. Powder Technol. **295**, 234–244 (2016). https://doi.org/10.1016/j.powtec.2016.03.037
3. Marchisio, D.L., Barresi, A.A.: Investigation of soot formation in turbulent flames with a pseudo-bivariate population balance model. Chem. Eng. Sci. **64**(2), 294–303 (2009). https://doi.org/10.1016/j.ces.2008.10.020
4. Marchisio, D.L., Fox, R.O.: Solution of population balance equations using the direct quadrature method of moments. J. Aerosol Sci. **36**(1), 43–73 (2005). https://doi.org/10.1016/j.jaerosci.2004.07.009
5. Pollack, M., Salenbauch, S., Marchisio, D.L., Hasse, C.: Bivariate extensions of the extended quadrature method of moments (EQMOM) to describe coupled droplet evaporation and heat-up. J. Aerosol Sci. **92**, 53–69 (2016). https://doi.org/10.1016/j.jaerosci.2015.10.008
6. Singh, M., Kaur, G., De Beer, T., Nopens, I.: Solution of bivariate aggregation population balance equation: a comparative study. React. Kinet. Mech. Catal. **123**(2), 385–401 (2018). https://doi.org/10.1007/s11144-018-1345-9
7. Zhi, J., Jie, S., Zhiming, L.: Bivariate Taylor-series expansion method of moment for particle population balance equation in Brownian coagulation. J. Aerosol Sci. **114**, 94–106 (2017). https://doi.org/10.1016/j.jaerosci.2017.08.010
8. Zucca, A., Marchisio, D.L., Vanni, M., Barresi, A.A: Validation of bivariate DQMOM for nanoparticle processes simulation. AIChE J. **53**(4), 918–931 (2007). https://doi.org/10.1002/aic.11125

Application of Computational Mechanics in Emerging Areas

Application of Novel Radial Thresholding Method for the Segmentation of Different Phases from X-Ray Microtomographic Images of Concrete

Vishwas Chhimpa, Subhasis Pradhan, and Sudhirkumar V. Barai

1 Introduction

In general, concrete can be segmented into three major components, that is, voids, aggregates and mortar. Aggregates and air voids are surrounded by the mortar matrix. The mechanical and morphological characteristics of these components govern the macro-level performance of concrete. The distribution of aggregates and voids acts as a key factor for characteristic strength, crack propagation and long-term performance (durability) of concrete [1].

X-ray microtomography (XRT) is an advanced non-destructive imaging technique using which very high-resolution images of cross-sectional and three-dimensional (3D) internal structures of a concrete specimen can be analysed. However, to obtain the morphological description of concrete, it is necessary to segment the three components effectively of generated XRT images. Generally, a binarization process is used to segregate the three discrete components.

Tashman et al. [2] used IPBasic sub-programming language of Visual Basic to write the macros for the image analysis and study the horizontal and vertical distribution of air voids in asphalt mixes by applying image analysing techniques on X-ray computed tomography images of the same. Gallucci et al. [3] analysed the three-dimensional images obtained from synchrotron microtomography to quantify the anhydrous cement content, porosity and connectivity and tortuosity of pore network of cement paste. It was observed that the variation in spatial resolution and contrast resolution of the scanned images due to the age of the concrete influence the determination of the degree of pore connectivity. Zelelew and Papagiannakis [4] discussed an automated digital image processing (DIP) algorithm called volumetric-based global

V. Chhimpa · S. Pradhan · S. V. Barai (✉)
Indian Institute of Technology Kharagpur, Kharagpur, West Bengal 721302, India
e-mail: skbarai@civil.iitkgp.ac.in

S. K. Saha and M. Mukherjee (eds.), *Recent Advances in Computational Mechanics and Simulations*, Lecture Notes in Mechanical Engineering,
https://doi.org/10.1007/978-981-15-8315-5_47

minima (VGM) thresholding algorithm for processing asphalt concrete (AC) XRT images. Zelelew et al. [5] presented a comparative evaluation of two image segmentation techniques: adaptive enhancement-based thresholding algorithm (AETA) and volumetric-based thresholding algorithm (VTA) for processing asphalt concrete microstructure images obtained from X-ray computed tomography. Kapitány and Barsi [6] proposed an efficient method to determine the presence of dense objects (aggregates) inside a composite material (asphalt and concrete core samples) of the computed tomography slices using fast Fourier transformation. Fonseca and Scherer [7] presented a method to quantify the volume, spacing and size distribution of the air bubbles in mortar or concrete using automatic image analysis. Otsu's method of thresholding was adopted by Fonseca and Scherer to process the binary images. Hashemi et al. [8] discussed the difficulties in segmenting the various phases in X-ray tomographic images because of the presence of noise in the obtained tomographic representation. They used simultaneous region growing method to reconstitute the three-dimensional segmented image of granular materials.

2 Research Significance

It is a challenge to segregate the three components of concrete accurately with less computational effort. The present study deals with addressing this problem. In this context, the images are obtained by scanning the concrete specimen in X-ray microtomography. The objective is to segregate the three phases with less computation effort and quantify the same.

3 Background

3.1 X-Ray Microtomography

Beer-Lambert law which depicts a relationship between the absorption of light and the properties of the material through which it is passing is the fundamental principle of X-ray microtomography (XRT). XRT is functionally identical to the computed tomography (CT) scan especially used in medical science. In X-ray microtomography, the sample is irradiated with X-ray beams, and a cross-sectional image or a sequence of such images is generated by using mathematical models. Basically, the 2D X-ray absorption images are obtained by relative rotation between the sample and the X-ray source.

3.2 Bit Depth of an Image

A digital image is a collection of pixels array. Each pixel stores the information regarding the brightness/darkness (tone) of the image at that portion in the form of an integer, known as the grey value of the pixel. This grey value ranges from 0 to 255 (2^8 tonnes) for an 8-bit image and 0–65535 (2^{16} tonnes) for a 16-bit image and represents different shades of grey. For an 8-bit image, a grey value of 0 represents the pure black colour (tone of grey); on the other hand, a grey value of 255 represents the pure white colour.

3.3 Filtering/De-noising

De-noising/filtering is the removal of unwanted noisy pixels from an image to restore the original image. Pre-processing of the raw images is an essential step before the analysis of the images. However, the uncontrolled pre-processing may result in loss of necessary information from the image. There are number of filtering techniques; such as mean filtering, median filtering and Gaussian filtering. The principle of the median filter is to run through each pixel entry, replacing each pixel value with the median of the neighbouring entries. The property of median filtering allows preserving the edges while removing noise, which is the primary reason for its use.

3.4 Threshold Grey Value

In an image, the grey value at a point is a representative of the material density at that point. Grey value thresholding is the simplest of the methods for image segmentation. Threshold grey value is a characteristic grey value such that all the pixels having grey value lower than its value will be characterized as one phase.

3.5 Beam Hardening

Beam hardening is the phenomenon occurring when an X-ray beam comprising polychromatic energies passes through an object, resulting in selective attenuation of lower energy photons. The effect is conceptually similar to a high-pass filter, in that only higher energy photons are left to contribute to the beam, and thus mean beam energy is increased. The low-energy X-ray photons are attenuated more easily, and the remaining high-energy photons are not attenuated as easily. As a result, regions interior to an object are traversed with higher energy X-rays than regions towards the edge, making the edges effectively more attenuating than interiors.

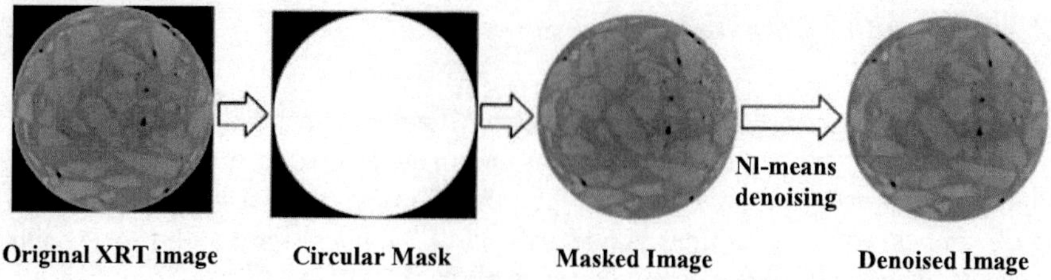

Fig. 1 Pre-processing steps of raw image

4 Methodology

The X-ray microtomography test was performed in an environmentally controlled chamber of 'GE Phoenix Vltomelx s', which is an industrial high-resolution computed tomography and X-ray system. The cylindrical concrete specimen was placed on a rotating stage. The X-ray source was set at a voltage of 150 kV and a tube current of 100 μA. For image analysis, two-dimensional 8-bit greyscale images of the scanned specimens were collected. In the present study, the images of the bottom to top plane were considered for the analysis. Hence, the collected images are circular in shape.

4.1 *Pre-processing of the Raw Images*

The sample raw image consists of various noise and background which are not important for the segregation of phases. Thus, pre-processing plays an important role in analysis of image. But sometimes pre-processing can remove some necessary information, so we should apply it bare minimum for less possible error. The following steps are followed in the pre-processing to obtain the denoised image and the steps are illustrated in Fig. 1.

(a) **Cropping**: Original size of raw image consisting of 5000 × 3452 pixels is then reduced to 600 × 600 pixels image of circular section.
(b) **Background Removal**: The original XRT image contains unnecessary edges and surface voids. To remove these, a mask of small diameter is taken, which is manually calculated and used to create a perfect masked sample image.
(c) **Noise Removal**: A non-local means (Nl-means) denoising filter is used to remove the unwanted noise from the masked image.

4.2 *Void Segregation*

After pre-processing, the source images are converted to several binary images by thresholding the source image with thresholds starting at minimum threshold. These thresholds are incremented by threshold step until maximum threshold to successfully segregate all the voids. So, the first threshold is (minimum threshold), the second is (minimum threshold + threshold step), the third step is (minimum threshold + 2 × threshold step), and so on. Moreover, while determining the threshold, a histogram of the de-noised image is plotted and any sudden jump in the number of pixels for certain greyscale value is analysed manually. In each binary image, the connected white pixels are grouped together and called as binary blobs. The centres of the binary blobs in the binary images are computed, and blobs located closer than minimum distance between blobs are merged. The centres and radii of the new merged blobs are computed and returned.

4.3 Aggregates Segregation

Aggregates cannot be segregated using simple thresholding. Thresholding for the aggregates required the knowledge of distribution of beam hardening effect. The following methods are applied earlier for the segregation of aggregates from the mortar;

- Using the greyscale value correction by employing the following equation:

$$GV^{new}(r) = \frac{GV(r) - GV^{p}(r)}{GV^{a}(r) - GV^{p}(r)} \tag{1}$$

- Using machine learning algorithm to segregate mortar and aggregate pixels.

Both methods consist of collection of training sample for better efficiency. Greyscale value correction method will have to take care of changed greyscale values of selected pixels of aggregates and mortar if sample is different. Similarly, for machine learning algorithm, a huge collection of training data is required for proper working in different sample images. The versatility decreases with the use of training data from images of same sample.

With the current rise and demand of speed in industrial field, we require a method which will provide efficiency in different types of samples with less computational time. So, in the present study, a method known as contrast threshold is used, which works on the principle similar to our own eyes to detect and distinguish the phases in one go, that is, the change in contrast. The contrast threshold can be used even when the filter is passed through radially, which will reduce the requirement of computational power while moving the filter through row-by-row. However, the efficiency of aggregate detection will be higher in latter case due to the inefficient segregation at outskirts of radius as rotation will skip $X^c \times r$ pixels, where X is the least angle of rotation (in radian) and r is the radial distance of a point from centre.

A graph of greyscale vs radius is observed in one specific direction (Fig. 2). It shows some sudden jumps which represent a sudden change of contrast, that

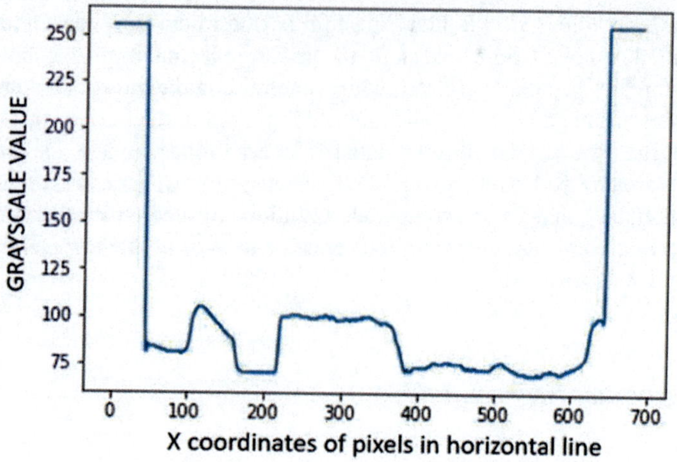

Fig. 2 Greyscale distribution in a horizontal line through a sample image

is, a phase change. The changes are detected using a simple function: $|GV_A(n) - GV_M(n-1)| - L\rangle 0$, where L is an average jump value which is taken 12 in our case. GV_A is the grey value of aggregate and GV_M is the average greyscale value of mortar found from the present study. The initial value of GV_M is considered as 60.

If the aforementioned condition satisfies, then a phase change occurred and the pixels are detected as aggregate until other change exists. As a change of 1 degree is taken per cycle then the total run cycle will be $300 \times 300 \times 450$ which is smaller than $600 \times 600 \times 450$. This proved that only small computational time is required than previous methods. But efficiency will be lower; hence we used the row-by-row moving with contrast threshold filter.

5 Results and Discussion

The void content was estimated by segregating the voids using simple threshold method, as discussed in Sect. 4.2. The detected voids in the pre-processed image can be seen in Fig. 3, where the accuracy in capturing the size and shape of the voids by the present method is justified. From this study the threshold grey value to detect all the voids was found to be 76. The void content using threshold method is estimated to be 0.91% for the 28 days cured specimen. The result obtained from VGStudio MAX 2.2 commercial software [9] is 0.96%. The difference in detecting the voids by the present method is about 5.5%, which infers the appropriateness of the used method.

The contrast threshold method discussed in Sect. 4.3 was applied in the radial direction to segregate aggregates and mortar. The detected aggregates, mortar and voids are shown in false colour in Fig. 4 for a sample image. The aggregate content

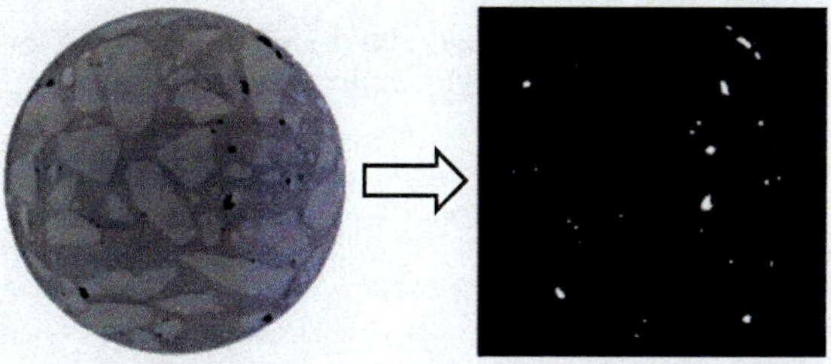

Fig. 3 Detected voids in binary image

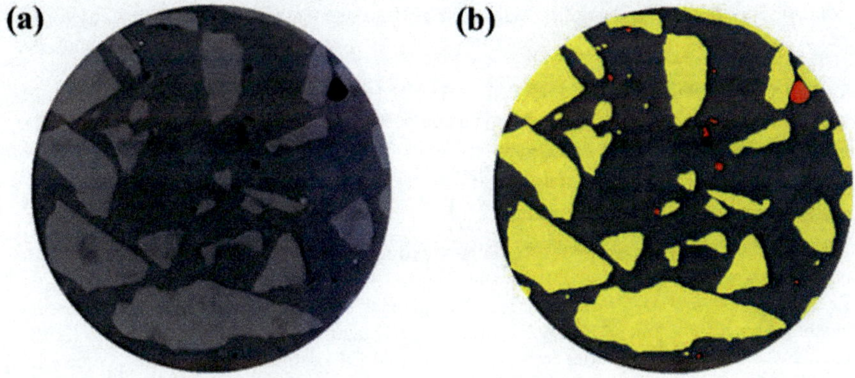

Fig. 4 **a** Original image, **b** detected aggregates and voids

estimated in this method is 49.19%. The error in determining the aggregate quantity is approximately 0.6% as compared to the actual aggregate content.

6 Conclusions

The primary objective of this study was to segregate three phases, that is, voids, aggregate and mortar of concrete using XRT images. First, the XRT images were pre-processed to avoid the presence of any noise. Subsequently, data is used to detect voids using simple threshold-based methodology and the detected voids content exhibited a good correlation with the results obtained from the commercial software VGStudio MAX 2.2. The segregation of aggregates and mortar by this method is not reliable owing to the association of beam hardening effect. In this context, a contrast thresholding method is proposed in the present study. The developed contrast threshold technique detected the aggregates to a decent level of accuracy. Consequently, the segregation of all the three phases of concrete in the XRT images using thresholding approach and novel method of contrast and radial threshold techniques is achieved with satisfactory accuracy. The study will be helpful to investigate further on the durability aspects of concrete.

Acknowledgements The authors gratefully acknowledge the financial support provided for the project on "Sustainable and Cost Effective Housing using Recycled Aggregate Based Concrete" under the mega project on Future of Cities by MHRD, Government of India.

References

1. Stamati, O., Roubin, E., Andò, E., Malecot, Y.: Phase segmentation of concrete X-ray tomographic images at meso-scale: validation with neutron tomography. Cement Concr. Compos. **88**, 8–16 (2018)
2. Tashman, L., Masad, E., D'Angelo, J., Bukowski, J., Harman, T.: X-ray tomography to characterize air void distribution in superpave gyrat compacted specimens. Int. J. Pavement Eng. **3**(1), 19–28 (2002)
3. Gallucci, E., Scrivener, K., Groso, A., Stampanoni, M., Margaritondo, G.: 3D experimental investigation of the microstructure of cement pastes using synchrotron X-ray microtomography (μCT). Cem. Concr. Res. **37**, 360–368 (2007)
4. Zelelew, H.M., Papagiannakis, A.T.: A volumetrics thresholding algorithm for processing asphalt concrete X-ray CT images. Int. J. Pavement Eng. **12**, 543–551 (2011)
5. Zelelew, H.M., Almuntashri, A., Agaian, S., Papagiannakis, A.T.: An improved image processing technique for asphalt concrete X-ray CT images. Road Mater. Pavement Des. **14**, 341–359 (2013)
6. Kapitány, K., Barsi, Á.: Fourier transformation-based CT correction method for geomaterial core samples. J. Mater. Civ. Eng. **28**, 1–5 (2016)
7. Fonseca, P.C., Scherer, G.W.: An image analysis procedure to quantify the air void system of mortar and concrete. Mater. Struct. **48**, 3087–3098 (2015)
8. Hashemi, M.A., Khaddour, G., François, B., Massart, T.J., Salager, S.: A tomographic imagery segmentation methodology for three-phase geomaterials based on simultaneous region growing. Acta Geotech. **9**, 831–846 (2014)
9. VGStudio Max 2.2. https://www.volumegraphics.com/en/products/vgstudio-max.html

Design of a Nonlinear Energy Harvesting Dynamic Vibration Absorber

Soumi Bhattacharyya and Shaikh Faruque Ali

1 Introduction

Structural vibrations are undesirable from safety and serviceability considerations. Several research works have been carried out over the past decades by engineers on different vibration control techniques such as isolation, passive energy dissipation devices, and active control systems to safeguard structures [1]. To ensure proper application of these technologies, an efficient round the clock structural health monitoring system plays a key role in which sensors are deployed all through the structure, sometimes at remote, inaccessible locations, which are exposed to extreme environmental conditions. Presently, wireless sensors work on battery power which is hazardous to the environment and also their replacement at remote locations is not cost-effective. One cost-effective and environment friendly way out of this problem is tapping power from ambient energy resources and using it for powering sensors which is known as energy harvesting [2].

Harvesting energy from the ambient vibration has garnered prominence since the last decade owing to its advantages in powering the wireless sensors [3, 4]. Three basic vibration energy harvesting mechanisms have been reported in the literature, namely, the electromagnetic [3, 5], electrostatic [6], and piezoelectric transduction [4, 7]. This paper focuses on the piezoelectric harvesters (PEHs) in which piezoelectric material is used to transform the mechanical strain, generated by ambient vibration around the power harvesting device, into electrical charge. This method of piezoelectric transduction has received much attention because of its simplicity in design and ability to have higher power densities [8, 9]. PEHs are best suitable for high-frequency application and where voltage is a primary concern.

S. Bhattacharyya (✉) · S. F. Ali
Indian Institute of Technology Madras, Chennai 600036, Tamil Nadu, India
e-mail: soumibhttchr86@gmail.com

S. K. Saha and M. Mukherjee (eds.), *Recent Advances in Computational Mechanics and Simulations*, Lecture Notes in Mechanical Engineering,
https://doi.org/10.1007/978-981-15-8315-5_48

Piezoelectric energy harvesting exploiting linear devices has been investigated widely and explicit expressions for optimal parameters are available in the literature [8–10]. One of the major drawbacks of linear energy harvesters is that they are efficient only when the excitation frequency is around the resonance frequency [11]. However, harvesters with a broadband or adaptive response [10, 12] and multiple harvesters [13, 14] are likely to be beneficial in case of mistuned condition or nonharmonic excitations. In broadband energy harvesters, nonlinear structural systems are used to maximize the harvested energy over a wide range of excitation frequencies [9, 10, 12, 15], which is under study in this paper.

In literature, detailed investigations have been carried out on structural control as well as energy harvesting. However, very less number of studies have been found on nonlinear controller-harvesting combined system. The main objective of this paper is to design a nonlinear device, which can control the vibration of a primary single degree of freedom (SDOF) system as well as generate energy from ambient vibration. For this purpose, a device is designed combining a conventional nonlinear dynamic vibration absorber (DVA) and the mechanism of piezoelectric transduction. DVA is a well-established passive vibration control device that undergoes large displacements in order to dissipate the energy from the primary structure. This large displacement is utilized to generate energy by using piezoelectric material. To be specific, in this paper, a nonlinear energy harvesting dynamic vibration absorber (EHDVA) is proposed and designed to control the vibration of a base-excited SDOF primary structure. The modeling of the SDOF-EHDVA system and the time-domain formulation of the same are done. A detailed numerical study with nondimensional equations of motions is carried out. The important design parameters are identified and the optimal values of those parameters are evaluated. The design procedure for EHDVA system is elaborated for an example SDOF primary system. Finally, the performance of the designed EHDVA system is evaluated numerically.

2 Modeling of SDOF-Nonlinear EHDVA System and Formulation

The mathematical model for the SDOF-nonlinear DVA system with piezoelectric harvester is shown in Fig. 1, where the mass, stiffness, and viscous damping of the primary SDOF system are represented by m_0, k_0 and c_0, respectively. The EHDVA system of equivalent mass m_h and equivalent viscous damping c_h is considered to be attached with the primary system by a nonlinear hard spring which is modeled as Duffing oscillator. The linear and nonlinear components of the spring force are represented as k_{h1} and k_{h2}, respectively, and the force-displacement relationship is described by $F_s = k_{h1}x + k_{h2}x^3$, $k_{h1} > 0$ and $k_{h2} > 0$. The piezoelectric element of capacitance C_p and load resistance R_L is attached to the nonlinear DVA system. The coupling between electrical and mechanical part of the harvester is represented by θ. $x_0(t)$, and $x_h(t)$ are the absolute displacements of the primary SDOF and

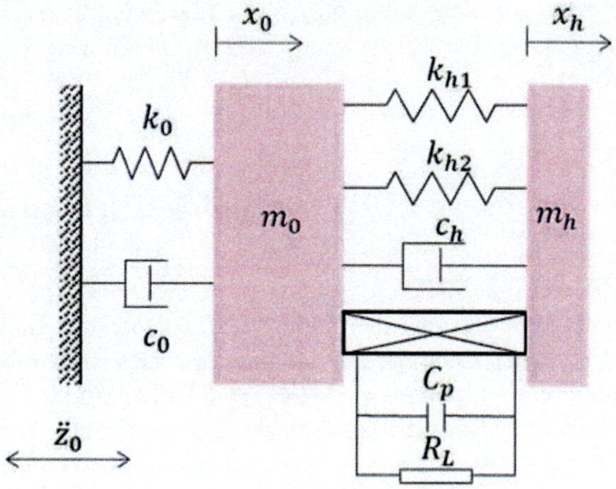

Fig. 1 Schematic diagram of the energy harvesting dynamic vibration absorber attached to a host vibrating system (primary structure)

DVA system, respectively, considering the primary system to be subjected to a base acceleration of $\ddot{z}(t)$. The voltage across the load resistor is represented by v.

The equations of motion of the primary SDOF-EHDVA system subjected to base acceleration $\ddot{z}_0$ are as follows:

$$m_0(\ddot{x}_0 + \ddot{z}_0) + c_0\dot{x}_0 + k_0x_0 - c_h(\dot{x}_h - \dot{x}_0) - k_{h1}(x_h - x_0) - k_{h2}(x_h - x_0)^3 = 0 \quad (1)$$

$$m_h(\ddot{x}_h + \ddot{z}_0) + c_h(\dot{x}_h - \dot{x}_0) + k_{h1}(x_h - x_0) + k_{h2}(x_h - x_0)^3 - \theta v = 0 \quad (2)$$

$$\dot{v} + v/(R_LC_p) + \theta(\dot{x}_h - \dot{x}_0)/C_p = 0 \quad (3)$$

A sinusoidal base excitation of displacement amplitude a, force amplitude F and frequency ω rad/s is considered. Introducing nondimensional coefficients, damping ratio of the primary system, $\zeta_0 = c_0/2m_0\omega_0$, natural frequency of the primary system, $\omega_0 = \sqrt{k_0/m_0}$, damping ratio of the DVA system, $\zeta_h = c_h/2m_h\omega_h$, natural frequency of the linear part of the nonlinear DVA system, $\omega_h = \sqrt{k_{h1}/m_h}$, mass ratio, $\mu = m_h/m_0$, tuning ratio, $\nu = \omega_h/\omega_0$, nonlinear coefficient, $\alpha = k_{h2}/k_{h1}$, nondimensional frequency, $\Omega = \omega/\omega_0$, nondimensional time, $\tau = \omega_0 t$, $f = F/k_0$, $q_1 = x_0/f$, $q_2 = x_h/f$ and $q_3 = v\theta/F$, the following equations are obtained.

$$q_1'' + 2\zeta_0q_1' + q_1 - 2\zeta_h\mu\nu(q_2' - q_1') - \nu^2\mu(q_2 - q_1) - \alpha f^2\mu\nu^2(q_2 - q_1)^3 = \cos(\Omega\tau) \quad (4)$$

$$q_2'' + 2\zeta_h\nu(q_2' - q_1') + \nu^2(q_2 - q_1) + \alpha f^2\nu^2(q_2 - q_1)^3 - q_3/\mu = \cos(\Omega\tau) \quad (5)$$

$$C_pR_L\omega_0q_3' + q_3 + R_L\theta^2\omega_0(q_2' - q_1')/k_0 = 0 \quad (6)$$

Here, the derivatives of q_1, q_2, and q_3 with respect to the nondimensional time τ are presented as q_1', q_1'', q_2', q_2'', and q_3', respectively. Load resistance is evaluated by $R_L = 1/(\omega_h C_p)$.

3 Nonlinear EHDVA—Tuning Criterion

To design the EHDVA system, it is necessary to know the effect of nonlinearity on DVA responses and the output voltage generated. In order to obtain the responses of DVA-harvester system, a frequency sweep study is carried out by numerical simulation considering $\alpha = 0.8$, $k_{h1} = 0.0187$ N/m [16], $\omega_h = 1$ rad/s, and $\zeta_h = 0.02$. The formulation presented in Sect. 2 is modified for DVA-harvester system and used for this study. The nondimensional equations of motion for EHDVA system are as follows:

$$q_2'' + 2\zeta_h q_2' + q_2 + \alpha f_h^2 q_2^3 - q_3 = \cos(\Omega_h \tau_h) \tag{7}$$

$$C_p R_L \omega_h q_3' + q_3 + R_L \theta^2 \omega_h q_2' / k_{h1} = 0 \tag{8}$$

where $f_h = F_h/k_{h1}$, nondimensional frequency, $\Omega_h = \omega_0/\omega_h = 1/\nu$ and nondimensional time, $\tau_h = \omega_h t$. F_h is amplitude of force at the base of EHDVA induced by motion of the primary SDOF system which is considered as 0.25 N.

A d_{31} macro fiber composite (MFC) of type M8557-P2 is considered as piezoelectric material for this study. The values of the parameters of DVA system and MFC used for the simulation study are $C_p = 342.93$ nF, width of MFC patch $b = 0.064$ m, Young's modulus $E_y = 15.857 \times 10^9$ N/m^2, piezoelectric strain coefficient $d_{31} = 170 \times 10^-12$ C/N, distance between MFC bottom surface to neutral axis of supporting beam $h_m = 0.0002$ m. The value of electromechanical coupling is evaluated by $\theta = bE_y d_{31}(2h_m + 0.0003)/2$ [17].

The amplitude of the nondimensional displacement of EHDVA is plotted over a range of nondimensional excitation frequency, Ω, as shown in Fig. 2. From Fig. 2, the hardening effect of the Duffing type nonlinear spring is clearly visible. For forward sweep the jump is occurring at $\Omega_h = 6$, and for reverse sweep the jump frequency is $\Omega_h = 4$ roughly. It is also observed that the displacement response is higher over a wide range of frequency, namely, between $\Omega_h = 0.6$ and $\Omega_h = 3.6$. Higher displacement of EHDVA ensures more harvested energy and better control. Hence, to implement this nonlinear EHDVA system to a primary SDOF system, an equivalent range of tuning ratio from $\nu = 0.278$ to $\nu = 16.667$ can be adopted as $\Omega_h = \omega_0/\omega_h = 1/\nu$ (see Eqs. 4 and 7). It is also to be noted that, this nonlinear EHDVA provides the flexibility in selection of tuning ratio with the primary SDOF structure and unlike the linear DVA it will work in mistuned condition.

The displacement time history of nonlinear EHDVA is plotted in Fig. 3 considering $\Omega_h = 1$ and compared with that of nonlinear DVA. The presence of the nonlinearity

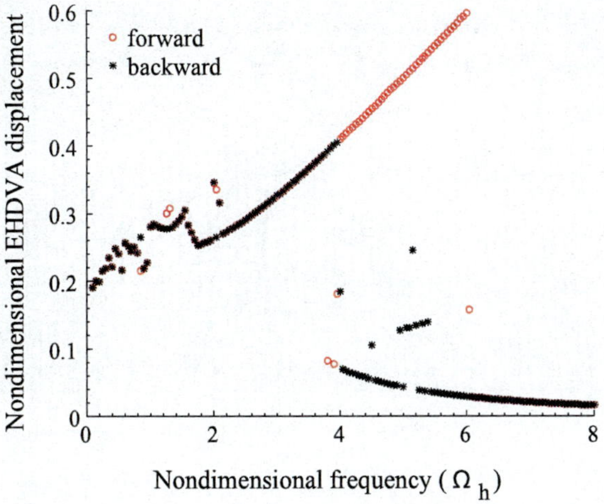

Fig. 2 Amplitudes of nondimensional displacement of EHDVA over a range of nondimensional excitation frequency (Ω_h)

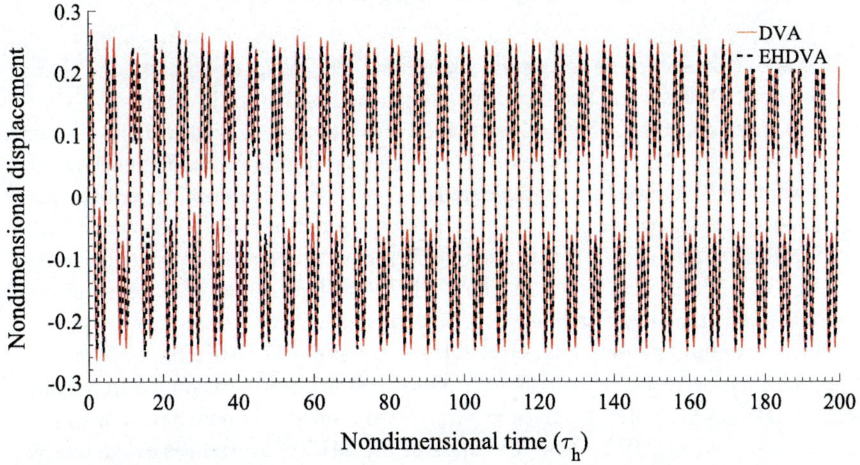

Fig. 3 Nondimensional time histories of nondimensional displacement responses of nonlinear EHDVA and nonlinear DVA

is observed from Fig. 3 for both the cases of EHDVA and DVA. It is also observed that after reaching the steady state the response amplitude of DVA is higher than that of EHDVA. This is the effect of an additional damping induced due to the presence of the harvesting system. The time history for the actual power generated by the EHDVA system is presented in Fig. 4 which shows that power of a steady-state amplitude 0.11 W is achieved for the given parametric values.

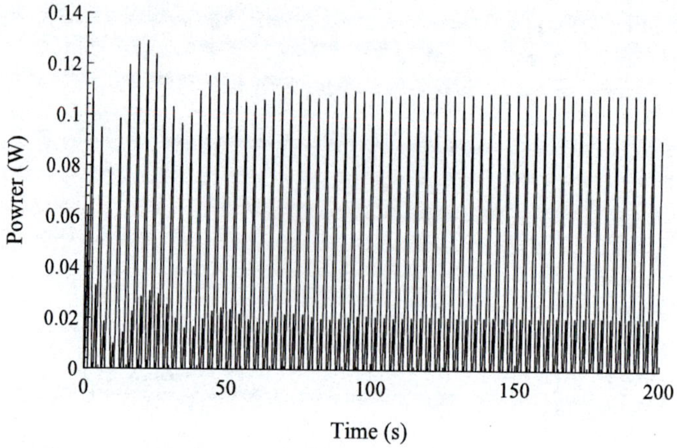

Fig. 4 Time history of power generated by nonlinear EHDVA

4 Design and Performance Evaluation of EHDVA

The focus of this section is to design a nonlinear EHDVA, which can control the vibration of a SDOF system and also generate power. For this study, a primary SDOF system of mass $m_0 = 1$ kg, natural frequency $\omega_0 = 1$ rad/s, and equivalent viscous damping ratio $\zeta_0 = 0.01$ is considered. The force amplitude of sinusoidal excitation at the base of the SDOF system is considered as $F = 0.05$ N. The control performance of EHDVA system is evaluated in terms of percent reduction in displacement of primary SDOF structure, which is the ratio of difference between steady-state response amplitude of uncontrolled and controlled SDOF system to steady-state response amplitude of uncontrolled SDOF system. To design the EHDVA for a SDOF system, first the value of tuning ratio, ν, is selected from the range obtained from Fig. 2 as per the required output responses of EHDVA (described in Sect. 3). Then the frequency of the EHDVA is evaluated from $\omega_h = \nu\omega_0$. As the linear component of the spring force, k_{h1}, is fixed as 0.0187 N/m, the mass of the EHDVA system is evaluated by $m_h = k_{h1}/\omega_h^2$. Hence the mass ratio is $\mu = m_h/m_0$. The values of these parameters and the responses of the SDOF-EHDVA system is evaluated for a range of ν values, obtained in Sect. 3, using the formulation presented in Sect. 2.

The variation in percent reduction in displacement of primary SDOF system with the variation in ν value are presented in Fig. 5, which reveals the presence of an optimum tuning ratio, $\nu_{opt} = 0.7$, where the response reduction has the maximum value of 87.95%. This is a satisfactory performance provided by the designed EHDVA as a vibration absorber. At $\nu = \nu_{opt}$, the value of the mass ratio is $\mu = 0.038$ and the mass of the EHDVA system is 0.038 kg.

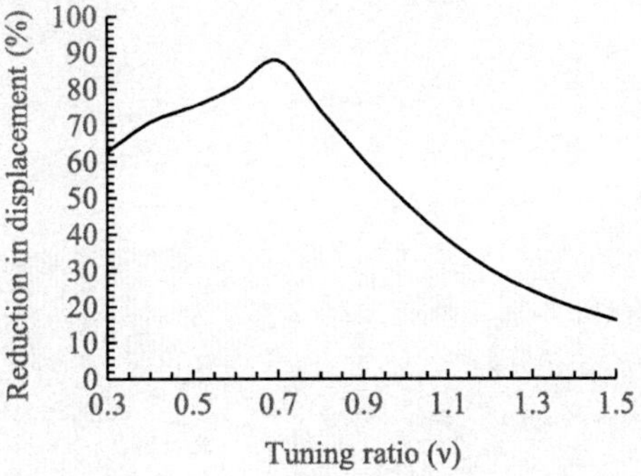

Fig. 5 Variation in percent reduction in displacement of SDOF system over a range of tuning ratio (ν)

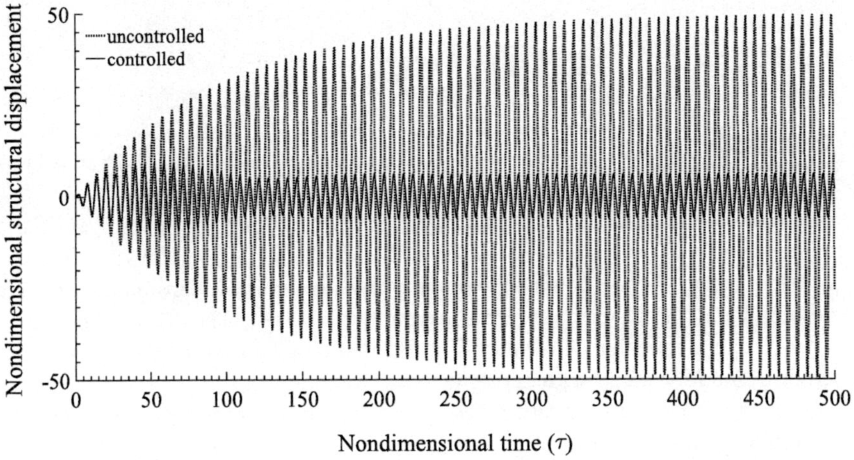

Fig. 6 Nondimensional time histories of uncontrolled and controlled nondimensional displacement of primary SDOF system

For these optimum values of the design parameters, the time history analysis is carried out for SDOF-EHDVA system. The uncontrolled and EHDVA-controlled displacement time histories of the SDOF primary system are presented in Fig. 6. A satisfactory controlling performance of the EHDVA is visible in Fig. 6. The time history for actual power generated by the EHDVA is presented in Fig. 7, from which it is observed that a steady-state power of amplitude 7.9 mW is generated by the EHDVA system.

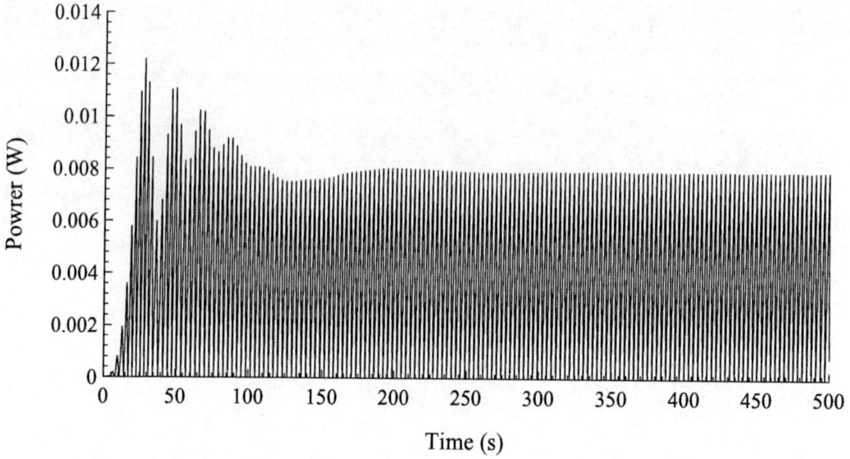

Fig. 7 Time history of power generated by EHDVA attached to primary SDOF system

5 Conclusions

A design procedure is described for a nonlinear EHDVA system to mitigate the vibration of a SDOF system while harvesting electrical energy from the vibration of the primary system. A time-domain method is proposed for nondimensional analysis and design for the coupled electromechanical nonlinear system. The SDOF-nonlinear EHDVA system is modeled considering a hardening type Duffing nonlinearity. The numerical simulation study with harmonic base excitation shows the effect of nonlinearity in the responses of the EHDVA system which is used to harvest a higher value of generated power over a broader range of excitation frequency. The effect of the additional damping provided by the harvesting system attached to the DVA is also observed. Tuning ratio is found to be the most important design parameter for EHDVA. A sine sweep study with nondimensional frequency excitation provides the basis of selecting the value of tuning ratio from a broadband response of EHDVA system. However, a sensitivity study on tuning ratio shows that while designing, the optimal tuning ratio should be evaluated from the performance aspect of EHDVA to control the vibration of primary system. The procedure to obtain the values of other important design parameters such as mass ratio and mass of EHDVA are described. Representative time histories of designed nonlinear EHDVA attached to primary SDOF system show that a displacement reduction of 87.95% could be achieved while generating a steady-state power of amplitude 7.9 mW. Hence, the performance of a designed nonlinear EHDVA system is satisfactory from both control and harvesting aspects.

References

1. Housner, G.W., Bergman, L.A., Caughey, T.K., Chassiakos, A.G., Claus, R.O., Masri, S.F., Skelton, R.E., Soong, T.T., Spencer, B.F., Yao, J.T.P.: Structural control: past, present, and future. J. Eng. Mech. **123**, 897–971 (1997)
2. Chongfeng, W., Xingjian, J.: A comprehensive review on vibration energy harvesting: modelling and realization. Renew. Sustain. Energy Rev. **74**, 118 (2017)
3. Williams, C.B., Yates, R.B.: Analysis of a micro-electric generator for microsystems. Sens. Actuators, A **52**, 8–11 (1996)
4. Anton, S.R., Sodano, H.A.: A review of power harvesting using piezoelectric materials (2003–2006). Smart Mater. Struct. **16**, R1–R21 (2007)
5. Jones, P.G., Tudor, M.J., Beeby, S.P., White, N.M.: An electromagnetic, vibration powered generator for intelligent sensor system. Sens. Actuators, A **110**, 244–349 (2004)
6. Mitcheson, P., Milao, P., Start, B., Yeatman, E., Holmes, A., Green, T.: MEMS electrostatic micro power generator for low frequency operation. Sens. Actuators, A **115**, 523–529 (2004)
7. Sodano, H.A., Park, G., Inman, D.J.: A review of power harvesting from vibration using piezoelectric materials. Shock Vib. Dig. **36**, 197–205 (2004)
8. Erturk, A., Inman, D.J.: Piezoelectric Energy Harvesting. Wiley, UK (2011)
9. Ali, S.F., Friswell, M.I., Adhikari, S.: Piezoelectric energy harvesting with parametric uncertainty. Smart Mater. Struct. **19**(10), 105010 (2010)
10. Adhikari, S., Friswell, M.I., Inman, D.J.: Piezoelectric energy harvesting from broadband random vibrations. Smart Mater. Struct. **18**(11), 115005 (2009)
11. Daqaq, M.: Response of a uni-modal Duffing type harvesters to random force excitations. J. Sound Vib. **329**, 3621–3631 (2010)
12. Friswell, M.I., Ali, S.F., Adhikari, S., Lees, A.W., Bilgen, O., Litak, G.: Nonlinear piezoelectric vibration energy harvesting from an inverted cantilever beam with tip mass. J. Intell. Mater. Syst. Struct. **23**(3), 1505–1521 (2012)
13. Ferrari, M., Ferrari, V., Guizzetti, M., Marioli, D., Taroni, A.: Piezoelectric multi-frequency energy converter for power harvesting in autonomous micro-systems. Sens. Actuators, A **142**, 329–335 (2008)
14. Malaji, P., Ali, S.F.: Magneto-mechanically coupled electromagnetic harvesters for broadband energy harvesting. Appl. Phys. Lett. **111**, 083901 (2017)
15. Ali, S.F., Adhikari, S., Friswell, M.I., Narayanan, S.: The analysis of piezomagnetoelastic energy harvesters under broadband random excitations. J. Appl. Phys. **109**(7), 074904 (2011)
16. Li, L., Cui, P.: Novel design approach of a nonlinear tuned mass damper with duffing stiffness. J. Eng. Mech. **143**(4), (2017)
17. Erturk, A., Inman, D.J.: A distributed parameter electromechanical model for cantilevered piezoelectric energy harvesters. J. Vib. Acoust. **130**, (2008)

Harvesting Energy from a Series of Harvesters

Mohammad Reyaz Ahmad Vali and Shaikh Faruque Ali

1 Introduction

The idea of energy harvesting is to extract energy from the ambient sources that are available in various forms [1]. In this paper, focus is on mechanical energy harvesting, where energy is harvested from a mechanical system that undergoes vibration. Vibrational energy is omnipresent in nature and it is abundant in nature. It can be tapped easily for harvesting energy. The harvested energy can be used for low-powered self-sustaining sensors replacing battery-powered sensors for various wireless operations such as structural health monitoring, environmental control, and military applications [2]. The maintenance costs related to replacement, recharge of batteries of battery-powered sensors and manpower required for their maintenance are eliminated by making use of this type of energy harvesting. Vibrational energy harvesting can be achieved by means of electromagnetic, electrostatic, and piezoelectric transductions [3–6]. This manuscript looks into a study related to electromagnetic energy harvester but nonetheless the study is a generic one and can be very well extended to piezoelectric harvester or any set of energy harvesters.

A lot of early research on energy harvesting was on linear harvesting devices [7, 8]. Linear harvesting devices can generate maximum power only at resonance. A slight variation or mismatch between the ambient vibrating frequency and the natural frequency of the harvester can reduce the power harvested to a large extent [9]. Broadband energy harvesting gained importance because of this limitation. Broadband energy harvesting is achieved using various means [10]. Broadband is expected when the harvester show nonlinearity with variation of excitation frequency [11–14]. Mistuning of the harvesters is another way of broadening harvester operating frequency [15–17]. The system shows multiple peaks in its frequency response curve

M. Reyaz Ahmad Vali (✉) · S. Faruque Ali
Indian Institute of Technology Madras, Chennai 600036, India
e-mail: am18s014@smail.iitm.ac.in

S. K. Saha and M. Mukherjee (eds.), *Recent Advances in Computational Mechanics and Simulations*, Lecture Notes in Mechanical Engineering,
https://doi.org/10.1007/978-981-15-8315-5_49

due to different natural frequencies of harvesters. This principle is used in the present work by testing the system to different ranges of mistuning percentages and then finding out the bandwidths and peak power. Response of mistuned structure is observed to be better than those of perfectly tuned structure. When series of pendulums are considered and length of each pendulum is identical, natural frequency of each pendulum is same. Mistuning is where it is expected the pendulums are with same natural frequency, but due to operational reasons they are slightly varying and not largely varying. This gives rise to broadband energy harvesting.

Malaji et al. [15, 16] proposed parametric study on effect of mistuning on a 2 pendulum system. Malaji et al. [15, 16] did the necessary groundwork and established through experiments that mistuning is indeed an important factor to be considered in analysis of multiple harvesters. In [17], parametric effect on frequency bandwidth and magnitude of harvested power is studied through experiments and simulations. Simulation is restricted to 20 pendulums and experiments are restricted to 5 pendulums. It is shown through experiments and simulations that strong mechanical coupling between harvesters enhances frequency bandwidth of operation compared to a set of independent harvesters. It is also shown that the maximum total power harvested by a set of harvesters saturates after a particular number of harvesters.

In this work, 100 pendulums are considered in a set of harvesters. Responses of the harvester set in four different configurations are studied. Frequency response functions of voltage and power are plotted, analyzed, and compared. In this analysis, special focus is given to mistuned cases where harvester lengths are randomly generated such that they form a Gaussian distribution. Studies are conducted for different values of standard deviation, coupling stiffness. Plots representing the mean and standard deviation of bandwidths are obtained.

This paper is organized as follows. Section 2 describes the different arrangements in which the system is studied and gives mathematical equations governing different arrangements of the system. Section 3 describes various simulation results obtained based on numerical study. Finally conclusion of the present study is described in Sect. 4.

2 Different Arrangements of the Harvesters

Figure 1 shows the schematic of the system which has been considered for study in this manuscript. It consists of a series of electromagnetic pendulum harvesters connected to the same source of excitation. Each pendulum harvester has a magnet attached to its free end and an electric coil is placed beneath it. Further, each pendulum is connected with another pendulum by a spring of stiffness K at a distance of a from the hinge of the pendulum. Support excitation generates a relative motion between the coil and the magnet attached with pendulum and the change in magnetic field due to pendulum motion generates electricity in the coils. Voltage is received across each coil through a load resistor (R) connected to it. In Fig. 1, X_g denotes amplitude of support motion which has been assumed to be harmonic in the current study.

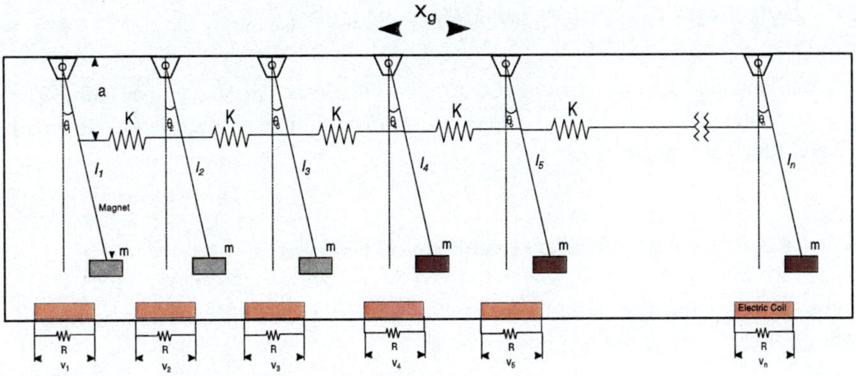

Fig. 1 Schematic arrangement of n pendulum harvesters

To generate higher power, an array of tuned electromagnetic pendulum (lengths are same) harvesters are used to generate more electrical power compared to the standalone harvester. In practice, manufacturing a set of tuned harvesters is difficult. Further due to operational non-similarity tuning may get lost. There may be power loss from the designed power due to loss in tuning. To understand the performance of tuned, mistuned, coupled, and uncoupled systems, four different configurations are considered for numerical study. They are

case(a): Series of independent tuned harvesters: We get case (a) by making the stiffness, K of the springs in between the pendulums to be zero and keeping the length of all pendulums same. So, $K = 0, l_1 = l_2 = \cdots = l_n$.

case(b): Series of independent mistuned harvesters: We get case (b) by making the stiffness, K of the springs in between the pendulums to be zero and keeping the length of all pendulums different. So, $K = 0, l_1 \neq l_2 \neq \cdots \neq l_n$.

case(c): Series of interacting tuned harvesters: We get case (c) by making the stiffness, K of the springs in between the pendulums to be non-zero value and keeping the length of all pendulums same. So, $K \neq 0, l_1 = l_2 = \cdots = l_n$.

case(d): Series of mistuned interacting harvesters: We get case (d) by making the stiffness, K of the springs in between the pendulums to be non-zero value and keeping the length of all pendulums different. So, $K \neq 0, l_1 \neq l_2 \neq \cdots \neq l_n$.
Where $l_1, l_2, \ldots, l_n$ denotes the length of pendulums starting from left in the configuration.

2.1 Equations of Motion for Different Cases

The mathematical models for the above mentioned cases are developed here. The magnetic interaction among the neighbouring pendulums are assumed to be small and are, therefore, neglected.

2.1.1 Case(a): Series of Independent Tuned Harvesters

As the system is uncoupled, equations of motion are independent. The equation of motion for each pendulum can be written as

$$ml_i^2\ddot{\theta}_i + cl_i^2\dot{\theta}_i + mgl_i \sin\theta_i + ml_i\ddot{x}_g \cos\theta_i = 0; \quad i = 1, 2, \ldots, n \tag{1}$$

where m is mass of the magnet and n is the number of pendulums. $l_i = l_1, l_2, l_3, \ldots, l_n$ are the length of the pendulums. In this case, length of pendulums are equal (*i.e.,*. $l_1 = l_2 = l_3 = \ldots = l_n = l$). θ_i's are the angular displacement of pendulums. c is damping coefficient which consists of both structural and electromagnetic damping. g is acceleration due to gravity. $\ddot{x}_g$ denotes support acceleration.

2.1.2 Case(b): Series of Independent Mistuned Harvesters

The equations of motion for mistuned independent harvesters can be written as

$$ml_i^2\ddot{\theta}_i + cl_i^2\dot{\theta}_i + mgl_i \sin\theta_i + ml_i\ddot{x}_g \cos\theta_i = 0; \quad i = 1, 2, \ldots, n \quad l_1 \neq l_2 \neq \cdots \neq l_n; \tag{2}$$

2.1.3 Case(c): Series of Interacting Tuned Harvesters

As the system is coupled, the equations of motion are dependent. Equations of motion for each pendulum is written as

$$\begin{aligned} ml_1^2\ddot{\theta}_1 + cl_1^2\dot{\theta}_1 + mgl_1 \sin\theta_1 + ml_1\ddot{x}_g \cos\theta_1 + Ka^2 \cos\theta_1(\sin\theta_1 - \sin\theta_2) = 0 \\ ml_i^2\ddot{\theta}_i + cl_i^2\dot{\theta}_i + mgl_i \sin\theta_i + ml_i\ddot{x}_g \cos\theta_i + 2Ka^2 \cos\theta_i \sin\theta_i \\ +Ka^2 \cos\theta_i(\sin\theta_{i+1} - \sin\theta_{i-1}) = 0 \\ ml_n^2\ddot{\theta}_n + cl_n^2\dot{\theta}_n + mgl_n \sin\theta_n + ml_n\ddot{x}_g \cos\theta_n + Ka^2 \cos\theta_n(\sin\theta_n - \sin\theta_{n-1}) = 0 \\ i = 2, 3, \ldots, (n-1) \end{aligned} \tag{3}$$

Since it is a tuned case, the length of all pendulums are equal.

2.1.4 Case(d): Series of Mistuned Interacting Harvesters

In this case, the harvesters are mistuned but they are connected with springs in between such that an interaction is possible. As system is coupled through mechanical spring the governing differential equations of motion are coupled and are written as

$$
\begin{aligned}
ml_1^2\ddot{\theta}_1 + cl_1^2\dot{\theta}_1 + mgl_1 \sin\theta_1 + ml_1\ddot{x}_g \cos\theta_1 + Ka^2\cos\theta_1(\sin\theta_1 - \sin\theta_2) = 0 \\
ml_i^2\ddot{\theta}_i + cl_i^2\dot{\theta}_i + mgl_i \sin\theta_i + ml_i\ddot{x}_g \cos\theta_i + 2Ka^2\cos\theta_i \sin\theta_i \\
+Ka^2\cos\theta_i(\sin\theta_{i+1} - \sin\theta_{i-1}) = 0 \\
ml_n^2\ddot{\theta}_n + cl_n^2\dot{\theta}_n + mgl_n \sin\theta_n + ml_n\ddot{x}_g \cos\theta_n + Ka^2\cos\theta_n(\sin\theta_n - \sin\theta_{n-1}) = 0 \\
i = 2, 3, \ldots, (n-1); \quad l_1 \neq l_2 \neq \ldots \neq l_n
\end{aligned}
\tag{4}
$$

This is a mistuned case and the length of the pendulums are not same.

2.2 *Electrical Equation for the Harvester*

Under the external excitation at the support, the magnets oscillate over the electric coil and generate voltage due to change in magnetic flux. Using Faraday's law, the voltage generated in each electromagnetic harvester can be obtained as [15]

$$v_i = \zeta B_s L l_i \dot{\theta}; \quad i = 1, 2, 3, \ldots, n \tag{5}$$

where ζ denotes coil filling factor, B_s is residual magnetic flux, L represents coil length and l_i is length of pendulum (including magnet), $\dot{\theta}$ is the relative velocity between magnet and coil.

Power harvested from each harvester is given as

$$p_i = \frac{v_i^2}{R}; \quad i = 1, 2, 3, \ldots, n \tag{6}$$

where R is the coil resistance. Finally, the total harvested power through the series of electromagnetic harvester system is given as

$$P = \sum_{i=1}^{n} p_i \tag{7}$$

Table 1 Harvester parameters and their values used in simulation studies

Parameter	Value
Filling factor (ζ)	0.3
Residual magnetic flux (B_s)	0.6 T
Coil length (L)	7 m
Tuned pendulum length (l)	0.06 m
Damping coefficient (c)	0.006 N-s/m
Coil resistance (R)	160 Ω

3 Results and Discussions

This section provides results obtained from the numerical simulations. The parameters used in numerical simulations are given in Table 1. For simulations [18], 100 harvesters are considered in a set.

3.1 *Response of System to Harmonic Excitation*

Harmonic base excitation with frequency $(\omega) = 3.5\pi$, amplitude $(X_g) = 2.3$ mm (this amplitude is chosen such that it is physically realizable) is given to the system considered and the response of system is observed. System in all the four configurations (cases) discussed previously are considered. In tuned configurations, the length of each harvester considered is 0.06 m. In the mistuned configurations, 100 harvesters with Gaussian distribution of mean $(\mu) = 0.06$ m and standard deviation $(\sigma) = 5\%\ of\ \mu = 0.003$ m are randomly generated. Same set of pendulum lengths are used for both the mistuned cases (b) and (d). The time histories of 50th pendulum for different cases are as shown in Fig. 2. Displacement on the Y-axis is corresponding to rotational displacement. The response of case(a) and case(c) are same. As there is no mistuning of length in case(c), there will be no forces developed in the spring due to excitation, and response of system will be same as case(a). In Case(d), $K = 212.55$ N/m. It is chosen by considering non-dimensional quantity Kl/mg as 100 randomly. It can be seen from cases (b) and (d) that amplitude of response is slightly less in (d) when compared with (b). Similarly simulations are carried out for this system by increasing the value of K to 10 times for each step. It is seen that the amplitude of response of the system kept on decreasing with increase in stiffness value.

Some interesting aperiodic response came into picture when the excitation frequency is close to natural frequency of individual tuned harvester (we can also say at natural frequency of tuned independent harvester system). Response of 50th pendulum near-natural frequency for tuned configurations is shown in Fig. 3 with increasing value of stiffness. It can be seen from Fig. 3 that system shows periodic motion for

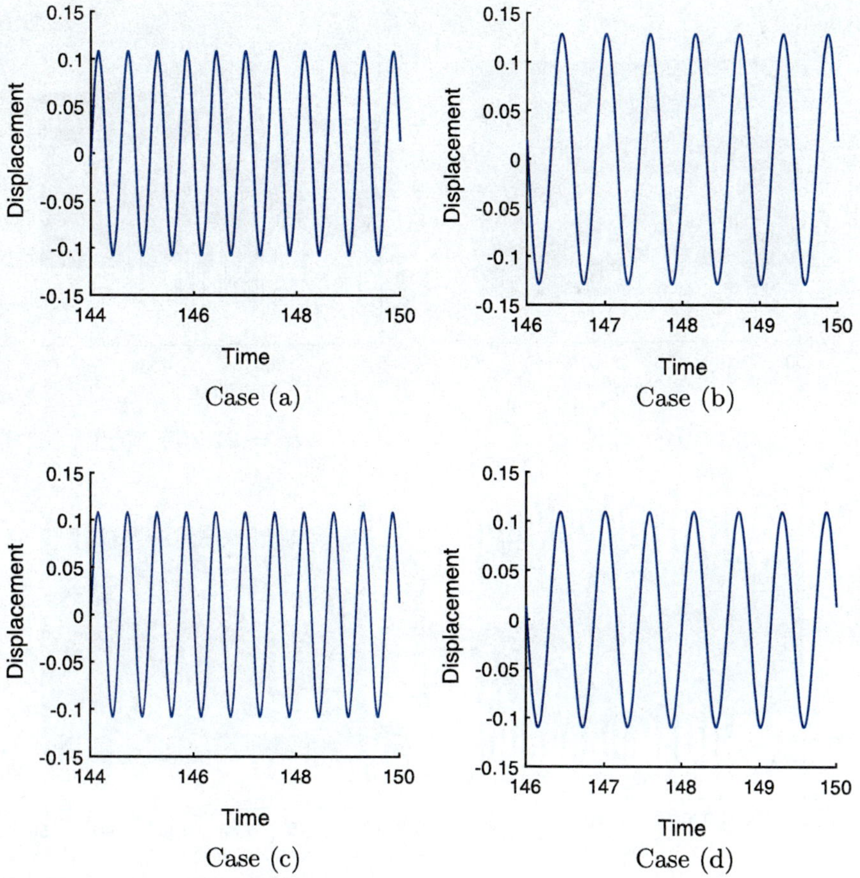

Fig. 2 Time histories of 50th pendulum for different cases (a), (b), (c), (d)

uncoupled configuration and aperiodicity kept on decreasing with increase in stiffness for coupled configuration. Phase portraits of 50th pendulum near-natural frequency for the mistuned configuration (case(d)) are shown in Fig. 4 with increasing value of K. It is seen from Fig. 4 that with increase in value of coupling stiffness, aperiodicity in the system kept on increasing.

3.2 FRF of Voltage and Power

The variation of amplitude of voltage and power with change in frequency ratio is discussed in this section to analyze the operating bandwidth and peak power. For mistuned cases, harvester lengths are generated as Gaussian distribution with $\mu = 0.06$ m and $\sigma = 5\%\ of\ \mu = 0.003$ m and for coupled cases stiffness is considered as

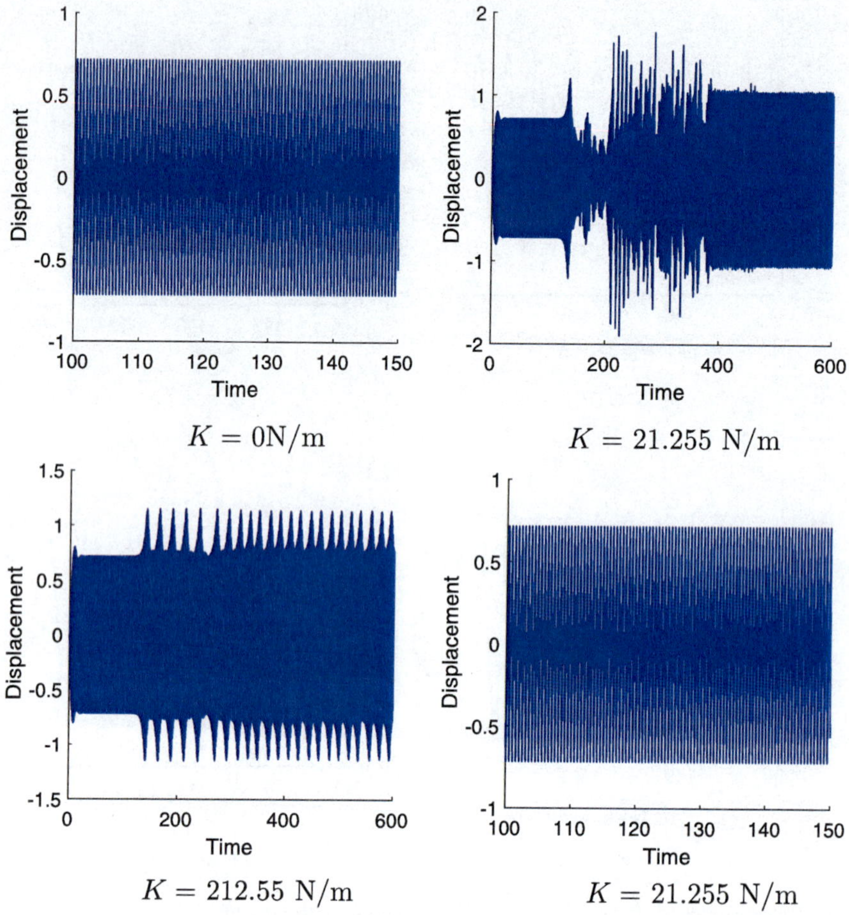

Fig. 3 Time histories for the tuned configuration at excitation frequency of 4π with increasing value of K

21.255 N/m (by taking $Kl/mg = 10$). Let ω_1 be denoted as natural frequency of each pendulum. In each case, harmonic sweep of frequency is swept from $0.8\,\omega_1$ to $1.2\,\omega_1$ Hz (because the amplitudes of response out of this range came out to be negligible). Variation of voltage amplitude of each harvester and total power amplitude of system is obtained by changing the frequency ratio ($\Omega = \omega/\omega_1$). ω is excitation frequency. Voltage FRF of 70th pendulum for each case is shown in Fig. 5. It can be seen that cases (a) and (c) will have same FRF because of same response for both cases and in case(d) plot is not smooth because of coupling effect.

Power FRF for each case is shown in Fig. 6. From Fig. 6, it can be observed that Peak power is more in tuned cases and bandwidth of FRF is more in mistuned cases.

When the system is excited close to natural frequency then tuned cases can be used for energy harvesting. When system is operating in band of frequency then mistuned

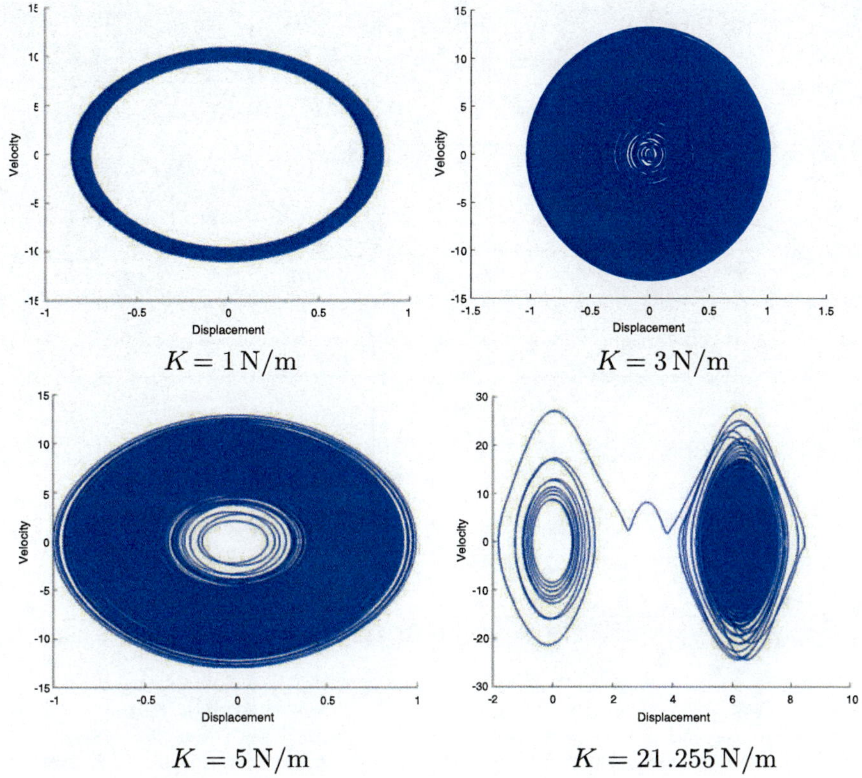

Fig. 4 Phase portraits for the mistuned configuration at excitation frequency of 4π with increasing value of K

cases are very useful. As ambient vibrations fall in frequency band, mistuned cases have to be explored for energy harvesting.

3.3 *Study of Bandwidths for Mistuned Cases*

In this section, sets of 100 harvesters are randomly generated such that their lengths follow Gaussian distribution with mean $(\mu) = 0.06\,\mathrm{m}$. Standard deviation (σ) is left as choice. $\sigma = 1\%\ of\ \mu = 0.0006\,\mathrm{m}$ and $\sigma = 5\%\ of\ \mu = 0.003\,\mathrm{m}$ are considered for this study. Choosing $\sigma = 1\%\ of\ \mu$, 30 such sets are created randomly. For each set, frequency is swept from $0.8\,\omega_1$ to $1.2\,\omega_1$ Hz and peak power, bandwidths (at 0.15W, 0.1W, 0.05W) are calculated. From this data, mean and standard deviation of peak power and bandwidths are calculated. Same procedure is repeated for $\sigma = 5\%\ of\ \mu$. With change in value of coupling stiffness these values changes. $\alpha = Kl/mg$ is the non-dimensional quantity used for considering different values of K. $\alpha = 0$, 0.1

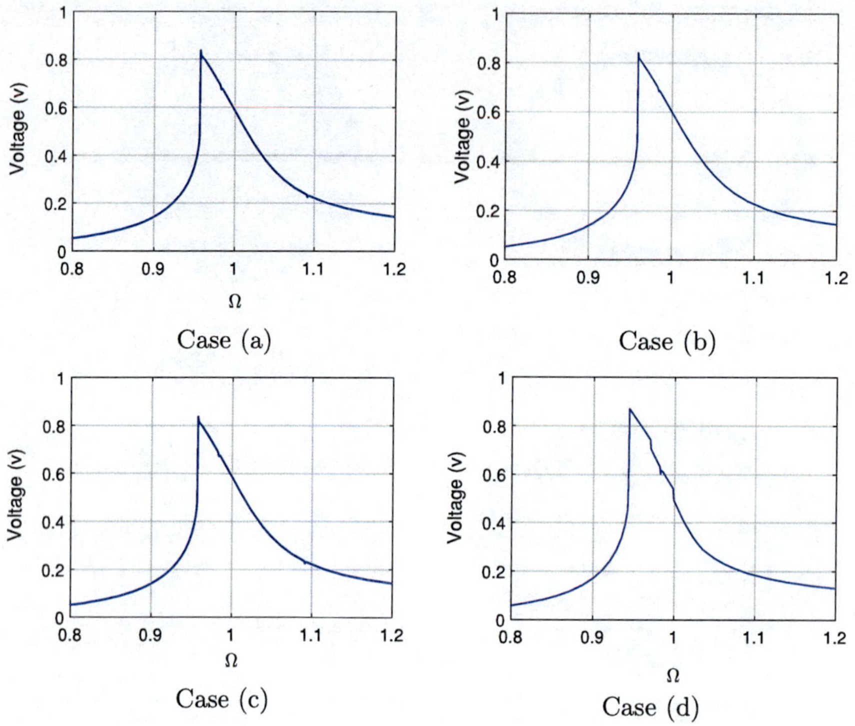

Fig. 5 Frequency Response Function (FRF) for voltage of 70th pendulum corresponding to different cases

(< 1), 10 (> 1) are considered for analysis. Now mean and deviations of bandwidth for different α values are plotted seperately for $\sigma = 1\%\, of\ \mu$ and $\sigma = 5\%\, of\ \mu$ in Fig. 7 and Fig. 8, respectively. In these figures, X-axis is normalized power which is ratio of power (P) and mean of peak power (P_{max}) at particular values of α and σ.

From these plots it can be seen that bandwidth is decreasing with increasing value of power level at particular value of α and also deviations in bandwidths is more for $\sigma = 5\%\, of\ \mu$ than $\sigma = 1\%\, of\ \mu$. Normalized mean peak power value is 1 and its bandwidth is 0 for all values of α and σ.

From these kind of plots, based on the power and bandwidth required for application, we can select the value of σ and K.

4 Conclusion

The system under four different configurations is mathematically modeled. Pendulums with tip magnets are considered as electromagnetic harvesters. Simulations are carried to know the response of the system and to know the power, voltage charac-

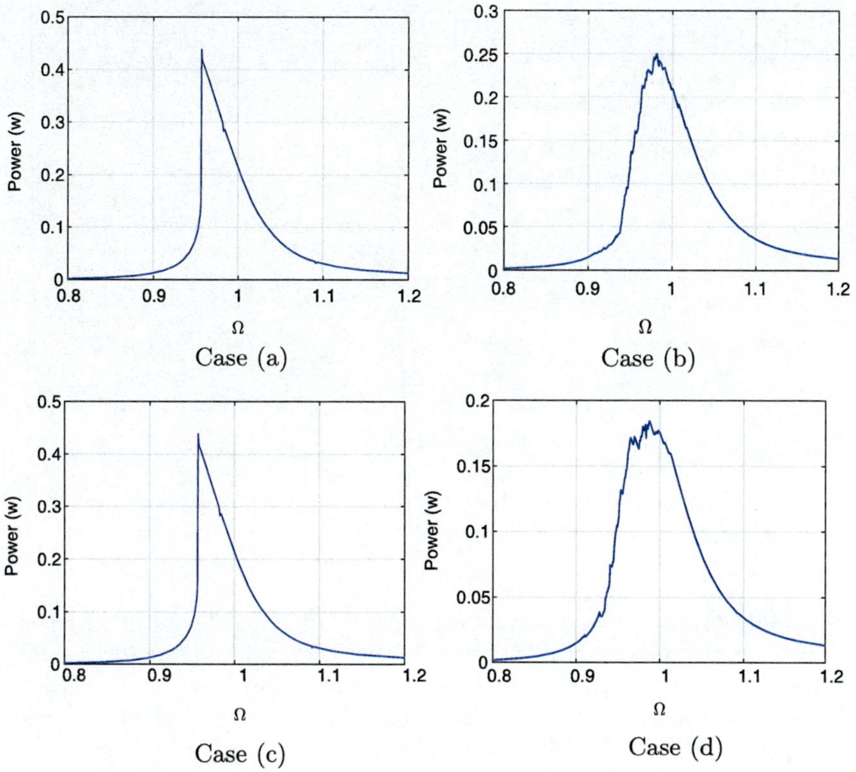

Fig. 6 Frequency Response Function (FRF) of total power for different cases

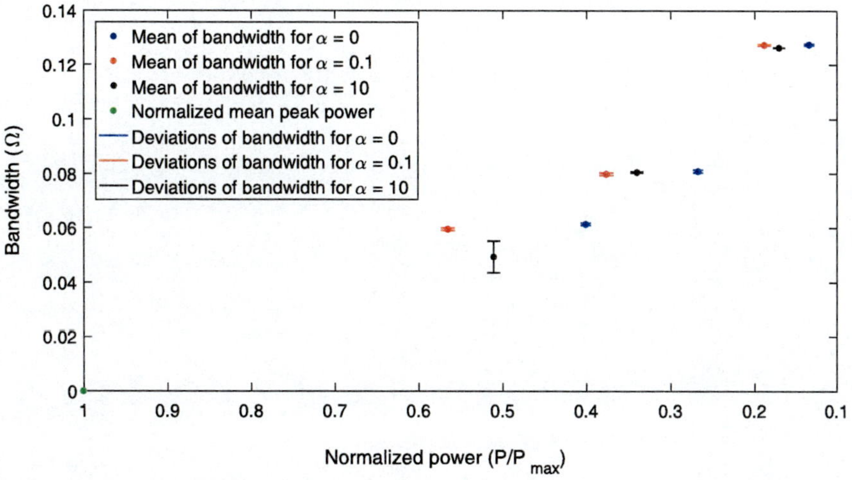

Fig. 7 Bandwidth variations at normalized power level for 0.15W, 0.1W, 0.05W for different values of α at $\sigma = 1\%\, of\ \mu$

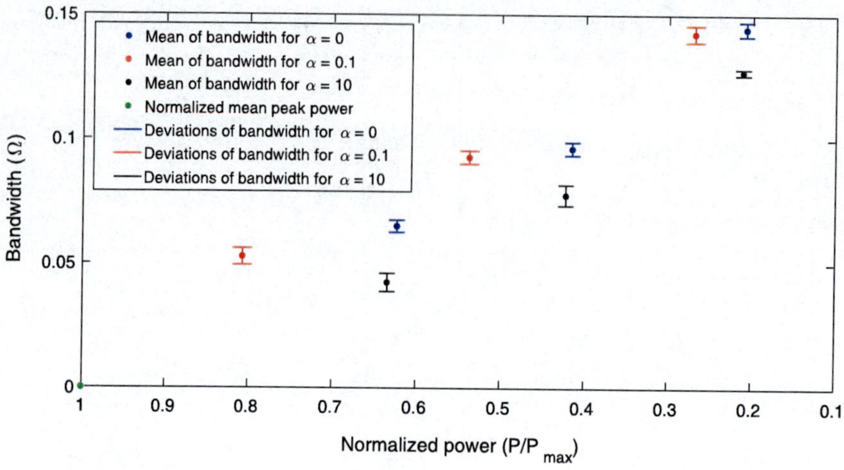

Fig. 8 Bandwidth variations at normalized power level for 0.15W, 0.1W, 0.05W for different values of α at $\sigma = 5\% of\ \mu$

teristics of the system under different configurations. It is found out that mistuned configurations give wider power bandwidths. Later mistuned cases are focused for energy harvesting by taking the harvester lengths such that they form Gaussian distribution with mean equal to tuned pendulum length. The mean and standard deviation of peak power, bandwidths of system are plotted for $\alpha = 0, 0.1, 10$ at 2 values of standard deviations. Similarly, this study can be exploited for various other values of σ and α to suit our need.

References

1. Bănică, M.A.: Energy harvesting from renewable energy sources. In: International Conference of Mechatronics and Cyber-Mixmechatronics, pp. 247–254. Springer, Cham (2019)
2. Roundy, S., Wright, P.K., Pister, K.S.: Micro-electric vibration-to-electricity converters, In: Proceedings of ASME International Mechanical Engineering Congress and Exposition, pp. 1–10. (2002)
3. Glynne-Jones, P., Tudor, M.J., Beeby, S.P., White, N.M.: An electromagnetic, vibration-powered generator for intelligent sensor systems. Sens. Actuators, A **110**(1–3), 344–349 (2004)
4. Mitcheson, P.D., Miao, P., Stark, B.H., Yeatman, E.M., Holmes, A.S., Green, T.C.: MEMS electrostatic micropower generator for low frequency operation. Sens. Actuators, A **115**(2–3), 523–529 (2004)
5. Erturk, A., Inman, D.J.: Piezoelectric energy harvesting. Wiley (2011)
6. Rajarathinam, M., Ali, S.F.: Energy generation in a hybrid harvester under harmonic excitation. Energy Convers. Manage. **155**, 10–19 (2018)
7. Williams, C.B., Yates, R.B.: Analysis of a micro-electric generator for microsystems. Sens. Actuators, A **52**(1–3), 8–11 (1996)
8. Hadas, Z., Kluge, M., Singule, V., Ondrusek, C.: Electromagnetic vibration power generator. In: 2007 IEEE International Symposium on Diagnostics for Electric Machines, Power Electronics and Drives, pp. 451–455. IEEE (2007)
9. Ali, S.F., Friswell, M.I., Adhikari, S.: Piezoelectric energy harvesting with parametric uncertainty. Smart Mater. Struct. **19**(10), 105010 (2010)
10. Tang, L., Yang, Y., Soh, C.K.: Toward broadband vibration-based energy harvesting. J. Intell. Mater. Syst. Struct. **21**(18), 1867–1897 (2010)
11. Kumar, K.A., Ali, S.F., Arockiarajan, A.: Piezomagnetoelastic broadband energy harvester: nonlinear modeling and characterization. Eur. Phys. J. Spec. Top. **224**(14–15), 2803–2822 (2015)
12. Kumar, K.A., Ali, S.F., Arockiarajan, A.: Magneto-elastic oscillator: modeling and analysis with nonlinear magnetic interaction. J. Sound Vib. **393**, 265–284 (2017)
13. Friswell, M.I., Ali, S.F., Bilgen, O., Adhikari, S., Lees, A.W., Litak, G.: Non-linear piezoelectric vibration energy harvesting from a vertical cantilever beam with tip mass. J. Intell. Mater. Syst. Struct. **23**(13), 1505–1521 (2012)
14. Stanton, S.C., McGehee, C.C., Mann, B.P.: Nonlinear dynamics for broadband energy harvesting: investigation of a bistable piezoelectric inertial generator. Physica D **239**(10), 640–653 (2010)
15. Malaji, P.V., Ali, S.F.: Analysis of energy harvesting from multiple pendulums with and without mechanical coupling. Eur. Phys. J. Spec. Top. **224**, 2823–2838 (2015)
16. Malaji, P.V., Ali, S.F., Adhikari, S., Friswell, M.I.: Analysis of harvesting energy from mistuned multiple harvesters with and without coupling. Procedia Eng. **144**, 621–628 (2016)
17. Malaji, P.V., Ali, S.F.: Broadband energy harvesting with mechanically coupled harvesters. Sens. Actuators, A **255**, 1–9 (2017)
18. MATLAB.: $version$ 9.1.0.441655($R2016b$). Natik, Massachusetts: The MathWorks Inc. (2016)

Design of a Flow Control Device Using a Special Class of Hybrid Symmetric Bistable Laminates in Clamped Boundary Condition

Aghna Mukherjee, Akash Mundwaik, Shaikh Faruque Ali, and Arunachalkasi Arockiarajan

1 Introduction

Traditionally, the buckling of structural elements has been considered detrimental to the design process. However, with the increase in our knowledge base and advancement in the computational capabilities, structures in the post-buckled regime have gained immense importance as design elements. For instance, Hampali et al. [1] developed a multi-stable shell element to design an assistive chair for the elderly. Daynes et al. [2] used a system of prestressed post-buckled laminates to design a bistable composite flap for wing morphing application. In this work, a special class of hybrid bistable laminates is explored for their potential as design elements in flow control devices.

In recent years multi-stable laminates have captured the attention of the research community because of their capability of assuming more than one equilibrium shapes, which can be switched reversibly by a mere application of an external force which is sufficient to cross the potential barrier between the stable states. Multi-stability in thin unsymmetric laminate was first observed and analyzed by Hyer et al. [3]. Ever since there has been a growing interest regarding these multi-stable laminates among the research community for their potential in morphing and energy harvesting applications. One important aspect of using these laminates is their integrability. The layup introduced by Hyer could not be used as a part of a compliant structure since it lost its multi-stable characteristics on integration. Toward that, Mattioni et al. [4] gave a layup that could be clamped at one end. Arrieta et al. [5] introduced a special class

A. Mukherjee (✉) · S. F. Ali · A. Arockiarajan
Indian Institute of Technology Madras, Chennai 600036, India
e-mail: aghna11@gmail.com

A. Mundwaik
Airbus Group India Private Limited, Bangalore 560048, India

S. K. Saha and M. Mukherjee (eds.), *Recent Advances in Computational Mechanics and Simulations*, Lecture Notes in Mechanical Engineering,
https://doi.org/10.1007/978-981-15-8315-5_50

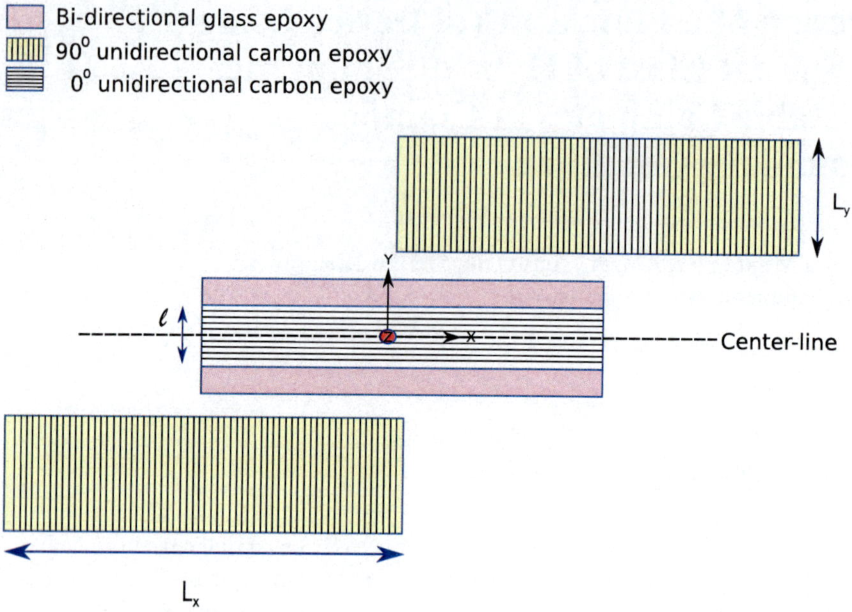

Fig. 1 Schematic diagram of the multi-sectioned hybrid bistable laminate layup

of variable stiffness multi-stable structural element that could be clamped at both ends. On similar lines, Li et al. [6] introduced a class of multi-sectioned metal hybrid symmetric laminates that had two symmetric curved equilibrium shapes. Later, it was shown these hybrid bistable symmetric laminates could be clamped at one end without losing bistability [7].

The motivation of this manuscript is to design a concept flow control device using multi-stable structural elements. A multi-sectioned symmetric layup is introduced that includes a combination of uni-directional carbon-epoxy plies and bi-directional glass epoxy layer, as shown in Fig. 1. A finite element analysis using ABAQUS® is done to show the existence of two stable equilibrium shapes, which is validated using a Rayleigh-Ritz based potential energy minimization scheme: a technique which was introduced in [3]. This special class of bistable laminates is used as a part of a flow control device. The load-displacement characteristics and the design space is then studied under the action of a free-flowing stream using XFoil and FEA software ABAQUS®. The potential of such bistable laminates in the design of a passive flow control device is highlighted in the study. The manuscript is divided into three sections. The first section details the special class of bistable laminates, which is used as the central member in the design. The subsequent section details the flow control device and the performance of the laminate under the action of a flowing stream. Finally, the manuscript ends with a conclusion.

2 Multi-section Bistable Structural Member

In this section, the multi-stable structural element is detailed. Li et al. [6] introduced a hybrid bistable symmetric laminate, which used a combination of aluminum and carbon epoxy. He achieved two stable states which could be integrated with the cantilever boundary condition. However, in using metals as a part of the layup, there is an issue of slippage [7, 8]. In this work, a hybrid layup is introduced which uses a combination of glass epoxy and carbon epoxy prepregs, as shown in Fig. 1. The layup used in this manuscript comprises of three sections: a central symmetric section having a layup $[90_2/0_2/90_2]_T$ and two outer sections symmetrically placed having a layup $[90_2/BD_1/90_2]_T$. The glass epoxy and the carbon epoxy prepregs use the same epoxy, so as to ensure that the bonding is good and there is minimum issue of delamination and slippage. The material properties of the bi-directional glass epoxy and the uni-directional carbon epoxy used in the layup are detailed in Table 1.

The proposed laminate is analyzed using the commercial finite element software ABAQUS®. To obtain the stable equilibrium shapes the laminates are subjected to a temperature field with a change of temperature of $\Delta T = 120$ °C. Since the layup is symmetric, the lamina settles into a flat, unstable equilibrium shape. To obtain the stable equilibrium states, the central node of the lamina is arrested, and a small load is applied at the four ends of the lamina. The loads are then removed in the subsequent steps. The lamina settles to one of the equilibrium shapes depending on the direction of the perturbation. Figure 2 shows the equilibrium shapes obtained using FEA. The equilibrium shapes obtained are validated against the ones obtained using Rayleigh-Ritz minimization of the potential energy function, where the potential energy is defined as

$$U_{tot} = \int_V \left(\begin{bmatrix} \epsilon^o \\ \kappa^o \end{bmatrix}^T \begin{bmatrix} A & B \\ B & D \end{bmatrix} \begin{bmatrix} \epsilon^o \\ \kappa^o \end{bmatrix} - \begin{bmatrix} \epsilon^o \\ \kappa^o \end{bmatrix}^T \begin{bmatrix} N^T \\ M^T \end{bmatrix} \right) dV - W_{ext} \quad (1)$$

Table 1 Properties of the carbon epoxy uni-directional (UD) and glass epoxy bi-directional (BD) lamina used in the layup

Lamina	E1 (GPa)	E2 (GPa)	ν_{12}	G_{12} (GPa)	α_1 (°C^{-1})	α_2 (°C^{-1})
Carbon epoxy (UD)	137.47	10.07	0.23	4.4	0.37e-6	24.91e-6
Glass epoxy 1 (BD)	29	29	0.189	4.75	14.31e-6	14. 31e-6
Glass epoxy 2 (BD)	22.2	22.2	0.192	4.41	19.78e-6	19.78e-6
Glass Epoxy 3 (BD)	21.3	21.3	0.166	3.19	20.47e-6	20.47e-6

Fig. 2 **a** The first equilibrium shape obtained using ABAQUS® in "free-free" boundary condition. **b** The second equilibrium shape obtained using ABAQUS® in "free-free" boundary condition

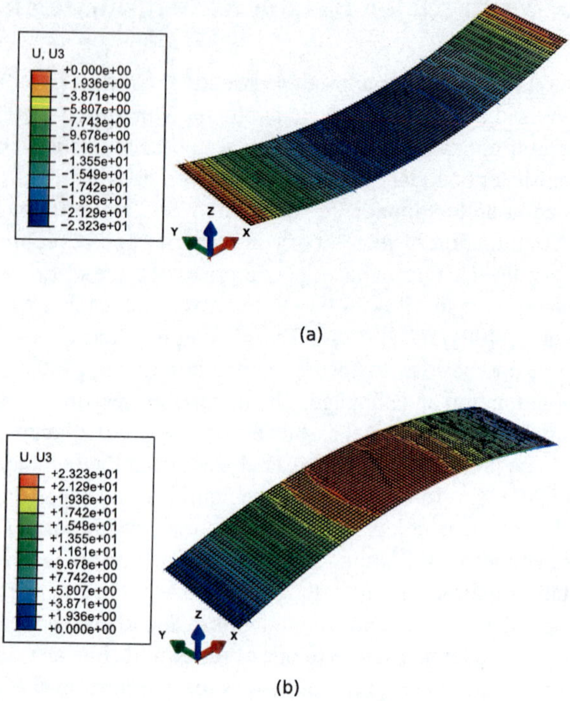

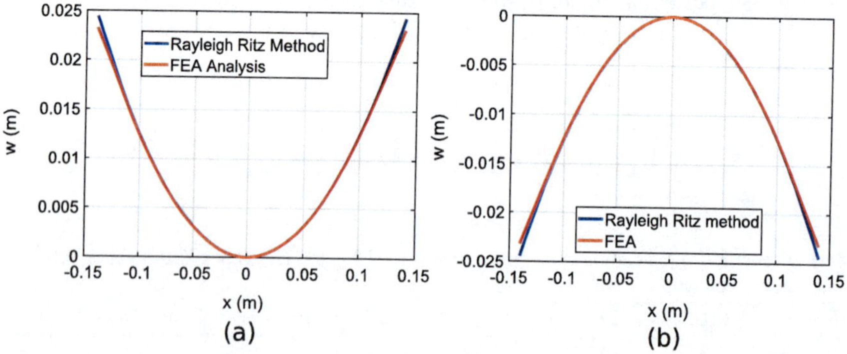

Fig. 3 **a** The center-line deflection of the first equilibrium shape obtained using FEA software ABAQUS® and Rayleigh-Ritz based analytical scheme [6]. **b** The center-line deflection of the second equilibrium shape obtained using FEA software ABAQUS® and Rayleigh-Ritz based analytical scheme

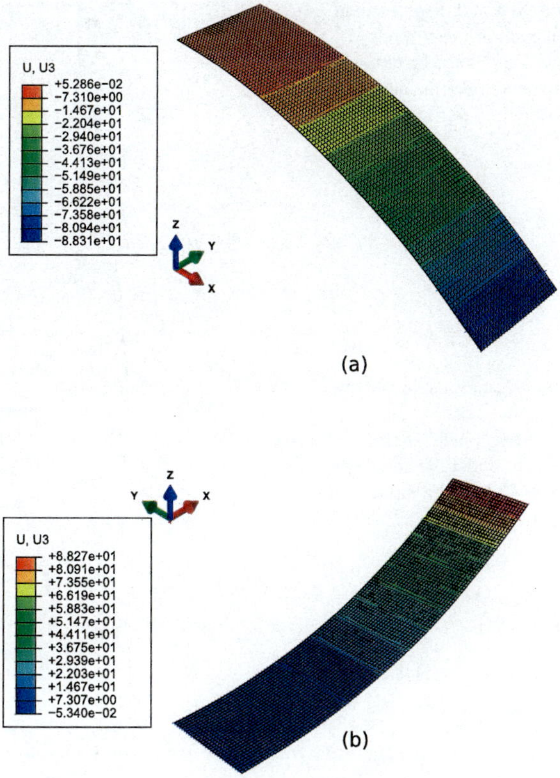

Fig. 4 **a** The first equilibrium shape obtained using ABAQUS® in "clamped" boundary condition. **b** The second equilibrium shape obtained using ABAQUS® in "clamped" boundary condition

where $[\epsilon^o]$ and $[\kappa^o]$ are mid-surface strains and curvatures and $[A]$, $[B]$, and $[D]$ are stiffness matrices [9]. $[N^t]$ and $[M^t]$ are the thermal force and moment vectors that capture the curing process. For the validation, using the Rayleigh-Ritz minimization scheme, the strain and the displacement fields are assumed as in Li et al. [6]. Figure 3 compares deflected profiles of the center-line obtained using the FEA analysis and Rayleigh-Ritz based analytical scheme. It can be seen that the results obtained using the two methods are in close agreement with each other.

To use these multi-stable laminates as a part of larger compliant structures, it is required to analyze them in clamped boundary conditions. Since the layup used in this manuscript is symmetric, after the curing step, the laminate assumes an unstable flat shape. In the subsequent steps, all the degree of freedom of one of the edges is restricted using "Encastre" option in ABAQUS®, and two-point test loads are applied at the endpoints of the other edge. On removing the test load, the structure assumes one of the equilibrium states depending on the direction of the test load. Figure 4 shows the stable shapes obtained in the clamped configuration.

Apart from the equilibrium shapes, another aspect to be studied for their deployment is the load-displacement characteristics. Figure 5 shows the load-displacement diagram for hybrid laminate where the choice of the bi-directional lamina is Glass

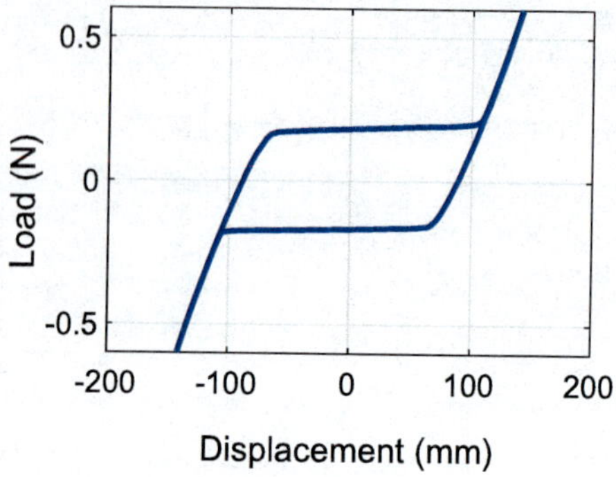

Fig. 5 Load-displacement diagram of the multi-sectioned bistable element under the action of a concentrated load

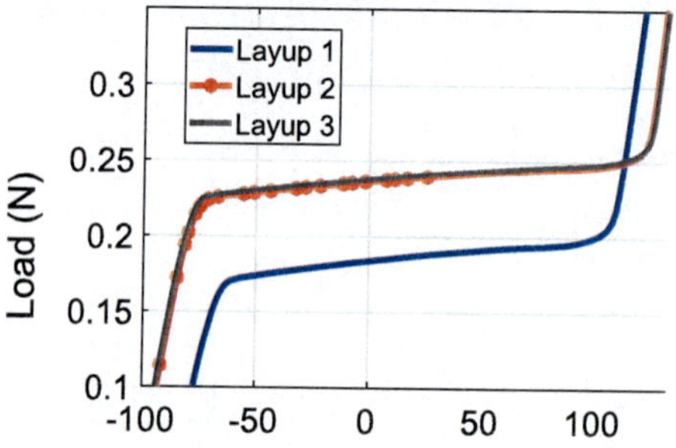

Fig. 6 Load-displacement diagram of the multi-sectioned bistable element under the action of a point load for the three layups

Epoxy 1 (Layup 1), whose properties are given in Table 1. It is observed that the load-displacement curve is symmetric about the x-axis. This is expected as the laminate is symmetric. Since the load-displacement curve is symmetric for the other laminates (i.e., one with the choice of bi-directional lamina being Glass Epoxy 2 (Layup 2) and other being Glass Epoxy 3 (Layup 3)), only one half load-displacement cycle is simulated using ABAQUS®. Figure 6 shows the load-displacement plots for the three layups. It can be observed that with the change in the layup, there is a substantial change in the snapping characteristics of these multi-stable elements. Note that in this study, a point load is applied at the central node of the far edge of the bistable plate.

3 Design of the Passive Flow Control Device

The schematics of the proposed flow control device is shown in Fig. 7. Figure 7a shows the flow device when the bistable member is in its first stable configuration. As the flow rate increases the pressure distribution on the laminate changes. Beyond a particular value of the flow rate and corresponding pressure distribution, the bistable lamina snaps from the first equilibrium shape to the second. In doing so, the flow stream gets another lower resistance pathway, which is redirected to the flow source,

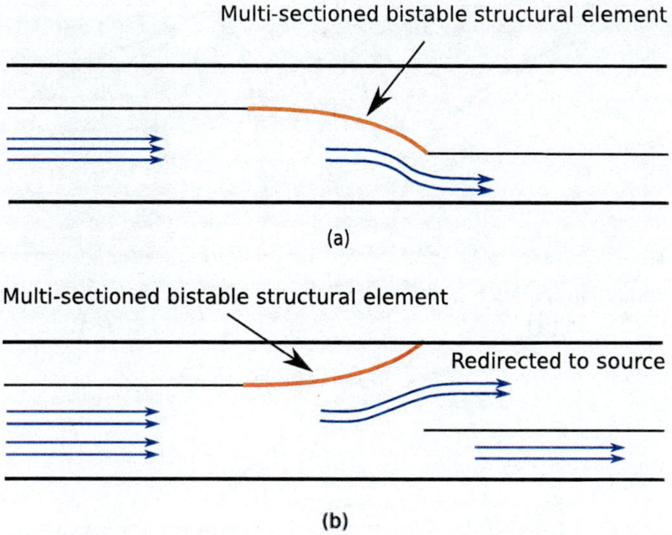

Fig. 7 Schematic of the adaptive flow control device. **a** The flow control device with the bistable element in the first equilibrium condition. **b** The flow control device with the bistable element in the second equilibrium condition

and the flow rate to the attached device is attenuated. The flow rate at which the bistable laminates snaps to the second equilibrium shape can be controlled by controlling the different design parameters of the laminate, i.e., the relative dimensions and the thickness of the individual layers.

To study the performance of these bistable members as part of the flow control device shown in Fig. 7, the pressure distribution on the laminate is obtained using Xfoil. In literature, Xfoil has been used to analyze airfoils, turbine blades, and flaps. For a given Reynolds number or the flow velocity, Xfoil is capable of evaluating the pressure distribution around an airfoil. In this work, the pressure distribution on the lower surface under the assumption of free stream for the configuration shown in Fig. 7a is evaluated for three different flow velocities using Xfoil. Figure 8 shows the pressure distribution on the lower face of the bistable element, under the assumption of viscous free flow, as evaluated using Xfoil, for three different flow velocities, i.e., 1, 5 and 10 m/s. It is observed that the pressure distribution for the different flow velocities is qualitatively similar, only changing in magnitude.

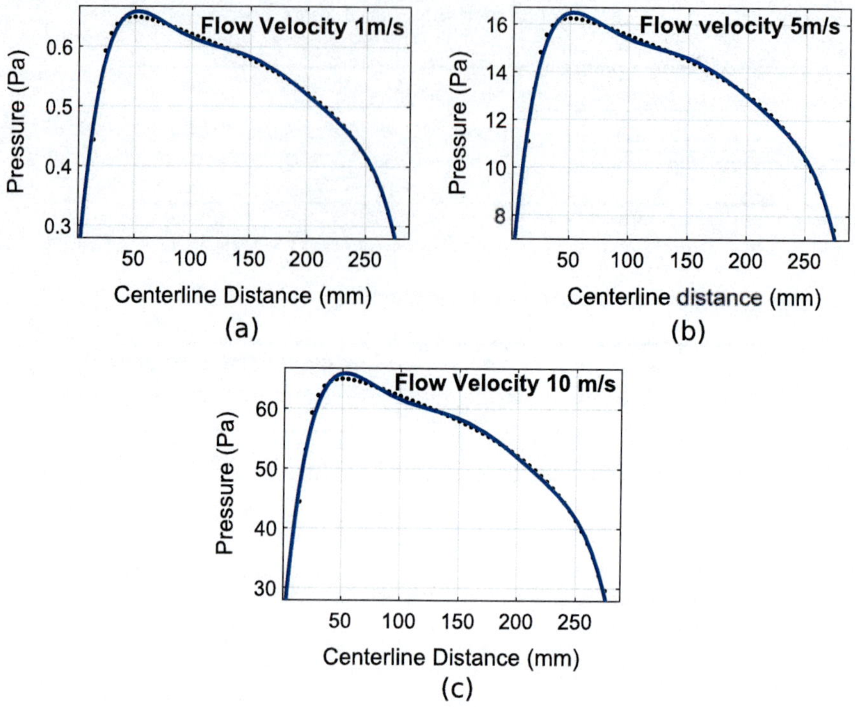

Fig. 8 **a** The pressure distribution along the surface of the bistable laminate for a flow velocity of 1 m/s. **b** The pressure distribution along the surface of the bistable laminate for a flow velocity of 5 m/s. **c** The pressure distribution along the surface of the bistable laminate for a flow velocity of 10 m/s

3.1 The Load-Displacement Characterization of the Bistable Element

In Sect. 2, it was shown that the multi-sectioned bistable structural element, which is used as a part of the flow control device has two equilibrium shapes. It was also shown that it is possible to snap back and forth from one state to another by a simple application of force. To analyze the performance of these special class of bistable laminates, the load-displacement characteristics of the laminate is obtained under the action of pressure distribution. The pressure distribution across the element under the action of a free stream was obtained in the previous section using the viscous flow analysis capabilities of Xfoil. Since the pressure distribution remains the same for the different flow velocities, a sixth-order polynomial function is used to fit the pressure distribution corresponding to the flow with the flow velocity of 1 m/s. To obtain the distribution corresponding to different flow velocities, a multiplication factor is used along with the sixth-order polynomial. Figure 9 shows the load-displacement diagram for the three layups introduced in the study.

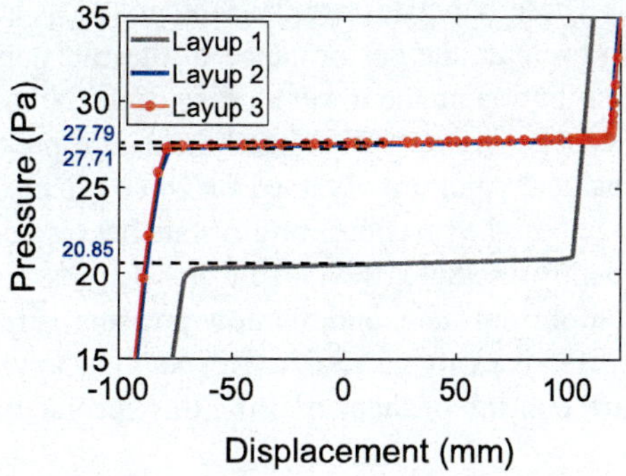

Fig. 9 Load-displacement diagram of the multi-sectioned bistable element under the action of pressure field for the three layups

It can be seen that depending upon the layup, the magnitude of the pressure distribution for which the laminate snaps to the other equilibrium state changes. Furthermore, comparing the pressure profile for the different flow rates and the snap-through pressure, the range of flow velocities for which the laminate retains the first equilibrium shape (refer Fig. 7) can be ascertained. For instance, corresponding to a flow rate of 5 m/s, the maximum pressure on the lower surface of the bistable element is 16.4 Pa, as shown in Fig. 8b. In Fig. 9, it is seen that the critical snapping pressure for an element with layup 1 subjected to a pressure distribution similar to Fig. 8b is 20.85 Pa. So layup one would not snap under a flow of 5 m/s. As the flow rate increases the maximum value of pressure profile increases and when it reaches a critical value of 20.85 Pa, the multi-sectioned bistable element snaps, hence reducing the flow rate, as a part of the fluid coming from the source is directed back to the source as shown in the schematic diagram Fig. 7b. Furthermore, it can be seen from Fig. 9 that depending on the layup, the critical snap-through pressure load changes, hence changing the critical flow velocity beyond which the flow control device releases pressure.

4 Conclusion

In this manuscript, a class of symmetric multi-sectioned hybrid laminates is introduced, which uses uni-directional carbon epoxy and bi-directional glass epoxy prepregs. When cured, these laminates exhibit two symmetric equilibrium shapes. The equilibrium shapes are then obtained using FEA in ABAQUS® which is validated against a Rayleigh-Ritz based numerical scheme in "free-free" boundary condition. Furthermore, it has been shown that the laminates retain its bistability when clamped along one of its shorter edges. The equilibrium shapes in clamped boundary conditions are then obtained using pseudo-dynamic FE analysis. The analysis is then extended to obtain the load-displacement characteristics of these laminates in the clamped boundary condition.

These laminates are then integrated as a part of a concept flow control device. To evaluate the performance of these bistable laminates numerically, the pressure distribution on the lower surface of the laminate is evaluated using Xfoil, under different flow rates. Subsequently, using the pressure distribution, the load-displacement characteristics are obtained for three selected layups. It is shown that the laminates maintain the first equilibrium state below a critical value of the flow rate and consequently, pressure. Beyond the critical pressure load, the laminate snaps to the second equilibrium state, and the flow pressure is released, as a part of the flowing fluid is carried back to the source. Hence it is shown that it is possible to achieve a passive flow control mechanism using this special class of integrable bistable laminates.

References

1. Hampali, S., Anoosha, P.S., Ananthasuresh, G.K.: An open-section shell designed for customized bending and twisting to ease sitting and rising in a chair. Mach. Mech. Rob. 427–439. Springer, Singapore
2. Daynes, S., Nall, S., Weaver, P., Potter, K., Margaris, P., Mellor, P.: On a bistable flap for an airfoil. In: 50th AIAA/ASME/ASCE/AHS/ASC Structures, Structural Dynamics, and Materials Conference 17th AIAA/ASME/AHS Adaptive Structures Conference 11th AIAA No, p. 2103
3. Hyer, M.W.: Calculations of the room-temperature shapes of unsymmetric laminates. J. Compos. Mater. **15**(4), 296–310 (1981)
4. Mattioni, F., Weaver, P.M., Friswell, M.I.: Multistable composite plates with piecewise variation of lay-up in the planform. Int. J. Solids Struct. **46**(1), 151–64 (2009)
5. Arrieta, A.F., Kuder, I.K., Waeber, T., Ermanni, P.: Variable stiffness characteristics of embeddable multi-stable composites. Compos. Sci. Technol. **16**(97), 12–8 (2014)
6. Li, H., Dai, F., Weaver, P.M., Du, S.: Bistable hybrid symmetric laminates. Compos. Struct. **1**(116), 782–92 (2014)
7. Mukherjee, A., Friswell, M.I., Ali, S.F., Arockiarajan, A.: Composite Structures. Elsevier (2020)
8. Daynes, S., Weaver, P.: Analysis of unsymmetric CFRP-metal hybrid laminates for use in adaptive structures. Compos. Part A: Appl. Sci. Manuf. **41**(11), 1712–8 (2010)
9. Jones, R.M.: Mechanics of Composite Materials. CRC Press (2014)

Dynamic Stability of Spent Fuel Trays Stack Submerged in Water Pool Incorporating Coupled Fluid Structure Interaction

Binu Kumar, R. S. Singh, O. P. Singh, G. R. Reddy, K. M. Singh, and N. Gopala Krishnan

1 Introduction

The spent fuel from nuclear reactors is stored in water pools (Fig. 1) which are provided with SS-lined thick RCC walls and floor. These spent fuel bundles are stored for 3–10 years for cooling and decay of thermal heat before reprocessing. The seismic behavior of these submerged freestanding fuel trays in spent fuel storage water pool (SFSWP) is highly nonlinear due to sliding, impact of trays stack and hydrodynamic effect of sloshing water. The stacks of trays are required to be stable during seismic event for safe storage. Earlier, only rigid, uncoupled or simplified methods were implemented to consider the hydrodynamic effect of water on submerged trays stack system.

The problem of computing the behavior of fluid in response to structural vibrations and vice versa has a wide range of applications. Earlier, a lot of researchers have attempted to model the fluid to study the slosh behavior of fluid in a moving container using acoustics wave, potential and Navier–Stokes equation. However, these studies are limited to rigid water tank, or with rigid fixed submerged structure. There is a limited study to understand the dynamic behavior of spent fuel storage pool containing trays stack kept freely on the floor with or without locators and to provide proper design procedures. Also, there is no provision in code [1] for earthquake analysis/design of stacks incorporating gap, friction, sliding and hydrodynamic effects. Efforts have been put only to study the seismic behavior of liquid containers and on

B. Kumar (✉) · R. S. Singh · O. P. Singh · G. R. Reddy · K. M. Singh
Bhabha Atomic Research Centre, Mumbai 400094, India
e-mail: binu@barc.gov.in

N. Gopala Krishnan
CSIR - SERC, Chennai 600113, India

S. K. Saha and M. Mukherjee (eds.), *Recent Advances in Computational Mechanics and Simulations*, Lecture Notes in Mechanical Engineering,
https://doi.org/10.1007/978-981-15-8315-5_51

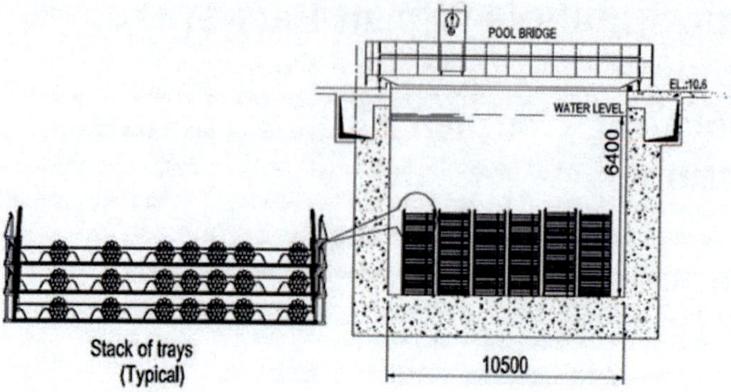

Fig. 1 Storage arrangement of spent fuel trays in fuel pool (sectional view)

the earthquake-resistant design methods of liquid storage tanks without submerged freestanding objects.

Furthermore, from years back to 1960, Houser [2, 3] and many other researchers have studied the behavior of 2D liquid tank under seismic condition, using mechanical spring-mass model. Mustafa [4] in his paper emphasized on the development of finite element (FE) formulation to study the sloshing of liquid in a partially filled rectangular container with rigid baffle under the horizontal base excitation and found a remarkable reduction in the hydrodynamic force by using baffles. Choun [5, 6] presented a method to determine sloshing frequencies and mode shapes in a rectangular tank with an internal rigid block of arbitrary size by using the linear potential wave theory and studied the effect of rigid fixed block on slosh pressure distribution. Similarly, Mitra [7] developed 2D pressure-based finite element computer code to analyze the slosh dynamics of a two-dimensional partially filled rigid container with bottom submerged rigid block and found that the height and width of the submerged structure have very strong influence on the slosh dynamics. Koh [8] and Abramson [9] presented finite element procedure for slosh dynamics analysis of water tank without any submerged block. Nayak [10] used Galerkin-based two-dimensional fully nonlinear finite element model for investigation of the seismic behavior of partially filled rigid rectangular liquid tank with submerged internal components incorporating Mixed Eulerian–Lagrangian (MEL) method. Furthermore, Liu [11] concentrated on the seismic response of spent fuel rack including phenomena like sliding, overturning, and impact between rack and spent fuel pool wall, however, water is modeled using added mass principle which undermines the exact slosh behavior. Furthermore, Muthumani [12] and many researchers have investigated the problem experimentally. Hence, it can be observed that coupled stability analysis of stacks incorporating gap, friction, sliding and hydrodynamic effects is not well understood and analyzed. For understanding the behavior of submerged stacks, a simplified methodology was adopted by Kumar [13–15] in which the rigid and fixed stack boundary was considered for estimating hydrodynamic pressure and the estimated pressure is applied along with seismic motion for coupled analysis. 20–27%

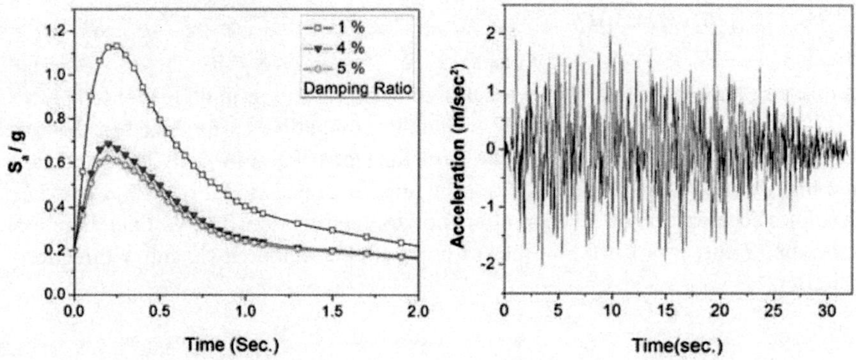

(a) Response spectra (b) Horizontal Earthquake time history

Fig. 2 Earthquake response of SFSWP site

differences were observed in trays displacement compared to experimental values. The study has been carried out for different site-specific time histories and different peak ground acceleration (PGA). It is also observed in the study that 0.2 g PGA site-specific seismic time history has the most catastrophic effect as compared to other seismic time history. Hence in the present paper 0.2 g PGA seismic time history is considered for the analysis.

Structural stability of pool, stability of stacks and structural integrity of fuel bundle are of major concerns during earthquake. Sloshing behavior (sloshing height) and the hydrodynamic pressure exerted on the wall of a spent fuel storage pool is different when the stacks of fuel bundle are present and when they are not. The spent fuel storage pool structure should assure the safety of the stored spent fuels and its structural integrity against design earthquake load. In addition, it should prevent toppling, undesired sliding of the stacked trays and the overflow of contaminated cooling water over the working area by providing sufficient freeboard. Hence for the safety, a good understanding and accurate estimation of sloshing characteristics is a must which will be helpful for calculating the required freeboard, and for the estimation of hydrodynamic pressure on the pool wall and stored spent fuel. The present paper is being emphasized to study the behavior of submerged freestanding sliding stacks of fuel bundle using coupled ALE numerical model. The numerical analysis and modeling of the cases considered are discussed in detail. It is found that convective parts of hydrodynamic pressure induce more instability to the freestanding trays stack system. The results like, slosh frequency, displacement of tank and slosh height are in very close agreement with experimental results and discussed in detail.

2 Input Time History

As discussed above, Fig. 2 shows the design basis seismic ground response spectra (DBSGRS) of 0.2 g PGA for SFSWP site and its compatible earthquake time history is generated from DBSGRS as per the guidelines of IEEE-344 [16]. The generated time history is used for base excitation of finite element model of SFSWT. It has maximum acceleration of 2.1 m/s^2 along horizontal direction. The vertical direction excitation (Z-direction) is 0.66 times of horizontal component (X and Y-direction) excitation.

3 Numerical Method

Arbitrary Lagrangian–Eulerian (ALE) based finite element (FE) model accounting gap, contact, and friction and sliding between trays and bottom surface with coupled hydrodynamic effect on the trays is developed. The water mass is simulated using Navier–Stokes equation. The numerical model of the two systems, that is, free-standing trays stack structure and contained water mass are solved simultaneously using coupled ALE method to resolve high mesh deformation issues which can lead to non-convergence in the solution. Firstly, the finite element model is validated with available codal provision and acoustic wave code [13]. The details of the numerical formulation and modeling are given below.

3.1 FSI-ALE Formulation

In FSI-ALE, the structure is modeled by Lagrangian approach while fluid domain is modeled by ALE approach. In ALE an arbitrary reference coordinate is introduced in addition to the Lagrangian and Eulerian coordinate. Equation 1 describes the material derivative w.r.t. the reference coordinate. On substituting a relation between material time derivative and reference configuration time derivative, ALE equations are formed.

$$\frac{\partial f(X_i, t)}{\partial t} = \frac{\partial f(x_i, t)}{\partial t} + w_i \frac{\partial f(x_i, t)}{\partial x_i} \tag{1}$$

where

X_i Lagrangian coordinate
x_i Eulerian coordinate
w_i Relative velocity = v-u
u Velocity of mesh
v Velocity of material

Equations 2, 3 and 4 represent the governing equations for mass, momentum and energy (E) conservation, respectively, used in ALE formulation.

$$\frac{\partial \rho}{\partial t} = -\rho \frac{\partial v_i}{\partial x_i} - w_i \frac{\partial \rho}{\partial x_i} \tag{2}$$

$$\rho \frac{\partial v_i}{\partial t} = \sigma_{ij,j} + \rho b_i - \rho w_i \frac{\partial v_i}{\partial x_i} \tag{3}$$

σ_{ij} is the stress tensor defined by $\sigma_{ij} = -p + \tau$, where τ is the shear stress from the material constitutive law and p is the pressure.

$$\rho \frac{\partial E}{\partial t} = \sigma_{ij} v_{i,j} + \rho b_i v_i - \rho w_i \frac{\partial E}{\partial x_i} \tag{4}$$

These ALE governing equations reduce to Eulerian governing equations by considering the velocity of reference configuration equal to zero, that is, $u = 0$ and hence relative velocity between the material and the reference configuration becomes equal to material velocity, w = v. The term containing relative velocity in the above equation is known as advective term, which represents the transport of the material across the mesh. In the two-way coupling, first the Lagrangian step is applied. In this step, mesh moves with the material and also the effect of internal and external forces on variation in velocity and internal energy is calculated. Eqs. 5 and 6 represent the equilibrium equations.

$$\rho \frac{\partial v_i}{\partial t} = \sigma_{ij,\,j} + \rho b_i \tag{5}$$

$$\rho \frac{\partial E}{\partial t} = \sigma_{ij} v_{i,j} + \rho b_i v_i \tag{6}$$

In Lagrangian step, no material deforms with the element, and the material does not cross the boundary of the element; hence mass is automatically conserved. In the second step (advection phase), flow of mass, momentum and energy are calculated. Explicit dynamics-based central difference scheme is used for time integration. For successive step, velocity and displacement are updated by the following Eqs. 7 and 8.

$$u^{n+1/2} = u^{n-1/2} + t \cdot M^{-1} \cdot (F_{\text{ext}} + F_{\text{int}}) \tag{7}$$

$$x^{n+1} = x^{n-1} + t u^{n+1/2} \tag{8}$$

where F_{int} represents the internal vector force and F_{ext} represents the external vector force associated with coupling force, body force and pressure boundary condition. M is diagonal lumped mass matrix. For sloshing problem with water in a moving tank where the velocity of the tank is time-dependent, the fluid mesh follows the tank and moves like a rigid mesh. With this ALE feature the mesh remains regular, and also the time step, which is dependent on mesh distortion, remains stable. This algorithm is generally used for problem where the fluid container falls, rotates or moves as in the present case. The integrity of fluid mesh remains intact and fluid mesh moves as a rigid mesh in the coordinate system attached to the container.

4 Numerical Modeling and Analysis

ALE based finite element modeling (FE) of SFWST containing trays stack with and without water mass is carried out. A typical FE model and mesh of SFWST containing freestanding trays stacks with or without water is shown in Figs. 3 and 4, respectively.

Each tray in a stack has been modeled using solid elements. The tray is made of SS 304L having density 8000 kg/m^3 and 2×10^{11} N/m^2 as modulus of elasticity. There are four locators in a tray. The gap between the locators and its corresponding holes have been modeled using gap element. Resting surface between bottom-most tray and pool floor has been simulated by defining contact elements. Contact element has the capability to simulate the friction and to capture the sliding of trays with respect to

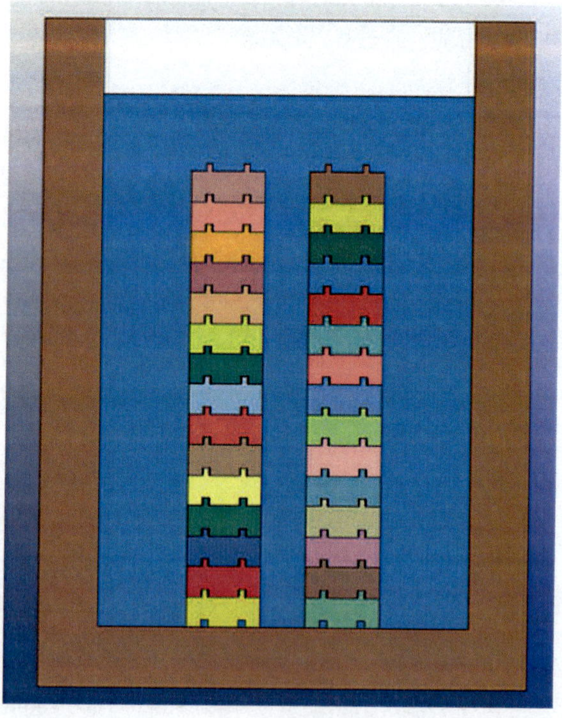

Fig. 3 ALE-FE model of SFSWT trays stack submerged in water

Fig. 4 FE model of trays stack in SFSWT without water

resting surface. The contact element is also used to simulate the contact and friction between the trays in a stack. The available horizontal gap (i.e. operational clearance of 3 mm) between the locator and the corresponding locating hole has also been simulated. The freestanding stack of 27 trays has been analyzed for 0.1 as friction coefficient (μ).

Firstly, Mesh convergence and optimization study with varying mesh size and time step for model has been carried out. Then after, FE model of SFWST without internal trays is validated with Housner [2] and in-house finite element fluid acoustic code (FEFAC) [13]. The comparisons of various parameters observed during analysis are shown in Table 1. Dynamic analysis is carried out for 0.2 g PGA time history.

Slosh frequency obtained using ALE-FE method has a value of 0.533 Hz, which is marginally varied by 5.06% with the analytical value calculated by Housner [2], and

Table 1 Comparison of results for SFSWF tank without stack

S. no.	Code	First sloshing frequency (Hz)	Slosh displacement (m)
1	Present ALE-FE analysis	0.533	0.267
2	FEAFAC code [13]	0.559	0.23
3	Housner [2]	0.56	0.29

a difference of 4.48% with the value obtained by acoustic code. Also the obtained maximum slosh displacement of 0.267 m differs by 8.61% from the Housner [2] analytical value of 0.29 m and by 16.08% from the FEAFAC code value of 0.23 m.

Modal analysis is performed to get slosh frequency and mode shape. It can be observed form Figs. 5 and 6 that analysis produces agreeable slosh mode shape with very less distortion of the mesh using ALE formulation.

Fig. 5 First slosh mode shape

Fig. 6 Third slosh mode shape

5 Results and Discussion

After validation of the present method with available codes and literature, ALE-based FE model of trays stack submersed in SFSWT is simulated and analyzed using site-specific seismic time history. The results obtained are compared with the experimental observation and from in-house FEAFAC code and are discussed in detail.

Figs. 7 and 8 show the slosh displacement of water at 10 and 20.30 s, respectively. Similar slosh behavior is observed during shake table experiment.

Table 2 shows the comparison of responses of trays stack system submerged in SFSWT. The response obtained is compared with shake table experiment and in-house FEAFAC code [13]. For 27 trays stack submerged in water, maximum displacement of 80.21 and 78.49 mm is observed in shake table test and ALE-FE analysis, respectively. Therefore, only 2.14% difference is observed between experimental and ALE-FE method trays stack displacement results.

Hence, it is observed that the ALE-FE analysis gives more accurate displacement with respect to in-house FEAFAC code. It is also found from numerical analysis and experiment that maximum 4.8 and 5 mm lifting is observed for 0.2 PGA seismic excitation, respectively. Hence trays will not come out from beneath trays guides. Although the tray stack has maximum relative displacement of 80.21 mm w.r.t.

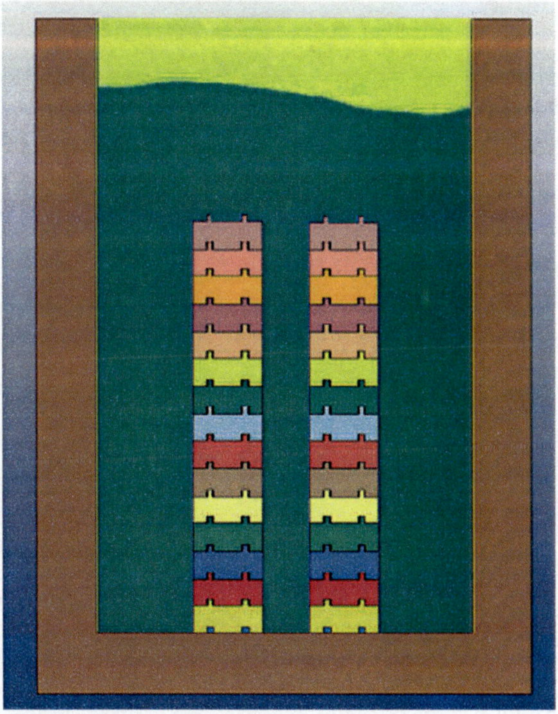

Fig. 7 Slosh displacement at 10 s

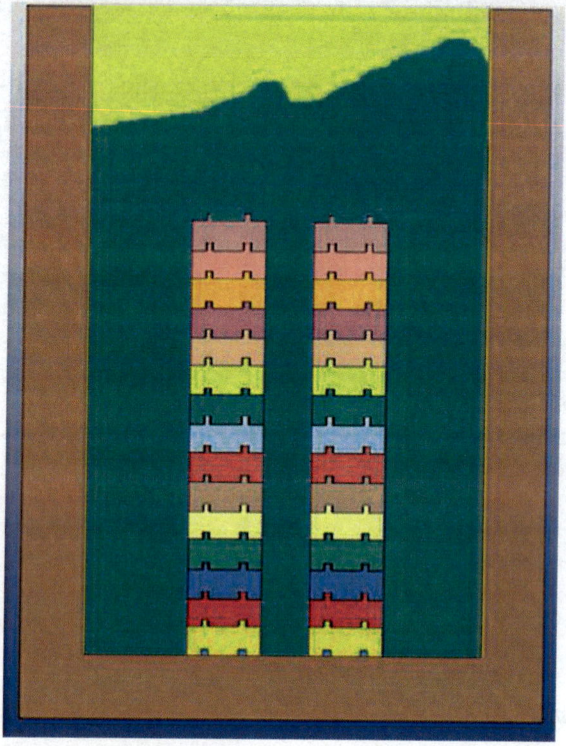

Fig. 8 Slosh displacement at 20.30 s

Table 2 Comparison of responses of trays stacks and tank for 0.2 PGA seismic excitation

S. no.	Response	Experiment	FEAFAC code	ALE-FE analysis
1	Maximum slosh height (m)	0.82	0.62	0.76
2	Frequency of stacks (Hz)	1.43	1.21	1.34
3	Frequency of tank (Hz)	16.00	–	16.10
4	Slosh frequency (Hz)	0.61	0.55	0.57
5	Hydrodynamic pressure (Pa) at the top	7454.12	6179.70	7360.34
6	Relative displacement at 27th tray of stack (mm) w.r.t. tank	80.21	69.89	78.49
7	Maximum relative vertical lifting of tray (mm)	5	–	4.8

tank wall and the gap between the two trays stack is 150 mm, but as trays stack move in the same direction, hence both does not strike each other. Furthermore, experimentally evaluated sloshing frequency of water is around 0.61 Hz for 27 trays stack submerged in water, which is in very close agreement with both ALE-FE and in-house numerical results. The first mode frequency of the water tank is 16.10 Hz, which is approximately the same as observed in the experiment. The natural frequency for 27 trays stack test setup is equal to 1.43 and 1.23 Hz, respectively, for trays stack system without water and with water condition, respectively. The reduction in response frequency of stacks is due to coupling effect of water on trays stacks.

The various stack height is also analyzed using ALE-FE method and the results of tray relative displacement is compared with the shake table experimental data and in-house FEAFAC code. Figure 9 shows the maximum relative displacement of 10, 14 and 27 trays stack under earthquake excitation of 0.2 g peak acceleration. It is also observed that maximum displacement of top tray in a stack increases with increase of stack height. It can be understood that displacement of trays are highly nonlinear and results from ALE-FE matches very close to the shake table experiment results. Therefore, it may be concluded that the ALE-FE method gives results very close to the shake table experiment observations.

Maximum 7.31% difference in slosh height is observed between the experimental and the ALE-FE method. Also, the ALE-FE method is capable of finding various essential parameters like convective and impulsive frequencies, mode shape, hydrodynamic pressure and stack displacement for design of pool and stacking patterns. However, the mesh size and time step has major influence on convergence and

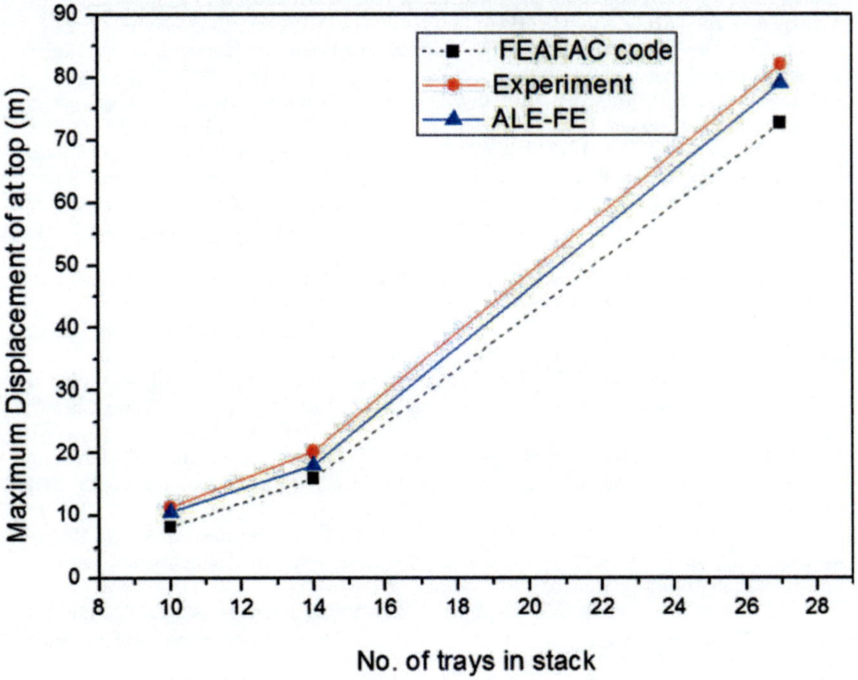

Fig. 9 Comparison of maximum displacement of stacks under earthquake excitation of 0.2 g peak acceleration

stability of results, which requires high computational time and processing. Considering high computational process requirement, the in-house FEAFAC code requires very little computational time and gives results with maximum 25% of difference with the shake table experimental observation. Furthermore, in the present study the interaction of bundles is not considered. Hence, bundles interaction with trays during seismic event is required to be investigated in future to predict bundle failure modes and behavior.

References

1. Jaiswal, O.R.: Review of Code Provisions on Seismic Analysis of Liquid Storage Tanks. Document No. IITK-GSDMA-EQ04-V1.0, Earthquake Codes IITK-GSDMA Project on Building Codes (2003)
2. Housner, G.W.: Dynamic pressures on accelerated fluid containers. Bull. Seismol. Soc. Am. **47**, 15–35 (1957)
3. Housner, G.W.: The dynamic behavior of water tanks. Bull. Seismol. Soc. Am. **53**, 381–387 (1963)
4. Mustafa, A.: Finite Element Analysis of Sloshing in Liquid-Filled Containers. Cairo University, Cairo (2006)
5. Choun, Y.S., Yun, C.B.: Sloshing characteristics in rectangular tanks with a submerged block. J. Comput. Struct. **61**(3), 401–413 (1996)
6. Choun, Y.S., Yun, C.B.: Sloshing analysis of rectangular tanks with a submerged structure by using small-amplitude water wave theory. J. Earthq. Eng. Struct. Dyn. **28**, 763–783 (1999)
7. Mitra, S., Sinhamahapatra, K.P.: Slosh dynamics of liquid-filled containers with submerged components using pressure-based finite element method. J. Sound Vib. **304**, 361–381 (2007)
8. Koh, H.M., Kim, J.K., Park, J.H.: Fluid structure interaction analysis of 3-D rectangular tanks by a variationally coupled BEM-FEM and comparison with test results. J. Earthq. Eng. Struct. Dyn. **27**, 109–124 (1998)
9. Abramson, H.N.: The Dynamic Behavior of Liquids in Moving Containers. NASA SP-106, Washington, DC (1966)
10. Nayak, S.K., Biswal, K.C.: Nonlinear seismic response of a partially-filled rectangular liquid tank with a submerged block. J. Sound Vib. **368**, 148–173 (2016)
11. Liu, Y., Lu, D., Wang, Y., Liu, H.: The sliding and overturning analysis of spent fuel storage rack based on dynamic analysis model. Sci. Technol. Nuclear Install. **2016**, 1–11 (2016)
12. Muthumani, K., Lakshmanan, N., Iyer, N.R.: Dynamic Response of Liquid Tanks during Near Fault Earthquakes. CSIR-SERC, CSIR-Structural Engineering Research Centre, Chennai (2012)
13. Kumar, B., Singh, O.P., Reddy, G.R., Nair, K.N.S.: Experimental and numerical analysis of stacked spent fuel trays submerged in water pool subjected to earthquake loading. SMiRT **21**(240), 1–8 (2011)
14. Kumar, B., Singh O.P., Reddy, G.R., Singh, K.M., Agarwal, K., Gopalakrishnan, N.: Hydrodynamic effect on stability of tray stacks incorporating simplified fluid structure analysis. Structural Engineering Convention (SEC-2016), 1–6 (2016)
15. Kumar, B., Singh, O.P., Reddy, G.R., Singh, K.M., Agarwal, K., Gopalakrishnan, N.: Influence of slosh dynamics on stability of stack of spent fuel trays submerged in water pool under seismic event. SEE-2018, 1–8 (2018)
16. IEEE Standard 344-2004: Recommended Practice for Seismic Qualification of Class 1E Equipment for Nuclear Power Generating Stations (2004)

Frequency Domain Based Robust Flutter Analysis of Swept Back Wing Using μ Method

A. Arun Kumar and Amit Kumar Onkar

1 Introduction

Aeroelastic flutter is defined as the mutual interaction between elastic forces, inertial forces, and aerodynamic forces caused by the unsteady fluid flow around the flexible structure. It is a dynamic aeroelastic phenomenon which can lead to catastrophic structural failure. Further, it is highly sensitive to small variations in the structural and aerodynamic parameters. Since the flutter computational analysis is based on assumptions and approximations, there exist inherent differences between the computational model and the actual flight vehicle. Also, the parameters involved in the computational analysis such as material properties, flight conditions, etc., may be subjected to errors and uncertainties. Thus computational flutter predictions may not be totally reliable and hence conservative safety margins are applied to the analysis results [1]. Also, confident results from the computations are necessary for reliable flutter clearance.

Traditionally, variations and uncertainties associated with the flutter computational model are analyzed using the nominal flutter analysis techniques by enumerating all possible combinations of uncertainties [2]. The drawback of this procedure is that it requires enormous computational hours and also the possibility of missing the worst-case combination [3]. In the present work, robust flutter analysis using $\mu - k$ method [4] is adopted to study the worst-case flutter boundary of sweptback wing in the presence of various structural and aerodynamic uncertainties. This method is based on the application of structured singular value (μ) technique (from the control theory) into aeroelastic analysis. In this method, uncertain flutter analysis is posed as the robust stability problem with variations in the uncertain parameters. Linear

A. Arun Kumar · A. K. Onkar (✉)
Structural Technological Division, CSIR - National Aerospace Laboratories, Bengaluru 560017, India
e-mail: aeroamit@nal.res.in

S. K. Saha and M. Mukherjee (eds.), *Recent Advances in Computational Mechanics and Simulations*, Lecture Notes in Mechanical Engineering,
https://doi.org/10.1007/978-981-15-8315-5_52

Fig. 1 LFT representation of an uncertain system

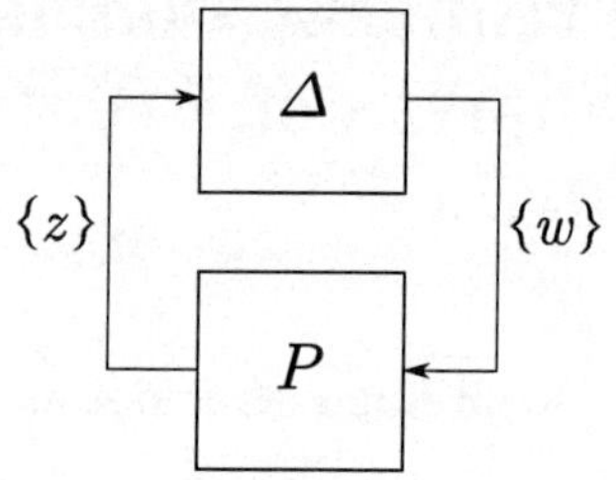

Fractional Transformation (LFT) representation and structured singular value (μ) analysis form the basis for this approach. The maximum value of $\mu < 1$ determines the robust stability of the aeroelastic system under prescribed uncertainty set [4]. Initially, the method is validated for the 3DOF airfoil system under various parametric uncertainties. Later, the method is applied to AGARD 445.6 sweptback wing to study its worst-case flutter characteristics in the presence of structural and aerodynamic uncertainties.

2 Theoretical Background

2.1 μ Method

Robust flutter analysis deals with the stability of an aeroelastic system subjected to uncertainties. Its aim is to quantify the gap between the computational flutter predictions resulting from the nominal aeroelastic model (model without uncertainties) and uncertain aeroelastic model in which the entire uncertainty set is contemplated. From the given family of uncertain aeroelastic models, robust flutter analysis predicts the model having the worst-case stability characteristics.

The theory behind the robust flutter analysis based on the structured singular value was introduced by Lind [5] for robust aeroservoelastic analysis in 1997. μ measures the robustness of a linear, stable system with respect to specified uncertainties within it. For a given bounded family of uncertain linear systems, μ finds the system having the worst-case stability characteristics. μ method has the advantage of utilizing both theoretical models and actual flight data for the determination of robust flutter boundary.

The application of μ method requires the representation of uncertain aeroelastic system in LFT form. In this representation, the uncertain aeroelastic model is separated into nominal and uncertain parts and then interconnected into LFT feedback loop as shown in Fig. 1. In the figure, P contains the nominal aeroelastic model and terms associated with uncertainties. The uncertainty operator, $\boldsymbol{\Delta}$, contains the uncertainties associated with the inaccuracies present in the model. The uncertainty operator is a norm bounded set, which, with the help of weighting matrices, becomes unity norm bounded given by [6]:

$$\boldsymbol{\Delta} = \{\Delta : \|\Delta_{\infty}\| \leq 1\} \tag{1}$$

Now, to measure the robust stability of the system P with respect to $\boldsymbol{\Delta}$ set, structured singular value (μ) is used whose definition is given as [6]:

$$\mu(P) = \frac{1}{\min\limits_{\Delta \in \boldsymbol{\Delta}}\{\bar{\sigma}(\Delta) : \det(I - P\Delta) = 0\}} \tag{2}$$

with $\mu = 0$ if no $\Delta \in \boldsymbol{\Delta}$ exists such that $\det(I - P\Delta) = 0$. The system P is robustly stable for the set $\boldsymbol{\Delta}$ if $\mu(P) < 1$. This condition is tested for the uncertain aeroelastic model by computing Eq. (2) on a grid of frequency range of interest. The velocity at which $\mu(P) \approx 1$ is called the robust flutter velocity. Thus, the robust flutter boundary for an aeroelastic system can be computed using this stability criterion.

2.2 Governing Equations

The governing aeroelastic equations of motion in dimensional Laplace variable s is given by [1]:

$$\left(\boldsymbol{M}s^2 + \boldsymbol{C}s + \boldsymbol{K} - q\,\boldsymbol{Q}(\bar{s})\right) X(s) = 0 \tag{3}$$

where $\boldsymbol{M}$, $\boldsymbol{C}$, and $\boldsymbol{K}$ represent mass, damping, and stiffness matrices, respectively, in the generalized coordinates of size $n \times n$ where n is the number of modes considered, q is the dynamic pressure, $s = i\omega$ is the dimensional Laplace variable, and $X(s)$ is the vector of generalized coordinates of size n. $\boldsymbol{Q}(\bar{s})$ represents generalized aerodynamic forces and $\bar{s} = ik$ represents the non-dimensional Laplace variable. Here, k and ω are defined as the reduced frequency and circular frequency, respectively, which are related by $k = \omega b/U$ where b is the reference semi-chord and U is the freestream velocity.

For the robust flutter studies, uncertainties need to be introduced in the parameters of the nominal aeroelastic system. Thus the structural and aerodynamic properties of the nominal aeroelastic system are introduced with parametric uncertainties whose definition is given as follows [4]:

$$\boldsymbol{M} = \boldsymbol{M}_0 + \boldsymbol{V}_M\boldsymbol{\Delta}_M\boldsymbol{W}_M \tag{4}$$

$$\boldsymbol{C} = \boldsymbol{C}_0 + \boldsymbol{V}_C\boldsymbol{\Delta}_C\boldsymbol{W}_C \tag{5}$$

$$\boldsymbol{K} = \boldsymbol{K}_0 + \boldsymbol{V}_K\boldsymbol{\Delta}_K\boldsymbol{W}_K \tag{6}$$

$$\boldsymbol{Q} = \boldsymbol{Q}_0 + \boldsymbol{V}_Q\boldsymbol{\Delta}_Q\boldsymbol{W}_Q \tag{7}$$

where $\boldsymbol{V}$'s and $\boldsymbol{W}$'s are the scaling matrices and the subscript 0 represents nominal value of the system parameters.

By introducing the above uncertainties in the nominal aeroelastic system given by Eq. (3), the following uncertain flutter equation is obtained.

$$\left(\boldsymbol{M}_0 s^2 + \boldsymbol{C}_0 s + \boldsymbol{K}_0 - q\,\boldsymbol{Q}_0\right)X(s) + \left(\boldsymbol{V}_M \boldsymbol{\Delta}_M \boldsymbol{W}_M s^2 + \boldsymbol{V}_C \boldsymbol{\Delta}_C \boldsymbol{W}_C s + \boldsymbol{V}_K \boldsymbol{\Delta}_K \boldsymbol{W}_K - q\boldsymbol{V}_Q \boldsymbol{\Delta}_Q \boldsymbol{W}_Q\right)X(s) = 0 \quad (8)$$

For the application of μ method to find the robust flutter velocities, the uncertain flutter equation needs to be converted into Linear Fractional Transformation (LFT) representation. This is achieved by separating the uncertainties from the nominal system and then reintroduce it in a feedback manner. Let the input signals to the uncertainty matrix is given by:

$$z_M = \boldsymbol{W}_M X(s) \quad (9)$$

$$z_C = \boldsymbol{W}_C X(s) \quad (10)$$

$$z_K = \boldsymbol{W}_K X(s) \quad (11)$$

$$z_Q = \boldsymbol{W}_Q X(s) \quad (12)$$

The above equation in vector form can be written as:

$$z = \boldsymbol{W} X(s) \quad (13)$$

By defining the uncertainty matrix $\boldsymbol{\Delta}$ as:

$$\boldsymbol{\Delta} = \begin{bmatrix} \boldsymbol{\Delta}_M & 0 & 0 & 0 \\ 0 & \boldsymbol{\Delta}_C & 0 & 0 \\ 0 & 0 & \boldsymbol{\Delta}_K & 0 \\ 0 & 0 & 0 & \boldsymbol{\Delta}_Q \end{bmatrix} \quad (14)$$

The output signals from the uncertainty matrix are given by:

$$\boldsymbol{w} = \boldsymbol{\Delta} z \quad (15)$$

where $\boldsymbol{w} = \left[w_M \; w_C \; w_K \; w_Q\right]^T$. By using Eq. (15) in Eq. (8), the LFT representation of the uncertain aeroelastic system in compact form is then given by:

$$\boldsymbol{P}_0 X = \boldsymbol{V}\boldsymbol{w} \quad (16)$$

where $\boldsymbol{P}_0$ is the nominal flutter matrix given by:

$$\boldsymbol{P}_0 = \left[\boldsymbol{M}_0 s^2 + \boldsymbol{C}_0 s + \boldsymbol{K}_0 - q\,\boldsymbol{Q}_0\right] \quad (17)$$

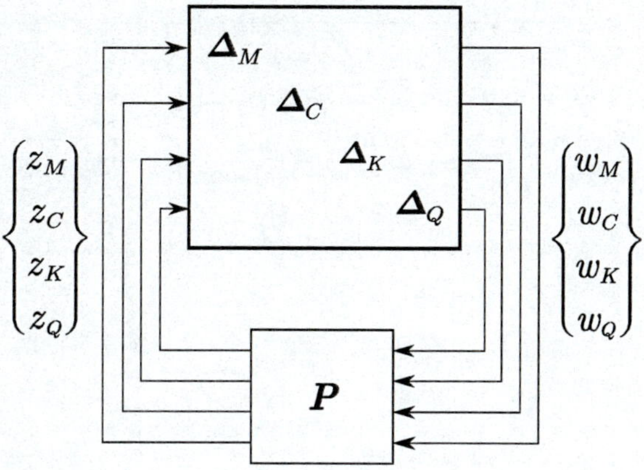

Fig. 2 LFT representation of the uncertain flutter equation

and V is the total perturbation transfer function matrix given by:

$$V = \left[-V_M s^2,\ -V_D s,\ -V_K,\ qV_Q\right] \tag{18}$$

Using Eqs. (13) and (16), the output signals from the nominal flutter matrix is given by:

$$z = W P_0^{-1} V w = P w \tag{19}$$

where P is the flutter loop transfer function matrix which relates the signals w and z. Figure 2 shows the LFT representation given by Eqs. (15) and (19). Robust flutter analysis is then performed by computing the structured singular values (μ) of the above-generated LFT model at various grid points in the frequency range of interest. Robust Control Toolbox of MATLAB [7] is used for this purpose. The velocity at which the peak value of $\mu \approx 1$ is considered to be the worst-case flutter velocity.

3 Results and Discussions

In the present work, two aeroelastic problems are considered for the robust flutter studies, namely, a 3DOF airfoil system [1] for validation and AGARD 445.6 swept-back wing for application [8]. The analysis and results for each of these cases are discussed in the following sections.

3.1 3DOF Airfoil System (Validation Case)

Figure 3 shows a 3DOF airfoil system along with its geometric, structural, and aerodynamic descriptions. The motion of the airfoil is described by the plunge h, the

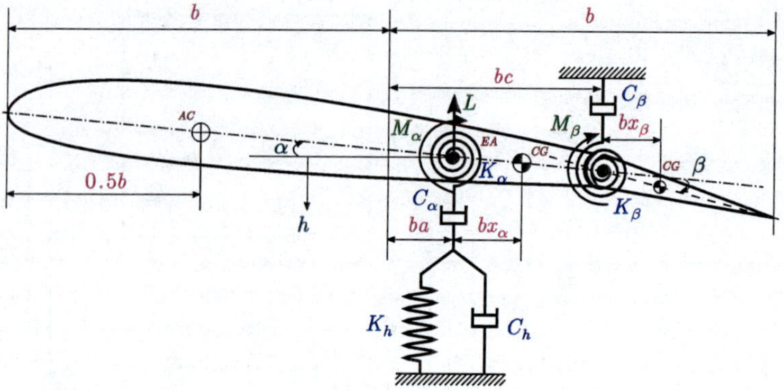

Fig. 3 A 3DOF airfoil system

pitch α, and the flap angle β. h is measured positive downward, α is measured positive nose up from the x-axis and β is measured positive down from the airfoil reference chord. K_h, K_α, K_β represent plunge, pitch, and torsional stiffness, respectively, C_h, C_α, C_β represent plunge, pitch, and torsional damping, respectively. b is the semi-chord of the airfoil, a is the non-dimensional distance between the mid-chord and the elastic axis, x_α is the non-dimensional distance between the elastic axis and the center of gravity, c is the location of the flap hinge point from the airfoil mid-chord, x_β is the distance of the flap CG from the flap hinge point. L, M_α and M_β are the aerodynamic lift, airfoil pitching moment, and flap hinge moment, respectively. The expressions for mass, stiffness, damping, and aerodynamic matrices are given in [1].

The values of various parameters used in the present analysis are [1]: $b = 1$ m; $a = 0.4$; $c = 0.6$; $S = 2b^2$; $m_s = 153.94$ kg/m; $x_\alpha = 0.2$; $x_\beta = -0.025$; $K_h = 8.66 \times 10^5$ N; $K_\alpha = 3.85 \times 10^5$ N; $K_\beta = 3.85 \times 10^5$ N, and $\rho = 1.225$ kg/m^3. The dimensionless radius of gyration of the airfoil section r_α and of the control surface section r_β are 0.497 and 0.0791, respectively.

3.1.1 Nominal Flutter Analysis

Nominal flutter analysis is carried out using the *p-k* method implemented in MATLAB. The theoretical description of this method is given in [9]. The mass, stiffness, and aerodynamic matrices are computed using the parameters defined above. For the velocity range of interest, the state matrix of the *p-k* method is solved for each mode by monitoring the convergence of the reduced frequency. This results in the converged root for each mode from which coupled frequency and damping will be computed for a given velocity. The velocity at which the computed damping of a particular mode becomes positive from negative indicates nominal flutter velocity.

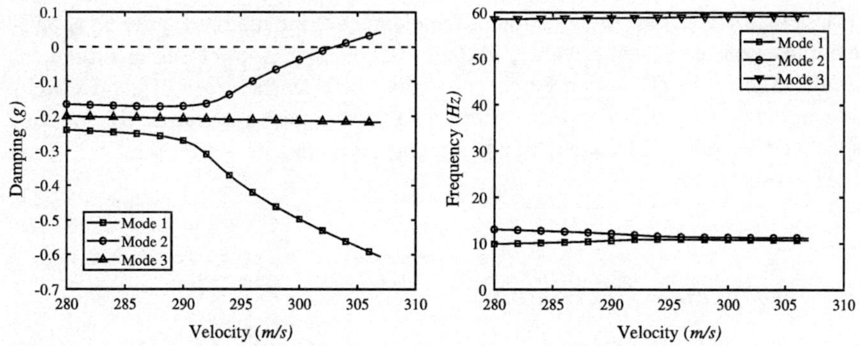

Fig. 4 Variation of damping and frequency for various modes with velocity for 3DOF airfoil system

Table 1 Nominal flutter comparison

	Flutter velocity (m/s)	Flutter frequency (Hz)
Present	303	11.25
Ref. [1]	303.3	11.25

Figure 4 shows the damping and frequency trends for each mode for a range of velocities. At $U = 303$ m/s, zero crossing is observed in the damping curve of Mode 2 thus indicating the occurrence of flutter. From the V-f graph, the coalescence of plunge and pitch mode can be clearly observed.

Table 1 shows the comparison of the present computed flutter velocities with the literature. From the comparison, it can be seen that there exists a good match between the present result and those given in Ref. [1].

3.1.2 Robust Flutter Analysis

For the robust flutter validation studies, three cases are considered as reported in [1], namely, structural uncertainty alone, aerodynamic uncertainty alone, and the combination of both. A sub-critical velocity of 270 m/s is chosen for all the cases as reported in [1]. The validation results for each case will be discussed in the following paragraphs.

3.2 *Structural Uncertainty*

In this case, uncertainties are assumed in the mass and stiffness matrices of the 3DOF system. The percentage of uncertainties assigned are [1]: $M_{11} = 10\%$, $M_{22} = 10\%$, $K_{22} = 10\%$, $M_{12} = 5\%$, and $K_{11} = 5\%$, where the subscripts i, j in both mass and stiffness matrices denote the location of the element in the matrix. Figure 5a shows the comparison of present μ values in the presence of real structural uncertainties with those given in [1]. It can be observed that there exists a overall good match between the present and literature results. The present peak μ value 1.398 and the respective frequency 72 Hz are in good agreement with the literature values 1.38 and 72.2 Hz, respectively.

3.3 Aerodynamic Uncertainty

In this case, uncertainties are prescribed in the aerodynamic matrix computed using Theodorsen's method. Complex form of uncertainty is assumed for the terms Q_{12}, Q_{21} and Q_{22} of the AIC matrix. These complex uncertainties are defined as a disk in the complex plane whose nominal values form the center of the disk and the radius of the disk equals 10% of the magnitude of the nominal values [1]. Figure 5b shows the comparison of the computed μ values with those given in [1]. There exists an overall good match between the two results.

3.4 Structural and Aerodynamic Uncertainties

Here, the uncertainties defined in the previous two cases are combined, resulting in the mixed real/complex uncertainties for the μ calculations. Figure 5c shows the comparison of the present μ values with those given in [1] and a good match is observed between them.

3.5 AGARD 445.6 Sweptback Wing (Application Case)

In the previous section, μ method was successfully validated for 3DOF airfoil system with the literature results. In this section, the method is applied to find the worst-case flutter velocities for AGARD 445.6 wing in various flow regimes. Figure 6 gives the geometric details of the AGARD wing. The wing has a NACA 65A004 airfoil cross section. The other geometric properties are 45° quarter chord sweep, aspect ratio = 1.65, and taper ratio = 0.66. Various test models are available in [8] but in the present work, weakened wing model is chosen for the uncertainty study. The analysis conditions considered for the study are given in Table 2.

Initially, the free vibration studies are carried out to verify the computed natural frequencies with the available experimental results. For this purpose, 3D finite element model is generated for the AGARD wing in MSC Nastran and the orthotropic material properties are assigned.The properties of the material are as follows: ρ = 381.98 kg/m^3, E_x = 3.1511 GPa, $E_y = E_z$ = 0.4162 GPa, $\nu_{xy} = \nu_{yz} = \nu_{zx}$ = 0.31 and

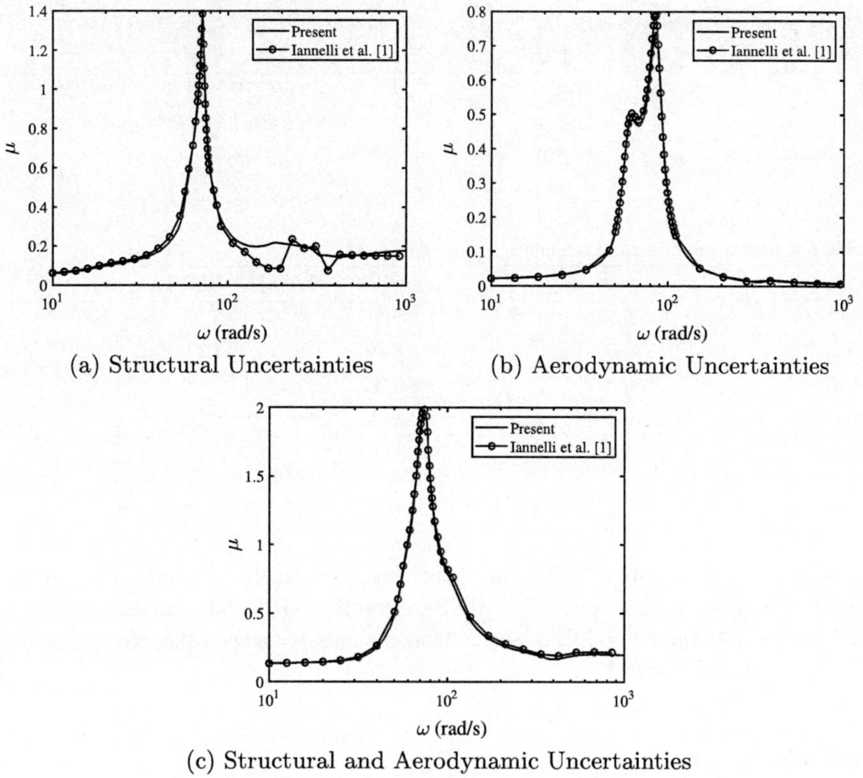

Fig. 5 Comparison of μ values in the presence of various uncertainties between present computation and Iannelli et al. [1] for 3DOF airfoil system at 270 m/s

Fig. 6 AGARD 445.6 wing

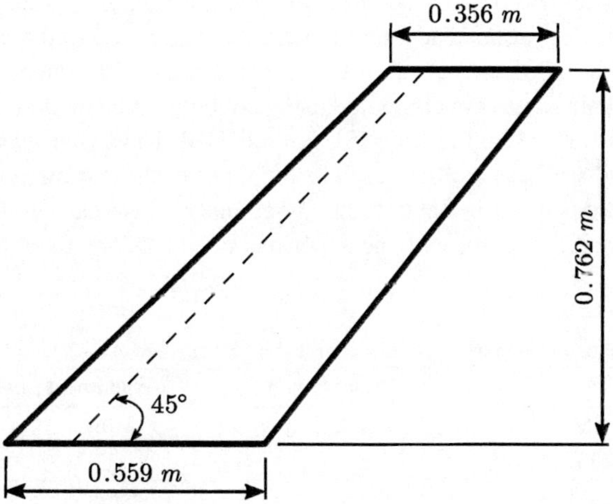

Table 2 Analysis conditions [8]

Cases	Mach No.	Density (kg/m^3)
1	0.499	0.4278
2	0.678	0.2082
3	0.901	0.0995

Table 3 Comparison of natural frequencies of AGARD 445.6 wing

Modes	Present (Hz)	Experiment (Hz) [8]	Difference (%)
Mode 1	9.611	9.5992	0.12
Mode 2	40.0	38.1650	4.81
Mode 3	50.257	48.3482	3.95
Mode 4	96.284	91.5448	5.18
Mode 5	124.852	118.1132	5.71

$G_{xy} = G_{yz} = G_{zx} = 0.4392$ GPa. The modal mass and stiffness matrices are then extracted from MSC Nastran using DMAP subroutines [10]. The matrices are then used to calculate the natural frequencies whose comparison with the experimental values are given in Table 3.

3.5.1 Nominal Flutter Analysis

Nominal flutter analysis is then carried out to compute the flutter velocities of AGARD wing in the absence of uncertainties. The analysis is carried out in the modal coordinates using the first five fundamental modes of vibration. For the computation of unsteady aerodynamics, a list of reduced frequencies is chosen ranging from 0.001 to 1. At all the considered reduced frequencies, the generalized aerodynamic forces are computed using doublet lattice method (DLM) and extracted using DMAP subroutines of MSC Nastran [10]. These matrices along with the modal mass and stiffness are then used as the input to the *p-k* method. For all the analysis conditions given in Table 2, the flutter analyses are carried out and the computed flutter velocities along with the experimental values are given in Table 4.

Table 4 Comparison of nominal flutter velocities of AGARD 445.6 wing

Mach	Present (m/s)	Experiment (m/s) [8]	Difference (%)
0.499	184.23	172.46	6.83
0.678	253.0	231.37	9.35
0.901	320.42	296.69	8

Table 5 Comparison of robust flutter velocities of AGARD 445.6 wing

Mach	Present (m/s)	Experiment (m/s) [8]	Difference (%)
0.499	173.2	172.46	0.43
0.678	236	231.37	2.0
0.901	272	296.69	−8.32

It can be observed that there exists a considerable amount of difference between the flutter velocities computed using *p-k* method and the experimental results. The variation of damping and frequency with respect to velocity for all the cases are shown in Fig. 7.

3.5.2 Robust Flutter Analysis

For the robust flutter studies, the differences observed in the first five natural frequencies between the experiment and the present computations are introduced as uncertainties in the modal stiffness matrix of the aeroelastic equation of motion. The respective percentages are 0.24, 9.85, 8.05, 10.62, and 11.74%. To model the uncertainties in aerodynamics, the following percentages are assumed for generalized aerodynamic forces for each Mach number: 1% for 0.499 case, 2% for 0.678 case, and 5% for 0.901 case. The increase in the percentage error values as Mach number increases is due to the fact that, at high Mach numbers, the flow becomes nonlinear and hence DLM accuracy falls short. The μ analysis is then carried out by including both aerodynamic and structural uncertainties for all the Mach numbers considered and the respective worst-case velocities are computed based on the criteria $\mu \approx 1$. Figure 8 shows the variation of μ with frequency computed at robust flutter velocity for all the considered Mach numbers given in Table 2.

Table 5 shows the comparison of the computed robust flutter velocities with the experimental results. It can be observed that, by the introduction of structural and aerodynamic uncertainties, the robust flutter velocities move toward the experimental results for all the Mach numbers considered. Also, the % difference between the robust velocity and experimental values is less at low subsonic Mach numbers and increases with increase in Mach number. This difference can be brought down by defining the uncertainty magnitudes associated with the aerodynamics using CFD and/or experimental results.

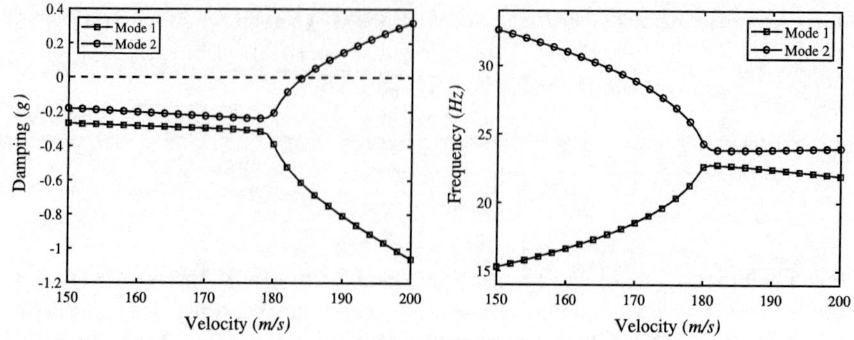

(a) $U_{flutter} = 184.23\ m/s,\ Mach = 0.499$

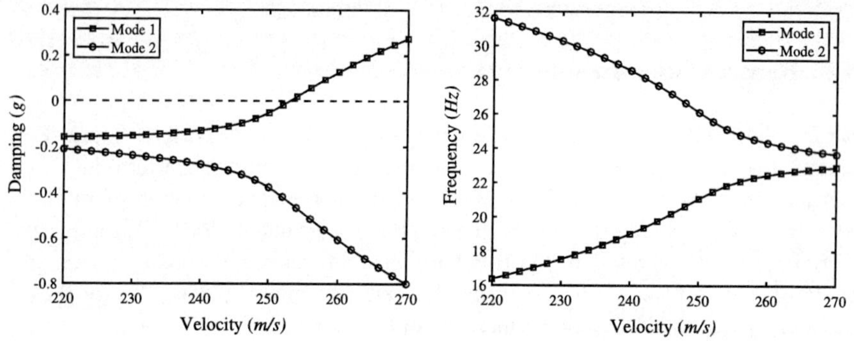

(b) $U_{flutter} = 236\ m/s,\ Mach = 0.678$

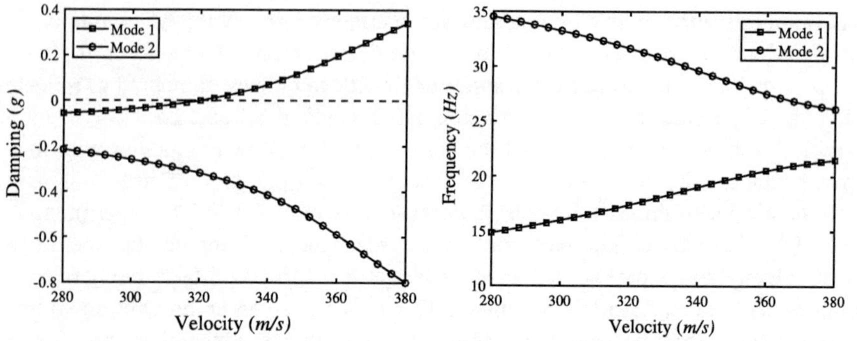

(c) $U_{flutter} = 272\ m/s,\ Mach = 0.901$

Fig. 7 Variation of damping and frequency for different modes with velocity for AGARD 445.6 wing at various Mach numbers

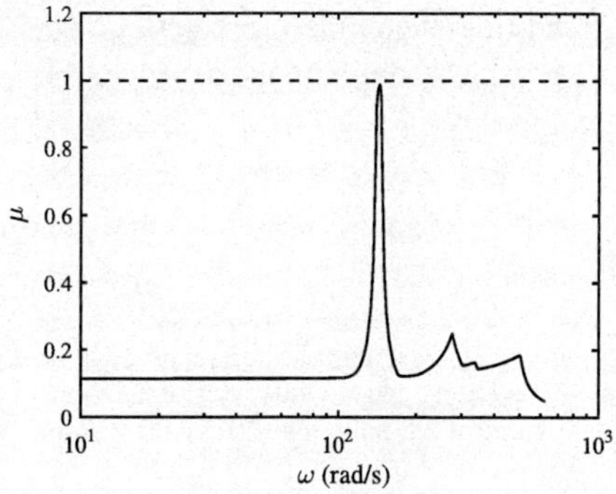

(a) $U_{rob} = 173.2\ m/s,\ Mach = 0.499$

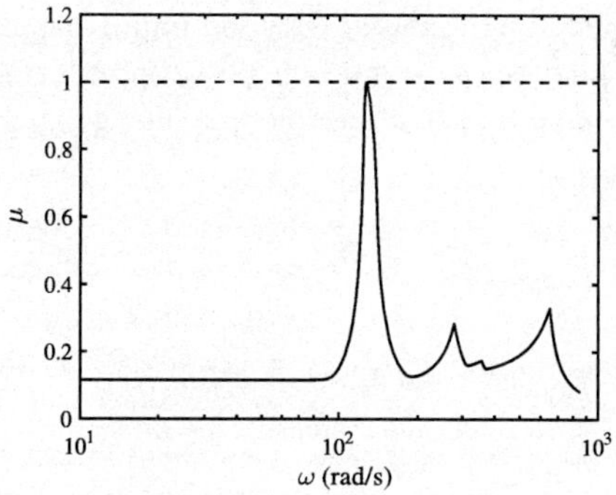

(b) $U_{rob} = 236\ m/s,\ Mach = 0.678$

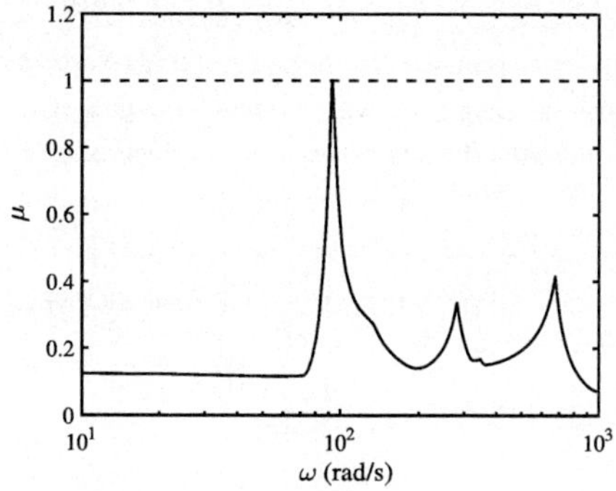

(c) $U_{rob} = 272\ m/s,\ Mach = 0.901$

Fig. 8 μ bound for AGARD 445.6 wing in the presence of structural and aerodynamic uncertainties at various Mach numbers

4 Conclusions

In this study, the robust flutter analysis using structured singular value (μ) method is successfully carried out to determine the worst-case flutter speed for AGARD 445.6 sweptback wing. Initially, a 3DOF airfoil system is robustly evaluated at a sub-critical velocity of 270 m/s for various structural and aerodynamic uncertainties and the present μ values are successfully validated with the available literature results. The method is then applied to AGARD wing by considering structural (resulting from the differences observed in the first five natural frequencies) and aerodynamic uncertainties. From the μ analysis, it is observed that the present robust flutter velocities are quite close to the experimental values as compared to nominal flutter velocities at all the Mach numbers studied. Thus by incorporating uncertainties in the aeroelastic system, there is a good improvement in the flutter boundary prediction. The predictions in the present work can be further improved by defining the aerodynamic uncertainties in a rigorous way using experiment or CFD data.

References

1. Iannelli, A., Marcos, A., Lowenberg, M.: Aeroelastic modelling and stability analysis: a robust approach to flutter problem. Int. J. Robust Nonlinear Control **28**, 342–364 (2018). https://doi.org/10.1002/rnc.3878
2. Yuting, D., Chao, Y.: Methods and advances in the study of aeroelasticity with uncertainties. Chin. J. Aeronaut. **27**(3), 461–474 (2014). https://doi.org/10.1016/j.cja.2014.04.016
3. Zhigang, W., Chao, Y.: A new approach for aeroelastic robust stability analysis. Chin. J. Aeronaut. **21**(5), 417–422 (2008). https://doi.org/10.1016/S1000-9361(08)60054-0
4. Borglund, D.: The $\mu - k$ method for robust flutter solutions. J. Aircraft **41**(5), 1209–1216 (2004). https://doi.org/10.2514/1.20190
5. Lind, R., Brenner, M.: Robust Aeroservoelastic Stability Analysis. Springer, London (1999)
6. Lind, R., Brenner, M.: Incorporating flight data into a robust aeroelastic model. J. Aircraft **35**(3), 470–477 (1998). https://doi.org/10.2514/2.2320
7. Balas, G.J., Doyle, J.C., Glover, K., Packard, A., Smith, R.: μ—Analysis and Synthesis Toolbox User's Guide. The MathWorks Inc., Natick, MA (1996)
8. Yates, Jr., E.C.: AGARD Standard Aeroelastic Configurations for Dynamic Response I—Wing 445.6. NASA-TM-100492, Hampton, USA (1987)
9. Aeroelastic Analysis User's Guide, MSC Nastran (2017)
10. DMAP Programmer's Guide, MSC Nastran (2017)

Optimal Parameters Identification of Quarter Car Simulink Model for Better Ride Comfort and Road Holding

N. P. Puneet, Abhinandan Hegale, Hemantha Kumar, and K. V. Gangadharan

1 Introduction

With advancement in technology each and every moment of time, human tendency to lean toward comfort zone has been seen in recent decades. Hence, every invention has been witnessing day-by-day modification in the field of research. Automobile industry is one among those sectors which is truly working for 'more than need' of a person. The happiness of a passenger inside a vehicle has been termed 'ride comfort' in vehicle industries. This is generally done by isolating the passenger from road disturbances. One main element which is essential to serve this purpose is damper. A very smooth damper can provide better ride comfort. But the intension of providing the ride comfort alone may result in losing the road holding of a vehicle [1]. This created a new challenge in the vehicle suspension design, and a greater level of research is going on in this field.

Els et al. [1] with the help of three case studies explained the opposite extremes of ride comfort and road handling in terms of spring and damper characteristics. The experiments were conducted on/off road vehicles and proposed semi-active suspension to sustain compromising nature of ride comfort and road handling. An attempt to control the vibration transferred from the road disturbance to the passenger in the cabin was made by Roumy et al. [2]. Experiments were conducted in a quarter car test rig equipped with an active damper integrated with control logic. A multibody dynamics simulation code (Vehsim2d) was linked to leap frog optimization algorithm by Naude and Snyman [3] in a constrained problem to optimize the parameters of passive suspension system. Results indicated the changes in performance with variation in terrain. Sun [4] made an attempt to optimize the suspension system

N. P. Puneet (✉) · A. Hegale · H. Kumar · K. V. Gangadharan
National Institute of Technology Karnataka, Surathkal, Mangaluru 575025, India
e-mail: puneetnp12@gmail.com

S. K. Saha and M. Mukherjee (eds.), *Recent Advances in Computational Mechanics and Simulations*, Lecture Notes in Mechanical Engineering,
https://doi.org/10.1007/978-981-15-8315-5_53

design suitable for rough pavement surfaces. Walking beam suspension system was employed to demonstrate the vehicle suspension design. A passive vehicle suspension system was used to analyze the ride comfort of a vehicle by Mostaani et al. [5]. Experiments considered different road roughness conditions and different vehicle velocities. The study used design of experiment (DOE) method to optimize the conditions for better ride comfort employed on 7 DOF vehicle model. Reddy et al. [6] used genetic algorithm (GA) in optimizing the parameters of quarter car suspension system for better ride comfort. Quarter car was modeled using Matlab/Simulink. Response surface method was used to model the equation for ride comfort and was used as objective function for GA. Mitra et al. [7] used design of experiment and response surface method to determine the optimal setting of spring and damper for better ride comfort. Quarter car Simulink model was used for simulation purpose and the Box-Behnken method of DOE was used in the study. Shirahatt et al. [8] made a simulation study of a passenger car model and used genetic algorithm to optimize the control parameters of a suspension design. These results were again compared with the results of simulated annealing (SA) technique. This comparative study between active and passive suspension systems showed the upper hand of active system in providing better ride comfort. Gangadharan et al. [9] presented a study on ride comfort evaluation of railway coach both analytically and experimentally. Ride comfort was evaluated based on Sperling's ride index and ISO 2631. Response surface method and Taguchi method had been used to find the optimal parameter and effective parameter, respectively, by Reddy et al. [10]. The simulation using Matlab/Simulink was conducted for finding the optimal seat suspension parameters using quarter car model. Gurubasavaraju et al. [11] used response surface method to understand the variation of responses against various parameters in magneto-rheological damper and optimized the magneto rheological fluid particle concentration for greater damping force.

In this work, an attempt has been made to characterize the commercially available passive damper of a light motor vehicle and to implement this damper data in the quarter car simulation model. Many previous works have seen quarter car simulation by assuming the value of damping coefficient and other parameters. But in this work, value of actually available damper after characterization has been implemented to know the effectiveness of this damper. Also, multiobjective optimization has been done for better ride comfort and road holding using response surface optimization.

2 Experimental Setup and Procedure

Experiments were conducted on a commercially available passive damper of a light motor vehicle using damper testing machine (DTM) and is shown in Fig. 1. The damper testing machine (make-HEICO) with a hydraulic capacity of 10 kN was used for conducting experiments. This machine consists of potentiometer for displacement measurement and a load cell for measuring force. The measured data can be acquired through a data acquisition system (DAQ) and the input conditions can be varied using MOOG controller. A desktop computer is also available for monitoring purpose.

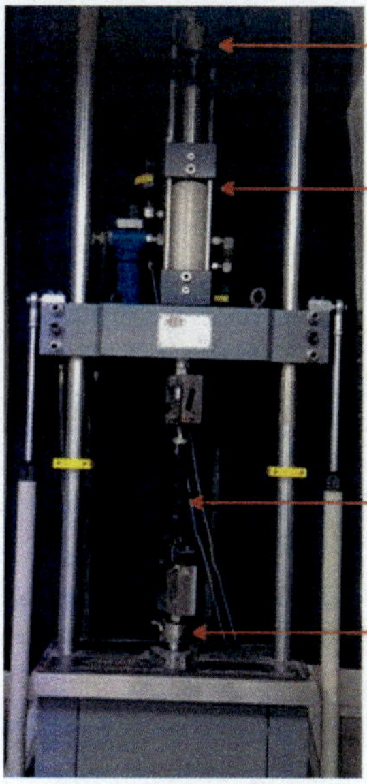

Fig. 1 Damper testing machine for damper characterization

2.1 Characterization of Passive Damper

In this work, a passive damper of a light motor vehicle was used for conducting experiments. The characterization tests were performed using damper testing machine by choosing three distinct frequency and amplitude conditions. The experiments were conducted for frequencies of 0.5, 1 and 1.5 Hz and each with variation in displacement amplitudes of 10, 15 and 20 mm, respectively. The characterization tests with combination of each frequency and amplitude provide force–displacement curve. Area enclosed by the force–displacement curve gives energy dissipated per cycle and this concept can be utilized to arrive at damping coefficient for a particular cycle by using Eq. 1.

$$E = \pi C\omega A^2 \tag{1}$$

where E = Energy dissipated per cycle (Nm), ω = Frequency (rad/s), C = Damping coefficient of damper (Ns/m), A = Amplitude (m).

Equation 1 helps to calculate the damping coefficient of damper in a particular cycle. The combination of three frequencies and amplitudes provides nine cycles and hence nine damping coefficients. Among these, three values, that is, low, medium and high values of damping coefficients were selected for the design of experiments and hence to be used in quarter car model using Matlab Simulink.

2.2 Design of Experiment (DOE)

Currently, design of experiments (DOE) is one of the salient techniques used by several researchers before conducting the actual experiments and this can provide number of experiments to be carried out based on DOE method preferred. The time-saving characteristic of the DOE made it popular where the number of input parameters involved is more. In this work, central composite design (CCD) has been used to design the experiments by using Minitab software. In this work, controlling parameters considered are sprung mass (kg), vehicle velocity (m/s), damping coefficient (Ns/m) and the spring stiffness (N/m). To design the experiment, three levels of each parameter have been considered in this work which could be able to provide 31 sets of experiments, as per CCD design. The level of parameters and their magnitude along with DOE table are discussed in later section. These 31 sets of experiments are conducted with the help of quarter car model using Matlab Simulink model by varying the parameters in each set.

2.3 Quarter Car Model

Quarter car model is exclusively used to understand the effect of certain vehicle parameters on the four-wheeler by simply using 1/4th of the vehicle model. A quarter car model involves sprung and unsprung mass, spring, damper and tire. Quarter car model can be mathematically represented with the help of Newton's second law of motion and are represented by Eqs. 2 and 3.

$$m_2\ddot{y}_2 - c_2(\dot{y}_2 - \dot{y}_1) - k_2(y_2 - y_1) = 0 \tag{2}$$

$$m_1\ddot{y}_1 - c_2(\dot{y}_2 - \dot{y}_1) - k_2(y_2 - y_1) + c_1\left(\dot{y}_1 - \dot{R}_d\right) + k_1(y_1 - R_d) = 0 \tag{3}$$

where m_1 = unsprung mass, m_2 = sprung mass, R_d = road displacement, k_1 = tire stiffness, k_2 = spring stiffness, c_1 = tire damping coefficient, c_2 = damping coefficient of damper, y_1 = unsprung mass displacement, y_2 = sprung mass displacement.

In this work, a sine profile has been considered as the road input to be given for quarter car model with amplitude of 6 cm and 10 m wavelength.

A relation between wavelength (υ), vehicle velocity (v) and angular frequency (ω) can be used to input the vehicle speed in quarter car model and is given in Eq. 4.

$$\omega = 2\pi v/\upsilon \tag{4}$$

With the help of all the mentioned control parameters, a quarter car model has been constructed using Matlab Simulink for simulation and is depicted in Fig. 2.

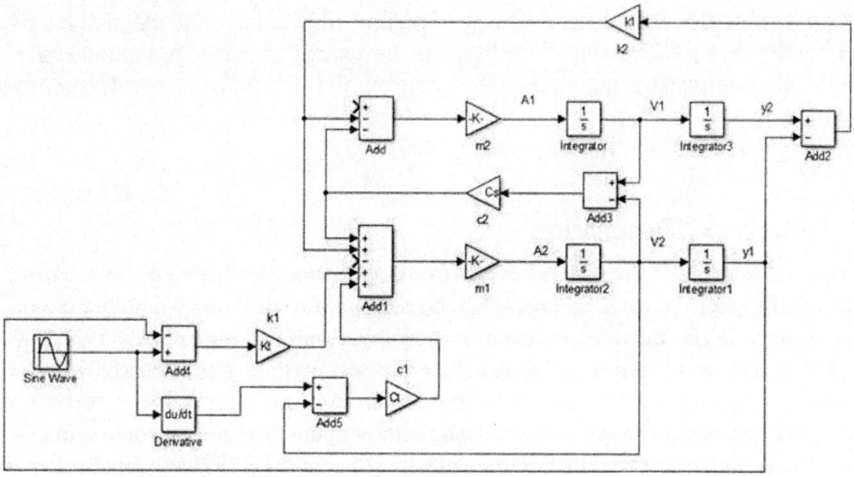

Fig. 2 Quarter car simulink model

A total of 31 experiments were carried out using the quarter car Simulink model as per DOE to evaluate ride comfort and road holding in every case. In this work, the standard ISO 2631-1:1997 is applied to evaluate ride comfort and road holding. Referring to this standard, ride comfort can be quantified as RMS acceleration of sprung mass. Also, the relative displacement between the unsprung mass and road can be used to quantify road holding.

2.4 ANOVA and Response Surface Optimization

Response surface method (RSM) is an important statistical tool used to understand the relationship between the control variables and responses. Obtaining an optimal surface with the help of designed experiments is a basic idea behind RSM. Analysis of variance (ANOVA) is another useful statistical tool to check the significance of the result by separating the total variance. ANOVA helps in the estimation of interaction between parameters and in identifying the desirable parameter. This also explores the significance and influence of one parameter over the response. One of the important tests in ANOVA is '*p-test*' or probability test. For 95% confidence level, if *p-value* is lesser than or equal to 0.05, then the null hypothesis will be rejected for total population and the parameter is said to significant and if $p > 0.05$, then the parameter is not significant.

To optimize the response, it is necessary to set the boundaries with upper and lower limits of control variables. The goal for response can be set in terms of 'response maximization' or 'target the response' or 'response minimization'. Based on the problem definition and requirements the goals can be set for single response or multiple responses. In this problem, it is desired to minimize the magnitudes of ride comfort and road holding, by optimizing the control variables. A multiobjective optimization approach is required in this case to minimize the quantity of ride comfort and road holding together.

3 Results and Discussion

The results from passive damper characterization for different amplitudes and frequencies, DOE, quarter car simulation using Simulink model, ANOVA and multiobjective optimization using response surface method are discussed in this section.

In the first stage, a passive damper of a light motor vehicle (which is commercially available) was characterized in the damper testing machine (DTM). Frequency condition for experiments were 0.5, 1 and 1.5 Hz with each frequency subjected to amplitude variation of 10, 15 and 20 mm, respectively. The nature of force–displacement curve is shown in Fig. 3a–c.

The characteristic curve for force–displacement data provides information on damping force achieved at different cycles, but more importantly explores the energy dissipation per cycle. The energy dissipation is nothing but area covered by the curve. The damping coefficient can be estimated using Eq. 1. The damping coefficient calculated for each cycle is given in Table 1.

Thus, the damper characterization provides useful information on damping force achieved in different cycles, along with damping coefficient. If this experimental data be used with some other parameters, it may provide the behavior of vehicle suspension system under certain conditions. Hence, an attempt has been made to utilize these damping coefficient values into the quarter car simulation model. Control parameters considered in the simulation were sprung mass (kg), vehicle velocity (m/s), damping coefficient (Ns/m) and spring stiffness (N/m). For the simulation purpose, three values of damping constants have been considered from Table 2. These three values are nothing but least-medium-high values among all the damping constant values (1348, 1718 and 2079, respectively). Different control parameters and their levels considered for the simulation are given in Table 2.

Other influencing parameters like unsprung mass, tire stiffness and tire damping coefficients are also needed for simulation in quarter car and these parameters were maintained constant for all the test runs. The values maintained for these parameters were unsprung mass = 55 kg, tire stiffness = 254,800 N/m and tire damping coefficient = 12050 Ns/m.

The factors and levels mentioned in Table 2 were used to form design of experiment and central composite design (CCD) was used to design the experiment using Minitab software which developed 31 sets of experiments. The control variables as mentioned in Table 2 were set in the quarter car simulation model using Matlab/Simulink. The output of each experiment was RMS acceleration of sprung mass (which is quantitative measure of ride comfort) and RMS relative displacement between unsprung

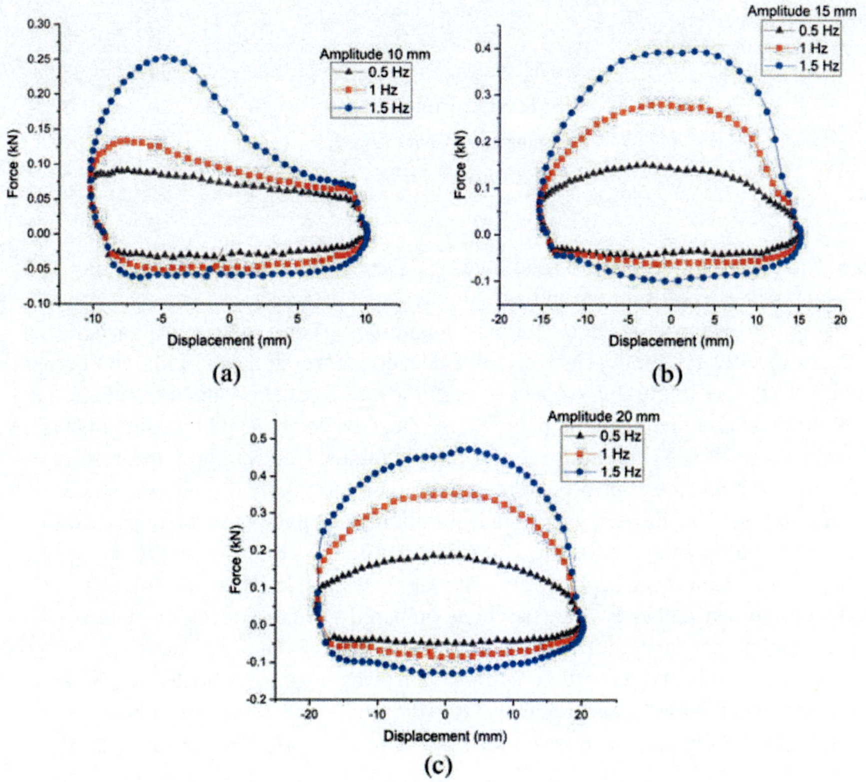

Fig. 3 Load displacement curves for **a** 10 mm, **b** 15 mm and **c** 20 mm amplitudes for different frequencies

Table 1 Damping coefficients for different cycles

S. no.	Amplitude (mm)	Frequency (Hz)	Damping coefficient (Ns/m)
1	10	0.5	1943.952
2	10	1	1347.979
3	10	1.5	1406.172
4	15	0.5	2079.294
5	15	1	1800.863
6	15	1.5	1747.215
7	20	0.5	1850.28
8	20	1	1718.524
9	20	1.5	1588.812

Table 2 Variables for quarter car model and their levels

	Level 1	Level 2	Level 3
Sprung mass (kg)	280	330	380
Vehicle velocity (m/s)	10	20	30
Damping coefficient (Ns/m)	1348	1718	2079
Spring stiffness (N/m)	24,000	28,000	32,000

mass and the road (to measure road holding). Design of experiment based on CCD and the result of each simulation has been tabulated in Table 3.

Using the response (ride comfort and road holding) and the control variables, a relationship between them can be established using regression equations. But before doing so, it is very important to check the significance level of parameters, interaction between them and the desirable parameters. This can be achieved by using analysis of variance (ANOVA) for both the responses. Tables 4 and 5 show the results of ANOVA, for both ride comfort and road holding, respectively.

As mentioned in the earlier section, the individual parameter or interaction of parameters with p value < 0.05 are said to be significant. The closeness of the model with the actual values can be identified with the R^2 values. In this work, full quadratic model has shown higher R^2 values when compared to linear or linear+ interaction or square+ interaction model.

The R^2 value observed for full quadratic value was 97.05%. Adjusted R^2 (R^2 adj.) was observed as 94.46% and predicted R^2 (R^2 pred.) was observed to be 86.83%. Based on this, a regression equation for ride comfort was developed and is given in Eq. 5.

$$\begin{aligned} RC = & -2.6 + 0.933 * Vel - 0.0018 * Mass - 0.000082 * Stiff \\ & - 0.00034 * Damp - 0.01833 * Vel * Vel + 0.000054 * Mass * Mass \\ & - 0.001365 * Vel * Mass + 0.000009 * Vel * Stiff \\ & + 0.000093 * Vel * Damp - 0.000003 * Mass * Damp \end{aligned} \tag{5}$$

where RC = Ride comfort, Vel = velocity, Mass = sprung mass, Stiff = spring stiffness and Damp = damping coefficient.

In a similar manner, ANOVA was performed over road holding also, the results of which are given in Table 5 which provided the significance parameters and terms over road holding model.

When checked for effectiveness of the model, the R^2 for the road holding was found to be 94.73%. Similarly, adjusted R^2 was found to be 90.12% and prediction R^2 was 76.7%. Based on this, similar to ride comfort model, a full quadratic regression model was developed for road holding also. This is given in Eq. 6, where RH is road holding.

$$\begin{aligned} RH = & -0.0045 + 0.000891 * Vel + 0.000041 * Mass \\ & - 0.000002 * Damp - 0.000022 * Vel * Vel \end{aligned}$$

Table 3 Design of experiment and simulation result

S. no.	Sprung mass (kg)	Vehicle velocity (m/s)	Damping coefficient (Ns/m)	Spring stiffness (N/m)	Ride comfort (m/s^2)	Road holding (m)
1	330	20	1718	28,000	6.3202	0.007
2	380	30	1348	32,000	5.1998	0.0047
3	330	20	1718	28,000	6.3202	0.007
4	380	20	1718	28,000	5.2919	0.0066
5	280	20	1718	28,000	7.7177	0.0075
6	330	20	2079	28,000	6.3675	0.0073
7	330	20	1718	28,000	6.3202	0.007
8	280	30	1348	32,000	7.5152	0.0054
9	380	10	2079	32,000	3.0698	0.0048
10	330	20	1718	28,000	6.3202	0.007
11	330	20	1718	28,000	6.3202	0.007
12	380	10	1348	24,000	3.9121	0.0059
13	330	10	1718	28,000	3.0697	0.0042
14	280	30	1348	24,000	5.9888	0.0045
15	280	30	2079	32,000	8.4513	0.0071
16	380	30	1348	24,000	4.2271	0.0041
17	380	30	2079	32,000	5.9886	0.0064
18	330	20	1348	28,000	6.3552	0.0068
19	330	20	1718	28,000	6.3202	0.007
20	330	20	1718	24,000	5.3867	0.006
21	330	20	1718	28,000	6.3202	0.007
22	380	10	1348	32,000	3.3685	0.0052
23	330	20	1718	32,000	7.4074	0.0083
24	380	30	2079	24,000	5.2128	0.0059
25	280	10	1348	24,000	3.0817	0.0037
26	280	10	2079	32,000	2.6126	0.0032
27	380	10	2079	24,000	3.2008	0.0049
28	330	30	1718	28,000	6.0038	0.0055
29	280	10	1348	32,000	2.7526	0.0034
30	280	30	2079	24,000	7.2401	0.0063
31	280	10	2079	24,000	2.7532	0.0033

Table 4 ANOVA for ride comfort

Source	df	Adj SS	Adj MS	*f*-value	*p*-value
Model	14	83.4916	5.9637	37.55	0.000
Linear	4	49.6406	12.4102	78.14	0.000
Vel	1	43.5496	43.5496	274.21	0.000
Mass	1	4.1475	4.1475	26.11	0.000
Stiff	1	1.5975	1.5975	10.06	0.006
Damp	1	0.3460	0.3460	2.18	0.159
Square	4	22.3617	5.5904	35.20	0.000
Vel × *Vel*	1	8.7188	8.7188	54.90	0.000
Mass × *Mass*	1	0.0474	0.0474	0.30	0.592
Stiff × *Stiff*	1	0.0019	0.0019	0.01	0.913
Damp × *Damp*	1	0.0001	0.0001	0.00	0.979
2-way interactions	6	11.4290	1.9048	11.99	0.000
Vel × *Mass*	1	7.4504	7.4504	46.91	0.000
Vel × *Stiff*	1	1.9813	1.9813	12.48	0.003
Vel × *Damp*	1	1.8484	1.8484	11.64	0.004
Mass × *Stiff*	1	0.0891	0.0891	0.56	0.465
Mass × *Damp*	1	0.0591	0.0591	0.37	0.550
Stiff × *Damp*	1	0.0007	0.0007	0.00	0.948

$$- 0.000001 * Vel * Mass + 0.000000 * Stiff * Damp \qquad (6)$$

It is a general assumption that the model can be said to be valid if the adjusted R^2 value is closer to R^2 of the model and this is true in this case also, both in terms of ride comfort and road holding. Hence, these models can be considered to establish relationship between the responses (ride comfort and road holding) and the control variables.

3.1 Results of Optimization

As explained by Els et al. [1], certain parameters seek opposite nature of settings when considering ride comfort and road holding. Hence it is very difficult to design the suspension variables so that both ride comfort and road holding meet the passenger satisfaction. Therefore, an attempt has been made in this work to optimize the quarter car suspension parameters, for greater ride comfort and road holding simultaneously. This is nothing but a multiobjective approach. This was achieved through response

Table 5 ANOVA for road holding

Source	df	Adj SS	Adj MS	f-value	p-value
Model	14	0.000056	0.000004	20.54	0.000
Linear	4	0.000011	0.000003	13.46	0.000
Vel	1	0.000007	0.000007	36.15	0.000
Mass	1	0.000001	0.000001	4.77	0.044
Stiff	1	0.000001	0.000001	4.31	0.054
Damp	1	0.000002	0.000002	8.59	0.010
Square	4	0.000034	0.000008	43.34	0.000
Vel × *Vel*	1	0.000013	0.000013	64.08	0.000
Mass × *Mass*	1	0.000000	0.000000	0.00	0.996
Stiff × *Stiff*	1	0.000000	0.000000	0.14	0.717
Damp × *Damp*	1	0.000000	0.000000	0.00	0.985
2-way interactions	6	0.000012	0.000002	9.97	0.000
Vel × *Mass*	1	0.000006	0.000006	28.21	0.000
Vel × *Stiff*	1	0.000001	0.000001	5.11	0.038
Vel × *Damp*	1	0.000005	0.000005	25.87	0.000
Mass × *Stiff*	1	0.000000	0.000000	0.32	0.580
Mass × *Damp*	1	0.000000	0.000000	0.21	0.651
Stiff × *Damp*	1	0.000000	0.000000	0.12	0.729

surface optimization. The intension of the optimization was to minimize the magnitude of ride comfort and road holding. The response optimization plot is shown in Fig. 4.

The response optimization plot provided the optimal condition for all the four control variables for 'compromised' minimum conditions of ride comfort and road holding. It can be observed that the desirability value achieved was 1 for road holding and it was close to unity in case of ride comfort. The optimum conditions indicated, and the corresponding response values are tabulated in Table 6.

The optimum conditions indicate that combined minimization of ride comfort and road holding preferably require minimum vehicle velocity and maximum damping coefficient value. The other parameters such as sprung mass and spring stiffness were intermediate values of provided input magnitude. To check the validity of the result, the optimum values were substituted back in the quarter car Simulink model. The result of this simulation and the deviation of this result with the result of response optimization is given in Table 7 in terms of percentage error.

The results in Table 7 indicate that the regression model for both ride comfort and road holding obtained using response surface method efficiently served the purpose. The percentage error between the predicted and validation through simulation is well within the acceptable limit. Even though RSM is one of the old statistical methods

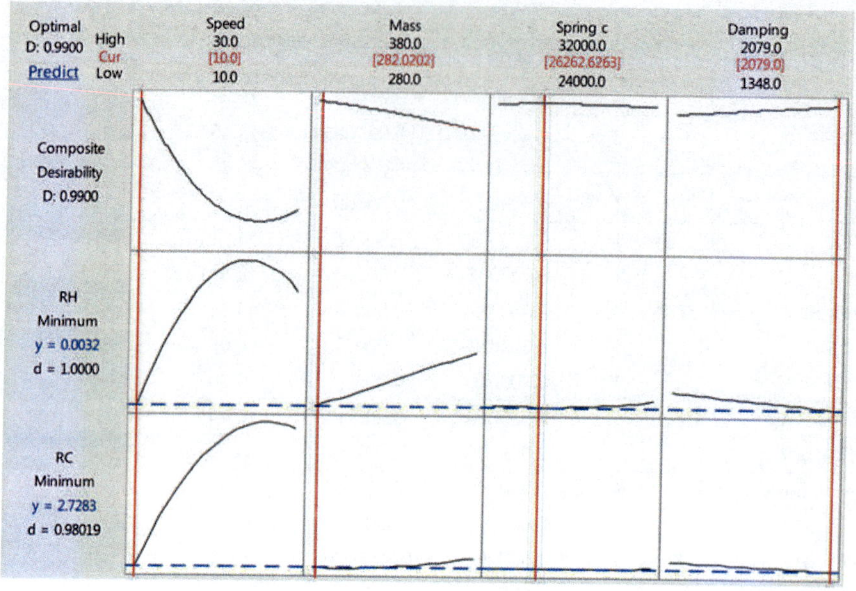

Fig. 4 Result of RSM optimization

Table 6 Results of optimization

Optimum conditions				Response output	
Sprung mass	Velocity	Damping constant	Spring stiffness	Ride comfort	Road holding
282.0202	10	2079	26262.62	2.7283	0.0032

Table 7 Result of validation simulation

	RSM optimization	Validation test	% error
Ride comfort	2.7283	2.7280	0.01
Road holding	0.0032	0.0033	3.03

and considering the fact that the study is involving multiobjective optimization, the method was able to model the conditions well.

4 Conclusion

The current work explores the dynamic characterization of the passive damper of a light motor vehicle, implementation of the damper characteristics into quarter car simulation model and multiobjective optimization for better ride comfort and road holding. In the first stage, commercially available passive damper was characterized using damper testing machine (DTM) for three different frequency and amplitude conditions. The force–displacement data was utilized to estimate the damping coefficient values with the help of energy equation of a damper. The damping coefficient values thus obtained were further utilized in the quarter car simulation model in Matlab/Simulink. Control variables such as sprung mass, vehicle velocity and spring stiffness were also considered for the quarter car simulation. The set of simulation was decided based on CCD method of design of experiment. The simulation was performed in the quarter car model to understand the ride comfort and road holding for the considered variables. The effect of each variable over the response was analyzed using ANOVA and regression equations were modeled with good R^2 value. Proceeding further, an attempt was done to optimize the control variables for minimum magnitudes of ride comfort and rod holding using response surface optimization. The optimized conditions were validated by quarter car simulation for the optimum values. Very low deviation of the validation result with the optimization result indicated the effectiveness of response surface method in modeling the conditions, in this work.

Acknowledgments The authors are thankful to IMPRINT Project No. IMPRINT/2016/7330 titled with "Development of Cost Effective Magnetorheological (MR) Fluid Damper in Two wheelers and Four Wheelers Automobile to Improve Ride Comfort and Stability" funded by Ministry of Human Resource Development and Ministry of Road Transport and Highways, Govt. of India.

References

1. Els, P.S., Theron, N.J., Uys, P.E., Thoresson, M.J.: The ride comfort vs. handling compromise for off-road vehicles. J. Terramech. **44**(4), 303–317 (2007)
2. Roumy, J.G., Boulet, B., Dionne, D.: Active control of vibrations transmitted through a car suspension. Int. J. Veh. Auton. Syst. **2**(3/4), 236–254 (2004)
3. Naudé, A.F., Snyman, J.A.: Optimisation of road vehicle passive suspension systems. Part 2. Qualification and case study. Appl. Math. Model. **27**(4), 263–274 (2003)
4. Sun, L.: Optimum design of "road-friendly" vehicle suspension systems subjected to rough pavement surfaces. Appl. Math. Model. **26**(5), 635–652 (2002)
5. Mostaani, S., Singh, D., Firouzbakhsh, K., Ahmadian, M.T.: Optimization of a passive vehicle suspension system for ride comfort enhancement with different speeds based on DOE method. In: Proceedings of the International Joint Colloquiums on Computer Electronics Electrical Mechanical and Civil (CEMC'11), pp. 149–154 (2011)
6. Reddy, M.S., Rao, S.S., Vigneshwar, P., Akhil, K., RajaSekhar, D.: An intelligent optimization approach to quarter car suspension system through RSM modeled equation. In: Proceedings of First International Conference on Information and Communication Technology for Intelligent Systems, pp. 97–108, Springer, Cham (2016)
7. Mitra, A.C., Patil, M.V., Banerjee, N.: Optimization of vehicle suspension parameters for ride comfort based on RSM. J. Inst. Eng. (India) Ser. C **96**(2), 165–73 (2015)
8. Shirahatti, A., Prasad, P.S., Panzade, P., Kulkarni, M.M.: Optimal design of passenger car suspension for ride and road holding. J. Braz. Soc. Mech. Sci. Eng. **30**(1), 66–76 (2008)
9. Gangadharan, K.V., Sujatha, C., Ramamurti, V.: Experimental and analytical ride comfort evaluation of a railway coach. In: Proceeding of SEM ORG IMAC XXII Conference, pp. 1–15 (2004)

10. Reddy, M.S., Vigneshwar, P., RajaSekhar, D., Akhil, K., Reddy, P.L.: Optimization study on quarter car suspension system by RSM and Taguchi. In: Proceedings of the International Conference on Signal, Networks, Computing, and Systems, pp. 261–271, Springer, New Delhi (2016)
11. Gurubasavaraju, T.M., Kumar, H., Arun, M.: Evaluation of optimal parameters of MR fluids for damper application using particle swarm and response surface optimisation. J. Braz. Soc. Mech. Sci. Eng. **39**(9), 3683–3694 (2017)

Parametric Study of Lithium-ion Battery Module for Electric Vehicle Application

R. Umesh, Basant Singh Sikarwar, Ayush Goyal, and Sachin Singh Gautam

1 Introduction

Electric and hybrid vehicles (EVs and HVs) are playing a pivotal role in curbing and tackling the rising fuel prices and pollution caused by gasoline-powered vehicles [1]. EVs are considered to be one of the emerging markets in the automotive industry to meet the world demand for clean alternative transportation [2]. Lithium-ion (Li-ion) batteries are considered as one of the emerging alternative sources to power electric and hybrid vehicles. Li-ion cells have many advantages due to its, (1) high specific energy, (2) high power density, and (3) high nominal voltage and low discharge cycle [3]. However, Li-ion batteries have performance issues due to heating problems that are not desired for practical application [4]. The desired operating temperature range for Li-ion batteries is $0 - 40\,^{\circ}\mathrm{C}$ [5]. In practical application, if the non-uniform temperature profile exists within the battery module, it causes a severe effect on the battery performance and the vehicle. If the overheating and non-uniform temperature of Li-ion ions is not controlled, it can further cause failure and degradation of Li-ion cells. Therefore, a thermal management system is needed, which can maintain the desired operating temperature in the battery module for optimal battery performance and prevent it from further degradation.

R. Umesh · S. S. Gautam (✉)
Indian Institute of Technology Guwahati, Guwahati 781039, Assam, India
e-mail: ssg@iitg.ac.in

B. S. Sikarwar
Department of Mechanical Engineering, Amity University UP, NOIDA 301303, Uttar Pradesh, India

A. Goyal
Texas A&M University Kingsville, Kingsville, TX 78363, USA

S. K. Saha and M. Mukherjee (eds.), *Recent Advances in Computational Mechanics and Simulations*, Lecture Notes in Mechanical Engineering,
https://doi.org/10.1007/978-981-15-8315-5_54

The different thermal management system has been used to cool the Li-ion battery module such as (1) thermal control using water cooling, (2) air cooling, and (3) phase change material thermal cooling [6, 7]. Thermal cooling using air has many added advantages over another cooling method due to its simple design, low cost, and easy maintenance [6]. Various techniques have been used by researchers to model the battery, such as the finite element method [8, 9], artificial neural network [10], and CFD models [11–13]. The present work is based on CFD-based thermal modeling of Li-ion cells to study its thermal characteristics.

The thermal performance of Li-ion cells can be affected by different boundary conditions such as inlet velocity, ambient temperature, and duct sizes (inlet and outlet). In the open literature, Wang et al. [14] investigated the thermal profile of different array arrangements using forced air convection methods. They studied the thermal profile of different configuration for inlet velocity 1 m/s, and 25 °C ambient temperature. However, their numerical results did not give further insight into the thermal profile of Li-ion cells at different boundary conditions such as different ambient temperatures and inlet velocities. Lu et al. [15] numerically investigated the thermal characteristics of densely packed 252 cylindrical Li-ion cells. They studied the effect of the flow path, airflow rates on temperature uniformity of battery pack with 15 and 59 vents. However, they limited the duct size to 12 mm (inlet and outlet) for 15 and 59 vents, which could not predict the temperature distribution for different duct sizes.

To best of authors' knowledge in the open literature, very few studies have been carried out to study the thermal profile of 3×8 Li-ion cell arrangement, and limited boundary conditions have been applied. Hence, the motivation of the present work is to provide innovative thermal management using a forced air convection method for 3×8 Li-ion cell arrangement. This can further provide insight into thermal profiles of 3×8 Li-ion cells, which can aid in designing an efficient battery management system for the given arrangement.

The objective is to study the thermal profile of 3×8 Li-ion cells at different boundary conditions. This paper focuses on the thermal modeling of 3×8 Li-ion cells rather than the electrochemical process. In the open literature, Wang et al. [14] limited the CFD model of 3×8 Li-ion cell arrangement to one specific ambient temperature, i.e., (25 °C) and inlet velocity ($V = 1$ m/s). This paper investigates the 3×8 Li-ion cell arrangement over different ambient temperature (T_{am}), inlet velocity (V), and duct sizes. The 3×8 Li-ion cell arrangement is analyzed over the different inlet and out duct sizes to understand the effect of duct sizes on thermal profiles of the given arrangement.

The rest of the paper has been organized as follows. It gives the details of the design of the 3×8 Li-ion cell arrangement. Section 3 describes the mathematical models developed for the given arrangement. Section 4 presents the results and discussion of the numerical results of the given arrangement. Finally, the conclusion of the paper is summarized in Sect. 5.

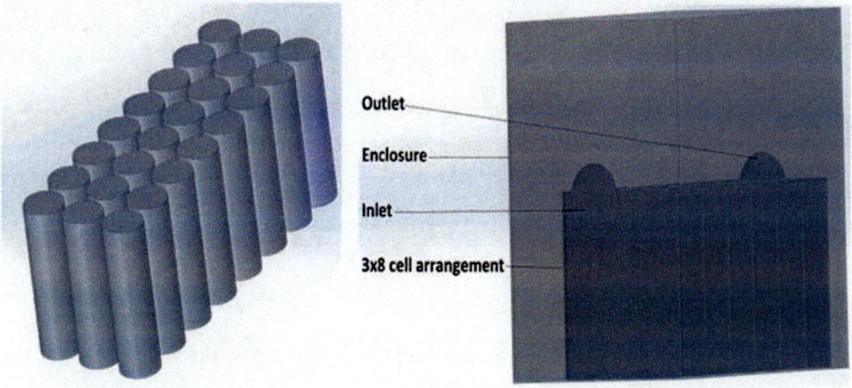

(a) 3×8 Li-ion cell pack (b) 3×8 Li-ion cell pack in the enclosure

Fig. 1 Schematic of 3 × 8 Li-ion cell arrangement with inlet and outlet duct size of 20 mm

2 The Geometry of 3 × 8 Li-ion Cell Arrangement

Li-ion cells have been closely stacked in 3 × 8 cell arrangement (see Fig. 1) with a 2 mm inter-cell distance between each cell. When forced air convection method is employed, the air passed through the 2 mm gaps between individual cells and walls of the battery module. The distance between the ends of the right and left side of the cells is 20 mm, and the top is 35 mm, where the overall height of the battery module is 100 mm. The battery module consists of an enclosure and closely stacked Li-ion cells. The enclosure houses the Li-ion cells. The inlet and outlet duct size for 3 × 8 cell arrangement is 5 mm. Three different duct sizes, i.e., 10, 15, 20 mm, have been designed to study the thermal profile of 3 × 8 cell arrangement at 25 °C ambient temperature.

3 Model Formulation

3.1 Governing Equations

The governing equations of the battery module during discharge are expressed as [14]

Continuity equation

$$\nabla \vec{\upsilon} = 0 \tag{1}$$

Momentum equation

$$\frac{\partial \vec{v}}{\partial t} + (v.\nabla)\vec{v} = \frac{1}{\rho}(-\nabla p + \mu \nabla^2 \vec{v}) \quad (2)$$

Energy equation

$$\rho c_p \left(\frac{\partial E}{\partial t} + v_x \frac{\partial y}{\partial x} + v_y \frac{\partial E}{\partial y} + v_z \frac{\partial E}{\partial z} \right) = k\left(\frac{\partial^2 E}{\partial x^2} + \frac{\partial^2 E}{\partial y^2} + \frac{\partial^2 E}{\partial z^2}\right) \quad (3)$$

where ∇ is the gradient operator, ρ is the density (kg/m^3), p is the pressure (Pa), μ is the viscosity (Pas).

The set of governing equations was solved numerically using ANSYS FLUENT.

3.2 *CFD Model*

In this paper, the thermal profile of 3 × 8 Li-ion cell arrangement is numerically simulated using ANSYS FLUENT 19. The total number of elements in the meshing of the 3 × 8 Li-ion arrangement is 400119. The boundary condition used to model Li-ion cells is velocity inlet, pressure outlet, and the no-slip boundary condition at the walls. The cooling air is assumed to be steady, incompressible, and laminar due to the low flow velocity and short characteristics length ($Re < 2300$). ABS plastic is used as a battery material. Each cell in a 3 × 8 arrangement is considered as a homogenous cylinder with an internal heat source, which is a lumped model for a single cell. Various assumptions made in the study are

1. The cell material is considered to be homogenous.
2. The cell is considered to be isotropic.
3. The heat generated within the Li-ion cell is assumed to be uniformly distributed.
4. Thermal conductivity (K) and other thermo-physical properties are assumed to be constant.

The thermo-physical properties of Li-ion cells are summarized in Table 1.

The SIMPLE algorithm [16] is used to solve the momentum and energy equation of the computational unit.

Table 1 Thermo-physical properties of 3 × 8 Li-ion cells

Parameter	Value
Density (kg/m^3)	2100
Heat transfer coefficient (W/m^2 K)	10

4 Results and Discussion

In this section, the numerical result of the 3 × 8 cell arrangement is presented. The effect of different parameters, such as ambient temperature, inlet velocity, and duct size on the current model, is discussed in detail. Ambient, maximum, average, optimal temperature, and inlet velocity is denoted as T_{am}, T_{max}, $\bar{T}$, T_{opt}, and V, respectively.

The 3 × 8 cell arrangement is investigated for two different velocities, i.e., 1 and 2 m/s. 3 × 8 Li-ion cells are analyzed for five different ambient temperature viz. −25, −15, 0, 15, and 25 °C. A detailed observation is carried out to analyze the performance of 3 × 8 Li-ion cells for their ability to work in optimal temperature ranges at different ambient temperatures. Thermal performance of 3 × 8 Li-ion cells is judged on (1) maximum temperature (T_{max}) of the individual cells, and battery pack, and (2) the optimal temperature profile (T_{opt}) (0–40 °C) of the 3 × 8 battery pack.

4.1 Validation of Numerical Simulation

The numerical model is validated by comparing it with the results of Wang et al. [14] for a 3 × 8 cell arrangement. The model is analyzed for inlet velocity 1 m/s, and T_{am} = 25 °C for 2 mm inter-cell distance.

The maximum temperature and average temperature obtained from the current numerical study is presented in Table 2, along with the results of Wang et al. [14]. It can be seen that the numerical result from the current model closely matches the results from Wang et al. [14].

4.2 Thermal Analysis of 3 × 8 *Cell Arrangement*

Figure 2 compares the thermal profile of the 3 × 8 cell arrangement at 25 °C ambient temperature. It can be observed that for inlet velocity 1 m/s, there is uneven temperature distribution within the battery module. Li-ion cells at extreme ends operate above the optimal temperature range. As the inlet velocity increases to 2 m/s, the average and individual cell temperatures come below 40 °C. All cells operate within the optimal temperature range of 0–40 °C.

Table 2 Validation of the current CFD model

CFD model	T_{max} (°C)	$\bar{T}$ (°C)
3 × 8 arrangement (Wang et al. [14])	43.4	36.0
3 × 8 arrangement (Current model)	43.5	36.2
% error in the 3 × 8 arrangement	0.22	0.55

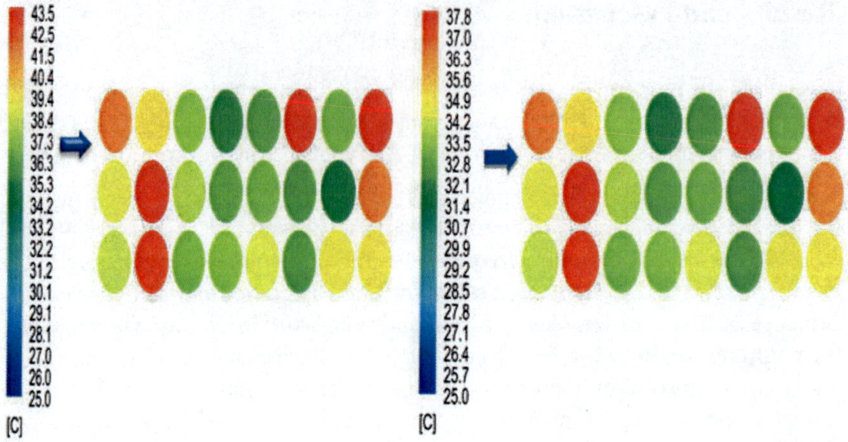

(a) $V = 1$ m/s (b) $V = 2$ m/s

Fig. 2 Thermal profile of 3 × 8 cell arrangement at $T_{am} = 25\,°C$

At 15 °C ambient temperature, it can be observed (see Fig. 3) that for inlet velocity 1 m/s, all cells work within the optimal temperature range. As the inlet velocity increases to 2 m/s, it further cools down the cells and brings the maximum temperature T_{max} to 29.4 °C. All the cells at 15 °C ambient temperature work within the optimal temperature range.

From Fig. 4, it can be observed that a similar trend is followed to 15 and 25 °C ambient temperature. The maximum temperature (see Fig. 4a) is 27.4 °C for an inlet velocity of 1 m/s. As the inlet velocity increases to 2 m/s, the maximum temperature

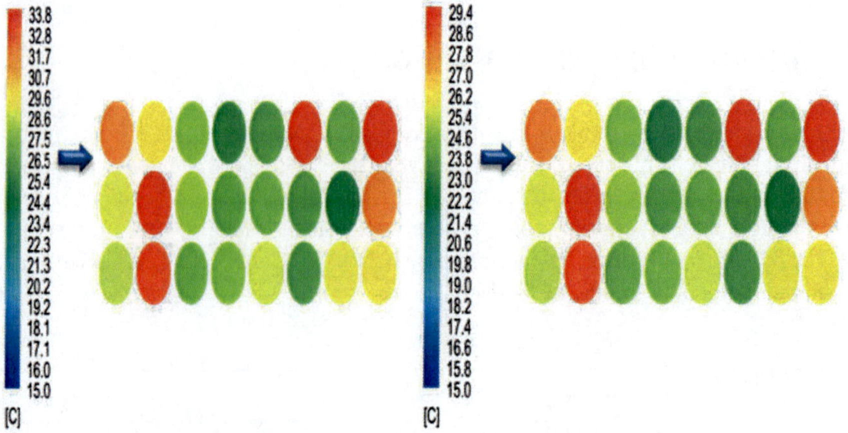

(a) $V = 1$ m/s (b) $V = 2$ m/s

Fig. 3 Thermal profile of 3 × 8 cell arrangement at $T_{am} = 15\,°C$

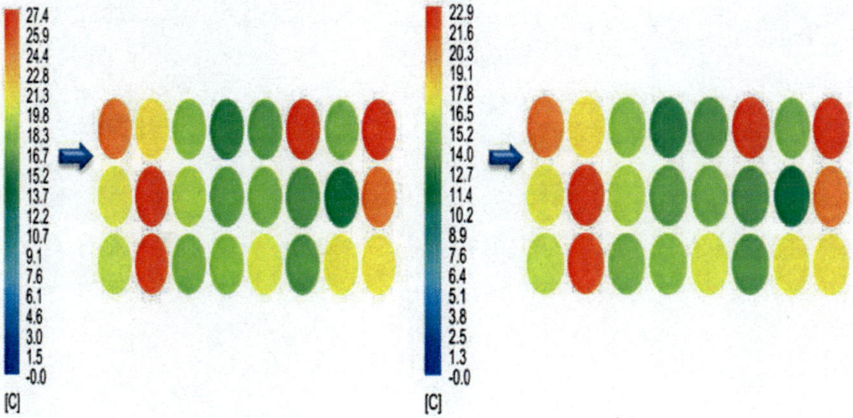

(a) $V = 1$ m/s (b) $V = 2$ m/s

Fig. 4 Thermal profile of 3 × 8 cell arrangement at $T_{am} = 0\,°C$

decreases to 22.9 °C. This further cools the cells and helps to operate in ideal operating temperatures.

Based on the above analysis (see Figs. 5 and 6), it can be observed that for lower ambient temperature −15 and −25 °C, all cells operate below the optimal temperature range of 0–40 °C. From Fig. 5a, b, it can be observed that as the inlet velocity to 2 m/s the maximum temperature T_{max} increases to 6 °C. However, all the cells at −15 °C ambient temperature operate below the optimal temperature range.

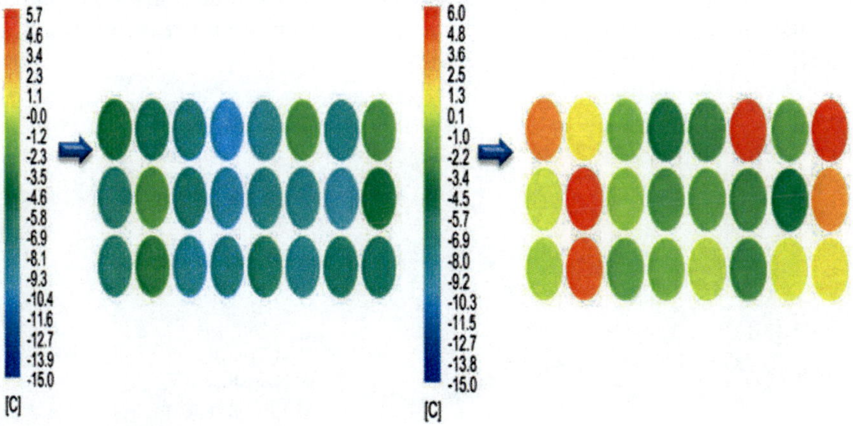

(a) $V = 1$ m/s (b) $V = 2$ m/s

Fig. 5 Thermal profile of 3 × 8 cell arrangement at $T_{am} = -15\,°C$

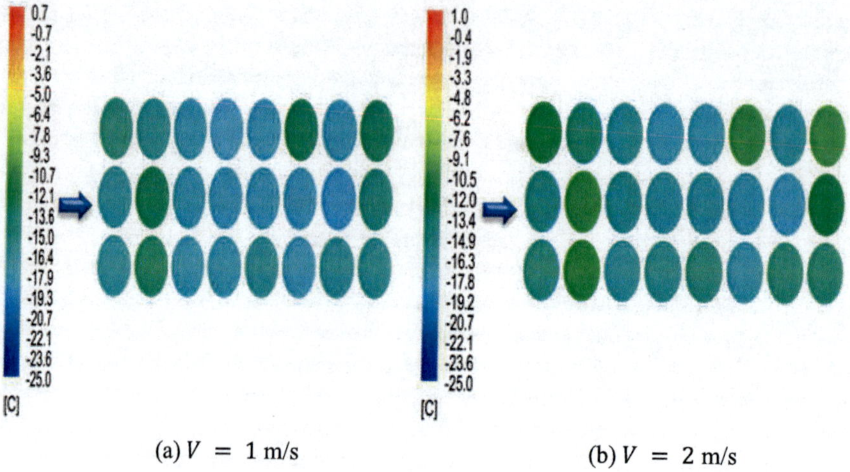

Fig. 6 Thermal profile of 3 × 8 cell arrangement at $T_{am} = -25\,°C$

The same trend is observed at −25 °C ambient temperature (see Fig. 6); all cells operate below the ideal temperature range. Inlet velocity has no significant effect on the thermal profile of the 3 × 8 cell arrangement.

Figure 7 shows the plot of maximum and minimum temperature range of 3 × 8 cell arrangement at different ambient temperatures. It can be observed that at a higher ambient temperature range, all the cells operate within the optimal temperature range. However, at lower temperature ranges, all cells operate below the ideal operating ranges.

It can be observed from simulation results (see Fig. 8) that duct size (inlet and outlet) has an impact on the maximum and average temperature range of 3 × 8

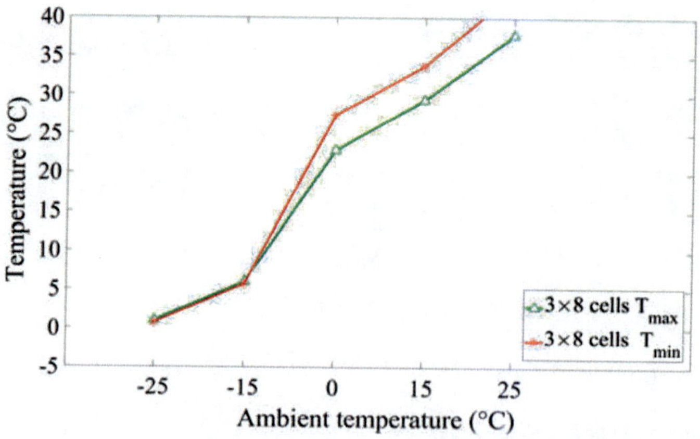

Fig. 7 The plot of temperature vs. ambient temperature of 3 × 8 cell arrangement

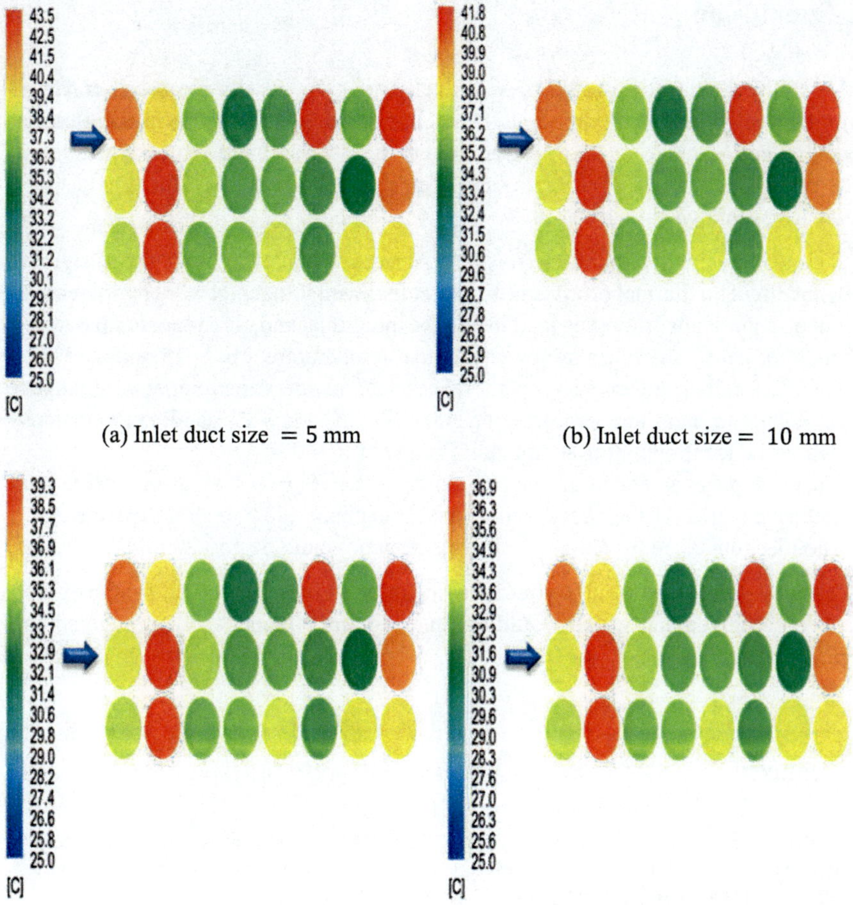

(a) Inlet duct size = 5 mm

(b) Inlet duct size = 10 mm

(c) Inlet duct size = 15 mm

(d) Inlet duct size = 20 mm

Fig. 8 Thermal profile of different inlet and outlet duct sizes of 3 × 8 cell arrangement at T_{am} = −25 °C

cell arrangement. As the duct size increases to 10 mm, the maximum temperature decreases to 41.8 °C. As the duct size increases, it is observed (see Fig. 5d) that maximum temperature decreases, indicating that bigger the duct size, better the cooling of Li-ion cells.

5 Conclusion

In this study, the thermal profile of 3 × 8 cell arrangement with forced air convection method is investigated numerically to optimize the performance of the battery module. The influence of various parameters such as ambient temperature, inlet velocity, and duct size on of 3×8 array arrangement has been studied. The conclusions drawn are as follows:

1. At lower ambient temperature, it is observed that the inlet velocity has no significant effect on thermal profiles of Li-ion cells. Even if the inlet velocity increases but no significant difference is found on the individual and maximum temperature range of cells. Therefore, at lower ambient temperature, i.e., −15 and −25 °C, the 3 × 8 cell arrangement does not meet the ideal operating temperature range.
2. At higher ambient temperature ranges, i.e., 0, 15, and −25 °C all cells operate within the ideal operating temperature range of 0–40 °C.
3. Duct size plays an essential role as it controls the volume of air going inside the battery module. As the duct size increases, the maximum temperature decreases. The ideal duct size for a 3 × 8 cell arrangement would be 15–20 mm.

The above-presented results would provide insights into the thermal profile of the 3 × 8 cell arrangement, which would aid in designing efficient array arrangements for practical application.

References

1. Neelakrishnan, S., Rudramoorthy, R., Amjad, S.: Review of design considerations and technological challenges for successful development and deployment of plug-in hybrid electric vehicles. Renew. Sustain. Energy Rev. **14**(3), 1104–1110 (2010)
2. Pruteanu, A., Scarlatache, V.-A., Ciobanu, R.C., Viziteu, G.: Thermal study of a superior lithium-ion polymer battery. In: World Energy System Conference–WESC (2012)
3. Ritchie, A., Howard, W.: Recent developments and likely advances in lithium-ion batteries. J. Power Sources **162**, 809–812 (2006)
4. Xing, Y., Miao, Q., Tsui, K.-L., Pecht, M.: Prognostics and health monitoring for lithium ion battery. In: Proceedings of 2011 IEEE International Conference on Intelligence and Security Informatics, pp. 242–247. IEEE (2011)
5. Xing, Y., He, W., Pecht, M., Tsui, K.L.: State of charge estimation of lithium-ion batteries using the open-circuit voltage at various ambient temperatures. Appl. Energy **113**, 106–115 (2014)
6. Xu, X.M., He, R.: Research on the heat dissipation performance of battery pack based on forced air cooling. J. Power Sources **240**, 33–41 (2013)
7. Park, H.S.: A design of air flow configuration for cooling lithium-ion battery in hybrid electric vehicles. J. Power Sources **239**, 30–36 (2013)
8. Yeow, K., Thelliez, M., Teng, H., Tan, E.: Thermal analysis of a Li-ion battery system with indirect liquid cooling using finite element analysis approach. SAE Int. J. **1**(1), 65–78 (2012)
9. Huo, Y., Rao, Z., Liu, X., Zhao, J.: Investigation of power battery thermal management by using mini-channel cold plate. Energy Convers. Manage. **89**, 387–395 (2015)
10. He, F., Li, X., Ma, L.: Combined experimental and numerical study of thermal management of battery module consisting of multiple Li-ion cells. Int. J. Heat Mass Transfer **72**, 622–629 (2014)

11. Ling, Z., Wang, F., Fang, X., Gao, X., Zhang, Z.: A hybrid thermal management system for lithium ion batteries combining phase change materials with forced-air cooling. Appl. Energy **148**, 403–409 (2015)
12. Li, X., He, F., Ma, L.: Thermal management of cylindrical batteries investigated using wind tunnel testing and CFD simulation. J. Power Sources **238**, 395–402 (2013)
13. Panchal, S., Khasow, R., Dincer, I., Agelin-Chaab, M., Fraser, R., Fowler, M.: Numerical modeling and experimental investigation of a prismatic battery subjected to water cooling. Numer. Heat Trans. Part A Appl. **71**(6), 626–637 (2017)
14. Wang, T., Tseng, K.J., Zhao, J.Y., Wei, Z.B.: Thermal investigation of lithium-ion battery module with different cell arrangement structures and forced air cooling strategies. Appl. Energy **134**, 229–238 (2014)
15. Lu, Z., Meng, X.Z., Wei, L.C., Hu, W.Y., Zhang, L.Y., Jin, L.W.: Thermal management of densely-packed EV battery with forced air cooling strategies. Energy Procedia **88**, 682–688 (2016)
16. Patankar, S.: Numerical Heat Transfer and Fluid Flow. CRC Press (1980)

An Efficient Implementation of Lanczos Method for Dynamic Analysis of Launch Vehicle Structures

P. Deepak, P. V. Anil Kumar, and R. Neetha

1 Introduction

FEAST is a Finite Element (FE)-based structural analysis software developed by structures group of VSSC. Figure 1, shows the architecture of FEAST software with the capabilities of its pre- and post-processor [1] and a solver module called Substructure and Multi-Threaded (SMT). The SMT solver is designed based on domain decomposition of the FE model and implemented based on multi-threaded programming to solve large order problems by exploiting multi-core system architecture.

The substructure method [2] is employed for solving both static and dynamic problems of structural analysis by reducing the number of unknown variables in the solution process. A parallel direct solver for static problems of structural analysis has efficient sparse storage schemes, to reduce the number of arithmetic operations, during the factorization of coefficient stiffness matrix at substructure level. A multi-threaded implementation scales the parallelism of substructure computation based on number of CPU cores of multi-core system.

1.1 Eigenvalue Problems in Structural Dynamics

The initial eigenvalue problem (Unshifted vibration) for structural dynamics is

$$([K] - \lambda[M])\phi = 0 \tag{1}$$

P. Deepak (✉) · P. V. Anil Kumar · R. Neetha
Vikram Sarabhai Space Centre, ISRO Thiruvananthapuram, Thiruvananthapuram 695021, India
e-mail: p_deepak@vssc.gov.in

S. K. Saha and M. Mukherjee (eds.), *Recent Advances in Computational Mechanics and Simulations*, Lecture Notes in Mechanical Engineering,
https://doi.org/10.1007/978-981-15-8315-5_55

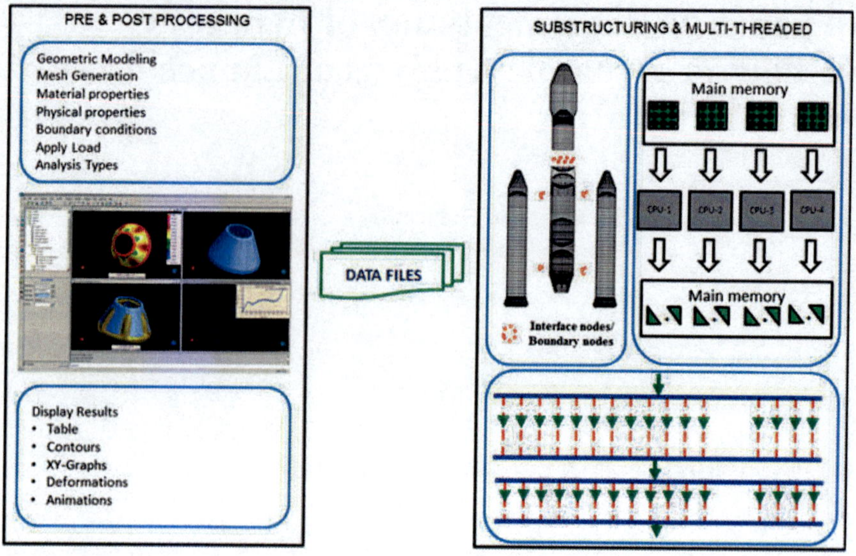

Fig. 1 Software architecture of $FEAST^{SMT}$ software

$[K]$, $[M]$ represents global stiffness and mass matrices. λ represents eigenvalues and $\{\phi\}$ represents eigenvectors. For the shifted eigenvalue problems, the Eq. (1) is represented by an initial shift value λ_0.

$$([K] - \lambda_0[M] - (\lambda - \lambda_0)[M])\phi = 0$$

$$([\overline{K}] - (\lambda - \lambda_0)[M])\phi = 0 \tag{2}$$

where $[\bar{K}] = ([K] - \lambda_0[M])$.

For buckling problems, the mass matrix $[M]$ is replaced by a differential stiffness matrix $[K^d]$.

$$([K] - \lambda[K^d])\phi = 0 \tag{3}$$

The general form of inverse eigenvalue problem is written as

$$[B]\{X\} = \Lambda[D]\{X\} \tag{4}$$

These eigenvalue problems of Eqs. 1 2, and 3) are converted into a symmetric inverse form represented in Eq. (4). The symmetric stiffness matrix $[K]$ is factorized in $[L]^{-1}\{d\}[L]^{-T}$ form. The $[B]$, $[D]$, $\{X\}$, and Λ are represented as shown in Table 1.

Table 1 Inverse eigenvalue problem definitions

Problem type	[B]	[D]	{X}	Λ
Shifted vibration	$[M][L]^{-T}\{d\}[L]^{-1}[M]$	$[M]$	$\{\phi\}$	$1/(\lambda - \lambda_0)$
Unshifted vibration	$[L]^{-1}\{d\}[L]^{-T}$	$[I]$	$[L]^T\{\phi\}$	$-1/\lambda$
Buckling	$[L]^{-1}\{K^d\}[L]^{-T}$	$[I]$	$[L]^T\{\phi\}$	$-1/\lambda$

1.2 Eigenvalue Problems in Structural Dynamics

The basic Lanczos recurrence algorithm is a transformation process to a reduced tridiagonal form. The algorithm truncates the tridiagonalization process and provides approximations to the eigenpairs (eigenvalues and eigenvectors) of the original matrix [3]. The tridiagonal reduction is an automatic matrix reduction scheme whereby the eigen solutions in the neighborhood of a specified point in the eigen spectrum can be accurately extracted from a tridiagonal eigenvalue problem whose order is much lower than that of the full problem. Specifically, the order, m of the reduced problem is never greater than.

$$m = 2 * q + 10 \tag{5}$$

where q is the desired number of accurately computed eigenvalues. Thus, the intrinsic power of the method lies in the fact that the size of the reduced eigenvalue problem is of the same order of magnitude as the number of desired roots, even though the discretized system model may possess thousands of degrees of freedom. Tridiagonal reduction method employs only a single initial shift of eigenvalues and hence usually requires only one matrix decomposition, namely global stiffness matrix $[K]$. The Lanczos tridiagonal reduction method is implemented in $FEAST^{SMT}$ for real eigenvalue analysis as typified by structural vibration (Shifted and Un-shifted) and buckling problems.

The Recurrence algorithm is based on forward and backward passes on factorized global stiffness matrix to perform the inverse iteration operations. The reduced tridiagonal matrix is of the form:

$$[A] = \begin{bmatrix} a_{1,1} & d_2 & & & & \\ d_2 & a_{2,2} & d_3 & & & \\ & d_3 & a_{3,3} & d_4 & & \\ & & . & . & . & \\ & & & . & . & d_m \\ & & & & d_m & a_{m,m} \end{bmatrix}$$

The mth order eigenvalue problem $[A]\{y\} = \bar{\Lambda}\ \{y\}$ is solved for eigenvalues $\bar{\Lambda}$ and eigenvectors $\{y\}$ using QR-algorithm.

1.3 Multi-frontal Method

In multi-frontal method, numerical factorization is reduced to a series of partial factorization operations on dense frontal matrices. The frontal matrices are generally formed by assembling the stiffness associated with each nodal Degree of Freedom (DOF). They are composed of three main components: diagonal factors L_B, off-diagonal factors L_{OFF}, and Schur complement S. Once the partial factorization is completed for a frontal matrix, only S is required for the subsequent partial factorization steps. After the partial factorization of a frontal matrix, diagonal and off-diagonal factors are not required until the triangular solution. Figure 2 shows the assembly tree structure used for multi-frontal-based factorization.

2 Proposed Method

To handle large order problems, Lanczos recurrence algorithm discussed in Sect. 1.2 is modified based on multi-frontal-based partial factorization method 1.3. The existing implementation of Lanczos method requires both stiffness and mass matrices to be fully assembled to perform factorization, followed by forward and backward substitution. For this purpose, a new matrix data structure for unassembled elemental matrices is created. Mass matrix of each element is created and kept on a symmetric matrices list.

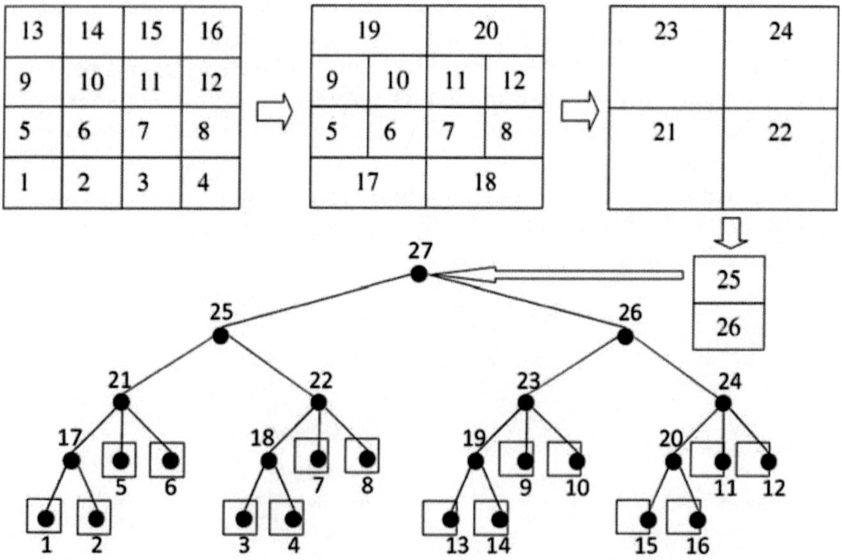

Fig. 2 Assembly tree structure used for multi-frontal method

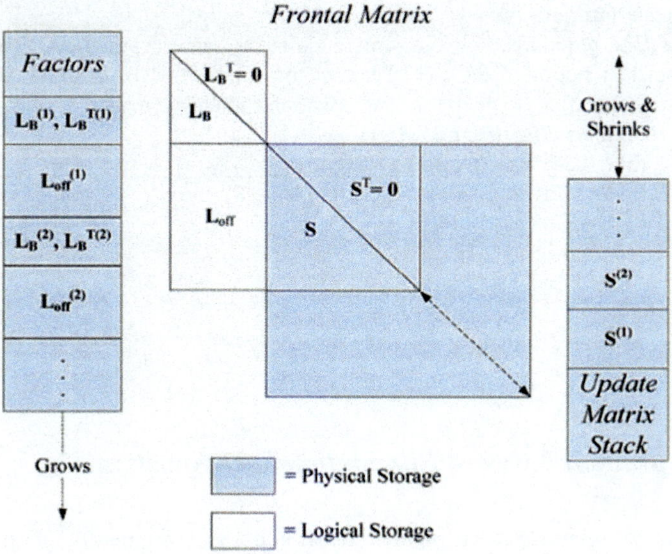

Fig. 3 Matrix data structure used for proposed multi-frontal-based partial factorization

The stiffness is arranged as an assembly tree of frontal matrices discussed in Sect. 1.3. Factorization of eliminated nodes and update matrix operations of remaining nodes on each assembly tree level are performed on multiple threads. This ensures computation on multiple tree branches simultaneously on frontal matrices. The following subsections explain the partial factorization of stiffness matrix with multi-threaded execution, setting up unassembled mass matrices, and implementation of parallel multiplication of unassembled mass matrices with a vector. Figure 3 shows the matrix data structure used for storing and propagating frontal matrices.

2.1 *Multi-threaded Frontal Factorization*

The partial factorization operations are carried out on frontal matrices L_B, L_{OFF}, and S, explained in Sect. 1.3, using BLAS and LAPCK kernels. Multiple threads are started at each level while computing partial factors of the frontal matrices.

Step 1. Set **current node** = **root**
Step 2. Lauch a new thread on each tree node at the same depth (Level), starting from leaf nodes of the tree. Perform the following BLAS and LAPACK kernel operations to get partial factorization and update matrix.
(a) $L_B = L_B^{-1}\{DPOTRF(L_B)\}$
(b) $L_{OFF} = L_{OFF} * L_B^{-T}\{DTRSM(L_{OFF}, L_B)\}$
(c) $S = S$â€“$L_{OFF} * L_{OFF}^T\{DSYRK(L_{OFF}, S\}$
(d) **Synchronize** threads
Step 3. Level = Level - 1
Step 4. **repeat** step 2 and step 3 **until** Level = 0
Step 5. **Continue**

2.2 *Independent Forward Elimination/back Substitution*

Forward elimination process starts from the leaf nodes and progress is subsequently on parent nodes until it reaches the root tree node. The implementation is done recursively starting from root of the tree.

Step 1. **Level** = maximum tree depth
Step 2. **repeat** for each child node of **current node**
(a) **current node** = child node
(b) Execute steps: **2,3,4,5,6,7** and **8**
until all child nodes are processed
Step 3. **Assemble** $\{F_e\}$, the eliminated DOFs from the original RHS vector.
Step 4. **repeat** for each child nodes of the current node
(a) **Compute** the local indices of eliminated and remaining nodes
(b) **Assemble** $\{F_e\}$ and $\{F_r\}$ with RHS vectors of corresponding child node
until all child nodes are processed
Step 5. **Load** $\{L_B$ and $L_{OFF}\}$
Step 6. **Execute** the following BLAS and LAPACK kernel operations for forward elimination.
(a) $F_e = L_B^{-1} * F_e\{DTRSV(L_B, F_e)\}$
(b) $F_r = F_r - L_{OFF} * F_e\{DGEMV(F_r, L_{OFF}, F_e)\}$

Step 7. **Update** RHS vectors using $\{F_e\}$ and $\{F_r\}$
Step 8. **Release** $\{L_B$ and $L_{OFF}\}$
Step 9. **Continue**

Backward substitution process starts at the root node and continues subsequently on each child nodes until all the tree nodes are processed. The implementation is explained in the following steps.

Step 1. Set **current node = root**
Step 2. **Load** $\{F_e\}$ from the updated RHS vector
Step 3. **If** current node is **not** root **then**
(a) **Load** $\{D_e$ and $D_r\}$ of parent node from the updated RHS vector
(b) **Assemble** RHS vector $\{F_r\}$ from $\{D_e$ and $D_r\}$
end If
Step 4. **Load** $\{L_B$ and $L_{OFF}\}$
Step 5. **Execute** the following BLAS and LAPCK operations for back substitution.
(a) $F_e = F_e - L_{OFF}^T * F_r\{DGEMV(F_e, L_{OFF}, F_r)\}$
(b) $F_e = L_B^{-T} * F_e\{DTRSV(F_e, L_B)\}$

Step 6. **Release** $\{L_B$ and $L_{OFF}\}$
Step 7. **repeat** for each child nodes of the current node
(a) **Current node** = child node
(b) **Execute** steps: **3,4,5,6,7** and **8**
until all child nodes are processed
Step 8. **Update** original RHS vector with $\{F_e\}$
Step 9. **Continue**

2.3 *Parallel Multiplication of Mass Matrices with a Vector*

In order to avoid memory-related issues, mass matrix is set up as a list of elemental symmetric matrices, with associated DoF indexing. The mass multiplication with vector data structure is performed with unassembled mass matrices. To speed up the process, the multiplication is carried out in multiple threads, spanning the list into groups. Figure 4 shows the grouped data ready for multi-threaded execution on independent CPU cores. *N* indicates number of CPU cores, which is also equal to number of sets. The multi-threaded computation requires creation of number of sets, which is dynamically determined by number of CPU cores. Each set data is represented by a data structure, which is formed by a reference to input vector, a temporary output vector, a list of elemental mass matrices, and its indices.

3 Results and Conclusion

For performance comparison, Lanczos solver is first used with the coefficient matrix decomposed with another direct solver based on Cholesky method. The solution time of the proposed method is compared with the Cholesky version of Lanczos implementation for satellite and Reusable Launch Vehicle (RLV) structural components.

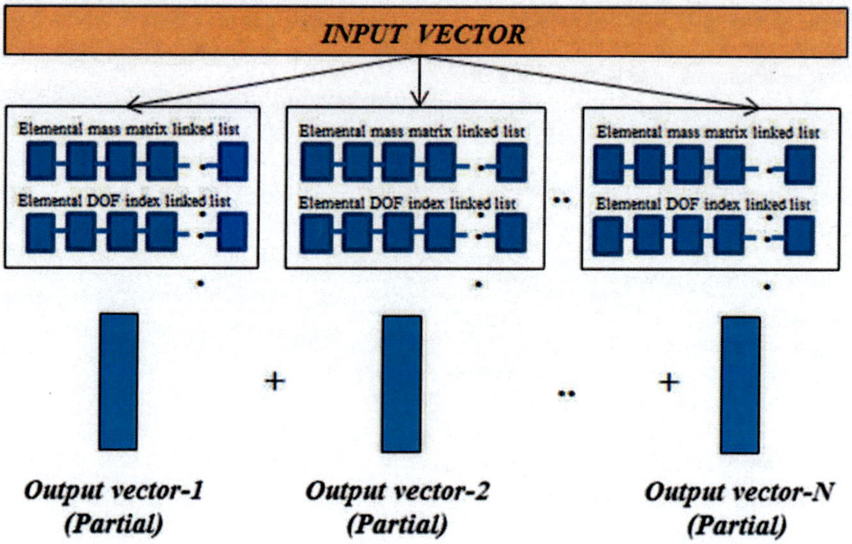

Fig. 4 Multi-threaded multiplication of unassembled mass matrix

Figure 5 shows the free-vibration results of a typical satellite structural component, with model size. The multi-frontal-based eigenvalue extraction procedure is implemented only for solution of interface variables.

Figure 6 shows a comparison of number of solution variables and solution time taken for different number of substructure super elements. The number of solution

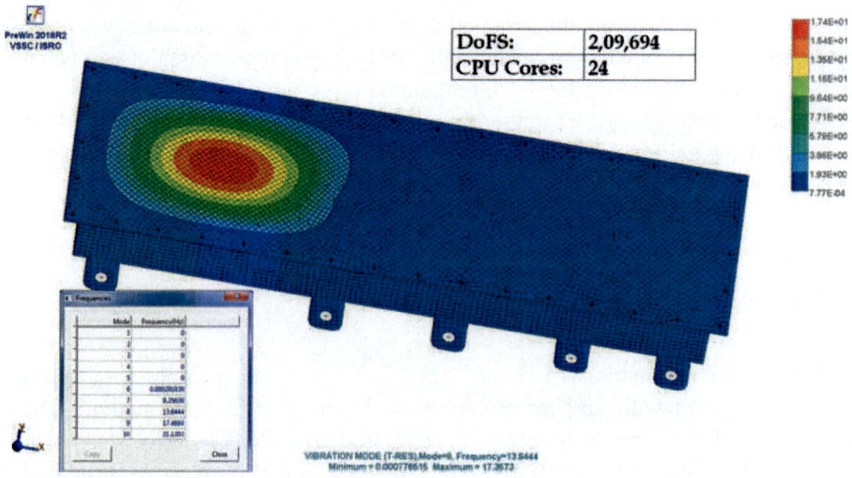

Fig. 5 FE model of a typical satellite structure

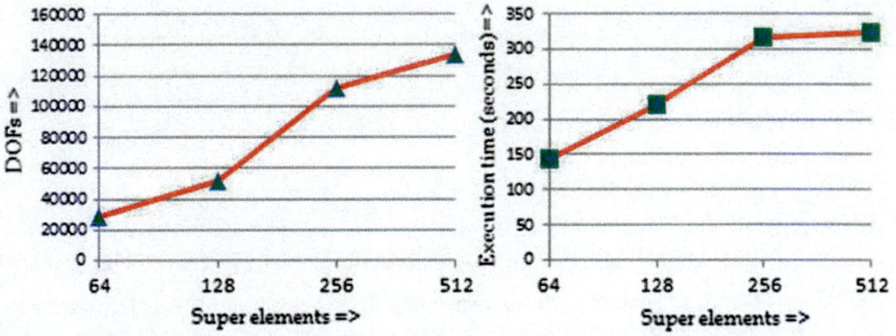

Fig. 6 Comparison of satellite structural components against number of super elements

variable increases as the number of super elements. However, the solution time depends on the model used and its connectivity information.

Figure 7 shows the solution performance of a reusable launcher structure. Table 2 shows improvement in execution time for extracting mode shapes using Lanczos method. Performance of satellite component FE model is approximately 4.62 times. RLV model was unsolvable using Cholesky factorization, is solved using multi-frontal factorization scheme in nearly 12 min.

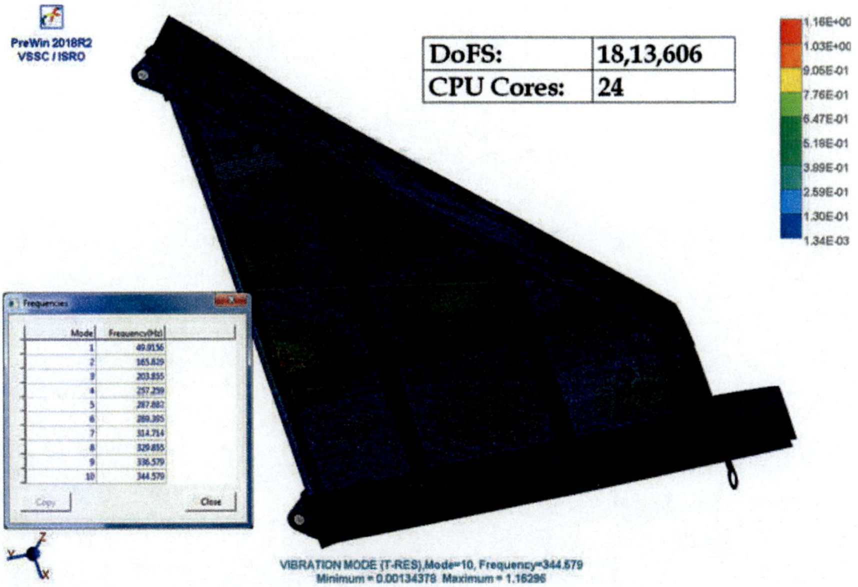

Fig. 7 FE model of a typical launch vehicle structure

Table 2 Inverse eigenvalue problem definitions

Satellite component		RLV component	
Multi-frontal	Cholesky	Multi-frontal	Cholesky
156 s	721.544 s	721 s	—

References

1. Anil kumar P.V., Raj Thilak, T.J., Sivasubramonian, B.: Development of an object oriented pre and post processing software for finite element analysis. In: International Conference on computer aided engineering, CAE 07, pp. 661–667
2. Karypis, G., Kumar V.: METIS "A Software Package for Partitioning Unstructured Graphs, Partitioning Meshes, and Computing Fill-Reducing Orderings of Sparse Matrices" Version 4.0. University of Minnesota (1998)
3. Newman, M., Flanagan, P.F.: Eigen value extraction in NASTRAN by the tridiagonal reduction (FEER) method–Real eigen value anlaysis